Lecture Notes in Computer Science 10178

Commenced Publication in 1973
Founding and Former Series Editors:
Gerhard Goos, Juris Hartmanis, and Jan van Leeuwen

More information about this series at http://www.springer.com/series/7409

Selçuk Candan · Lei Chen
Torben Bach Pedersen · Lijun Chang
Wen Hua (Eds.)

Database Systems for Advanced Applications

22nd International Conference, DASFAA 2017
Suzhou, China, March 27–30, 2017
Proceedings, Part II

 Springer

Editors

Selçuk Candan
Arizona State University
Tempe - Phoenix, AZ
USA

Lei Chen
Hong Kong University of Science
  and Technology
Hong Kong
China

Torben Bach Pedersen
Aalborg University
Aalborg
Denmark

Lijun Chang
University of New South Wales
Sydney, NSW
Australia

Wen Hua
The University of Queensland
Brisbane, QLD
Australia

ISSN 0302-9743   ISSN 1611-3349   (electronic)
Lecture Notes in Computer Science
ISBN 978-3-319-55698-7   ISBN 978-3-319-55699-4   (eBook)
DOI 10.1007/978-3-319-55699-4

Library of Congress Control Number: 2017934640

LNCS Sublibrary: SL3 – Information Systems and Applications, incl. Internet/Web, and HCI

Printed on acid-free paper

This Springer imprint is published by Springer Nature
The registered company is Springer International Publishing AG
The registered company address is: Gewerbestrasse 11, 6330 Cham, Switzerland

Preface

It is our great pleasure to welcome you to DASFAA 2017, the 22nd edition of the International Conference on Database Systems for Advanced Applications (DASFAA), which was held in Suzhou, China, during March 27–30, 2017.

The long history of Suzhou City has left behind many attractive scenic spots and historical sites with beautiful and interesting legends. The elegant classical gardens, the old-fashioned houses and delicate bridges hanging over flowing waters in the drizzling rain, the beautiful lakes with undulating hills in lush green, and the exquisite arts and crafts, among many other attractions, have made Suzhou a renowned historical and cultural city full of eternal and poetic charm. Suzhou is best known for its gardens: the Humble Administrator's Garden, the Lingering Garden, the Surging Wave Pavilion, and the Master of Nets Garden. These gardens weave together the best of traditional Chinese architecture, painting, and arts. Suzhou is also known as the "Venice of the East." The city is sandwiched between Taihu Lake and Grand Canal. A network of channels, criss-crossed with hump-backed bridges, give Suzhou an image of the city on the water.

We were delighted to offer an exciting technical program, including three keynote talks by Divesh Srivastava (AT&T Research), Christian S. Jensen (Aalborg University), and Victor Chang (Xi'an Jiaotong University and Liverpool University), one 10-year best paper award presentation; a demo session with four demonstrations; two industry sessions with nine paper presentations; three tutorial sessions; and of course a superb set of research papers. This year, we received 300 submissions to the research track, each of which went through a rigorous review process. Specifically, each paper was reviewed by at least three Program Committee (PC) members, followed by a discussion led by the PC co-chairs. Several papers went through a shepherding process. Finally, DASFA 2017 accepted 73 full research papers, yielding an acceptance ratio of 24.3%.

Four workshops were selected by the workshop co-chairs to be held in conjunction with DASFAA 2017. They are the 4th International Workshop on Big Data Management and Service (BDMS 2017), the Second Workshop on Big Data Quality Management (BDQM 2017), the 4th International Workshop on Semantic Computing and Personalization (SeCoP), and the First International Workshop on Data Management and Mining on MOOCs (DMMOOC 2017).

The workshop papers are included in a separate volume of the proceedings also published by Springer in its *Lecture Notes in Computer Science* series.

The conference received generous financial support from Soochow University. The conference organizers also received extensive help and logistic support from the DASFAA Steering Committee and the Conference Management Toolkit Support Team at Microsoft.

We are grateful to the general chairs, Karl Aberer, EPFL, Switzerland, Peter Scheuermann, Northwestern University, USA, and Kai Zheng, Soochow University, China, the members of the Organizing Committee, and many volunteers, for their great support in the conference organization. Special thanks also go to the DASFAA 2017 local Organizing Committee: Zhixu Li and Jiajie Xu, both from Soochow University, China. Finally, we would like to take this opportunity to thank the authors who submitted their papers to the conference and the PC members and external reviewers for their expertise and help in evaluating the submissions.

February 2017

K. Selçuk Candan
Lei Chen
Torben Bach Pedersen

Organization

General Co-chairs

Karl Aberer — EPFL, Switzerland
Peter Scheuermann — Northwestern University, USA
Kai Zheng — Soochow University, China

Program Committee Co-chairs

Selçuk Candan — Arizona State University, USA
Lei Chen — HKUST, Hong Kong, SAR China
Torben Bach Pedersen — Aalborg University, Denmark

Workshops Co-chairs

Zhifeng Bao — RMIT, Australia
Goce Trajcevski — Northwestern University, USA

Industrial/Practitioners Track Co-chairs

Nicholas Jing Yuan — Microsoft, China
Georgia Koutrika — HP Labs, USA

Demo Track Co-chairs

Meihui Zhang — Singapore University of Technology and Design, Singapore
Wook-Shin Han — POSTECH University, South Korea

Tutorial Co-chairs

Katja Hose — Aalborg University, Denmark
Huiping Cao — New Mexico State University, USA

Panel Chair

Xuemin Lin — University of New South Wales, Australia

Proceedings Co-chairs

Lijun Chang — University of New South Wales, Australia
Wen Hua — University of Queensland, Australia

Publicity Co-chairs

Bin Yang	Aalborg University, Denmark
Xiang Lian	University of Texas-Pan American, USA
Maria Luisa Sapino	University of Turin, Italy

Local Organization Co-chairs

Zhixu Li	Soochow University, China
Jiajie Xu	Soochow University, China

Steering Committee Liaison

Xiaofang Zhou	University of Queensland, Australia

Conference Secretary

Yan Zhao	Soochow University, China

Webmaster

Yang Li	Soochow University, China

Program Committee

Amr El Abbadi	University of California at Santa Barbara, USA
Alberto Abello	UPC Barcelona, Spain
Divyakant Agrawal	University of California at Santa Barbara, USA
Marco Aldinucci	University of Turin, Italy
Ira Assent	Aarhus University, Denmark
Rafael Berlanga Llavori	University Jaume I, Spain
Francescho Bonchi	ISI Foundation, Italy
Selcuk Candan	Arizona State University, USA
Huiping Cao	New Mexico State University, USA
Barbara Catania	Università di Genova, Italy
Qun Chen	Northwestern Polytechnical University, China
Reynold Cheng	University of Hong Kong, SAR China
Wonik Choi	Inha University, South Korea
Gao Cong	Nanyang Technological University, Singapore
Bin Cui	Peking University, China
Lars Dannecker	SAP, Germany
Hasan Davulcu	Arizona State University, USA
Ugur Demiryurek	University of Southern California, USA
Francesco Di Mauro	University of Turin, Italy
Curtis Dyreson	Utah State University, USA
Hakan Ferhatosmanoglu	Bilkent University, Turkey

Dieter Pfoser	George Mason University, USA
Evaggelia Pitoura	University of Ioannina, Greece
Silvestro Poccia	University of Turin, Italy
Weixiong Rao	Tongji University, China
Matthias Renz	George Mason University, USA
Oscar Romero	UPC Barcelona, Spain
Florin Rusu	University of California Merced, USA
Simonas Saltenis	Aalborg University, Denmark
Maria Luisa Sapino	University of Turin, Italy
Claudio Schifanella	University of Turin, Italy
Cyrus Shahabi	University of Southern California, USA
Jieying She	HKUST, SAR China
Hengtao Shen	University of Queensland, China
Yanyan Shen	Shanghai Jiao Tong University, China
Alkis Simitsis	HP Labs, USA
Shaoxu Song	Tsinghua University, China
Yangqiu Song	HKUST, SAR China
Xiaoshuai Sun	University of Queensland, Australia
Letizia Tanca	Politecnico di Milano, Italy
Nan Tang	Qatar Computing Research Institute, Qatar
Egemen Tanin	University of Melbourne, Australia
Junichi Tatemura	Google, USA
Christian Thomsen	Aalborg University, Denmark
Hanghang Tong	Arizona State University, USA
Yongxin Tong	Beihang University, China
Panos Vassiliadis	University of Ioannina, Greece
Sabrina De Capitani Vimercati	University of Milan, Italy
Bin Wang	Northeastern University, China
Wei Wang	National University of Singapore, Singapore
Xin Wang	Tianjin University, China
John Wu	Lawrence Berkeley Lab, USA
Xiaokui Xiao	Nanyang Technological University, Singapore
Xike Xie	University of Science and Technology of China, China
Jianliang Xu	Hong Kong Bapatist University, SAR China
Jeffrey Xu Yu	Chinese University of Hong Kong, China
Xiaochun Yang	Northeastern University, China
Bin Yao	Shanghai Jiao Tong University, China
Hongzhi Yin	University of Queensland, Australia
Man Lung Yiu	Hong Kong Polytechnic, SAR China
Yi Yu	National Institute of Informatics, Japan
Ye Yuan	Northeastern University, China
Meihui Zhang	Singapore University of Technology and Design, Singapore
Wenjie Zhang	University of New South Wales, Australia
Ying Zhang	University of Technology Sydney, Australia

Zhengjie Zhang   Advanced Digital Sciences Center, Singapore
Xiangmin Zhou   RMIT, Australia
Yongluan Zhou   University of Southern Denmark, Denmark
Lei Zhu   University of Queensland, Australia
Esteban Zimanyi   Université Libre de Bruxelles, Belgium
Andreas Zufle   George Mason University, USA

Industry Track Program Committee

Akhil Arora   Xerox Research Centre, India
Jie Bao   Microsoft, China
Senjuti Basu Roy   UW Tacoma, USA
Neil Zhenqiang Gong   Iowa State University, USA
Defu Lian   University of Electronic Science and Technology,
                             China
Qi Liu   University of Science and Technology of China, China
Alkis Simitsis   HP Labs, USA
Kostas Stefanidis   University of Tampere, Finland
Lu-An Tang   NEC Lab, USA
Fuzheng Zhang   Microsoft, China
Hengshu Zhu   Baidu, China

Demo Track Program Committee

Jinha Kim   Oracle Labs, USA
Xuan Liu   Baidu, China
Yanyan Shen   Shanghai Jiao Tong University, China
Yongxin Tong   Beihang University, China

Additional Reviewers

Jinpeng Chen   Beihang University, China
Xilun Chen   Arizona State University, USA
Yu Cheng   Turn Inc., USA
Alexander Crosdale   UC Merced, USA
Tiziano De Matteis   University of Pisa, Italy
Vasilis Efthymiou   University of Crete, Greece
Roberto Esposito   University of Turin, Italy
Yixiang Fang   The University of Hong Kong, SAR China
Christian Frey   LMU Munich, Germany
Yash Garg   Arizona State University, USA
Concorde Habineza   George Mason University, USA
Jiafeng Hu   The University of Hong Kong, SAR China
Zhiyi Huang   UC Merced, USA
Zhipeng Huang   The University of Hong Kong, SAR China
Shengyu Huang   Arizona State University, USA

Contents – Part II

Graph and Network Data Processing

Spatial Databases

Real Time Data Processing

Big Data (Industrial)

Social Networks and Graphs (Industrial)

Demos

Contents – Part I

Trajectory and Time Series Data Processing

Data Mining

Query Processing and Optimization (I)

Text Mining

Recommendation

Security, Privacy, Senor and Cloud

Social Network Analytics (I)

Tutorials

Map Matching and Spatial Keywords

HIMM: An HMM-Based Interactive Map-Matching System

Xibo Zhou[1]([✉]), Ye Ding[2], Haoyu Tan[2], Qiong Luo[1], and Lionel M. Ni[3]

[1] Department of Computer Science and Engineering,
The Hong Kong University of Science and Technology,
Kowloon, Hong Kong
{xzhouaa,luo}@ust.hk
[2] Guangzhou HKUST Fok Ying Tung Research Institute,
The Hong Kong University of Science and Technology,
Kowloon, Hong Kong
{yeding,haoyutan}@ust.hk
[3] University of Macau, Zhuhai, China
ni@umac.mo

Abstract. Due to the inaccuracy of GPS devices, the location error of raw GPS points can be up to several hundred meters. Many applications using GPS-based vehicle location data require map-matching to pre-process GPS points by aligning them to a road network. However, existing map-matching algorithms can be limited in accuracy due to various factors including low sampling rates, abnormal GPS points, and dense road networks. In this paper, we propose the design and implementation of HIMM, an **H**MM-based **I**nteractive **M**ap-**M**atching system that produces accurate map-matching results through human interaction. The main idea is to involve human annotations in the matching process of some elaborately selected error-prone points and to let the system automatically adjust the matching of the remaining points. We use both real-world and synthetic datasets to evaluate the system. The results show that HIMM can significantly reduce human annotation costs comparing to the baseline methods.

Keywords: Map-matching · Interactive system · Trajectory

1 Introduction

With the ubiquity of location sensing technologies in a wide range of location-based devices such as vehicle GPS navigators and mobile phones, large amounts of trajectory data have been collected from different sources. These data have been utilized by various location-based services such as route recommendation, traffic control, and location-based social networks. A trajectory consists of a sequence of location points with latitudes, longitudes, and time-stamps. In practice, the location information of a trajectory are imprecise due to measurement noises and sampling errors [5]. It is therefore necessary to perform *map-matching* [5] by aligning the observed location points to the road networks

© Springer International Publishing AG 2017
S. Candan et al. (Eds.): DASFAA 2017, Part II, LNCS 10178, pp. 3–18, 2017.
DOI: 10.1007/978-3-319-55699-4_1

in a digital map so that these position data can be sufficiently accurate for trajectory-based applications.

The fundamental difficulty of map-matching is that raw trajectory data typically do not consist the actual paths of moving objects, especially when the location information is collected passively. Without the ground truth, it is difficult to train or evaluate any map-matching algorithms. Hence, in order to collect the ground truth, it is necessary to involve human contributions such as driving a vehicle along the road network and collecting the raw trajectory data along with the corresponding path manually. However, the cost of such methods is quite high. Hence, in this paper, we propose an interactive system called HIMM to process the raw trajectories to reduce the cost of generating the ground truth.

The main process of the system is to interactively select raw trajectory points for human annotators to match, and the challenge is in how to facilitate the interactive map-matching process. A naïve approach would be to simply throw all the sample points on a trajectory onto the road network, and then ask the annotator to drag each point to the correct road segment. However, it is tedious and at times challenging for the annotator to find a proper candidate road segment for each point of the trajectory. A more effective and user-friendly approach would be the following: (1) generate an initial path using certain map-matching algorithms, and display the path on the digital map along with the original trajectory; (2) ask the annotator to drag each mismatched point to the correct road segment. Furthermore, given the feedbacks of an annotator, the interactive system could keep updating the path in display after each human annotation. Unfortunately, to the best of our knowledge, none of the existing map-matching algorithms is able to utilize the feedbacks of annotators.

In this paper, we propose a novel interactive map-matching algorithm that takes the feedbacks of annotators to improve the matching result. Although such an interactive system can help an annotator to easily adjust a single point, the total annotation cost of a trajectory may still be high, because in order to pick and confirm the exact points that are mismatched, the annotator may check a large portion of the points, which could be up to the entire trajectory. As a result, it is desirable that the interactive map-matching system provides some guidance recommending potentially mismatched points for the annotator to check. Such a strategy of posing queries to the annotator can reduce the annotation cost, which is a key research issue in active learning and crowd-sourcing [12]. However, due to the complexity of map-matching algorithms as well as the input trajectories and the road network, existing query selection strategies are not suitable for the interactive map-matching task. Therefore, we design efficient strategies to pose queries for the interactive map-matching algorithm.

The contributions of this paper lie in the following aspects: (1) we propose a novel system framework for interactive map-matching. It is a general framework that combines human efforts with algorithms in an iterative manner to achieve high map-matching accuracy; (2) we design a new HMM (Hidden Markov Model)-based map-matching algorithm that can take an arbitrary number of human annotations into consideration. To the best of our knowledge, this is the first map-matching algorithm whose accuracy can be largely enhanced by the input of human knowledge; (3) we propose different query selection strategies to

effectively reduce the number of points that are required to be manually annotated. Compared with traditional approaches, our query selection strategies can
reduce the number of queries by up to 44%; and (4) we use both real world and
synthetic trajectory datasets to perform experiments and analyze the empirical results. The results demonstrate that HIMM can significantly reduce human
annotation cost.

In the remainder of this paper, we first discuss related work in Sect. 2, and
then introduce the problem definitions and the framework of our system in
Sect. 3. The map-matching algorithms and the query selection strategies are
described in Sects. 4 and 5, respectively. We evaluate our system in Sect. 6, and
conclude the paper in Sect. 7.

2 Related Work

Map-Matching. Existing map-matching algorithms can be categorized into
three types: geometric algorithms, topological algorithms, and statistical algorithms. Geometric algorithms [6] utilize spatial information to find local matches
for each point of the trajectory, thus the accuracy is highly affected by the measurement noises. Topological algorithms [2] consider both the connectivity and
contiguity of the road network as well as the geometric information, but the
accuracy is reduced when the sampling rate of the trajectory is low. Statistical
algorithms make use of advanced statistical models such as Kalman filter [10],
particle filter [7], and HMM [8,9,14], to find the global optimal path for the
trajectory. These algorithms are less sensitive to measurement noise and sampling rate, but the time complexity is high. To the best of our knowledge, none of
the existing map-matching algorithms takes feedbacks from human annotators to
improve accuracy or provides interactive mechanisms to facilitate map-matching.
More details are shown in Sect. 4.2.

Active Learning. Many learning tasks face a situation where unlabeled data
are easy to obtain but annotation is costly [12]. Active learning aims to minimize
the annotation cost by querying the most informative instances in the unlabeled
dataset. Although the scope of active learning is broad, most of the methods
are not suitable to the map-matching problem. For example, graph-based active
learning [1] focuses on using graph-based metrics to define the informativeness
of instances and querying the most informative instances. These approaches utilize link information with node-specific features or partial network structures to
improve the classification accuracy. Different from graph-based active learning,
the points of a trajectory in this paper are not part of the graph, but have
certain mapping relations with the edges of the graph. Another example is the
active learning algorithms for structured prediction tasks [13], which ignore the
annotation cost of a single structured object, but query the instances with the
highest joint uncertainty or utility. However, the map-matching task for a single
trajectory is costly, which cannot be disregarded. To the best of our knowledge, none of the existing active learning frameworks is designed for interactive
map-matching.

3 Overview

3.1 Preliminary

Definition 1 (Road Segment). *A road segment e is a directed polyline between two road intersections v_i and v_j, and there is no other road intersection on e. We denote $v_i \in e$ and $v_j \in e$.*

Definition 2 (Road Network). *A road network is a weighted directed graph $G = (V, E)$, where V is a set of road intersections (or vertices), and E is a set of road segments (or edges). The weight of a road segment is represented by its properties.*

A moving object is only allowed to travel on the road segments within the road network.

Definition 3 (Trajectory). *A trajectory T is a sequence of location points sampled from the GPS device of a moving object, denoted as $T = (p_1, p_2, \cdots, p_n)$. We say $p_i \in T$ for $i = 1, \cdots, n$ and $|T| = n$.*

A location point is represented by its latitude and longitude. The sampled location points on a trajectory may not be the actual locations of the moving object due to measurement inaccuracy.

Definition 4 (Path). *A path $P = (e_1, e_2, \cdots, e_n)$ is a sequence of road segments where e_i and e_{i+1} are connected for $i = 1, 2, \cdots, n-1$. Two road segments e_i and e_j are connected if there exists some intersection v such that $v \in e_i$ and $v \in e_j$.*

Definition 5 (Match). *Given a trajectory T and a road network $G = (V, E)$, a match $m_{i,j} = \langle p_i, e_j \rangle$ where $p_i \in T$ and $e_j \in E$ specifies point p_i was sampled when the object was moving on road segment e_j.*

Definition 6 (Map-Matching Query). *Given a trajectory T and a road network $G = (V, E)$, a map-matching query $Q(T, G)$ finds a path P, such that each point $p_i \in T$ is matched to exactly one road segment $e_j \in E$. The resulting set of matches is denoted as $M = \{\langle p_1, e_{j_1} \rangle, \cdots, \langle p_n, e_{j_n} \rangle\}$.*

Definition 7 (Interactive Map-Matching Query). *Given a trajectory T, a road network $G = (V, E)$, and a set of matches M' conducted by the annotator, an interactive map-matching query $Q(T, G, M')$ finds a new path P, such that each point $p_i \in T$ is matched to exactly one road segment $e_j \in E$. The resulting set of matches is denoted as M, where $M' \subseteq M$.*

3.2 Framework

Figure 1 shows the workflow of *HIMM*, our interactive map-matching system. First, an annotator requests a trajectory T to perform the map-matching task.

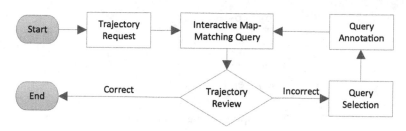

Fig. 1. The workflow of HIMM.

If T is not map-matched before, HIMM automatically generates a path for T using our interactive map-matching algorithm with $M' = \emptyset$. Then, HIMM plots the trajectory T along with the path onto the digital map for the annotator to review. In each iteration, if the annotator considers that the trajectory is not correctly map-matched, an unlabeled point p is selected for the annotator to review using our query selection strategy. If the annotator considers that p is not correctly matched, the annotator marks a correct match for p; otherwise, the annotator leaves the match as is. During the task, HIMM maintains a set of matches M' that are specified by the annotator. After receiving the feedbacks from the annotator, HIMM adds the match of p to M', and then performs an interactive map-matching query with M' to complete the iteration. Finally, if the annotator considers that all points are correctly map-matched, the map-matching task for the trajectory T terminates.

HIMM contains two major components shown in Fig. 1: (1) an interactive map-matching algorithm that takes the feedbacks of the annotator and automatically adjusts the map-matching results; and (2) a query selection strategy that recommends potentially mismatched points for the annotator to review. The details are introduced in Sects. 4 and 5, respectively.

4   Interactive Map-Matching

4.1   Map-Matching Model

As shown in Fig. 2, we model a map-matching query as a hidden Markov model, which is one of the most suitable models in this area [9,15]. Given a map-matching query $Q(T, G)$, trajectory T represents an observation sequence, where each point $p_i \in T$ is an observation, and each candidate road segment $e_{i,j} \in E$ represents a hidden state of p_i. The total number of points of a trajectory is denoted by n, where $n = |T|$, and the total number of hidden states of each point p_i is denoted by r, where $r = |E|$.

For each point p_i, each state $e_{i,j}$ has an *emission probability* denoted as $\Pr(e_{i,j}|p_i)$, which represents the likelihood of p_i being observed if the vehicle is on road segment $e_{i,j}$. A higher emission probability is associated to p_i if $e_{i,j}$ is closer to p_i, and the emission probability follows a Gaussian distribution of positioning measurement noise [9]:

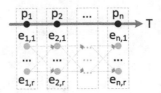

Fig. 2. The map-matching model of HIMM.

$$\Pr(e_{i,j}|p_i) = \frac{1}{\sqrt{2\pi\delta}} e^{-\frac{1}{2}\left(\frac{\text{pdist}(e_{i,j}, p_i)}{\delta}\right)^2} \tag{1}$$

where δ is the standard deviation of the positioning measurement noise, and pdist$(e_{i,j}, p_i)$ is the *minimum perpendicular distance* [3] between $e_{i,j}$ and p_i.

For each pair of consecutive points (p_i, p_{i+1}), each pair of candidate states $(e_{i,j_i}, e_{i+1,j_{i+1}})$ associated with them has a *transition probability* denoted as $\Pr(e_{i,j_i}, e_{i+1,j_{i+1}}|p_i, p_{i+1})$, which represents the likelihood for a vehicle moving from $e_{i,j}$ to $e_{i+1,j_{i+1}}$. e_{i,j_i} and $e_{i+1,j_{i+1}}$ are more likely to be matched to p_i and p_{i+1}, respectively, if the driving distance along e_{i,j_i} and $e_{i+1,j_{i+1}}$ from p_i to p_{i+1} is closer to the great circle distance between p_i and p_{i+1}; and the transition probability follows an exponential distribution:

$$\Pr(e_{i,j_i}, e_{i+1,j_{i+1}}|p_i, p_{i+1}) = \frac{1}{\beta} e^{-\frac{|\text{cdist}(p_i, p_{i+1}) - \text{route}(p_i, p_{i+1})|}{\beta}} \tag{2}$$

where β is the rate parameter [9], cdist(p_i, p_{i+1}) is the great circle distance between p_i and p_{i+1}, and route(p_i, p_{i+1}) is the driving distance along e_{i,j_i} and $e_{i+1,j_{i+1}}$ from p_i to p_{i+1}.

4.2 Interactive Map-Matching Algorithm

Recall that a map-matching query $Q(T, G)$ finds a path P, such that each point $p_i \in T$ is matched to exactly one road segment $e_j \in E$. Hence, the objective of the map-matching algorithm is to find a sequence of hidden states $P = (e_{1,j_1}, e_{2,j_2}, \cdots, e_{n,j_n})$ with the maximum joint probability $\Pr(P)$, where:

$$\Pr(P) = \prod_{i=1}^{n} \Pr(e_{i,j_i}|p_i) \times \prod_{i=1}^{n-1} \Pr(e_{i,j_i}, e_{i+1,j_{i+1}}|p_i, p_{i+1}) \tag{3}$$

Traditional hidden Markov model uses the Viterbi algorithm [11] to find the optimal solution, denoted as P^*. The Viterbi algorithm uses dynamic programming to quickly find the state sequence that maximizes $\Pr(P^*)$ in a recursive manner. Hence, if the annotator specifies a match $\langle p_i, e_{i,k}\rangle$ where $e_{i,k} \notin P^*$, the Viterbi algorithm will ignore such feedback of the annotator. Therefore, the traditional Viterbi algorithm cannot utilize the feedbacks of an annotator.

We propose an interactive map-matching algorithm based on the Viterbi algorithm to utilize the feedbacks of the annotator. The recursive formulation (a.k.a., *forward formulation* [11]) is defined as:

$$
C(i,j) = \begin{cases} O(i,j) & \langle p_i, e_{i,k} \rangle \in M', k = j \\ 0 & \exists \langle p_i, e_{i,k} \rangle \in M', k \neq j \\ \Pr(e_{i,j}|p_i) \times O(i,j) & \not\exists \langle p_i, e_{i,k} \rangle \in M' \end{cases} \qquad (4)
$$

where $1 \leq k \leq r$, and:

$$
O(i,j) = \begin{cases} 1 & i = 1 \\ \max_{1 \leq k \leq r} C(i-1,k) \Pr(e_{i-1,k}, e_{i,j}|p_{i-1}, p_i) & i > 1 \end{cases} \qquad (5)
$$

In Formula (4), $C(i,j)$ represents the highest value of the probabilities of state sequences $P_i = (e_{1,j_1}, e_{2,j_2}, \cdots, e_{i,j_i})$ for the first i observations $T_i = (p_1, p_2, \cdots, p_i)$ that have $e_{i,j}$ as the final state.

The recursion terminates when the last observation is processed. The optimal state sequence that results in $C(i,j)$ can be retrieved reversely from the last hidden state that results in the maximum $C(i,j)$ in each step through the following formulation (a.k.a. *backward formulation* [11]):

$$
e_{i,j_i} = \begin{cases} \arg\max_{1 \leq j \leq r} C(i,j) & i = n \\ \arg\max_{1 \leq j \leq r} \dfrac{C(i+1,j)}{\Pr(e_{i,j}, e_{i+1,j_{i+1}}|p_i, p_{i+1})} & 1 \leq i < n \end{cases} \qquad (6)
$$

Similar to the Viterbi algorithm, our interactive map-matching algorithm uses dynamic programming [11] to quickly find the optimal state sequence in a recursive manner. When the algorithm calculates the local optimal probability $C(i,j)$ for each hidden state $e_{i,j}$ as the final state for the first i observations $T_i = (p_1, p_2, \cdots, p_i)$, it first checks whether p_i is manually matched by the annotator. If so, the algorithm prunes all candidate hidden states of p_i except $e_{i,j}$, which is chosen by the annotator (i.e., $\langle p_i, e_{i,j} \rangle \in M'$) by modifying the emission probability $\Pr(e_{i,j}|p_i)$ to 1, and all other emission probabilities $\Pr(e_{i,k}|p_i)$ to 0, where $1 \leq k \leq r$ and $k \neq j$. Otherwise, the emission probability $\Pr(e_{i,j}|p_i)$ is set via Formula (1). This way, all $C(i,k)$ where $1 \leq k \leq r$ and $k \neq j$ are 0, and the backward formulation is guaranteed to select the state sequence that contains $e_{i,j}$ with respect to $C(i,j)$.

5   Query Selection Strategy

A good query selection strategy is critical to effectively guide the annotator by picking the points that are likely to be mis-matched. In this section, we propose four query selection strategies for comparison.

5.1 Distance-Based Strategy

A commonly used query selection strategy in active learning is uncertainty sampling. The basic idea is to query the instance whose label is the least certain. In the map-matching problem, the most straight-forward factor that reflects the uncertainty of a trajectory point p_i is the distance distribution from p_i to its candidate road segment set E_i. Consider the example shown in Fig. 3(a), p_1 and p_3 are clearly closer to e_1 and e_7, respectively. Thus, there is no need to check p_1 or p_3 since their labels are almost certain. However, the distances between p_2 and $e_2/e_3/e_6$ are quite similar, which makes p_2 the most uncertain point to be labeled.

A general strategy of uncertainty measurement in information theory is Shannon entropy [12], which represents the average amount of information generated by a probability distribution. As described in Sect. 4.1, the emission probability distribution of the candidate state set E_i of p_i reflects the distance distribution from p_i to each $e_{ij} \in E_i$. Thus, a distance-based strategy defines the uncertainty $H(p_i)$ of p_i as the Shannon entropy of the emission probability distribution of the candidate state set E_i of p_i. More specifically,

$$H(p_i) = - \sum_{j=1}^{r} \Pr(e_{ij}|p_i) \log(\Pr(e_{ij}|p_i)) \tag{7}$$

where r is the number of candidate states of p_i.

Based on Formula (7), given a trajectory T, in each iteration of the interactive map-matching process, the next point recommended for the annotator is:

$$p' = \arg\max_{p_i \in T} H(p_i) \tag{8}$$

After the annotator checks p_i, the system modifies $H(p_i)$ to $-\infty$, so that p_i will not be checked again until the interactive map-matching process terminates. Hence, the time complexity of the distance-based strategy is $O(1)$ with respect to the number of points on the trajectory n.

5.2 Confidence-Based Strategy

In the distance-based strategy, the emission probability distribution of a trajectory point p_i only considers the local information of each point on the trajectory. Consider the example shown in Fig. 3(b). The distances between p_2 and e_2/e_4 are similar, whereas p_3 is closer to e_5 than e_3. According to the distance-based strategy, p_2 has a higher uncertainty than p_3. However, if we consider the entire trajectory, p_3 is the point that mostly likely to be checked, because the optimal path differs a lot if p_3 is matched to e_3 or e_5. On the other hand, p_2 is not likely to be matched to e_4 considering the topological information ($(e_1, e_4, e_8, e_5, e_3)$ vs. (e_1, e_2, e_3)). Hence, in order to utilize the connectivity and contiguity of the road segments along each trajectory point, we define the *confidence* $\Pr(\langle p_k, e_{k,l} \rangle)$ for

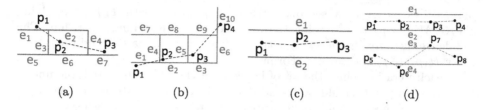

Fig. 3. Examples of the query selection strategy scenarios.

each candidate match $\langle p_k, e_{k,l} \rangle$ of p_k, and evaluate the uncertainty of p_k according to the Shannon entropy of the confidence distribution among all candidate matches for p_k. More specifically, given a trajectory T and a match $\langle p_k, e_{k,l} \rangle$, we generate a set of matches $M_{k,l}$ and a path $P_{k,l}$ by applying $Q(T, G, \{\langle p_k, e_{k,l} \rangle\})$ using the interactive map-matching algorithm. Thus,

$$\Pr(\langle p_k, e_{k,l} \rangle) = \Pr(P_{k,l}) = \prod_{i=1}^{n} \Pr(e_{i,j_i} | p_i) \times \prod_{i=1}^{n-1} \Pr(e_{i,j_i}, e_{i+1,j_{i+1}} | p_i, p_{i+1}) \quad (9)$$

where each $e_{i,j_i} \in P_{k,l}$.

Based on Formula (9), the uncertainty $H(p_i)$ of p_i is:

$$H(p_i) = -\sum_{j=1}^{r} \Pr(\langle p_i, e_{i,j} \rangle) \log(\Pr(\langle p_i, e_{i,j} \rangle)) \quad (10)$$

where r is the number of candidate states of p_i.

The next point recommended in each iteration is similar to the distance-based strategy with the uncertainty computed in Formula (10). The time complexity is $O(1)$ with respect to the number of points on the trajectory n.

5.3 Dynamic Confidence-Based Strategy

In the confidence-based strategy, the confidence for each candidate match of a point is defined under the assumption that all the other points of the trajectory are not map-matched. Consider the example shown in Fig. 3(c), where p_1, p_2 and p_3 are close to both e_1 and e_2. According to the confidence-based strategy, the probabilities of the paths that pass either e_1 or e_2 are similar, thus the uncertainty of p_1, p_2 and p_3 are similar. If there is another point of the trajectory that is wrongly matched but has a lower uncertainty, p_1, p_2 and p_3 will all be checked. However, if we have confirmed that p_2 is matched to e_2, the labels of p_1 and p_3 are no longer uncertain. Hence, the confidence-based method cannot prune the case when the match of a point is constrained by other points of the trajectory.

To deal with such situations, we propose a dynamic confidence-based strategy utilizing the set of labeled matches M' maintained by the system. More specifically, given a trajectory T, a set of labeled matches M', and a match

$\langle p_k, e_{k,l} \rangle$, we generate a set of matches $M_{k,l}$ and a path $P_{k,l}$ by applying $Q(T, G, M' \cup \{\langle p_k, e_{k,l} \rangle\})$ using the interactive map-matching algorithm. The confidence measure $\Pr(\langle p_k, e_{k,l} \rangle)$ is then defined as Formula (9), and the uncertainty $H(p_i)$ of p_i is defined as Formula (10).

In each iteration, since the set of labeled matches M' is updated, the uncertainty of each point should be re-calculated via Formula (10). Hence, the time complexity of the dynamic confidence-based strategy $O(n \times r)$, where n is the number of points on the trajectory, and r is the number of hidden states of each point. In practice, we restrict r in order to ensure high efficiency, which will be explained in Sect. 6.1. The next point recommended in each iteration is similar as the confidence-based strategy.

5.4   Stability-Based Strategy

Another strategy of uncertainty measurement is *stability*. Based on our observation, if the match of a point is frequently influenced by other points on the trajectory (i.e., the match of this point is not *stable*), the uncertainty of this point is often high. Consider the example shown in Fig. 3(d), p_1 has a similar probability to be matched to e_1 or e_2, and p_5 has a similar probability to be matched to e_3 or e_4. Therefore, the uncertainty of p_1 and p_5 are similar. However, since p_6 is much closer to e_3 while e_7 is much closer to e_4, there must be a large measurement noise for either of these two points. Hence, the match of p_5 is very unstable if either p_6 or p_7 is removed from the trajectory. On the contrary, p_2, p_3, p_4 are all slightly closer to e_1. Thus the match of p_1 is much more stable than p_5. In this case, since p_1 and p_5 belong to the same trajectory, the priority of checking p_5 is higher than p_1, because there is a higher probability that there exists a major measurement noise for the points around p_5.

In order to define the stability of a point, we first define the *influence* between two points. Given a trajectory $T = (p_1, p_2, \cdots, p_n)$ and two points $p_a, p_b \in T$, we first generate a set of matches M for T using our interactive map-matching algorithm. Next we generate another trajectory T_b omitting p_b, so that $T_b = (p_1, p_2, \cdots, p_{b-1}, p_{b+1}, \cdots, p_n)$, and then similarly obtain M_b for T_b. Suppose $\langle p_a, e_a \rangle \in M$, we denote that p_a is influenced by p_b as $p_a \prec p_b$ if and only if $\langle p_a, e_a \rangle \notin M_b$. Given a point $p_i \in T$, the stability $H(p_i)$ of p_i is:

$$S(p_i) = |D_{p_i}| \tag{11}$$

where D_{p_i} is the set of points that have no influence on p_i. More specifically, $D_{p_i} = \{p_1, p_2, ..., p_k\}$, where $p_j \in T$ and $p_i \nprec p_j$ for all $p_j \in D_{p_i}$.

Based on Formula (11), given a trajectory T, in each iteration of the interactive map-matching process, the next point recommended is:

$$p' = \arg \min_{p_i \in T} S(p_i) \tag{12}$$

The stability-based strategy is efficient for selecting problematic points. However, since $S(p_i)$ is defined as a cardinality rather than a probability, it is possible that the points of a trajectory have the same $S(p_i)$. In this case, the stability-based method is not able to determine an efficient order for these points. To deal with this case, we dynamically switch to dynamic confidence-based strategy in each iteration if two points have the same $S(p_i)$. Hence, the time complexity of the stability-based strategy is between $O(n)$ and $O(n \times r)$. Similar to dynamic confidence-based strategy, we restrict r to ensure high efficiency.

6 Evaluation

6.1 Experiment Setup

Experiment Environment. The experiments are conducted on a Linux server with a CPU of Intel Core i5-4590 and 8 GB memory. The operating system is Ubuntu 14.04, and the code is written in Python 2.7.6.

Road Network. In our experiments, the road network data is provided by the government of a large city in China and consisted of 25,613 intersections and 36,451 road segments. There are no direction information thus all the road segments are bi-directional.

Synthetic Trajectory Data. We build a trajectory generator to generate synthetic trajectories with the following parameters: (1) the number of points n_p on the trajectory, (2) the number of road segments n_e covered by the trajectory, and (3) the standard deviation δ of the positioning *measurement noise*.

The trajectory generator selects a starting point $p_1 \in e_1$ and an ending point $p_{n_p} \in e_{n_e}$ from an n_e-hop random path $P = (e_1, e_2, \ldots, e_{n_e})$, and then computes the *distance interval* Δ between two consecutive points:

$$\Delta = \frac{\text{route}(p_1, p_{n_p})}{n_p - 1} \tag{13}$$

where $\text{route}(p_1, p_{n_p})$ is the driving distance between p_1 and p_{n_p}. Starting from p_1, the trajectory generator derives the locations of the remaining points along P such that $\text{route}(p_1, p_{i+1}) = \text{route}(p_1, p_i) + \Delta$. Finally, the generator adds a Gaussian distributed measurement noise to each point with δ so that $T = (\delta(p_1), \delta(p_2), \ldots, \delta(p_{n_p}))$.

In our experiments, we study the impact of different parameters including: (1) the *number of points* on a trajectory; (2) the *initial accuracy* of a trajectory in terms of the number of points that are matched to the correct road segments by comparing $Q(T, G, \emptyset)$ with P; (3) the *sampling rate* of a trajectory which is represented by n_e given a fixed average driving speed; and 4) the standard deviation of the *measurement noise*.

Real Trajectory Data. Our real world trajectory data contains 154 million records of 15,231 vehicles for 26 days [4].

Evaluation Metrics. To evaluate the effectiveness of our interactive map-matching algorithm, we simulate the work flow of HIMM with the assumption that each query annotation is correct and the time of trajectory review is trivial. In order to reduce the response time, for each point on the trajectory, we empirically reserve the top 10 nearest road segments as its candidate set. In our experiments, it is sufficient to produce a high percentage of the correct matches within this range. The correct road segment is also added into the candidate set in case it is excluded.

We use two metrics to evaluate the efficiency of our query selection strategies: (1) the response time for a query selection strategy; and (2) the execution time of an interactive map-matching task for a single trajectory. In addition, we define three metrics to evaluate the effectiveness of our query selection strategies. Given a trajectory T along with the initial path generated by HIMM, the number of mis-matched points is denoted as $\zeta(T)$. After the map-matching task for T is terminated, the total number of points that are reviewed by the annotator is denoted as $\eta(T)$, and the total number of points that are corrected by the annotator is denoted as $\psi(T)$. The three metrics are defined as follows: (1) ***cost ratio*** $\mathtt{CR} = \eta(T)/|T|$ representing the review cost of the interactive map-matching task; (2) ***selection accuracy*** $\mathtt{SA} = \psi(T)/\eta(T)$ representing the accuracy of selecting mis-matched points; and (3) ***true negative rate*** $\mathtt{TNR} = \psi(T)/\zeta(T)$ representing the ratio of the corrections conducted by the annotator rather than the interactive map-matching algorithm. A lower $\mathtt{CR}$ indicates fewer iterations for a map-matching task; a higher $\mathtt{SA}$ indicates a higher rate of selecting mis-matched points; and a lower $\mathtt{TNR}$ indicates more points are automatically corrected by our interactive map-matching algorithm.

6.2   Experiment Results

The experiments are conducted on both synthetic and real data. For each data set, we apply all the query selection strategies proposed in this paper: *distance-based strategy* (DIST), *confidence-based strategy* (CONF), *dynamic confidence-based strategy* (D-CONF), and *stability-based strategy* (STAB); as well as two baselines: *sequential strategy* (SEQ) and *random strategy* (RAND), where the annotator checks each point along the trajectory in a sequential and random order, respectively.

Efficiency. In order to evaluate the scalability of our query selection strategies, we conduct experiments on 8 groups of trajectories whose numbers of points range from 10 to 80. The initial accuracy is fixed within 60–70%. The sampling rate and measurement noise are fixed to 1.5 min and 101.04 m, respectively.

Figure 4(a) shows the impact of number of points on response time. Consistent with the analysis in Sects. 5.3 and 5.4, the response time of D-CONF and STAB increases linearly with the number of points. Figure 4(b) shows the impact of number of points on TNR, where TNR drops at first when the number of points grows, but rises after the number of points reaches 50. Therefore,

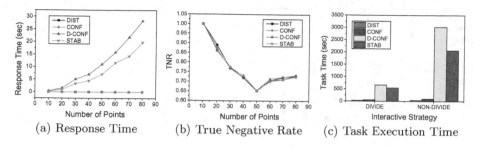

(a) Response Time   (b) True Negative Rate   (c) Task Execution Time

Fig. 4. Performance of query selection strategies.

HIMM achieves the best performance when the number of points on the query selection trajectory is 50. Hence, for long trajectories, the most efficient strategy is to divide them into sub-trajectories with 50 points each, and perform an interactive map-matching task for each sub-trajectory. In order to show the effectiveness of this dividing strategy, we generate a group of trajectories consisting of 100 points each, and then compare the average execution time of the interactive map-matching task for each trajectory with or without using the dividing strategy. As a result, Fig. 4(c) shows that dividing long trajectories into sub-trajectories significantly reduces the task time.

Effectiveness on Synthetic Data. To study the impact of initial accuracy, we generate 5 groups of trajectories with 5 categories of initial accuracy: 50–60%, 60–70%, 70–80%, 80–90%, and 90–100%. The sampling rate and measurement noise are fixed to 1.5 min and 101.04 m respectively. To study the impact of sampling rate, we generate 3 groups of trajectories with 3 categories of sampling rates: 0.5, 1.5, and 4.5 min. The initial accuracy and measurement noise are fixed to 70–80% and 11.23 m respectively. To study the impact of measurement noise, we generate 3 groups of trajectories with 3 categories of measurement noises: 11.23, 33.68, and 101.04 m. The initial accuracy and sampling rate are fixed to 60–70% and 0.5 min respectively.

The experiment results on synthetic trajectory data are shown in Fig. 5. In general, the performance (in terms of CR and SA) of our query selection strategies (DIST, CONF, D-CONF, and STAB) achieve a much higher efficiency than the two baseline strategies (SEQ and RAND). Among our query selection strategies, the performance of D-CONF is better than the two global strategies (DIST and CONF), and STAB outperforms the other three strategies. Compared with the two baseline strategies, CR reduced by our query selection strategies is up to 44%, and SA is improved up to 24%.

Moreover, the TNR results in Fig. 5 show that the percentage of mis-matched points that are automatically corrected by the interactive map-matching algorithm during human annotation is up to 59%, which indicates a significant reduction of the annotation cost.

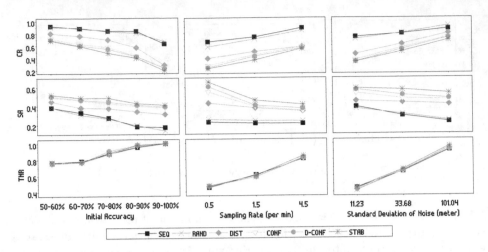

Fig. 5. Effectiveness of query selection strategies on the synthetic trajectory data.

Next, we discuss the impact of each parameter on the performance of our query selection strategies. Firstly, we observe that the gaps of CR and SA between baseline strategies and our query selection strategies enlarge when initial accuracy grows. This indicates that our query selection strategies are more effective in picking out wrongly matched points when the initial accuracy is high. Meanwhile, TNR decreases when the initial accuracy falls for all the query selection strategies. This indicates that the ratio of the automatic corrections triggered by human annotation rises when the initial accuracy is low, which also saves the annotation cost. In conclusion, HIMM can reduce $\eta(T)$ no matter the initial map-matching accuracy is low or high.

Secondly, we observe that a larger sampling rate or measurement noise will hurt the performance both in CR and SA for all the query selection strategies. However, compared with the baseline strategies, our query selection strategies are more sensitive to sampling rate, but less sensitive to measurement noise. This is because a larger sampling rate reduces the topological correlations between points, thus the advantage of our query selection strategies is less effective. In contrast, a larger measurement noise only increases the deviation of each point within its local area rather than the topological information, thus the advantage of our query selection strategies remains. As a result, our query selection strategies outperform the baseline strategies in most of the cases.

Effectiveness on Real Data. Since the cost of a manual map-matching task is very high, due to the limit of time, we manually processed 200 trajectories with 50 points each. We use HIMM to annotate these trajectories, and record the resulting paths as the ground truth. For the experiments on real trajectory data, based on our statistics, the initial accuracy is 89% on average; the sampling rate ranges from 30 s to 5 min; and the measurement noise is around 33.68 m.

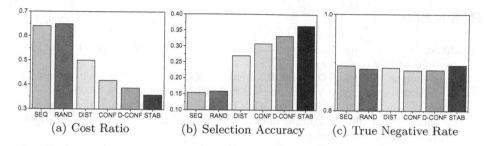

(a) Cost Ratio   (b) Selection Accuracy   (c) True Negative Rate

Fig. 6. Effectiveness of query selection strategies on the real trajectory data.

The experiment results on real trajectory data are shown in Fig. 6. It is clear that the effectiveness of our query selection strategies are much higher than the baseline strategies in terms of both CR and SA. Similar to the experiments on synthetic trajectory data, the performance of STAB is the best, which reduces 29% of CR and improves 21% of SA compared with baseline strategies. Moreover, 12% mis-matched points are automatically corrected by the interactive map-matching algorithm, which is a satisfactory result for such a high initial accuracy.

In general, the performance of HIMM on the real trajectory data is similar to that on the synthetic trajectory data, which indicates that HIMM achieves a satisfactory performance on a wide range of trajectories, and significantly reduces the annotation cost.

7   Conclusion

In this paper, we propose an interactive map-matching system called HIMM for the annotators to perform effective interactive map-matching tasks. We design and implement an interactive map-matching algorithm that can be improved by manual annotations, and propose four different query selection strategies to reduce the costs of interactive map-matching tasks. We conduct intensive experiments on both synthetic and real trajectory data. The results show that our query selection strategies achieve a satisfactory performance. In this paper, we only consider single annotators for interactive map-matching tasks, and it could be further discussed when multiple annotators and crowd-sourcing are introduced, which will be our future work.

Acknowledgments. This work is supported in part by NSFC Grant 61300030 and the National Key Basic Research and Development Program of China (973) Grant 2014CB340304.

References

1. Bilgic, M., Mihalkova, L., Getoor, L.: Active learning for networked data. In: Proceedings of ICML, pp. 79–86 (2010)
2. Brakatsoulas, S., Pfoser, D., Salas, R., Wenk, C.: On map-matching vehicle tracking data. In: Proceedings of VLDB, pp. 853–864. VLDB Endowment (2005)
3. Ding, Y., Liu, S., Pu, J., Ni, L.M.: HUNTS: a trajectory recommendation system for effective and efficient hunting of taxi passengers. In: Proceedings of MDM, vol. 1, pp. 107–116. IEEE (2013)
4. Ding, Y., Zheng, J., Tan, H., Luo, W., Ni, L.M.: Inferring road type in crowd-sourced map services. In: Bhowmick, S.S., Dyreson, C.E., Jensen, C.S., Lee, M.L., Muliantara, A., Thalheim, B. (eds.) DASFAA 2014. LNCS, vol. 8422, pp. 392–406. Springer, Cham (2014). doi:10.1007/978-3-319-05813-9_26
5. Jagadeesh, G., Srikanthan, T., Zhang, X.: A map matching method for GPS based real-time vehicle location. J. Navig. **57**(03), 429–440 (2004)
6. Karimi, H.A., Conahan, T., Roongpiboonsopit, D.: A methodology for predicting performances of map-matching algorithms. In: Carswell, J.D., Tezuka, T. (eds.) W2GIS 2006. LNCS, vol. 4295, pp. 202–213. Springer, Heidelberg (2006). doi:10.1007/11935148_19
7. Liao, L., Patterson, D.J., Fox, D., Kautz, H.: Learning and inferring transportation routines. Artif. Intell. **171**(5), 311–331 (2007)
8. Lou, Y., Zhang, C., Zheng, Y., Xie, X., Wang, W., Huang, Y.: Map-matching for low-sampling-rate GPS trajectories. In: Proceedings of SIGSPATIAL, pp. 352–361. ACM (2009)
9. Newson, P., Krumm, J.: Hidden Markov map matching through noise and sparseness. In: Proceedings of SIGSPATIAL, pp. 336–343. ACM (2009)
10. Pink, O., Hummel, B.: A statistical approach to map matching using road network geometry, topology and vehicular motion constraints. In: Proceedings of ITSC, pp. 862–867. IEEE (2008)
11. Rabiner, L.R.: A tutorial on hidden Markov models and selected applications in speech recognition. Proc. IEEE **77**(2), 257–286 (1989)
12. Settles, B.: Active learning literature survey, vol. 52, no. 55–66, p. 11. University of Wisconsin, Madison (2010)
13. Settles, B., Craven, M.: An analysis of active learning strategies for sequence labeling tasks. In: Proceedings of Empirical Methods in Natural Language Processing, pp. 1070–1079. Association for Computational Linguistics (2008)
14. Wang, G., Zimmermann, R.: Eddy: an error-bounded delay-bounded real-time map matching algorithm using HMM and online Viterbi decoder. In: Proceedings of SIGSPATIAL, pp. 33–42. ACM (2014)
15. Xue, A.Y., Qi, J., Xie, X., Zhang, R., Huang, J., Li, Y.: Solving the data sparsity problem in destination prediction. VLDB J. **24**(2), 219–243 (2015)

HyMU: A Hybrid Map Updating Framework

Tao Wang, Jiali Mao, and Cheqing Jin[✉]

School of Data Science and Engineering,
School of Computer Science and Software Engineering,
East China Normal University, Shanghai, China
{toy_king,jlmao1231}@stu.ecnu.edu.cn, cqjin@sei.ecnu.edu.cn

Abstract. Accurate digital map plays an important role in mobile navigation. Due to the ineffective updating mechanism, existing map updating methods cannot guarantee completeness and validity of the map. The common problems of them involve huge computation and low precision. More importantly, they scarcely consider inferring new roads on sparse unmatched trajectories. In this paper, we first address the issue of finding new roads in sparse trajectory area. On the basis of sliding window model, we propose a two-phase hybrid framework to update the digital map with inferred roads, called HyMU, which takes full advantage of line-based and point-based strategies. Through inferring road candidates for consecutive time windows and merging the candidates to form missing roads, HyMU can even discover new roads in sparse trajectory area. Therefore, HyMU has high recall and precision on trajectory data of different density and sampling rate. Experimental results on real data sets show that our proposal is both effective and efficient as compared to other congeneric approaches.

1 Introduction

With the widespread use of onboard navigators and smart phones, the accuracy of navigation map has aroused universal concern. An inaccurate road map with disconnected and misaligned roads may make the experienced drivers get lost and even cause traffic accidents. Essentially, the accuracy and completeness of a digital map depend on whether road information is updated timely and effectively. However, such a task is difficult to achieve due to two factors, one is the rapid development of road construction, and the other is the ineffective map updating mechanism. Specifically, rapid construction of roads has increased great difficulties to timely update of digital map. Massive amount of roads all over the world change every year. According to the reports by the Ministry of Transport of China[1], 4,500 km of new expressways will be built in China, and 29 road projects will be pushed forward in Shanghai in 2016. While at the same time, existing techniques cannot guarantee the timeliness of map updating. The commercial map companies update digital map by periodically conducting geological survey of the entire road network. To cut down the overall cost, survey

---

[1] http://www.chinahighway.com/.

© Springer International Publishing AG 2017
S. Candan et al. (Eds.): DASFAA 2017, Part II, LNCS 10178, pp. 19–33, 2017.
DOI: 10.1007/978-3-319-55699-4_2

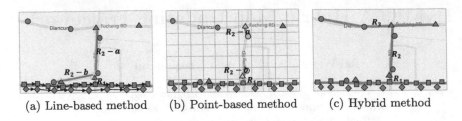

(a) Line-based method   (b) Point-based method   (c) Hybrid method

Fig. 1. An example of point-based and line-based method

period is quite long and thus the map updating rate lags far behind the construction of new roads. Alternative mechanism is to adopt the crowdsourced map project to generate customized map (e.g. OpenStreetMap), but it largely depends on the geographic data directly provided by volunteers. As a result, the amount of users and even the editing skills of users greatly influence the quality of map updating. As mentioned above, it is desirable to devise a low-cost but high reliable map updating mechanism.

Huge amount of trajectory data of vehicles can be applied to update the map. Recently, a few researches have been done in map updating with trajectories [13,16,18], and they can be grouped into two classes: line-based strategy (e.g., CrowdAtlas [16]) and point-based strategy (e.g., Glue [18] and COBWEB [13]). To be specific, the former is to infer the missing roads for a given map based on clustering considerable volume of unmatched trajectory segments, and the latter on massive unmatched trajectory points. These methods still face a series of problems, such as high computational overhead, low accuracy of inferred roads, and bad timeliness of map updating, etc. Moreover, line-based strategy has poor performance in processing low-sampling data (sampling interval longer than 30 s [18]), because it may infer the roads with false directions when the line segments cross over several roads. Although point-based methods can overcome this issue, they easily infer some short road segments rather than long roads due to the low coverage caused by point-based clustering. As shown in Fig. 1(a), two consecutive sampling points that are located on two roads are connected as a line segment, and accordingly an incorrect road R_2-b is inferred by line-based strategy. Though point-based strategy solves the above deficiencies, it infers two short road segments, R_2-a and R_2-b, instead of a long road that covers them, as illustrated in Fig. 1(b). Thus, the inferred roads in Fig. 1(a) and (b) are incorrect. To improve the inferring accuracy and obtain the ideal result in Fig. 1(c), it necessitates a hybrid framework to integrate virtues of both line-based and point-based strategies.

Furthermore, the aforementioned map updating mechanisms focus on discovering the missing roads on trajectory data of dense areas. They usually define a threshold of minimum clustering quantity standard, and cluster unmatched trajectory line segments (or points) to infer new roads only when satisfying a specific threshold. Hence, for the top road region with sparse positional points in Fig. 1(a) and (b), both line-based and point-based strategies cannot infer the road R_3 in Fig. 1(c). Actually, we can obtain two insights from the observation of trajectories. When the new roads first come into service, relatively few vehicles will drive along them and thus the track data are more sparse than that of normal roads. Distinct from noisy data, sparse trajectories appear on such roads in many days, i.e., the amount of trajectories will not increase tremendously in a short time period. If simply lowering the threshold of aforementioned methods, noisy data may also be clustered and some incorrect roads can be inferred. Thus, both methods are not tailored to inferring new roads in sparse trajectory area. Given the two insights above, on the basis of sliding window model, we propose a two-phase road inferring framework, including candidate generation and missing roads inferring, called HyMU. Additionally, we employ a hybrid scheme to enhance the accuracy of map updating by integrating line-based and point-based strategies. Specifically, the contributions of this paper are summarized below.

- We first address the issue of new roads inferring on sparse trajectory data to improve the overall inferring precision.
- Based on the sliding window model, we take full advantage of line and point-based strategies, and propose a two-phase hybrid framework to update the map, called HyMU.
- We compare our proposal with other congeneric approaches by conducting substantial experiments on real data sets. Experimental results show that HyMU method has good inferring performance on trajectory data under different sampling rate and density.

The remainder of this paper is organized as follows. Section 2 reviews the most related work. In Sect. 3, the preliminary concepts are introduced and the problem is defined formally. In Sect. 4, we outline and analytically study the details of HyMU framework. In Sect. 5, a series of experiments are conducted on real datasets to evaluate our proposal. Finally, we briefly conclude this article in Sect. 6.

2   Related Work

In this section, we briefly conduct a systematic review over the related work in two relevant areas: map inference and map updating.

2.1   Map Inference

Based on the track data set or satellite images, map inference aims to infer the entire road map. Image processing technology is mainly applied to infer map from satellite images [10,12]. But it is costly to obtain high-resolution satellite images data. Therefore, most researches on map inference are based on the trajectories of vehicles and they can be divided into three classes: K-means [1,5,11], KDE algorithm [2,4], and trace merging algorithm [3,8]. Nevertheless, most approaches have poor performance in handling trajectory data with excessive random noise, nonuniform distribution, and uneven sampling rate. Additionally, they are too time-consuming to fit online map inference.

2.2 Map Updating

Compared with time overhead of inferring the whole map, simply adding or modifying the roads for a given map is more realistic. Map updating methods are to discover missing roads to update a given map based on unmatched trajectories, including CrowdAtlas [16], Glue [18] and COBWEB [13]. CrowdAtlas consists of four stages: trajectory clustering, centerline fitting, connection and iteration. When the number of unmatched trajectory segments reaches the specified quantity criterion, CrowdAtlas implements clustering and polyline fitting functions to generate the centerlines that represent new roads. However, CrowdAtlas may infer new roads with false directions when unmatched trajectory segments cross over two or more roads, as illustrated in Fig. 1(a). In addition, CrowdAtlas obtains poor accuracy when dealing with low sampling rate data. To improve the precision of inferred roads on low sampling rate trajectory data, Glue clusters the unmatched trajectory points to infer new roads. Similarly, COBWEB organizes the GPS points using a Cobweb data structure and reduces the vertices and edges from Cobweb to generate Road-Tree, and finally finishes map updating. Nevertheless, both Glue and COBWEB cluster unmatched trajectory points and easily infer incomplete road segments instead of intrinsically long roads, as illustrated in Fig. 1(b). Moreover, all the above mentioned approaches cannot infer the new roads based on sparse trajectory data. Thus, it necessitates devising a map updating mechanism with high precision and noise tolerance to online infer the new roads on trajectory data of various sampling rate and density.

3 Problem Definition

In this section, we introduce preliminary concepts, and formally define the problem of map updating upon trajectory data.

A complete digital map contains road type, geometry, turn restriction, speed limit, etc. We aim to find the roads that have not been marked on the map. The road network G that corresponds to the map is defined as follows.

Definition 1 (Road Network). *A road network is denoted by a graph $G = (V, E)$, where V is a set of vertexes and E refers to a set of edges. Each edge $e \in E$ represents a road segment.*

To infer the missing roads, we need to cope with the continuously arrived trajectories. The trajectory of an object that consists of a series of points is defined below.

Definition 2 (Trajectory). *The trajectory of a moving object, denoted as Tr, consists of a sequence of points, $(p_1, t_1), (p_2, t_2), \cdots$, where p_i is the position at t_i. Such records arrive in chronological order, i.e., $\forall i < j$, $t_i < t_j$. A trajectory segment is a line segment between two adjacent trajectory points, which is denoted as $Ts = (p_i, p_{i+1})$.*

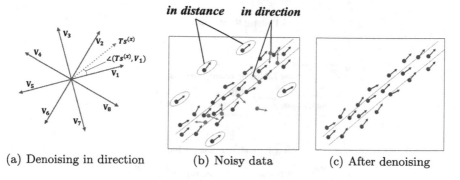

(a) Denoising in direction   (b) Noisy data   (c) After denoising

Fig. 2. An example of denoising in distance and direction (Color figure online)

Trajectory data are collected in real-time with massive scale. In order to describe the portions of trajectories in different time periods, we employ the sliding window model, and a trajectory in a time window is denoted as Tw. Given a window size N, the window range at timestamp t_0 is $(t_0, t_0 + N)$. Hereafter, we infer the missing road candidates based on the trajectories in each time window.

Due to different resolutions of various GPS-enabled equipments and city canyon surrounded by high-rise buildings, trajectory data are noisy. According to our observation, noisy data often behave abnormally in direction or distance relative to its neighborhood. The neighborhood of a trajectory segment $Ts^{(x)}$ is defined as follows.

Definition 3 (Trajectory Segment Neighborhood). *Given a trajectory segment $Ts^{(x)}$, a distance threshold th_{dis}, and a set of trajectory segments TS, if we denote $dist(Ts^{(x)}, Ts^{(y)})$ as the shortest Euclidean distance between any two points in two line segments, the neighborhood of $Ts^{(x)}$ is defined as follows:*

$$Nd(Ts^{(x)}) = \{Ts^{(y)} \in TS | dist(Ts^{(x)}, Ts^{(y)}) \leq th_{dis}\}$$

Correspondingly, the neighborhood of a trajectory point p_i is denoted as $Nd(p_i)$, which represents the set of points that their distances to p_i are within a distance threshold th_{dis}. Subsequently, we define noisy trajectory segment as below.

Definition 4 (Noisy Trajectory Segment). *Given a trajectory segment $Ts^{(x)}$ and the directions' distribution of its surrounding segments $U(Ts^{(x)})$, $Ts^{(x)}$ is noisy if $Nd(Ts^{(x)})$ is empty or the direction of $Ts^{(x)}$ does not tally with the top-k most popular directions of its surrounding segments.*

We take the starting point of $Ts^{(x)}$ as center and the length of $Ts^{(x)}$ as radius of a circle, and generate a region. For example, in Fig. 2(a), we divide the region into 8 pieces representing 8 sector [6]. The distribution is represented as below.

$$U(Ts^{(x)}) = (C_1, C_2, C_3, C_4, C_5, C_6, C_7, C_8)$$

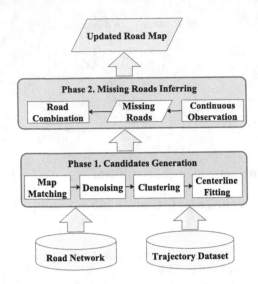

Fig. 3. The framework of HyMU

where C_i records the number of trajectory segments of $Nd(Ts^{(x)})$ that belong to the ith direction. Considering that each road usually has at least two lanes with opposite directions, we decide whether $Ts^{(x)}$ is a noisy trajectory segment by calculating whether $Ts^{(x)}$ belongs to the top two most popular directions. For example, in Fig. 2(a), for a trajectory segment $Ts^{(x)}$, we can determine which direction the trajectory segment $Ts^{(x)}$ belongs to according to the angle range between $Ts^{(x)}$ and V_1, denoted as $\angle(Ts^{(x)}, V_1)$.

Besides, each inferred road is represented by a road centerline.

Definition 5 (Road Centerline). *A road centerline, denoted as Rc, is represented by a polyline. It consists of a sequence of continuous positional points,* $(p_1, p_2, \ldots, p_n)$, *where p_i is the geographical position.*

Finally, we summarize the problem as below.

Given a road network G and a set of trajectories in different time periods, our goal is to infer the missing roads as early as possible, and then update the road network G by using the inferred missing roads.

4   Framework

In this section, we introduce a novel framework, which is called Hybrid Map Updating (HyMU). HyMU is to identify missing roads based on trajectory data. As shown in Fig. 3, HyMU is mainly composed of two phases: *candidates generation* and *missing roads inferring*. During the first phase, we obtain the unmatched trajectories in each time window by map matching, distance denoising and direction denoising. Then, through clustering and centerline fitting on

---

Algorithm 1. Candidate Generation

   Input: A trajectory set TwS in current time window
   Output: Line-based candidate set RC_l and point-based candidate set RC_p
1 $RC_l \leftarrow \emptyset$; $RC_p \leftarrow \emptyset$; //line-based and point-based candidate set
2 $TuS \leftarrow \emptyset$; //unmatched trajectory segment set
3 **foreach** *trajectory* $Tw^{(i)}$ *in* TwS **do**
4 | $Tu \leftarrow MapMatching(Tw^{(i)})$; //unmatched trajectory segments
5 |__ $TuS \leftarrow TuS \cup Tu$;
6 **foreach** *trajectory segment* $Ts^{(x)}$ *in* TuS **do**
7 | **if** $Ts^{(x)}$ *is noisy trajectory segment* **then**
8 |__ |__ $TuS \leftarrow TuS \setminus \{T_s^{(x)}\}$;
9 $CS_l \leftarrow LClustering(TuS)$; //line-based clustering
10 $CS_p \leftarrow PClustering(TuS)$; //point-based clustering
11 **foreach** *cluster* $CS_l^{(i)}$ *in* CS_l **do**
12 |__ $RC_l \leftarrow RC_l \cup CLFitting(CS_l^{(i)})$; //line-based candidate generation
13 **foreach** *cluster* $CS_p^{(j)}$ *in* CS_p **do**
14 |__ $RC_p \leftarrow RC_p \cup CLFitting(CS_p^{(j)})$; //point-based candidate generation
15 **return** RC_l and RC_p;

---

the unmatched and denoised trajectories, we derive the road candidates in each time window. During the second phase, we combine the candidates of multiple time windows via continuous observation. When the number of hybrid candidates related to a certain road reaches the threshold k, they will be merged to form a missing road. Finally, through road combination, we update the road network with inferred roads. Note that our hybrid framework integrates the advantages of line-based and point-based strategies, including high coverage and greater precision of inferred roads.

4.1 Candidate Generation

As shown in Algorithm 1, the candidate generation phase involves *map matching* (at lines 3–5), *denoising* (at lines 6–8), *clustering* (at lines 9–10) and *centerline fitting* (at lines 11–14). First, the trajectories in each time-window are matched with the road network to obtain unmatched trajectories. Then, after denoising, the denoised and unmatched trajectories are grouped into clusters using both line-based and point-based clustering methods. Finally, each cluster is fitted into a polyline that represents a road candidate through centerline fitting.

Map Matching. The purpose of map matching is to match the GPS trajectories to the right roads. Commonly used map matching can be divided into two categories: incremental approach [9,15], which aims to select the best matching candidate only on the basis of the preceding observations; global methods [14,17], which is to observe the entire series to select the best candidate. The Fast Viterbi [17], one of the most popular map matching methods, has been

---

Algorithm 2. LClustering

Input: Trajectory segment set TuS, a threshold th_c
Output: Cluster set CS

1 $CS \leftarrow \emptyset$; $l \leftarrow 1$;
2 **foreach** *unvisited segment* $Ts^{(i)}$ *in* TuS **do**
3 $\quad$ Mark $Ts^{(i)}$ as visited;
4 $\quad$ **if** $Nd(Ts^{(i)}) > th_c$ **then**
5 $\quad\quad$ $C_l \leftarrow \{Ts^{(i)}\}$; $Q \leftarrow \emptyset$; $Q.enqueue(Ts^{(i)})$;
6 $\quad\quad$ **while** Q *is not empty* **do**
7 $\quad\quad\quad$ $Ts^{(x)} \leftarrow Q.dequeue()$;
8 $\quad\quad\quad$ **foreach** *segment* $Ts^{(y)}$ *in* $Nd(Ts^{(x)})$ **do**
9 $\quad\quad\quad\quad$ Mark $Ts^{(y)}$ as visited;
10 $\quad\quad\quad\quad$ **if** $Nd(Ts^{(y)}) > th_c$ **and** $Ts^{(x)}$ *and* $Ts^{(y)}$ *are similar* **then**
11 $\quad\quad\quad\quad\quad$ $Q.enqueue(Ts^{(y)})$;
12 $\quad\quad\quad\quad$ **if** $Ts^{(y)}$ *does not belong to any cluster* **then**
13 $\quad\quad\quad\quad\quad$ $C_l \leftarrow C_l \cup \{Ts^{(y)}\}$;
14 $\quad\quad$ $CS \leftarrow CS \cup \{C_l\}$; $l \leftarrow l + 1$;
15 **return** CS;

---

adopted by most of map updating methods (e.g. CrowdAtlas and Glue) due to its excellent performance. Likely, the *MapMatching* function in Algorithm 1 (at line 4) also implements Viterbi, and derives unmatched trajectory segments by selecting candidates with the maximal weight after calculating the candidate positions within a certain radius. Finally, we will obtain a set of unmatched trajectory segments.

Denoising. GPS samples often have a few noisy data of position or direction. To improve the accuracy of inferred missing roads, denoising process is required to reduce the noisy samples. For example, there are a few noisy points (in red) in Fig. 2(b). First, the red circled points can be removed through distance denoising because they are far from most of its surrounding points. Subsequently, as the red track points in Fig. 2(b) are significantly different from most of its surrounding points in directions, they are removed by direction denoising [6]. Specifically, we search the nearby segments of each trajectory segment, and compare the direction of it with its neighboring segments. Then, we identify a noisy trajectory segment according to the significant gap between its direction and most of its surrounding segments' direction. The denoising result after distance and direction denoising is shown in Fig. 2(c).

Clustering. After map matching and denoising, the unmatched trajectory segments need to be clustered to infer the road candidates. To enhance the accuracy of inferred missing roads, we combine both line-based clustering (*LClustering*) and point-based clustering (*PClustering*). The point-based clustering takes two endpoints of all trajectory segments as input, while the line-based cluster takes trajectory segments as input. In Algorithm 2 (*LClustering*), each trajectory segment is initialized as a cluster once the number of its similar trajectory segments is greater than a specific threshold, i.e., $N_d(p_i) > th_c$ (at lines 2–5). The similar trajectory segment is defined below.

Definition 6 (Similar Trajectory Segment). *Given two trajectory segments* $T_s^{(x)}$ *and* $T_s^{(y)}$, *a distance threshold* th_{dis} *and a direction threshold* th_{dir}, $T_s^{(x)}$ *and* $T_s^{(y)}$ *are two similar trajectory segments if the distance between* $Ts^{(x)}$ *and* $Ts^{(y)}$ *is smaller than* th_{dis}, *and the angle between* $Ts^{(x)}$ *and* $Ts^{(y)}$ *is less than* th_{dir}.

Then, for each segment $Ts^{(x)}$ in one cluster and each segment $Ts^{(y)}$ in $Nd(Ts^{(x)})$, if they are similar and $Nd(Ts^{(y)}) > th_c$, we add $Ts^{(y)}$ into queue. If $Ts^{(y)}$ does not belong to any cluster, it should also be added into the cluster of $Ts^{(x)}$ (at lines 7–13). The *PClustering* approach also divides the input points into several clusters according to the similar criterion. The directions of two endpoints of a segment can be seen as the direction of the segment. Due to space limitations, we omit the detail of *PClustering*.

Centerline Fitting. The centerline fitting step aims to generate the centerlines to represent road candidates. Since a cluster that consists of the trajectory points or segments may belong to the same road candidate, we need to fit a centerline to represent a road candidate. For the clustering results of former stage, we use the sweeping line method in [7] to realize the centerline fitting process. The *CLFitting* function in Algorithm 1 takes trajectory points or segments as input, and generates the road candidates (at lines 11–14). Finally, we obtain line-based candidates and point-based candidates.

4.2   Missing Roads Inferring

In this phase, we group road candidates belonging to the same road based on two kinds of road candidates, RC_l and RC_p, generated in Algorithm 1. To be specific, if k road candidates (at least one line-based candidate and one line-based candidate) are located on the same road, they are merged to infer a missing road. After that, we connect the inferred roads with existing roads in network. Therefore, the missing roads inferring phase is composed of two steps, including *continuous observation* and *road combination*.

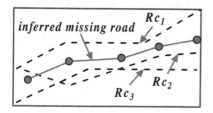

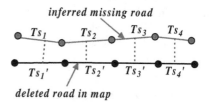

Fig. 4. An example of missing road generated in the MBR

Fig. 5. An example of two similar road centerlines

Algorithm 3. Continuous Observation

    Input: Road candidate sets RC_p and RC_l, a threshold k $(k \geqslant 3)$
    Output: A missing road set $RS = \{R_1, R_2, ..., R_m\}$
1 $RS \leftarrow \emptyset; i \leftarrow 1;$
2 **foreach** *unvisited candidate* $Rc^{(i)}$ *in* $RC_p \cup RC_l$ **do**
3     $Z \leftarrow Rc^{(i)} \cup \{Rc^{(j)} \mid Rc^{(j)} \in RC_p \cup RC_l, Rc^{(i)}$ and $Rc^{(j)}$ are similar$\}$;
4   Mark all road candidates in Z as visited;
5   **if** $|Z| \geqslant k$ **and** *at least one point-based candidate in Z* **then**
6         $R_i \leftarrow CLFitting(Z)$;
7         $RS \leftarrow RS \cup \{R_i\}; i \leftarrow i + 1;$

8 **return** RS;

Continuous Observation. As mentioned in Sect. 1, since the sparse trajectories in one time window may be confused with noise, the road candidates derived from them may imply wrong missing roads. To improve the precision, we propose a continuous observation approach to infer the missing roads based on the candidates of multiple time windows. To be specific, as show in Algorithm 3, we collect the candidates of consecutive time windows so far and take them as input. First, we divide all road candidates according to the roads which they belong to (at line 3). When the number of road candidates exceeds a predefined threshold k $(k \geqslant 3)$ and at least one point-based candidate is involved, we can identify a missing road. Next, we fit them into a missing road by invoking $CLFitting$ function (at line 6).

For example, there are three similar road candidates (e.g. the black polyline) co-exist in Fig. 4. As the number of road candidates reaches the predefined threshold $(k = 3)$, we combine them to generate a new road centerline to represent a missing road. To be specific, given two road candidates $Rc^{(x)}$ and $Rc^{(y)}$, if $\exists Ts^{(i)} \in Rc^{(x)}, Ts^{(j)} \in Rc^{(y)}$, and $Ts^{(i)}$ and $Ts^{(j)}$ are similar, we take $Rc^{(y)}$ as one of the similar road candidates of $Rc^{(x)}$ and take them as candidates of the same road. In Fig. 4, if the number of road candidates reaches k $(k = 3)$, a missing road will be inferred through centerline fitting. If $k = 4$, we continue to observe the road candidates in the following time windows until the number of road candidates belonging to the same road reaches 4.

Road Combination. After inferring missing roads, we update the existing road network by connecting the inferred roads to the existing neighboring roads. Given an inferred missing road $Rc^{(x)}$, we try to find a road $Rc^{(y)}$ in the road network such that the $Rc^{(y)}$ is close to one of the endpoints of $Rc^{(x)}$ (e.g. smaller than 20 m). If such $Rc^{(y)}$ exists, we update the existing road network by connecting $Rc^{(x)}$ and $Rc^{(y)}$.

5   Experimental Evaluation

We conduct substantial comparison experiments on real data sets to evaluate the performance of HyMU. Specifically, we compare HyMU with line-based method

(CrowdAtlas [16]) and point-based method (Glue [18]) to verify the superiority of HyMU. Our codes, written in Java, are conducted on a PC with 16 GB RAM, Intel Core CPU 3.2 GHz i7 processor, and the operating system is Windows 10.

5.1 Evaluation Method

In order to ensure fairness, we randomly select an area on the existing map and remove some road segments from this region. The goal is to verify whether the deleted road segments can be inferred by different map updating methods. Evaluation criteria includes *Precision*, *Recall* and *F-measure* [8,18]. Let *truth* denote the deleted roads, *inferred* denote all inferred road segments, and *tp* denote the correctly inferred roads. Accordingly, we use *len(truth)*, *len(inferred)* and *len(tp)* to represent the length of all the deleted roads, the inferred roads and the correctly inferred roads respectively. Then, *Precision*, *Recall* and *F-measure* can be calculated as follows.

$$Precision = \frac{len(tp)}{len(inferred)} \qquad Recall = \frac{len(tp)}{len(truth)}$$

$$\textit{F-measure} = \frac{2 \times Precision \times Recall}{Precision + Recall}$$

As shown in Fig. 5, the deleted roads and their corresponding inferred missing roads are split into small segments with fixed length. Then, *tp* can be denoted as below.

$$tp = \{si(Ts^{(x)}, Ts^{(y)}) | \forall Ts^{(x)} \in inferred, \ \forall Ts^{(y)} \in truth\}$$

The function $si(Ts^{(x)}, Ts^{(y)})$ returns $Ts^{(x)}$ if $Ts^{(x)}$ and $Ts^{(y)}$ are similar. Otherwise, it returns *null*.

5.2 Data Sets and Map

We use two real data sets to evaluate the effectiveness of HyMU method, including a taxi trajectory data set of 2015 in *ShanghaiOpen Data Apps*[2] (hereafter termed *Taxi*2015) and a high-sampling Shanghai taxi data set in 2013 (hereafter termed *Shanghai*2013). In addition, we choose an open source map *OpenStreetMap(OSM)*[3] as our map data.

*Taxi*2015 contains the GPS logs of taxis from Apr. 1 to Apr. 30, 2015. It involves about 10,000 trajectories every day (about 115 million points). Each GPS log, represented by a sequence of time-stamped points, contains Vehicle ID, Time, Longitude and Latitude, Speed, etc.

*Shanghai*2013 contains the GPS logs of taxis in 2 days (from Oct. 1 to Oct. 2). It involves about 50,000 trajectories every day (about 107 million points). The average sampling rate of the objects is about 60 s. Besides, each GPS log, represented by a sequence of time-stamped points, contains Vehicle ID, Time, Longitude and Latitude, Speed, etc.

---

[2] http://soda.datashanghai.gov.cn/.
[3] http://wiki.openstreetmap.org/.

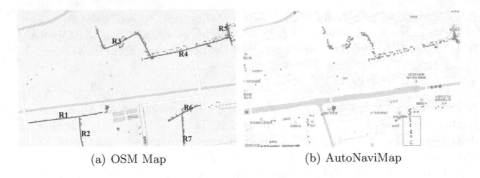

(a) OSM Map   (b) AutoNaviMap

Fig. 6. Visualization result of HyMU on $Taxi2015$

5.3   Effectiveness Evaluation

Results for Taxi2015. We first implement HyMU on $Taxi2015$ to infer about 150 road segments that haven't been described in OSM map. The visualization result is shown in Fig. 6(a), where the red lines represent the missing roads detected by HyMU. As compared to the roads in $AutoNaviMap$[4] (as shown in Fig. 6(b)), we can find that six roads (R_1–R_6) are correctly inferred by HyMU. This verifies the high precision of our proposal. In addition, we infer the road R_7 that is not marked on $AutoNaiveMap$, which further confirms the superiority of HyMU in discovering missing roads on sparse trajectory data.

Results for Shanghai2013. We compare HyMU with CrowdAtlas and Glue on $Shanghai2013$ and randomly select a test area consists of 19 road segments (from North Zhang Yang Road, through Wuzhou Avenue and Shenjiang Road, to Jufeng Road). Firstly, to verify the robustness of HyMU, we evaluate sensitivity of parameters (th_{dis}, th_{dir} and k) on $Shanghai2013$, as illustrated in Fig. 7. After tuning them repeatedly, we find that HyMU achieves the best performance on $Shanghai2013$ when $th_{dis} = 20\,m$, $th_{dir} = \frac{\pi}{6}$ and $k = 3$. Secondly, we further evaluate HyMU, CrowdAtlas and Glue by varying the sampling interval from 40 s to 160 s. As shown in Fig. 8, we find that Glue has the best precision because the point-based strategy will not infer the missing roads with wrong direction. But it does not take into account inferring roads on sparse region, which result in a lower recall rate. By contrast, HyMU combines the advantage of line-based and point-based strategies. It attains almost the same precision as Glue, and the highest recall as well as F-measure. Thirdly, we evaluate the performance of HyMU, CrowdAtlas and Glue under various data volume, as shown in Fig. 9. As data volume becomes larger, we observe that the precision, recall and F-measure value of HyMU increases accordingly, and the precision approaches Glue. Hence, HyMU has a good scalability. Additionally, we observe that HyMU has higher recall than other methods in all situations, which demonstrates that it is capable of inferring missing roads on sparse trajectory data.

---

[4] http://ditu.amap.com/.

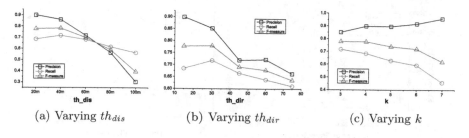

(a) Varying th_{dis} (b) Varying th_{dir} (c) Varying k

Fig. 7. Performance of HyMU under different parameters on *Shanghai*2013

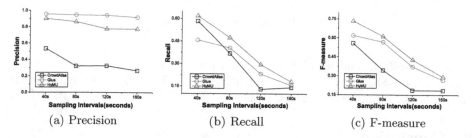

(a) Precision   (b) Recall   (c) F-measure

Fig. 8. Performance comparison under various sampling intervals on *Shanghai*2013

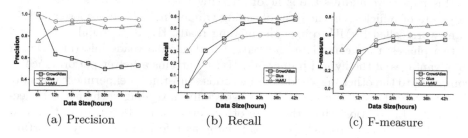

(a) Precision   (b) Recall   (c) F-measure

Fig. 9. Performance comparison under various data volume on *Shanghai*2013

5.4 Efficiency Evaluation

Next, we assess the efficiency of HyMU by comparison with CrowdAtlas and Glue on *Shanghai*2013. As shown in Fig. 10(a), HyMU run faster than the other two methods with the increase of trajectory data. It indicates that HyMU is more efficient than other map updating methods. GLUE, by contrast, is extremely time-costing, due to the cost on calculating direction of each point. Additionally, we evaluate the efficiency of HyMU by varying the time window size N. Figure 10(b) shows the processing time comparison when N is set to 3 h, 6 h and 21 h respectively. When $N = 6$ h, the execution time is the smallest. This is due to that massive amount of data in a time window requires to be denoised and clustered which is quite time-consuming if the time window size is large. Conversely, when time window size is small, we need to deal with too many road candidates, which is also time-consuming. So the appropriate window size is 6 h

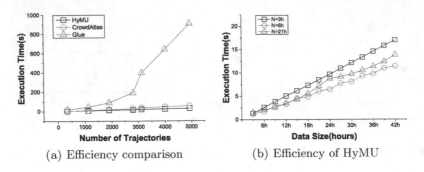

(a) Efficiency comparison   (b) Efficiency of HyMU

Fig. 10. Efficiency evaluation

on *Shanghai*2013, and we also use this optimal value to execute effectiveness evaluation on *Shanghai*2013. Consequently, HyMU is efficient and effective to infer the missing roads for a given map.

6   Conclusion

In this paper, we address the issue of inferring missing roads on sparse trajectory data of vehicles. On the basis of sliding window model, we propose a hybrid framework called HyMU to infer the missing roads. HyMU is mainly composed of two phases: road candidates generation and missing roads inferring. Owing to advantages of the hybrid framework, HyMU attains a better performance as compared to the other map updating methods. Substantial experimental results demonstrate the superiority of HyMU especially in dealing with sparse trajectory data. In addition, since there are other forms of road changes in the road network (e.g. blocked roads). Such road changing information is very important in navigation applications. In the future work, we proceed to study how to detect road changes, and provide real-time traffic information to users to enable route planning.

Acknowledgement. Our research is supported by the National Key Research and Development Program of China (2016YFB1000905), NSFC (61370101, 61532021, U1501252, U1401256 and 61402180), Shanghai Knowledge Service Platform Project (No. ZF1213).

References

1. Agamennoni, G., Nieto, J.I., Nebot, E.M.: Robust inference of principal road paths for intelligent transportation systems. IEEE Trans. Intell. Transp. Syst. **12**(1), 298–308 (2011)
2. Biagioni, J., Eriksson, J.: Map inference in the face of noise and disparity. In: SIGSPATIAL, pp. 79–88 (2012)
3. Cao, L., Krumm, J.: From GPS traces to a routable road map. In: GIS, pp. 3–12 (2009)

4. Davies, J.J., Beresford, A.R., Hopper, A.: Scalable, distributed, real-time map generation. IEEE Pervasive Comput. **5**(4), 47–54 (2006)
5. Edelkamp, S., Schrödl, S.: Route planning and map inference with global positioning traces. In: Computer Science in Perspective, Essays Dedicated to Thomas Ottmann, pp. 128–151 (2003)
6. Ge, Y., Xiong, H., Zhou, Z., Ozdemir, H.T., Yu, J., Lee, K.C.: Top-eye: top-k evolving trajectory outlier detection. In: CIKM, pp. 1733–1736 (2010)
7. Lee, J., Han, J., Whang, K.: Trajectory clustering: a partition-and-group framework. In: SIGMOD, pp. 593–604 (2007)
8. Liu, X., Biagioni, J., Eriksson, J., Wang, Y., Forman, G., Zhu, Y.: Mining large-scale, sparse GPS traces for map inference: comparison of approaches. In: KDD, pp. 669–677 (2012)
9. Mazhelis, O.: Using recursive bayesian estimation for matching GPS measurements to imperfect road network data. In: International IEEE Conference on Intelligent Transportation Systems, pp. 1492–1497 (2010)
10. Mokhtarzade, M., Zoej, M.J.V.: Road detection from high-resolution satellite images using artificial neural networks. Int. J. Appl. Earth Obs. Geoinf. **9**(1), 32–40 (2007)
11. Schrödl, S., Wagstaff, K., Rogers, S., Langley, P., Wilson, C.: Mining GPS traces for map refinement. Data Min. Knowl. Discov. **9**(1), 59–87 (2004)
12. Seo, Y., Urmson, C., Wettergreen, D.: Exploiting publicly available cartographic resources for aerial image analysis. In: SIGSPATIAL, pp. 109–118 (2012)
13. Shan, Z., Wu, H., Sun, W., Zheng, B.: COBWEB: a robust map update system using GPS trajectories. In: UbiComp, pp. 927–937 (2015)
14. Thiagarajan, A., Ravindranath, L., Balakrishnan, H., Madden, S., Girod, L.: Accurate, low-energy trajectory mapping for mobile devices. In: NSDI (2011)
15. Velaga, N.R., Quddus, M.A., Bristow, A.L.: Developing an enhanced weight-based topological map-matching algorithm for intelligent transport systems. Transp. Res. Part C Emerg. Technol. **17**(6), 672–683 (2009)
16. Wang, Y., Liu, X., Wei, H., Forman, G., Chen, C., Zhu, Y.: CrowdAtlas: self-updating maps for cloud and personal use. In: MobiSys, pp. 27–40 (2013)
17. Wei, H., Wang, Y., Forman, G., Zhu, Y., Guan, H.: Fast viterbi map matching with tunable weight functions. In: SIGSPATIAL, pp. 613–616 (2012)
18. Wu, H., Tu, C., Sun, W., Zheng, B., Su, H., Wang, W.: GLUE: a parameter-tuning-free map updating system. In: CIKM, pp. 683–692 (2015)

Multi-objective Spatial Keyword Query with Semantics

Jing Chen[1], Jiajie Xu[1(✉)], Chengfei Liu[2], Zhixu Li[1], An Liu[1],
and Zhiming Ding[3]

[1] Department of Computer Science and Technology,
Soochow University, Suzhou, China
20164227012@stu.suda.edu.cn, {xujj,zhixuli,anliu}@suda.edu.cn
[2] Faculty of SET, Swinbourne University of Technology, Melbourne, Australia
cliu@swin.edu.au
[3] Department of Computer Science and Technology,
Beijing University of Technology, Beijing, China
zmding@bjut.edu.cn

Abstract. Multi-objective spatial keyword query finds broad applications in map services nowadays. It aims to find a set of objects that can cover all query objectives and are reasonably distributed in spatial. However, existing approaches mainly take the coverage of query keywords into account, while leaving the semantics behind the textual data to be largely ignored. This limits us to return those rational results that are synonyms but morphologically different. To address this problem, this paper studies the problem of multi-objective spatial keyword query with semantics. It targets to return the object set that is optimum regarding to both spatial proximity and semantic relevance. We propose an indexing structure called LIR-tree, as well as two advanced query processing approaches to achieve efficient query processing. Empirical study based on real dataset demonstrates the good effectiveness and efficiency of our proposed algorithms.

1 Introduction

Spatial keyword query is widely used in location based service (LBS) systems to recommend users the needed services or places to visit. The study on this topic has attracted a great deal of attention. Existing methodologies mainly study the efficient retrieval of spatial web objects that can best match the query in terms of both spatial and textual relevances. The spatial keyword query itself sometimes has multiple objectives, which may lead to none or few objects that can fully cover all keywords in query. To address this problem, [5] returns a group of objects that can cover all required keywords with reasonable spatial distribution. But the keyword match cannot help us to find out those objects with highly related semantics but low similarity in spellings, such as *market* and *Wal-Mart*. This limitation motivates us to investigate other approaches to capture the semantic relatedness to multi-objective spatial keyword queries.

© Springer International Publishing AG 2017
S. Candan et al. (Eds.): DASFAA 2017, Part II, LNCS 10178, pp. 34–48, 2017.
DOI: 10.1007/978-3-319-55699-4_3

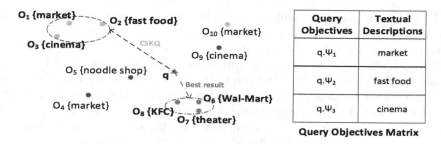

Fig. 1. Distribution of Spatial Web Objects

Example 1. Figure 1 shows an example with ten spatial web objects, each has a geographical location and a set of keywords. A user issues a query with three objectives described by *market*, *fast food* and *cinema* respectively. By using traditional methods [7,9] to process each objective in query independently, the objects $\{O_4, O_5, O_9\}$ are returned because of the spatial and textual similarities to query. Alternatively by using the collective spatial keyword querying method [5], the search engine tends to return a more qualified result such as $\{O_1, O_2, O_3\}$, because they are coherent in spatial and been close to the query together. However if we check the semantics of query objectives more carefully, instead of $\{O_1, O_2, O_3\}$, we can easily observe that $\{O_6, O_7, O_8\}$ is the set of objects that should be returned, because they are best matched in spatial, and all objectives in the query can be fully matched in semantics. The key issue is how to take the semantics into account and process the query efficiently.

To represent the semantics of spatial web objects and query objectives, we can apply powerful tools in the field of machine learning, such as probabilistic topic model or word embedding. By using them on textual descriptions, query objectives (e.g., *market* in q of Fig. 1) and spatial web objects are represented as high dimensional vectors called topic distributions in semantic space. A topic distribution indicates the semantic relevance between a textual description and a latent topic, and accordingly, the similarity between an object and an objective in query can be measured on top of their topic distributions. In this way it is possible to find the collective object set that can satisfy all query objectives while coherent in spatial and close to the query point.

While the incorporation of semantics helps us to return more meaningful feedbacks, the query processing becomes more challenging and time-consuming for three main reasons: firstly, finding the optimal result (the subset according to spatial and semantic similarity) is an NP complete problem, which cannot be solved in a polynomial time; secondly, existing spatial keyword indices, such as IR-tree [4], cannot be directly used to organize the information of spatial web objects because of its difficulties in representing their topic distributions regarding to semantics. Last but not the least, the high dimensionality of vector (topic distribution in semantics) deteriorates the pruning effectiveness in query processing due to the large dead space.

To address all above difficulties, we propose a novel query processing mechanism that has good efficiency and precision. To ensure the pruning effect in semantic space, we take advantage of the locality sensitive hashing (LSH) to hash the objects by their high dimensional topic distributions. Each bucket is understood as a semantic tag, and the LSH mechanism ensures that objects in the same bucket to have consistent semantic meanings. We design a candidate bucket set oriented searching mechanism to reduce the search space. It retrieves and compares local result for each candidate bucket set, and finally derives a result in global optimum. In addition, a more efficient approach is proposed to avoid checking all candidate bucket sets while ensuring high accuracy of the result. The main contributions of the paper can be briefly summarized as follows:

- We formalize a probabilistic topic model based similarity measure between a multi-objective query and a set of objects;
- We design a semantic hashing based algorithm by applying LSH index structure, so that collective objects can be derived by making use of the collective spatial keyword querying technologies.
- We propose a novel mechanism that can start from a good result directly, and then guide us to improve the result while ensuring the accuracy by distance based replacement strategy.
- We conduct an extensive experiment analysis based on real spatial databases and make the comparisons with baseline algorithm, and then demonstrate the efficiency of our method.

2   Preliminaries and Problem Definition

In this section, we introduce some preliminaries about probabilistic topic model and then formalize the problem of this paper.

2.1   Probabilistic Topic Model

Probabilistic topic model is a well-known technique on theme interpretation and document classification. In this paper, we apply one of the most frequently used probabilistic topic models, i.e. the *Latent Dirichlet Allocation* (LDA) model to understand the semantic meanings of textual descriptions. In LDA, each latent topic, or topic in short, is a feature that represents a semantic meaning. By carrying out statistical analysis on the large amount of textual descriptions, the LDA model automatically derives the semantic relevance of a textual description to all latent topics, known as topic distribution defined as follows:

Definition 1 (*Topic Distribution*). Given a textual description W, a topic distribution derived from LDA is a high dimensional vector that describes the semantic relevance between the textual description and each latent topic. We use TD_W to denote the topic distribution of W over finite latent topics, and a component $TD_W[i]$ indicates the relevance between W and the i_{th} latent topic.

Table 1. Topic distributions of textual descriptions

Textual descriptions	Topics				
	Exercise	Movie	Drink	Shop	Food
market (in O_1, O_{10})	0.09	0.09	0.09	0.64	0.09
fast food (in O_2)	0.04	0.04	0.16	0.04	0.72
cinema (in O_3, O_9)	0.07	0.72	0.07	0.07	0.07
noodle shop (in O_5)	0.07	0.07	0.07	0.07	0.72
Wal-Mart (in O_6)	0.07	0.07	0.07	0.72	0.07
theater (in O_7)	0.04	0.84	0.04	0.04	0.04
KFC (in O_8)	0.03	0.03	0.03	0.03	0.88

Example 2. Table 1 shows the LDA interpretation on all the spatial web objects in Fig. 1. Each tuple in Table 1 is a topic distribution over five topics. Each component is the relevance between the textual description and a specific topic, for example, $TD_{market}[1] = 0.09$ means the relevance between *market* and *exercise* is 0.09. We can learn from Table 1 that *cinema* has high coherence with *theater* due to $TD_{cinema}[2] = 0.72$ and $TD_{theater}[2] = 0.84$ while in contrast that *KFC* is distinct to *cinema* because of $TD_{KFC}[2] = 0.03$.

2.2 Problem Definition

A spatial web object is a place of interest in LBS systems, and it is formalized as $o = (o.\lambda, o.\psi)$ where $o.\lambda$ is the position of o and $o.\psi$ is the textual information for describing o. A user issues a multi-objective query $q = (q.\lambda, q.\Psi)$, where $q.\lambda$ represents a geographical location, and $q.\Psi$ is a set of query objectives which are textual descriptions for describing an activity intention. In the rest of this paper, we simply use *objects* to represent *spatial web objects*.

Definition 2 (*Spatial Distance*). The objects in the result set are supposed to be not only close to query, but also close with each other. We thus follow collective spatial keyword query [5] and measure the spatial distance D_S to range [0,1] from a query q to an object set O as follow:

$$\mathcal{D}_S(q, O) = \beta \times max_{o_i \in O} \ (s_d(q, o_i))$$
$$+ (1 - \beta) \times max_{o_i, o_j \in O}(s_d(o_i, o_j)) \quad (1)$$

where $\beta \in [0, 1]$ is a user-specified weight parameter, $s_d(q, o_i) = \frac{2}{1+e^{||q.\lambda, o_i.\lambda||}} - 1$ is the normalized spatial distance between query q and object o_i by sigmoid function. The spatial measure allows us to find a set of objects close to query and have spatial coherence with each other. That means, the objects are rationally distributed in spatial when $\mathcal{D}_S(q, O)$ is small.

Definition 3 (*Semantic Distance*). Semantic distance $\mathcal{D}_T$ between a query q and an object set O can be measured on top of their topic distributions through

LDA. By calculating the distance between high dimensional vectors, we define $\mathcal{D}_T$ to range [0,1] by using the sigmoid function as follow:

$$\mathcal{D}_T(q, O) = \sum_{q.\Psi_i \in q.\Psi} min_{o_j \in O}(d_T(q.\Psi_i, o_j)) \tag{2}$$

such that,

$$d_T(q.\Psi_i, o_j) = \frac{2}{1 + e^{-\sqrt{\Sigma(TD_{q.\Psi_i}[z] - TD_{o_j}[z])^2}}} - 1 \tag{3}$$

where $\mathcal{D}_T(q, O) \in [0, 1]$. It is obvious that when semantic distance is smaller, the query q and a correspond object o are more relevant in semantics.

Definition 4 (*Distance*). By combining spatial distance $\mathcal{D}_S(q, O)$ and semantic distance $\mathcal{D}_T(q, O)$, we define the distance $Dist(q, O)$ of query q and object set O in Equation below.

$$Dist(q, O) = \alpha \times \mathcal{D}_S(q, O) + (1 - \alpha) \times \mathcal{D}_T(q, O) \tag{4}$$

where $\alpha \in [0, 1]$ is a user-specified weight parameter that balance spatial distance and semantic distance.

Problem Statement. Given an object set O and a query $q = (q.\lambda, q.\Psi)$, the multi-objective spatial keyword query (MoSKQ, in short) in this paper aims to return a subset O' of objects $O(O' \subset O, |O'| \leqslant |q.\Psi|)$, such that $\forall O'' \subset O$, $Dist(q, O') \leqslant Dist(q, O'')$.

3   Baseline Algorithm

In this section, we propose a baseline algorithm which seeks to find the optimal result within a subspace incrementally. A lower bound and an upper bound are used to stop the searching process in the middle if possible.

Starting from a search region centered at the query q with a radius r, we execute an exhaustive search to get the best object set $R(|R| \leqslant |q.\Psi|)$ which minimizes the distance to the query according to *Definition* 4. If needed, we enlarge the search radius $r = r + \Delta r$ and search all combinations of objects in this region to find a best object set R'. During this process, a set S is used to store all the solutions found. In this process, we dynamically maintain an upper bound $\mathcal{UB} = min_{R \in O}(Dist(q, R))$ and a lower bound $\mathcal{LB} = \alpha \times \beta \times (\frac{2}{1+e^{-r}} - 1)$ which equals to part of spatial distance.

During the process, if $\mathcal{UB} \leqslant \mathcal{LB}$ or the search radius extends to the most distant object, the algorithm terminates and returns the best group in the solution set S, because the spatial distance of all unprocessed objects are no less than the distance to query of the found result object sets. However, this algorithm may require an exhaustive search sometimes because the bound is relative loose. Therefore more efficient approaches are required to find the results.

4   Semantic Hashing Based Algorithm

In this section, we propose a novel solution called semantic hashing based algorithm (SH-based algorithm in short) to speed up the querying process. In Sect. 4.1, we introduce the details of the LSH based indexing structure. Section 4.2 plots the search algorithm over the index.

4.1   Index Structure

In this subsection, we devise a new index, namely LIR-tree, based on LSH and IR-tree. As is known, LSH [3, 8, 17] is a method widely used for similarity search in high dimension. We first utilize LSH to preprocess all objects in the dataset, i.e., hash the object into buckets based on their topic distributions. Every bucket in LSH can be regarded as the tag of the semantic meanings of the objects in this bucket. That is to say, the objects in the same bucket are considered to be similar in semantics. In this way, all the objects in the dataset derive the bucket ids that they are hashed into, which makes the semantic similarity search possible based on the bucket ids. Then we use IR-tree [6, 16, 19] to organize the objects according to their geographical locations and corresponding bucket ids for a given query with specified location and bucket ids.

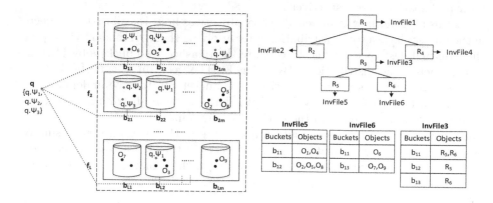

Fig. 2. An example of LIR-tree

LSH part. The LSH is a well-known index scheme for high-dimensional similarity search with the basic idea to use a family of locality-sensitive hashing functions to map the objects into the same buckets with high probability. LSH hash families have the property that objects close to each other have a higher colliding probability than those far apart, which is determined by different distance measure functions. In this paper, we use the hash family proposed by Datar et al. [8] based on p-stable distributions [12], which is defined as:

$$h\left(p\right) = \lfloor \frac{a \times p + b}{W} \rfloor \tag{5}$$

where a is a random topic distribution vector, W represents the width of the hash function, b is a random variable belongs to $[0, W]$. All the objects in the dataset are divided into corresponding buckets based on their topic distributions. Each bucket can be considered as the semantic tag of the objects in this bucket and the objects in the same bucket have high proximity in semantics. We record the geographical location and the bucket ids that every object in dataset are hashed into.

IR-tree part. The IR-tree part of LIR-tree is similar to the conventional inverted R-tree, except that we store the inverted list of the buckets of the objects derived by LSH, rather than the keywords that describe the objects. All the objects in the dataset are organized using the R-tree according to their geographical location. Since the objects also have the bucket ids that they are hashed into, we build the inverted list of the R-tree node in a bottom up fashion. The inverted list of both leaf node and non-leaf node includes the *buckets* and the *objects* that are hashed to this bucket.

4.2 Search Algorithm

In this subsection, we propose a search algorithm that prunes the search space effectively over the proposed index. The prune process is complished on topic layer and spatial layer respectively.

Let us consider how to match all query objectives first. Recall that all objects in the dataset have a topic distribution after applying the LDA model to interpret their textual descriptions. The objects are then hashed into buckets by LSH on top of their topic distributions. By using LSH, objects in a same bucket are supposed to be consistent in semantics, each bucket can thus be understood as a semantic tag. Given a query q, we can derive a topic distribution for each objective in q, and then hash the query objectives into the LSH buckets in the same way to objects. By taking advantage of the LSH structure, we can simply evaluate if an object can match a query objective if they share a same semantic tag (i.e. in a same LSH bucket), and accordingly, the semantic distance can be rewritten to:

$$d_T(q.\Psi_i, o) = \begin{cases} 0 & \exists B_{ij} : q.\Psi_i \in B_{ij} \land o \in B_{ij} \\ \infty & otherwise \end{cases} \qquad (6)$$

where q and o are the query and the object respectively. Equation 6 means that an object can be matched to a query objective in semantics if they share at least one LSH bucket, and they have no semantic relevance otherwise.

On top of the LSH based semantic distance defined above, the next issue is to justify if a given object set is semantically relevant to all query objectives. Conceptually, the relevance requires us to find an object from the set to share a same bucket for each query objective (having a same semantic tag). To define the semantic relevance more clearly, we further define the concept of *candidate bucket set* as follows.

Definition 5 (*Candidate bucket set*). A candidate bucket set is a smallest unit of buckets to ensure the relevance to a query about its objectives. Given a query q, a candidate bucket set cbs satisfy the following two requirement: (1) containment. A cbs contains at least one bucket in bucket set of each query objective $q.\Psi_i \in q.\Psi$ such that $BS(q.\Psi_i) \cap cbs \neq \varnothing$, where $BS(q.\Psi_i)$ is the set of buckets which $q.\Psi_i$ is hashed to; (2) minimum. The above condition fails for each subset of these buckets, i.e., $cbs' \subsetneq cbs$.

The candidate bucket set ensures that all query objectives can be matched. Given a set of objects O, it can be returned if the union of buckets containing an object in O can cover a candidate bucket set of query q. In Fig. 2, $\{b_{1M}, b_{21}, ..., b_{L2}\}$ is a candidate bucket set which covers all query objectives, but $\{b_{12}, b_{22}, ..., b_{L2}\}$ is not a candidate bucket set which covers only part of query objectives $\{q.\Psi_1, q.\Psi_2\}$.

Here we describe the searching mechanism. The target of SH-based algorithm search is to find the object set such that: (1) its related bucket set covers at least one candidate bucket set to meet all query objectives; (2) the distance to query is the minimum. We use a candidate bucket set oriented searching mechanism. For each candidate bucket set, if each bucket is regarded as a keyword (denoting a semantic tag), our problem can be transfered to the well studied collective spatial keyword query [5]. We apply the Top-Down Search algorithm in [5] to obtain the best object set for the given candidate bucket set (line 7). The basic idea of the Top-Down Search algorithm is to perform a best-first search on the IR-tree to find the covering node sets, such that some objects from these nodes can constitute a group to cover all required buckets in the set. We process the covering node set with the lowest distance to query to find covering node sets from their child nodes. While reaching a covering node set consisting of leaf nodes, a group of objects with the lowest distance to query can be found by performing an exhaustive search (lines 6–10). The Top-Down Search algorithm return the exact result set and it is invoked $L^{|q.\Psi|}$ times according to Lemma 1.

Lemma 1. *There are L hash functions and query q have $|q.\Psi|$ query objectives. The SH-based Algorithm would retrieve collective objects for $L^{|q.\Psi|}$ times.*

Proof. Assuming that there are L hash functions, each query objective $q.\Psi_i$ can be put into L buckets as a bucket set $BS(q.\Psi_i)$ after hashing. The quantity of this bucket set is L, namely $|BS(q.\Psi_i)| = L, i \in [1, |q.\Psi|]$. Then the $|q.\Psi|$ query objectives correspond to $|q.\Psi|$ bucket sets. All candidate bucket sets produced by Cartesian product $BS(q.\Psi_1) \times BS(q.\Psi_2) \times ... \times BS(q.\Psi_{|q.\Psi|})$ and obviously the size of these sets is $L^{|q.\Psi|}$. Therefore, this query step will be repeated $L^{|q.\Psi|}$ times. Lemma 1 can be proven. ∎

This method avoids the worst situation in which the whole search region needs to be scrutinized. The search process is subject to at most $BS(q.\Psi_1) \times ... \times BS(q.\Psi_{|q.\Psi|})$ candidate bucket sets, each of them calls for a Top-Down Search whose time complexity is $|q.\Psi| - |N| + 1$, where N represents the number of nodes which cover the query keywords and each node contributes at least one

---

Algorithm 1. *SH* based Search Algorithm

Input: IR-tree ir, query q, λ
Output: a set of objects O
1 $O \leftarrow \varnothing$;
2 $CBS \leftarrow BS(q.\Psi_1) \times BS(q.\Psi_2)... \times BS(q.\Psi_{|q.\Psi|})$;
3 **for** *each cbs in CBS* **do**
4   O' and $Dist(q, O') \leftarrow$ TopDownSearch(q, cbs);
5   **if** $Dist(q, O') < Dist(q, O)$ **then**
6        $Dist(q, O) = Dist(q, O')$;
7        $O = O'$;

8 **return** O and $Dist(q, O)$;

---

object to the final result [5]. The time complexity of SH-based Algorithm is $L^{|q.\Psi|} \times (|q.\Psi| - |N| + 1)$ which cannot be solved in polynomial time either. The computational overhead will rapidly increase when the number of query objectives grows.

5   Distance Based Replacement Algorithm

This section presents a novel strategy called Distance Based Replacement (DBR) Algorithm, which starts at the SH-based algorithm but aims to find a high quality result more efficiently. Instead of taking every possible candidate bucket set as input, the DBR algorithm randomly sample a number of candidate bucket sets and derive a result based on SH-based algorithm. Then it aim to improve the result by replacement iteratively. Besides, the DBR Algorithm takes both spatial layer and topic layer into consideration, and LIR-tree is used to accelerate query processing as well.

We randomly sample some candidate bucket sets to obtain object sets by invoking SH-based Algorithm and choose the object set with the minimum distance to query from these object sets before replacement. Objects in this set are replaced individually and iteratively until a stable object set is found according to Lemma 2. A stable object set is obtained by iteratively replacing object until no reduction can be achieved on distance to query. The critical operation is that how to set the standard for replacement. In the procedure of replacement, an object in initial object set replaced by an object which beyond this set and has farther $\mathcal{MG}$ than others. The important concept *marginal gain* $\mathcal{MG}$ defined as

$$\mathcal{MG}(o_i, o_j) = Dist(q, O) - Dist(q, O') \qquad (7)$$

where $O = \{o_1, .., o_i, .., o_n\}$ is the initial object set and $O' = \{o_1, .., o_j, .., o_n\}$ represents a replaced object set where o_i is replaced by o_j. The object o_i should be replaced by o_j while $\max_{o_j \in O}(\mathcal{MG}(o_i, o_j))$ and $\mathcal{MG}(o_i, o_j) > 0$. The replacement strategy terminates when no object can be found to touch positive marginal gain, i.e., for each object o'_i in object set, $\forall o'_j \in O, \mathcal{MG}(o'_i, o'_j) < 0$. The stable object set is final result of DBR Algorithm. And this result may have lower distance to query than the object set found by SH-based Algorithm.

Lemma 2. *Given dataset and bucket sets related to query objectives. The stable object set can be found through distance based replacement strategy.*

Proof. We assume that the object set after DBR Algorithm is not stable. Thus there must exist an object o_i in this object set which can be replaced by another object o_j and $\mathcal{MG}(o_i, o_j) > 0$. According to terminating condition, replacement will be continued until the stable object set is found. Therefore, the object set must be stable. Lemma 2 has been proved. ∎

Algorithm 2. Distance based Replacement Search Algorithm

Input: IR-tree ir, query q, λ
Output: a set of objects V
1 $V \leftarrow \varnothing$;
2 $CBS \leftarrow BS(q.\Psi_1) \times BS(q.\Psi_2)... \times BS(q.\Psi_{|q.\Psi|})$;
3 $Somecbs \leftarrow$ randomSome(CBS);
4 V and $Dist(q, V) \leftarrow getMinimum(q, Somecbs)$;
5 **for** *object o_i in V* **do**
6 $\quad$ **for** *object o_j in O* **do**
7 $\quad\quad$ $(o'_i, o'_j) = argmax(\mathcal{MG}(o_i, o_j))$;
8 $\quad\quad$ $V = $Replace($o_i, o'_j$);
9 return V;

The pseudocode is shown in Algorithm 2. Firstly, we obtain all candidate bucket sets CBS which is the cartesian product of bucket sets (line 2). Next, we randomly select some candidate bucket sets (line 3). Then we apply Top-Down Search Algorithm to query q and each $cbs(cbs \in Somecbs)$ and obtain the object set V with minimum distance $\mathcal{D}ist(q, V)$ (line 4). Then, we compute $\mathcal{MG}(o_i, o_j)$ to find an object o_j with the maximum marginal gain and apply DBR strategy to replace it until finding a stable object set (line 5–8). Through DBR strategy, the problem of time consuming caused by messive candidate bucket sets has been solved. Assuming there are n objects in datasets and the stable object set is found after m replacement, the time complexity of DBR algorithm is $O(n \times m)$.

6 Experiment Study

In this section, we conduct extensive experiments on real datasets to evaluate the performance of our proposed algorithms.

6.1 Experiment Settings

We create the real datasets by using the online check-in records of Foursquare within the areas of New York City. Each record contains the user ID, venue with geographical location (place of interest) and the tips written in English. We put the records belonging to the same object to form textual descriptions of the objects, and the textual descriptions for each place are interpreted into a probabilistic topic distribution by the *LDA* model. The number of objects in this dataset is 206,097 in sum.

Table 2. Default values of parameters

Parameter	Default value	Description
$|q.\Psi|$	3	Number of query objectives
α	0.5	Weight factor for distance
β	0.5	Weight for spatial distance
L	100	Number of hash table
M	8	Number of *LSH* function
t	50	Number of latent topics

We compare the query time cost of proposed algorithms respectively. The default values for parameters are given in Table 2. All algorithms are implemented in Java and run on a PC with a 2-core Intel CPU at 2.5 GHz and 4 GB memory.

6.2 Performance Evaluation

(1) Comparisons of proposed methods

We evaluate the efficiency and accuracy of the proposed methods by varying the parameters in Table 2.

Effect of $|q.\Psi|$. We investigate the effect of $|q.\Psi|$ on the efficiency and accuracy of the proposed algorithms. As shown in Fig. 3, the DBR algorithm has much less time consumption than both the baseline and the SH-based algorithms. The query time of the SH-based algorithm and the baseline algorithm are exponentially increasing with the growth of the number of query objectives, while in contrast the query time for the DBR algorithm increases smoothly. Within our expectations, the baseline and DBR algorithms have similar accuracy performance, better than the SH-based algorithm in all $|q.\Psi|$ settings.

Effect of α. We study the effect of weight factor α which varying from 0 to 1. As shown in Fig. 4, all algorithms including the baseline algorithm, the SH-based algorithm and the DBR algorithm have a marginal increase on query time when α grows. The DBR algorithm has the best time performance while the baseline algorithm always takes most of the time to complish search processing. The DBR algorithm and the baseline algorithm outperform the SH-based algorithm in accuracy. And the distances to query for these algorithms are smoothly increase with the grows of value of α.

Effect of β. We investigate the performance of these algorithms when the threshold of weight factor β for spatial distance is varying. Figure 5 shows the results of our experiment. With the increase of β, these algorithms have the same increasing trend on query time. On the other hand, the baseline algorithm and the DBR algorithm always outperform the SH-based algorithm in accuracy when β varies from 0 to 1.

Effect of t. We proceed to examine the effect of number of latent topics by ploting query time and distance to query. As Fig. 6 shown, query time for these algorithms will increase as t grows. Besides, the SH-based algorithm and the baseline algorithm need more query time than the DBR algorithm. The DBR algorithm achieves high accuracy compared with other algorithms and the distances to query for these algorithms are almost unaffected by the increase in the number of topics.

(2) Evaluations of LSH parameters

The performance of the SH-based Algorithm and the DBR Algorithm is mainly influenced by LIR-tree. LIR-tree can be measured by parameters hash tables L and hash families M. We will tune these parameters to evaluate the performance of LIR-tree in sequence in this part.

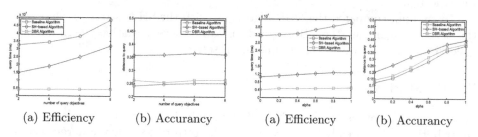

(a) Efficiency   (b) Accurancy   (a) Efficiency   (b) Accurancy

Fig. 3. Effect of $|q.\Psi|$ **Fig. 4.** Effect of α

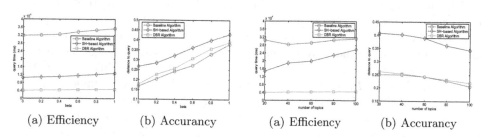

(a) Efficiency   (b) Accurancy   (a) Efficiency   (b) Accurancy

Fig. 5. Effect of β **Fig. 6.** Effect of t

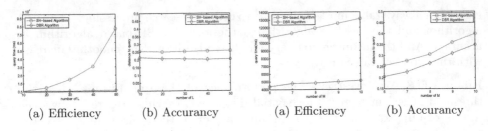

(a) Efficiency   (b) Accurancy   (a) Efficiency   (b) Accurancy

Fig. 7. Effect of L **Fig. 8.** Effect of M

Effect of L. Parameter L denotes the number of hash tables. Intuitively, a larger L indidates more information provided by the LIR-tree, which facilitates the accuracy of object sets. To achieve higher quality with lower time, we vary L from 10 to 50. As shown in Fig. 7, the DBR algorithm takes much less query time than the SH-based algorithm with the growth of the topics. This can be explained by the fact that the SH-based algorithm avoids computing for all candidate bucket set. Compared to the SH-based algorithm, the DBR algorithm always obtains result with less distance to query when L varies from 10 to 50.

Effect of M. A group of experiments are conducted to evaluate the performance of LIR-tree under different M, which is a fundamental parameter for constructing a hash table. We vary M from 6 to 10 with L fixed as 30. Figure 8 indicates that the query time goes up as M increases. And similar to Effect of L, the DBR algorithm transcends the SH-based algorithm in both efficiency and accuracy as we expected.

To sum up, compared the DBR algorithm with the SH-based algorithm and the baseline algorithm, the DBR algorithm can achieve a relatively high quality collective object set within short time in all settings.

7   Related Work

Spatial keyword query has been intensively studied in previous decades. Many contributions have already been made in the literature to support different types of spatio-textual querying. Some efforts are made to support the Spatial Keyword Boolean Query (SKBQ) [7,9,10,20,24] that requires exact keywords match, which may lead few or no results to be found. To overcome this problem, lots of work have been done to support the Spatial Keyword Approximate Query (SKAQ) [14,16,19,21–23], which ensures the query results are no longer sensitive to spelling errors and conventional spelling differences. Many novel indexing structures are proposed to support efficient processing on $SKBQ$ and $SKAQ$, such as IR-tree [7], IR^2-tree [9], MHR-tree [19], $S2I$ [16], etc. Numerous work studies the problem of spatial keyword query on *why-not questions* [18], *continuous querying* [1], *interactive querying* [23], *pub/sub system* [11,15], etc. Specifically, [20] addresses a more challenging problem on spatial keyword top-k queries, where some known object is unexpectedly missing from a result; and [18] investigates a novel problem, namely, continuous top-k spatial keyword queries on road networks; [13] eliminates the requirement of users to explicitly specify their preferences between spatial proximity and keyword relevance by enhancing the conventional queries with interaction; towards multi-objective query, [5] studies the $CSKQ$ problem to retrieve a group of objects that can collectively cover all keywords in query, while having the minimum distance cost.

But as far as we know, none of those existing approaches can retrieve spatial objects that are semantically relevant but morphologically different which are collectively cover all user-supplied keywords. The probabilistic topic models are statistical methods to analyze the words in documents and to discover the themes that run through them, how those themes are connected to each other, with no prior annotations or labeling of documents been required. Based on topic models, it is possible to measure the relevance of a testual description with regrad to a theme, as well as the relevance between different textual descriptions. The most classical topic models includes LDA [2], Dynamic Topic Model, etc. However, this method cannot find the object set that can satisfy all query objectives collectively and distributed in spatial rationally. Therefore, we investigate the topic model based collective spatial keyword querying to recommend users collective spatial objects that have both high spatial and semantic similarities to query.

8 Conclusion and Future Work

This paper committed to the problem of retrieving a group of spatial web objects more effectively and reasonably by converting keywords matching to topic distribution. The probabilistic topic model is utilized to interpret the textual descriptions attached to spatial objects and query objectives into topic distributions. To find the object set that can satisfy all query objectives collectively and distributed in spatial rationally, we propose an indexing structure to combine the spatial and semantic information effectively, as well as a searching algorithm to achieve efficient query processing. Extensive experimental results on real datasets demonstrate the efficiency of our proposed method. In the future, it will be interesting to consider the possible spatial dynamics of users and investigate the problem of continuous multi-objective spatial keyword querying with semantics.

Acknowledgement. This work was partially supported by Chinese NSFC project under grant numbers 61402312, 61232006, 61402313, 61572336, 61502324, 61572335, and Australia Research Council discovery projects under grant numbers DP140103499, DP160102412.

References

1. Barbieri, D.F., Braga, D., Ceri, S., Valle, E.D., Grossniklaus, M.: C-SPARQL: SPARQL for continuous querying. In: WWW, pp. 1061–1062 (2009)
2. Blei, D.M., Ng, A.Y., Jordan, M.I.: Latent dirichlet allocation. J. Mach. Learn. Res. **3**, 993–1022 (2003)
3. Buhler, J.: Efficient large-scale sequence comparison by locality-sensitive hashing. Bioinformatics **17**(5), 419–428 (2001)

4. Cao, X., Cong, G., Jensen, C.S.: Retrieving top-k prestige-based relevant spatial web objects. PVLDB **3**(1), 373–384 (2010)
5. Cao, X., Cong, G., Jensen, C.S., Ooi, B.C.: Collective spatial keyword querying. In: SIGMOD, pp. 373–384 (2011)
6. Chen, Y.-Y., Suel, T., Markowetz, A.: Efficient query processing in geographic web search engines. In: SIGMOD Conference, pp. 277–288 (2006)
7. Cong, G., Jensen, C.S., Wu, D.: Efficient retrieval of the top-k most relevant spatial web objects. PVLDB **2**(1), 337–348 (2009)
8. Datar, M., Immorlica, N., Indyk, P., Mirrokni, V.S.: Locality-sensitive hashing scheme based on p-stable distributions. In: Symposium on Computational Geometry, pp. 253–262 (2004)
9. De Felipe, I., Hristidis, V., Rishe, N.: Keyword search on spatial databases. In: ICDE, pp. 656–665 (2008)
10. Jiang, H., Zhao, P., Sheng, V.S., Xu, J., Liu, A., Wu, J., Cui, Z.: An efficient location-aware top-k subscription matching for publish/subscribe with Boolean expressions. In: Navathe, S.B., Wu, W., Shekhar, S., Du, X., Wang, X.S., Xiong, H. (eds.) DASFAA 2016. LNCS, vol. 9643, pp. 335–350. Springer, Cham (2016). doi:10.1007/978-3-319-32049-6_21
11. Huiqi, H., Liu, Y., Li, G., Feng, J., Tan, K.-L.: A location-aware publish, subscribe framework for parameterized spatio-textual subscriptions. In: ICDE, pp. 711–722 (2015)
12. Indyk, P., Motwani, R.: Approximate nearest neighbors: towards removing the curse of dimensionality. In: STOC, pp. 604–613 (1998)
13. Jin, J., Szekely, P.: Interactive querying of temporal data using a comic strip metaphor. In: IEEE VAST, pp. 163–170 (2010)
14. Li, F., Yao, B., Tang, M., Hadjieleftheriou, M.: Spatial approximate string search. IEEE Trans. Knowl. Data Eng. **25**(6), 1394–1409 (2013)
15. Li, G., Wang, Y., Wang, T., Feng, J.: Location-aware publish, subscribe. In: KDD, pp. 802–810 (2013)
16. Rocha-Junior, J.B., Gkorgkas, O., Jonassen, S., Nørvåg, K.: Efficient processing of top-k spatial keyword queries. In: Pfoser, D., Tao, Y., Mouratidis, K., Nascimento, M.A., Mokbel, M., Shekhar, S., Huang, Y. (eds.) SSTD 2011. LNCS, vol. 6849, pp. 205–222. Springer, Heidelberg (2011). doi:10.1007/978-3-642-22922-0_13
17. Slaney, M., Casey, M.: Locality-sensitive hashing for finding nearest neighbors. IEEE Sig. Process. Mag. **25**(2), 128–131 (2008)
18. Tran, Q.T., Chan, C.-Y.: How to conquer why-not questions. In: SIGMOD, pp. 15–26 (2010)
19. Yao, B., Li, F., Hadjieleftheriou, M., Hou, K.: Approximate string search in spatial databases. In: ICDE, pp. 545–556 (2010)
20. Zhang, C., Zhang, Y., Zhang, W., Lin, X.: Inverted linear quadtree: efficient top k spatial keyword search. In: ICDE, pp. 901–912 (2013)
21. Zheng, B., Yuan, N.J., Zheng, K., Xie, X., Sadiq, S., Zhou, X.: Approximate keyword search in semantic trajectory database. In: ICDE 2015, pp. 975–986 (2015)
22. Zheng, K., Huang, Z., Zhou, A., Zhou, X.: Discovering the most influential sites over uncertain data: a rank-based approach. IEEE TKDE **24**(12), 2156–2169 (2012)
23. Zheng, K., Han, S., Zheng, B., Shang, S., Jiajie, X., Liu, J., Zhou, X.: Interactive top-k spatial keyword queries. In: ICDE 2015, pp. 423–434 (2015)
24. Ding, Z., Xu, J., Yang, Q.: SeaCloudDM: a database cluster framework for managing and querying massive heterogeneous sensor sampling data. J. Supercomput. **66**(3), 1260–1284 (2013)

Query Processing and Optimization (II)

RSkycube: Efficient Skycube Computation by Reusing Principle

Kaiqi Zhang[1], Hong Gao[1], Xixian Han[1], Donghua Yang[2(✉)], Zhipeng Cai[3],
and Jianzhong Li[1]

[1] School of Computer Science and Technology,
Harbin Institute of Technology, Harbin, China
{zhangkaiqi,honggao,lijzh}@hit.edu.cn, hanxixian@gmail.com
[2] Academy of Fundamental and Interdisciplinary Sciences,
Harbin Institute of Technology, Harbin, China
yang.dh@hit.edu.cn
[3] Department of Computer Science, Georgia State University, Atlanta, USA
zcai@gsu.edu

Abstract. Over the past years, the skyline query has already attracted wide attention in database community. In order to meet different preferences for users, the skycube computation is proposed to compute skylines, or cuboids, on all possible non-empty dimension subsets. The key issue of computing skycube is how to share computation among multiple related cuboids, which classified into *sharing strict space dominance* and *sharing space incomparability*. However, state-of-the-art algorithm only leverages *sharing strict space dominance* to compute skycube. This paper aims to design a more efficient skycube algorithm that shares computation among multiple related cuboids. We first propose a set of rules named *identical partitioning* (*IP*) for constructing a novel structure VSkyTree. Moreover, we present the reusing principle, which utilizes both *sharing strict space dominance* and *sharing space incomparability* by reusing VSkyTree on parent cuboids to compute child cuboids. Then, in *top-down* fashion, we design an efficient skycube computation algorithm RSkycube based on the reusing principle. Our experimental results indicate that our algorithm RSkycube significantly outperforms state-of-the-art skycube computation algorithm on both synthetic and real datasets.

Keywords: Skyline · Skycube · Space partitioning

1 Introduction

The skyline query is coined out by Börzsönyi *et al.* in [1]. In the past several years, many researchers have paid much attention to it, and it has becomes an important preference query in multi-criteria decision making applications. For a multi-dimensional dataset, the skyline query returns a set of interesting points, which are not dominated by any other point in the dataset. Specifically, given two points p and q, if p is not worse than q in all dimensions and better than q in at least one dimension, then p dominates q.

© Springer International Publishing AG 2017
S. Candan et al. (Eds.): DASFAA 2017, Part II, LNCS 10178, pp. 51–64, 2017.
DOI: 10.1007/978-3-319-55699-4_4

However, it is common that users have many different preferences. For a dataset in an online hotel booking system with 4 dimensions, *i.e.*, hotel (*price, distance, room number, star level*). Some users may focus on *price* and *distance*. And others could pay attention to *room number* and *star level*. Different people may have different preferences (*i.e.*, dimension subsets). In order to meet all possible preferences, Yuan *et al.* propose the concept of skycube [10]. For a d-dimensional dataset, skycube computes $2^d - 1$ skylines, or cuboids, on all possible non-empty dimension subsets.

Skycube computation has recently received considerable attention. [9] emphasizes updating skycube in dynamic environment. Kailasam *et al.* [2] focus on utilizing bitmap index for skycube computation. [6–8] concentrate on using skycube to construct *skyline group lattice* and *decisive subspace*. The most relevant work with us is [5,10].

The key issue of computing skycube is how to share computation among multiple related cuboids as much as possible. Yuan *et al.* [10] propose two skycube computation algorithms, Bottom-Up Skycube (BUS) and Top-Down Skycube (TDS). BUS adopts *bottom-up* fashion, which computes a cuboid by utilizing their child cuboids. In *top-down* fashion, TDS takes parent cuboids as input and avoids accessing entire dataset. With *top-down* fashion, TDS is reported to be faster than BUS, especially in large or high-dimensional data [10].

Based on *top-down* fashion, Lee *et al.* [5] propose a new skycube algorithm QSkycube exploiting SkyTree [4], which is a space partitioning technique. However, we observe that there is still much room for optimization. Sharing strategies on space partitioning is divided into *sharing strict space dominance* and *sharing space incomparability*. It aims to reduce duplicate computation in skycube computation. *Sharing strict space dominance* can straightly filter out some non-skyline points and reduce the amount of input. And *sharing space incomparability* can reduce duplicate comparisons between incomparable points. While QSkycube only leverages *sharing strict space dominance* to compute skycube. Because it is necessary for QSkycube to reconstruct SkyTree to compute child cuboids, which makes *space incomparability relationship* between points disappeared.

This paper aims to design a more efficient skycube algorithm that shares computation among multiple related cuboids. We first propose a set of rules named *identical partitioning* (*IP*) for constructing a novel structure VSkyTree. Moreover, we present the reusing principle, which utilizes both *sharing strict space dominance* and *sharing space incomparability* by reusing VSkyTree on parent cuboids to compute child cuboids. Then, in *top-down* fashion, we design an efficient skycube computation algorithm RSkycube based on the reusing principle. Our experimental results indicate that our algorithm RSkycube significantly outperforms state-of-the-art skycube computation algorithm on both synthetic and real datasets.

Our contributions can be summarized as follows:

- We analyze the room of optimization for the existing skycube computation algorithms.

- We propose a set of rules named *identical partitioning* (*IP*) for constructing a novel structure VSkyTree.
- we present the reusing principle, which utilizes both *sharing strict space dominance* and *sharing space incomparability* by reusing VSkyTree on parent cuboids to compute child cuboids.
- We design an efficient skycube computation algorithm RSkycube based on the reusing principle.
- We evaluate our proposed algorithm RSkycube by comparing it with state-of-the-art skycube algorithm in dimensionality and cardinality on both synthetic and real datasets.

The rest of this paper is organized as follows. Section 2 introduces some definitions about skycube. Section 3 proposes a set of rules named *identical partitioning* (*IP*) for constructing a novel structure VSkyTree and present a reusing principle which utilizes both sharing strategies to compute child cuboids. Section 4 designs an efficient skycube algorithm RSkycube based on the reusing principle. Section 5 evaluates experimentally our proposed algorithm with state-of-the-art skycube algorithm on both real and synthetic datasets. Finally, our conclusion is summarized in Sect. 6.

2   Preliminaries

For easy explanation, we present some notations used throughout this paper. Given a dataset S with d dimensions, $\mathcal{D}$ represents the full dimension set. *i.e.*, $\mathcal{D} = \{D_0, D_1, ..., D_{d-1}\}$. And the domain space for $\mathcal{D}$ is denoted by $\mathbb{S}^{\mathcal{D}}$, where $\mathbb{S}_i^{\mathcal{D}}$ is the domain on dimension D_i. Actually, S is a subset of $\mathbb{S}^{\mathcal{D}}$. For any point p in S, we denote the value of point p on ith dimension by $p(D_i)$, abbreviated as $p(i)$, where $i \in [0, d\text{--}1]$. Without loss of generality, we assume that lower values are better for all users on all dimensions.

Given two points p and q in the dataset S on $\mathcal{D}$, p *dominates* q on dimension subset $\mathcal{U}$, where $\mathcal{U} \subseteq \mathcal{D}$, denoted by $p \succ_P^{\mathcal{U}} q$ iff $\forall D_i \in \mathcal{U}$, $p(i) \leq q(i)$ and $\exists D_i \in \mathcal{U}$, $p(i) < q(i)$. Otherwise $p \nsucc_P^{\mathcal{U}} q$. If $p \nsucc_P^{\mathcal{U}} q$ and $q \nsucc_P^{\mathcal{U}} p$, then p and q are incomparable on $\mathcal{U}$. The skyline query on $\mathcal{U}$ returns all the points that are not dominated by any other points on $\mathcal{U}$, denoted by $SKY_{\mathcal{U}}(S) = \{p \in S \mid \forall q \in S, q \nsucc_P^{\mathcal{U}} p\}$.

Definition 1 (Skycube and Cuboid). *Given a dataset S on $\mathcal{D}$, the skycube of S consists of all skyline results on $2^d - 1$ non-empty dimension subsets, denoted by $SKYCUBE(S) = \{SKY_{\mathcal{U}}(S) \mid \mathcal{U} \subseteq \mathcal{D}, \mathcal{U} \neq \emptyset\}$, where $SKY_{\mathcal{U}}(S)$ is called the cuboid on $\mathcal{U}$, abbreviated as cuboid $\mathcal{U}$.*

Actually, to meet users having different preferences, Skycube computes skyline results on all possible non-empty dimension subsets. Skycube can be visualized as a lattice as shown in Fig. 1(b) for 3-dimensional datasets. For two cuboids A on $\mathcal{U}$ and B on $\mathcal{V}$, if $\mathcal{U} \subset \mathcal{V}$ and $|\mathcal{V} - \mathcal{U}| = 1$, then A is a child of B and B is a parent of A. For example, Fig. 1(c) gives the skycube computation result for the dataset in Fig. 1(a).

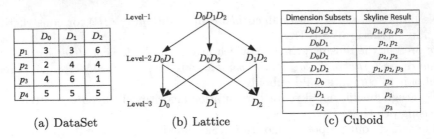

Fig. 1. Skycube

To explore sharing strategies, we first assume *distinct value condition*, which is also adopted in [5,10] and the general case beyond the assumption is also resolved in them. This assumption is that $\forall p, q \in S, \forall D_i \in \mathcal{D} : p(i) \neq q(i)$.

Based on *distinct value condition*, for two dimension subsets $\mathcal{U}$ and $\mathcal{V} \subseteq \mathcal{D}$, if $\mathcal{U} \subseteq \mathcal{V}$, then $SKY_{\mathcal{U}}(S) \subseteq SKY_{\mathcal{V}}(S)$. It means that the non-skyline points on $\mathcal{V}$ must be non-skyline points on $\mathcal{U}$. For example in Fig. 1(c), $SKY_{D_{01}}(S) \subseteq SKY_{D_{012}}(S)$. Therefore, $SKY_{\mathcal{V}}(S)$ can be took as input for computing $SKY_{\mathcal{U}}(S)$, which is called *top-down* fashion. The correctness is proved in [5,10].

3   Sharing Strategies by Reusing Principle

3.1   Sharing Strategies on Space Partitioning

We first introduce the space partitioning technique, given a dataset S on $\mathcal{D}$, $\mathbb{S}^{\mathcal{D}}$ is the whole d-dimensional space, for any point $\hat{p}$ in $\mathbb{S}^{\mathcal{D}}$, $\hat{p}$ divides $\mathbb{S}^{\mathcal{D}}$ into 2^d disjoint subspaces $\mathbb{S}^{\mathcal{D},\hat{p}} = \{\mathbb{S}_0, \mathbb{S}_1, ..., \mathbb{S}_{2^d-1}\}$. For any subspace $\mathbb{S}_b$, the subscript b indicates the address of $\mathbb{S}_b$, represented by a d-bit bitmap vector B. $\forall p \in \mathbb{S}_B$, $\forall i \in [0, d-1]$, $p(i)$ satisfies the following condition:

$$\begin{cases} p(i) < \hat{p}(i), & if\ B[i] = 0, \\ p(i) \geq \hat{p}(i), & if\ B[i] = 1. \end{cases}$$

Obviously, $\hat{p}$ also divides S into 2^d subsets, each of them belongs to one of subspaces. Figure 2(a) gives a 3-dimensional partitioning. If $\mathbb{S}_{000}$ is not empty, then the points in $\mathbb{S}_{111}$ must be dominated by any one in $\mathbb{S}_{000}$ and straightly pruned.

Definition 2 (Dominance on Spaces). *Given two subspaces $\mathbb{S}_B$ and $\mathbb{S}_{B'} \in \mathbb{S}^{\mathcal{D},\hat{p}}$, $\mathbb{S}_B$ dominates $\mathbb{S}_{B'}$ on dimension set $\mathcal{U} \subseteq \mathcal{D}$, denoted by $\mathbb{S}_B \succ_S^{\mathcal{U}} \mathbb{S}_{B'}$, if $B \mid B' = B'$, where "$\mid$" is bitwise or operator. Otherwise $\mathbb{S}_B \not\succ_S^{\mathcal{U}} \mathbb{S}_{B'}$. In addition, $\mathbb{S}_B$ strictly dominates $\mathbb{S}_{B'}$ on $\mathcal{U}$, if $\forall D_i \in \mathcal{U}$, $B[i] = 0$ and $B'[i] = 1$.*

Given two subspace $\mathbb{S}_B$ and $\mathbb{S}_{B'} \in \mathbb{S}^{\mathcal{D},\hat{p}}$, if $\mathbb{S}_B \not\succ_S^{\mathcal{U}} \mathbb{S}_{B'}$ and $\mathbb{S}_{B'} \not\succ_S^{\mathcal{U}} \mathbb{S}_B$, we call that $\mathbb{S}_B$ and $\mathbb{S}_{B'}$ are space incomparability.

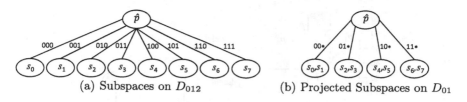

(a) Subspaces on D_{012} (b) Projected Subspaces on D_{01}

Fig. 2. Subspaces and projected subspaces

Theorem 1. *Given two subspaces $\mathbb{S}_B$ and $\mathbb{S}_{B'} \in \mathbb{S}^{\mathcal{D},\widehat{p}}$, if $\mathbb{S}_B \not\succ_S^{\mathcal{U}} \mathbb{S}_{B'}$, $\mathcal{U} \subseteq \mathcal{D}$, then $\forall p \in \mathbb{S}_B$ and $\forall q \in \mathbb{S}_{B'}$, $p \not\succ_P^{\mathcal{U}} q$.*

Theorem 1 has already been proved in [11]. Given any two points $p \in \mathbb{S}_B$ and $q \in \mathbb{S}_{B'}$, if $\mathbb{S}_B$ and $\mathbb{S}_{B'}$ are space incomparability on $\mathcal{D}$, then p and q are incomparable each other on $\mathcal{D}$.

A point $\widehat{p}$ partitions 3-dimensional space into 2^3 subspaces $\{\mathbb{S}_{000}, \mathbb{S}_{001}, \mathbb{S}_{010}, \mathbb{S}_{011}, \mathbb{S}_{100}, \mathbb{S}_{101}, \mathbb{S}_{110}, \mathbb{S}_{111}\}$ as shown in Fig. 2(a). The projections of these subspaces on D_{01} are $\{\mathbb{S}_{00*}, \mathbb{S}_{01*}, \mathbb{S}_{10*}, \mathbb{S}_{11*}\}$ in Fig. 2(b), symbol "*" represents any binary value. There still exist two relationships between the projected subspaces: *strict space dominance* and *space incomparability*. For example, $\mathbb{S}_{00*}$ and $\mathbb{S}_{11*}$ are *strict space dominance* on D_{01}, $\mathbb{S}_{01*}$ and $\mathbb{S}_{10*}$ are *space incomparability* on D_{01}.

Sharing strategy on space partitioning for computing skycube is that sharing the two relationships from parent cuboid to compute child cuboids. It aims to reduce duplicate computation in skycube computation. *Sharing strict space dominance* can straightly filter out some non-skyline points and reduce the amount of input. For example, $\mathbb{S}_{001}$ and $\mathbb{S}_{110}$ are *strict space dominance* on D_{01}, the points in $\mathbb{S}_{110}$ must be dominated by any point in $\mathbb{S}_{001}$ on D_{01}, therefore, for the computation of cuboid D_{01}, the points in $\mathbb{S}_{110}$ can be straightly pruned without new comparisons. *Sharing space incomparability* can reduce duplicate comparisons between incomparable points. For example, for any $p \in \mathbb{S}_{011}$ and $q \in \mathbb{S}_{100}$, we can conclude that p and q can not dominate each other on D_{01} without duplicate comparisons.

3.2 SkyTree

It needs three steps to construct SkyTree using space partitioning. First step, select a skyline point $\widehat{p}$ as root of SkyTree to partition space into 2^d subspaces $\{\mathbb{S}_{0...00}, \mathbb{S}_{0...01}, ..., \mathbb{S}_{1...11}\}$. Since $\widehat{p}$ is a skyline point, $\mathbb{S}_{0...00}$ must be empty and all points in $\mathbb{S}_{1...11}$ must be dominated by $\widehat{p}$. $\mathbb{S}_{1...11}$ need not be considered in subsequent processing. The other $2^d - 2$ subspaces are organized as the child nodes of $\widehat{p}$. Second step, for any $\mathbb{S}_B$ and $\mathbb{S}_{B'}$ in the above $2^d - 2$ subspaces, if $\mathbb{S}_B \succ_S^{\mathcal{D}} \mathbb{S}_{B'}$, that is the points in $\mathbb{S}_B$ may dominate the points in $\mathbb{S}_{B'}$, then the dominated points in $\mathbb{S}_{B'}$ need to be filtered out. Third step, for each non-empty subspace, recursively repeat above two steps.

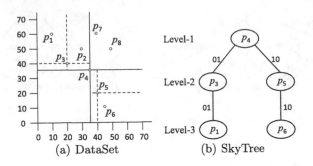

Fig. 3. SkyTree

Given a dataset in Fig. 3(a), first select skyline point p_4 as root to divide 2-dimensional space into 2^2 subspaces $\{\mathbb{S}_{00}, \mathbb{S}_{01}, \mathbb{S}_{10}, \mathbb{S}_{11}\}$, p_1, p_2, $p_3 \in \mathbb{S}_{01}$, p_5, $p_6 \in \mathbb{S}_{10}$, and p_7, $p_8 \in \mathbb{S}_{11}$. p_7 and p_8 are dominated by p_4 and straightly pruned. Since $\mathbb{S}_{01} \not\succ^{\mathcal{D}}_S \mathbb{S}_{10}$ and $\mathbb{S}_{10} \not\succ^{\mathcal{D}}_S \mathbb{S}_{01}$, no point needs to be filtered out. Then, select skyline point p_3 to recursively partition $\mathbb{S}_{01}$ into 2^2 subspaces. In order to distinguish $\mathbb{S}_{01}$ and partitioned subspaces by it, the four subspaces are denoted by $\{\mathbb{S}_{0100}, \mathbb{S}_{0101}, \mathbb{S}_{0110}, \mathbb{S}_{0111}\}$. After partitioned by p_3, $p_1 \in \mathbb{S}_{0101}$, $p_2 \in \mathbb{S}_{0111}$, p_2 is dominated by p_3 and pruned. Select skyline point p_5 to recursively partition $\mathbb{S}_{10}$ into 2^2 subspaces $\{\mathbb{S}_{1000}, \mathbb{S}_{1001}, \mathbb{S}_{1010}, \mathbb{S}_{1011}\}$, $p_6 \in \mathbb{S}_{1010}$. Finally, each leaf node has only one point, SkyTree is completely constructed as shown in Fig. 3(b).

QSkycube utilizes SkyTree to compute skycube in *top-down* fashion. Given $\mathcal{U} \subset \mathcal{V} \subseteq \mathcal{D}$ and $|\mathcal{V} - \mathcal{U}| = 1$, QSkycube utilizes SkyTree $\mathcal{T}_{\mathcal{V}}$ on $\mathcal{V}$ to compute $\mathcal{T}_{\mathcal{U}}$, which contains cuboid $\mathcal{U}$. Firstly, it exploits *sharing strict space dominance* to straightly filter out some non-skyline points to reduce duplicate comparisons. Then it traverses $\mathcal{T}_{\mathcal{V}}$ to get the remaining points and use them to construct $\mathcal{T}_{\mathcal{U}}$. However, we observe that there is still much room for optimization. To guarantee the partitioning points of SkyTree are skyline points, QSkycube must reconstruct SkyTree, which makes *space incomparability relationship* between points disappeared. Therefore, QSkycube only adopts *sharing strict space dominance* to accelerate skycube computation.

In order to utilize both sharing strategies to compute $\mathcal{T}_{\mathcal{U}}$ from $\mathcal{T}_{\mathcal{V}}$, $\mathcal{T}_{\mathcal{U}}$ should be straightly derived from $\mathcal{T}_{\mathcal{V}}$, instead of being reconstructed completely.

3.3   Sharing Strategies by Reusing VSkyTree

We first propose a novel structure VSkyTree based on *identical partitioning (IP)* rules. Then, we utilize both sharing strategies by reusing VSkyTree to compute child cuboids.

The difference between SkyTree and VSkyTree is how to determine partitioning points. Given any subspace $\mathbb{S}_B = [(l_0, h_0),..., (l_{d-1}, h_{d-1})]$, $\forall D_i \in \mathcal{D}$, $\mathbb{S}^i_B = \{p(i) \mid \forall p \in S, l_i \leq p(i) \leq h_i\}$, l_i and h_i are boundaries of $\mathbb{S}_B$ on D_i,

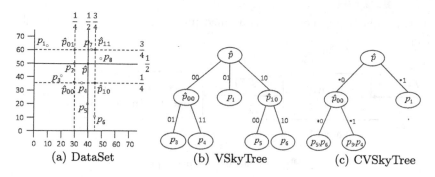

Fig. 4. VSkyTree and CVSkyTree

M_i is the median of $\mathbb{S}_B^i$. The partitioning point $\widehat{p}$ of $\mathbb{S}_B$ is $(M_0, M_1, ..., M_{d-1})$. Actually, the partitioning points based on this method may be not in dataset S, therefore we call them virtual partitioning points. This method of selecting virtual partitioning points is call *identical partitioning* (IP) rules.

The construction of VSkyTree is similar to that of SkyTree. First step, determine the partitioning point $\widehat{p}$ based on above *IP* rules. $\widehat{p}$ partitions space into 2^d subspaces $\{\mathbb{S}_{0...00}, \mathbb{S}_{0...01}, ..., \mathbb{S}_{1...11}\}$. If $\mathbb{S}_{0...00}$ is not empty, all the points in $\mathbb{S}_{1...11}$ must be dominated by any point in $\mathbb{S}_{0...00}$ and $\mathbb{S}_{1...11}$ need not be considered in subsequent processing. These subspaces are organized as the child nodes of $\widehat{p}$. Second step, for any $\mathbb{S}_B$ and $\mathbb{S}_{B'}$ in the above subspaces, if $\mathbb{S}_B \succ_S^{\mathcal{D}} \mathbb{S}_{B'}$, that is the points in $\mathbb{S}_B$ may dominate the points in $\mathbb{S}_{B'}$, then the dominated points in $\mathbb{S}_{B'}$ need to be filtered out. Third step, for each subspace with more than one point, recursively repeat above two steps.

Given a dataset in Fig. 4(a), first select the virtual partitioning point $\widehat{p}$ based on above *IP* rules as root, $\widehat{p}(0) = p_5(0)$, $\widehat{p}(1) = p_2(1)$, $p_5(0)$ and $p_2(1)$ are medians of the projections of all points in whole space on D_0 and D_1, respectively. $\widehat{p}$ divides 2-dimensional space into 2^2 subspaces $\{\mathbb{S}_{00}, \mathbb{S}_{01}, \mathbb{S}_{10}, \mathbb{S}_{11}\}$, p_3, $p_4 \in \mathbb{S}_{00}$, p_1, $p_2 \in \mathbb{S}_{01}$, p_5, $p_6 \in \mathbb{S}_{10}$ and p_7, $p_8 \in \mathbb{S}_{11}$. $\mathbb{S}_{00}$ is not empty, p_7 and p_8 are dominated by any point in $\mathbb{S}_{00}$ and straightly pruned. Since $\mathbb{S}_{00} \succ_S^{D_{01}} \mathbb{S}_{01}$ and $\mathbb{S}_{00} \succ_S^{D_{01}} \mathbb{S}_{10}$, p_1, p_2, p_5 and p_6 need to be compared with p_3 and p_4, p_2 is dominated by p_3 and pruned. Then, select the virtual partitioning point $\widehat{p}_{00}$ of subspace $\mathbb{S}_{00}$ based on above *IP* rules. $\widehat{p}_{00}$ recursively partition $\mathbb{S}_{00}$ into 2^2 subspaces $\{\mathbb{S}_{0000}, \mathbb{S}_{0001}, \mathbb{S}_{0010}, \mathbb{S}_{0011}\}$, $p_3 \in \mathbb{S}_{0001}$, $p_4 \in \mathbb{S}_{0011}$. Since $\mathbb{S}_{0001} \succ_S^{D_{01}} \mathbb{S}_{0011}$, p_4 is compared with p_3 and not dominated by it. Select the virtual partitioning point $\widehat{p}_{10}$ of $\mathbb{S}_{10}$ to recursively partition $\mathbb{S}_{10}$ into 2^2 subspaces $\{\mathbb{S}_{1000}, \mathbb{S}_{1001}, \mathbb{S}_{1010}, \mathbb{S}_{1011}\}$, $p_5 \in \mathbb{S}_{1000}$, $p_6 \in \mathbb{S}_{1010}$. Since $\mathbb{S}_{1000} \succ_S^{D_{01}} \mathbb{S}_{1010}$, p_6 is compared with p_5 and not dominated by it. Finally, each leaf node has only one point, VSkyTree is completely constructed as shown in Fig. 4(b).

Given SkyTree $\mathcal{T}_\mathcal{V}$ on $\mathcal{V}$, we can combine the points in $\mathcal{T}_\mathcal{V}$ to generate a combined VSkyTree, named CVSkyTree. For example in Fig. 4, $\widehat{p}_{00}(1) = \widehat{p}_{10}(1)$, when we only consider the partitioning on D_1, $\widehat{p}_{00}$ and $\widehat{p}_{10}$ can be combined together. Similarly, p_5 and p_6 are combined into $\mathbb{S}_{*0*0}$, p_3 and p_4 are combined into $\mathbb{S}_{*0*1}$.

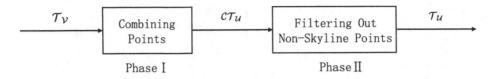

Fig. 5. The reusing principle

We present the reusing principle, which utilizes both *sharing strict space dominance* and *sharing space incomparability* by reusing VSkyTree on parent cuboids to compute child cuboids as shown in Fig. 5. Given VSkyTree $\mathcal{T}_\mathcal{V}$ on $\mathcal{V}$ and VSkyTree $\mathcal{T}_\mathcal{U}$ on $\mathcal{U}$, where $\mathcal{U} \subset \mathcal{V}$ and $|\mathcal{V} - \mathcal{U}| = 1$. $\mathcal{T}_\mathcal{V}$ and $\mathcal{T}_\mathcal{U}$ are took as input and output, respectively. The basic principle of reusing VSkyTree is utilizing $\mathcal{T}_\mathcal{V}$ to compute $\mathcal{T}_\mathcal{U}$. The principle is divided into two phases: Phase I mainly aims to combine the points of $\mathcal{T}_\mathcal{V}$ into CVSkyTree $\mathcal{CT}_\mathcal{U}$ on $\mathcal{U}$. $\mathcal{CT}_\mathcal{U}$ can share both *strict space dominance* and *space incomparability* from $\mathcal{T}_\mathcal{V}$. And the cost of combining points is low. Phase II focuses on filtering out non-skyline points in $\mathcal{CT}_\mathcal{U}$, it is similar to the *second step* in the process of constructing VSkyTree. The $\mathcal{CT}_\mathcal{U}$ after being filtered is the output result $\mathcal{T}_\mathcal{U}$. Finally, the cuboid $\mathcal{U}$ can be obtained by traversing $\mathcal{T}_\mathcal{U}$.

4 Algorithms

We design an efficient skycube computation algorithm RSkycube, which takes advantage of both sharing strategies by reusing principle. RSkycube computes skycube in *top-down* fashion, which utilizes parent cuboid to compute child cuboids.

Algorithm 1 depicts the pseudo code of RSkycube algorithm. We conduct $D.size$ iterations in *top-down* fashion as shown in line 1. And each iteration computes the cuboid for these dimension sets with same size in line 2–8. For the first iteration in line 3–4, function ComputeVSkyTree computes cuboid on full dimension set $\mathcal{D}$ and organizes them as VSkyTree according to IP rules and space partitioning technique. For any other dimension subset $\mathcal{U} \subset \mathcal{D}$, as shown in line 6, first selecting a parent dimension set $\mathcal{V}$ of $\mathcal{U}$. Then, according to reusing principle, utilizing $\mathcal{T}_\mathcal{V}$ to compute $\mathcal{T}_\mathcal{U}$, which contains the cuboid $\mathcal{U}$. The reusing principle has two phases as shown in line 7–8, function CombineVSkyTree implements Phase I, it aims to combine the points in $\mathcal{T}_\mathcal{V}$ to generate $\mathcal{CT}_\mathcal{U}$, which can share *strict space dominance* and *space incomparability* from $\mathcal{T}_\mathcal{V}$. Owing to the limitation of space, we omit the detail of CombineVSkyTree. The detailed process of combining the points is introduced in Section. Phase II mainly focuses on filtering out all non-skyline points. After filtering, all the points in $\mathcal{CT}_\mathcal{U}$ is the cuboid $\mathcal{U}$.

Algorithm 1. RSkycube(S, $\mathcal{D}$)

Input: A dataset S on dimension set $\mathcal{D}$
Output: The *SKYCUBE* result of S on $\mathcal{D}$

1: **for** *iteration* $\leftarrow |\mathcal{D}|$ **to** 1 **do**
2:   **for** $\forall \mathcal{U} \subseteq \mathcal{D}$ and $|\mathcal{U}| = iteration$ **do**
3:     **if** *iteration* $= |\mathcal{D}|$ **then**
4:       $\mathcal{T}_{\mathcal{U}} \leftarrow$ **ComputeVSkyTree**(S, $\mathcal{D}$).
5:     **else**
6:       select $\mathcal{V} \supset \mathcal{U}$ and $|\mathcal{V}| = |\mathcal{U}| + 1$.
7:       $\mathcal{CT}_{\mathcal{U}} \leftarrow$ **CombineVSkyTree**($\mathcal{T}_{\mathcal{V}}$, $\mathcal{U}$, $\mathcal{V}$). // Phase I of Reusing Principle
8:       $\mathcal{T}_{\mathcal{U}} \leftarrow$ **FilterCVSkyTree**($\mathcal{CT}_{\mathcal{U}}$, $\mathcal{U}$). // Phase II of Reusing Principle
9:     insert skyline points in $\mathcal{T}_{\mathcal{U}}$ into *SKYCUBE*(S).
10: **return** *SKYCUBE*(S).

Algorithm 2. ComputeVSkyTree(S, $\mathcal{D}$)

Input: A dataset S on dimension set $\mathcal{D}$
Output: The VSkyTree on $\mathcal{D}$

1: $max \leftarrow 2^d - 1$.
2: $L[0, max] \leftarrow \{\}$.
3: Select virtual partitioning point as root $\widehat{p}$.
4: **for** $\forall p \in S$ **do**
5:   $i \leftarrow$ Partition($\widehat{p}$, p)
6:   $L[i]$.Add(p).
7: **for** $i \leftarrow 0$ **to** max **do**
8:   **if** $L[i]$.Size() > 0 **then**
9:     **for** $\forall \mathcal{T}_{tmp} \in \widehat{p}.child$ and $\mathcal{T}_{tmp}.space \succ_S^{\mathcal{U}} \mathbb{S}_i$ **do**
10:       **Filter**($\mathcal{T}_{tmp}$, $L[i]$).
11:     $\mathcal{T} \leftarrow$ **ComputeVSkyTree**($L[i]$, $\mathcal{D}$).
12:     $\widehat{p}$.AddChild($\mathcal{T}$).
13: **return** *SKYCUBE*(S).

Algorithm 2 depicts the pseudo code of ComputeVSkyTree algorithm. The pseudo code of ComputeVSkyTree algorithm, as shown in Algorithm 2, accords with the processing of constructing VSkyTree. It aims to compute VSkyTree $\mathcal{T}_{\mathcal{D}}$ on $\mathcal{D}$. $\mathcal{T}_{\mathcal{D}}$ contains all the skyline points on cuboid $\mathcal{D}$. As well as it can be input in next iteration for computing its child cuboids. First step, in line 3–6, according to *IP* rules, selecting a virtual partitioning point $\widehat{p}$. Using $\widehat{p}$, all the points in S are partitioned into 2^d subsets, which are stored in $L[0, max]$. For example, $L[i]$ contains all the points in subspace $\mathbb{S}_i$. Second step, for any non-empty subset $L[i]$, as shown in line 8–10, the points in $L[i]$ need to be compared with the points in $\mathcal{T}_{tmp}$, where $\mathcal{T}_{tmp}.space \succ_S^{\mathcal{U}} \mathbb{S}_i$. And the non-skyline points in $L[i]$ dominated by any one in $\mathcal{T}_{tmp}$ need to be pruned. Third step, recursively partition subset $L[i]$ and add subtree corresponding to $L[i]$ as a child of $\widehat{p}$.

Algorithm 3 gives the pseudo code of function FilterCVSkyTree, which is the Phase II of reusing principle. It is a recursive function and aims to filter out

---

Algorithm 3. FilterCVSkyTree($\mathcal{CT}_\mathcal{U}$, $\mathcal{U}$)

Input: A CVSkyTree $\mathcal{CT}_\mathcal{U}$ on $\mathcal{U}$

1: **for** $\forall \mathcal{CT} \in \mathcal{CT}_\mathcal{U}.child$ **do**
2:     **for** $\forall \mathcal{CT}_{tmp} \in \mathcal{CT}_\mathcal{U}.child$ and $\mathcal{CT}_{tmp}.space \succ_S^\mathcal{U} \mathcal{CT}.space$ **do**
3:         **Filter**($\mathcal{CT}_{tmp}$, $\mathcal{CT}$).
4:     **FilterVSkyTree**($\mathcal{CT}$).

---

all non-skyline points in CVSkyTree $\mathcal{CT}_\mathcal{U}$ using itself. Line 3 implies that the points in subtree $\mathcal{CT}$ only need to be compared with the points in $\mathcal{CT}_{tmp}$, where $\mathcal{CT}_{tmp}.space \succ_S^\mathcal{U} \mathcal{CT}.space$. It accelerates the efficiency of filtering.

5 Experimental Evaluation

We conduct the evaluation for our proposed algorithm RSkycube in dimensionality and cardinality, by comparing it with the naive method and state-of-the-art skycube computation algorithm QSkycube [5] on both synthetic and real datasets. The experimental results indicate that our algorithm RSkycube significantly outperforms other skycube computation algorithms.

5.1 Experimental Settings

According to the instructions in [1], we generate two types of synthetic datasets that are Independent (IND) and Anti-correlated (ANT), respectively. On the synthetic datasets, dimensionality d is from 5 to 9 and cardinality is from 200 K to 1000 K. By default, they are 8 and 200 K, respectively. The domain of all values in points is [0, 1000]. For real datasets, ColorMoments[1] and IPUMS (see Footnote 1) are collected. They have 9 dimensional 68404 data points and 6 dimensional (selected from 23) 74954 data points, respectively. Colormoments is related with the image features and IPUMS describes unweighted PUMS census data from the Los Angeles and Long Beach areas in the year 1980. All algorithms are implemented by C++ languages and run on Intel Core-i7 CPU at 3.6 GHz, with 32 GB of RAM.

We regard *response time* and DC as the performance parameter. DC is the number of comparisons on dimensions per point for skycube computation, and computation formula is shown as follows.

$$DC = \frac{Total\ number\ of\ dimension\ comparisons}{The\ number\ of\ points}$$

Skycube computation is a compute-intensive operation, and main operation is the comparisons among points. Similar measure DT has been widely utilized in prior work [3,4]. There are two kinds of dominance tests: (1) Only check one point p whether is dominated by another point q or not. It does not require

---

[1] The data set is collected from kdd.ics.uci.edu.

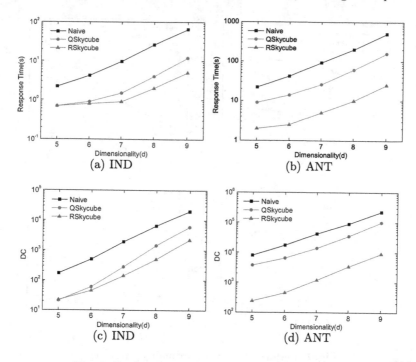

Fig. 6. Performance over dimensionality variation

to compare all dimensions when p is not dominated. (2) Dominance tests in space partitioning, it needs to compare all dimensions. For example, $p = (1, 3, 5)$ and $q = (3, 1, 2)$. For the first case, it only compares one dimension D_0 and indicates that p is not dominated by q. For the second case, over three dimensions comparisons, p locates in $\mathbb{S}_{011}$ $w.r.t$ q. Different conditions adopt corresponding cases. In order to unify above two conditions, we first present DC measure. It is applicable for the two cases. Apparently, DC is more persuasive than DT. Furthermore, DC as the measure of performance, can avoid the difference of implementation of all developers.

We compare our algorithm RSkycube with Naive and QSkycube. Naive straightly computes all cuboids without any sharing strategy. It adopts BSkytree [4] as the skyline algorithm for cuboid computation. QSkycube is state-of-the-art skycube algorithm. It develops *sharing strict space dominance* by exploiting SkyTree.

5.2 Scalability

Varing Dimensionality. This section evaluates the effect of above three algorithms on dimensionality as shown in Fig. 6. The size of dataset is 200 K and dimensionality varies from 5 to 9. IND and ANT represent independent and anti-correlated datasets, respectively. RSkycube and QSkycube have no performance difference on IND when dimensionality is 5, because almost points are

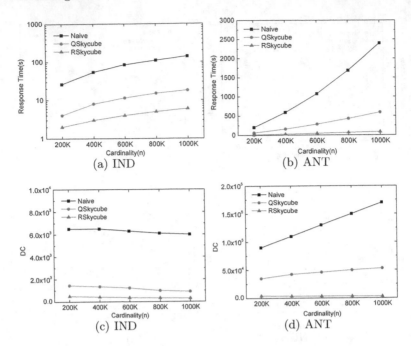

Fig. 7. Performance over cardinality variation

non-skyline points such that *sharing strict space dominance* is dominant. In other conditions on both kinds of datasets, our proposed algorithm RSkycube is greatly better than Naive and QSkycube. The skycube computation cost on ANT is larger than that on IND, because there are more skyline points on ANT. In addition, the advantage of our proposed algorithm RSkycube for the others on ANT is rather larger than that on IND, since that the skyline points on ANT are more uniform in subspaces than that on IND, which accelerates the reusing principle in RSkycube.

Varing Cardinality. This section evaluates the effect about cardinality as depicted in Fig. 7. The dimensionality of dataset is 8 and cardinality varies from 200 K to 1000 K. RSkycube outperforms the other algorithms in every cardinality. Similar with dimensionality, our algorithm RSkycube has a stronger applicability than others on ANT. We observe that the three methods all decrease for DC with cardinality increasing on IND. This is because the density of skyline decreases as cardinality augments on IND. However, ANT is not.

5.3   The Performance on Real Datasets

This section evaluates the effect of all algorithms on real datasets ColorMoments and IPUMS as shown in Table 1. We observe that our proposed RSkycube is better than Naive and QSkycube on both datasets.

Table 1. Performance on real datasets.

Algorithm	ColorMoments	IPUMS
	n = 68,040; d = 9	n = 74,954; d = 6
Naive	17 s $(DC = 18815)$	8.3 s $(DC = 10961)$
QSkycube	1.4 s $(DC = 650)$	0.3 s $(DC = 58)$
RSkycube	0.8 s $(DC = 342)$	0.2 s $(DC = 41)$

6   Conclusion

This paper studied efficient skycube computation. We analyzed two sharing strategies *sharing strict space dominance* and *sharing space incomparability*. And observed that there is still much room for optimization for state-of-the-art skycube algorithm. We first propose a set of rules named *identical partitioning (IP)* for constructing a novel structure VSkyTree. Moreover, we present the reusing principle, which utilizes both *sharing strict space dominance* and *sharing space incomparability* by reusing VSkyTree on parent cuboids to compute child cuboids. Then, in *top-down* fashion, we design an efficient skycube computation algorithm RSkycube based on the reusing principle. Our experimental results indicate that our algorithm RSkycube significantly outperforms state-of-the-art skycube computation algorithm on both synthetic and real datasets.

Acknowledgments. This work was supported in part by the Key Research and Development Plan of National Ministry of Science and Technology under grant No. 2016YFB1000703, the National Natural Science Foundation of China under grant Nos. 61402130, 61272046, U1509216, the Natural Science Foundation of Heilongjiang Province, China under grant No. F201317.

References

1. Börzsönyi, S., Kossmann, D., Stocker, K.: The skyline operator. In: Proceedings of the 17th International Conference on Data Engineering, Heidelberg, Germany, 2–6 April 2001, pp. 421–430 (2001)
2. Kailasam, G.T., Lee, J., Rhee, J., Kang, J.: Efficient skycube computation using point and domain-based filtering. Inf. Sci. **180**(7), 1090–1103 (2010)
3. Khalefa, M.E., Mokbel, M.F., Levandoski, J.J.: Skyline query processing for incomplete data. In: Proceedings of the 24th International Conference on Data Engineering, ICDE, Cancún, México, 7–12 April 2008, pp. 556–565 (2008)
4. Lee, J., Hwang, S.: Bskytree: scalable skyline computation using a balanced pivot selection. In: EDBT Proceedings of the 2010 13th International Conference on Extending Database Technology, Lausanne, Switzerland, 22–26 March 2010, pp. 195–206 (2010)
5. Lee, J., Hwang, S.: Qskycube: efficient skycube computation using point-based space partitioning. PVLDB **4**(3), 185–196 (2010)

6. Pei, J., Fu, A.W., Lin, X., Wang, H.: Computing compressed multidimensional skyline cubes efficiently. In: Proceedings of the 23rd International Conference on Data Engineering, ICDE 2007, The Marmara Hotel, Istanbul, Turkey, 15–20 April 2007, pp. 96–105 (2007)
7. Pei, J., Jin, W., Ester, M., Tao, Y.: Catching the best views of skyline: a semantic approach based on decisive subspaces. In: Proceedings of the 31st International Conference on Very Large Data Bases, Trondheim, Norway, 30 August–2 September 2005, pp. 253–264 (2005)
8. Raïssi, C., Pei, J., Kister, T.: Computing closed skycubes. PVLDB **3**(1), 838–847 (2010)
9. Xia, T., Zhang, D.: Refreshing the sky: the compressed skycube with efficient support for frequent updates. In: Proceedings of the ACM SIGMOD International Conference on Management of Data, Chicago, Illinois, USA, 27–29 June 2006, pp. 491–502 (2006)
10. Yuan, Y., Lin, X., Liu, Q., Wang, W., Yu, J.X., Zhang, Q.: Efficient computation of the skyline cube. In: Proceedings of the 31st International Conference on Very Large Data Bases, Trondheim, Norway, 30 August–2 September 2005, pp. 241–252 (2005)
11. Zhang, S., Mamoulis, N., Cheung, D.W.: Scalable skyline computation using object-based space partitioning. In: Proceedings of the ACM SIGMOD International Conference on Management of Data, SIGMOD 2009, Providence, Rhode Island, USA, 29 June–2 July 2009, pp. 483–494 (2009)

Similarity Search Combining Query Relaxation and Diversification

Ruoxi Shi[(✉)], Hongzhi Wang, Tao Wang, Yutai Hou, Yiwen Tang,
Jianzhong Li, and Hong Gao

Harbin Institute of Technology, Harbin, China
{shiruoxi,wangzh,lijzh,honggao}@hit.edu.cn,
isabeltang147@gmail.com, atma.hou@gmail.com,
yt6789299@163.com

Abstract. We study the similarity search problem which aims to find the similar query results according to a set of given data and a query string. To balance the result number and result quality, we combine query result diversity with query relaxation. Relaxation guarantees the number of the query results, returning more relevant elements to the query if the results are too few, while the diversity tries to reduce the similarity among the returned results. By making a trade-off of similarity and diversity, we improve the user experience. To achieve this goal, we define a novel goal function combining similarity and diversity. Aiming at this goal, we propose three algorithms. Among them, algorithms genGreedy and genCluster perform relaxation first and select part of the candidates to diversify. The third algorithm CB2S splits the dataset into smaller pieces using the clustering algorithm of k-means and processes queries in several small sets to retrieve more diverse results. The balance of similarity and diversity is determined through setting a threshold, which has a default value and can be adjusted according to users' preference. The performance and efficiency of our system are demonstrated through extensive experiments based on various datasets.

1 Introduction

Similarity search that finds objects with distance larger than a given similarity threshold or within a certain distance threshold with the query in a dataset has a wide range of applications, such as web page detection, entity linking and protein identification [13–15].

Recently, the quality of similarity search results has attracted more attention. The result quality is often measured in two dimensions.

One is the number of results. Too few results provide insufficient results to the user, while too many results are inefficient to display and impossible for users to explore. When too few results are returned, the query has to be relaxed to obtain more results. For example, wrong or fuzzy input may cause few searching results if the keyword is "Briatney". The searching engine will obtain more results by correcting the keyword to "Britney", which is the name of a famous singer.

The other is the diversification of results, which is the quantitative description of the variety of elements in the result set. A good search engine attempts to provide

© Springer International Publishing AG 2017
S. Candan et al. (Eds.): DASFAA 2017, Part II, LNCS 10178, pp. 65–84, 2017.
DOI: 10.1007/978-3-319-55699-4_5

various kinds of information within limited number of results. In web search engines and recommendation systems, query result diversification helps counteract the over-specialization problem in which the retrieved results are too homogeneous to meet users' needs [3, 16, 17].

During the similarity search process, these two dimensions are correlative and should be balanced. We use an example to illustrate this point. Consider the scenario of searching for commodities in an e-commerce site. The best search result is to show users abundant but not redundant commodities. These commodities meet the requirement of user input and meanwhile, different enough to one another. Similarly, such technique can also be applied to information retrieval, image search and some other areas [28, 29].

Such requirements bring challenges to query processing to obtain high-quality results. To find the most diverse elements has been proved to be an NP-complete problem [20], and one optimal solution leads to incredible time and space cost, especially on massive and complex datasets.

Even though many query relaxation and result diversification approaches have been proposed, they fail to balance result number and diversification. Relaxation techniques in [21, 22] only perform relaxation when the query result is empty, and the result number is uncontrollable. Moreover, the similarity among the results gets high due to the relaxation. Algorithms in [3–5] return k diverse neighbors, and most of them have a two-step of candidate-filter selection based on greedy selection. Due to the facts that the optimality of greedy selection is not guaranteed, and the result quality of candidate-filter algorithms is greatly affected by the quality of the candidate set, a bad candidate set may lead to worse results.

In this paper, we attempt to obtain a proper number of results with high diversity. We control the result number within a range $[k_{min}, k_{max}]$ instead of a fixed integer k in previous studies [7, 19]. In practice, the lower bound is often given by the user, while the upper bound is limited to the result display interface or the user's ability of exploring the result. With the consideration of relaxation, we define the problem with the measure combining diversity and similarity and develop various algorithms based on this measure.

For different scenarios, we develop three algorithms to solve this problem. gen-Greedy is based on greedy selection strategy, with high efficiency, and is more applicable for a frequently changing dataset. Based on multiple sequence alignment, genCluster costs more time than genGreedy, but more stable. The third algorithm CB2S is based on cluster analysis and machine learning. It is designed to achieve high efficiency aiming at complex and massive datasets.

The contributions of this paper are as follows.

- We study efficient query processing with the consideration of both result number and diversification. As far as we know, this is the first paper considering both of these dimensions.
- To achieve the goal, we design a novel measure of query result quality combining similarity and diversity. We develop three efficient algorithms for different scenarios.

- For efficient query processing, we develop a string vectorization strategy and iterative query processing strategy to speed up the search process.
- We tested our approaches on various real datasets. Extensive experimental results show that when returning similar diversity with existing algorithms, our approach provides a proper number of results. The runtime comparison shows that our approaches are more efficient.

The rest of the paper is organized as follows. Problem definition is discussed in Sect. 2. Sections 3, 4 and 5 describes our three searching approaches in detail. Our experimental results are presented in Sect. 6, and we conclude our paper in Sect. 7.

2 Problem Definition

In this section, we define the problem by defining the quality of query results integrating similarity between query and result, the number of results and diversification. For simplification, we focus on string and use edit distance [23] as the similarity measure. Our approaches can also be adapted to other applications such as semantic or image similarity search with minor changes on the search criteria.

We denote the dataset as $DS = \{s_1, s_2, \ldots\ldots, s_n\}$, the query as q, the given threshold of distance as ε and the given result number range $[k_{min}, k_{max}]$.

Definition 1. Given a query q and a result set S, the similarity between q an S is defined as the average distance between q and all elements in S, i.e.

$$argSim(S, q) = \frac{1}{|S|} \sum_{s_i \in S} Dis(q, s_i)$$

where $Dis(q, s_i)$ is the distance between q and s_i.

In this paper, we adopt content-based diversity based on edit distance, since the other two kinds (intent based diversity and novelty based diversity) are mainly used for semantic analysis [2]. The definition is as follows.

Definition 2. Given a result set S, the diversification of S is defined as follows.

$$argDiv(S, q) = \frac{2}{k(k - 1)} \sum_{s_i, s_j \in S} Dis(s_i, s_j)$$

Intuitively, the goal of query processing is to minimize the similarity distance and maximize the diversity. However, these two dimensions are correlative. To balance these two dimensions, we use a coefficient λ and define the objective function of the query process as follows.

Definition 3. Given a trade-off parameter of similarity and diversity, coefficient λ ($\lambda \in [0,1]$), the objective function $F(S, q)$ for a result set S is as follows.

$$F(S,q) = \lambda \, argDiv(S,q) + (1 - \lambda)\,(-argSim(S,q))$$

In this definition, the trade-off parameter λ can be determined by users. Hence the inner structure of returned set is flexible, i.e. a small λ leads to more relevant results while a large λ leads to more diverse results. Also, this parameter can be also decided by analyzing various datasets through model building or sampling like [20].

We chose the form $F(S,q)$ in three reasons. First, $F(S,q)$ is an efficient and effective assessment since it combines similarity and diversity into one expression and uses an adjustable parameter to balance these two dimensions. Furthermore, $F(S,q)$ increases with the growth of $argDiv(S,q)$ and drops with the growth of $argSim(S,q)$, which excellently reflects our aim at finding the most diverse results which are also similar to query q. Finally, this formula is simple and the computation cost is small.

According to this definition, the query processing algorithm works for a given query q and range $[k_{min}, k_{max}]$, to retrieve a result set S with $|S| = k \in [k_{min}, k_{max}]$ and maximize $F(S,q)$. According to [20], even when $\lambda = 0$, this problem is NP-Complete. Thus, we attempt to design efficient heuristic algorithms in the following sections.

3 genGreedy

Intuitively, the proposed problem can be solved by two steps, generating sufficient candidates through relaxation and greedy selection. Based on this framework, we develop the query processing algorithm genGreedy.

This algorithm has two phases, candidate generation and diversification filter.

3.1 Candidate Generation

Candidate generation phase first generates k results with the highest relevance with the query, which will be used for further selection. If the result number of an accurate query is smaller than k, relaxation is performed.

To ensure the number of final results, k should be large enough, while to achieve high efficiency in the diversification filter phase and select the results similar enough to q, k should not be too large. Hence in the relaxing process we make k dependent on k_{min} and k_{max}, $k \in [(\lambda+1)k_{min}, (\lambda+1)k_{max}]$, $\lambda \in [0,1]$, as mentioned in Sect. 2. Thus, $[k, 2\,k]$ strings are retrieved in the phase of relaxation, and in selecting phase, we pick $1/(\lambda+1)$ of the candidates since we enlarge the number constraint by $\lambda+1$ in relaxation.

To obtain the results which are the most similar to q, we develop an iterative algorithm for candidate generation phase. In this algorithm, the query is relaxed iteratively from the one most similar string to q to those different ones until total k results are obtained with the relaxed queries.

That is, if insufficient results are obtained through q in the first round, then a greater threshold is used for query relaxation to retrieve results within difference ε with q. Initially, ε is set to 1 to retrieve the results within distance smaller than 1 with q. If such relaxation does not return sufficient results, ε is relaxed to a larger value.

The pseudo code is shown in Algorithm 1. In this algorithm, for efficiency issues q-gram and inverted index [24] are adopted. We first initialize min_com with 0 and the output set $rlxResult$ with $\emptyset$ (Line 1–2). Line 3–12 describe the iterative process. In Line 3, we start the iteration until the result number equals or exceeds $(\lambda + 1) * k_{min}$. During each iteration, we turn to next string if the current string s is already in $rlxResult$ (Line 5–6). In each iteration, we set the value of min_com, which is the minimum number of same grams that two strings should contain. Considering that the similarity of each result cannot be guaranteed to be within ε if only one step of q-gram approach is used. Hence, we add a verification step in Line 8–9. The results that pass verification are added into the result set in Line 9. In Line 10, we check the number of results. The program jumps out the loop and returns set $rlxResult$ (Line 13) if $|rlxResult| = (\lambda + 1) * k_{max}$ is satisfied. If there are insufficient results, we enlarge ε (Line 12) to perform the next round of searching.

Note that the computation of set similarity for the q-gram set of each string is inefficient, we involve inverted list to accelerate the process. The details will be discussed in Sect. 6.

Algorithm 1. Relaxation

Input:
   q: query string; k_{min}, k_{max}: minimum and maximum bound
   λ: tradeoff parameter; ε: threshold of edit distance

Output:
   $rlxResult$: set selected from dataset

1. $min\_com \leftarrow 0$
2. $|rlxResult| \leftarrow \emptyset$
3. **while** $(|rlxResult| < (\lambda + 1) * k_{min})$ **do**
4.   **for** (each string s in $dataset$) **do**
5.   **if** s in $rlxResult$
6.   **continue**
7. $min\_com \leftarrow q.length + 2 - 1 - \varepsilon * 2$
8.   **if** $|s.\text{grams} \cap q.\text{grams}| > min\_com$ && $Dis(q, s) < \varepsilon$
9. $rlxResult.add(s)$
10.   **if** $|rlxResult| = (\lambda + 1) * k_{max}$
11.   **break**
12. $\varepsilon \leftarrow \varepsilon + 1$
13. **return** $rlxResult$

The complexity of Algorithm 1 is $O(k_{min}N)$, where k_{min} is the minimum bound of the result number, and N is the size of dataset DS.

3.2   Diversification Filter

In the diversification filter phase, we select top $(1/\lambda + 1) * rlxResult|$ strings that make the greatest contribution to result diversity. Since the diversity of a set is measured through *argDiv* in Sect. 2, we define the contribution that a string t makes to the final result set S, denoted by $DD_t(S)$, as follows.

Definition 4. Given two strings s_i and s_j in dataset S, the edit distance between them is $Dis(s_i, s_j)$. The contribution is computed as the sum of each distance between t and any other string s, which is denoted as $DD_t(S) = \sum_{s \in S}^n Dis(s, t)$.

We accelerate the selection by pruning the strings that are not diverse enough according to a prune function $F(\sigma, \Omega) = \Omega \frac{1}{\sigma * |rlxResult|} \sum_{s \in rlxResult, t \in samSet}^{|rlxResult|} Dis(s, t)$ where σ and Ω are the parameters of pruning, $\sigma \in (0, 0.5)$ with default value of 0.25, $\Omega \in (0.5, 1)$ with default value of 0.75. The sample set of *rlxResult* is *samSet* with size of $\sigma * |rlxResult|$. σ decides how many elements *samSet* contains. In order to guarantee the number of query results, we use parameter Ω to control pruning number. And Ω can be adjusted by users according to the preferences. Higher σ and Ω increase the accuracy but decrease the efficiency of algorithm, and vice versa.

Algorithm 2. greedy

Input:
> *ED_matrix*: two-dimensional matrix string edit distances between each pair of strings
> *rlxResult*: strings generated by Algorithm Relaxation ; λ: tradeoff parameter

Output:
> *S*: final result set

1. **for** each candidate c_i in *rlxResult* **do**
2. $\qquad DD_{c_i}(rlxResult) \leftarrow \sum_{j=0, j \neq i}^{n} ED\_matrix_{ij}$
3. calculate $F(\sigma, \Omega)$
4. **for** each candidate c_i in *rlxResult* **do**
5. $\qquad$ **if** $DD_{c_i}(rlxResult) < F(\sigma, \Omega)$
6. $\qquad\qquad rlxResult.\text{remove}(c_i)$
7. $S \leftarrow \text{mergeSort}(rlxResult)$
8. **return** S

The pseudo code is shown in Algorithm 2. Initially, for each candidate c_i in *rlxResult*, we calculate how much contribution c_i makes for *rlxResult* by $DD_{c_i}(rlxResult)$ in Line 1 and 2. In Line 3, we calculate $F(\sigma, \Omega)$ for pruning in Line 5. Candidates with $DD_{c_i}(rlxResult)$ lower than $F(\sigma, \Omega)$ are considered to have a too low diversity and removed from *rlxResult* (Line 6). After this pruning, we sort the candidates by $DD_{node}(rlxResult)$ and return top $(1/\lambda + 1) * |rlxResult|$ results as S (Line 7 and 8).

The time complexity of Algorithm 2 is $O(k \log k)$, in which $k = |rlxResult|$. Since the cost of merge sort is $O(k \log k)$ and that of one loop is $O(k)$, the total complexity of genGreedy is $O(kN + k \log k)$.

This algorithm is simple and efficient without heavy preprocessing cost. Thus, it is suitable for scenarios with frequently changing datasets. As shown in Sect. 6, this algorithm could generate a good result set efficiently in most conditions, especially when dealing with complex and massive datasets. However, this method is not stable when datasets are too small. To remedy the shortage, we present a more stable approach genCluster in the next section.

4 genCluster

genCluster is presented to solve the unstable problem of genGreedy. As a trade-off, it relatively costs more time than genGreedy. Hence, it is more applicable for scenarios when the result quality requirement is more important than query runtime restriction.

To make the algorithm more stable, we cluster the candidates in *rlxResult* based on multiple sequence alignment. Such idea is inspired by the method of multiple sequence alignment in bioinformatics [25], which finds genetic relation among series of DNA or proteins. We apply such idea to find similarity connection among strings. First of all, we make pairwise alignment to create a distance matrix. Thereafter, a guide tree is built by applying clustering algorithms. Then a motif string is created by a method of scoring. Finally, strings far away from the motif in edit distance are picked out to maximize the diversity. This algorithm can also be divided into two parts, relaxing (described in Sect. 3 hence we will not repeat here) and clustering.

4.1 Definitions

Before discussing the specific steps of this algorithm, we first introduce two concepts, substitution matrix and score function. After accomplishing multiple sequence alignment on strings, in order to obtain the motif sequence, which is considered to be the center to have the closest edit distance to all sequences, we define substitution matrix as follows.

Definition 5. Given a group of m sequences, $\alpha = \{A_1, A_2, \ldots, A_s, \ldots, A_m\}$. A substitution matrix is a group of sequences $\alpha' = \{A_1', A_2', \ldots, A_s', \ldots, A_m'\}$ generated by changing A_s to A_s' by enlarging every A_s in α to the same length with place holders filling in the unmatched blanks. That is to say, all sequences in the matrix have a same length.

For each sequence A_s which has not been enlarged in α, we fill its i-th character in in A_s' if it matches the i-th character in the enlarged sequences in α'. Otherwise, we fill this i-th position in A_s' with a place holder. Figure 1 shows the process of transformation. The left figure shows the original α with sequences of various lengths, and the right one shows α', in which sequences are extended to the same length.

The method of obtaining the substitution matrix is as follows.

First, each pair of leaf nodes is compared and scored using a scoring matrix. Global optimization of dynamic programming algorithm is used in this process [30]. As for the comparison among clusters, actually, it is the comparison among groups of the multiple

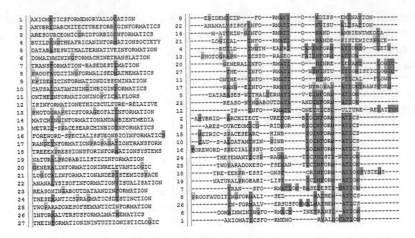

Fig. 1. Substitute matrix transformation

sequences which have already been compared. Until all sequences are processed, we obtain the substitution as a result.

To obtain the motif sequence for further selection, we need to score the sequences in the substitute matrix through a score function. This function computes scores of each kind of characters in each column of substitute matrix.

Definition 6. Score function, also called penalty function, is used to score the sequence alignment and generate the substitute matrix α' of sequences α and then, create motif according to α'. The basis of score function is the scoring matrix, usually obtained by hamming distance. Higher mark represents a higher similarity among sequences.

$$dH(a, b) = \begin{cases} 0, & if\ a = b \\ 1, & else \end{cases}$$

In addition, one of the popular methods of computing scores is called *SP* (sum of pairs) standard. Take the following sequences as an example.

$$C1 = - - -gttag$$

$$C2 = acag - - - g$$

$$C3 = -cagttag$$

If the bit of one sequence matches with the other one, it is marked 1, otherwise marked 0. Its mark is deducted by 1 during the inserting process. Considering the example above, we can easily find that the score of comparison between C1 and C2 is −4. Then, comparison of C1 and C3 is scored 3, and that of C2 and C3 is 0. Thus, the total score of the multiple sequences is −4 + 3 + 0 = −1.

Hereby, we select the character with the highest score to fill in the corresponding position of the motif sequence. That is to say, we find characters which appear most frequently in every column to create the motif.

Generally, creating motif in this method makes a good result, except some undesirable situation when the selected sequences are far away from the motif sequence but close to each other, which negatively affects the diversity. Hence, we use $F(S, q)$ to measure the result quality. When meeting with unsatisfying results, technique mentioned in [18] is applied, where several profiles sequences or sub-sequences are used to create a more accurate motif with a cost of longer runtime. Fortunately, such special situations seldom happen in our experiment. In fact, according to triangle inequality, i.e. the sum of the length of two edges is larger than the length of the third in a triangle, two sequences cannot be too similar if they are both far away from the motif.

4.2 Description of Cluster

In this section, we propose the specific cluster algorithm to solve the unstable problem of genGreedy. This algorithm tries to create a motif sequence of $rlxResult$ and select results with the farthest distance to the motif for diversification.

Algorithm 3. cluster

Input:

   ED_matrix: two-dimensional matrix string edit distances between each pair of strings
   $rlxResult$: strings generated by Algorithm Relaxation; λ: tradeoff parameter

Output:

   S: final result set

1. set initialization $(rlxResult)$
2. **for** each set in Set **do**
3.   branchLen$(set\ A, set\ B)$ ← Dis(A, B)
4. **while** $(|set| != 1)$ **do**
5.   Find set_j and set_k with min length of branchLen
6.   conbine set_j, set_k to new set_z
7.   **for** each set_t in Set **do**
8.   branchLen(set_z, set_t) ← $\frac{1}{2}$ (Dis(set_t, set_j) + Dis(set_t, set_k))
9. $Submatrix$ ← treeSetTranstoSubstituteMatrix
10. $motif$ ← score$(Submatrix)$
11. **for** each $sequence_i$ in $Submatrix$ **do**
12. $sequence_i.dis$ ← Dis $(sequence_i, motif)$
13. S ← MergeSort$(sequence_i.dis)$
14. **Return** S

We first treat each string as a set among which the branch length is initialized by ED_matrix. After that, we perform search for each set in SET to find two sets set_j and set_k that has the minimum branch length. These two sets are merged into set_z, and the branch lengths are updated by average distance from set_j and set_k to the other sets.

Hence, a phylogenetic tree is built, in which closer nodes are more similar to each other. From the leaves, we start to compare the nodes. For each time, we choose two closest nodes to be added to the substitution matrix and build up this matrix by iterative processing. After that, by applying the score function, we obtain the motif sequence, which is considered to be the center with the closest edit distance to all sequences. After that, we sort the sequences by the distance to motif and return $1/(1 + \lambda)*|$ *rlxResult*| items as the final result.

In Algorithm 3, we initialize the branch lengths among sets by edit distance among strings (Line 1 to Line 3). Line 4–Line 8 are the iterative process of building a phylogenetic tree. When two sets are merged into one, we update the branch lengths of the new set in Line 8 by computing the average of two old sets, $\frac{1}{2}\left(\text{Dis}\left(set_t, set_j\right) + \text{Dis}(set_t, set_k\right)$. In Line 9, we transfer our tree into a substitution matrix by multiple sequence alignment and use the score function to find the motif (Line 10). Thereafter, we calculate the distance among sequences to motif (Line 12) and return those with the farthest distance (Line 13 and 14). In this algorithm, the time complexities of 'while' loop, merge sort and other lines are $O(k^2 \log k)$, $O(k \log k)$ and $O(k)$, respectively. Thus, the total time complexity is $O(k^2 \log k)$, in which k is the size of *rlxResult*.

5   CB2S

genGreedy and genCluster perform well in some cases. However, they still have disadvantages. genGreedy is unstable for greedy selection, and genCluster costs more time. Motivated by this, we develop a novel algorithm CB2S (Cluster-Based String Search), which is stable and meanwhile, efficient. This method combines query relaxation with diversification in one iterative process instead of two separate steps of picking candidates and filtering. Therefore, it eliminates the exceptional situations that bad candidates lead to terrible results. Meanwhile, one iterative process helps to reduce algorithm runtime. Hence, the efficiency of CB2S outperforms these algorithms and is high especially when meeting with massive datasets.

Basically, this algorithm reduces the searching space by cluster analysis in advance and then searches several clusters to retrieve results that fit our requirement. This method involves a complex pretreatment process (described in Sect. 5.1) of cluster analysis based on *kmeans* algorithm [26]. In Sect. 5.2, we discuss the details of the searching process based on *knn* algorithm [27], which is used for string classification in our approach. Given some clusters of classified strings and an unclassified string t, KNN tells which cluster t belongs to according to a training set by uniform random sampling from classified strings. For efficiency issues of clustering and classification, strings are vectored before searching.

During the whole process, we first separate dataset into clusters, and treat the sampled data of these clusters as the training set. Given a query q, the training set and KNN algorithm vaguely classify q to one of the clusters. This cluster is considered as the search center. The search starts from the center cluster, and the search space spreads to neighbor clusters to diversify the results. This process iterates until we obtain enough number of results.

5.1 Data Pretreatment

The pretreatment process of CB2S aims at splitting a large dataset into small clusters and generating a complete graph by treating clusters as vertexes and distances among clusters as edges. The process is divided into the following steps. First, strings are changed to vectors. This step is necessary since using feature vectors to conduct cluster and classification work is more efficient [9]. Second, the dataset is separated into clusters by *kmeans* algorithm. In this step, similar strings are clustered in the same cluster, and strings in different clusters are less alike. After that, we generate the complete graph of clusters by calculating the distance from the center of one cluster to the others'. Strings are more similar if their clusters are near, vice versa. The details of these steps are shown below.

Vectorization. In the area of machine learning, text is transferred into feature vectors to perform text mining [9]. In our work, we transfer strings into vectors and classify strings by cluster analysis on vectors. Since the representation of a string has a strong impact on the accuracy of a learning system, various techniques are proposed to fit the need of various systems [10, 11]. Word stems work well on strings. Hence, similar strings have a closer space vector. For example, the distance between the vector of "computer" and "computing" should be smaller than that between "computer" and "apple" because "computer" and "computing" are mapped in the same stem. In our paper, we use the method of vectorization presented in [9] and feature selection technique proposed in [8]. Before this process, the long strings are segmented and some end words or stop words such as "the" and "a" are removed.

Establishment of clusters. We cut the whole searching space into smaller pieces and search in only small clusters to reduce the cost. Given the feature vectors of long strings obtained by vectorization, we use algorithm *K-means* to cluster them into M categories. In our approach, the number of categories (M) is determined according to the sizes of datasets. The size of each cluster is controlled to no more than 64 MB, for it is the default capacity of block storage in a distributed system, considering a future optimizing work of running our algorithm on distributed platforms. We create a distance matrix by calculating the distances among the cluster centers, i.e., the median of the set. Through the process, we obtain the complete graph of M clusters.

5.2 The Searching Process

During search, we first vaguely determine which category the query input q belongs to, and this category will be considered as the center set for further searching. During this process of classification, we use KNN algorithm for it does not need any evaluation parameters [27]. After that, the distance matrix among clusters is checked and sorted. A set list is returned according to the ascending order of distances between center set and the other clusters.

After determining the center set, the iterative process of retrieving enough items is shown in Fig. 2. We first focus the searching range on center set, adding the string whose vector is the closest to that of query q. This process continues until the objective

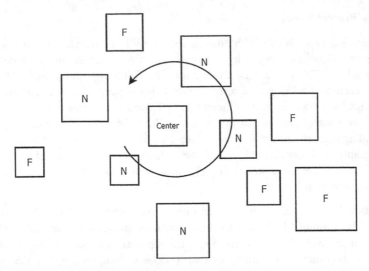

Fig. 2. Searching process of CB2S

function $F(S, q)$ starts to decrease, which means that the quality of result set starts to drop. Hence, we need to switch to next category to get items with better quality. The iteration ends when the number of results satisfies user's requirement.

The pseudo code of the algorithm is shown in Algorithm 4. We search at least $(1 - \lambda)k_{min}$ strings for similarity and at least λk_{min} strings for diversity. We first use KNN algorithm to decide the center set (Line 1). In Line 2 we check the distances among sets and prune sets that are far from the center set. From Line 3–13, we search iteratively until $|S| < k_{min}$. For each round, we add the closest string to S and check the change of $F(q, S)$ (Line 7). When $F(q, S)$ decreases, or when $|S| > (1 - \lambda)k_{min}$ is satisfied, which means that we already have enough similar strings, the searching space switches to the next cluster (Line 9). We check the number of results and jump out the loop when $|S| > = k_{max}$ to finish searching. Otherwise, we update $F(q, S)$ and continue the search processing.

We apply pruning technique when determining the search space of CB2S algorithm. The first step utilizes k NN algorithm to find the set of search center, to which the string of query input is vaguely classified. To reduce the search space, we abandon some of the separated sets with the lowest possibility to contain the final search result. Consider that we have M sets in total, a parameter σ is used to prune $(1 - \sigma) *M$ sets that have the farthest distance to the center set. This question is converted into SSSP (single source shortest path) problem which pick s $\sigma *M$ closest sets to the center set. σ is changeable according to various datasets and the user input of k_{min} and k_{min}. No matter how large a dataset is, the real search space includes only several clusters. For a large dataset, we set a small σ. And we set a relatively large one for small datasets to ensure enough but not redundant search space.

Algorithm 4. CB2S

Input:

       q:query string; k_{min}, k_{max}:minimum and maximum bound; λ: tradeoff parameter
       ε: threshold of edit distance; Vec_DS: vector set of Dataset DS

Output:

       S: final result set

1. centerSet $\leftarrow$ KNN(q, Vec_DS)
2. setList $\leftarrow$ Prune(centerSet)
3. **while**$(|S| < k_{min})$ **do**
4.   **for** $(set = centerSet; set_{num} < setList.length;)$ **do**
5.   add closest string s to S
6.   delete s from set
7. $temp \leftarrow$ update $(F(q,S))$
8.   **if** $temp < F(q,S)|\ |S| > (1 - \lambda)k_{min}$
9.   set.switch
10.   **else if** $|S| >= k_{max}$
11.   **return** S
12.   **else**
13. $F(q,S) \leftarrow temp$

Suppose that we have N strings in the whole dataset DS, and DS is cut into M small categories. The training set used to decide the center set is N' which is much smaller than N. (In our work, we sample 5% content of each dataset randomly to do training task.) Thus The time complexity of CB2S is $O(kM + N')$, in which k is the number of searching results. After applying the technique of pruning, the complexity is reduced to $O(\sigma kM + N')$, where $\sigma < 1$ and is a changeable threshold of pruning.

6 Experiments

In this section we evaluate the performance of our methods of optimizing the search results, and compare them with some other methods from previous papers [12].

6.1 Setup

We use four datasets from different domains, including Computer Conference, Information of Mammal, Protein and Random Sequence. Conference set is extracted from DBLP, containing names of journals and conferences respectively. The datasets Protein and Mammal are available on UNIPROT. Finally, the Random Sequence is generated by ourselves, containing strings made of random combinations of letters specially used to evaluate the runtime of the algorithms. The specific statistics of all datasets are summarized in Table 1. The statistics show that the strings in Mammal and Conference are shorter than the other two datasets, usually with lengths of 50–100 characters. On the contrary, Protein and Random consist of relatively long strings.

Table 1. Dataset characteristics

Dataset	Number of items	Max length	Average length
Conference	2199	125	89
Protein	10000	2163	465
Mammal	50000	142	73
Random	150194	572	277

All of our experiments are performed on a PC with quad-core, 64-bit, 1.7 GHz CPU and 4 GB memory. Apart from the preprocessing of CB2S which is written in Python (containing transferring strings to vectors and classification of the query input), the other parts of this system are implemented with c++. The operating system is Windows 7. Comparisons were made among our methods and two previously presented algorithms dealing with a similar problem. Performances of these five algorithms are measured through the value of $F(S, q)$ and the number of results. Efficiency is measured through runtime and the impacts of some parameters are tested by variable controlling.

6.2   Preprocessing Time

Establishing index. Before doing similarity query, inverted table needs to be established to shorten the query time. The cost of our inverted index is $O(n^2)$. This experiment is tested on three real datasets and one synthetically generated dataset. The dataset description and the time used to build inverted table are shown in Table 2. Such is offline time, for the inverted table is only built once every time when a new dataset was read, which means that when processing other queries, we use the same inverted table.

By using such index, query processing can be very fast. We make comparison of processing query by traditional method and by using our index, the histogram Fig. 4(a) with time unit of millisecond illustrates that when doing similarity query, the runtime of using index only costs about one-third time when the dataset is not very big and the advantage can be more obvious on more complex datasets. Even though we add the query time together with time of building inverted table, the total time is still much shorter than that of the traditional method.

Table 2. Time of establishing inverted table

Dataset	Number of strings	Time of establishing inverted table (s)
Conference	2199	0.09245
Protein	10000	0.64021
Mammal	50000	0.94835
Random	150194	3.78483

Vector transformation and dataset cluster. In algorithm CB2S we perform vectorization by calling the python interface 'word2vec' provided by Spark 2.0.0. with Hadoop 2.7. The development tool is pyCharm. After getting access to the vectors, we use the interface of *k-means* in MATLAB to separate datasets into clusters. The runtime of transformation in four datasets is shown below. Although the vectorization costs some time, this process does not need to be run for a second time if there is not any change in the searching space. Also, vector matrix does not need to rebuild when new items are added. Just update the vector matrix by calling the interface of increment (Table 3).

Table 3. Preprocessing time of CB2S

Dataset	Number of strings	Time of vectorization (s)	Time of clustering (s)
Conference	2199	5.373	0.810
Protein	10000	24.694	1.383
Mammal	50000	11.025	1.278
Random	150194	41.78483	1.735

6.3 Impact of Parameters

In this section, we tested the performance of our query processing algorithms considering two parameters that might influence the final object function $F(S,q)$, including the trade-off threshold λ, the value of average of k_{min} and k_{max}. When making analysis of $F(S,q)$, we can clearly find the associated relationship between $F(S,q)$ and λ. However, when it comes to k, k influences *argDiv* and *argSim*. Hence, k influences $F(S,q)$ but not directly. We fix k to see how $F(S,q)$ changes with the change of λ. After that, we do the same to k. This part is tested in dataset Random, for this dataset is more well-distributed without too much special or extreme data. The default value of λ is 0.5 and $k_{min} = 25$, $k_{max} = 55$. We set $\varepsilon = 30$ and changes of $F(S,q)$ of three algorithms are shown as below.

From Fig. 3, we see an increasing trend of three algorithms, which is just as what we expected. The goal function $F(S,q) = \lambda argDiv(S,q) + (1 - \lambda)(-argSim(S,q)) = \lambda(argDiv(S,q) + argSim(S,q)) - argSim(S,q)$, when $\lambda = 0.5$, $F(S,q) = 0.5(argDiv - argSim)$. Both *argDiv* and *argSim* rise when k gets larger. However, *argDiv* grows faster than *argSim*, thus leads to the final trend.

The result of $F(S,q)$ changing with λ is relatively similar. $\lambda = 0$ means traditional similarity query without considering inner diversity, while a higher value of λ tries to involve more diversity. When $\lambda = 1$, only diversity is taken into consideration ignoring the distance to query input. Although the line of genCluster slightly decreased from $\lambda = 0.6$ to $\lambda = 0.7$, the increasing trends of three algorithms are obvious.

6.4 Comparisons

In this section, we compared the performance of our relaxation-diversify algorithms with two other algorithms swap and comGreedy presented in previous paper [12].

Although they are used to solve a different problem of document mining, when changing the document to be processed into dataset of strings, processing query in datasets and diversifying the query results can get a result set similar to our approaches. Thus, we choose these two algorithms for comparison.

In this experiment, we set $k_{min} = 25$ and $k_{max} = 55$ to do query with our algorithms and use $\varepsilon = 40$ as the initial edit distance threshold. The datasets used for comparison are mentioned in Sect. 6.1. We use runtime to evaluate the efficiency and the value of object function $F(S, q)$, the number of results to estimate their performance.

Efficiency. The runtime of five algorithms is tested on dataset Mammal, for Mammal is well-distributed and large enough. The string length covers from 40–200 and does not have too much special data. We set λ, ε and k the default values mentioned in the last section, and change the data size to see how runtime changes. The efficiency shows in Fig. 4(b) and the time unit is second.

From the figure we observe that CB2S runs fastest and comGreedy is the slowest one. CB2S puts more time on preprocessing which makes a contribution to its fast speed. Especially when running on larger datasets, the advantage of CB2S is more obvious for it effectively reduces the searching space. With the growth of data size, the runtime does not increase too much. Behind CB2S, the efficiencies of genGreedy, genCluster and swap are similar. ComGreedy is the slowest algorithm due to its complexity. The speed can also be proved by comparing the algorithm complexity of

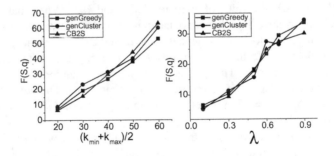

Fig. 3. Impacts of parameters

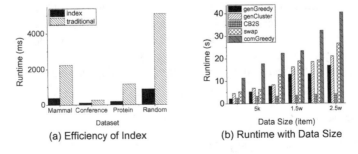

(a) Efficiency of Index   (b) Runtime with Data Size

Fig. 4. Runtime comparison

these five methods. Complexities of CB2S, genGreedy, swap, genCluster and com-Greedy are $O(\sigma kM + N')$, $O(kN + k\log k)$, $O(Nk\log k)$, $O(Nk)$ and $O(Nk)$, respectively. CB2S wins in complexity. Although the comGreedy also possesses a low complexity, it takes longer time to run for multiple passes [12].

Performance. We tested the performance through measuring the value of $F(S, q)$ and the number of results. We change the threshold of edit distance to query a specific string. Relevance of each algorithm is obtained by calculating the average of edit distance between result items and the user input, and the diversity is calculated by inner edit distance between each pair of strings. In four datasets the results appear similar in some extent.

Figure 5 illustrates the generally increasing trend of five algorithms. On different datasets, the lines fluctuate at some thresholds of ε. Such conditions happen when the items added to result set are not good enough but still has to be added to fit the requirement of returning k_{min} to k_{max} results. The gaps among algorithms are not very significant.

From Fig. 6, we observe that the numbers of items in the result set generated by genCluster and genGreedy and CB2S are obviously more than the other two, nearly double in Conference, Protein and Radom. genGreedy and genCluster usually return the most results in five algorithms while CB2S is just a little fewer. These three algorithms fit the requirement of range from 25–55 items. We can find that comGreedy is not very stable according to variable datasets. Also, when the threshold of edit distance is small, the results returned by swap and comGreedy can be too few. In this section, our algorithms perform better.

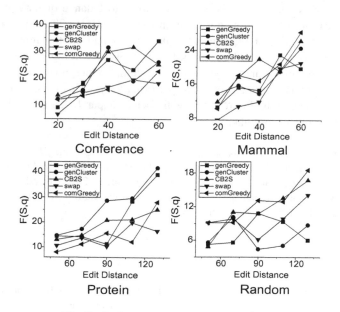

Fig. 5. Performance comparison of $F(S, q)$

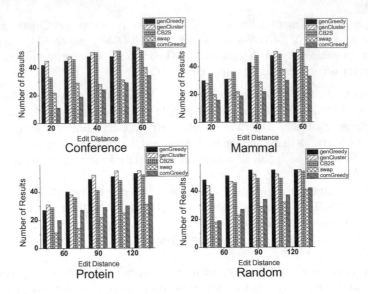

Fig. 6. Performance comparison

7   Conclusion

To obtain high-quality results, this paper combines query relaxation and result diversification. We develop a new measure for such combination. To process query efficiently, we propose three algorithms, genGreedy, genCluster and CB2S, for different scenarios. As far as we know, this is the first work to balance query relaxation and result diversification. We evaluate our work on various datasets. The experiment shows that when providing similar relevance to query input and similar inner diversity in result set, our algorithms relax the result set to a proper size at high speed. genGreedy and genCluster are simple and effective without complex preprocessing. CB2S needs some time to do preprocessing work, but performs very well in speed, especially in large searching space. Our future work will focus on parallelizing CB2S on a distributed system to achieve higher efficiency.

Acknowledgments. This paper was partially supported by NSFC grant U1509216, 61472099, National Sci-Tech Support Plan 2015BAH10F01, the Scientific Research Foundation for the Returned Overseas Chinese Scholars of Heilongjiang Province LC2016026 and MOE–Microsoft Key Laboratory of Natural Language Processing and Speech, Harbin Institute of Technology. Hongzhi Wang is the corresponding author of this paper.

References

1. Li, C., Lu, J., Lu, Y.: Efficient merging and filtering algorithms for approximate string searches. In: IEEE International Conference on Data Engineering (2008)
2. Zheng, K., Wang, H.: A survey of query result diversification. Knowl. Inf. Syst. **50**, 1–36 (2016)

3. Ziegler, C.N., Mcnee, S.M., et al.: Improving recommendation lists through topic diversification. Promontory Press (1974)
4. Drosou, M., Pitoura, E.: DisC diversity: result diversification based on dissimilarity and coverage. In: Proceedings of the Vldb Endowment (2013)
5. Agrawal, R., Gollapudi, S., Halverson, A., et al.: Diversifying search results. In: ACM International Conference on Web Search & Data Mining (2009)
6. Deng, D., Li, G., Feng, J.: A pivotal prefix based filtering algorithm for string similarity search. SIGMOD (2014)
7. Jain, A., Sarda, P., Haritsa, J.R.: Providing diversity in K-Nearest neighbor query results. In: Dai, H., Srikant, R., Zhang, C. (eds.) PAKDD 2004. LNCS (LNAI), vol. 3056, pp. 404–413. Springer, Heidelberg (2004). doi:10.1007/978-3-540-24775-3_49
8. Yang, Y., Pedersen, J.O.: A comparative study on feature selection in text categorization. In: Advances in Information Sciences & Service Sciences (2012)
9. Joachims, T.: Text categorization with support vector machines: learning with many relevant features. In: Proceedings of European Conference (1998)
10. Kim, J.D., Ohta, T., Tateisi, Y., et al.: GENIA corpus–semantically annotated corpus for bio-text mining. Bioinformatics 19, 180–182 (2003)
11. Larsen, B., Aone, C.: Fast and effective text mining using linear-time document clustering. In: KDD-ACM (1999)
12. Yu, C., Lakshmanan, L., Amer-Yahia, S.: It takes variety to make a world: diversification in recommender systems. In: EDBT (2009)
13. Haveliwala, T.H., Gionis, A., Klein, D., et al.: Evaluating strategies for similarity search on the web. In: International Conference on World Wide Web (2010)
14. Zheng, J.G., Howsmon, D., Zhang, B., et al.: Entity linking for biomedical literature. BMC Med. Inform. Decis. Making 15, S4 (2015)
15. Gish, W., States, D.J.: Identification of protein coding regions by database similarity search. Nat. Genet. 3(3), 266–272 (1993)
16. Drosou, M., Pitoura, E., et al.: Search result diversification. In: Proceedings of the National Academy of Sciences (2010)
17. Vee, E., Srivastava, U.: Efficient computation of diverse query results (2008)
18. Jones, C., Pevzner, P.: An Introduction to Bioinformatics Algorithms, pp. 97–100. MIT Press, Cambridge (2004)
19. Santos, L., et al.: Combine-and-conquer: improving the diversity in similarity search through influence sampling. In: ACM Symposium on Applied Computing (2015)
20. Santos, L.F.D., Oliveira, W.D., Ferreira, M.R.P.: Parameter-free and domain-independent similarity search with diversity. In: SSDBM (2013)
21. Mirzadeh, N., Ricci, F., Bansal, M.: Supporting user query relaxation in a recommender system. In: Bauknecht, K., Bichler, M., Pröll, B. (eds.) EC-Web 2004. LNCS, vol. 3182, pp. 31–40. Springer, Heidelberg (2004). doi:10.1007/978-3-540-30077-9_4
22. Zhou, X., Gaugaz, J.: Query relaxation using malleable schemas. In: ACM SIGMOD (2007)
23. Wagner, R.A., Lowrance, R.: The string-to-string correction problem. J. ACM 21(1), 168–173 (1974)
24. Zhang, Z., Hadjieleftheriou, M.: Bed-tree: an all-purpose index structure for string similarity search based on edit distance. In: SIGMOD (2010)
25. Thompson, J.D.: CLUSTAL W: improving the sensitivity of progressive multiple sequence alignment through sequence weighting position-specific gap penalties and weight matrix choice. Nucleic Acids Res. 22(22), 4673–4680 (1994)
26. Hartigan, J.A., Wong, M.A.: A K-Means clustering algorithm. Appl. Stat. 28, 100–108 (1979)

27. Han, E.H., Karypis, G.: Text categorization using weight adjusted k-Nearest neighbor classification. In: Pacific-Asia Conference on Knowledge Discovery and Data Mining (2001)
28. Vargas, S., Castells, P.: Explicit relevance models in intent-oriented information retrieval diversification. In: International ACM SIGIR Conference on Research & Development in Information Retrieval (2012)
29. Sun, F., Wang, M., Wang, D., et al.: Optimizing social image search with multiple criteria: relevance, diversity, and typicality. Neurocomputing **95**, 40–47 (2012)
30. Yang, J., Hu, G.: Computational biology: methods and applications for the analysis of biological sequences (2010). www.sciencep.com

An Unsupervised Approach for Low-Quality Answer Detection in Community Question-Answering

Haocheng Wu[1], Zuohui Tian[2], Wei Wu[3], and Enhong Chen[1(✉)]

[1] University of Science and Technology of China, Hefei, China
ustcwhc@outlook.com, cheneh@ustc.edu.cn
[2] Harbin Institute of Technology, Harbin, China
zuohuitian@gmail.com
[3] Microsoft Research, Beijing, China
weiwu@microsoft.com

Abstract. Community Question Answering (CQA) sites such as Yahoo! Answers provide rich knowledge for people to access. However, the quality of answers posted to CQA sites often varies a lot from precise and useful ones to irrelevant and useless ones. Hence, automatic detection of low-quality answers will help the site managers efficiently organize the accumulated knowledge and provide high-quality contents to users. In this paper, we propose a novel unsupervised approach to detect low-quality answers at a CQA site. The key ideas in our model are: (1) most answers are normal; (2) low-quality answers can be found by checking its "peer" answers under the same question; (3) different questions have different answer quality criteria. Based on these ideas, we devise an unsupervised learning algorithm to assign soft labels to answers as quality scores. Experiments show that our model significantly outperforms the other state-of-the-art models on answer quality prediction.

Keywords: Community question answering · Answer quality evaluation

1 Introduction

In the last decade, many community question answering (CQA) sites such as Yahoo! Answers and Baidu Knows have emerged and accumulated a large number of questions, answers, and users. The quality of answers may be high in the sense that the answers are precise and useful. However, it may be low in the sense that the answers are irrelevant to the topic and thus useless. It becomes an important problem how to detect low-quality answers in order to improve the experience of user when he browses a question and its answers (a QA thread).

One way to improve user experience in CQA is to provide the best answer for each QA thread. Many studies have been conducted along this line [7,18,21,23]. However, we have found that many non-factoid questions do not have single best answers, especially those asking for reasons, instructions or opinions. Therefore,

© Springer International Publishing AG 2017
S. Candan et al. (Eds.): DASFAA 2017, Part II, LNCS 10178, pp. 85–101, 2017.
DOI: 10.1007/978-3-319-55699-4_6

selection of one best answers may not satisfy the user needs in such cases. In the meantime, the existence of low-quality answers can seriously decrease user satisfaction. Therefore, there is a clear need to detect low-quality answers, which is the problem we want to address in this paper.

There are three main challenges in low-quality answer detection. (1) Manually labeled datasets are costly and hard to obtain, which is one of the bottlenecks of existing methods, since they are mainly based on supervised learning. (2) Existing methods usually focus on relevance measures between questions and answers. However, a low-quality answer usually uses the same key words with the question but talks on totally different and irrelevant topics. In this case, it is very hard to say that they are useless answers. (3) The answer quality criteria may vary on different questions. For example, for non-factoid questions long answers are quite common, while for factoid questions short answers are often sufficient. Existing supervised learning methods usually favor long answers because long answers tend to have good human judged labels in training datasets.

Table 1 shows an example[1] of a question with a low-quality answer. We find that most answers can answer the question and the high-quality ones (A3–A6) are quite similar. While, A1 is marked as the best answer in Yahoo! Answers, but it makes an impolite joke and has no similarities in terms of content to other answers. Thus, we can pick A1 as low-quality answer from the others by comparing the differences among them. From this example, we can infer that low-quality answers are usually outliers from all the other answers.

In this paper, we propose a new method for low-quality answer detection, on the basis of unsupervised learning. There are three assumptions in our method: (1) Most answers are normal answers and only a few answers are of low-quality. (2) Low-quality answers can be found by checking whether they are significantly different from its peer-answers, i.e. the other answers under the same QA thread. (3) Different questions should have different answer quality criteria.

Our method takes three solutions based on the three assumptions to tackle the three challenges. (1) Inspired by outlier detection algorithms, we propose a novel unsupervised optimization approach to detect low-quality answers by minimizing the data variance and maximizing the number of normal answers. (2) We incorporate a set of features which can capture the content differences between the answer and its peer-answers. (3) We apply the optimization model on each question individually, rather than on all questions at once.

We conduct experiments with three datasets. We make use of two benchmark datasets for answer quality prediction: an English dataset with 3,229 questions and 20,162 answers, and an Arabic dataset with 1,700 questions and 8,501 answers. We also label a third dataset sampled from Yahoo! Answer with 636 questions and 3,723 answers to test the performances on the popular CQA site. Experimental results on three datasets show that our method significantly outperforms other state-of-the-art methods.

Our contributions in this paper are of three-fold: (1) a proposal for an unsupervised optimization model for low-quality answer detection; (2) a proposal of

---

[1] https://answers.yahoo.com/question/index?qid=20090408172834AArbCtu.

Table 1. An example of a question thread.

Question	
What 2 colors make green?	
Description	
I was painting a mission so I needed green paint for the grass but I run out	
Answers	
A1*:	You know there is something called google right?
A2:	If you are mixing pigments it is blue and green. Things work differently on a computer and in some photo stuff. If you are mixing paint/pigments lookup RBG color wheel
A3:	Yellow, blue, mix em together
A4:	Blue and yellow
A5:	Blue and yellow, but more yellow than blue
A6:	Blue and yellow?

*: A1 is marked as the best answer.

using a set of features for capturing content differences among peer-answers; (3) empirical verification of the efficacy of the proposed method on two benchmark datasets and another dataset from a popular CQA site.

2   Related Work

Although there have been many studies on CQA, to our best knowledge, no work has been done aiming at detecting low-quality answers based on the content differences among answers in a QA thread before. The related work can be broadly categorized into three threads.

Answer Quality Prediction. In previous studies, there are no commonly agreed definitions of answer quality in CQA. Jeon et al. [7] define that a good answer tends to be relevant, informative, objective, sincere and readable. Sakai et al. [20] propose a new evaluation methods based on graded-relevance metrics. Recently, two workshops of answer quality prediction are hold for SemEval-2015 & 2016 Task 3 [15,16]. The organizers publish large manually judged English and Arabic datasets where answers are labeled in three levels: good, potential useful, and bad. And bad answers are subdivided into four categories: Irrelevant, Dialogue, Non-English and Others. We conduct our experiments on the two datasets published in SemEval-2015 Task 3.

Despite the lack of agreement on the answer quality definitions, researchers have made great progress in answer quality prediction these years. In the beginning, many works focus on the methods based on non-textual features. Jeon et al. [7] use the maximum entropy approach and kernel density estimation to predict answer quality scores based on statistics of question and answer.

Shah et al. [21] propose a supervised method based on additional user information, which is considered as one of the state-of-the-art methods who use only non-textual features. Recently, a workshop in SemEval-2015 Task 3 [15] starts to target on semantically oriented solutions using rich language representations. Tran et al. [23] and Nicosia et al. [18] win on English and Arabic datasets respectively by using supervised methods with various of lexical, syntactic and semantic similarity measures. We take the methods of Shah et al. [21], Tran et al. [23] and Nicosia et al. [18] as three baselines in this paper.

Review Spam Detection. Review spam detection are also related to our work. Crawford et al. [4] categorize review spams into three groups in their survey: untruthful reviews, reviews only on brands and non-reviews. By representing a review using a set of features of reviews, reviewers and products, classification techniques are used to assign spam labels to reviews [8,11].

However, our work on low-quality answer detection is clearly different from review spam detection on three aspects. (1) Target: our work only detects irrelevant and useless answers. While as suggested by Crawford et al. [4], a good review spam detection system should be able to identify whether a review is fake or untruthful. (2) Feature: a bunch of features representing the relevance between question-answer pair can be used in our work. While review comments can only refer to product names, thus features may heavily rely on review content. (3) Candidates: a question only have seven answers on average (See Sect. 4), while a popular product may have thousands of review comments.

Anomaly Detection. Anomaly detection (also known as outlier detection), referring to the problem of finding patterns in data that do not conform to expected behavior, is also related to our work. Hodge et al. [6] indicate in their survey that techniques in unsupervised mode do not require training data, and thus are most widely applicable on this problem. Unsupervised methods make the implicit assumption that normal instances are far more frequent than anomalies [3]. Based on this assumption, many optimization models are proposed based on minimizing a customized loss function of data variance, where variables are classification labels of 0(anomaly) or 1(normal), and parameters are features. And techniques like soft labels [12] and gradient descent method [25] are proved useful when solving the optimization problems.

Our work still makes a difference with classical anomaly detection methods on the scope of similarities between instances. In anomaly detection, instances usually share lower similarities. For example, in review spam detection, reviews are often on a wide range of topics. While in CQA, questions, especially factoid questions, usually have very specific information needs, which narrow the topics and contents of answers. Thus, similarities in answers tend to be higher than in reviews. We take advantage of this characteristic in our optimization model.

3   Our Method of Low-Quality Answer Detection

In this section, we first formally present the problem of low-quality answer detection, and then illustrate our three key assumptions, and then devise a unsupervised method, and finally propose a set of features.

3.1   Problem

Given a question q and a set of its n answers $\{a_1, a_2, \cdots, a_n\}$. Each answer a_i is represented by an m-dimensional feature vector $\mathbf{x}_i = \{x_{i1}, x_{i2}, \ldots, x_{im}\}^\top$. All answers $\{a_i\}$ construct a $m \times n$ feature matrix $\mathbf{X} = \{\mathbf{x}_1, \mathbf{x}_2, \ldots, \mathbf{x}_n\}$.

Our goal is to learn a label vector $\mathbf{y} = \{y_1, y_2, \ldots, y_n\}^\top$ with $y_i \in \{0, 1\}$, where $y_i = 0$ means the corresponding answer a_i is low-quality, and $y_i = 1$ means a_i is a normal one and should be kept.

3.2   Assumptions

The first assumption is based on the observations in Table 1. In fact, most unsupervised anomaly detection methods have the same implicit assumption [3]. And the labeled datasets in Table 3 verify that it is true. And it helps to construct the second factor of our loss function in Formula (1).

Assumption 1. *Most answers under a question are normal ones and only a few of them are low-quality answers.*

Secondly, a question usually has specific information needs, which makes answers tend to have similar content. Then if an answer is significantly different with its peer-answers, it is likely to be low-quality. The second assumption helps to construct the first factor of Formula (1), and also inspires us to design features to capture the content differences between an answer and its peer-answers.

Assumption 2. *Whether an answer is low-quality or not can be known by checking its peer-answers.*

Moreover, different questions should judge answers in different quality criteria. For example, non-factoid questions favor long answers, and general questions expect yes/no, and factoid questions accept short noun-phrase, etc. Based on this observation, we have the third assumption, and apply our unsupervised model on each question instance, rather than on overall questions like supervised models.

Assumption 3. *Questions should have different answer quality criterion.*

3.3 Method

We propose an unsupervised learning approach to detect low-quality answers by minimizing the data variance and maximizing the number of kept answers.

Let $\overline{\mathbf{X} \cdot \mathbf{y}}$ denotes average weighted vector $\frac{1}{n} \sum_{i=1}^{n} y_i \cdot \mathbf{x}_i$. According to Assumption 3, we consider the optimization problem for each question instance:

$$\underset{\mathbf{y}=\{y_i\}}{\operatorname{argmin}} \frac{1}{mn} \sum_{i=1}^{n} ||y_i \cdot \mathbf{x}_i - \overline{\mathbf{X} \cdot \mathbf{y}}||^2 - \frac{\alpha}{n} \sum_{i=1}^{n} y_i \tag{1}$$

$$\text{s.t. } y_i \in \{0,1\}, 1 \leqslant i \leqslant n$$

Where $\frac{1}{mn} \sum_{i=1}^{n} ||y_i \cdot \mathbf{x}_i - \overline{\mathbf{X} \cdot \mathbf{y}}||^2$ comes from Assumption 2, representing the data variance averaged on feature count. It will lead answers to have same labels, and those significantly different with other answers to have label 0. And $-\frac{\alpha}{n} \sum_{i=1}^{n} y_i$ comes from Assumption 1, helping to maximize number of answers with label 1, and α denotes the trade-off for the number of kept answers.

Formula (1) is novel in anomaly detection methods [12]. As described in Sect. 2 and Assumption 2, questions have specific information needs thus narrow topics, then answers tend to be more similar than instances in classical anomaly detection, such as reviews. We use variance with average weighted vector: $||y_i \cdot \mathbf{x}_i - \overline{\mathbf{X} \cdot \mathbf{y}}||^2$, instead of using variance with prediction: $||f(\mathbf{x}_i) - y_i||^2$, or variance with average vector: $||\mathbf{x}_i - \overline{\mathbf{x}}||^2$. In this way, the negative influence of low-quality answers is removed by labeling them 0, and the similarities between high-quality answers are highlighted since their feature values are similar.

This is an 0-1 programming problem, and thus it is NP-hard. To solve it, we adopt the soft label technique [12] and soften the label constraints to interval [0, 1]. By this means, it becomes a probabilistic constraint solving problem, and the learned distribution can represent the probabilities to be high-quality answers. By denoting as $\mathcal{L}(\{y_i\})$, we have a new optimization problem:

$$\underset{\{y_i\}}{\operatorname{argmin}} \mathcal{L}(\{y_i\}) = \underset{\{y_i\}}{\operatorname{argmin}} \frac{1}{mn} \sum_{i=1}^{n} \sum_{j=1}^{m} (y_i \cdot x_{ij} - \frac{1}{n} \sum_{k=1}^{n} y_k \cdot x_{kj})^2 - \frac{\alpha}{n} \sum_{i=1}^{n} y_i \tag{2}$$

$$\text{s.t. } 0 \leqslant y_i \leqslant 1, 1 \leqslant i \leqslant n$$

We employ a coordinate descent method to solve Problem (2) by taking y_i as a variable and fix other labels in each iteration. We calculate the partial derivative of $\mathcal{L}(\{y_i\})$ with respect to y_i, and set the result to be 0, we have

$$\frac{\partial \mathcal{L}}{\partial y_i} = \frac{2(n-1)}{mn^2} ||\mathbf{x}_i||^2 \cdot y_i - \frac{2}{mn^2} \sum_{k=1, k \neq i}^{n} \mathbf{x}_i^\top \cdot \mathbf{x}_k \cdot y_k - \frac{\alpha}{n} = 0$$

$$\text{s.t. } 0 \leqslant y_i \leqslant 1$$

By defining the solution as $\hat{y}_i$, we have:

$$\hat{y}_i = \frac{2 \sum_{k=1, k \neq i}^{n} \mathbf{x}_i^\top \cdot \mathbf{x}_k \cdot y_k + \alpha mn}{2(n-1)||\mathbf{x}_i||^2}$$

Then the optimal solution of Problem (2) is:

$$y_i^* = \begin{cases} 0, & \text{if } \hat{y}_i < 0, \\ \hat{y}_i, & \text{if } 0 \leqslant \hat{y}_i \leqslant 1, \\ 1, & \text{if } \hat{y}_i > 1. \end{cases}$$

The procedure ends when the Euclidean distance of two label vectors in consecutive iterations is less than ϵ or iteration number exceeds N^2. The final $\{y_1^*, y_2^*, \ldots, y_n^*\}$ represents the possibilities of being normal answers. A threshold $\mu \in [0, 1]$ is used for classification, if $y_i^* < \mu$, then a_i is a low-quality answer.

3.4   Features

Given a question q, and its description d and n answers $\{a_1, a_2, \cdots, a_n\}$, we have 173 features for each answer a_i. We choose these features because they represent nearly all aspects of a question-answer pair.

Table 2 clusters them into five groups. Group 1,2 and 3 are widely used and proved to be effective [18, 21, 23]. And Group 4 is a simple expansion of Group 3. We propose the last group, which seems to be new for answer quality prediction, as far as we know, although the idea is quite simple. All features are normalized to interval of $[0, 1]$ in our unsupervised models.

Question Features: features with prefix "Q" (denoted as $\{f_1(q)\}$, and its feature index set is denoted as F_Q) are obtained from statistics of question's and asker's information. Although all a_i share the same feature values (normalized 0 or 1) in this group, $\{f_1(q)\}$ can still be proved to be useful: in Formula (2), if t answers are classified as 1, the contribution of $\{f_1(q)\}$ in the first factor is:

$$\frac{1}{mn} \sum_{i=1}^{n} \sum_{j \in F_Q} (y_i \cdot x_{ij} - \frac{1}{n} \sum_{k=1}^{n} y_k \cdot x_{kj})^2 = \frac{t(n-t)}{mn^2} \sum_{j \in F_Q} x_{1j}^2 \tag{3}$$

When $\sum_{j \in F_Q} x_{1j}^2 \neq 0$, Formula (3) have minimum value when $t = 0$ or $t = n$. And influenced by $-\frac{\alpha}{n} \sum_{i=1}^{n} y_i$, t will approach n.

Answer Features: features (denoted as $\{f_2(a)\}$) with prefix "A" are obtained from the statistics of answer a_i and the answerer's information.

Question-to-Answer Features: features (denoted as $\{f_3(q, a)\}$) with prefix "QA" are obtained from contents of question-answer pair. The relevance of q and a_i are measured by various similarities on lexical, syntactic and semantic levels. Many of the state-of-the-art methods focus on this part.

Description-to-Answer Features: features (denoted as $\{f_4(d, q, a)\}$) in DA_repeat and QDA_repeat are obtained from contents of question, description and answer. DA_repeat take the same feature calculation methods in $\{f_3(q, a)\}$ to measure the relevance between d and a_i. QDA_repeat does the same way by concatenating question and description.

---

[2] We set $\epsilon = 0.00001$ and $N = 200$ in our experiments.

Table 2. Features of low-quality answer detection

Feature	#*	Description	[23]	[18]	[21]
Q_len	2	Word # of q and d			○
Q_category	1	The q's category. This feature is discretized to c boolean dimensions, where c equals to # of categories	○		
Q_ans#	1	# of answers for the q			○
Q_u_post#	2	# of total questions and answers of the asker	○		○
Q_u_other	5	Other features of the asker: points, level, # of best answers, # of resolved questions and star count			○
A_len	1	Length of the a_i			○
A_rank	1	Reciprocal rank of the a_i in the answer list			○
A_symbol	2	Two booleans to identify whether the a_i contains some special strings (question marks, laugh symbols), and words which are only frequent in bad answers	○		
A_u_same	1	A boolean to indicate whether answerer is the asker	○		
A_u_post#	2	# of total questions and answers of the answerer	○		○
A_u_other	5	Other feature of the answerer: points, level, # of best answers, # of resolved questions and star count			○
QA_word	1	Cosine similarity of bag-of-word vectors of qa pair	○		
QA_ngram	20	5 similarity measures for n-grams($n \in \{1, 2, 3, 4\}$) of qa pair: greedy string tiling [24], long common subsequence, Jaccard index, word containment [13] and cosine similarity		○	
QA_pos	1	Cosine similarity of bag-of-POS [22] tags vectors of qa pair		○	
QA_noun	1	Cosine similarity of bag-of-noun vectors of qa pair. Noun are words containing "NN" in POS tags	○		
QA_tfidf	2	Sum of tf-idf [17] scores in answer collection of intersect subset of unigrams/bigrams between q and a_i		○	
QA_dep	1	Cosine similarity of bag-of-word-dependency vectors of qa pair. We parse sentences to dependency trees [10] and regard dependency arcs (like "pre:buy-for") as words	○		
QA_meteor	1	Alignment score from Meteor Toolkit [5] between q and a	○		
QA_lda	1	Cosine similarity of LDA [2] topic vectors of qa pair	○		
QA_w2v	1	Alignment score between q word vectors and to a_i word vectors from pre-trained word2vec model [14]. (See details in [23])	○		
QA_trans	1	Translation probability [1] from q to a by utilizing pre-trained translation model. (See details in [23])	○		
DA_repeat	30	For each method in $\{f_3(a, q)\}$, get $f_3(d, a_i)$ as feature			
QDA_repeat	30	For each method in $\{f_3(a, q)\}$, get $f_3(q + d, a_i)$ as feature			
AP_repeat	60	For each method in $\{f_3(a, q)\}$, get $\max_{p \neq i}(f_3(a_i, a_p))$ and $\frac{1}{n-1}\sum_{p \neq i}(f_3(a_i, a_p))$ as features			

*: The second column represents the actual feature count.

Answer-to-Peer Features: features (denoted as $\{f_5(a, a_p)\}$) in AP_repeat are obtained from answer a_i and its peer-answers $\{a_p | 1 \leqslant p \leqslant n, p \neq i\}$. They consist of two basic values: maximum and average of the similarities between a_i and peer-answers $\{a_p\}$ by repeating the feature methods in $\{f_3(q, a)\}$. According to Assumption 2, we propose these features to capture the differences between bad answers and peer-answers.

The time complexity of our method consists of two parts: feature calculation and coordinate descent method. In feature calculation, the time complexity is $O((l_Q + l_d)l_a mn^2)$, where l_Q, l_d, l_a the maximum length of questions, descriptions and answers, respectively, and m, n are the feature count and answer count, respectively. In coordinate descent method, it is $O(Nmn^2)$, where N is the maximum iteration count. How to improve the efficiency of the process is still an interesting topic for future research that we will not address at this time. One possibility is to calculate the features off-line and store them in database.

4   Experiment

4.1   Experimental Setup

Datasets Preparation. We conduct experiments on three datasets: two public datasets from a workshop for SemEval-2015 Task 3[3] and our labeled dataset from Yahoo! Answers, and their statistics are given in Table 4.

Qatar Corpus: The workshop provides an large English dataset from Qatar Living website. Each question contains a description, several answers and aliases of askers and answerers. Each answer is labeled with one of six labels: "Good", "Potential", "Irrelevant", "Dialogue", "Non-English", "Other". The last four are regarded as "Bad" answers. See Table 3 for labeling guidelines and distributions. The task provides a split: Train, Dev and Test, for training model, tuning parameters and testing performances, respectively.

Fatwa Corpus: The workshop also provides an Arabic dataset from The Fatwa website. Each question has five answers, some of them are carefully answered by knowledgeable scholars in Islamic studies, while some are answers to other questions. Each answer is labeled with one of three labels: "Good, "Potential" and "Irrelevant". The task also provides a split: Train, Dev and Test.

Yahoo Corpus: In order to evaluate our methods on popular CQA site, we also label a dataset from Yahoo! Answer. Specifically, we crawl 6.4M questions associated with descriptions, answers and users' information[4] from Yahoo! Answer using a public API[5]. We then sampled 636 questions for labeling. Two expert labelers are invited to give each answer one of six labels with guideline in Table 3. If the judges disagree on an answer, we invited a third expert to make the final decision. We provide a split by ratio 2:1: Train and Test, for tuning parameters by cross validation, and testing performances, respectively.

---

[3] http://alt.qcri.org/semeval2015/task3.
[4] Features of Q_u_other and A_u_other in Table 2 are only traceable in Yahoo dataset.
[5] http://developer.yahoo.com/answers.

Table 3. Labeling guideline and label distribution

Label	Description	Qatar	Fatwa	Yahoo
Good	The answer directly responds to the question with relevant and useful content	49.3%	20.2%	66.3%
Potential	The answer is potentially useful to the question	10.0%	22.5%	11.3%
Bad	The answer is bad or irrelevant that the asker does not expect to receive	40.7%	57.3%*	22.4%
- Irrelevant	The answer is totally irrelevant to the question	17.9%	-	12.5%
- Dialogue	The answer does not directly respond to the question but hold an irrelevant chat, such as expressing gratitude or asking questions	22.3%	-	8.0%
- Non-English	Irrelevant non-English answer	0.5%	-	1.5%
- Other	Other irrelevant answer, such as advertisements	0.0%	-	0.4%

*: Bad answers in Fatwa are manually added as noise data, so the proportion is large.

Preprocessing. We illustrate the preprocessing method on each corpus. (1) We prepare an initial collection $\{q, d, \{a_i\}\}$ consisting of questions with descriptions and answers, where words are stemmed and stopwords are removed. (2) We prepare a concatenated collection $\{qd, \{a_i\}\}$ by concatenating q and d. (3) We prepare a document collection of qd and a_i to train a LDA model and a Word2vec model. (4) We prepare a mapping collection where qd is source and a_i is target to train a translation model.

The tools we used are: Stemmers(English/Arabic) in Lucene system[6] for word stemming, stopword lists(English/Arabic) from the Ranks website[7] for stopword removal, GibbsLDA++[8] for training LDA models, Word2vec tool[9] for training word2vec models, GIZA++[10] for training translation models, Stanford Tagger[11] and Parser [12] (English/Arabic) for part-of-speech tagging and dependency parsing, and Meteor System[13] (English/Arabic) for translation alignment score evaluation.

Evaluation Measure. We use four widely used measures to evaluate the performances: accuracy, precision, recall and F1-score.

---

[6] http://lucene.apache.org/.
[7] http://www.ranks.nl/stopwords.
[8] http://gibbslda.sourceforge.net/.
[9] https://code.google.com/archive/p/word2vec/.
[10] http://www.statmt.org/moses/giza/GIZA++.html.
[11] http://nlp.stanford.edu/software/tagger.shtml.
[12] http://nlp.stanford.edu/software/lex-parser.shtml.
[13] http://www.cs.cmu.edu/~alavie/METEOR/.

Table 4. Overview of three CQA datasets

	Qatar			Fatwa			Yahoo	
	Train	Dev	Test	Train	Dev	Test	Train	Test
# of question	2,600	300	329	1,300	200	200	419	217
# of description	2,599	300	329	1,300	200	200	367	175
# of answer	16,541	1,645	1,976	6,500	1,000	1,001	2,407	1,316
# of answer per question	6.3	5.5	6.0	5.0	5.0	5.0	5.7	6.1
# of good answer	8,069	875	997	1,300	200	215	1,582	888
# of potential answer	1,659	187	167	1,469	222	222	278	144
# of bad answer	6,813	583	812	3,731*	578*	564*	547	284
- # of irrelevant answer	2,981	269	362	-	-	-	305	160
- # of dialogue answer	3,755	312	435	-	-	-	196	101
- # of non-English answer	74	2	15	-	-	-	36	18
- # of other answer	3	0	0	-	-	-	10	5

*: Bad answers in Fatwa are manually added as noise data, so the number is large.

Baselines and Our Methods. We consider four state-of-the-art baselines: Tran et al. [23] and Nicosia et al. [18] use feature-rich supervised models and win first place on the Qatar and the Fatwa datasets respectively, denoted as T^S and N^S. Method of Shah et al. [21] is another state-of-the-art method among supervised methods using only non-textual features, denoted as S^S. We create another supervised baseline by utilizing all above features plus DA_repeat and QDA_repeat features, denoted as A^S. Features of baselines are listed in Table 2.

To evaluate the effectiveness of answer-to-peer features, we create a supervised model by utilizing all features in Table 2, denoted as O^S.

Last, and most importantly, to evaluate the effectiveness of our unsupervised method, five comparison models are considered. They are the unsupervised models based on the same features from T^S, N^S, S^S, A^S and O^S, denoted as T^U, N^U, S^U, A^U and O^U respectively.

Parameter Tuning. There are three parameters in supervised models and two parameters in unsupervised models need to be tuned.

We train SVM classification models for supervised methods by the tool of SVMLight [9]. There are three parameters $\{c, j, b\}^{14}$. The ranges are: c in $\{0.001, 0.002, 0.005, 0.01, \cdots, 50\}$, j in $\{0.5, 1, 1.5, \cdots, 8\}$, and b in $\{1, 0\}$. For Qatar and Fatwa datasets, we train models on Train sets, and choose the combinations with best F1-score on Dev sets. And for Yahoo dataset, we conduct four-fold cross validation on Train set and choose the one with the best F1-score.

In unsupervised methods, there are two parameters: the tradeoff weight α and decision threshold μ. The ranges are: α in $\{0, 0.1, 0.2, \cdots, 1\}$, and μ in

---

[14] c: trade-off between training error and margin. j: cost-factor of training errors difference between positive and negative examples. b: use biased hyperplane or not.

Table 5. Low-quality answer detection results on Qatar, Fatwa and Yahoo datasets

Corpus	Measures	S^S[21]	S^U	N^S[18]	N^U	T^S[23]	T^U	A^S	A^U	O^S	O^U
Qatar	Precision	50.0	49.6	54.3	55.0	64.6	69.8*	64.7	70.1*	67.0#	**73.6***
	Recall	76.5	72.7	71.2	67.8	79.2	71.7	80.3	72.3	**80.5**	75.3
	F1-score	60.5	57.7	61.6	60.7	71.2	70.7	71.6	71.2	73.1#	**74.4**
	Accuracy	58.9	56.2	63.5	64.0	73.6	75.6*	73.9	75.9*	75.7#	**78.8***
Fatwa	Precision	56.7	56.8	80.9	79.4	84.3	84.7	84.5	84.8	87.3#	**89.4***
	Recall	84.9	**91.8***	87.8	90.4*	88.3	90.4*	89.3	89.9	90.1	90.5
	F1-score	68.0	70.2*	84.2	84.6	86.3	87.5	86.8	87.3	88.7#	**90.0**
	Accuracy	54.9	56.0*	81.4	81.4	84.1	85.4*	84.7	85.2	87.0#	**88.6***
Yahoo	Precision	51.4	54.0*	47.6	49.3	59.8	65.7*	63.5	69.1*	66.4#	**73.6***
	Recall	64.1	56.2	66.2	52.0	76.2	67.4	78.8	73.2	**79.2**	75.5
	F1-score	57.1	55.1	55.4	50.6	67.0	66.5	70.3	71.1	72.2#	**74.5***
	Accuracy	79.2	80.2	77.0	78.1	83.8	85.4	85.6	87.1	86.9	**88.9***

Bold: the highest performance in terms of the measure.
*,#: statistically significant improvement of our models (two-sided sign-test, $p < 0.05$).

$\{0, 0.01, 0.02, \cdots, 1\}$. We choose the combinations with best F1-scores on Dev sets for Qatar and Fatwa and on Train set for Yahoo.

4.2 Main Results

Table 5 shows the results on Qatar(English), Fatwa(Arabic) and Yahoo(English). The experimental results show that: (1) Our unsupervised model O^U outperforms all baseline methods on three datasets for nearly all metrics. (2) And our supervised model O^S outperforms other supervised methods on three datasets for all metrics. Most of the improvements are statistically significant by two-sided sign-test ($p < 0.05$). The results indicate that our methods are effective for low-quality answer detection.

4.3 Analysis Based on Models

We investigate the main reasons of the improvements of our unsupervised method.

(1) Supervised models do not take advantage of the fact that only minority are low-quality answers. Therefore, more answers are tended to be classified as bad answers. That is why supervised models often have high recall but low precision on this problem. This is particularly serious when the question is short. Specifically, for short questions, feature values are usually small on relevance measures. Then answers will have low SVM score and tend to be classified as bad answers. For example, we observe that T^S misclassifies all six correct answers under question "Write 5/2 as a percent?", since all the answers do not have common words with the question. While our unsupervised method T^U only misclassifies one, which increases the precision.

(2) Answers in supervised models share the same quality criteria. For example, it can be inferred from the labeled datasets that longer answers tend to be

"Good". Thus, the supervised models have a bias on short answers. This bias causes misclassification when short answers are also acceptable. For example, for question "What is your favorite poptart flavor?", six in eight are one-word-answer "strawberry". We observe that T^S misclassifies them to be bad. While our unsupervised method T^U applies on each question instance, and is able to capture the high similarities between these answers. T^U results in the safest solution by assigning them all "Good", which increases the precision again.

4.4   Analysis Based on Features

In Table 5, our unsupervised model O^U effectively improves supervised model O^S, while others with less features are not the same. For example, A^U is only comparable with A^S, and S^U is even worse than S^S on Qatar and Yahoo. Since feature utilization is the only difference between unsupervised models, an interesting question is: what features are effective or useless in unsupervised models?

(1) We study on the effective features for unsupervised models. Notice that by bringing in extra answer-to-peer features, both O^S and O^U outperform A^S and A^U, and O^U is even better then O^S, while A^U and A^S have similar performances. We investigate the main reason of effectiveness of answer-to-peer features.

Many baselines focus on the relevance between question and answer, but overlook the difference between answer and its peer-answers. In fact, normal answers in a question usually share words that bad answers do not have. For example, for question "Where can I buy carrot cake?", most normal answers contain the same shop names but do not contain "carrot cake". While one bad answer expresses his dislike on carrot cake, which is useless to the question but contains "carrot cake". A^S gives out totally opposite wrong classifications. On the contrary, answer-to-pear features in O^S provide more clues for identifying normal answers, then less normal answers are predicted to be bad, which increases the precision indeed. Moreover, O^U uses the strategy of operating on each question instance, then the effects of answer-to-peer features are larger in a single question than in the whole dataset, which makes O^U has better performances than O^S.

(2) Then, we investigate on the useless features for unsupervised model. A feature is useless if it is uncorrelated with human labels. Specifically, we assign 2, 1 and 0 to "Good", "Potential" and "Bad" answers, respectively. By this means, each feature has two vectors in a dataset, one is the vector of feature values, the other one is the assigned label vector, both dimensions are the size of dataset. We then calculate the Pearson correlation coefficient for the two vectors to represent the correlation between a feature and human labels.

Figure 1[15] (left) counts the features according to correlation coefficients. All values turn out to be in range of $[-0.3, 0.3]$. We divide them into five groups by thresholds $\{-0.1, -0.01, 0.01, 0.1\}$. We find that: 20% of features are highly positive correlated with quality, such as QA_w2v, QDA_trans, QDA_w2v, etc.; 25% are median positive, such as QA_tfidf, DA_lda, QDA_meteor, etc.; 4% are highly

---

[15] To save space we only report the results on Qatar dataset. The results in terms of Fatwa and Yahoo have similar trends.

negative, such as A_u_same, A_len; 20% are median negative, which include most of answer-to-peer features. While 32% have near-zero coefficient, we treat them as uncorrelated with human labels, such as A_rank, A_symbol, etc.

To test the effectiveness of uncorrelated features, we conduct another experiment for O^S and O^U by removing uncorrelated features, denoted as O^S_- and O^U_-. Figure 1 (right) shows their performances on Qatar dataset: O^S_- drops apparently from O^S in supervised scope, while O^U_- drops slightly from O^U in unsupervised scope. There are two reasons for these changes. Firstly, features are normalized in unsupervised methods but not normalized in supervised methods. Therefore, those features uncorrelated in unsupervised models may be correlated in supervised models. Secondly, features have global influences in supervised models, since all feature weights will change if any feature is removed. While uncorrelated features may have limited influences in unsupervised models since it is operating on each question instance. Take feature A_symbol as an example. In fact, only a few answers with special symbols and words inside are influenced by this feature.

4.5   Analysis Based on Data Sources

It is interesting to notice that, Fatwa dataset has more bad answers than other answers (see Table 3), which seems inconsistent with Assumption 1, but its performance scores are even higher than in Qatar and Yahoo (see Table 5). In fact, the Fatwa dataset is quite different. As we described in Sect. 4.1, bad answers in Fatwa are manually inserted as noise data. Thus, organizers can create as many bad ones as they want. That is why Fatwa has more bad answers. Moreover, all answers in Fatwa are carefully answered to the original questions. This means they are normal originally. Therefore, it is easier to distinguish a bad answer from other answers in Fatwa since they have totally different topics.

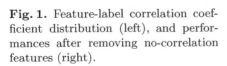
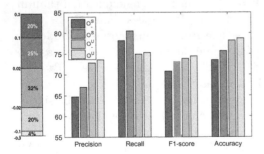
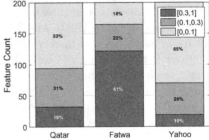

Fig. 1. Feature-label correlation coefficient distribution (left), and performances after removing no-correlation features (right).

Fig. 2. Distributions of divergence of feature distributions between bad answer and normal answers.

In order to confirm our guess, we investigate the divergences between bad and normal answers on each dataset. Specifically, (1) for each feature, we get the min and max among all answers. (2) Then we split the region $[min, max]$ into 100 slots, and count bad/normal answers on each slots, and then divide by total bad/normal answer count to get a discrete distribution. Thus, for each feature we have a distribution of bad answers and a distribution of normal answers. (3) We calculate a Jensen-Shannon divergence for the two distributions. (4) Finally, we count divergences in three slots $[0, 0.1]$, $(0.1, 0.3]$ and $[0.3, 1]$.

Figure 2 shows the results on three datasets. 61% of features in Fatwa have big divergences ($[0.3, 1]$), which is much larger than Qatar's 16% and Yahoo's 10%. It means that in Fatwa bad answers are more different from normal answers, making it easier to detect bad answers. Meanwhile, 65% of features in Yahoo have small values ($[0, 0.1]$). It means that in Yahoo bad answers are more ambiguous with normal answers, making it more difficult to detect bad answers. That explains why Yahoo has the lowest performance scores.

4.6 Analysis Based on Answer Labels

We study the effectiveness on different answer types. Models of A^S, O^S and O^U are used to see the gradual changes, where O^S uses extra answer-to-peer features to improve A^S, and O^U uses unsupervised model to improve O^S. We study on Qatar dataset since it is the largest one with full labels.

Figure 3 shows the number of correct and wrong predictions on each answer type[16]. (1) It is interesting to notice that O^S improves A^S mainly by reducing false predicted "Good" answers. It indicates that answer-to-peer features help to reduce false detections in supervised model, which also confirms the statements in Sect. 4.4. (2) It is also interesting that the "Potential" answers have exactly fifty-percent precision, well proving that they are potentially useful or useless. (3) O^U classifies less answers to be low-quality based on Assumption 1. Therefore, precision on "Good" answers is significantly improved, while the recall of "Irrelevant" answers also drops. (4) The performances on "Dialogue" answers is steady since some features are too strong, such as A_u_same, A_u_symbol, etc. For example, nearly all answers are labeled by "Dialogue" if their answerers are identical with the askers, or they contain special words like "thank".

4.7 Analysis Based on User Experience

As we discussed in Sect. 1, the target of detecting low-quality answers is to improve the user experience when he browses a question page. A user usually browse answers from top-ranked to lower-ranked on question page. A top-ranked low-quality answer will reduce the user experience.

We conduct a re-ranking experiment to study on the improvement of user experience. Specifically, (1) re-rank answers according quality scores by descending order. (2) then evaluate user experience by precision at top positions

---

[16] "Non-English" and "Other" answers are categorized into "Irrelevant" answers.

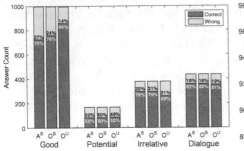

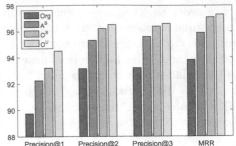

Fig. 3. Prediction results of A^S, O^S and O^U methods on each answer type.

Fig. 4. Performances of re-ranking for improving user experience in browsing.

(Precision@1,2,3) and mean reciprocal rank (MRR) [19] methods. We report the results on Yahoo since the original rank is obtainable only on Yahoo! Answer.

Figure 4 shows the result of original rank (denoted as "Org") and the re-ranking results of three different models A^S, O^S and O^U. All Precision@1,2,3 and MRR are improved by answer quality prediction methods compared with original results, which means some low-quality answers are correctly removed from the top positions by re-ranking methods. Moreover, the better method in classification generates better results in re-ranking. For example, O^U is better than O^S on classification in Table 5, and O^U is also better than O^S on re-ranking in Fig. 4. It is because performances of classification and re-ranking are highly correlated.

5 Conclusion

In this paper, we have investigated the problem of low-quality answer detection in community question and answering. We propose an unsupervised learning method based on three assumptions that most answers under a question are normal ones, and low-quality answers are different from other answers under the same question, and questions have different quality criteria. We propose a set of features to describe the difference from answers by taking advantage of the state-of-the-art methods. We empirically study the efficacy of the proposed unsupervised learning method as well as supervised methods on three datasets, including two benchmark datasets. The evaluation results show that our unsupervised method can significantly improve the supervised method.

Acknowledgements. This research was partially supported by grants from the National Key Research and Development Program of China (Grant No. 2016YFB1000904), the National Science Foundation for Distinguished Young Scholars of China (Grant No. 61325010), the National Natural Science Foundation of China

(Grant No. 61672483), and the Fundamental Research Funds for the Central Universities of China (Grant No. WK2350000001).

References

1. Berger, A., et al.: Bridging the lexical chasm: statistical approaches to answer-finding. In: SIGIR 2000 (2000)
2. Blei, D.M., et al.: Latent Dirichlet allocation. In: NIPS 2001 (2001)
3. Chandola, V., et al.: Anomaly detection: a survey. ACM Comput. Surv. **41**(3) (2009)
4. Crawford, M., et al.: Survey of review spam detection using machine learning techniques. J. Big Data **2**(1), 23 (2015)
5. Denkowski, M.J., Lavie, A.: Meteor universal: language specific translation evaluation for any target language. In: EACL 2014 (2014)
6. Hodge, V.J., Austin, J.: A survey of outlier detection methodologies. Artif. Intell. Rev. **22**(2), 85–126 (2004)
7. Jeon, J., et al.: A framework to predict the quality of answers with non-textual features. In: SIGIR 2006 (2006)
8. Jindal, N., Liu, B.: Review spam detection. In: WWW 2007, pp. 1189–1190 (2007)
9. Joachims, T.: Learning to Classify Text Using Support Vector Machines - Methods, Theory, and Algorithms. Kluwer/Springer, New York (2002)
10. Klein, D., Manning, C.D.: Accurate unlexicalized parsing. In: ACL 2003 (2003)
11. Li, F., et al.: Learning to identify review spam. In: IJCAI 2011 (2011)
12. Liu, W., et al.: Unsupervised one-class learning for automatic outlier removal. In: CVPR 2014 (2014)
13. Lyon, C., et al.: Detecting short passages of similar text in large document collections. In: EMNLP 2001, pp. 118–125 (2001)
14. Mikolov, T., et al.: Efficient estimation of word representations in vector space. CoRR, abs/1301.3781 (2013)
15. Nakov, P., et al.: Semeval-2015 task 3: answer selection in community question answering. In: SemEval@NAACL-HLT 2015 (2015)
16. Nakov, P., et al.: Semeval-2016 task 3: community question answering. In: SemEval@NAACL-HLT 2016, pp. 525–545 (2016)
17. Nallapati, R.: Discriminative models for information retrieval. In: SIGIR 2004 (2004)
18. Nicosia, M.Q., et al.: QCRI: answer selection for community question answering - experiments for arabic and english. In: SemEval@NAACL-HLT 2015 (2015)
19. Radev, D.R., et al.: Evaluating web-based question answering systems. In: LREC's 2002 (2002)
20. Sakai, T., et al.: Using graded-relevance metrics for evaluating community QA answer selection. In: WSDM 2011 (2011)
21. Shah, C., Pomerantz, J.: Evaluating and predicting answer quality in community QA. In: SIGIR 2010 (2010)
22. Toutanova, K., et al.: Feature-rich part-of-speech tagging with a cyclic dependency network. In: HLT-NAACL (2003)
23. Tran, Q.H., et al.: JAIST: combining multiple features for answer selection in community question answering. In: SemEval@NAACL-HLT 2015 (2015)
24. Wise, M.J.: YAP3: improved detection of similarities in computer program and other texts. In: SIGCSE 1996, pp. 130–134 (1996)
25. Xia, Y., et al.: Learning discriminative reconstructions for unsupervised outlier removal. In: ICCV 2015 (2015)

Approximate OLAP on Sustained Data Streams

Salman Ahmed Shaikh[1]([✉]) and Hiroyuki Kitagawa[2]

[1] Center for Computational Sciences, University of Tsukuba, Tsukuba, Japan
salman@kde.cs.tsukuba.ac.jp
[2] Faculty of Engineering Information and Systems,
University of Tsukuba, Tsukuba, Japan

Abstract. Many organizations require detailed and real time analysis of their business data for effective decision making. OLAP is one of the commonly used methods for the analysis of static data and has been studied by many researchers. OLAP is also applicable to data streams, however the requirement to produce real time analysis on fast and evolving data streams is not possible unless the data to be analysed reside on memory. Keeping in view the limited size and the volatile nature of the memory, we propose a novel architecture *AOLAP* which in addition to storing raw data streams to the secondary storage, maintains data stream's summaries in a compact memory-based data structure. This work proposes the use of piece-wise linear approximation (PLA) for storing such data summaries corresponding to each materialized node in the OLAP cube. Since the PLA is a compact data structure, it can store the long data streams' summaries in comparatively smaller space and can give approximate answers to OLAP queries.

OLAP analysts query different nodes in the OLAP cube interactively. To support such analysis by the PLA-based data cube without the unnecessary amplification of querying errors, inherent in the PLA structure, many nodes should be materialized. However, even though each PLA structure is compact, it is impossible to materialize all the nodes in the OLAP cube. Thus, we need to select the best set of materialized nodes which can give query results with the minimum approximation errors within the given memory bound. This problem is NP-hard. Hence this work also proposes an optimization scheme to support this selection. Detailed experimental evaluation is performed to prove the effectiveness of the use of PLA structure and the optimization scheme.

1 Introduction

With the increase of stream data sources, such as sensors, GPS, micro blogs, e-business, etc., the need to aggregate and analyze stream data is increasing. Many applications require instant decisions exploiting the latest information from the data streams. For instance, timely analysis of business data is required for improving profit, network packets need to be monitored in real time for identifying network attacks, etc. Online analytical processing (OLAP) is a well-known and useful approach to analyse data in a multi-dimensional fashion, initially

© Springer International Publishing AG 2017
S. Candan et al. (Eds.): DASFAA 2017, Part II, LNCS 10178, pp. 102–118, 2017.
DOI: 10.1007/978-3-319-55699-4_7

given for disk-based static data (we call it *traditional OLAP*). For the effective OLAP analysis, the data is converted into a multi-dimensional schema, also known as star schema. The data in star schema is represented as a data cube, where each cube cell contains measure across multiple dimensions. A user may be interested in analysing data across different combination of dimensions or examining different views of it. These are often termed as OLAP operations and to support these operations, data is organized as lattice nodes.

A number of solutions have been proposed for OLAP analysis on data streams in the near past [1–3]. The requirement to produce real time OLAP analysis on fast and evolving data streams is not possible unless the data to be analysed reside on primary memory. However the size of the primary memory is limited and it is volatile. Therefore we need an in-memory compact data structure that can quickly answer user queries in addition to a non-volatile backup of the data streams. Hence we propose a novel architecture $AOLAP$ (Approximate Stream OLAP), which in addition to storing raw data streams to the secondary storage, maintains data stream's summaries in a compact memory-based data structure. This work proposes the use of piece-wise linear approximation (PLA) for storing such data summaries corresponding to each materialized node in the OLAP cube. PLA can store the long data streams' summaries in comparatively smaller space on the primary memory and can give approximate answers to OLAP queries. It provides an impressive data compression ratio and answers user queries with max error guarantees, and has been studied by many researchers [4–7].

When performing OLAP analysis, users usually request different lattice nodes. Generally only a few nodes are materialized while the other requested nodes are computed on ad-hoc basis, because the materialization of all the nodes is memory expensive. On the other hand, materialization of too less nodes require a lot of ad-hoc computation which is computationally (time) expensive. The optimization of the space-time trade-off in traditional OLAP, when choosing the lattice nodes to materialize is a NP-hard problem [8]. Selecting the lattice nodes to materialize for PLA-based stream OLAP is a space-error trade-off in addition to the space-time trade-off. Since PLA can maintain long data stream summaries on memory and in the most cases can answer queries from it, where the computation is extremely fast, space-error trade-off is more significant among the two trade-offs in the context of PLA-based stream OLAP. Hence this work also proposes an optimization scheme which selects the η (user defined parameter) lattice nodes to materialize such that the overall querying error is minimized. We support the contributions of this work with the following real-world example.

Example 1. A big retail chain collects sales quantities of their stores at the granularity of individual product, store location and promotion (under which the product is sold) dimensions which arrive every minute as an infinite time series data stream. The top management is interested in analysing the sales in real time to avoid shortage of products' supply in any city or state. In addition, the top management is interested in finding the recent past advertisement strategies and/or promotional campaigns for specific products and brands to decide future advertisement budget/strategy and to execute promotional campaigns.

It is not possible to perform such analysis in real-time if all the data to be analysed must be fetched from secondary storage. Keeping in view the importance and demand of real-time analysis, a small degree of error in query results may be tolerated as a trade-off for timely analysis. ∎

Our contributions in this work can be summarized as follows:

- A novel architecture $AOLAP$, which in addition to storing raw data streams to the secondary storage, maintains data streams summaries in a compact memory-based data structure.
- Use of the PLA structure to compactly maintain the stream OLAP cubes.
- An optimization scheme to select the lattice nodes to materialize which can minimize the querying error.
- Detailed experimental evaluation to prove the effectiveness of the use of PLA structure for the materialized nodes and the optimization scheme.

The rest of the paper is organized as follows: Sect. 2 reviews essential concepts. Section 3 discusses the related work. In Sect. 4, PLA-based sustained storage is presented. In Sect. 5, the proposed AOLAP architecture and query processing over PLA-based storage are presented. The estimation of querying error is presented in Sect. 6. The proposed optimization scheme is presented in Sect. 7. The effectiveness of our contributions is experimentally evaluated in Sects. 8 and 9 concludes this paper and discusses future directions.

2   Essential Concepts

2.1   Piecewise Linear Approximation (PLA)

PLA is a method of constructing a function to approximate a single valued function of one variable in terms of a sequence of linear segments [9]. Precisely, let S be a time series of discrete data points (t_i, x_i), where $i \in [1, n]$, t_i is the i-th timestamp, x_i is the i-th value and we wish to approximate x_i with a piecewise linear function $f(t_i)$, using a small number of segments such that the error $|f(t_i) - x_i| \le \epsilon$, where ϵ is a user defined error parameter. The goal is to record only the successive line segments, and not the individual data points, to reduce the overhead of recording entire time-series.

Authors in [9] proposed an online algorithm to construct such an f having the minimum number of line segments. For completeness, the algorithm is described in Algorithm 1. It takes a data point $p = (t_i, x_i)$ and an error parameter ϵ. Let P be the set of points processed so far, the algorithm maintains the property that all the points in P can be approximated with a line segment within ϵ. If $P \cup \{p\}$ can be approximated with a line segment then it is added to P, else the points in P are output as a line segment and a new line segment is started with the point p.

Example 2. Consider a retail chain time series with the dimensions and a business fact of Example 1. It is a series of a 5-tuple $< t, p, s, m, x >$; the timestamp (minute) (t), product (p), store (s), promotion (m) and the sales quantity (x).

$(1, p_1, s_1, m_1, 48)$, $(1, p_2, s_1, m_1, 48)$, $(2, p_1, s_1, m_1, 43)$, $(2, p_2, s_1, m_1, 64)$, $(3, p_1, s_1, m_1, 60)$, $(3, p_2, s_1, m_1, 73)$, $(4, p_1, s_1, m_1, 75)$, $(4, p_2, s_1, m_1, 58)$, $(5, p_1, s_1, m_1, 35)$, $(5, p_2, s_1, m_1, 87)$, $(6, p_1, s_1, m_1, 52)$, $(6, p_2, s_1, m_1, 7)$, $(7, p_1, s_1, m_1, 95)$, $(7, p_2, s_1, m_1, 2)$, ...

Algorithm 1. PLA

Input: data point p, error parameter ϵ
  /*P: Set of points processed so far.*/
1: **if** $P \cup \{p\}$ can be approximated with a
   line segment within ϵ **then**
2:     Add p to P;
3: **else**
4:     **return** a line segment approximat-
       ing P;
5:     Set $P \leftarrow \{p\}$;
6: **end if**

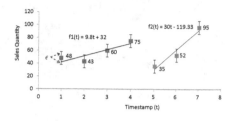

Fig. 1. Data points approximated by PLA segments

Assuming $\epsilon = 10$, the tuples for the dimension keys p_1, s_1 and m_1 in the above time series can be approximated by the following piecewise function.

$$f_{p_1, s_1, m_1}(t) = \begin{cases} 9.8t + 32 & 1 \leq t \leq 4 \\ 30t - 119.33 & 5 \leq t \leq 7 \\ \dots \end{cases}$$

Figure 1 shows the PLA segments of the $f_{p_1, s_1, m_1}(t)$. $f1(t)$ and $f2(t)$ are the PLA segments formed by the tuples for timestamps $1 \leq t \leq 4$ and $5 \leq t \leq 7$, respectively. The accurate sales quantities are shown in the figure for illustration only, while the approximate sales quantities can be obtained from the PLA segments. Note that when using PLA, we only maintain PLA segments (slopes and intercepts) in memory and not the actual data points, resulting in data size reduction. ∎

2.2 Online Analytical Processing (OLAP)

OLAP is a technique for interactive analysis over multidimensional data. For efficient OLAP analysis, the underlying database schema is usually converted into a partially-normalized star schema. The data in star schema is represented as a data cube consisting of several dimension tables and a fact table. Dimension tables contain descriptive attributes, while the fact tables contain business facts called measures and foreign keys referring to primary keys in the dimension tables. Some of the dimension attributes are hierarchically connected. A number of dimension hierarchies compose a cube lattice, where each node corresponds to different combination of attributes at different hierarchy levels and an edge between two nodes represents a subsumption relation between them. Hence nodes in a lattice are combinations of dimension attributes and represent OLAP queries.

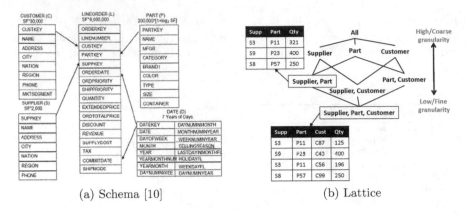

(a) Schema [10]   (b) Lattice

Fig. 2. Star schema benchmark

For instance, consider the star schema benchmark [10] shown in Fig. 2a with a fact table *LINEORDER* and four dimension tables, *PART, CUSTOMER, SUP-PLIER* and *DATE*. Attributes *Quantity, ExtendedPrice, OrdTotalPrice, Discount, Revenue*, etc. of the *LINEORDER* are the business facts, while *CustKey, PartKey* and *SuppKey* are foreign keys of *CUSTOMER, PART* and *SUPPLIER* dimensions respectively. Additionally, each dimension table contains hierarchical relationship among some of its attributes. For example, *SUPPLIER* dimension contains hierarchy among the following attributes: $City->Nation->Region$. If we consider interaction of the *PART, CUSTOMER* and *SUPPLIER* dimensions only (without considering their internal hierarchies), the corresponding lattice is given by Fig. 2b. In the figure, the nodes with the border are materialized and the associated tables show their tuples. Once an OLAP lattice has been generated, users can register queries and apply OLAP operations to it. The queries registered to non-materialized nodes are computed from the materialized nodes on ad-hoc basis.

3   Related Work

3.1   Compact Data Structures and Approximate Querying

Compact data structures have long been utilized to summarize voluminous and velocious data streams and answer queries from them approximately. H. Elmeleegy et al. in [4] proposed two PLA-based stream compression algorithms, swing filters and slide filters, to represent a time-varying numerical signal within some preset error value. The PLA line segments in the swing filter are connected whereas mostly disconnected in the slide filter. The slide filter proposed in their work is almost similar to the one proposed by O'Rourke in [9].

Zhewei et al. in [7] proposed sketching techniques that support historical and window queries over summarized data. The data summary is maintained using the count-min sketch and the AMS sketch and the persistence is achieved by utilizing PLAs. Their work can provide persistence for counters only and can support point, heavy hitter and join size queries. [6] presented an online algorithm to optimize the representation size of the PLA for streaming time-series data. A PLA function f can be constructed using either only continuous (joint) line segments or only disjoint line segments. To optimize the size of f, the authors gave an adaptive solution that uses a mixture of joint and disjoint PLA segments and they named it *mixed-type* PLA.

Wavelet is also a famous technique which is often used for hierarchical data decomposition and summarisation. The technique proposed in [11] can effectively perform the wavelet decomposition with maximum error metrics. However, since the technique uses dynamic programming, it is computationally expensive. Therefore it cannot be used effectively for the data streams, which require one-pass methodology in linear time. [12] proposed a method for one-pass wavelet synopses with the maximum error metric. [12] shows that by using a number of intuitive thresholding techniques, it is possible to approximate the technique discussed in [11]. However, wavelet summarization can have a number of disadvantages in many situations as many parts of the time series may be approximated very poorly [12]. [13] used a sampling approach to answer OLAP queries approximately, however they did not consider lattice nodes materialization issue as we do. [2] compared different summarization methods on data streams and proved that the PLA is the best data summarization technique as far as querying error is concerned. Hence we propose the use of PLAs to summarize the data streams in this work. None of the above work considered the use of compact data structure for the OLAP as we do in this work.

3.2 Stream OLAP and View Maintenance

OLAP has been intensively studied by database researchers. [14] proposed a systematic study of the OLAP node and index-selection problem. Authors in [8] investigated the issue of nodes materialization when it is expensive to materialize all nodes. They presented a greedy algorithm that determines a good set of nodes to materialize. However, these work can only deal with static data.

One of the earliest work on stream OLAP was given by J. Han et al. [1]. They proposed an architecture called *StreamCube* to facilitate OLAP for streams. In order to reduce the query response time and the storage cost, StreamCube keeps the distant data at coarse granularity and very new data at fine granularity and pre-computes some OLAP queries at coarser, intermediate, and finer granularity levels. However, their work does not use compact data structures, therefore can not be used to maintain long data histories. Furthermore older data in their work is only available at coarser granularity, thus limiting the range of queries.

Phantoms are intermediate queries to accelerate user queries. Zhang et al. [15] proposed the use of phantoms to reduce the overall cost (processing and data transfer cost) within very limited memory of a network interface card. Although their work can reduce aggregation query cost, but is not capable of answering ad-hoc OLAP queries. M. Sadoghi et al. in [3] presented a lineage-based data store that combines real-time transactional and analytical processing within an engine with the help of the lineage-based storage. However their focus is storage architecture and not the core OLAP. Ahmad et al. [16] presented *viewlet transforms*, which materializes a query and a set of its higher-order deltas as views resulting in a reduced overall view maintenance cost by trading space.

In contrast to the above works, this work proposes a compact data structure based stream OLAP, capable of maintaining sustained data stream summaries and answering user queries approximately with maximum error guarantees.

4   PLA-Based Sustained Storage

The PLA is a compact data structure and can be used for sustained in-memory data summaries. The term sustained in this work corresponds to the long data summaries that PLA can accommodate by approximating several data points with a segment. Thus the main idea of our proposal is to store time series data points as PLA line segments for all the OLAP lattice nodes that need to be materialized, to reduce the overhead of recording complete time-series. This paper assumes that only the business facts arrive as time-series stream, while the dimensions are not treated as stream as they are updated less frequently.

Let S be a time-series data stream consisting of tuples $(t_i, k_{1i}, k_{2i}, ..., k_{di}, m_i)$, where t_i is a timestamp, $i \in [1, n]$ and $t_i \leq t_{i+1}$, $k_{1i}, k_{2i}, ..., k_{di}$ constitute a d-dimensional key and m_i is a business fact or measure. To keep the discussion simple, this work assumes that one tuple for every key combination arrives at each timestamp, however it is easily extendible for the general case. Recall that PLA approximates data points using a piece-wise linear function, such that the error between the approximated and the actual data point is within the user-defined error parameter, ϵ. For the data points in S, we wish to approximate m_i using a piece-wise linear function $f_{k_{1i},...,k_{di}}(t_i)$, such that the $|f_{k_{1i},...,k_{di}}(t_i) - m_i| \leq \epsilon$. The PLA needs to be maintained for each d-dimensional key. This paper, like most of the previous work that discussed this problem in an online setting [4,6,17,18], assumes L_∞-metric for the error computation. This is due to the fact that other error computations are not suitable for online algorithms as they require sum of errors over the entire time series.

The above approach would result in a sustained PLA-based storage for each key. The number of line segments required for each PLA and the cost of a PLA line segment computation depend on the choice of error parameter ϵ. Larger ϵ would result in a smaller number of line segments but larger line segment computation cost and approximation error and vice versa. Also note that for multiple measures, multiple PLA structures need to be maintained per key.

5   Architecture and Query Processing

5.1   AOLAP Architecture

This section presents the proposed Approximate Stream OLAP (AOLAP) architecture, shown in Fig. 3, that enables users to obtain approximate answer of their

OLAP queries. Given the dimension information, the number of dimensions to materialize (η) and utilizing the proposed optimization scheme (Sect. 7), the AOLAP system selects the η nodes to materialize. For each materialized node, the AOLAP system maintains a PLA structure discussed in Sect. 4.

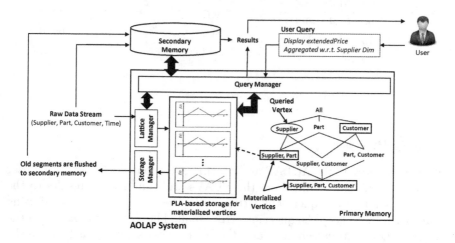

Fig. 3. AOLAP Architecture

As the time series data arrive, the *Lattice Manager* calls the PLA algorithm (Algorithm 1) for each materialized node and update the corresponding PLA structures (hereafter materialized node is called *PLAV*). In Fig. 3, lattice nodes within rectangular boundaries represents PLAVs. The node at the lowest granularity, i.e., the node (*Supplier, Part, Customer*) in the figure, is always materialized to enable the AOLAP system to answer all possible user queries. The raw data stream is also stored in some non-volatile storage to avoid permanent data loss in case of system failure and to enable users to obtain accurate answers of their queries if needed.

In contrast to data stream, primary memory is finite. Since the users are interested in analysing recent data more frequently than the old or historical data, the *Storage Manager* flushes the old PLA segments to the secondary storage once they reach the memory limits or as specified by end user. These segments may be used to answer the historical queries to avoid computing the results from raw data stream, which is computationally expensive. Since the data is compact, this flushing may be done periodically rather than continuously or when the system is not overloaded by user queries. This also makes the system durable as in the case of system crash, the old segments are not permanently lost while the very new segments, not yet flushed to the secondary storage, can be reconstructed from the raw data stream available in the non-volatile storage.

Table 1. Querying PLA

Timestamp (t_i)	m_i	$\hat{m}_i$	$m_i - \hat{m}_i$
t_1	48	41.8	6.2
t_2	43	51.6	-8.6
t_3	60	61.4	-1.4
t_4	75	71.2	3.8
t_5	35	30.67	4.33
t_6	52	60.67	-8.67
t_7	95	90.67	4.33

Table 2. OLAP Operations on Table 1 data

OLAP Operations	$m_{1,7}$	$\widehat{m_{1,7}}$
MAX	95	90.67
MIN	35	30.67
SUM	408	408.01
AVG	58.286	58.287

5.2 Query Processing

The *Query Manager* in the AOLAP architecture is responsible for accepting user queries, computing the results from the PLAVs and sending the results to the end user. Since a user can query any lattice node, the results are generated using the nearest PLAV to keep the querying error small. The *Lattice Manager*, on the request from *Query Manager*, generates the query results and sends them to the *Query Manager*. For example in Fig. 3, the user query (*Supplier*), represented by oval boundary, can be answered using the PLAV (*Supplier, Part*).

OLAP queries over data streams generally involve aggregation operations over current, historical or some window data. Typical OLAP aggregation operations include SUM, MAX, MIN, AVG, etc. Users may also be interested in analysing raw facts across multiple dimensions. To answer a historical window query for a key k or any combination of keys from d-dimensional key for time range $(t', t]$, find the recorded measures $\hat{m}_i$ for all $t_i \in (t', t]$ as an approximation of m_i and perform the requested aggregate operations to obtain $\widehat{m_{t',t}}$. Let the average length of a PLA line segment in terms of timestamp is l, then the cost of finding a measure $\hat{m}_i$ can be given by $\frac{n}{l}$, where n is the length of stream.

Example 3. Once again consider the time series and piecewise function $f_{p_1,s_1,m_1}(t)$ of Example 2. Now we would like to query $f_{p_1,s_1,m_1}(t)$ segments for the following OLAP aggregation operations: MAX, MIN, SUM, AVG, where $1 \leq t \leq 7$.

Table 1 shows the accurate (m_i) and approximate $(\hat{m}_i)$ measures of the Example 2 time series for keys p_1, s_1 and m_1. The approximate measures are obtained from the PLA-based storage. Table 2 lists the OLAP aggregation operations performed on $\hat{m}_i$. It is interesting to note that the exact and the approximate measures for the OLAP operations SUM and AVG are quite similar, although they are computed from several approximate measures. This is due to the mutual cancellation of + and - errors in the individual approximate values. ∎

6   Querying Error

In order to select the optimal PLAVs (nodes to materialize), an estimation of the overall querying error is needed, i.e., the aggregated querying error of all the lattice nodes, which forms the basis of our optimization problem presented in Sect. 7. In a d-dimensional lattice, there exist 2^d nodes [19]. Let $V = \{v_1, ...v_{2^d}\}$ denotes the set of all the lattice nodes. The overall querying error is computed by taking into consideration the set of nodes chosen for materialization (V_m) and the number of rows in each node ($|v_i|$). Since the number of rows in a node is not known beforehand, it is estimated using the domain size of dimension attributes.

Consider two nodes $v_i \in V$ and $v_j \in V_m$, then $v_i \preceq v_j$ shows the dependence relationship between the queried node (v_i) and the materialized node (v_j), that is, a query can be answered from a materialized node if the queried node is dependent on the materialized node. Since a query can be answered from more than one materialized nodes, we choose the nearest node which can minimize the fraction $\frac{|v_j|}{|v_i|}$, as the larger $\frac{|v_j|}{|v_i|}$ results in the amplification of the overall querying error. The overall querying error can be expressed as:

$$\epsilon. \sum_{v_i \in V} min_{\forall v_j \in V_m | v_i \preceq v_j} \left(\frac{|v_j|}{|v_i|} \right) \tag{1}$$

Note that in the Eq. 1, the fraction $\frac{|v_j|}{|v_i|}$ depends on the number of rows in the materialized and the querying nodes. By choosing the smaller fraction we actually choose the node v_j with the smaller number of rows, that is, we need to aggregate a less number of rows to answer a query, resulting in smaller processing time and querying error as each row may contributes to the querying error.

7   Optimization Scheme

The PLA-based sustained storage discussed in Sect. 4 can be used to materialize only one lattice node. Since an OLAP lattice contain several nodes and during analysis a user may request any node, a baseline approach is to materialize all the lattice nodes. However, the baseline approach may results in prohibitively large number of nodes to materialize (2^d), specially when the number of dimensions is high, which is extremely memory costly. In the following we propose an optimization algorithm to solve this issue. Additionally we consider the reference frequency (f_i), the frequency with which a lattice node is queried by end-users, of each lattice node in the computation of querying error. Nodes or queries with the low reference frequencies contribute less to the overall querying error and vice versa. Hence the overall querying error considering the reference frequencies can be expressed as:

$$\epsilon. \sum_{v_i \in V} min_{\forall v_j \in V_m | v_i \preceq v_j} \left(\frac{|v_j|}{|v_i|} . f_i \right) \tag{2}$$

7.1   Optimization Problem

Choosing which lattice nodes to materialize for the PLA-based stream OLAP is a space-error trade-off in addition to the space-time trade-off of traditional OLAP. However the focus of this work is only the space-error trade-off which is more significant in the context of PLA-based stream OLAP and is a NP-hard problem. Hence we propose a greedy optimization algorithm to find the optimal solution. Here we assume that the number of nodes to be materialized (η) is provided and the reference frequencies of the lattice nodes is known.

Optimization Problem: Given the number of nodes to materialize, η, and the reference frequency of each lattice node, $\mathbf{f} = \{f_1, f_2, ..., f_{2^d}\}$, materialize the nodes that can minimize the overall querying error.

7.2   Greedy Optimization Algorithm

Having introduced the optimization scheme, we are ready to present the proposed optimization algorithm (Algorithm 2). The algorithm takes as input a set of lattice nodes (V), the finest node (v_f), the number of nodes to materialize (η), the PLA error parameter (ϵ) and the reference frequencies ($\mathbf{f}$). The algorithm outputs a set of nodes to materialize (V_m). Note that the node at the finest granularity, (v_f), is always chosen to materialize because it contains data at the most granular level and therefore can answer all the queries, however answering coarser level queries from v_f results in the amplification of querying error. Therefore the proposed greedy algorithm finds η nodes to materialize, besides v_f, such that the overall querying error (Eq. 2) is minimized.

In the algorithm, the inner *for* loop (Lines 5–13) computes the overall querying error for each candidate node v_j in $V \setminus V_m$ using Eq. 2 (Line 7). Lines 8–11 keeps track of the current best candidate node. At the end of the inner *for* loop, the best candidate node is selected and added to the set of materialized nodes (Line 14). This loop is executed η times to select the η best nodes to materialize (outer *for* loop). At the end of the algorithm, the set of the best nodes to materialize V_m is returned (Line 17).

8   Experiments

8.1   Experimental Setup

Environment: For the sake of experiments a prototype system corresponding the AOLAP architecture is developed in C++. The experiments are performed on one of the node of HP BladeSystem c7000 with Intel Xeon (ES-2650 v3 @ 2.3 GHz) processor and 6 GB RAM running Ubuntu 14.10 OS.

Data: We used TPC-H[1] benchmark for experiments, well-known for OLAP analysis. However its schema is modified according to the Star Schema Benchmark (SSB)

---

[1] TPC-H. http://www.tpc.org/tpch/.

Algorithm 2. Greedy Optimization Algorithm

Input: V: a set of lattice nodes, v_f: the finest node, η: the number of nodes to materialize, **f**: a set of reference frequencies corresponding to nodes in V, ϵ

Output: V_m: a set of nodes to materialize

1: $V_m \leftarrow \{v_f\}$;
2: $V_c \leftarrow V \setminus V_m$; {$V_c$: List of candidate nodes}
3: **for** $i = 1$ to η **do**
4:     $\Delta_{min} \leftarrow \infty$;
5:     **for each** $v_j \in V_c$ **do**
6:         $V_m \leftarrow V_m \cup \{v_j\}$;
7:         $\Delta \leftarrow QueryingError(V, V_m, \epsilon, \mathbf{f})$;
        {Δ: Overall querying error computed for current V_m using Eq. 2}
8:         **if** $\Delta < \Delta_{min}$ **then**
9:             $\Delta_{min} \leftarrow \Delta$;
10:          $v_{min} \leftarrow v_j$;
11:         **end if**
12:         $V_m \leftarrow V_m \setminus v_j$;
13:     **end for**
14:     $V_m \leftarrow V_m \cup \{v_{min}\}$;
15:     $V_c \leftarrow V_c \setminus \{v_{min}\}$;
16: **end for**
17: **return** V_m;

---

[10] as shown in Fig. 2a. The *LINEORDER* fact table contains 6,000,000 tuples and the dimension tables, *PART*, *CUSTOMER* and *SUPPLIER* contains 200,000, 30,000 and 2,000 tuples respectively. We considered the following dimensional hierarchies. *CUSTOMER*: Custkey $->$ Nation $->$ Region, *SUPPLIER*: Suppkey $->$ Nation $->$ Region, PART: PARTKEY, where NATION and REGION contain 25 and 5 unique tuples, respectively. The hierarchical lattice of the dimensions contain 32 nodes.

The time series is generated by identifying $10\,K$ unique dimension keys combinations in the *LINEORDER* fact table and feeding them repeatedly to the system. In order to avoid the repetition of fact values, we only fed the dimension keys repeatedly. The fact values are repeated after every 6,000,000 tuples (which is the size of the fact table) and are quite non-uniform. We selected this business fact to show the usability of the PLA on non-uniform data, as the PLA results in low compression ratio on non-uniform data while high compression ratio on uniform data. The system time is used as the time series timestamp.

Comparative Methods: To evaluate the effectiveness of the proposed optimization scheme, we compared it with the following methods: **(1) Random:** The lattice nodes to materialize are chosen randomly. **(2) Frequency:** The lattice nodes with high reference frequencies are chosen for materialization.

We used the following five ways to assign reference frequencies to lattice nodes to cover different types of use cases in various applications.

– **Rand:** Frequencies are assigned randomly within [0, 1] range.

- **AllHigh:** High frequencies are assigned randomly within [0.8, 1] range.
- **AllLow:** Low frequencies are assigned randomly within [0, 0.2] range.
- **CoarseHigh:** Higher frequencies are assigned to coarser aggregation levels.
- **FineHigh:** Higher frequencies are assigned to finer aggregation levels.

8.2   Experimental Evaluation

Experimental evaluation is subdivided into measuring the memory space utilization and querying error percentage. The evaluation is done for the worst case SUM operation, i.e., we aggregated the absolute querying error values. Unless otherwise stated, the following default parameter values are used in the experiments: $\eta = 6$, $\epsilon = 3\%$ (the value of ϵ is set as the percentage of the maximum value in the fact table) and frequency method = Rand. Each experiment is performed 5 times and their average values are reported in the graphs.

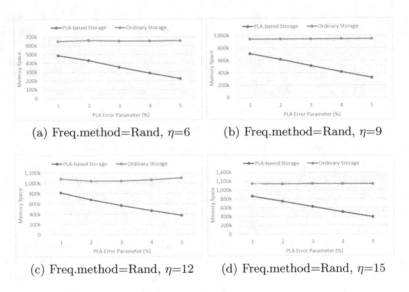

(a) Freq.method=Rand, η=6   (b) Freq.method=Rand, η=9

(c) Freq.method=Rand, η=12   (d) Freq.method=Rand, η=15

Fig. 4. Average memory usage for different η

Memory Space Utilization. To evaluate the effectiveness of the PLA, we compared the memory space consumed when using PLA-based storage to that of ordinary storage (which stores the actual data points) in Figs. 4 and 5. The storage space is measured in terms of the number of PLA segments for the PLA-based storage and the number of data points for the ordinary storage. Since a PLA segment requires twice memory space than a data point, we divided the total number of data points by a factor of 2 to keep the comparison fair.

The average amount of memory consumed by the PLA-based storage decreases with the increase in PLA error-parameter (ϵ), as can be observed from Fig. 4. This is because as the ϵ increases, a PLA segment can approximate a larger number of data points thereby reducing the number of line segments required by the PLA-based storage, which results in reduced memory space consumption. In most of the cases in Fig. 4, the memory space used by the PLA-based storage is upto 3 times less than the ordinary

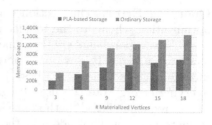

Fig. 5. Effect of varying η on space (Freq. method = Rand, $\epsilon = 3\%$)

storage for $\epsilon = 4\%$ and higher. This proves that the use of PLA for the materialization of lattice nodes can significantly reduce the memory consumption.

We also measured the memory space consumption by varying the number of materialized lattice nodes (η) as shown in Fig. 5. As η increases, the memory space consumption of both the PLA-based storage and the ordinary storage increases because we need to store data at increased number of aggregations levels. However the memory consumption of the PLA-based storage is lower than that of ordinary storage for all the η values. Note that we used highly non-uniform data values for the experiments, where it is difficult for the PLA algorithm to approximate the larger number of data points with one line segment. For the uniform time series data, for instance hourly temperature values or stock price data, the PLA-based storage is expected to be far more advantageous.

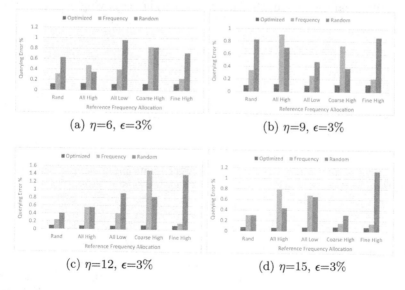

Fig. 6. Querying error percentage for different frequency allocation methods

Querying Error. This section compares the querying error of the proposed optimization scheme to the frequency and the random methods.

Firstly, experiments are performed for different frequency allocation methods as shown in Fig. 6. It is evident from the Figs. 6a, b, c and d that the greedy optimization scheme selects the best η nodes to materialize, resulting in the least querying error. Note the use of logarithmic scale on the y-axis. On the other hand, frequency based method gives priority to the nodes with high frequencies to materialize, while random method randomly chooses η nodes to materialize, however both the comparative methods results in higher querying error. Furthermore in Fig. 6, the proposed optimized scheme results in the similar querying error for all the reference frequency allocations, because it always chooses the nodes that minimizes the querying error. Additionally the frequency method performs best for the frequency allocation approach *FineHigh* because it causes most of the nodes at finer granularity to materialize, which are at the middle or finer level of the lattice. When the nodes from these levels are materialized, coarser level queries can be answered from them leading to the reduction of the big querying error, that may otherwise results when the coarser level nodes need to be answered from the most finer level node. On the other hand, the random method behaves randomly for all the frequency allocations because of the reasons discussed above.

Next we perform experiments by varying η. Increasing η reduces the querying error as can be observed from Fig. 7. Here again the optimization scheme performs the best. Furthermore when using the proposed optimized scheme, we do not need to materialize many nodes to get the results with acceptable querying error. For instance in Fig. 7, out of the total 32 nodes, materialization of 9 or 12 nodes can significantly reduces the querying error, hence saving a lot of memory.

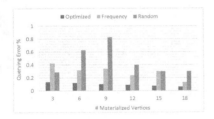

Fig. 7. Varying # materialized nodes (Freq. method = Rand, $\epsilon = 2\%$)

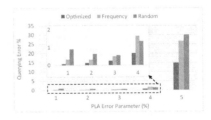

Fig. 8. Varying PLA error parameter ϵ (Freq. method = Rand, $\eta = 6$)

Finally experiments are performed by varying ϵ in Fig. 8. Increasing ϵ slightly increases the querying error which is mainly observable from the bars of the optimized scheme and the frequency method. However, here again the random method results in random querying errors for each ϵ value due to the random selection of the lattice nodes to materialize. Moreover, the querying error increases significantly for $\epsilon = 5\%$, because for higher ϵ, the PLA algorithm approximates a larger number of data points with a single PLA segment, possibly with higher absolute error values. Thus resulting in higher querying error.

9 Conclusion and Future Work

In this work we propose a novel architecture Approximate Stream OLAP (*AOLAP*) for maintaining time series data streams summaries, corresponding to each materialized lattice node, in a compact memory-based data structure, in addition to storing raw data streams to the secondary storage. We used *piece-wise linear approximation* as an in-memory compact data structure which can answer user queries approximately. In addition, we propose an optimization scheme to select the η lattice nodes to materialize, such that the overall querying error, caused by the approximation, is minimized. Experiments prove that the PLA-based storage can significantly reduce the memory consumption for a small cost of querying error and the nodes selected by the optimization algorithm to materialize can minimize the overall querying error. In the future we plan to extend this work to incorporate dependence relation between lattice nodes so that the number of PLA structures need to be maintained can be further reduced.

Acknowledgment. This research was partly supported by the program "Research and Development on Real World Big Data Integration and Analysis" of the RIKEN, Japan.

References

1. Han, J., Chen, Y., et al.: Stream cube: an architecture for multi-dimensional analysis of data streams. Distrib. Parallel Databases **18**(2), 173–197 (2005)
2. Duan, Q., Wang, P., Wu, M.X., Wang, W., Huang, S.: Approximate query on historical stream data. In: Hameurlain, A., Liddle, S.W., Schewe, K.-D., Zhou, X. (eds.) DEXA 2011. LNCS, vol. 6861, pp. 128–135. Springer, Heidelberg (2011). doi:10.1007/978-3-642-23091-2_12
3. Sadoghi, M., Bhattacherjee, S., Bhattacharjee, B., Canim, M.: L-store: a real-time OLTP and OLAP system. In: CoRR (2016)
4. Elmeleegy, H., Elmagarmid, A.K., Cecchet, E., Aref, W.G., Zwaenepoel, W.: Online piece-wise linear approximation of numerical streams with precision guarantees. Proc. VLDB Endow. **2**(1), 145–156 (2009)
5. Xie, Q., Zhu, J., et al.: Efficient buffer management for piecewise linear representation of multiple data streams. In: ACM CIKM, pp. 2114–2118 (2012)
6. Luo, G., Yi, K., Cheng, S.W., Li, Z., Fan, W., He, C., Mu, Y.: Piecewise linear approximation of streaming time series data with max-error guarantees. In: 2015 IEEE 31st ICDE, pp. 173–184 (2015)
7. Wei, Z., Luo, G., Yi, K., Du, X., Wen, J.-R.: Persistent data sketching. In: ACM SIGMOD 2015, pp. 795–810 (2015)
8. Harinarayan, V., Rajaraman, A., Ullman, J.D.: Implementing data cubes efficiently. In: ACM SIGMOD, pp. 205–216 (1996)
9. O'Rourke, J.: An on-line algorithm for fitting straight lines between data ranges. Commun. ACM **24**(9), 574–578 (1981)

10. O'Neil, P., O'Neil, E., Chen, X., Revilak, S.: The Star Schema Benchmark and Augmented Fact Table Indexing. In: Nambiar, R., Poess, M. (eds.) TPCTC 2009. LNCS, vol. 5895, pp. 237–252. Springer, Heidelberg (2009). doi:10.1007/978-3-642-10424-4_17

11. Garofalakis, M., Kumar, A.: Deterministic wavelet thresholding for maximum-error metrics. In: ACM PODS (2004)

12. Karras, P., Mamoulis, N.: One-pass wavelet synopses for maximum-error metrics. In: PVLDB, pp. 421–432 (2005)

13. De Rougemont, M., Cao, P.T.: Approximate answers to OLAP queries on streaming data warehouses. In: Proceedings of the Fifteenth International Workshop on Data Warehousing and OLAP, pp. 121–128 (2012)

14. Talebi, Z.A., Chirkova, R., Fathi, Y., Stallmann, M.: Exact and inexact methods for selecting views and indexes for OLAP performance improvement. In: EDBT (2008)

15. Zhang, R., Koudas, N., Ooi, B.C., Srivastava, D., Zhou, P.: Streaming multiple aggregations using phantoms. VLDB J. **19**(4), 557–583 (2010)

16. Koch, C., Ahmad, Y., Kennedy, O., Nikolic, M., Nötzli, A., Lupei, D., Shaikhha, A.: DBToaster: higher-order delta processing for dynamic, frequently fresh views. VLDB J. **23**(2), 253–278 (2014)

17. Lazaridis, I., Mehrotra, S.: Capturing sensor-generated time series with quality guarantees. In: ICDE, pp. 429–440 (2003)

18. Olston, C., Jiang, J., Widom, J.: Adaptive filters for continuous queries over distributed data streams. In: ACM SIGMOD, pp. 563–574 (2003)

19. Gray, J., Chaudhuri, S., et al.: Data cube: a relational aggregation operator generalizing group-by, cross-tab, and sub-totals. Data Min. Knowl. Discov. **1**(1), 29–53 (1997)

Search and Information Retrieval

Hierarchical Semantic Representations of Online News Comments for Emotion Tagging Using Multiple Information Sources

Chao Wang[1], Ying Zhang[1(✉)], Wei Jie[2], Christian Sauer[2], and Xiaojie Yuan[1]

[1] College of Computer and Control Engineering,
Nankai University, Tianjin, People's Republic of China
{wangchao,zhangying,yuanxiaojie}@dbis.nankai.edu.cn
[2] School of Computing and Engineering, University of West London, London, UK
{wei.jie,christian.sauer}@uwl.ac.uk

Abstract. With the development of online news services, users now can actively respond to online news by expressing subjective emotions, which can help us understand the predilections and opinions of an individual user, and help news publishers to provide more relevant services. Neural network methods have achieved promising results, but still have challenges in the field of emotion tagging. Firstly, these methods regard the whole document as a stream or bag of words and can't encode the intrinsic relations between sentences. So these methods cannot properly express the semantic meaning of the document in which sentences may have logical relations. Secondly, these methods only use semantics of the document itself, while ignoring the accompanying information sources, which can significantly influence the interpretation of the sentiment contained in documents. Therefore, this paper presents a hierarchical semantic representation model of news comments using multiple information sources, called Hierarchical Semantic Neural Network (HSNN). In particular, we begin with a novel neural network model to learn document representation in a bottom-up way, capturing not only the semantics within sentence but also semantics or logical relations between sentences. On top of this, we tackle the task of predicting emotions for online news comments by exploiting multiple information sources including the content of comments, the content of news articles, and the user-generated emotion votes. A series of experiments and tests on real-world datasets have demonstrated the effectiveness of our proposed approach.

Keywords: Emotion tagging · Hierarchical semantic representation · Multiple information sources · Neural network

1 Introduction

Due to the development of the internet, the past decades have witnessed an explosive growth in different types of web services such as blogs, forums, social networks and online news services. Among these various types of web services,

S. Candan et al. (Eds.): DASFAA 2017, Part II, LNCS 10178, pp. 121–136, 2017.
DOI: 10.1007/978-3-319-55699-4_8

online news has been an important type of information that attracts billions of users to read and actively respond by making comments. Users often express subjective emotions like sadness, happiness and surprise in their comments. Extracting these emotions contained in the comments can help us understand the preferences and perspectives of users, and help online news publishers to provide users with more personalized services. Therefore, an automatic emotion tagging method for online news comments is strongly desirable.

Emotion tagging is a fundamental problem in the research area of opinion mining and sentiment analysis, which has attracted much attention in information retrieval and natural language processing communities [14,17]. The emotion tagging problem can be formulated as a multi-classification problem, which calls for identifying multiple emotion categories (e.g., happiness, sadness and angry, etc.) from user-generated content including product reviews, posts on blogs or social networks, comments in forums or comments in online news services.

The dominating approaches usually utilize machine learning algorithms to build a classifier with hand-crafted features. Since the performances of traditional machine learners are heavily dependent on feature representations [5], deep learning methods become more and more popular recently due to the ability to learn discriminative features from data automatically.

Despite the achievement of neural network approaches, there still are some challenges. Firstly, how to encode the intrinsic relations between sentences in the semantic meaning of a document. This is important for emotion tagging because relations such as causality and contrast have great influence on determining the meaning of a document. However, existing studies usually fail to effectively capture the intrinsic relations, since sentences influence the semantic meaning equally whether they are before or after the adversatives. Secondly, these methods only use semantics of the document itself, while ignoring the accompanying information sources, which can have significant influence on interpreting the sentiment of the document. In the news comment scenario, the comments are users' response to the news articles, thus the emotions of the comments are influenced by the content of the news articles obviously. Moreover, many online news websites provide a emotion voting service through which users can share their emotions after reading news articles. These user-generated emotion votes can naturally provide guidance for assigning emotion tags to comments.

Therefore, this paper presents a hierarchical semantic representation model of news comments using multiple information sources, called Hierarchical Semantic Neural Network (HSNN). Firstly we bring in a novel neural network model to learn a hierarchical semantic representation of documents which encodes not only the semantics between words in a sentence but also the relations between sentences. Further, we combine the representations of multiple information sources including the comments, the news articles and the user-generated emotion votes together and introduce a novel classification method utilizing this hierarchical semantic representation in order to improve the result of emotion predicting and tagging. A series of experiments and tests on real-world datasets have demonstrated that our HSNN demonstrated good performances in emotion tagging compared with a selection of baseline models.

The remainder of this paper is organized as follows. Section 2 gives a brief overview of some state-of-the-art research on emotion tagging and makes discussions regarding the differences between our work and previous works. In Sect. 3, we present our proposed approach HSNN including hierarchical semantic representation model of document and semantic representations using multiple information sources with their classifiers utilized. Experiments are shown in Sect. 4. We end the paper with conclusions and an outlook on future work.

2 Related Work

Emotion tagging has become an important subtask of opinion mining and sentiment analysis [14], which aims at identifying the emotion tag of a document (e.g., review of products [24–26], news article [1,2,12,21], news comment [29,30]). For a general survey, please refer to [17]. This paper focuses on emotion tagging for comments of online news.

Many machine learning techniques have been applied on sentiment classification, such as unsupervised learning techniques (e.g., [26]), supervised learning techniques (e.g., [18]) and semi-supervised learning techniques (e.g., [22]). Many studies now focus on designing an effective feature schema. On this basis, relevant features can be extracted and classifiers like SVM could be used to classify each text into emotion categories. Other than these methods using only words to classify text, prior works [1,2,21] asserted it is arguable that emotions should be linked to specific topics instead of a single keyword, and proposed emotion-topic models by incorporating a intermediate layer of emotion into LDA. Moreover, in Li's method [12], documents are not treated equally and influence the prediction at different levels, in order to reduce the impact of noisy documents. The weakness of the aforementioned methods are obvious. They regarded the document as a bag of words, and didn't take semantics of the document into account, while the sentiment of the document have close ties to the semantic meaning. At the meantime, some other studies analyse the emotion present in documents by considering semantics. Zhang et al. [28] brought in a Conditional Random Fields based model which take the context into account to encode the reviews. and mined the sentiment polarity to the products. Tang et al. [24,25] constructed a neural network model, which modelled user-comment and product-comment consistencies and rated numeric scores to products accordingly. Inspired by word embedding, [15,16] presented a batch of methods by using both local and global semantics to improve the performance on sentiment analysis. Differing from the aforementioned approaches, this paper presents a neural network model capturing both the semantics within sentence and relations between sentences to learn hierarchical document representation. Thus we can make full use of the semantic information to predict the sentiment of the documents.

On another hand, these methods only use the information of the document itself, while ignoring the accompanying information sources, which can significantly influence the interpretation of the sentiment contained in documents. This paper uses heterogeneous information sources to analyse the sentiment.

To the best of our knowledge, the only work on emotion tagging for news comments is Zhang's prior works [29,30], which used a fixed combination strategy to merge heterogeneous information sources, and employed traditional machine learning method to tag emotions for the comments of news. Our work differs from Zhang's work since we build our model based on artificial neural networks, instead of traditional machine learning. In addition, Zhang only uses two kinds of information sources, while we use more.

3   Hierarchical Semantic Neural Network

We now state the emotion tagging problem as follows: Given a set of users' comments on news along with the news articles and user-generated emotion votes of the news, we should identify the emotion tags of individual comments.

Furthermore, we formulate the problem setting as follows: Given a collection of comments C and a collection of news articles D, each $c \in C$ has its $d \in D$ which means c is made by a user after reading d. We also have a predefined emotion set $E = \{e_1, e_2, e_3, \cdots, e_K\}$ from which we assign emotion tag for each comment. Afterwards each news article d is accompanied by user-generated emotion votes $M_d = \{\mu_1, \mu_2, \mu_3, \cdots, \mu_K\}$ where $\mu_k \in \mathbb{R}$ is the count of votes over emotion e_k. On the top of this, we cast the emotion tagging problem into a multi-class classification problem that we classify a comment c into one emotion tag e_k of the emotion set, according to the content of the comment itself, the content of its news article d and the emotion votes M_d of d.

The problem involves three issues. First, we develop a hierarchical semantics representations model of the document, according to not only the contextual relations within sentence, but also the intrinsic relations between sentences to encode the semantic meaning of document. Second, we reconstruct comments as a combination of the semantic representations of the comment itself, its news article and the user-generated emotion vote of the news. Then we use this representation as feature to classify and assign emotion tags to comments.

3.1   Hierarchical Semantic Representation Model of the Document

We introduce our proposed hierarchical semantic representation model of the document in this section, which computes fixed length continuous vector representations for documents of variables length.

Words are the basic components of sentences, and sentences constitute documents structurally and semantically. The principle of compositionality [7] states that the meaning of a longer expression (e.g., a sentence or a document) comes from the meanings of its constituents and the rules used to combine them. Thus our method to compute the document representation can be divided into two steps. We first model sentence semantic representations by producing continuous sentence vectors from word vectors/representations. Then we use sentence semantic representations to get the final document semantic representations.

3.1.1   Sentence Semantic Representation.

In order to model the sentence semantic representation, word embedding [3] is innovated to represent each word. According to word embedding, each word is represented as a low dimensional, continuous and real-valued vector, all of which are stored in a matrix $L \in \mathbb{R}^{dim \times |V|}$, where dim is the dimension of word vectors and V is the vocabulary. The word embedding can be initialized randomly from a uniform distribution and learned as a parameter at the some time with the training of a neural network [10,23], or be pre-trained from text corpus with embedding learning algorithms [16,19,24]. We employ the latter method using $word2vec$[1] to make better use of semantic and grammatical associations of words.

After that, we apply a modified convolutional neural network (CNN) to compute representations of sentences. CNN are a state-of-the-art semantic model from sentiment classification and emotion tagging [10,11,24], and it can learn fixed length vectors for sentences of varying length, according to the words order in a sentence and doesn't depend on an external parse tree.

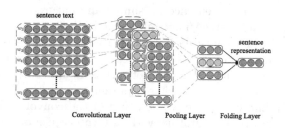

Fig. 1. Convolutional neural network for sentence semantic representation

Figure 1 shows the overview of our sentence method to capture the sentence semantic representation. The first lookup layer mapping words into low-dimensional vectors. The next layer performs convolutions over the embedded word vectors using filters with multiple sizes of windows. Next, we average-pool and average-fold the outputs of the convolutional layer into the representation.

We use different convolutional filters with different window widths to capture local semantics of various granularities to generate the sentence representation, which have been proven effective for sentiment classification and emotion tagging. For example, a convolutional filter with a window width of 3 essentially captures the semantics of a sentence in the perspective of trigram. In this paper, we use three different convolutional filters with widths of 3, 4 and 5 to encode the semantics of trigrams, 4-grams and 5-grams in a sentence.

Formally, given a sentence consisting of n words denoted as $\{w_1, w_2, w_3, \ldots, w_n\}$, l_{cf} is the window width of a convolutional filter cf, W_{cf} and b_{cf} is the shared parameters of linear layers of this filter. Each word w_i in the sentence is mapped to its word embedding $we_i \in \mathbb{R}^{dim}$ through word a embedding matrix $L \in \mathbb{R}^{dim \times |V|}$, where dim is the dimension of word embedding. The input of

---

[1] https://code.google.com/archive/p/word2vec/.

a linear layer is the concatenation of l_{cf} word embeddings in the window of this filter, which is denoted as $I_{cf} = [we_i; we_{i+1}; \ldots; we_{i+l_{cf}-1}] \in \mathbb{R}^{dim \cdot l_{cf}}$. The output of a linear layer is shown as follows:

$$O_{cf} = tanh\left(W_{cf} \cdot I_{cf} + b_{cf}\right), \tag{1}$$

where $W_{cf} \in \mathbb{R}^{l_{ocf} \times dim \cdot l_{cf}}$, $b_{cf} \in \mathbb{R}^{l_{ocf}}$, l_{ocf} is the length of the output of this convolutional layer, $tanh$ is the hyperbolic tangent to increase the non-linear property without affecting the receptive fields of the convolution.

Afterwards, we feed all the outputs of a convolutional filter into an average pooling layer to capture the overall semantics. Then we use an average fold layer to merge the outputs of different filters to get the final sentence representation.

3.1.2 Document Semantic Representation.

Next, we introduce our method to generate a document representation from the obtained sentence vectors, utilizing long-short term memory model (LSTM).

Given a set of vectors of sentences, a simple and natural strategy to form a text vector is taking the average/max/min value of the sentence vectors as text vector. Obviously it can't capture complex relations such as causality and contrast between sentences since it totally ignores the order and logical relationship of sentences. Using convolutional neural network is an alternative to model local relations using its convolution with shared parameters partly. But this capability is considerably limited by the window size of the convolutional filter. The main idea behind recurrent neural network is to make use of sequential information of sentences. RNN is called recurrent because it performs the same task for every element of a sequence, with the output being depended on the previous computations. This helps it to encode the relations between sentences in long sequences, even if the two related sentences are far from each other in theory. Unfortunately, RNN suffers from gradient vanishing or exploding [4], which means gradients may grow or decay exponentially over long sequences. This makes it nearly impossible to model long-distance correlations in a sequence.

To solve this problem, we use a modified long-short term memory model. The transition function of LSTM used in this paper is shown as follows:

$$f_t = \delta\left(W_f \cdot [h_{t-1}; x_t] + b_f\right), \tag{2}$$

$$i_t = \delta\left(W_i \cdot [h_{t-1}; x_t] + b_i\right), \tag{3}$$

$$\tilde{C}_t = \delta\left(W_C \cdot [h_{t-1}; x_t] + b_C\right), \tag{4}$$

$$C_t = f_t \odot C_{t-1} + i_t \odot \tilde{C}_t, \tag{5}$$

$$h_t = \delta\left(W_h \cdot [h_{t-1}; x_t] + b_h\right) \odot tanh(C_t), \tag{6}$$

where x_t is the input vector of LSTM at the t-th step, in this section it's the t-th sentence semantic representation. f_t, i_t, W_f, W_i, b_f, b_i adaptively forget and update the information of hidden vector and input vector, W_C and b_C form the candidate vector, h_{t-1} is the hidden vector which represents the history status and maintains the accumulated knowledge of previous $t - 1$ step, C_{t-1} and

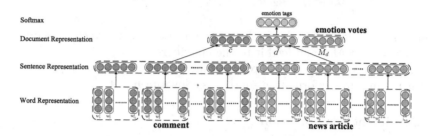

Fig. 2. Classical LSTM

Fig. 3. Avg LSTM

Fig. 4. The overview of our hierarchical semantic representation model using multiple information sources. w_j^i is the j-th word in the i-th sentence, l_i is the length of the i-th sentence, n and m are the numbers of sentences in the comment and the news article.

$\tilde{C}_t$ represent the old cell state and new candidate vector respectively at the t-th step, W_h and d_h help to update the hidden vector from the old hidden vector, input vector and cell state vector. As a side note, $\odot$ is element-wise multiplication of two vector, which means two vectors are multiplied element by element.

In classical LSTM [8], the last hidden vector is regarded as the text representation as shown in Fig. 2. In this paper, we make a further extension called Avg LSTM by using the average of all hidden vectors as text representation. Thus we can take considerations of the differences of semantics and sentiment relations between sentences and with different historical granularities (Fig. 3).

3.2 Hierarchical Semantic Representation Using Multiple Information Sources

In this subsection, we introduce three heterogeneous information sources to mine sentiment of user comments, which are content of comments, content of news articles, and user-generated emotion votes of news articles.

The first and second information sources are hierarchical semantic representation of the contents of the comments and contents of the news articles. The comments are users' response to the news articles, thus the emotions of the comments are directly influenced by the content of the news articles. So we take contents of the news articles into consideration. For modelling the semantics of the comments and the news articles, we embed the content of comment c and news article d as continuous vectors $\bar{c} \in \mathbb{R}^{dim_c}$ and $\bar{d} \in \mathbb{R}^{dim_d}$ using the hierarchical semantic representation model in Sect. 3.1, where dim_c and dim_d are dimensions of the comment vector and the news vector respectively.

The last information source is derived from the emotion votes of the news articles generated by users. When tagging emotion for each comment, we can follow the normalized user-generated emotion votes of the news article to which the comment belongs. How likely the comment c of news article d will be tagged by emotion e_i according to the information of emotion votes can be denoted by $\bar{\mu}_i$. Hence we reconstitute the votes vector M_d through normalization as follows:

$$\bar{M}_d = \{\bar{\mu}_1, \bar{\mu}_2, \bar{\mu}_3, \cdots, \bar{\mu}_K\}, \bar{\mu}_i = \frac{\mu_i}{\sum_{j=1}^{K} \mu_j}. \tag{7}$$

On this basis, we define the final semantic representation vector of comment c as $rep(c) = \left[\bar{c}; \bar{d}; \bar{M}_d\right]$ and feed it to the classifier.

At last, an overview of our proposed hierarchical semantic representation model using multiple information sources is shown in Fig. 4.

3.3 Sentiment Classification

In this section, we use hierarchical semantic representations using multiple information sources as discriminative features instead of handcrafted features which are used in traditional machine learning.

As shown in Sect. 3.2, the hierarchical semantic representation is the concatenation of semantic representation of comment, semantic representation of news article and continuous representation of user-generated emotion votes. On top of this, we introduce a *Softmax* classifier to transform the feature representations into conditional probabilities which can be interpreted as the probabilities of comments to be assigned into each emotion tag.

Given the i-th comment c_i in the corpus, the conditional probability that the comment should be associated with emotion $e_k(k = 1, 2, \cdots, K)$ within the set of emotion tags can be calculated as probability values with a *softmax* function.

$$P(e_k|c_i) = P(e_k|rep_i) = \frac{exp(\omega_k^T rep_i)}{\sum_{e_j \in E} exp(\omega_j^T rep_i)} \tag{8}$$

where c_i is the i-th comment, rep_i is the input hierarchical semantic representation feature of c_i, E is the set of emotion categories, ω is the matrix that transforms representation rep_i into a real-valued vector with dimension of $|E|$, ω_j is the combination parameter for each term with emotion e_j.

Afterwards, we train the model in a supervised way, where each comment in the training corpus is accompanied with its ground truth emotion tag. We introduce the cross-entropy error between ground truth sentiment distribution and predicted sentiment distribution as the objective loss function as follows:

$$J(\theta) = -\sum_{c \in C} \sum_{e \in E} P^g(e|c) \cdot log\left(P(e|c)\right), \tag{9}$$

where $c \in C$ is a comment, e is a emotion in the set of emotion categories E, $P(e|c)$ is the predicted distribution, $P^g(e|c)$ is the ground truth sentiment distribution with the same dimension of E, in which only the dimension corresponding to the ground truth is set to 1, and the others are set to 0.

We feed the cross-entropy error loss function into the back propagation algorithm to update the whole set of parameters of $\theta = [W_{cf}, b_{cf}, W_f, b_f, W_i, b_i, W_C,$ $b_C, W_h, b_h, \omega]$ with stochastic gradient descent.

In this paper, we didn't enforce L2 norm constraints on parameters, instead we employ dropout [20, 27] as a regularization method to reduce overfitting. The main idea of dropout is bringing in random removal of some units in a neural network during training, but keeping all of them during testing. Dropout involves a hyper parameter p, which means individual units are either "dropped out" of the network with the probability $1 - p$ or kept with the probability p in each iteration, so that a reduced network is left to be trained in each iteration and the removed units keep their original weights.

Specifically, for the CNN part in this paper, before we feed I_{cf} into the convolutional layer, we add a dropout mask vector to the input vector to produce a dropout-modified input vector $\hat{I}_{cf}$ which is formulated as follows:

$$\hat{I}_{cf} = I_{cf} \odot m, \tag{10}$$

$$m_{(i)} \sim Bernoulli(p), \tag{11}$$

where m is the dropout mask with the same dimension of I_{cf}, and $m_{(i)}$ is the i-th element of m. Note that, m keeps changing for every I_{cf}. For the LSTM part, the hidden vector is also converted into a dropout-modified form similarly.

4   Experiment

In this section, we first introduce our experimental settings including the datasets used, evaluation metrics and baseline algorithms, then we present the experimental results with analysis and discussion.

4.1   Dataset

We collected the most-viewed news articles with their comments and user emotion votes in 6 months of 2011 from the *Society* channel of Sina News[2] and the *Entertainment* channel of QQ News[3]. We only use these Chinese datasets since we have not found similar services in English yet, but the proposed model is language independent. We randomly sampled news articles with their top-20 popular comments[4] and user-generated emotion votes as our training and testing datasets, which are referred as the Sina dataset and the QQ dataset respectively in the following pages. There are 5,185 comments, 369 news articles and 83,634 emotion votes in the Sina dataset, and 5,414 comments, 372 news articles and 993,089 emotion votes in the QQ dataset. Each comment is accompanied by its corresponding news articles and emotion votes of the news articles.

---

[2] http://news.sina.com.cn/society/.
[3] http://ent.qq.com/.
[4] If the number of comments was under 20, then we took all of them.

Table 1. The statistics of labeled comments of datasets.

Emotion	Number	Proportion
Sina dataset		
Touched	905	17.45%
Sympathetic	614	11.84%
Bored	336	6.48%
Angry	1,752	33.79%
Amused	408	7.87%
Sad	654	12.61%
Surprised	196	3.78%
Fervent	320	6.17%
QQ dataset		
Happy	1,619	29.90%
Touched	139	2.57%
Sympathetic	641	11.84%
Angry	1,639	30.27%
Amused	563	10.40%
Sad	355	6.56%
Surprised	85	1.57%
Anxious	373	6.89%

For the purpose of performance evaluation, emotion labels in both datasets are manually annotated. In Sina News and QQ News, even though users can tag articles with built-in emotion categories, the tag-systems are independent from the commenting systems so a tag cannot be paired with a specific comment. Thus we cannot utilize users tags as labels, instead, we just borrow the built-in emotion categories as predefined emotion categories in the annotating task. Due to the substantial laboring efforts, each dataset is annotated by only three annotators. The detailed statistic of labelled comments on the 8 emotions in Sina and QQ dataset are shown in Table 1. To test the annotating quality, 100 comments are randomly sampled from each dataset and a reviewer (not the annotator) annotated them blindly from the original labels. The number of consistent labels are 91 for the Sina dataset and 94 for the QQ dataset.

4.2   Evaluation Metrics

In this paper, we apply two measures to compare the performances:

1. **Mean Reciprocal Rank (MRR).** Given a comment $c \in C$ with its ground truth emotion tag $\hat{e}_c$ and the predicted emotion ranking list L_c of c, let

$rank_{L_c}(\hat{e}_c)$ be the position of $\hat{e}_c$ in L_c, MRR can be denoted as follows:

$$MRR = \frac{1}{|C|} \sum_{c \in C} \frac{1}{rank_{L_c}(\hat{e}_c)}.$$ (12)

2. **Accuracy (*Accu@m*).** Given a comment $c \in C$ with its ground truth emotion tag $\hat{e}_c$ and the predicted emotion ranking list $L_c@m$ including top-m emotions in L_c, $accu_c@m$ can be defined as follows:

$$accu_c@m = \begin{cases} 1, \hat{e}_c \in L_c@m \\ 0, \hat{e}_c \notin L_c@m \end{cases}.$$ (13)

and $Accu@m$ for the entire dataset is $Accu@m = \sum_{c \in C} accu_c@m / |C|$.

4.3 Baseline Methods

We compared the proposed HSNN with the following methods for emotion tagging with 10-fold cross validation on the two datasets.

1. In SVM+n-grams, we used bag-of-n-grams of comments as features and trained SVM classifier with LIBLINEAR [6].
2. WE, namely Word-Emotion method [21], is a generative model based on emotional dictionaries. It first builds the word-level and topic-level emotion dictionaries, then uses them to predict the emotions of given comments.
3. In RPWM, or Reader Perspective Weighted Model [12], comments are not treated equally and influence the prediction at different levels.
4. Standard CNN [11] and LSTM [8] are also implemented as baseline methods which are state-of-the-art technologies for semantics and sentiment analysis. Note that we used three convolutional filters with widths of 3, 4 and 5 for standard CNN as the same as our proposed HSNN.
5. Content-based Model (CM) [29] builds a supervised fixed combination classification model and uses traditional machine learning methods to predict emotions for the comments.
6. Finally, HSNN is our proposed model.

4.4 Comparison to Baselines

The first set of experiments in this section is conducted to evaluate the performance of our proposed HSNN in comparison to the baseline methods using only the content of comments. Experimental results are shown in Table 2.

We can see that the SVM classifiers are very strong, which are almost the strongest among all baselines even though they nearly don't catch any linguistic information when the value of n is small. But with the increase of n, the bag-of-n-grams features become more and more sparse especially the comments part, since there are too few words in the comments. For example, the feature dimensions of unigrams, bigrams and trigrams on QQ dataset are 12,574, 90,687 and 158,741.

Table 2. Performances of emotion tagging using single information source.

	Sina dataset				QQ dateset			
	MRR	Accu@1	Accu@2	Accu@3	MRR	Accu@1	Accu@2	Accu@3
SVM+unigrams	0.6298	0.4455	0.6308	0.7615	0.6153	0.4256	0.6130	0.7549
SVM+bigrams	0.5901	0.4057	0.5734	0.6990	0.6028	0.4081	0.6094	0.7331
SVM+trigrams	0.5497	0.3528	0.5237	0.6627	0.5477	0.3084	0.5913	0.7118
WE	0.5687	0.3650	0.5587	0.7052	0.5340	0.3365	0.5077	0.6395
RPWM	0.5347	0.3356	0.4973	0.6512	0.5438	0.3638	0.5156	0.6206
Standard CNN	0.6166	0.4225	0.6668	0.7642	0.6326	0.4400	0.6172	0.7797
Standard LSTM	0.6414	0.4384	0.6856	0.7909	0.6833	0.4455	0.6317	0.8082
CM	0.6577	0.4838	0.6716	0.7810	0.6558	0.4907	0.6535	0.7636
HSNN_{CC}	**0.6841**	**0.5293**	**0.7478**	**0.8232**	**0.7046**	**0.4967**	**0.7077**	**0.8525**

This is also the reason why the performance of SVM with trigrams is the worst among three SVMs. We try to reduce the dimensions of features by only picking up emotion terms, but the performance shows no noticeable improvement.

WE is effective since it uses emotional dictionaries to predict the emotions of given comments. However, it only models comments as bag of words and doesn't take the semantic information of comments into account. RPWM is an improvement of WE, since it (1) jointly models emotions and topics by LDA, (2) calculates emotional entropy as document weights to reduce the impact of the noisy comments on the prediction. However, the results show no obvious improvement, we assume this is due to the fact that there is no significant difference between comments in the datasets used in this paper.

CM utilizes emotions terms[5] in the comments as features, and feeds them into a L2 regularization model. Since CM only takes considerations of the terms which are more likely to convey the emotions, it has a obviously better performance than the aforementioned baselines. From the comparison between CM and WE/RPWM, we also can tell that discriminative models usually have better performances and accuracies than generative models, which is proved in [9,13].

Standard CNN and LSTM outperform the vast majority of baseline methods significantly since they model the local semantics within the comment, from which we can tell that compositionality is important to understand the semantics and sentiment. However, there is still some room for improvement as long as the complex semantics, like the relations between sentences, are not captured well.

HSNN_{CC}, which is our proposed model with single information source of content of comments, has an outstanding performance over all baseline methods, since it models not only the semantics within each sentence with modified CNN but also the relations between sentences with Avg LSTM. This gives HSNN the capability to model the complex semantics in documents. In addition, comparing HSNN_{CC} with Standard CNN and LSTM, we can tell that the logical

---

[5] Emotion terms can be extracted by several lexical resources developed for these tasks, such as NTU Sentiment Dictionary and Hownet.

Table 3. Performances of HSNN with different information sources.

	Sina dataset				QQ dateset			
	MRR	Accu@1	Accu@2	Accu@3	MRR	Accu@1	Accu@2	Accu@3
HSNN_{CC}	0.6841	0.5293	0.7478	0.8232	0.7046	0.4967	0.7077	0.8525
HSNN_{CN}	0.6013	0.3732	0.4766	0.6725	0.5997	0.3575	0.4538	0.6732
HSNN_{UEV}	0.6019	0.3969	0.5299	0.6836	0.5995	0.3596	0.5077	0.7089
HSNN_{CC+CN}	0.6831	0.5232	0.7499	0.8357	0.7017	0.4986	0.7218	0.8557
HSNN_{CC+UEV}	0.7290	0.5859	0.8049	0.8713	0.7537	0.5403	0.7252	0.8947
HSNN_{CN+UEV}	0.6791	0.5105	0.7277	0.8515	0.6823	0.4860	0.6916	0.8291
HSNN_{CC+CN+UEV}	**0.7505**	**0.5905**	**0.8066**	**0.8904**	**0.7639**	**0.5605**	**0.7443**	**0.9049**

relations between sentences do help understanding the sentiment and semantics of the whole comment positively.

Statistical significance tests have been conducted. HSNN_{CC} outperforms other methods with a confidence level of 0.95 on all datasets.

4.5   Effect of Multiple Information Sources

The second set of experiments is conducted to (1) find out whether every information source would be helpful for emotion tagging for comments, and (2) evaluate the performance of HSNN with different information sources.

HSNN_{CC}, HSNN_{CN} and HSNN_{UEV} are our proposed models with single information source, either of the comments, the news article or the user-generated emotion votes. HSNN_{CC+CN}, HSNN_{CC+UEV} and HSNN_{CN+UEV} are models with two information sources. HSNN_{CC+CN+UEV} is our integrated proposed model with all three information sources. Finally, from Table 3, we can see that HSNN_{CC} achieves the best performance compared to the other two single information source HSNN, which indicates the comments is more reliable and effective to predict the emotion of comments, since the comment is our object for emotion tagging obviously. HSNN_{UEV} is the second best, which means that the user-generated emotion votes are more useful than the news articles. This may be because the user emotion votes convey users' sentiments after reading the news articles more directly.

It can also be seen that HSNNs with multiple information sources generally outperform HSNNs with single information source, which shows that combining different information sources is more effective than using only one specific source of information. Furthermore, every information source is more or less helpful to understand the semantics and sentiment of comments.

Finally, HSNN_{CC+CN+UEV} yields the best performances, which clearly demonstrates that utilizing all three information sources is more effective than using only one or two specific sources of information. We can tell that each source of information provides a different perspective on emotion tagging, and respectively helps the model to achieve better prediction accuracy.

Statistical significance tests have been conducted. HSNN_{CC+CN+UEV} outperforms all other methods with a confidence level of 0.95 on all datasets.

4.6   Effect of Dropout

The final set of experiments is conducted to explore the effect of dropout.

Since a common value of dropout rate is $p = 0.5$ in practice [20], we designate it as our baseline, and effects of different dropout rates are measured by changes in Acc@1 compared with 0.5. Experimental data show that changes in MRR and Accu@2,3 have the same trend and a similar curve as Accu@1, so we only show the changes in Acc@1, which are shown in Fig. 5.

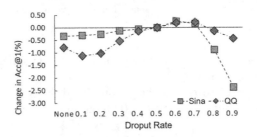

Fig. 5. Effect of dropout rate.

From Fig. 5 we can see that non-zero dropout rates can improve the performance of emotion tagging at a range from 0.1 to 0.7, depending on datasets, which is consistent with the conclusions of previous research work [27]. In this paper, we choose $p = 0.6$ as our dropout rate during the experiment.

5   Conclusions and Future Work

In this paper, we proposed a novel methodology, namely Hierarchical Semantic Neural Network (HSNN), for emotion tagging for online news comments. Specifically, we developed a novel hierarchical semantic representation model to learn a semantic representation of a document based on both the semantics within a sentence and the relations between sentences. We also proposed a novel classification method utilizing the hierarchical semantic representation of multiple information sources. In this approach we use the information of not only the comment but also the accompanied news article and the user-generated votes to improve the classification accuracy of emotion tagging. The experimental results show that our approach outperforms the traditional approaches.

For possible future research, there are several assumptions, such as the improvement of HSNN to reorient the model to cross-domain and cross-language online news comments emotion tagging problem, or modelling the reading habits and emotional tendencies of individual users to improve the prediction accuracy.

Acknowledgement. This work is partially supported by National Natural Science Foundation of China under Grant No. 61402243 and National 863 Program of China under Grant No. 2015AA015401. This work is also partially supported by Tianjin Municipal Science and Technology Commission under Grant No. 16JCQNJC00500 and No. 15JCTPJC62100.

References

1. Bao, S., Xu, S., Zhang, L., Yan, R., Su, Z., Han, D., Yu, Y.: Joint emotion-topic modeling for social affective text mining. In: 2009 Ninth IEEE International Conference on Data Mining, pp. 699–704. IEEE (2009)
2. Bao, S., Xu, S., Zhang, L., Yan, R., Su, Z., Han, D., Yu, Y.: Mining social emotions from affective text. IEEE Trans. Knowl. Data Eng. **24**(9), 1658–1670 (2012)
3. Bengio, Y., Ducharme, R., Vincent, P., Jauvin, C.: A neural probabilistic language model. J. Mach. Learn. Res. **3**, 1137–1155 (2003)
4. Bengio, Y., Simard, P., Frasconi, P.: Learning long-term dependencies with gradient descent is difficult. IEEE Trans. Neural Netw. **5**(2), 157–166 (1994)
5. Domingos, P.: A few useful things to know about machine learning. Commun. ACM **55**(10), 78–87 (2012)
6. Fan, R.E., Chang, K.W., Hsieh, C.J., Wang, X.R., Lin, C.J.: Liblinear: a library for large linear classification. J. Mach. Learn. Res. **9**, 1871–1874 (2008)
7. Frege, G.: Sense and reference. Philos. Rev. **57**(3), 209–230 (1948)
8. Hochreiter, S., Schmidhuber, J.: Long short-term memory. Neural Comput. **9**(8), 1735–1780 (1997)
9. Jordan, A.: On discriminative vs. generative classifiers: a comparison of logistic regression and naive Bayes. Adv. Neural Inf. Process. Syst. **14**, 841 (2002)
10. Kalchbrenner, N., Grefenstette, E., Blunsom, P.: A convolutional neural network for modelling sentences. arXiv preprint arXiv:1404.2188 (2014)
11. Kim, Y.: Convolutional neural networks for sentence classification. arXiv preprint arXiv:1408.5882 (2014)
12. Li, X., Rao, Y., Chen, Y., Liu, X., Huang, H.: Social emotion classification via reader perspective weighted model. In: Proceedings of the 30th AAAI Conference on Artificial Intelligence (2016)
13. Liang, P., Jordan, M.I.: An asymptotic analysis of generative, discriminative, and pseudolikelihood estimators. In: Proceedings of the 25th International Conference on Machine Learning, pp. 584–591. ACM (2008)
14. Liu, B.: Opinion mining and sentiment analysis. In: Web Data Mining, pp. 459–526. Springer, Heidelberg (2011)
15. Maas, A.L., Daly, R.E., Pham, P.T., Huang, D., Ng, A.Y., Potts, C.: Learning word vectors for sentiment analysis. In: Proceedings of the 49th Annual Meeting of the Association for Computational Linguistics: Human Language Technologies, vol. 1, pp. 142–150. Association for Computational Linguistics (2011)
16. Mikolov, T., Sutskever, I., Chen, K., Corrado, G.S., Dean, J.: Distributed representations of words and phrases and their compositionality. In: Advances in Neural Information Processing Systems, pp. 3111–3119 (2013)
17. Pang, B., Lee, L.: Opinion mining and sentiment analysis. Found. Trends Inf. Retrieval **2**(1–2), 1–135 (2008)
18. Pang, B., Lee, L., Vaithyanathan, S.: Thumbs up?: sentiment classification using machine learning techniques. In: Proceedings of the Conference on Empirical Methods in Natural Language Processing, vol. 10, pp. 79–86. Association for Computational Linguistics (2002)
19. Pennington, J., Socher, R., Manning, C.D.: Glove: global vectors for word representation. In: Proceedings of the Conference on Empirical Methods in Natural Language Processing, vol. 14, pp. 1532–1543 (2014)

20. Pham, V., Bluche, T., Kermorvant, C., Louradour, J.: Dropout improves recurrent neural networks for handwriting recognition. In: Proceedings of the 14th International Conference on Frontiers in Handwriting Recognition, pp. 285–290. IEEE (2014)
21. Rao, Y., Lei, J., Wenyin, L., Li, Q., Chen, M.: Building emotional dictionary for sentiment analysis of online news. World Wide Web **17**(4), 723–742 (2014)
22. Sindhwani, V., Melville, P.: Document-word co-regularization for semi-supervised sentiment analysis. In: 2008 Eighth IEEE International Conference on Data Mining, pp. 1025–1030. IEEE (2008)
23. Socher, R., Perelygin, A., Wu, J.Y., Chuang, J., Manning, C.D., Ng, A.Y., Potts, C.: Recursive deep models for semantic compositionality over a sentiment treebank. In: Proceedings of the Conference on Empirical Methods in Natural Language Processing, pp. 1631–1642. Association for Computational Linguistics, October 2013
24. Tang, D., Qin, B., Liu, T.: Document modeling with gated recurrent neural network for sentiment classification. In: Proceedings of the Conference on Empirical Methods in Natural Language Processing, pp. 1422–1432 (2015)
25. Tang, D., Qin, B., Liu, T.: Learning semantic representations of users and products for document level sentiment classification. In: Proceedings of the 53rd Annual Meeting of the Association for Computational Linguistics and the 7th International Joint Conference on Natural Language Processing, pp. 1014–1023 (2015)
26. Turney, P.D.: Thumbs up or thumbs down?: semantic orientation applied to unsupervised classification of reviews. In: Proceedings of the 40th Annual Meeting on Association for Computational Linguistics, pp. 417–424. Association for Computational Linguistics (2002)
27. Wu, H., Gu, X.: Towards dropout training for convolutional neural networks. Neural Netw. **71**, 1–10 (2015)
28. Zhang, K., Xie, Y., Yang, Y., Sun, A., Liu, H., Choudhary, A.: Incorporating conditional random fields and active learning to improve sentiment identification. Neural Netw. **58**, 60–67 (2014)
29. Zhang, Y., Fang, Y., Quan, X., Dai, L., Si, L., Yuan, X.: Emotion tagging for comments of online news by meta classification with heterogeneous information sources. In: Proceedings of the 35th International ACM SIGIR Conference on Research and Development in Information Retrieval, pp. 1059–1060. ACM (2012)
30. Zhang, Y., Zhang, N., Si, L., Lu, Y., Wang, Q., Yuan, X.: Cross-domain and cross-category emotion tagging for comments of online news. In: Proceedings of the 37th International ACM SIGIR Conference on Research and Development in Information Retrieval, pp. 627–636. ACM (2014)

Towards a Query-Less News Search Framework on Twitter

Xiaotian Hao[(⊠)], Ji Cheng, Jan Vosecky, and Wilfred Ng

Department of Computer Science and Engineering,
Hong Kong University of Science and Technology,
Clear Water Bay, Kowloon, Hong Kong
{xhao,jchengac,jvosecky,wilfred}@cse.ust.hk

Abstract. Twitter enables users to browse and access the latest news-related content. However, given user's interest in a particular news-related tweet, searching for related content may be a tedious process. Formulating an effective search query is not a trivial task. And due to the often small size of smart phone screens, instead of typing, users always prefer click-based operations to retrieve related content. To address these issues, we introduce a new paradigm for news-related Twitter search called *Search by Tweet(SbT)*. In this paradigm, a user submits a particular tweet which triggers a search task to retrieve further related tweets. In this paper, we formalize the SbT problem and propose an effective and efficient framework implementing such a functionality. At the core, we model the public Twitter stream as a dynamic graph-of-words, reflecting the importance of both words and word correlations. Given an input tweet, our framework utilizes the graph model to generate an implicit query. Our techniques demonstrate high efficiency and effectiveness as evaluated using a large-scale Twitter dataset and a user study.

1 Introduction

Twitter has already become a popular platform to browse recent news updates all around the world. However, the way Twitter users interact with content that interests them is still fairly limited.

Given that users have seen an interested news-related tweet (e.g., outbreak of the Ebola epidemic), they may wish to find more content related to the topic (e.g., potential spread of Ebola in the US). The user has the option to identify related Twitter accounts (e.g., BBC News) and browse to find relevant tweets, which is a tedious process. Alternatively, the user may perform a Twitter search which results in another set of problems.

First, it may be difficult to identify an effective search query that retrieves relevant content. A general query may yield low precision while a long and specific query may yield few or no search results.

Second, we observe that smart phone users always prefer click-based interactions instead of typing due to the small screen size of the device.

To solve those problems, our goal in this work is therefore to allow a user to search for information about a specific news-related tweet without formulating an

© Springer International Publishing AG 2017
S. Candan et al. (Eds.): DASFAA 2017, Part II, LNCS 10178, pp. 137–152, 2017.
DOI: 10.1007/978-3-319-55699-4_9

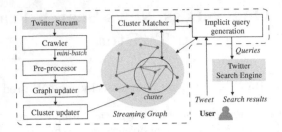

Fig. 1. A conceptual perspective of the SbT framework

explicit search query. In this new paradigm, called *Search by Tweet(SbT)*, users are enabled to select a particular news-related tweet and provided automatically retrieved tweets related to such news topic. The conceptual perspective of the SbT framework is illustrated in Fig. 1.

We model Twitter content as a dynamic graph of words. The graph is continuously updated by monitoring the Twitter public stream [1]. Given an input user selected tweet, we map it onto the graph. We use weights of individual words and correlations among words to select a set of keywords as the *implicit query*. We also detect word clusters in the graph, which correspond to popular topics discussed in Twitter. For input tweets that correspond to a cluster, we generate an *implicit query* based on both the tweet's words and related words in the cluster. Finally, the generated *implicit query* is utilized to perform a search task using a Twitter search engine.

The main contributions of this paper are as follows:

- We develop a novel dynamic graph model to capture words and word correlations in Twitter streams.
- We develop a new streaming graph clustering method, which efficiently discovers dense clusters in the graph-of-words. Each cluster corresponds to a popular topic in Twitter.
- We propose a novel technique to match an input tweet to a word cluster.

2   Related Work

Microblog search. Prior work on traditional query-driven Twitter search incorporates tweet-query relevance, query expansion [15] and temporal aspects of tweets [6,16]. However, such traditional approaches are insufficient in many real-life scenarios.

Query expansion and suggestion. Recent approaches on *query expansion* include random walks on a word graph [8] and entity queries produced using auxiliary knowledge bases [5]. For long query documents, frequency-based models may be used to produce *query suggestions* [10]. However, these techniques are not readily applicable to the Search by Tweet problem, which aims at eliminating the need of an explicit query.

Topic detection from document streams. Detection of hot topics from social media collections has been an active research area in recent years [13, 14, 18, 19]. We mention two important approaches to topic detection in recent years: burstiness analysis and streaming graph clustering.

Burstiness. Burstiness of a term is commonly used for detecting "hot" or "trending" topics [11, 13, 18]. Apart from detecting emerging and hot topics, burstiness is also employed in news clustering [9] and IR [13].

Streaming graph clustering. To model the semantics in document streams, a graph-based representation is commonly adopted [2–4, 14, 19]. Streaming graph clustering may then be applied to detect hot topics in the graph. Some existing approaches to streaming graph clustering may be adapted to our scenario, however they suffer from various drawbacks. Aggarwal et al. [3] do not consider edge deletions, which are common in our setting. [4] assumes a complete graph, which is not a reasonable assumption for social media streams.

In summary, we are not aware of any previous work on the *Search by Tweet* problem proposed in this paper. Our work is inspired by the advances in the mentioned areas and we develop a framework that integrates a streaming graph model, burstiness-aware weighting and implicit query generation.

3   Search by Tweet(SbT) Framework

3.1   Preliminaries

Definition 1 *(Implicit Query). An implicit query iQ is a search query automatically generated from an input tweet T. The implicit query concisely represents the tweet's topic for the purpose of performing a search task. Formally, iQ is represented as a set of words $\{w_1, \ldots, w_n\}$.*

Definition 2 *(Search by Tweet(SbT)). Given an input tweet T, the goal of SbT is to produce an implicit query iQ based on the most recent semantics extracted from the Twitter platform.*

An important assumption in the SbT paradigm is that microblog users are interested in *recent* content. The techniques in this paper are tailored towards search tasks focusing on recent data (cf. Sect. 3.3, a time decay function is utilized to promote recency).

3.2   Graph Model of Twitter Content

The first step in our framework is to construct a dynamic graph-of-words, referred to as a *base graph*. This graph is then utilized to generate implicit queries based on an input tweet. The main requirement of the base graph BG is to capture important keywords occurring in Twitter and correlations among them.

Formally, let $BG = (V, E)$ denote a weighted graph, where each node $v \in V$ corresponds to a word and an edge $e(u, v) \in E$ represents correlation of node u with node v. Let further $w(v)$ and $w(u, v) = w(e)$ denote the weight of node v and edge $e(u, v)$, respectively. $w(\cdot)$ can also be viewed as a k-dimensional weight vector, thus associating k weights with each node or edge. The details of weighting functions are presented in Sect. 3.3.

Dynamic Updating. A graph update involves the processing of words (i.e., nodes) and word-pairs (i.e., edges) in the arriving tweets. To tackle the large volume and fast rates of tweets, the following techniques are employed to achieve efficient updates.

Mini-batches. To alleviate overhead caused by the fast arrival of tweets, we perform tweet caching. When the cache $\mathbb{C}$ reaches its capacity $|\mathbb{C}|$, we send its contents as a mini-batch to update BG.

Lazy updates. When updating BG at time τ_i, only nodes and edges occurring in $\mathbb{C}$ are added or updated. A similar approach is used to update edges.

Probation buffers. The amount of unique words in the Twitter stream grows rapidly. This would result in an ever-increasing size of BG and seriously affect efficiency. To manage this issue, we adopt a *probation buffer* approach.

When processing a mini-batch, we initialize an empty word buffer $\mathbb{B}_w$ and edge buffer $\mathbb{B}_e$ with thresholds θ_w and θ_e. When a mini-batch is processed, only words with $\mathbb{B}_w(w) > \theta_w$ and edges with $\mathbb{B}_e(u, v) > \theta_e$ will enter the graph.

Graph Maintenance. To remove out-of-date content from the graph, we employ a maintenance procedure at regular time intervals. Let δ be the length of the time intervals. For each node and edge in BG, we maintain its latest update time τ_{upd}. For the maintenance, first, we remove all nodes and edges, for which the difference of τ_{upd} from the current time τ_i exceeds δ. Second, the frequency weight of each node $w \in V$ decays over time, as discussed in Sect. 3.3. For all remaining nodes in BG, we check the node weight at current time τ_i and remove all nodes with a weight below θ_w. Additionally, every 24 h we calculate the average daily frequency of each word in BG.

3.3   Weights of Nodes and Edges

Node Weights. In traditional long documents, word importance is typically based on term frequency or inverse document frequency. But tweets are short and the number of them is rapidly growing. This inspires new metrics of word importance, such as *burstiness* and node correlations.

Self-based Weights

Frequency. Frequency weight $fr(v)$ is based on the occurrences of v in a mini-batch. After a mini-batch is processed at time τ, the new frequency of word v

is interpolated with the value lastly stored in BG at time $\tau_{upd}(v)$, i.e., $fr(v)' = \lambda fr(v)^{(\tau_{upd})} + (1 - \lambda)fr(v)^{(\tau_i)}$.

To promote recency, we adjust the word frequency using an exponential time decay function, i.e. $\lambda = e^{-\rho \cdot |\tau_i - \tau_{upd}|}$. At query time τ_i, we may obtain the adjusted frequency of v using $\tau_{upd}(v)$: $fr(v, \tau_i) = fr(v) * e^{-\rho \cdot |\tau_i - \tau_{upd}|}$.

Long-term average daily frequency. For each node v in BG, we monitor its long-term average daily frequency, represented as a Gaussian (μ_v, σ_v). We perform lazy counting of node frequencies, only processing nodes that occurred in a mini-batch. The average daily frequency is only calculated on-demand by keeping track of updates during the current day. In 24-hour intervals, we examine the last update time and use the current day frequency to update (μ_v, σ_v).

Burstiness. To reflect the "trending" behaviour of a word, we calculate a bursti-ness score based on its average daily frequency (μ_v, σ_v). We adopt the z-score as the measure of burstiness, $bu(v) = \frac{df(v) - \mu_v}{\sigma_v}$, where $df(v)$ is the frequency of v in the current day. As the daily frequency of each word is only updated in 24-hour intervals, we provide the most recent *estimate* of the current day fre-quency $df(v)$. The *estimated current day frequency* is calculated as $df_{estim}(v) = \alpha df_{curr}(v) * \omega + (1 - \alpha)df_{prev}(v)$, where $t \in [0, 1]$ is relative time since midnight, $\omega = \frac{1}{t}$ and $\alpha = t$. After substituting α, ω into the above equation, we obtain

$$df_{estim}(v) = df_{curr}(v) + (1 - t)df_{prev}(v). \tag{1}$$

Connectivity-based weights. For a node v, the following connectivity-based weights are calculated based on its incident edges $e \in E(v)$:
(i) Degree ($deg(v)$), (ii) Sum of edge correlations ($sumCo(v)$), (iii) Sum of fre-quencies of incident edges (log) ($sumFr(v)$) and (iv) Adjusted avg. edge corre-lation $adjCo(v) = \frac{1}{|E(v)|} \sum_{e \in E(v)} co(e) * fr(e)$.

Edge Weights. (i) *Frequency.* Calculation and updates of edge frequency $fr(u, v)$ proceed similarly to node frequency (cf. Sect. 3.3). (ii) *Correlation.* Cor-relation is defined as $co(u, v) = P(u|v) = \frac{P(u,v)}{P(v)}$.

3.4   Implicit Query Generation

Given an input tweet T, we generate implicit queries from BG as follows. First, we map words in T onto BG. Second, we extract a set of candidate queries at various query lengths. Third, candidate queries are ranked based on node and edge weights. Finally, we select the k highest-ranked queries as the final implicit queries. In this section, we describe ranking of queries of different lengths.

1-Word Query Ranking. To rank individual words from tweet T, we directly use their node weights in BG (cf. Sect. 3.3).

2-Word Query Ranking. Given tweet T, we extract all word pairs (u, v), for which we find a matching edge in BG. We utilize edge weights (cf. Sect. 3.3) and their linear combination to rank candidate 2-word queries.

3 or More Words Query Ranking. We first generate all 3 (or more) words combinations $w_1, w_2...w_i$ from T, such that each $w_i \in V$ and each $(w_i, w_j) \in E$. The following weighting functions (or linear combinations thereof) may be used to rank 3 (or more) words queries.

- **Definition:** $E.fr = \sum fr(w_i, w_j)$, **Definition:** $E.co = \sum co(w_i, w_j)$.
- **Definition:** $N.fr = \sum fr(w_i)$, **Definition:** $N.co = \sum avgCo(w_i)$.

Global Query Ranking. To produce an overall ranking of queries across different lengths, we propose a heuristic function that considers both the query length and the weights of nodes and edges. Given a candidate query Q, we assign a Global Query Score GQS as follows.

$$GQS(Q) = \left(1 + \frac{\alpha \sum_{(u,v) \in Q} w(u,v)}{1 + |\{(u,v) \in Q\}|}\right) \frac{\sum_{v \in Q} w(v)}{|\{v \in Q\}|}, \tag{2}$$

where $w(v)$ is node weight and $w(u,v)$ is edge weight. Parameter α controls the importance of edge weights, thus influencing the query length preference. A higher value of α results in queries with more edges to be promoted.

3.5   Clustering for Implicit Query Generation

In this section, we detect dense clusters in the graph and utilize such clusters to obtain queries with context knowledge that not included in the input tweet. In SbT framework, popular topics in Twitter are exhibited as dense clusters in BG. Our clustering approach is tailored towards the dynamic nature of BG. First, we select high-weight nodes and edges from BG into a *core graph* CG. Second, we use triangles in CG as the building blocks of clusters. The approach is shown to be highly effective (Sect. 4).

Core Graph from Base Graph. The *core graph* CG is an undirected subgraph constructed from the base graph BG by selecting a subset of *core nodes*. A *core node* should exhibit two properties: (1) strong correlation with other nodes, indicating a strong semantic relationship, and (2) frequent co-occurrence with other nodes, indicating a high popularity. We collectively refer to these properties as *node importance*, defined as $\varphi(v) = \sum_{e \in E_{in}(v)} co(e) * \frac{1}{|E_{in}(v)|} \sum_{e \in E_{in}(v)} co(e)$. where $E_{in}(v)$ is the set of incoming edges to v.

We may now form core graph CG by inserting all core nodes and all edges between them. $CG = (V_C, E_C)$, where $V_C = \{v | v \in BG \wedge \varphi(v) > \vartheta_w\}$, $E_C = \{(u,v) | u \in V_C \wedge v \in V_C \wedge \varphi(u,v) > \vartheta_e\}$, where $\varphi(u,v) = co(u,v)fr(u,v)$ and ϑ_w is a threshold.

Finally, we prune all core nodes, which have less than 2 neighbours in CG. The pruning operation will be further explained in Sect. 3.5.

Streaming Cluster Detection. As a core concept for efficient cluster detection, we exploit the triangle property in CG. We define a *cluster* as follows.

Definition 3. *A cluster C is a subgraph of CG that satisfies the following conditions: (i) C consists of one or more triangles $\Delta(u, v, w)$, (ii) All triangles in C must have an edge in common, (iii) Any clusters C_i, C_j with a common edge will be merged.*

To account for the dynamic nature of the graph, our culstering algorithm should support the following maintenance operations:

Core Node Addition. The operation AddCoreNode(v) proceeds as follows:

- Add v to CG and for each neighbour $u \in N_{CG}(v)$: AddCoreEdge(u, v)

Core Edge Addition. Each time an edge is added, we apply procedure AddCoreEdge(u, v). Assuming that u and v are already in CG, it proceeds as follows:

- For each $w \in N_{CG}(u) \cap N_{CG}(v)$: FoundTriangle(u, v, w)

Each time a triangle is found, it will either become the basis of a new cluster, or existing clusters will be merged.

Core Node Deletion. Operation RemoveCoreNode(v) proceeds as follows:

- For each edge (v, w) adjacent to v: RemoveCoreEdge(v,w)
- Remove v from CG

Core Edge Deletion. The removal of an edge from CG requires additional checks related to the triangle property of clusters. Figure 2 illustrates an example of the cases. In case 1, the removal of the edge (u, v) results in the splitting of a cluster. But in case 2, the additional edge enables the cluster to hold even after the removal of edge (u, v).

Tweet-to-Cluster Matching. Given an input tweet T, we may consult CG and identify one or more clusters related to T. We propose a matching algorithm to determine a correct cluster assignment.

First, we identify a candidate cluster C if the number of matching words in T and C exceeds a minimum threshold (set to 2 in our work). Second, we extract a number of features from each matching word and construct a feature vector $f(T, C) = \langle f_1, ..., f_n \rangle$. Third, we pass $f(T, C)$ to a classifier to determine the matching result.

To train the classifier, we build a training dataset of tweet-cluster pairs and manually assign a relevance score to each pair. Although our method initially requires human input, we note that the obtained classifier is applicable to any new tweets or clusters. The following features are used.

Features of matching nodes (44 features). We compute the following features for each matching word and aggregate using $\{max, min, sum, avg\}$:

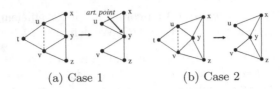

(a) Case 1   (b) Case 2

Fig. 2. Cases after removing edge (u, v)

(i) No. of matching nodes, (ii) Self-based properties (cf. Sect. 3.3), (iii) Connectivity based properties on BG (cf. Sect. 3.3), (iv) Connectivity based properties on matched cluster C.

Features of edges among matching nodes (8 features). The following features are aggregated using $\{max, min, sum, avg\}$: Edge frequency (log) ($Edge.fr$), correlation ($Edge.co$).

Approximate matching (2 features). Apart from analyzing words from T that exactly match C, we may utilize the remaining words in T as additional contextual evidence. We denote all tweet's words except already matched words as T_a and calculate the following features: (i) Average edge correlation of words in T_a and C, (ii) Average edge frequency (log) of all T_a and C.

Implicit Query from Cluster. After matching input tweet T to cluster C, we can generate implicit queries from C. However, query generated from relatively large clusters may contain words not related to T. To address this issue, we propose two approaches.

1. *Tweet-independent cluster query.* All words in C are considered.
2. *Tweet-dependent cluster query.* In this approach, the choice of words from cluster C is influenced by input tweet T. Specifically,
   (a) We consider nodes in C that have an edge to words in T, denoted as C_T^1.
   (b) For each node $v \in C_T^1$, we obtain its *tweet-dependent weight* based on its correlation to words $u \in T$, $w_{C,T}(v) = \max_{(u,v)}\{corr(u, v)\}$. Weight $w_{C,T}(v)$ is then used in conjunction with any node weighting function.

Both approaches above lead to the selection of a set of *candidate queries* from cluster C. Selection and ranking of candidate queries proceeds in the same manner as described in Sect. 3.4.

4   Experiments

4.1   Datasets

Background Twitter Dataset. We collect a Twitter dataset by monitoring the Twitter Streaming API [1] between March 1 and October 10, 2014. In total, our dataset contains over 307 million English tweets.

We utilize data between March 1-September 30 2014 to measure long-term daily average words frequency (cf. Sect. 3.3). The remainder (October 1–10) is used for constructing our graph model and cluster detection.

News-Related Tweet Pools. We first gather a list of 100 most influential news accounts in Twitter[1]. For each account, we then crawl all tweets published from October 1-10-2014 using the Twitter REST API [1]. The reason we use those tweets to build the experiment pool is that the "quality" of them are comparatively higher (with less typos and oral English expression etc.) This makes it more convenient for users to finish the user study and give more reasonable results. Note that the proposed SbT framework will not just gather information from a fix set of users but the whole twitter stream.

Pool-R: Random News Tweets. For each day in our 10-day period, we randomly sample 2 tweets from each news account. In total, we obtain 200 tweets per day and 2,000 tweets in total.

Pool-P: News Tweets from Popular Topics. The popularity of news stories varies dramatically, from large-scale news (e.g., "Ebola") to small-scale stories. For *Pool-P* we select tweets that refer to popular topics. In our framework, popular topics are identified using our cluster detection algorithm (cf. Sect. 3.5). As a result, 200 tweets are identified to construct *Pool-P*.

4.2   Graph Model Construction

Our graph model is built over a 10-day period of our background Twitter dataset[2] (cf. Sect. 4.1). We use the following parameter settings in our experiments: $\delta = 1$ day, $\rho = 0.3$, $\theta_w = 5$, $\theta_e = 5$, $\vartheta_w = 10$, $\vartheta_e = 10$. We test different thresholds in preliminary experiments. Based on the experiment result, we set threshold here not high as we just want to filter out rare words and edges. Unless otherwise stated, we set mini-batch size to 100,000 tweets.

Baseline Graph Model. As a baseline for comparison with our graph model and clustering techniques, we implement a streaming graph model by Agarwal et al. [2]. The model constructs a dynamic, undirected and unweighted graph of words from a document stream and discovers density-based clusters. Clustering is performed by selecting an "active graph" of highly important nodes and detecting short-cycles as the basis of clusters. We utilize this model as a baseline both in our scalability evaluation and when evaluating implicit query generation.

Scalability. We evaluate the running time of our graph updating and clustering algorithms (cf. Sects. 3.2 and 3.5). First, we vary the size of the mini-batch (cf. Sect. 3.2) between 2,000 and 100,000 tweets and run our model on 500,000 tweets. Within intervals of 100,000 tweets, we calculate the average time to process 1,000 tweets. Figure 3(a) presents the timing results. We detect around 1,000 English tweets per minute arriving via the Twitter Streaming API [1] and that is about

---

[1] http://memeburn.com/2010/09/the-100-most-influential-news-media-twitter-accounts/.

[2] As pre-processing steps, we remove reply-tweets, user names and stopwords. All hashtags are retained and no stemming is applied.

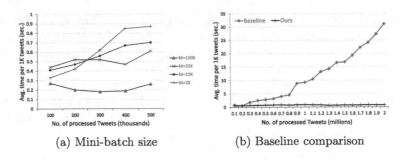

(a) Mini-batch size   (b) Baseline comparison

Fig. 3. Scalability evaluation

(a) "Ebola" (Oct 1)   (b) "Clean India" (Oct 2)   (c)   "England   football
                                                            team" (Oct 9)

Fig. 4. Example clusters

one percent of the total volume of Twitter Stream. The timing results thus demonstrate that our framework is able to process the real-time public stream while the news-related content is only a part of it.

To compare the scalability of our framework against the baseline [2], we apply both methods to a dataset of 2 million tweets. We follow the parameterization of the baseline as presented in [2] and set the mini-batch size in our framework to 10,000. Figure 3(b) presents the timing results. The results show that the running time of the baseline increases as more data is processed. By our analysis, this is caused by the node deletion and edge maintenance strategy in [2]. In summary, our experiments show that our framework is suitable for handling streaming datasets of a large scale.

Cluster Detection. Using our background dataset, we process a 10-day period of tweets. After each day, we store a snapshot of all discovered clusters. Figure 4 presents examples of clusters discovered using our method.

We now present an example to illustrate how a cluster is leveraged to generate an implicit query from an input tweet. As input tweet, we select the following tweet from *Pool-P*: *"Modi will pick up a broom and sweep the streets for nationwide "Clean India" campaign [URL]"*. Using our base graph *BG*, we may generate implicit queries from the words in the tweet. However, the tweet can also be matched to the "Clean India" cluster in Fig. 4(b). Using the cluster, we generate the following queries: *#mycleanindia* and *swachh bharat* (i.e., the indian name for the "Clean India" campaign).

4.3   Tweet-to-Cluster Matching

In this section, we evaluate our proposed method for assignment of tweets to clusters (cf. Sect. 3.5). We use simple word matching to generate 1000 candidate "tweet-cluster" pairs, based on a 2 matching words threshold. Then we invite four independent language experts to manually label each "tweet-cluster" pair as true or false. A "true" label is only applied if the tweet directly corresponds to the cluster's topic.

For the classification task, we use logistic regression. As evaluation metrics, we use precision, recall and F-measure. Table 1 reports the average classification results using 10-fold cross-validation. Note that the "optimal" feature set is generated by a wrapper feature selector [12]. The selected set contains 5 features[3]. This optimal feature set is then employed in our tweet-to-cluster matching task.

Table 1. Tweet to cluster classification

Features	Prec	Recall	F
NumMatches >= 2	0.539	1.000	0.700
NumMatches >= 3	0.871	0.467	0.608
Node-based	0.817	0.816	0.816
Connectivity-graph	0.721	0.720	0.720
Connectivity-cluster	0.750	0.746	0.747
Edges among matching nodes	0.756	0.739	0.738
Approximate matching	0.629	0.626	0.626
Optimal (5 features)	**0.896**	**0.828**	**0.861**
All (52 features)	0.840	0.838	0.839

4.4   Implicit Query Evaluation

Implicit Query Ratings. In order to evaluate implicit queries generated by our framework, we aim to collect user ratings for each query generated from an input tweet T. The maximum query length here is restricted to 3 words. The reason for setting the restriction is the number of queries generated without any restriction is too large for user evaluation. Shown by a small-scale pre-processing experiment, queries with length longer than 3 performs worse than shorter ones. Longer queries always provide just a few or even no new information.

The experiment result is reasonable based on the following arguments: (i) The average length of Tweets tends to be short after pre-processing. Hence it is easy for long queries to cause over-fitting problem; (ii) According to [17], most queries in microblogs tend to be short.

Then we start the evaluation by generating a set of queries with maximum length 3-word for each tweet. For each tweet, we generate queries up to 3 words

---

[3] Features: burst_sum, graphAvgCorr_min, clustDeg_min, clustRelDeg_sum, clustSumCorr_max.

Table 2. Query rating dataset statistics

Raters	103		Avg	Med
Ratings	16,829	Queries per tweet	12.5	13
Rated tweets	416	Ratings per tweet	3.2	3
... from Pool-R	214	Tweets per rater	12.9	16
... from Pool-P	202			
Tweet-query pairs	5,142			

with 3 queries of each length. Similarly, we generate queries from a cluster matching the tweet, with max. 2 queries of each length from the cluster. As mentioned in Sect. 3.5, we generate two types of queries from a cluster (i.e., tweet-dependent and independent). We also match the tweet against *baseline clusters* (cf. Sect. 4.2) and obtain 2 queries at each length of $\{1, 2, 3\}$ words.

Additional Baselines. To compare our methods with other applicable techniques, we extract *hashtags* and *named entities* identified by a NE tagger [7].

We conduct a user study with 103 university students from various disciplines. The user needs to choose tweets from the web-based interface and then rates each query as 2 ("matches the tweet's topic"), 1 ("somewhat matches") or 0 ("does not match"). In total, each user may rate queries from 20 tweets (10 from each pool). The statistics of the obtained ratings are shown in Table 2. After this step, we have 68% of queries with $rating < 1$ and 32% with $rating \in [1, 2]$.

Evaluation Methodology. We formulate implicit query generation as a ranking task: given an input tweet, the most relevant queries should be ranked at the top positions. For each tweet, we compare ranked queries by a method $\mathcal{M}$ against the human ratings. For precision calculation, we consider queries with $avgRating \geq 1$ as 'relevant'[4]. The evaluation scores across all tweets are then averaged using a *weighted average*. As the number of raters who chose each tweet varies, we assign a higher weight (i.e., higher confidence) to tweets chosen by more raters.

We employ 3 metrics for ranking evaluation: (1) *Precision@k*: we calculate precision at $k = \{1, 2, 3\}$. (2) *Mean Average Precision (MAP)*: this metric provides an overall score across all ranking positions, considering all 'relevant' queries in the ground truth. Thus, MAP can be interpreted as both *precision* and *recall* of method $\mathcal{M}$. (3) *Normalized Discounted Cumulative Gain (NDCG)@k*: we calculate NDCG at $k = \{1, 2, 3\}$.

Results. Overall ranking results are presented in Fig. 5 using tweets from *Pool-P*. We present (a) results using all rated tweets, (b) results using only tweets

---

[4] Thus, 96.4% of tweets have at least one 'relevant' query. Among these tweets, on average 4.1 out of 12.5 queries are 'relevant'.

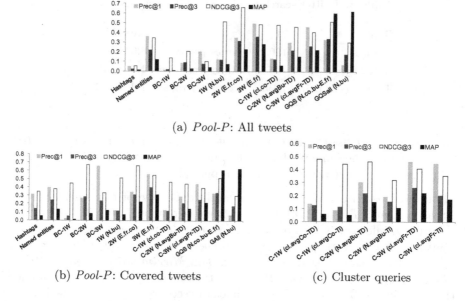

(a) *Pool-P*: All tweets

(b) *Pool-P*: Covered tweets

(c) Cluster queries

Fig. 5. Implicit query ranking evaluation

covered by the respective method. A method is regarded as *covering* tweet T if it produced at least 1 query from T. Due to space limitations, we only present the best result of each method and the used weights.

For *Pool-P*, the highest precision among all tweets is achieved by 3-word queries from our framework, followed by 3-word queries from our clusters (prefixed "C-3W"). Among the top-3 positions, 2-word queries perform the best by NDCG. When considering the retrieval of all queries, GQS has the highest MAP score, while maintaining a high NDCG@3. This indicates that our global query ranking effectively ranks queries of different lengths. Baseline methods do not perform as well on our datasets. Hashtags only cover around 16% of tweets. Named entities cover over 90% of tweets, however they retrieve a small portion of relevant queries, as shown by MAP. Baseline clusters (prefixed "B") fail to cover many tweets (cf. Sect. 4.1).

Regarding *Pool-R*, we note that our clusters only cover 2.5% of the rated tweets and baseline clusters cover 0%, thus we omit these results.

As the last step, we study implicit query generation from clusters (cf. Sect. 3.5) in *Pool-P*. Figure 5(c) shows the results of both tweet-dependent and tweet-independent queries at each query length. The results suggest that (1) longer cluster-based queries are more effective, and (2) tweet-dependent queries outperform queries independently selected from a cluster.

4.5   Twitter Search Results Evaluation

Twitter Search Results Ratings. We also set ah experiment to collect user rating for each search result from the implicit queries. Regarding the previous

Table 3. Twitter search results rating dataset statistics

Raters	21		Avg	Med
Ratings	1303	Queries per tweet	8.54	9
Rated tweets	139	Ratings per tweet	1.1	1.03
Tweet-query pairs	1,191	Tweets per rater	7.3	8.67

evaluation result, we use the set of queries generated from each tweet in *Pool-P*. Other settings are the same to that of the implicit query evaluation experiment but only 2 queries are generated for each type and length.

Additional Baselines. We extract one additional type of phrases from each tweet as baseline: *Twitter-LDA* (LDA) which is proposed by [20] to address the short and informal nature of tweets. We use the program implemented by the authors of [20], with parameters: $Number of topics = 50$, $alpha_g = 0.5$, $beta_word = 0.01$, $beta_b = 0.01$, $gamma = 20$ and $Maximum iterations = 100$.

We conduct a user study with 21 university students from various disciplines in this experiment. The user needs to choose tweets from the web-based interface and then rates each set of search results as 2 ("matches the tweet's topic and provide new information"), 1 ("somewhat matches or provides limited new information") or 0 ("does not match or provides no new information").

The statistics of the obtained ratings are shown in Table 3. For each tweet-query pair, we combine all ratings by different reviewers as the average. After this step, we have 39.7% of queries with $rating < 1$ and 60.3% with $rating \in [1, 2]$.

Evaluation Methodology. For the purpose of this evaluation, we formulate Twitter search by implicit query as a ranking task: given an input tweet, the best search result set should be ranked at the top positions. For each tweet, we therefore compare ranked search result sets by a method $\mathcal{M}$ against the human ratings. The other settings, including the calculation of evaluation score and ranking metrics, are exactly the same to the former experiment in Sect. 4.4.

Results. Overall ranking results are presented in Fig. 6 using tweets from *Pool-P* which are *covered* by the respective method. Among the top-3 positions, 2-word queries perform the best according to NDCG@2. When considering the retrieval of all queries, GQS has the highest MAP score, while maintaining a high NDCG@3. This indicates that our global query ranking effectively ranks queries of different lengths.

Baseline methods do not perform well on our datasets. Hashtag still suffers from low coverage problem. LDA is a commonly recognized effective text mining tool, but the Twitter-LDA only out performs our framework in precision@1 with small advantage, and our framework outperforms Twitter-LDA in all the other metrics. Besides, Twitter-LDA does not support real-time update operations. Thus, although Twitter-LDA is powerful, it is not suitable for the streaming application scenario required by SbT.

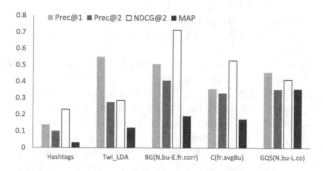

Fig. 6. Twitter search results ranking evaluation

5 Conclusion

In this paper, we propose a novel paradigm for Twitter search, referred to as Search by Tweet(SbT). Rather than formulating a search query, a user may trigger a search task by a single click to select a news-related tweet. The presented framework is both efficient and effective in generating implicit queries from input tweets, as demonstrated in our experiments. Moreover, our work opens up new directions to further refine the techniques supporting SbT. For example, we may study how to automatically generate the best amount of queries, in order to cover all aspects of the input tweet.

References

1. Twitter API (2016). https://dev.twitter.com/docs
2. Agarwal, M.K., Ramamritham, K., Bhide, M.: Real time discovery of dense clusters in highly dynamic graphs: identifying real world events in highly dynamic environments. In: VLDB (2012)
3. Aggarwal, C.C., Zhao, Y., Yu, P.S.: On clustering graph streams. In: SIAM, pp. 478–489 (2010)
4. Angel, A., Sarkas, N., Koudas, N., Srivastava, D.: Dense subgraph maintenance under streaming edge weight updates for real-time story identification. In: VLDB (2012)
5. Dalton, J., Dietz, L., Allan, J.: Entity query feature expansion using knowledge base links. In: SIGIR (2014)
6. Efron, M., Golovchinsky, G.: Estimation methods for ranking recent information. In: SIGIR (2011)
7. Finkel, J.R., Grenager, T., Manning, C.: Incorporating non-local information into information extraction systems by Gibbs sampling. In: ACL (2005)
8. Gao, J., Xu, G., Xu, J.: Query expansion using path-constrained random walks. In: SIGIR (2013)
9. He, Q., Chang, K., Lim, E.P.: Using burstiness to improve clustering of topics in news streams. In: ICDM (2007)
10. Kim, Y., Croft, W.B.: Diversifying query suggestions based on query documents. In: SIGIR (2014)

11. Kleinberg, J.: Bursty and hierarchical structure in streams. In: KDD (2002)
12. Kohavi, R., John, G.H.: Wrappers for feature subset selection. Artif. Intell. **97**(1–2), 273–324 (1997)
13. Lappas, T., Arai, B., Platakis, M., Kotsakos, D., Gunopulos, D.: On burstiness-aware search for document sequences. In: KDD, p. 477 (2009)
14. Lee, P., Lakshmanan, L.V., Milios, E.E.: Incremental cluster evolution tracking from highly dynamic network data. In: ICDE (2014)
15. Massoudi, K., Tsagkias, M., Rijke, M., Weerkamp, W.: Incorporating query expansion and quality indicators in searching microblog posts. In: Clough, P., Foley, C., Gurrin, C., Jones, G.J.F., Kraaij, W., Lee, H., Mudoch, V. (eds.) ECIR 2011. LNCS, vol. 6611, pp. 362–367. Springer, Heidelberg (2011). doi:10.1007/978-3-642-20161-5_36
16. Miyanishi, T., Seki, K., Uehara, K.: Improving pseudo-relevance feedback via tweet selection. In: CIKM (2013)
17. Teevan, J., Ramage, D., Morris, M.R.: #TwitterSearch: a comparison of microblog search and web search. In: WSDM (2011)
18. Wang, C., Zhang, M., Ru, L., Ma, S.: Automatic online news topic ranking using media focus and user attention based on aging theory. In: CIKM, October 2008
19. Yuan, M., Wu, K.-L., Jacques-Silva, G., Lu, Y.: Efficient processing of streaming graphs for evolution-aware clustering categories and subject descriptors. In: CIKM (2013)
20. Zhao, W.X., Jiang, J., Weng, J., He, J., Lim, E.-P., Yan, H., Li, X.: Comparing twitter and traditional media using topic models. In: Clough, P., Foley, C., Gurrin, C., Jones, G.J.F., Kraaij, W., Lee, H., Mudoch, V. (eds.) ECIR 2011. LNCS, vol. 6611, pp. 338–349. Springer, Heidelberg (2011). doi:10.1007/978-3-642-20161-5_34

Semantic Definition Ranking

Zehui Hao[1], Zhongyuan Wang[2], Xiaofeng Meng[1(✉)], Jun Yan[2],
and Qiuyue Wang[1]

[1] School of Information, Renmin University of China, Beijing, China
{jane0331,xfmeng,qiuyuew}@ruc.edu.cn
[2] Microsoft Research Asia, Beijing, China
wzhy@outlook.com, junyan@microsoft.com

Abstract. Question answering has been a focus of much attention from academia and industry. Search engines have already tried to provide direct answers for question-like queries. Among these queries, "What" is one of the biggest segments. Since results excerpted from Wikipedia often have a coverage problem, some models begin to rank definitions that are extracted from web documents, including Ranking SVM and Maximum Entropy Context Model. But they only adopt syntactic features and cannot understand definitions semantically. In this paper, we propose a language model incorporating knowledge bases to learn the regularities behind good definitions. It combines recurrent neural network based language model with a process of mapping words to context-appropriate concepts. Using the knowledge learnt from neural networks, we define two semantic features to evaluate definitions, one of which is confirmed to be effective by experiments. Results show that our model improves precision a lot. Our approach has been applied in production.

Keywords: Definition ranking · Question answering · Recurrent neural network · Conceptualization

1 Introduction

Question Answering (QA), which provides direct answers instead of ten blue links for users' queries, has become a hot trend in web searching. Definition questions, like "What is bandy?", occupy more than 20% of query logs in QA systems [4]. In this paper, we focus on this type of queries.

Figure 1 shows that search engines have already offered definitions for some definienda (the terms being defined). However, these answers are a little coarse and narrow in coverage. Some are just excerpted from *Wikipedia*.

This research was partially supported by the grants from the National Key Research and Development Program of China (No. 2016YFB1000603, 2016YFB1000602); the Natural Science Foundation of China (No. 61532010, 61379050, 91646203, 61532016); Specialized Research Fund for the Doctoral Program of Higher Education (No. 20130004130001), and the Fundamental Research Funds for the Central Universities, the Research Funds of Renmin University (No. 11XNL010).

© Springer International Publishing AG 2017
S. Candan et al. (Eds.): DASFAA 2017, Part II, LNCS 10178, pp. 153–168, 2017.
DOI: 10.1007/978-3-319-55699-4_10

Fig. 1. Definitions excerpted from *Wikipedia*

To break the bottleneck of *Wikipedia*, some approaches begin to extract answers from unstructured texts in the web. A general pipeline consists of two steps. First, use human-defined rules to collect definition candidates automatically from web documents. Then, rank the candidates through a scoring system. Typically, the second step is based on co-occurrence frequency [8,23], scores learnt from Support Vector Machine (SVM) or Language Models (LMs) [4,24], and discriminant functions of Maximum Entropy Model [5]. These methods, even if some simple LMs, cannot analyze sentences thoroughly in semantics. The candidates they prefer are the ones that look like definitions, but may not give a precise description. For the second step here, which we call "definition ranking", unsatisfying results mainly come from the following challenges:

- **Semantic analysis:** The semantics in unstructured texts are complicated. It is hard to define semantic features for machines to tell good definitions from bad ones like human beings.
- **Language polysemy:** Most words denote more than one sense in natural language. Different senses often result in misunderstanding and increase the difficulty of ranking.
- **Text particularity:** Definitions are a special kind of texts. Traditional language models cannot be applied directly to analyze them.

With the effective practice of neural networks, recent studies on text understanding and information retrieval have begun to take advantage of implicit knowledge representation models. Based on the co-occurrence relation, the most successful method [9] maps words to real-number vectors in a low dimensional space. But these models pay little attention to other relations, such as the is-a relation, which plays an important role when defining a term.

To make up for above drawbacks, we take an explicit knowledge representation model [6,18,21] into consideration. It infers the most likely concept of a word in a specific context. In other words, it understands texts with the help of the is-a relation among words. Because each concept is a clear category name shared by a group of hyponyms in KBs, this model captures explicit semantics.

Good definitions usually contain the notion and major attributes of definienda. In general, the concepts of these words are quite related to the concepts of definienda, while the concepts of irrelevant words or trivial matters are not. **Our starting point is that good definitions should be sentences not only full of related words with definienda, but keeping the concepts of these words coherent.** Take the following definition candidates of "emphysema" as an example:

Candidate1. Emphysema is a disease associated with smoking, and usually manifests itself in patients after 50 years of age, affecting about two million Americans each year.

Candidate2. Emphysema is a progressive lung disease that primarily affects smokers and causes shortness of breath and difficulty breathing.

Candidate1 does not mention any symptom of "emphysema". Impacts, like "affecting about two million Americans each year", are less important when defining a disease. Candidate2 contains the notion and major symptoms without too much trivial details.

If we only rely on syntactic features or word-vector similarity, Candidate1 can still rank high. It has a definition format and seems longer than Candidate2. Moreover, "emphysema" is indeed related to some words in it, and is likely to appear with "Americans" frequently in web documents. But looking at a concept level, we will find that the concepts of some words in Candidate1 (like $\langle$Americans, $nation\rangle$[1]) do not relate to the concept of "emphysema" (*disease*) as closely as Candidate2. Section 5.4 explains how our model analyzes at a concept level in details.

In this paper, we propose a combined model to rank answers semantically for definition questions. Our contributions are in two aspects:

- We combine the neural network with the is-a relation through conceptualization and make them complement each other, so that our model is more suitable to handle definition sentences.
- We provide a semantic feature to judge definitions, which is learnt from both unstructured texts and KBs.

The rest of this paper is organized as follows. Section 2 introduces related works, including definition ranking and conceptualization. Section 3 gives some preliminaries about Recurrent Neural Network based Language Model and analyzes its drawbacks. Section 4 describes our model and defines semantic features. Section 5 explains the experiment framework and shows the results. We conclude the paper in Sect. 6.

2   Related Work

Many approaches have been proposed to rank definition candidates collected from the web. Xu, Licuanan, and Weischedel [23] learn the centroid vector (i.e., vector of word frequency) from *Wikipedia*. Definitions are ranked according to

---

[1] The angle brackets mean a word and its concept.

how closely the distribution of their descriptive words correlates with the centroid vector of definienda. Kaisser, Scheible, and Webber [8] employ a similar approach and use the frequency of descriptive words as their weights. These statistical methods only grasp the shallow co-occurrence relation among words. The performance falls into decrease for definienda not covered by *Wikipedia*.

Xu et al. [24] rank definitions through Ranking SVM [7,20]. The features they adopt are syntactic. In practice, sentence length often dominates the preference and reduces accuracy.

Language models are also used for definition ranking. Chen, Zhou, and Wang [4] compare the influence of three kinds of features, including unigrams, bigrams and biterms. However, these LMs ignore the problem of language polysemy. Only one representation for one word may lead to misunderstanding.

Figueroa and Atkinson [5] propose a Maximum Entropy Context (MEC) Model which performs better than centroid vector models and biterm LMs. It accounts for regularities across both positive and negative examples. Although some context indicators are merged to classify definienda, they still only adopt syntactic features finally.

A related work introduces an unsupervised reasoning process which determines context-appropriate concepts for words, called conceptualization [18,21]. For example, seeing "product of apple", we know that "apple" is a *company* or *brand* rather than *food* or *fruit*. Under the hypothesis that concepts of adjacent words should be coherent, Hua et al. [6] develop an efficient framework to perform human-like conceptualization. Given a word in a sentence, it first obtains a concept set $\{c_1, \ldots, c_n\}$ from the is-a relation among words. Then, it uses a weighted-vote algorithm to pick out the most appropriate concepts based on the co-occurrence relation among concepts. The two kinds of relation both come from KBs. Probabilistic knowledge in KBs enables probabilistic reasoning and helps understand sentences precisely. However, compared to the is-a relation, the offline co-occurrence relation is so large that the whole framework becomes too heavy.

3   Preliminary

Recurrent Neural Network based Language Model (RNNLM) [12] is a robust model in natural language processing. Although humans can use long contexts to understand sentences, classic N-gram models barely rely on several preceding words to predict the next word. To break the limitation, RNNLM maintains the hidden layer to make contextual information loop inside the network. It has an advantage of learning arbitrary-length history itself over other LMs.

Considering the task of definition ranking, our target is to let the LM learn the regularities behind good definitions. Despite semantic coherence, words in good definitions are usually organized in a regular way (which syntactic features also try to learn). A LM ignoring the order of words is not competent to handle this special kind of texts. Fortunately, RNNLM often shows an outstanding ability to comprehend sequential data [10,11,19]. This is why we choose it instead of

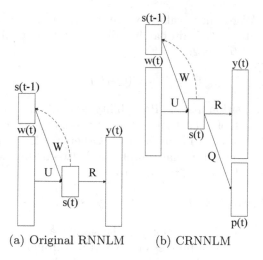

(a) Original RNNLM   (b) CRNNLM

Fig. 2. Model structure of RNNLM

other LMs which perform better on training word vectors or computing word similarity.

The structure of original RNNLM is shown in Fig. 2(a). The input vector $w(t)$ and the output vector $y(t)$ are both V-dimensional, where V is the vocabulary size. $w(t)$ represents the input word in time t and uses 1-of-V coding. $y(t)$ represents probability distribution over all the words. Its each dimension corresponds to the probability of a word to appear next, namely $P(w_i|w(t), s(t-1))$, where w_i is a word in vocabulary. The other vector in the input layer, $s(t-1)$, which represents contextual information, is copied from the hidden layer of last iteration. Vector $s(t)$ and $y(t)$ are computed as follows:

$$s(t) = F(Uw(t) + Ws(t-1)) \tag{1}$$

$$y(t) = G(Rs(t)) \tag{2}$$

where U, W and R on arrows are transition matrices. $F(z)$ is sigmoid activation function, and $G(z_m)$ is softmax function:

$$F(z) = \frac{1}{1 + e^{-z}}, \quad G(z_m) = \frac{e^{z_m}}{\sum_k e^{z_k}}$$

The objective of training RNNLM is to maximize the probability of words in the training corpus:

$$\sum_{t=1}^{T} \log P(w(t+1)|w(t), s(t-1)) \tag{3}$$

Despite its outstanding performance, RNNLM is often criticized for high computational complexity, especially the calculation between $s(t)$ and $y(t)$. One way to speedup is to factorize $y(t)$ as shown in Fig. 2(b), called CRNNLM (RNNLM with classes) [13]. Words in vocabulary are classified in advance according to their frequency. Vector $p(t)$ is probability distribution over all the classes, and is computed similarly to Eq. 2:

$$p(t) = G(Qs(t)) \qquad (4)$$

where Q is also a transition matrix. Probability estimation and backpropagation are done only on part of $y(t)$, that is the words sharing the same class with $w(t+1)$ (the real next word in the corpus). Other words are put aside for the moment. Therefore, computational complexity between the hidden and output layer is reduced. Word probability turns to:

$$P(w_i|w(t), s(t-1)) = P(w_i|p_i, s(t))P(p_i|s(t)) \qquad (5)$$

where w_i is a word which must belong to the same class p_i with $w(t+1)$.

We illustrate the training process of CRNNLM through the sentence "She is eating apple pies". We hypothesizes that "eating" is the current input word, and is the 100th word in the vocabulary. Hence, the 100th dimension of $w(t)$ is 1, and other dimensions are 0. Vector $s(t-1)$ is copied from the hidden layer of last iteration, namely when "is" is the input word. Vector $s(t)$ and $p(t)$ are obtained according to Eqs. 1 and 4. Since the next word is "apple", we check which words belong to the same class with it. Only the corresponding dimensions of these words, including "apple" itself, in $y(t)$ are computed. Each dimension of $p(t)$ is the probability of a class to appear next, namely $P(p_i|s(t))$ in Eq. 5. Each dimension of $y(t)$ is the conditional probability of a word, namely $P(w_i|p_i, s(t))$. After that, we do backpropagation based on the objective Function 3 to update the parameters in all transition matrices, namely U, W, R and Q. After reading all the sentences in the corpus, the training of these matrices is finished. Then, we can use the whole network to predict other sentences.

4   RNNLM Combined with Knowledge Bases

Despite the effectiveness of CRNNLM, there are some drawbacks in its structure. For the output layer, whether a word to be considered or not just depends on its frequency proximity to $w(t+1)$. The factorization method is a little arbitrary, making RNNLM lose much semantics.

For the input layer, since vector $w(t)$ uses 1-of-V coding, every column in matrix U is actually the implicit representation of a word in the vector space. $Uw(t)$ in Eq. 1 is adding the vector of the input word to the context. But words are polysemous. For sentences "product of apple" and "eating apple pie", the same vector of two "apple" is used. Different senses may disturb the training and predicting process.

Therefore, we propose two variants complementing CRNNLM with the is-a relation among words, making RNNLM adapt to the definition-ranking task naturally. Since the is-a relation comes from KBs, we call our model KRNNLM. Both of our variants conduct a reasonable method to factorize the output layer and implement learning-based conceptualization in the input layer. On the basis of KRNNLM-1, KRNNLM-2 adds a direct connection between the input and output layer, which strengthens the effect of conceptualization.

4.1   The First Variant (KRNNLM-1)

Factorization Based on Is-A Relation. As shown in Fig. 3(a), KRNNLM-1 replaces the class vector $p(t)$ with a concept vector $cout(t)$. Its each dimension equals to the probability of a concept given the context, namely $P(cout_j|s(t))$, where $cout_j$ is a concept in KBs. In the output layer, we estimate probability distribution first over all concepts ($cout(t)$), and then over the words sharing the same concepts with the next word (part of $y(t)$). Vector $cout(t)$ is obtained similarly to Eq. 2:

$$cout(t) = G(Qs(t)) \tag{6}$$

Word probability turns to:

$$P(w_i|w(t), s(t-1)) = P(w_i|C_{w_i}, s(t))P(C_{w_i}|s(t))$$
$$= P(w_i|C_{w_i}, s(t)) \sum_{j \in C_{w_i}} P(cout_j|s(t)) \tag{7}$$

where C_{w_i} denotes the concept set of w_i in KBs.

Vector $cout(t)$ here seems like $p(t)$ in CRNNLM. The difference between them does not lie in themselves, but how they influence the factorization of $y(t)$. In CRNNLM, words are classified according to frequency. When we update the network, it makes no sense to discard those words just because their frequency is not proximate to the next word.

In KRNNLM, words are classified according to the is-a relation in KBs. We think that words are more likely to appear in the next position if they belong to the same concepts with $w(t+1)$. These words are what we are really interested in when $time = t$. Besides, words can belong to more than one concept in KBs because of polysemy, so there is a summation in Eq. 7.

To understand our factorization method more clearly, we take "apple" as the next word $w(t+1)$ for example, and go through the computation process from the hidden layer to the output layer. It consists of two parts. For the concept part $cout(t)$, we use Eq. 6 to get the probability distribution of all concepts, including the concepts of "apple". For the word part $y(t)$, according to KBs, "apple" have four concepts, namely *company, brand, food* and *fruit*. Therefore, we only select words belonging to the four concepts to update (like "microsoft" and "orange") and discard others. This is how our model combines KBs to reduce computational complexity of original RNNLM without losing too much semantics.

Conceptualization in the Input Layer. In Fig. 3(a), KRNNLM-1 also adds a concept vector $cin(t)$ to the input layer. The objective function turns to:

$$\sum_{t=1}^{T} \log P(w(t+1)|w(t), cin(t), s(t-1)) \tag{8}$$

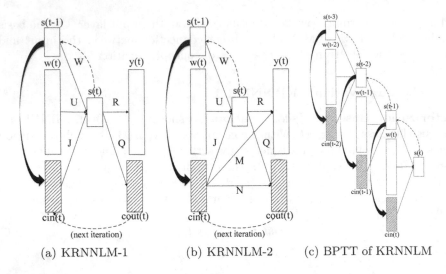

(a) KRNNLM-1   (b) KRNNLM-2   (c) BPTT of KRNNLM

Fig. 3. Structure of KRNNLM (The bold arrow means that there exists a conceptualization process.)

The hidden layer is computed as:

$$s(t) = F(Uw(t) + Jcin(t) + Ws(t-1)) \tag{9}$$

Similar to vector $cout(t)$, each dimension of $cin(t)$ equals to the probability of a concept given the context, namely $P(cin_j|s(t-1))$. Note that the context here is not $s(t)$ but $s(t-1)$. In fact, this value has already been estimated in the output layer when $time = t-1$, so we obtain $cin(t)$ from $cout(t-1)$ in last iteration. This operation is similar to the loop of $s(t)$. In Fig. 3(a), the dashed arrow means copying the concept vector in the output layer to the input layer of next iteration.

In practice, we do not directly copy $cin(t-1)$ to $cout(t)$. In order to decrease noise, we only retain concepts of the input word. The values of other concepts in $cin(t)$ are set to zero. In this way, $cin(t)$ represents the context-specified probability of linking the input word to each of its concepts. Specifically, each dimension of $cin(t)$ is computed as follows:

$$P(cin_j|s(t-1)) = \begin{cases} \frac{P(cout_j|s(t-1))}{\sum_{k \in C_w} P(cout_k|s(t-1))}, & j \in C_w \\ 0, & \text{else} \end{cases}$$

where cin_j is the j-th concept in $cint(t)$, and $cout_j$ corresponds to the same dimension in $cout(t-1)$. C_w denotes the concept set of the input word.

Vector $cin(t)$ here is intended to distinguish different senses of the input word in given contexts. Take "apple" as the input word $w(t)$ in sentence "eating apple pies" for example. Through Eq. 9, concept *food* and *fruit* are added to the hidden layer with larger probability than concept *company* and *brand*.

$Uw(t) + Jcin(t)$ represents not only the word "apple", but an "apple" with specific semantics. We do not assign a vector to each sense of a word like [3, 16], but achieve disambiguation through incorporating concepts to the vector space.

The process of determining $cin(t)$ is doing human-like conceptualization. Since concepts are added to the hidden layer with different probability, the whole model is embedded with semantics from a concept level. Meanwhile, compared to [6], we acquire the co-occurrence relation among concepts from the vector space rather than large offline data. That is why we call it learning-based conceptualization.

4.2   The Second Variant (KRNNLM-2)

The second variant is based on KRNNLM-1. As shown in Fig. 3(b), we connect $cin(t)$ to both the concept and word parts of the output layer. Even if the context vector $s(t)$ loses some information about concepts, the direct connection between $cin(t)$ and the output layer can compensate for it. Thus, the output layer becomes:

$$y(t) = G(Rs(t) + Mcin(t))$$
$$cout(t) = G(Qs(t) + Ncin(t))$$

4.3   Backpropagation Through Time

To ensure that the neural network can learn what information to store in the hidden layer itself, RNNLM usually conducts a different backpropagation, called Backpropagation Through Time (BPTT) [1,17]. As illustrated in Fig. 3(c), our two variants both implement BPTT between the input and hidden layer. The concept vector $cin(t)$ is recorded together with $w(t)$. Thus, the is-a relation from KBs loops with historical information in the neural network.

Through Sects. 4.2 and 4.3, our conceptualization in Sect. 4.1 strengthens its influence on the whole neural network.

4.4   Semantic Features

Here, we define two semantic features: the vector similarity between definienda and definitions (f-sim); the log-probability sum of words in definitions (f-obj). They can be derived from any variant of RNNLM, including CRNNLM and KRNNLM. We obtain the two features from trained models and compare their effects in Sect. 5.3.

f-sim. As analyzed in Sect. 3, every column in matrix U can be regarded as the vector of a word. Given a definiendum, we use its corresponding column u in U to represent it. Every definition is also represented by a vector s. We construct it through the traditional bag-of-words (BOW) approach, namely adding up the vectors of all the words in the sentence. After that, f-sim could be computed through any vector similarity between u and s. We use cosine similarity in the experiments.

f-obj. The log-probability sum of words is exactly the objective function of RNNLM, namely Function 3 for CRNNLM and Function 8 for KRNNLM. It tells how likely a series of words is to be a sentence. Moreover, since sentences in our corpus are all positive examples, our f-obj measures how much regularities of good definitions a given sentence correlates with. The larger f-obj is, the better a definition is.

4.5   Complexity

The complexity of a neural network is measured by its multiplications between transition matrices and vectors from each layer to the next one. For original RNNLM, the complexity of calculation from $s(t-1)$ to $s(t)$, namely $Ws(t-1)$, is $H \times H$. Similarly, the complexity of calculation from $s(t)$ to $y(t)$, namely $Rs(t)$, is $H \times V$. Because $w(t)$ uses 1-of-V coding, the complexity of from $w(t)$ to $s(t)$, namely $Uw(t)$, is $O(1)$. Hence, the complexity of original RNNLM is:

$$T(n) \propto O(H \times H + H \times V) \propto O(H \times V)$$

where H is the predefined dimensionality of $s(t)$ usually within 1000. As the vocabulary size, V is also the dimensionality of $y(t)$, which can reach several hundreds of thousands. Therefore, the complexity of RNNLM is dominated by the calculation from the hidden layer to the output layer. For CRNNLM:

$$T(n) \propto O(H \times P + H \times V')$$

where P is the predefined class size, and also the dimensionality of $p(t)$, usually within 1000. V' is the number of words sharing the same class with the next word. The value of V' depends on V and P, but often ranges from tens to tens of thousands. For both KRNNLM-1 and KRNNLM-2:

$$T(n) \propto O(H \times C + H \times V'')$$

where C is the concept size, and V'' is the number of words sharing the same concepts with the next word. The values of C and V'' are much smaller than V, although they depend on the knowledge bases and may be little larger than P and V'. Thus, our model also reduces the complexity of original RNNLM.

5   Experiments and Results

5.1   Framework

Our experiments consist of three steps in general:

Extract. The experiments are evaluated on a set of open-domain definienda randomly selected from Freebase [2]. Since our work focuses on ranking, we simply send these definienda to a search engine and extract their definition candidates from top three web documents according to the patterns: ... $\langle term \rangle$ is a|is the|is one of|is known as|stands for|, called|means ...

Label. All the candidates are annotated by three annotators with bad, indifferent, or good[2]. Bad means definitions do not refer to any general notion or attribute of definienda. Good means definitions contain the notion and major attributes of definienda. Indifferent means definitions mention some notions or attributes, but contain trivial matters. Definienda with less than five definitions are removed. As a result, we obtain 2,936 definienda with 417 definitions (bad: 1,386, indifferent: 927, good: 623). Note that these definitions are labeled for testing, not for training LMs. As an unsupervised model, RNNLM trains over corpus without any annotation, so do CRNNLM and KRNNLM.

Evaluate. The baseline method is [24] which only adopts syntactic features in Ranking SVM [7]. Besides, our model is compared with CRNNLM [13], MEC [5] and Skip-gram [9] model. MEC is designed for the definition-ranking task, while the other two are not. Skip-gram is a state-of-the-art LM to train word vectors and compute word similarity. We directly use the Word2Vec toolkit[3] to implement Skip-gram, and extend the RNNLM toolkit[4] to implement KRNNLM-1 and KRNNLM-2. We also add our two variants to the baseline method as a feature respectively. Table 1 explains our evaluation metrics.

Table 1. Evaluation metrics for definition ranking

Metrics	Explanation
Error rate	The ratio of mistaken ranked definitions. A definition is regarded as mistaken ranked when there is another definition with better label ranked behind it
Recall	The average ratio of good definitions in top-K candidates. K varies with each term, as it is the real number of good definitions
P@N	The ratio of definienda whose top-N candidates contain good definitions

5.2 Training Language Models

The corpus for training LMs (namely CRNNLM, Skip-gram, and our model) is composed of good definitions. In *Wikipedia*, the first sentence in every document is often a concise and accurate definition. We collect these sentences from Wikipedia Dumps[5] and filter out those obeying the patterns in Sect. 5.1. Finally, we construct a corpus having 1,343,249 good definitions (11M words, 237K vocabulary, words appearing less than 3 times is removed from vocabulary).

Any knowledge base providing the is-a relation among words can be adopted in our model. We use an open-domain KB, called Probase[6] [22], and aggregate

---

[2] If there is a contradiction among annotators, they will be asked to re-annotate the definition. If different opinions still exist, another two annotators will take part, and we will adopt the label given by most annotators.

[3] https://code.google.com/archive/p/word2vec/.

[4] http://www.rnnlm.org.

[5] https://dumps.wikimedia.org/.

[6] Probase data is available at http://probase.msra.cn/dataset.aspx.

all the concepts into 500 clusters. Besides, the size of the hidden layer in the four LMs is all set to 200. The BPTT step for CRNNLM and our model is 4.

5.3 Results

First, we examine whether f-sim and f-obj are capable to evaluate definitions. For the testing definitions in Sect. 5.1, we rank them according to their two features from trained CRNNLM and Skip-gram respectively. As a state-of-the-art model to train word vectors, Skip-gram has a similar matrix to matrix U of RNNLM. It can also provide f-sim for each definition. CRNNLM(f-sim), Skip-gram(f-sim) and CRNNLM(f-obj) in Table 2 show the results.

It is obvious that f-obj is more effective than f-sim, no matter f-sim from CRNNLM or Skip-gram. f-sim indeed measures the relevance between definienda and definitions, but does not reflect word sequence. Even if Skip-gram often performs best on finding related words or documents [9,14,15], it cannot be applied to the definition-ranking task directly. f-obj is derived from the whole neural network and thus takes advantage of the characteristic of RNNLM. It picks out definitions which not only contain relevant words but organize them as how positive examples in the training corpus do. The results confirm that definition ranking is not a task only requiring to find semantically related sentences. Therefore, for our two variants, we merely use them to calculate f-obj.

Table 2. Definition ranking results(%)

	Error rate	Recall	P@1	P@3
SVM	35.7	48.3	49.9	88.2
CRNNLM(f-sim)	55.1	31.5	30.4	75.0
Skip-gram(f-sim)	50.4	41.6	41.4	79.6
CRNNLM(f-obj)	33.8	53.3	51.8	89.8
MEC	32.7	58.1	54.7	92.2
KRNNLM-1(f-obj)	28.5	68.3	66.0	92.5
KRNNLM-2(f-obj)	26.2	70.9	68.1	92.5
KRNNLM-1(f-obj)+SVM	17.9	**74.6**	73.4	92.8
KRNNLM-2(f-obj)+SVM	**16.6**	74.2	**75.8**	**93.1**

We can see that only relying on one feature f-obj, CRNNLM and our two variants all perform better than the baseline method on P@1. Our model is also more accurate than MEC. After adding f-obj to the baseline method (KRNNLM-1(obj)+SVM and KRNNLM-2(obj)+SVM in Table 2), we get the best results. It indicates that no matter how many syntactic features the baseline method and MEC adopt, semantic analysis is the key to improve the precession. For MEC, the semantics complemented by context indicators are still not thorough enough. In contrast, our model provides more sufficient semantics for ranking.

In general, KRNNLM-1 and KRNNLM-2 surpass all compared models. Both CRNNLM and Skip-gram sidestep the problem of language polysemy. Since they do not distinguish different word senses, their training and prediction process is disturbed. To reduce this kind of disturbance, our model figures out word senses though determining their context-appropriate concepts with the help of the is-a relation from Probase. The coherence of these concepts meanwhile is used by our model to evaluate definitions, which exactly achieves our starting point described in the introduction.

Moreover, in KRNNLM-2, the direct connection between the input and output layer consolidates the effect of conceptualization and helps it get higher precision than KRNNLM-1. Though, KRNNLM-1 has less transition matrices, namely less parameters, than KRNNLM-2. For a neural network, less parameters mean smaller training corpus to reach a well-trained convergence. If there is not enough corpus, KRNNLM-1 can be a better choice.

Additionally, it is reasonable that the differences among all models are not that much on P@3. It is a quite loose standard compared to P@1. For terms having five definition candidates, a model can easily achieve 100% on P@3, as long as it puts one good definition in the top three. For real QA systems, P@1 is the most important metric since they usually offer only one answer to users.

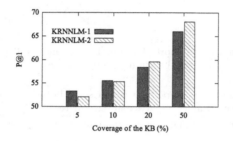

Fig. 4. P@1 with different coverage of the KB

In the above experiments of KRNNLM-1 and KRNNLM-2, about 50% words can find their concepts from Probase. To examine the influence of the KB, we change its coverage. Figure 4 shows that the larger the coverage is, the higher precision our model achieves. When only 5% words have concepts, P@1 on our two variants is very close to CRNNLM. Besides, we can see that the superiority of KRNNLM-2 becomes obvious with the increasing coverage.

5.4   Case Studies

We take "emphysema" in Sect. 1 for example to explain how our model analyzes definitions at a concept level. When we put Candidate1 into trained KRNNLM, four words can find their most appropriate concepts: ⟨disease, *disease*⟩, ⟨patients, *patient*⟩, ⟨Americans, *nation*⟩ and ⟨year, *time*⟩. These words, including "Americans" and "year", are likely to appear frequently with "emphysema" in corpus.

But concept *nation* and *time* are not coherent to the concept of "emphysema" (*disease*) as highly as the concepts of Candidate2, including ⟨lung disease, *disease*⟩, ⟨smokers, *patient*⟩ and ⟨breath, *physiological process*⟩.

Even though words are close to "emphysema" in the vector space, their context-appropriate concepts may not locate near *disease*. If users search for relevant sentences or relevant titles of web pages, Candidate1 is good enough. However, according to the analysis of our starting point in the introduction, Candidate2 should rank higher. Our model achieves this from a concept level.

Table 3 gives some other examples that KRNNLM-2 makes right judgements but CRNNLM does not. Inside one model, higher scores mean better definitions. Scores from different models are not comparable.

For the first term "bandy", Candidate1 has more relevant concepts (⟨team, *activity*⟩, ⟨sport, *activity*⟩, ⟨skaters, *athlete*⟩, ⟨sticks, *device*⟩, ⟨ball, *device*⟩, ⟨goal, *device*⟩) with it than Candidate2 (⟨sport, *activity*⟩, ⟨athletes, *athlete*⟩), even if the words themselves of Candidate1 may not appear as frequently as the words of Candidate2 in corpus. In other words, Candidate1 has more concepts close to the concepts of "bandy" (*sport*) in the vector space. Definition candidates of the other two terms in Table 3 are in a similar situation.

Table 3. Examples

Term	Definition	CRNNLM (f-obj)	KRNNLM-2 (f-obj)
Bandy	1. Bandy is a team winter sport played on ice, in which skaters use sticks to direct a ball into the opposing team's goal	0.3182	0.1214
	2. Bandy is the second most popular winter sport in the world based on the number of participating athletes	0.4938	0.0759
Honey	1. Honey is a sweet food made by bees foraging nectar from flowers	−0.1985	−0.1293
	2. Honey is adulterated if it has the addition of other sugars, syrups or compounds, making it cheaper to produce, or has many fructose contents in order to stave off	−0.3266	−0.1082
Diving	1. Diving is the sport of jumping or falling into water from a platform or springboard	0.0304	0.1182
	2. Diving is a separate sport in Olympic	0.3097	−0.0836

6   Conclusion

In this paper, we introduce a language model to do semantic definition ranking. It combines RNNLM with learning-based conceptualization and captures semantics from both unstructured texts and the is-a relation in knowledge bases. We define two semantic features derived from neural networks and confirm that the log-probability sum can evaluate definitions effectively. We compare our model with other approaches which only adopt syntactic features or sidestep the problem of language polysemy. Results indicate that semantic analysis is the key to improve the precision, and the incorporated knowledge base makes our model suit for the definition-ranking task better. Our approach has been applied in production.

References

1. Boden, M.: A guide to recurrent neural networks and backpropagation. Dallas Project Sics Technical report T Sics (2001)
2. Bollacker, K., Evans, C., Paritosh, P., Sturge, T., Taylor, J.: Freebase: a collaboratively created graph database for structuring human knowledge. In: Proceedings of the 2008 ACM SIGMOD International Conference on Management of Data, pp. 1247–1250. ACM (2008)
3. Chen, X., Liu, Z., Sun, M.: A unified model for word sense representation and disambiguation. In: EMNLP, pp. 1025–1035. Citeseer (2014)
4. Chen, Y., Zhou, M., Wang, S.: Reranking answers for definitional QA using language modeling. In: Proceedings of the 21st International Conference on Computational Linguistics and the 44th Annual Meeting of the Association for Computational Linguistics, pp. 1081–1088. Association for Computational Linguistics (2006)
5. Figueroa, A., Atkinson, J.: Maximum entropy context models for ranking biographical answers to open-domain definition questions. In: AAAI 2011, San Francisco, California, USA, August (2011)
6. Hua, W., Wang, Z., Wang, H., Zheng, K., Zhou, X.: Short text understanding through lexical-semantic analysis. In: International Conference on Data Engineering (ICDE) (2015)
7. Joachims, T.: Training linear SVMS in linear time. In: Proceedings of the 12th ACM SIGKDD International Conference on Knowledge Discovery and Data Mining, pp. 217–226. ACM (2006)
8. Kaisser, M., Scheible, S., Webber, B.L.: Experiments at the University of Edinburgh for the TREC 2006 QA track. In: TREC (2006)
9. Mikolov, T., Chen, K., Corrado, G., Dean, J.: Efficient estimation of word representations in vector space. arXiv preprint arXiv:1301.3781 (2013)
10. Mikolov, T., Deoras, A., Kombrink, S., Burget, L., Cernockỳ, J.: Empirical evaluation and combination of advanced language modeling techniques. In: INTERSPEECH, pp. 605–608, no. s1 (2011)
11. Mikolov, T., Deoras, A., Povey, D., Burget, L., Černockỳ, J.: Strategies for training large scale neural network language models. In: 2011 IEEE Workshop on Automatic Speech Recognition and Understanding (ASRU), pp. 196–201. IEEE (2011)
12. Mikolov, T., Karafiát, M., Burget, L., Cernockỳ, J., Khudanpur, S.: Recurrent neural network based language model. In: INTERSPEECH 2010, Makuhari, Chiba, Japan, 26–30 September 2010, pp. 1045–1048 (2010)
13. Mikolov, T., Kombrink, S., Burget, L., Černockỳ, J.H., Khudanpur, S.: Extensions of recurrent neural network language model. In: 2011 IEEE International Conference on ICASSP, pp. 5528–5531. IEEE (2011)
14. Mikolov, T., Sutskever, I., Chen, K., Corrado, G.S., Dean, J.: Distributed representations of words and phrases and their compositionality. In: Advances in Neural Information Processing Systems, pp. 3111–3119 (2013)

15. Mikolov, T., Yih, W.t., Zweig, G.: Linguistic regularities in continuous space word representations. In: HLT-NAACL, vol. 13, pp. 746–751 (2013)
16. Neelakantan, A., Shankar, J., Passos, A., McCallum, A.: Efficient non-parametric estimation of multiple embeddings per word in vector space. arXiv preprint arXiv:1504.06654 (2015)
17. Rumelhart, D.E.: Leaning internal representations by back-propagating errors. Nature **323**, 318–362 (1986)
18. Song, Y., Wang, H., Wang, Z., Li, H., Chen, W.: Short text conceptualization using a probabilistic knowledgebase. In: Proceedings of the Twenty-Second IJCAI-Volume Three, pp. 2330–2336. AAAI Press (2011)
19. Sutskever, I., Martens, J., Hinton, G.E.: Generating text with recurrent neural networks. In: Proceedings of ICML-11, pp. 1017–1024 (2011)
20. Vapnik, V.: The Nature of Statistical Learning Theory. Springer Science & Business Media, New York (2013)
21. Wang, Z., Zhao, K., Wang, H., Meng, X., Wen, J.R.: Query understanding through knowledge-based conceptualization. In: Proceedings of the Twenty-Fourth IJCAI (2015)
22. Wu, W., Li, H., Wang, H., Zhu, K.Q.: Probase: a probabilistic taxonomy for text understanding. In: Proceedings of the 2012 ACM SIGMOD International Conference on Management of Data, pp. 481–492. ACM (2012)
23. Xu, J., Licuanan, A., Weischedel, R.M.: TREC 2003 QA at BBN: answering definitional questions. In: TREC, pp. 98–106 (2003)
24. Xu, J., Cao, Y., Li, H., Zhao, M.: Ranking definitions with supervised learning methods. In: Special Interest Tracks and Posters of the 14th International Conference on World Wide Web, pp. 811–819. ACM (2005)

An Improved Approach for Long Tail Advertising in Sponsored Search

Amar Budhiraja[✉] and P. Krishna Reddy

FC Kohli Center in Intelligent Systems, IIIT-Hyderabad, Hyderabad 500032, India
amar.budhiraja@research.iiit.ac.in, pkreddy@iiit.ac.in

Abstract. Search queries follow a long tail distribution which results in harder management of ad space for sponsored search. During keyword auctions, advertisers also tend to target head query keywords, thereby creating an imbalance in demand for head and tail keywords. This leads to under-utilization of ad space of tail query keywords. In this paper, we have explored a mechanism that allows the advertisers to bid on concepts rather than keywords. The tail query keywords are utilized by allocating a mix of head and tail keywords related to the concept. In the literature, an effort has been made to improve sponsored search by extracting the knowledge of coverage patterns among the keywords of transactional query logs. In this paper, we propose an improved approach to allow advertisers to bid on high level concepts instead of keywords in sponsored search. The proposed approach utilizes the knowledge of level-wise coverage patterns to allocate incoming search queries to advertisers in an efficient manner by utilizing the long tail. Experimental results on AOL search query data set show improvement in ad space utilization and reach of advertisers.

Keywords: Data mining · Computational advertising · Coverage Patterns · Pattern mining · Sponsored search

1 Introduction

Sponsored search is one of the most dominant mediums to advertise on the web. In sponsored search, advertisers create ad campaigns and bid on keywords that they deem relevant to their product. For an incoming search query, advertisements from the ad campaigns containing the query keywords are shown along with the search results. If multiple advertisers demand to be shown on the same query's results page, they are ranked for the allocation of ad space. The ranking is based on multiple factors including the bid amount of the advertiser on the query keywords, relevance of ad content to the search query, Click-Through-Rate (CTR) and budget of the advertiser.

It has been established that search queries follow a long tail distribution of a small but fat head of frequent queries and a long-thin tail of infrequent queries [5,8]. Advertising on tail queries is challenging as tail queries are encountered rarely which makes them harder to interpret for sponsored search. Also, it has

© Springer International Publishing AG 2017
S. Candan et al. (Eds.): DASFAA 2017, Part II, LNCS 10178, pp. 169–184, 2017.
DOI: 10.1007/978-3-319-55699-4_11

been observed that during keyword auctions, advertisers tend to bid for the head keywords to reach more users. This creates a high demand for the head query keywords and little to no demand for the tail query keywords [5]. The long tail phenomenon also makes it quite difficult to capture the relevant keywords from the long tail. The above stated factors result in under-utilization of a significant amount of the ad space provided by tail queries in sponsored search which is identified as the research issue.

In this paper, we propose an approach to exploit the long tail of the search query keywords for sponsored search. We propose that instead of bidding on search query keywords, advertisers should bid upon high level concepts. The motivation for bidding on concepts is inspired from the trends in advertising on social media[1]. In social media advertising, advertisers target concepts beyond keywords such as photography, reading, travelling, lifestyle, etc. In sponsored search, bidding on concepts will result in capitalization of ad space of the tail queries as all the keywords would be considered based on the relevancy rather than frequency. Bidding on concepts instead of keywords would also ensure that the advertisers do not have to retrieve all the search keywords from the long tail.

In this paper, we propose an allocation mechanism for sponsored by considering concepts as bidding units rather than search keywords. We propose that during ad campaign creation, an advertiser is shown a taxonomy based on the content of the ad and is asked to select a concept in the shown taxonomy that seems to be the most relevant to the product. We propose an approximate allocation between the nodes of the taxonomy and the advertisers. To acknowledge the long tail phenomenon, we extract knowledge from search query logs using the notion of Coverage Patterns (CPs). In the literature, approaches to extract CPs have been proposed. Given a database of transactions, a CP is a set of items such that it covers a certain percentage of transactions having given overlap ratio [1,4]. By extending the notion of CPs, an effort has been made in the literature to propose allocation approach to improve the performance of Adwords [3] and display advertising by assuming that an advertiser requests a set of keywords [11]. In this paper, taking query logs and taxonomy as input, we propose a new approach to extract the knowledge of level-wise CPs and use the corresponding framework to allocate incoming queries to ads based on the high-level concepts requested by the advertisers. The proposed approach is compared against traditional sponsored search model. Experiments on the real world data set of AOL search query logs show the improvement in ad space utilization and reach of advertisers.

The remainder of this paper is organized as follows: in Sect. 2, we review the related work in the context of coverage patterns and long tail advertising in search engines; in Sect. 3 we discuss the background on coverage patterns and sponsored search; in Sect. 4 we discuss the basic idea followed by the proposed approach in Sect. 5; experiments are discussed in Sect. 6, followed by conclusions and a discussion on future work in Sect. 7.

---

[1] ads.twitter.com

2   Related Work

In the literature, challenges of tail queries in sponsored have been primarily addressed by means of query expansion [5–7]. In [6], the authors formulated a taxonomy based model to classify search queries, specifically tail queries. Organic clicks were used as blind feedback mechanism to learn the model. The authors explored its feasibility on search advertisement relevance. In another study [7], the authors expand search queries by adding multiple features including category of retrieved web pages and salient named entities. Furthermore, the authors propose an approach in [5] to expand tail queries in real time using an inverted index build from head and torso *expanded* queries. Using the expanded queries, the authors show improvement in ad retrieval.

Sponsored search has been also explored from the perspective of revenue optimization. In [9], it was modelled as an online bipartite matching problem such that advertisers are one set of disjoint vertices and queries are the other disjoint set. They developed an algorithm for advertisement allocation of incoming queries to optimize the revenue of the search engine. This bipartite approach is a high level architecture for Adwords, Google's sponsored search. A more detailed survey of the related literature [12] explains multiple models of bipartite graph matching with its context as Adwords, including algorithms from display ads and welfare maximization.

The model of coverage patterns has been proposed in the literature in the form of an apriori style approach proposed in [1] followed by a pattern growth approach in [4]. Coverage Patterns have been employed in improvement of delivering guaranteed contracts in display advertising [11] and in coverage of more advertisers in Adwords [3].

In this paper, we propose a framework to capitalize the long tail of search queries. We extend the bipartite model discussed in [9,12] into an end-to-end approach. The proposed approach is different from [5] as the authors propose to capture tail queries using a taxonomy by generalizing a query into a taxonomy node. However, the taxonomy was not exposed to the advertisers and was only employed internally whereas in this paper, we propose a mechanism to allow advertisers to bid on concepts by showing a taxonomy related to their ads. In [3], the authors used coverage patterns to group similar keywords but the model to group keywords was employed by abstracting similar keywords only to a single concept rather than a hierarchical relationship of taxonomy, as proposed in this paper. It should be noted that the previous approaches [3,5–7] have emphasized on keyword analysis or query expansion where in this paper we present an alternative approach of bidding on concepts rather than keywords in sponsored search.

3   Background: Sponsored Search and Coverage Patterns

In this section, we briefly explain the sponsored search framework and notion of coverage patterns.

3.1 Sponsored Search Background

The standard model for sponsored search is a bipartite model as shown in Fig. 1(a) such that each incoming query is matched to an advertiser based on certain constraints. These constraints are defined by multiple parameters including relevance score, bid of the advertiser on the query keywords and remaining budget of the advertiser.

The architecture of sponsored search for advertisement allotment has four main steps [12] as shown in Fig. 1(b):

1. **Analysis of Query:** In the first step, the query is analysed to extract important parameters such as session information to better serve advertisements.
2. **Retrieval of Relevant Advertisers:** Based on the query keywords and other query parameters learnt from Step 1, relevant advertisers are retrieved from ad campaigns which are to be considered for displaying alongside organic results.
3. **Bidding:** Due to competition among advertisers, incoming queries are allotted to advertisers through auctions such that advertisers bid for placing their ads on the query page. These bids can either be static or can be done in real time.
4. **Ranking of Advertisers:** Once the advertisers bid on a query, their bids are scaled according to a factor called *Quality Score*. The *Quality Score* is computed based on the parameters related to the respective advertisement. This includes expected *Click Through Rate (CTR)*, display URL's past *CTR*, quality of the landing page, remaining budget and advertisement/search relevance apart from several other parameters.

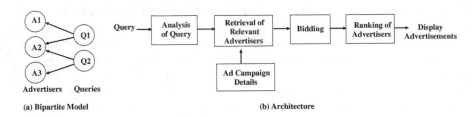

Fig. 1. Sponsored search: model and architecture

3.2 Coverage Patterns

In this section, we briefly explain about the notion of coverage patterns [1,4]. Let $W = \{w_1, w_2...w_n\}$ be set of web pages and D be a set of transactions such that each transaction T is a set of web pages $T \subseteq W$. X is a pattern of web pages such that $X \subseteq W$ and $X = \{w_p, ...w_q, w_r\}$ where $1 \leq p \leq q \leq r$. T^{w_i} denotes a set of transactions containing the web page w_i and its cardinality is denoted $|T^{w_i}|$.

The fraction of transactions containing a web page w_i is called as the *Relative Frequency* of w_i and is calculated as $RF(w_i) = \frac{|T^{w_i}|}{|D|}$. A web page is considered frequent if it has a relative frequency greater than the threshold value, *minRF*. Coverage Set of a pattern $X = \{w_p, ...w_q, w_r\}$, $CSet(X)$ is a set of all transactions that contain at least one web page from the patterns i.e. $CSet(X) = T^{w_p} \cup ...T^{w_q} \cup T^{w_r}$ such that $|T^{w_p}| > ... > |T^{w_q}| > |T^{w_r}|$. Coverage Support, $CS(X)$ is the ratio of size of $CSet(X)$ to size of D i.e., $CS(X) = \frac{|CSet(X)|}{|D|}$. Overlap ratio of a pattern X, $OR(X)$ is the ratio of number of transactions that are common between $X - w^r$ and w^r to the number of transactions in w_r i.e., $OR(X) = \frac{CSet(X-w^r) \cap CSet(w^r)}{CSet(w^r)}$.

Table 1. Sample Transactions

TID	1	2	3	4	5	6	7	8	9	10
Pages	{a,b,c}	{a,c,e}	{a,c,e}	{a,c,d}	{b,d,f}	{b,d}	{b,d}	{b,e}	{b,e}	{a,b}

A pattern is interesting if it has a high CS and low OR. A high CS value indicates more number of visitors and a low OR value means less repetitions amongst the visitors. Hence, a pattern is said to be interesting if $CS(X) > minCS(X)$, $OR(X) < maxOR$ and $RF(w^i) > minRF \ \forall \ w^i \in X$.

Example 1: To explain the notion of coverage patterns, we will consider a transactional database $|D|$ shown in Table 1. Let us assume the minRF to be 0.2, minCS to be 0.3 and maxOR to be 0.5. From Table 1, the number of transactions having a, T^a is 5, T^b is 7 and f, T^f is 1. So, RF for a is 0.5, for b is 0.7 and for f is 0.1, f will be removed. On the other hand, a and b satisfy the constraint of minRF and therefore, {b,a} is a candidate set. The order of items in a coverage pattern is in decreasing order of the relative frequency and hence, the pattern is {b,a} and not, {a,b}. The Coverage Set for {b,a} is {1,2,3,4,5,6,7,8,9,10} and $|CSet\{b,a\}|$ is 10. So, coverage support of {b,a} is $\frac{10}{10}$ is 1 which is greater than *minCS*. The transactions containing {b,a} together is {1,10} and $T^a = 5$, so the overlap ratio is $\frac{2}{5} = 0.4 < maxOR$ and hence, {b,a} is a coverage pattern. Thus, coverage patterns helps in extracting multiple sets of mutually exclusive subsets of items corresponding to coverage support and overlap ratio. In the literature, it has been demonstrated that coverage patterns can help in covering more advertisers and improve the diversity of viewers of individual ads [3,11].

4   Basic Idea

The long tail phenomenon of search queries makes them unpredictable for sponsored search which is identified as the research issue. Advertisers also tend to target head query keywords during keyword auctions in order to cover more eye balls. However, this leads to a high demand for head keywords while little to no demand for the tail keywords. This imbalance in demand results in underutilization of ad space of the tail keywords. Hence, an opportunity has been identified to capitalize this long tail of search query keywords.

We propose that for sponsored search, advertisers should bid upon high level concepts instead of specific keywords. Bidding on high level concepts will result in capitalization of the ad space of the tail keywords as the keywords would be considered to be allocated based on relevancy rather than frequency.

In the proposed approach, we achieve bidding on concepts such that an advertiser would be shown a taxonomy based on his/her advertisement content. The advertiser is then asked to select a node in the taxonomy which he/she deems the most relevant for the advertisement. For example, an advertiser like *Amazon.com* would be shown at taxonomy of *Shopping* and based on the advertisement, the advertiser can select the appropriate node. If the advertisement is of books, the advertiser would select the node *Books* in the *Shopping* taxonomy or if the ad is related to clothing, the advertiser would choose to bid upon *Clothing or Fashion*. Thus, we propose to add a middle layer of concepts through a taxonomy during the bidding process such that an advertiser would chose a concept which would ultimately translate to a set of keywords, compared to the present approach where the advertiser is responsible for selecting all the desired keywords (Fig. 2).

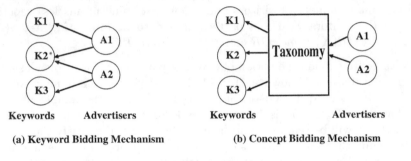

(a) Keyword Bidding Mechanism   (b) Concept Bidding Mechanism

Fig. 2. Sponsored search bidding: keywords based bidding and concept based bidding

We propose an estimated allocation model based on the concept bidding such that groups immediate children nodes of bidding node are allocated to advertisers. Allocation of only children nodes of the bidding node is done to ensure that the allocation mechanism should consider the amount of generalization requested by the advertiser. For example, an advertiser who chose to bid upon *Shopping* should not be allotted something like { *Outwear, Skirts, Shirts*} as he would like to show his ad to a larger audience consisting of *Books, Clothing, Electronics, etc.*

To create such combinations of children nodes, we employ the notion of Coverage Patterns (CPs) such that CPs are extracted from the query logs and a matching is performed between the CPs at each node and the corresponding advertisers. When a query is posed by a user to the search engine, it is classified into these concepts according to the taxonomy and the advertisers who have been allocated any of these concepts are eligible to be ranked for the query.

To address the issues of allocation for a multiple level taxonomy, we propose an approach to extract CPs with respect to the taxonomy followed by an allocation approach for advertisers using the extracted coverage patterns.

4.1 T-Cmine: Extraction of Coverage Patterns with Respect to a Taxonomy

In [2], an approach to extract generalized frequent patterns has been proposed. Similarly, we propose a methodology to extract coverage patterns involving the nodes of taxonomy by extending Cmine algorithm [4]. For a given transactional database $\mathcal{D}$ and the taxonomy $\mathcal{T}$ which relates the items of $\mathcal{D}$, we modify each transaction by appending the ancestors of each item in the transaction to the transaction. If we apply Cmine to this modified dataset several coverage patterns containing high-level as well as low level items would be extracted. Such patterns may not be useful for ad allocation. We are interested in the coverage patterns which contains the items at the same level and satisfy the following property.

$$CP = \{c \mid c \in (\mathcal{I} \cup \mathcal{T}) \ \& \ \forall \ c \ parent(c) = P\} \tag{1}$$

Here, CP is coverage pattern containing items c such that all items belong to the same parent P. To extract level-wise coverage patterns, we propose the T-Cmine algorithm which is as follows.

Algorithm 1. T-Cmine: Algorithm to extract Coverage Patterns with respect to a Taxonomy

---

Input: $\mathcal{D}$, dataset of transactions; $\mathcal{T}$, Taxonomy defined over items of $\mathcal{D}$;
Compute $\mathcal{D}*$ from $\mathcal{T}$ by appending ancestors to $\mathcal{D}$;
$TL_1 := \{frequent1 \ itemsets\}$;
$NO_1 := \{frequent1 \ itemsets\}$;
$C_2 := NO_1 \bowtie NO_1$;
$TL_2 :=$ Remove any patterns from C_2 which contain items other than sister nodes;
$TL_2 :=$ Remove any patterns from TL_k which do not satisfy $minCS, maxOR$ property;
$NO_2 :=$ Remove any patterns from TL_k which do not satisfy $maxOR$ property;
$k := 3$
while $TL_{k-1} \neq \phi$ **do**
    $C_k := NO_{k-1} \bowtie NO_{k-1}$;
    $TL_k :=$ Remove any patterns from TL_k which do not satisfy $minCS, maxOR$ property;
    $NO_k :=$ Remove any patterns from TL_k which do not satisfy $maxOR$ property;
end

The proposed algorithm takes the dataset $\mathcal{D}$ and a taxonomy $\mathcal{T}$ that defines the relationship between the items of the $\mathcal{D}$. The algorithm first adds ancestors of each item in a transaction to the transaction. Then, the first set of CPs (TL_1) is calculated by getting the frequent items for which relative frequency is greater than $minRF$. The same set (TL_1) is also considered as Non-Overlapping Patterns set (NO_1). Using the (NO_1), candidate-2 coverage patterns are computed in the same way as Cmine algorithm. We prune all the patterns which contains other than sister nodes as stated Eq. 1. From the pruned set, we extract patterns which satisfy both minCS and maxOR property which are the Coverage Patterns of length 2 (TL_2). In the next step, non-overlapping patterns (NO_2) are generated by sorting them in order of CS and removing any CPs which don't satisfy maxOR criteria. Note that the pruning step is only required at for $k = 2$ as once the patterns containing any non-sister nodes are removed, there will be no non-overlapping patterns that can be generated that contain non-sister nodes in a CP. From $k = 3$, for k^{th} iteration of the algorithm, first candidate CPs, C_k are generated by joining NO_{k-1} patterns. From C_k, any patterns which do not satisfy the minCS or maxOR are not considered to generate CPs of length k, TL_k. From C_k, patterns which do not satisfy the maxOR or contain non-sisters nodes are removed and the remaining are sorted according to coverage support to generate non-overlapping sets of items of length k, NO_k. It should be noted that OR follows a 'sorted' downward closure property [4], and hence, the item sets of candidate sets, C_k are sorted to obtain the corresponding non-overlapping sets NO_k. An example of the algorithm is also shown in Fig. 3.

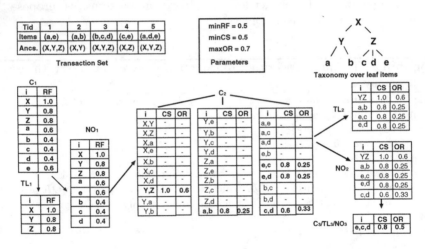

Fig. 3. Example 2: Example of T-Cmine

5   Proposed Approach

In this section, we discuss the proposed approach. In contrast to the sponsored search model of keyword based bidding, we proposed to add a middle layer of concepts during the bidding. Similarly, we also propose to add a middle layer to the allocation process such that when a user poses a query, it is first classified by a taxonomy into a set of nodes. For example, a query on *Harry Potter* would be classified into nodes *Shopping; Books; Fiction*. An advertiser who was allocated any of these concepts would be considered to be displayed on the query of *Harry Potter*. As compared to the standard sponsored search model of a bipartite graph between advertisers and queries as shown in Fig. 1 (a), we add a middle layer of CPs between search queries and advertisers as shown in Fig. 4 (a).

The sponsored search architecture has four major steps for query allocation to advertisers. The proposed architecture also has four major online steps for allocation of incoming queries to advertisers. But, in the proposed architecture, we also exploit the knowledge extracted from the query logs in the form of CPs. We discuss each step of the proposed architecture as follows.

1. **Query Analysis:** This step is same as the standard sponsored search architecture. But, we also extract the concepts of each incoming query. For example, if the query is *Harry Potter* which belongs to the taxonomy *Shopping* then, it's concepts would be *Shopping; Books; Fiction.*
2. **Retrieval of Relevant Advertisers:** Based on the concepts inferred from the query in the first step, we retrieve advertisers from the matching of CPs. (In the next part of this subsection, we show how this matching of CPs and advertisers is achieved.)
3. **Bidding:** This step is same as the standard sponsored search architecture.
4. **Ranking of Advertisers:** This step is same as the standard sponsored search architecture.

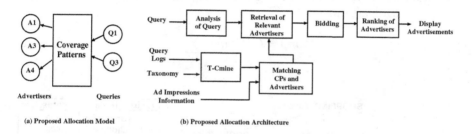

(a) Proposed Allocation Model   (b) Proposed Allocation Architecture

Fig. 4. Proposed sponsored search allocation model and architecture

5.1 Matching CPs and Advertisers

In this section, we explain how the matching between CPs and advertisers is achieved. It should be noted that while considering this approach we assume the CPM (Cost Per Mille) payment mechanism, which can be easily extended to CPC (Cost Per Click) mechanism [3]. The matching process has two main components:

1. **Extraction of CPs using T-Cmine**: This step takes input of the query logs and the taxonomy and extract CPs as explained in Sect. 4.1.
2. **Matching CPs and Advertisers**: In this step, we take the demands of the advertisers and the CPs extracted from query logs and perform a matching between the two. An allocation protocol has been proposed such that specialized requests are processed before generalized. The reason for doing a specialized-to-generalized allocation is to acknowledge that an advertiser who bids on a lower level in the taxonomy has less options of allocation compared to the advertiser who bids on a higher level. For example, an advertiser who

bids on the root node can be satisfied by any choice of children nodes. However, such an allocation poses a challenge where a coverage pattern containing a parent node has to be allocated given its descendants has been allocated to advertisers. Allocation at a node should take into account if any of its descendants have been allocated as coverage of a node is sum of coverage of its descendants. Hence, impressions of a node should be modified to take into consideration if any of its descendant nodes have been allocated to the advertisers. The necessary modification to a CP if any of its descendants have been allocated advertisers is to subtract the number of impressions allotted to the advertisers children of nodes contained in the respective CP.

Equation 2 captures the necessary changes required to a CP such for each node in the CP (denoted by k), count the impressions of allocated advertisers (denoted by j) of each descendant (denoted by i) and subtract it from total impressions of the CP. It should be noted that a coverage pattern is allocated to a set of advertisers if and only if it has enough impressions to satisfy the allocated advertisers. It may happen that advertisers are not allocated a coverage pattern if supply is greater than demand, and thus the following equation will never result in a negative value for the number of impressions of a coverage pattern.

$$CP.imp = CP.imp - \sum_k \sum_{ij} A_{ij} \tag{2}$$

Example Taxonomy

Ad ID	Node	Impressions
A1	Shopping	800
A2	Clothing	500
A3	Books	200
A4	Books	300
A5	Shopping	500

Table2: Example Impression Requests by Advertisers

Extracted CPs	Imp	Modified Imp
{Books, Clothing}	1400	400
{Electronics, Clothing}	1700	1200
{Books, Electronics}	1500	1000

Table1: Impressions provided by CPs beforeand after allocation at level 3

Ad Id	CPs
A1	{Books, Electronics}
A5	{Electronics, Clothing}

Table 3: Allocated CPs to Advertisers at Shopping

Fig. 5. Example allocation

Example 3: In Fig. 5, we show an example allocation. We consider the top two levels of a taxonomy to show and consider advertisers who bid on the first three levels. Each advertiser bids on a node and has a demand of certain impressions at that node. Assuming allocation was done at level two i.e. for *Electronics, Clothing* and *Books*, we will show how it will be done for *Shopping*. The node *Shopping* has three children and CPs pertaining to *Shopping* are shown in Table 1 of Fig. 5.

However, as we know that allocations have been done for advertisers who chose to bid upon *Books* and *Clothing*, we need to adjust the impressions provided by the CPs containing these two nodes. For example, the CP {*Book, Clothing*} has 1400 initial impressions, but some advertisers were already allocated *Books* and *Clothing* during allocations at lower level(s). Hence, those impressions need to be subtracted i.e. $1400 - (500 + 200 + 300) = 400$. Similarly, for {*Electronics, Clothing*}, the modified number of impressions is 1200 i.e. $1700 - 500$ and that of {*Books, Electronics*} is 100 i.e. $1500 - (200 + 300) = 1000$. In the next part of this section, the matching between CPs is performed considering the proposed modification.

A matching is performed with advertisers as one side of the bipartite and CPs as the other side. The matching is done at each node of the taxonomy where more than one advertisers choose to bid. In order to maximize the revenue, the matching should be performed in such a way that maximum number of impressions that can be provided by the coverage patterns should be allocated. We propose the matching as an optimization problem in the same respect such that the difference between the CPs and advertisers allocated to them should be minimal. For example, if an advertiser demands 100 impressions and there are two CPs with impressions 150 and 200 respectively, then we chose to allocate the CP with 150 impressions. A similar case can be made when the supply of CPs is 50 and 75 impressions and demand by the advertisers is 100 impressions, then the CP with 75 impressions is chosen. We frame the objective function of the matching on the same notion which is as follows. Equation 3a aims at minimizing the difference between the allocated advertising and the CPs. The objective function is such that for each advertiser Ad_{ij} who has been allocated the CP, CP_j the difference between the two is minimal. Equation 3b lays out the constraint such that the sum of impressions of allocated advertisers does not exceed the impression provided by the CP to avoid the objective from going negative.

$$MinZ = \sum_{level=d}^{0}(\sum_j |CP_j.Impressions - \sum_i^n(Ad_{ij}.Impressions)|) \quad (3a)$$
$$s.t \ \ CP_j.Impressions >= \sum_{i=1}^{n}(Ad_{ij}.Impressions) \quad (3b)$$

Continuing Example 3 from Fig. 5. From the last step, we have CPs whose impressions have been updated according to allocations at their descendants. We show how the allocation is to be done for the node *Shopping*. Two advertisers A_1 and A_5 chose to bid on the node *Shopping*. In the proposed approach, we decide to serve the advertisers on a first-come-first-serve basis. For ad A_1, we select the CP {*Books, Electronics*} because it has the lesser difference compared to the other node. It should be noted that now the number of impressions covered by CP {*Books, Electronics*} has been reduced to 200 as A_1 has been allotted to it. Next, we look at ad A_5 and we see that out of the three CPs, only {*Electronics, Clothing*} has enough impressions to satisfy the advertiser and after this allocation, the number of impressions covered by {*Electronics, Clothing*} reduces to 500. Through the example, we wanted to demonstrate how the proposed specialized-to-generalized allocation would work for advertisers who bid on *Shopping* considering a set of advertisers bid on children of *Shopping* and hence, the results for only A_1 and A_5 are shown. It should be noted that the matching between CPs and advertisers will be one-to-many as the number of impressions that can be covered by a CP is much large compared to demands of a single advertiser.

Considering the allocation done for Example 3, let us say a query related to the taxonomy is fired say, *Harry Potter*. As shown in Fig. 4, it will be first classified according to the taxonomy as *Shopping; Books; Fiction.* Advertisers who have been allotted a CP containing any of these nodes are considered for being displayed on this query's results page i.e. A_1, A_3, A_4 and A_5 would be considered to be displayed. The decision on who out these four would be shown and in which order will be decided by the ranking mechanism which includes their bids, remaining budget etc. (As stated earlier, ranking and bidding are independent of the proposed approach.)

6   Experiments

6.1   Dataset

For the experiments, we used the CABS120k08 [10] dataset which is a collection of search queries from the AOL500k dataset along with the documents clicked, document rank, timestamps and user id. The dataset models the web document as a unit. The data set also contains the classification of the clicked document according to a concept taxonomy of four levels. From the dataset, we extracted all the queries in the form: $< query, user-id, timestamp, concept\ taxonomy >$. Concept taxonomy present in the data is a four level taxonomy including the root node. Without loss of generality, we assumed that the search queries related to the documents also have the same category as the web document. The case where the same document had multiple categories, the first one was arbitrarily selected. After extracting queries, we extracted sessions of four most popular taxonomies – *Arts, Health, Society* and *Shopping* from the dataset that had more than a single query with at least two sub-concepts of the same concept in the same session. Each session is used a transaction to extract coverage patterns by T-Cmine as sessions form the logical boundary of searching. Table 2 shows the statistics of the extracted dataset.

6.2   Implementation Methodology

The standard sponsored search approach mentioned in [9] is compared with the concept based bidding approach. We simulate advertising demands randomly in terms of impressions for five sets of advertisers having 10, 20, 30, 40 and 50 advertisers. For the standard keyword bidding, a keyword is selected as the seed for each advertiser such that the probability of selection of a keyword as the seed is proportional to its frequency in the dataset, in order to mimic the advertising demand. Followed by selection of a seed keyword, all keywords from the dataset are selected to be in the advertiser's campaign for which the Wu-Palmer

Table 2. Search query dataset statistics

Taxonomy	Number of nodes	Sessions	Queries
Arts	48	7,107	15,317
Health	59	9,181	26,385
Society	68	6,471	13,223
Shopping	79	14,819	40,463
Total	254	37,578	95,388

similarity is more than 0.8. The number of requested impressions is randomly chosen between 100 and 1000. To simulate bid for each keyword, we consider the minimum bid as \$1.00, the maximum bid as \$10.00 and the actual bid for each keyword is considered as the function of its relative frequency between the minimum and maximum value. For the experimental setup, we assume the bid to be paid per hundred impressions instead of per 1000 impressions as in CPM model to analyse more number of requests. The bid amount here indicated how much the advertiser is willing to pay for 100 impressions. For the concept based bidding approach, bid of an advertiser on the concept is average of bids on all the keywords in his/her campaign.

6.2.1 Performance Metrics

Two performance metrics have been employed to compare the keywords based approach [9] and the proposed concept-based bidding approach.

To evaluate the utilization of ad space, we calculate the average number of unique Advertisements per Session (AS). It is calculated as the ratio of *Sum of Unique Advertisements of all Sessions (SUAS)* and *Number of Sessions with Advertisements (NSA)*. High value of AS indicates more utilization of a session, which in turn indicates covering of more advertisers.

$$AS = \frac{SU\ AS}{NSA} \tag{4}$$

We also measure the reach of each advertisement. Reach is defined as the number of users that view the ad. In this experiment, we consider reach of the ad with respect to the sessions instead of users as sessions define a logical boundary of tasks in search engines. To measure the reach, the value of Sessions per Advertisement (SA) is calculated which is the ratio of *Number of Unique Sessions for each Ad (NUSA)* to *Number of Advertisements (NA)*. A higher value of the metric implies the more number of unique eye balls and thus, increasing the chances of the advertisement being viewed by *diverse* users.

$$SA = \frac{N\ AS}{NA} \tag{5}$$

6.2.2   Results

Figure 6 reports the results with respect to ad space utilization. A fair improvement is observed in concept based bidding mechanism. Average improvement is 19.81% across all four taxonomies and all sets of advertisers. For individual taxonomies, average improvement for *Arts* is 18.33%, *Health* is 13.74%, *Society* is 17.29% *Shopping* is 29.86%. The improvement for *Shopping* show the highest improvement by a significant margin compared to the other three taxonomies. This is because for *Shopping* taxonomy average length of a session as well as distribution of nodes was higher compared to the other three taxonomies. Hence, it was possible to extract more interesting coverage patterns in the category of *Shopping*. These results align in the same way for the next performance metric as well.

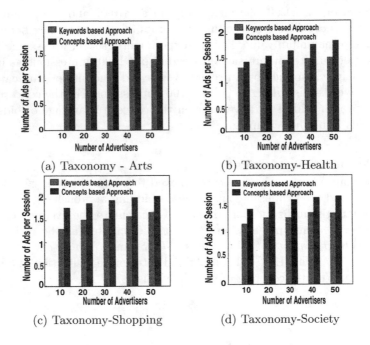

(a) Taxonomy - Arts

(b) Taxonomy-Health

(c) Taxonomy-Shopping

(d) Taxonomy-Society

Fig. 6. Performance with respect to utilization of ad space

Figure 7 shows the performance of two approaches with respect to reach of advertisements. An average improvement of 18% was observed. For individual taxonomies, improvement for *Arts* is 13.41%, *Health* is 14.83%, *Society* is 16.05% *Shopping* is 27.70%. The results for *Shopping* show significant improvements again because of the same reason as stated above.

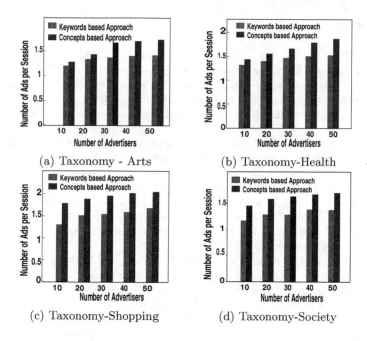

Fig. 7. Performance with respect to reach of advertisers

7   Conclusions and Future Work

In this paper, we address the issue of advertising on long tail search queries in search engines. We propose that advertisers should bid upon high level concepts represented by a taxonomy instead of search keywords during ad space auctions. To address the issues of inter-dependency of concepts on each other, we exploit search query logs and a taxonomy to extract level-wise coverage patterns. The corresponding architecture is used to perform allocation of incoming queries to advertisers for sponsored search. Experiments on a real world dataset of AOL search query logs show improvement in performance with respect to ad space utilization and reach of the advertisements.

As a part of future work, we plan to analyse what is the trade-off between relevance and bidding on concepts in terms of targeted advertising. Also, in this paper, we assumed that a taxonomy exists over search query logs. We plan to investigate how different taxonomies would suit the problem and if it is possible to build a taxonomy to suit sponsored search so to avoid the long tail phenomenon amongst the nodes of the taxonomy. We also intend to look at truthful auctions for concept-based bidding as the advertisers are targeting same keywords but using different concepts.

References

1. Srinivas, P.G., Reddy, P.K., Bhargav, S., Kiran, R.U., Kumar, D.S.: Discovering coverage patterns for banner advertisement placement. In: Tan, P.-N., Chawla, S., Ho, C.K., Bailey, J. (eds.) Advances in Knowledge Discovery and Data Mining. LNCS, vol. 7302, pp. 133–144. Springer, Heidelberg (2012)
2. Srikant, R., Agrawal, R.: Mining generalized association rules. Future Gener. Comput. Syst. **13**, 161–180 (1997)
3. Budhiraja, A., Reddy, P.K.: An approach to cover more advertisers in adwords. In: IEEE International Conference on Data Science and Advanced Analytics (DSAA 2015), 36678 2015, pp. 1–10 (2015)
4. Srinivas, P.G., Reddy, P.K., Trinath, A.V., Bhargav, S., Kiran, R.U.: Mining coverage patterns from transactional databases. J. Intell. Inf. Syst. **45**, 423–439 (2015)
5. Broder, A., Ciccolo, P., Gabrilovich, E., Josifovski, V., Metzler, D., Riedel, L., Yuan, J.: Online expansion of rare queries for sponsored search. In: International Conference on World Wide Web (2009)
6. Broder, A.Z., Fontoura, M., Gabrilovich, E., Joshi, A., Josifovski, V., Zhang, T.: Robust classification of rare queries using web knowledge. In: International ACM SIGIR Conference on Research and Development in Information Retrieval (2007)
7. Broder, A.Z., Ciccolo, P., Fontoura, M., Gabrilovich, E., Josifovski, V., Riedel, L.: Search advertising using web relevance feedback. In: Proceedings of the 17th ACM conference on Information and Knowledge management (2008)
8. Skiera, B., Eckert, J., Hinz, O.: An analysis of the importance of the long tail in search engine marketing. Electron. Commer. Res. Appl. **9**, 488–494 (2010)
9. Mehta, A., Amin, S., Umesh, V., Vijay, V.: Adwords and generalized online matching. J. ACM (JACM) **54**, 22 (2007)
10. Noll, M.G., Meinel, C.: The metadata triumvirate: Social annotations, anchor texts and search queries. In: IEEE/WIC/ACM International Conference on Web Intelligence and Intelligent Agent Technology, WI-IAT 2008, vol. 1, pp. 640–647 (2008)
11. Kavya, V.N.S., Reddy, P.K.: Coverage patterns-based approach to allocate advertisement slots for display advertising. In: Bozzon, A., Cudre-Maroux, P., Pautasso, C. (eds.) Web Engineering. LNCS, vol. 9671, pp. 152–169. Springer, Heidelberg (2016)
12. Mehta, A.: Online matching and ad allocation. Theor. Comput. Sci. **8**, 265–368 (2012)

String and Sequence Processing

Locating Longest Common Subsequences with Limited Penalty

Bin Wang[✉], Xiaochun Yang, and Jinxu Li

School of Computer Science and Engineering, Northeastern University,
Shenyang 110169, Liaoning, China
{binwang,yangxc}@mail.neu.edu.cn, lijinxu92@gmail.com

Abstract. Locating longest common subsequences is a typical and important problem. The original version of locating longest common subsequences stretches a longer alignment between a query and a database sequence finds all alignments corresponding to the maximal length of common subsequences. However, the original version produces a lot of results, some of which are meaningless in practical applications and rise to a lot of time overhead. In this paper, we firstly define longest common subsequences with limited penalty to compute the longest common subsequences whose penalty values are not larger than a threshold τ. This helps us to find answers with good locality. We focus on the efficiency of this problem. We propose a basic approach for finding longest common subsequences with limited penalty. We further analyze features of longest common subsequences with limited penalty, and based on it we propose a filter-refine approach to reduce number of candidates. We also adopt suffix array to efficiently generate common substrings, which helps calculating the problem. Experimental results on three real data sets show the effectiveness and efficiency of our algorithms.

Keywords: Longest common subsequence · Penalty score · Common substring

1 Introduction

The longest common subsequence (LCS) problem is a classic and well studied problem in computer science with extensive applications in diverse areas ranging from spelling error corrections to molecular biology. Especially in bioinformatics, LCS is the most important metric in all of local alignments, which are used for comparing primary biological sequence information, such as the amino-acid sequences of proteins or the nucleotides of DNA sequences. Locating LCS enables a researcher to compare a query sequence with a library or database of sequences, and identify library sequences that resemble the query sequence. The longest

---

This work is partially supported by the NSF of China for Outstanding Young Scholars under grant No. 61322208, the NSF of China under grant Nos. 61272178 and 61572122.

S. Candan et al. (Eds.): DASFAA 2017, Part II, LNCS 10178, pp. 187–201, 2017.
DOI: 10.1007/978-3-319-55699-4_12

common subsequence problem for two strings, is to find a common subsequence in both strings, having maximum possible length, where a subsequence of a string is obtained by deleting zero or more symbols of that string. In this paper, we require to get their matching regions for the strings.

The original version of LCS stretches a longer alignment between the query and the database sequence in the left and right directions, from the position where the exact match occurred. The extension does not stop until the accumulated threshold. However, in practice, the original LCS produces too many alignments, some of which are meaningless. For example, bio-scientists prefer to find matches of bio-sequences locally (i.e. within a small region). To tackle this problem, in this paper we propose a new version of LCS, called longest common subsequences with limited penalty, denoted LCSP. LCSP adopts a penalty threshold to maintain the same level of sensitivity for detecting sequence similarity.

The main challenges and contributions of this paper are listed as follows:

- In order to satisfy the requirement of real applications, we propose a new version of longest common subsequences with limited penalty score in Sect. 3. In order to be consistent with the alignment problem in bio-sequences, we adopt the flexible scoring scheme in bio-applications to quantify penalties in the LCS.
- Obviously, generating LCSs using dynamic programing and checking every generated LCS under the penalty threshold are time consuming. We propose an approach by concatenating common substrings to avoid the dynamic programming in Sect. 4. Furthermore, this could be help to generate small number of LCSs for checking using the penalty threshold. This algorithm can retrieve all correct results, and is thus an exact algorithm.
- The number of concatenated common substrings could be large, especially when the given strings are long. In order to reduce this number, we propose a filter-refine approach to further improve our algorithm, which can avoid useless concatenated common substrings and early terminate calculations in Sect. 5. We also in Sect. 6 to show how to efficiently find common substrings by constructing suffix array index structure.
- We conduct experimental evaluations on three real data sets with different alphabets, lengths, and distributions to test and analyze our algorithms in Sect. 7. The results demonstrate the effectiveness and efficiency of our proposed algorithms.

2   Related Work

A lot of research efforts have been made to design algorithms for string alignment, such as Needleman-Wunsch [14], Smith-Waterman [18], and their corresponding improvements OASIS [14], BWT-SW [11] and ALAE [21], all of which are based on dynamic programming. When conducting sequence comparison, these algorithms consider exact matching, as well as insertion, deletion and substitution, and assign a score scheme to these transformation operations. The goal of sequence matching is to find out the optimal matching, i.e. maximizing the number of matches and minimizing the number of spaces and mismatches. These algorithms usually suffer large space consumption, requiring a space complexity of $O(mn)$ [17], where m and n are the lengths of the two strings.

Although improvements have been made to reduce space complexity and enhance running efficiency using suffix array, they might return unsatisfactory results, which is caused by inappropriate score setting. Such a result typically acquires a decent score in their forepart, but confronts a score drop in the mid-part because of mismatches, and gets a relatively high score in the last part. As a consequence, these results usually end up with high overall scores, but are still unsatisfactory since their mismatched mid-parts are inconsistent with users' actual demands. Edit distance based approaches [8] retrieve dissimilar parts of the two strings, and restore them to the original strings, based on which the string similarity is evaluated. Other algorithms like BLAST [10] firstly acquire exact matched part of the two strings, then expand it to left/right, and form high-score matched sequences. In spite of their higher efficiency compared to dynamic programming based algorithms, they cannot guarantee retrieving all high-score segments without omission.

The classic dynamic programming solution to LCS problem, invented by Wagner and Fischer [19], has $O(mn)$ worst case running time. To reduce the space complexity, Hirschberg [4] provide an algorithm with $O(n)$ worst space cost, using a divide-and-conquer approach. The fastest known algorithm by Masek and Paterson [13] runs in $O(n^2/\log n)$ time. However, faster algorithms exist with complexities depending on special cases, such as when the input consists of permutations or when the output is known to be very long or very short. For example, Myers in [15] and Nakatsu et al. in [16] presented an $O(nB)$ algorithm, where the parameter B is the simple Levenshtein distance between the two given strings [12]. Hunt and Szymanski [3] studied the complexity of the LCS problem in terms of matching index pairs, i.e., they defined t to be the number of index-pairs (i, j) with $a_i = b_j$ (such a pair is called a match) and designed an algorithm that finds the LCS of two sequences in $O(t \log n)$ time. For a survey on the LCS problem see [2].

The rest of this paper is structured as follows: Sect. 3 elaborates the preliminaries of this paper and the problem definition; Sect. 4 proposes a basic approach for finding longest common subsequences with limited penalty. In Sect. 5, we analyze features of longest common subsequences with limited penalty, and based on it we propose a filter-refine approach to reduce number of candidates. And in Sect. 6 we adopt suffix array to efficiently generate common substrings, which helps calculating the problem. In Sect. 7 we present experimental results on real data sets to demonstrate the accuracy and time efficiency of the proposed technique. Then finally in Sect. 8, we conclude the paper.

3   Preliminaries and Problem Definition

Let Σ be an alphabet. For a string X of the characters in Σ, we use $|X|$ to denote the length of X, $X[i]$ to denote the i-th character of X (starting from 1), and $X[i \ldots j]$ to denote the substring from its i-th character to its j-th character.

A subsequence of a string is obtained by deleting zero or more symbols of that string. The longest common subsequence problem for two strings, is to find a common subsequence in both strings, having maximum possible length.

Definition 1. *Longest common subsequence (LCS). Given two strings* $X = X[1]X[2]\ldots X[m]$ *and* $Y = Y[1]Y[2]\ldots Y[n]$. *A subsequence* $X[i_1]X[i_2]\ldots X[i_r]$ *of* X *(* $0 < i_1 < i_2 < \ldots < i_r \le m$ *) is obtained by deleting* $m - r$ *symbols from* X. *A common subsequence of two strings* X *and* Y, *denoted* $cs(X,Y)$, *is a subsequence common to both* X *and* Y. *The longest common subsequence of* X *and* Y, *denoted* $lcs(X,Y)$ *or* $LCS(X,Y)$, *is a common subsequence of maximum length. We denote the length of* $lcs(X,Y)$ *by* $|lcs(X,Y)|$.

For example, for the two strings $X = \mathtt{traobcybgsfd}$ and $Y = \mathtt{tracycy}$ $\mathtt{raogsfdy}$, $lcs(X,Y) = \mathtt{tracygsfd}$. Based on the alignment of longest common subsequence, there exist three common substrings $\mathtt{tra}$, $\mathtt{cy}$, and $\mathtt{gsfd}$ along the alignment of $lcs(X,Y)$. We use (X^a, Y^b, l) to represent that $X[a \ldots a{+}l{-}1]$ and $Y[b \ldots b{+}l{-}1]$ share a common substring, where a and b are start positions of the matching substring in X and Y, respectively. In between every two adjacent common substrings, there is an uncommon substring pair $\langle X_i, Y_i \rangle$. We use penalty to evaluate the difference between all uncommon substrings in a longest common subsequence.

Definition 2. *Penalty of an LCS. Given two strings* X *and* Y, *let* $\langle X_1, Y_1 \rangle, \ldots, \langle X_k, Y_k \rangle$ *be the pairs of uncommon substrings in* $lcs(X,Y)$. *The penalty of* $lcs(X,Y)$ *is defined as:*

$$p(lcs(X,Y)) = \sum_{i=1}^{k} \alpha \cdot M(X_i, Y_i) + \beta \cdot S(X_i, Y_i), \tag{1}$$

where $M(X_i, Y_i)$ *and spaces* $S(X_i, Y_i)$ *represent the number of mismatches and spaces between* X_i *and* Y_i, *and* (α, β) *is the scoring scheme where* α *and* β *are penalty scores of a mismatch and a space, respectively.*[1]

For ease of presentation, we use Fig. 1 to show the penalties of different LCSs. From this figure, we can easily see that there are two alignments corresponding to the same LCS subsequence $lcs(X,Y) = \mathtt{tracygsfd}$. The first alignment consists of common strings $\mathtt{tra}$ with $(X^1, Y^1, 3)$, $\mathtt{cy}$ with $(X^6, Y^4, 2)$, and $\mathtt{gsfd}$ with $(X^9, Y^{11}, 4)$, and its penalty is $2\beta + (\alpha + 4\beta)$. The second alignment consists of $\mathtt{tra}$ with $(X^1, Y^1, 3)$, $\mathtt{cy}$ with $(X^6, Y^6, 2)$, and $\mathtt{gsfd}$ with $(X^9, Y^{11}, 4)$, and its penalty is $2\alpha + (\alpha + 2\beta)$. When both α and β equals 1, these two alignment have different penalties 7 and 5.

---

[1] Notice that, the definition of penalty score of LCS is different from edit distance even when $\alpha = \beta = 1$. The edit distance between two strings represents the *minimal* number of edit operations transforming from one string to another string, which does not guarantee to find an alignment with longest common subsequences as LCS does.

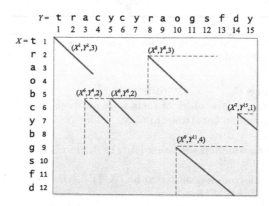

Fig. 1. Common substrings of $X = \texttt{traobcybgsfd}$ and $Y = \texttt{tracycyraogsfdy}$.

Problem Definition. The problem of longest common subsequence with limited penalty (a.k.a. LCSP) is to locate positions of every exact matching substrings along the longest common subsequences whose penalty is not greater than τ given two strings X and Y, and a penalty threshold τ, denoted $lcsp(X, Y, \tau)$.

4   A Basic Approach Based on Common Substrings

A straightforward approach of locating LCSP includes the following three steps: (i) calculate LCS using dynamic programming in $O(mn)$ time, where m and n are string lengths of the two given strings X and Y; (ii) get all possible alignments along the alignment of $lcs(X, Y)$ in $O(n \log n)$ (assuming $m \leq n$); and (iii) keep alignments whose penalty is not greater than the given penalty threshold. Therefore, the total cost is $O(mn)$.

Obviously, generating all LCSs firstly and then checking their penalties are time consuming. Now we propose our approach based on common substrings (We discuss how to get common substrings of X and Y in $O(n)$ time in Sect. 6). The basic idea of our approach is to start from the common substrings of X and Y since the final results must contain certain substring pair in the set of common substrings of X and Y. Then we concatenate the common substrings to get longer substring pairs of X and Y (lines 4–9), and verify each concatenated substring pair by calculating its penalty (lines 10–12). We call this baseline approach BASICLCSP (see Algorithm 1).

Reexamine the two strings $X = \texttt{traobcybgsfd}$ and $Y = \texttt{tracycyraogsfdy}$ and their common substrings $\texttt{tra}$ with $(X^1, Y^1, 3)$, $\texttt{rao}$ with $(X^2, Y^8, 3)$, $\texttt{cy}$ with $(X^6, Y^4, 2)$ and $(X^6, Y^6, 2)$, $\texttt{y}$ with $(X^7, Y^{15}, 1)$, and $\texttt{gsfd}$ with $(X^9, Y^{11}, 4)$. The algorithm BASICLCSP firstly puts these common substrings in a candidate set C_{set}. Secondly, it gets 10 concatenated substring pairs from the above common substrings, which are $\langle X[1 \ldots 7], Y[1 \ldots 5] \rangle$, $\langle X[1 \ldots 7], Y[1 \ldots 7] \rangle$, $\langle X[1 \ldots 4], Y[1 \ldots 10] \rangle$, $\langle X[1 \ldots 7], Y[1 \ldots 15] \rangle$, $\langle X[1 \ldots 12], Y[1 \ldots 14] \rangle$, $\langle X[6 \ldots 12], Y[4 \ldots 14] \rangle$, $\langle X[6 \ldots 7], Y[4 \ldots 15] \rangle$, $\langle X[6 \ldots 7], Y[6 \ldots 15] \rangle$,

---

Algorithm 1. BASICLCSP

Input: X and Y: Two strings; C: A set of common substrings; τ: A given penalty threshold

Output: $lcsp(X, Y, \tau)$

1   Common substrings $C_{set} \leftarrow$ CALCOMSTR(X, Y);
2   Rank strings in C_{set} in the order of their start positions in ascending order;
3   $k \leftarrow$ number of common substrings in C_{set};
4   **for** $i = 1; i < k; i + +$ **do**
5      $str_c \leftarrow$ the i-th common substring (X^a, Y^b, l_i) in C_{set};
6   **for** $j = i + 1; j \le k; j + +$ **do**
7   Let the j-th common substring be (X^c, Y^d, l_j);
8   **if** $a < c$ && $b < d$ **then**
9   Generate a candidate substring X' start from str_c and end at the j-th common string in C_{set};
10   **if** *penalty of* $X' \le \tau$ **then**
11               $Can \leftarrow X'$;
12               $str_c \leftarrow X'$;

13 **return** *the longest string in* Can;

$\langle X[2 \ldots 7], Y[8 \ldots 15] \rangle$, and $\langle X[2 \ldots 12], Y[8 \ldots 14] \rangle$. The algorithm keeps the concatenated substring as a candidate if its penalty $\le \tau$. Finally it returns the longest candidate as $lcsp(X, Y, \tau)$.

The algorithm BASICLCSP is correct. Any LCS of two strings X and Y must contain their common substrings and it must start from one common substring and end at another common substring. The time complexity of BASICLCSP is $O(k^2)$, where k is the number of common substrings of X and Y.

5   Reducing Number of Concatenated Common Substrings

The algorithm BASICLCSP enumerates all possible concatenated common substrings. Some of them will not generate the LCSP. To locate LCSP efficiently, we propose a *filter-refine* approach, called IMPROVEDLCSP. We first analyze the feature of LCSP, based on which we carefully prune those common substrings that could not generate LCSP. We propose one filtering in Sects. 5.1 and an early termination approach to avoid useless calculations in Sect. 5.2.

5.1   Avoiding Useless Concatenation of Common Substrings

property 1. Let A be an alignment of an LCSP of X and Y under the penalty threshold τ. For any two common substrings C_1 with (X^a, Y^b, l_1) and C_2 with (X^c, Y^d, l_2) in A ($a < c, b < d$). The penalty of the concatenated substring pair $\langle X[a, c+l_2-1], Y[b, d+l_2-1] \rangle$ must satisfy $p(X[a, c+l_2-1], Y[b, d+l_2-1]) \le \tau$.

Lower Bound of Penalty. Let C_1 with (X^a, Y^b, l_1) and C_2 with (X^c, Y^d, l_2) be two common substrings of strings X and Y, if there does not exist any common substring C with (X^e, Y^f, l_3) $(a < e < c, b < f < d)$, the lower bound of penalty of concatenating C_1 and C_2 is

$$LB(C_1, C_2) = \min(\alpha, \beta) \cdot \max(c - a - l_1, d - b - l_1). \qquad (2)$$

Theorem 1. *Two common substrings C_1 with (X^a, Y^b, l_1) and C_2 with (X^c, Y^d, l_2) cannot belong to the same alignment of an LCSP if there does not exist any common substring C with (X^e, Y^f, l_3) $(a < e < c, b < f < d)$ and $LB(C_1, C_2) > \tau$*

Proof. Assume C_1 and C_2 belong to the same alignment when Eq. 2 holds. Since there does not exist any common substring C with (X^e, Y^f, l_3) $(a < e < c, b < f < d)$, we let $\langle X_i, Y_i \rangle$ be the uncommon substring pair in between C_1 and C_2, then according to Eq. 1, we know $\alpha \cdot M(X_i, Y_i) + \beta \cdot S(X_i, Y_i) \leq \tau$. Since $\min(\alpha, \beta) \max(|X_i|, |Y_i|) \leq M(X_i, Y_i) + \beta \cdot S(X_i, Y_i)$, and $|X_i| = c - a - l_1$, $|Y_i| = d - b - l_1$, we can see the above assumption does not hold.

Based on Theorem 1, we can prune concatenated substrings that satisfy Eq. 2. For example, let $\alpha = 1, \beta = 1$. Given $\tau = 5$, it is useless to concatenate the two common substrings with $(X^6, Y^4, 2)$ and $(X^7, Y^{15}, 1)$ since $\min(\alpha = 1, \beta = 1) \cdot \max(7 - 6 - 2, 15 - 4 - 2) = 9 > \tau$. The same reason, we do not concatenate $(X^1, Y^1, 3)$ with $(X^7, Y^{15}, 1)$, and $(X^6, Y^6, 2)$ with $(X^7, Y^{15}, 1)$.

5.2   Early Termination of Calculations

Since we want to find the longest common subsequences with limited penalty, we prefer to check two common substrings with longest position distance. Therefore, instead of storing common substrings in an ordered set, we store them in a lattice such that we can use it to easily prune useless concatenation and early terminate the calculation of LCSP.

A Lattice Structure. We use a lattice to store all common substrings of X and Y. Each node in the lattice represents a common substring. Consider any two common substrings with (X^a, Y^b, l_1) and (X^c, Y^d, l_2). If $a < c$ and $b < d$, we call (X^a, Y^b, l_1) *dominates* (X^c, Y^d, l_2). Furthermore, if there does not exist any other common substring with (X^e, Y^f, l_3) such that $a < e < c$, $b < f < d$, then there is an edge from (X^a, Y^b, l_1) to (X^c, Y^d, l_2), we call (X^a, Y^b, l_1) *strictly dominates* (X^c, Y^d, l_2). We label the edge between any two common substrings with strictly dominate relationship using the penalty of its corresponding uncommon substring.

Figure 2 shows an example of the lattice for our running example. There is an edge in between $(X^1, Y^1, 3)$ and $(X^6, Y^4, 2)$, and the edge is labelled 2 since the penalty of concatenating these two common substrings is 2. Therefore, the penalty of concatenating strings along the path $(X^1, Y^1, 3)$, $(X^6, Y^4, 2)$, and $(X^7, Y^{15}, 1)$ is $2 + 7 = 9$.

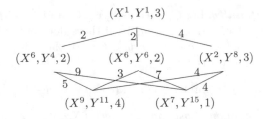

Fig. 2. A lattice of common substrings of $X = \texttt{traobcybgsfd}$ and $Y = \texttt{tracycyraogsfdy}$ when $\alpha = 1$ and $\beta = 1$.

Pruning Useless Concatenation Using the Lattice for Common Substrings. From the above example, we can see that the lattice structure can easily identify the strict dominate relationship between any two common substring pairs. When the summation of labels in a path is greater than τ, we do not need to concatenate common substrings along the path.

Algorithm 2 shows a pruning algorithm based on depth-first-search (DFS). For every search step, it adjusts the permitted penalty score so that we could avoid traversing those paths that could not generate LCSP.

Algorithm 2. PRUNE(L, τ)

Input: A lattice L for common substrings of X and Y, penalty threshold τ;
Output: A pruned lattice;

    // start from the root of L and traverse L using DFS
1 **if** L *is a single node* **then**
2   ⌊ return L;

3 **foreach** *node v pointed by the root r of L* **do**
4   **if** $edge(r,v) > \tau$ **then**
5   ⌊ remove the edge from r to v in L;
6   ⌊ PRUNE($L, \tau - edge(r,v)$);

Choosing a Good Calculation Order. In fact, we are only interested in the longest common subsequences whose penalty is not greater than τ, therefore, it is no need to calculate those common subsequences with shorter lengths.

Aiming at this target, we reorganize children of each node in the lattice by ranking their lengths in descending order. Then by using the DFS, the path with longest untraversed common strings will take precedence. When the first $lcsp(X, Y, \tau)$ is found, we are safe to early terminate all calculations since the later calculations can only generate a common subsequence with shorter length.

6   Efficiently Constructing Common Substrings

We can use dynamic programming to get all common substrings of X and Y in $O(mn)$ time. In order to accelerate this process, we can also use the suffix tree [20]

```
0 1 2 3 4  5  6 7 8 9 10 11
a b f a b #₁ a b e a b  #₂
```
(a) Strings with the text array.

i	$SA[i]$	SA^{-1}	$LCP[i]$	$text[i]$
0	5	5	0	1
1	11	9	0	2
2	3	11	0	1
3	9	2	2	2
4	6	6	2	2
5	0	0	2	1
6	4	4	0	1
7	10	8	1	2
8	7	10	1	2
9	1	3	1	1
10	8	7	0	2
11	2	1	0	1

(b) The SA and LCP array for T=abfab#₁abeab#₂.

Fig. 3. An example of SA array.

or suffix array [5]. Compared with the suffix tree, suffix array can be configured in linear time [9] and small space cost [7], so we consider the establishment of suffix array index structure.

Given a string X, its suffix array SA records the start positions of all the suffixes of one string. Since the suffixes are sorted lexicographically, $SA[i]$ is the start position of the i-th suffix based on the lexicographical order. The suffix array of string T, denoted as SA, is actually an array with integer from 1 to n, revealing the dictionary order of n suffixes. $T_{SA[i]}$ denotes the $SA[i]$th suffix $T[i \ldots n]$. The inverse SA^{-1} of suffix array is also an integer array, satisfying $SA^{-1}[SA[i]] = i$ ($1 \leq i \leq n$). Obviously, the inverse of suffix array can also be constructed in linear time.

LCP array is used for maintaining the length of the longest common prefix of two adjacent suffix in SA. Suppose we use $lcp(u, v)$ to denote the length of the longest common prefix of u and v, then $LCP[1] = 0$ and $LCP[i] = lcp(T_{SA[i-1]}, T_{SA[i]})$ where $2 \leq i \leq n$. Based on the suffix array and its inverse, this LCP array can be constructed in linear time [6].

Given two strings X and Y, we add #₁ and #₂ to their ends, respectively, forming a new string $T = X$#₁Y#₂. Suppose #₁ < #₂, and all characters in the string collection are larger than these separators according to dictionary order. We define the suffix array of the new string T and the related LCP array as the generalized augmented suffix array [1], which is consistent with the suffix arrays of X and Y. This approach can be done in $O(m + n)$ time.

For example, let X = abfab and Y = abeab, then T = abfab#₁abeab#₂, with their index starting from 0 as shown in Fig. 3(a). The SA array, SA^{-1} array and LCP array of T are all shown in Fig. 3. The position and length of the common substring in the outer matrix can be figured out according to the LCP array, which reduce the computation of mismatch in outer matrix. We can compute four common substrings ab with $(X^0, Y^6, 2)$, $(X^0, Y^9, 2)$, $(X^3, Y^6, 2)$, and $(X^3, Y^9, 2)$ in linear time, with the help of LCP array and the dynamic programming programming for common substrings of X and Y.

7   Experiments

In this section, we evaluated the effect of the different factors on the performance and used the following three data sets in the experiments.

- **Genome data set.** This data set contains human's first genome data, from which we randomly selected 1000 strings of various lengths as data strings. The query strings were generated similarly from mice genome data. We generated a query workload with 50 query strings.
- **DBLP data set.** We generated this data set from DBLP. It includes 1,632,442 papers, and each paper contains some of the properties of paper, such as title, author, abstract, and etc. We randomly selected 1000 strings of various lengths as data strings and randomly picked up 50 strings to construct a query workload.
- **AOL query log data set.** It contains the web pages from a large number of users Query records sorted by anonymous user IDs. Each record includes anonymous user ID, the contents of the query, and query time. The length of the records are from 20 and 100. We randomly chose 50 contents of queries to construct its query workload.

Our experimental results were run on Ubuntu (Linux) 13.10 with Interl (R) Core (TM) i7 CPU 870@2.93 GHZ 8 GB RAM. All the algorithms were implemented using GNU C++.

7.1   Evaluation of Effectiveness

We define *Locality Degree LD* to evaluate the effectiveness of LCSP as follows.

$$LD = \frac{avg(\sum l(lcs_p))}{avg(\sum l(lcs) - \sum l(lcs_p))},$$

where $l(lcs_p)$ represents the length of matching substrings generated by using LCSP, $l(lcs)$ represents the length of matching substrings generated by using LCS, and $avg(\cdot)$ is the average value. Notice that, we require that both LCSP and LCS generate the same longest common subsequences under the given penalty threshold.

Figure 4 shows the Locality Degree when increasing the penalty threshold ratio, which is the percentage of average data string length. We can see when getting the same longest common subsequence, LCSP prefers to find meaningful matching substrings with shorter lengths. Figure 4(a) shows the locality degree LD was very low, only less than 0.1 on DNA data set, which means that LCS

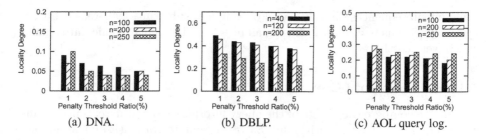

(a) DNA.　　　　　　　(b) DBLP.　　　　　　　(c) AOL query log.

Fig. 4. Effectiveness of LCSP.

generates much more meaningless results compared with LCSP. The locality degree on DBLP data set was less than 0.5, which was higher than it on DNA data set, since the selectivity of DBLP data was much less than the selectivity of DNA data. As the penalty threshold ratio increased, the locality degree on three data sets decreased since the smaller penalty threshold was, the more locality was required.

7.2 Comparison with Other Algorithms

We chose two state-of-the-art LCS algorithms DPA [2] and LIS [1], and modified them to support LCSP as discussed in Sect. 4. We call these modified LCS-based algorithms DPALCSP and LISLCSP. We compared the running time of our two algorithms (i.e. BASICLCSP and IMPROVEDLCSP) with these two LCS-based algorithms.

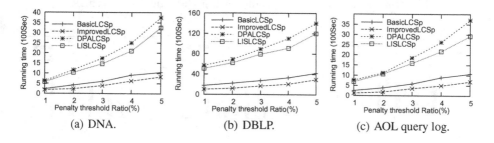

(a) DNA.　　　　　　　(b) DBLP.　　　　　　　(c) AOL query log.

Fig. 5. Comparison of different algorithms.

Figure 5 shows the comparison results of the four algorithms. We can see that the algorithm DPALCSP was the slowest algorithm. Both BASICLCSP and IMPROVEDLCSP algorithms ran much faster than LCS-based algorithms. When increasing the penalty threshold ratio, the running time increased. This is consistent with our expectation since more concatenated common substrings would be generated when the penalty threshold increases, incurring larger time cost for calculation. The running time of our algorithms kept more stable than both DPALCSP and LISLCSP since our algorithms can generate the longest common subsequences directly based on common substrings, whereas the LCS-based algorithms had to back chasing all possible alignments to calculate results, which were costly.

7.3 Evaluation of LCSP

In order to compare the effects of the query filtering among different algorithms, we define *Filtered Ratio* (FR) and *Early Terminate Ratio* (ETR) as follows. Filtered Ratio (FR) is the proportion that a number of pruned concatenated common substrings to the whole number of concatenated common substrings. Early Terminate Ratio (ETR) is the proportion that traversed nodes to the nodes in the lattice.

To measure the performance of algorithms, we take filtered ratio, early terminate ratio, and running time as the three metrics to evaluate pruning power, effect of early termination, and efficiency of our algorithms, respectively. It is obvious that a favored algorithm with high efficiency should have large filtered area ratio, larger early terminate ratio, and small running time. We use the scoring scheme $\alpha = 1$ and $\beta = 1$. We got similar results when varying α and β.

Pruning Power. We conducted experiments on strings with different lengths. The lengths of query string and data string were comparable. The detailed filtering ratios for three data sets are shown in Fig. 6. In Fig. 6(a), when the length of one string is fixed to 20×10^6, and the length of another string increases from 5×10^6 to 25×10^6, the filtering ratio is significant, increasing from 36.9% to 42.7%. Also, by comparing results on three different data sets in Fig. 6, we can see that the filtering ratios raise when increasing the lengths of strings.

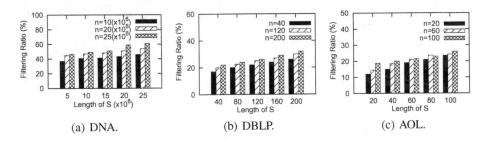

(a) DNA.   (b) DBLP.   (c) AOL.

Fig. 6. Pruning power.

Effect of Early Termination. The effect of early termination is related to string length, as well as the penalty threshold. Therefore we evaluated the impact of these two factors in this section.

Figure 7 shows how early terminate ratio would be affected with increasing string lengths. From Fig. 7, we can see that when the lengths of strings are generally comparable, the early terminate ratio increases with the increase of string length. The reason is that, for both strings, longer strings generally indicate larger probabilities of more common substrings.

Figure 8 shows the impact of the variance of penalty threshold ratios on the early termination. To be more specific, seen from Fig. 8, when string length is fixed, the larger the penalty threshold, the greater the early terminate ratio, especially for DBLP and ALO query log data sets (see Fig. 8(b) and (c)). It is because the larger the penalty threshold, the more the candidate starting from the same common substring, leading to a large number concatenated substrings, thus a larger early terminate ratio. Notice that, the early terminate ratio did not increase significantly when increasing the penalty threshold ratio since the distribution of frequencies for different substrings were similar, therefore, the number of concatenated substrings kept stable when increasing the penalty threshold ratio.

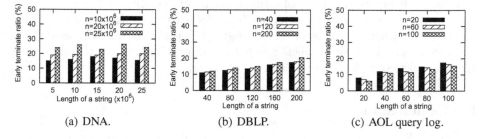

(a) DNA.   (b) DBLP.   (c) AOL query log.

Fig. 7. Effect of early termination with different string lengths.

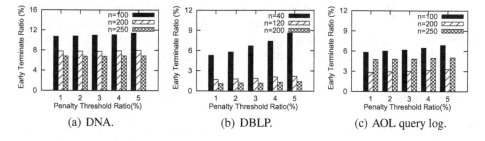

(a) DNA.   (b) DBLP.   (c) AOL query log.

Fig. 8. Effect of early termination with different penalty threshold.

Running Time. We also test the efficiency of our algorithms when varying the lengths of strings. Figure 9 reports the running time of IMPROVEDLCSP for different lengths of strings on DNA sequences when the penalty threshold $\tau = 2, 6, 10$, respectively. The results on DBLP and AOL query log data sets are similar.

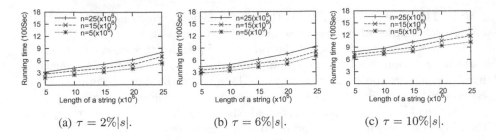

(a) $\tau = 2\%|s|$.   (b) $\tau = 6\%|s|$.   (c) $\tau = 10\%|s|$.

Fig. 9. The performance of IMPROVEDLCSP on DNA data set.

Figure 9(a) shows when $\tau = 2\%$ of string length, with the increase of the lengths of strings, the running time also increased from 259 s to 676 s. As can be seen in Fig. 9(b), when $\tau = 6\%$ of string length, with the increase of the lengths of strings, the running time increased from 380 s to 814 s. In Fig. 9(c), it can be seen that when $\tau = 10\%$ of string length, the running time increased from 722 s to 1,183 s. In a word, when τ is fixed, the running time of the algorithm IMPROVEDLCSP is linear to the lengths of strings.

8   Conclusion

In this paper, we propose a new problem, the longest common subsequence with limited penalty to get LCSs with good locality. We show that the existing LCS-based algorithms are not efficient since they have to back chasing alignments to do verifications. In order to avoid checking each generated LCS using the penalty threshold, we propose an approach based on common substrings. By improving the basic algorithm, we propose a filter-refine approach that can reduce the number of concatenated common substrings. It can efficiently prune useless concatenations of common substrings and early terminate calculations. Our experimental study demonstrate its effectiveness and efficiency.

References

1. Arnold, M., Ohlebusch, E.: Linear time algorithms for generalizations of the longest common substring problem. Algorithmica **60**(4), 806–818 (2011)
2. Bergroth, L., Hakonen, H., Raita, T.: A survey of longest common subsequence algorithms. In: Seventh International Symposium on String Processing and Information Retrieval, SPIRE 2000, A Coruña, Spain, pp. 39–48, 27–29 September 2000
3. Brodal, G.S., Kaligosi, K., Katriel, I., Kutz, M.: Faster algorithms for computing longest common increasing subsequences. In: Proceedings of the 17th Annual Symposium on Combinatorial Pattern Matching, CPM 2006, Barcelona, Spain, pp. 330–341, 5–7 July 2006
4. Hirschberg, D.S.: A linear space algorithm for computing maximal common subsequences. Commun. ACM **18**(6), 341–343 (1975)

5. Kärkkäinen, J., Sanders, P.: Simple linear work suffix array construction. In: Baeten, J.C.M., Lenstra, J.K., Parrow, J., Woeginger, G.J. (eds.) ICALP 2003. LNCS, vol. 2719, pp. 943–955. Springer, Heidelberg (2003). doi:10.1007/3-540-45061-0_73

6. Kasai, T., Lee, G., Arimura, H., Arikawa, S., Park, K.: Linear-time longest-common-prefix computation in suffix arrays and its applications. In: Proceedings of the 12th Annual Symposium on Combinatorial Pattern Matching, CPM 2001, Jerusalem, Israel, pp. 181–192, 1–4 July 2001

7. Kim, D.K., Sim, J.S., Park, H., Park, K.: Linear-time construction of suffix arrays. In: Proceedings of the 14th Annual Symposium on Combinatorial Pattern Matching, CPM 2003, Morelia, Michocán, Mexico, pp. 186–199, 25–27 June 2003

8. Knuth, D.E., Morris Jr., J.H., Pratt, V.R.: Fast pattern matching in strings. SIAM J. Comput. **6**(2), 323–350 (1977)

9. Ko, P., Aluru, S.: Space efficient linear time construction of suffix arrays. J. Discrete Algorithms **3**(2–4), 143–156 (2005)

10. Korf, I., Yandell, M., Bedell, J.A.: BLAST - An Essential Guide to the Basic Local Alignment Search Tool. O'Reilly, Sebastopol (2003)

11. Lam, T.W., Sung, W., Tam, S., Wong, C., Yiu, S.: Compressed indexing and local alignment of DNA. Bioinformatics **24**(6), 791–797 (2008)

12. Levenshtein, V.I.: Binary codes capable of correcting spurious insertions and deletions of ones. Probl. Inf. Transm. **1**(1), 817 (1965)

13. Masek, W.J., Paterson, M.: A faster algorithm computing string edit distances. J. Comput. Syst. Sci. **20**(1), 18–31 (1980)

14. Meek, C., Patel, J.M., Kasetty, S.: OASIS: an online and accurate technique for local-alignment searches on biological sequences. In: VLDB, pp. 910–921 (2003)

15. Myers, E.W.: An O(ND) difference algorithm and its variations. Algorithmica **1**(2), 251–266 (1986)

16. Nakatsu, N., Kambayashi, Y., Yajima, S.: A longest common subsequence algorithm suitable for similar text strings. Acta Inf. **18**, 171–179 (1982)

17. Overill, R.E.: Book review: "time warps, string edits, and macromolecules: the theory and practice of sequence comparison" by David Sankoff and Joseph Kruskal. J. Log. Comput. **11**(2), 356 (2001)

18. Smith, T.F., Waterman, M.S.: Identification of common molecular subsequences. J. Mol. Biol. **147**(1), 195–197 (1981)

19. Wagner, R.A., Fischer, M.J.: The string-to-string correction problem. J. ACM **21**(1), 168–173 (1974)

20. Weiner, P.: Linear pattern matching algorithms. In: 14th Annual Symposium on Switching and Automata Theory, Iowa City, Iowa, USA, pp. 1–11, 15–17 October 1973

21. Yang, X., Liu, H., Wang, B.: ALAE: accelerating local alignment with affine gap exactly in biosequence databases. PVLDB **5**(11), 1507–1518 (2012)

Top-k String Auto-Completion with Synonyms

Pengfei Xu$^{(\boxtimes)}$ and Jiaheng Lu

Department of Computer Science, University of Helsinki, Helsinki, Finland
{pengfei.xu,jiaheng.lu}@helsinki.fi

Abstract. Auto-completion is one of the most prominent features of modern information systems. The existing solutions of auto-completion provide the suggestions based on the beginning of the currently input character sequence (i.e. prefix). However, in many real applications, one entity often has synonyms or abbreviations. For example, "DBMS" is an abbreviation of "Database Management Systems". In this paper, we study a novel type of auto-completion by using synonyms and abbreviations. We propose three trie-based algorithms to solve the top-k auto-completion with synonyms; each one with different space and time complexity trade-offs. Experiments on large-scale datasets show that it is possible to support effective and efficient synonym-based retrieval of completions of a million strings with thousands of synonyms rules at about a microsecond per-completion, while taking small space overhead (i.e. 160–200 bytes per string). The implementation of algorithms is publicly available at http://udbms.cs.helsinki.fi/?projects/autocompletion/download.

1 Introduction

Keyword searching is a ubiquitous activity performed by millions of users daily. However, cognitively formulating and physically typing search queries is a time-consuming and error-prone process [3,6]. In response, keyword search engines have widely adopted auto-completion as a means of reducing the efforts required to submit a query. As users enter their query into the search box, auto-completion suggests possible queries the user may have in mind.

The existing solutions of auto-completion provide the suggestions based on the beginning of the currently input character sequence (i.e. prefix). Although this approach provides satisfactory auto-completion in many cases, it is far from optimal since it fails to take into account the semantic of users' input characters. There are many practical applications where syntactically different strings can represent the same real-world object [10]. For example, "Bill" is a short form of "William" and "Database Management Systems" can be abbreviated as "DBMS". These equivalence information suggests semantically similar strings that may have been missed by simple prefix based approaches. For instance, based on the DBLP dataset, when a user enters "Andy Pa" in the search box (see Fig. 1), the system should suggest "Andrew Palvo", because there is no record with the prefix "Andy Pa" and "Andy" is a nickname of "Andrew". Similarly, on an E-commerce

© Springer International Publishing AG 2017
S. Candan et al. (Eds.): DASFAA 2017, Part II, LNCS 10178, pp. 202–218, 2017.
DOI: 10.1007/978-3-319-55699-4_13

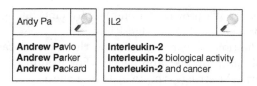

Fig. 1. Auto-completion with synonyms.

site, a user may type part of an abbreviation of a product name because she does not know the full name stored in a database. In a gene/protein database, one of the major obstacles that hinder the effective use is term variation [13], including acronym variation (e.g. "IL-2" and "interleukin-2"), and term abbreviation (e.g. "Ah receptor" and "Ah dioxin receptor"). *Therefore, this calls for auto-completion with synonyms to improve its usability and effectiveness.* For brevity we use "*synonym*" to describe any kind of equivalent pairs which may include synonym, acronym, nickname, abbreviation, variation and other equivalent expressions.

Often, when only a few characters of the lookup string have been entered, there are too many completions for auto completion to be useful. We thus consider a top-k synonym-based auto-completion strategy that provides the suggestion for the only top-k results according to predefined ranking scores and synonym sets. Given a large set of strings, an auto-completion system needs to be speedy enough to keep up with the user's key strokes. Meanwhile, we would like to fit all strings in the limited main memory. Hence, we need a both time-efficient and space-efficient data structure that enables us to return top-k completions without checking all the data in the synonyms set and the string collection.

In this paper, we propose three data structures to support efficient top-k completion queries with synonyms for different space and time complexity trade-offs: (i) *Twin tries (TT):* Two tries are constructed to present strings and synonym rules respectively in order to minimize the space occupancy. Each trie is a compact data structure, where the children of each node are ordered by the highest score among their respective descendants. Applicable synonym rules are indicated by pointers between two tries. An efficient top-k algorithm is developed to search both tries to find the synonym rules.

(ii) *Expansion trie (ET):* A fast lookup-optimized solution by integrating synonym rules with the corresponding strings. Unlike TT, ET uses a single expended trie to represent both synonym and string rules. Therefore, by efficiently traversing this trie, ET is faster than TT to provide top-k completions. Meanwhile ET often takes larger space overhead than TT, because ET needs to expand the strings with their applicable rules.

(iii) *Hybrid tries (HT):* An optimized structure to strike a good balance between space and time cost for TT and ET. We find a balance between lookup speed and space cost by judiciously selecting part of synonym rules to expand the strings. We show that given a predefined space constraint, the optimal selection of synonym rules is NP-hard, which can be reduced to a 0/1 knapsack problem with item interactions. We provide an empirically efficient heuristic algorithm by extending the branch and bound algorithm.

Large scale evaluation of search queries on three real datasets demonstrate the effectiveness of the proposed approaches. For example, on the US postal address dataset with 1M strings, the twin tries achieve a size of 160 bytes per string, which requires an average of only 5 ms to compute the top-10 completion on a simulated workload. In comparison, the expansion trie reduces the completion time to 0.1 ms, but increases the size to 200 bytes per string. The hybrid tries have a balanced performance, by achieving 1–2 ms per query, with the space overhead of 172 bytes per string. The implementation is available at http://udbms.cs.helsinki.fi/?projects/autocompletion/download.

2   Related Work

There is a plethora of literature on query auto-completion, especially in the field of information retrieval. We report here the results closest to our work. Readers may refer to a recent survey [2] for more comprehensive review.

Auto-completion with prefix matching can be easily implemented with a trie. Therefore, it is straightforward to extend trie to support top-k prefix matching. Li et al. [9] precompute and materialize the top-k completion of each possible word prefix and store them with each internal node of a trie. This requires a predetrmined k. Surajit et al. [3] provided solutions for error-tolerating auto-completion based on edit distance constraints, which during the lookup, maintains an *error* variable while searching for all possible candidate nodes. The collection of candidate strings are fetched by gathering strings stored under all leaf nodes under all candidates nodes. Xiao et al. [14] further extended the approach, by proposing a novel neighborhood generation-based algorithm, which can achieve up to two orders of magnitude speedup over existing methods for the error-tolerant query auto-completion problem. These solutions, however, are based on string similarity, which does not expose the semantic relations between words. For example, "iPod" and "iPad" have an edit distance only 1, but they should not be considered as the same word. In contrast, the edit distance between "DASFAA" and "International Conference on Database Systems for Advanced Applications" is big, but they refer to the same conference.

In [5], Hyvonen et al. proposed *semantic-based* auto-completion, which can include synonym-based, context-based and multilingual auto-completion. Unfortunately, this paper only mentions the concept of *semantic-based* auto-completion, but no algorithms are proposed. In this paper, we make the technical contribution by proposing space and time efficient algorithms to explore the synonym relations for top-k auto-completion.

Finally, synonym pairs can be obtained in many ways, such as existing dictionaries and synonyms mining algorithms [12]. Recently, Lu et al. [10] studied how to use the synonyms to improve the effectiveness of table joins. In this paper, with different research theme, we strike to use the synonyms to provide meaningful top-k auto-completion.

3   Preliminaries and Problem Description

In this section, we describe some of the data structures and primitives used in this paper and define our research problem.

Dictionary and synonym rule sets. Dictionary is a scored string set $\mathbb{D}$ in forms of pairs (s, r) where $s \in \Sigma^*$ is a string drawn from an alphabet Σ and r is an integer score. A synonym rule set $\mathbb{R}$ is a set of synonym pair. Let $\mathbb{R}$ denote a collection of synonym rules, i.e., $\mathbb{R} = \{r : lhs \rightarrow rhs\}$. A rule can be applied to s if lhs is a substring of s; the result of applying the rule is the string s' obtained by replacing the substring matching lhs with rhs. Given a string p, we say a replaced string p' from p, which is obtained from some non-overlapping substrings of p by applying the rules to generate new string p'. We can apply any number of synonym rules one after another. However, a token that is generated as a result of production application cannot participate in a subsequent production.

Problem description. Given a dictionary of strings and a collection of synonym rules, the goal is to suggest the top k strings with the highest scores with considerations of synonym pairs. Formally, we define the problem of top-k completion with synonyms as follows.

Problem 1 (Top-k completion with synonyms). Given a dictionary string $p \in \Sigma^*$, an integer k, and a synonym set $\mathbb{R}$, a top-k completion query in the scored dictionary string set $\mathbb{D}$ returns the k highest-scored pairs in $\mathbb{D}_p = \{s \in \mathbb{D} \mid p$ is a prefix of s, or there exists a replaced string p' of p using $\mathbb{R}$, such that p' is a prefix of $s.\}$ □

Example 1. See Fig. 1. Given three dictionary strings $\mathbb{D}$ including "Andrew Pavlo", "Andrew Parker" and "Andrew Packard" and one synonym rule $\mathbb{R} = \{$"Andy" $\rightarrow$ "Andrew"$\}$. If a user enters "Andy Pa". Then all three strings are returned as top-3 completions. Note that none of results can be returned based on the traditional prefix-based auto-completion. □

4   Twin Tries (TT)

A trie, or radix tree or prefix tree, is a tree data structure that encodes a set of strings, represented by concatenating the characters of the edges along the path from the root node to each corresponding leaf. All the descendants of a node have a common prefix of the string associated with that node, and the root is associated with the empty string. To encode the score and support top-k completions, we assign to each leaf node the score of the string it represents, while each intermediate node holds the maximum score among its descendants. We employ two tries, named *dictionary trie* ($\mathcal{T}_D$) and *rule trie* ($\mathcal{T}_R$), which hold all dictionary strings and the synonym rules, respectively. Moreover, we introduce *synonym links*, which are edges from $\mathcal{T}_R$ and pointing to corresponding nodes in $\mathcal{T}_D$. To support top-k lookups, each synonym link is assigned with an integer offset, denoted by *link.delta*, which equals to the length of *rule.lhs* minus length of $rule.rhs$. An example of the mentioned structure can be found in Fig. 2. Algorithm 1 gives the method for building TT.

We introduce a heuristic algorithm (see Algorithm 2) to find the best completion results that extends the search string. Specifically, starting from the root node, we iteratively search in the dictionary trie for any matching prefixes of search string. For the unmatched part, we look up in the rule trie for any possible synonym rules. If there are multiple link targets in this rule, we select the appropriate one by comparing the deepest locus node and the node prior to $rule.lhs$ (Line 18 in Algorithm 2). To support an efficient top-k lookup, we also introduce $node.depth$, which is the number of edges from $node$ to the trie's root node.

Example 2. Consider twin tries in Fig. 2. Assume that the search string is "abmp" and $k = 1$. The process for getting completion result with Algorithm 2 can be found in the following table[1]. "✓" and "×" represents a string is found or not found in corresponding trie: p_{r_a} in dictionary trie, p_{r_r} in rule trie.   □

Iter.	p_{r_a}	p_{r_r}	Note
1	Pop first element from queue: $m = \varepsilon$ (root of $\mathcal{T}_D$), $p_r = abmp$		
1.1	ε ✓	$abmp$ ×	ε is found in $\mathcal{T}_D$, but $abmp$ is not found in $\mathcal{T}_R$
1.2	a ✓	bmp ×	a is found in $\mathcal{T}_D$, but bmp is not found in $\mathcal{T}_R$
1.3	ab ✓	mp ✓	mp is found in $\mathcal{T}_R$. The target of its links are $\underline{c}$ and $ab\underline{c}$. $ab\underline{c}$ is the correct link target. Push it to queue
1.4	abm ×		Break loop
2	Pop first element from queue: $m = abc$, $p_r = \emptyset$		
2.1	abc ✓	$\emptyset$	Node $ab\underline{c}$ is a leaf, so add it to result set. p_{r_r} is empty, so push all children of $ab\underline{c}$ to queue (but it has no child).
3	The queue is empty. Therefore the final result is "abc"		

Complexity analyses. The worst-case time complexity of top-k auto-completion (Algorithm 2) is $O(pm + p^2n^2 + klm\log(klm) + st)$, where p is the length of search string, l is the maximum length of dictionary string, m and n is the maximum number of children per node in $\mathcal{T}_D$ and $\mathcal{T}_R$ respectively, s is the maximum number of links per rule and t is the maximum number of applicable rules per dictionary string.

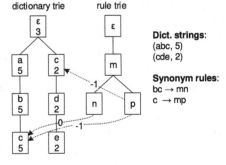

Fig. 2. TT example

---

[1] In this table, we use the denotation $ab\underline{c}$ to represent a node with label "c" with parent node labeled "b", in path $root - a - b - c$.

Algorithm 1. Generation of TT

Input: Set of dictionary strings ($\mathbb{D}$) and set of synonym rules ($\mathbb{R}$)
Output: Twin tries $\langle \mathcal{T}_D, \mathcal{T}_R \rangle$

1   for each of rules in $\mathbb{R}$, add its *rhs* to $\mathcal{T}_R$
2   **foreach** $s \in \mathbb{D}$ **do**
3   add s to $\mathcal{T}_D$
4   **foreach** $r \in \mathbb{R}$ **do**
5   **if** *r can be applied onto s* **then**
6          $f \leftarrow$ deepest locus node of r in $\mathcal{T}_R$
7   **foreach** $lo \in$ *all locus points of r on s* **do**
8            $l \leftarrow$ node from $\mathcal{T}_D$, which represents *r*.lhs in decendents of lo
9   f.links.add(l, *r*.lhs.length - *r*.rhs.length)   // (target, delta)
10   recursively set every score of every node in $\mathcal{T}_D$ to the maximum among its descendants
11   **return** $\langle \mathcal{T}_D, \mathcal{T}_R \rangle$

Specifically, we need to examine up to pm nodes in the dictionary trie to check whether a substring is from dictionary. We also need to lookup $(pn + \cdots + 2p + p)$ nodes in the rule trie in order to find possible synonym links. After we find one result, we need to scan upward l nodes on the way to root, which is totally $O(kl)$ time corresponding to k completions. As the algorithm inserts all nodes in corresponding path to the priority queue, we may add up to klm nodes, contributing an additional $O(klm \log(klm))$ term. Finally, $O(st)$ time is required to determining the correct synonym link.

5   Expansion Trie (ET)

In this section, we describe a compressed trie data structure to combine both dictionary and synonym strings into one trie, called Expansion Trie (ET).

The baseline algorithm is to generate a set of new strings by applying permutations of rules onto the dictionary strings, then add them to trie. The baseline algorithm has two problems: (i) Dictionary and synonym nodes are mixed together in the final trie, and thus it is hard to tell whether a string is from dictionary; (ii) the algorithm is extremely slow because the generation of permutations for all applicable rules in strings.

To address the above problems, we

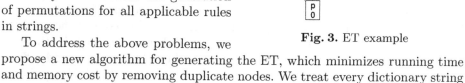

Fig. 3. ET example

Dict. strings:
d_1: (abc, 5)
d_2: (cde, 2)

Synonym rules:
r_1: bc → mn
r_2: c → mp

propose a new algorithm for generating the ET, which minimizes running time and memory cost by removing duplicate nodes. We treat every dictionary string

---

Algorithm 2. Top-k completions with TT

Input: Dictionary trie $\mathcal{T}_D$, rule trie $\mathcal{T}_R$, search string p and $k > 0$
Output: List of top-k completions C

1   $Q \leftarrow$ empty priority queue; $C \leftarrow$ empty priority list; Q.push($\langle$root node of $\mathcal{T}_D$, $0\rangle$)
2   **while** $Q \neq \emptyset$ **do**
3     $\langle m, i_{p_r} \rangle \leftarrow Q$.pop()   `// (current node, index of remaining p)`
4     $p_r \leftarrow p$.substring($0, i_{p_r}$)
5   **for** i from 0 to p_r.length **do**
6       $(p_{r_a}, p_{r_r}) \leftarrow p_r$.split($i$)
7       $l \leftarrow$ deepest locus node of p_{r_a} in descendants of node m
8   **if** l *is not found* **then** break the *for* loop
9   **else if** l *is a leaf node* **then**
10   C.add(full string of l)
11   **if** $|C| = k$ **then return** C
12   **if** p_{r_r} *is empty string* **then**
13   **foreach** $c \in l.children$ **do** Q.push($\langle c, i_{p_r} + i \rangle$)
14   **else**
15         $ns \leftarrow$ locus points of p_{r_r} in $\mathcal{T}_R$
16   **foreach** $n \in ns$ **do**
17   **foreach** $lk \in n.links$ **do**
18            $dest \leftarrow$ from $lk.target$, go up ($lk.depth + lk.delta$) levels
19   **if** l *and dest is the same node* **then**
20   Q.push($\langle lk.target, i_{p_r} + i + lk.target.depth \rangle$)
21
22 **return** C

---

as a unique path from root to leaf, while all its available synonym rules as "branches" that attached to it. At the end of each branch, a synonym link points it back to the dictionary string. Additionally, we set the score of every *synonym node* (new nodes introduced by expanding synonym rules on dictionary trie) to 0, because we do not give such suggestion. The pseudo-code of the proposed algorithm can be found in Algorithm 3.

Example 3. Given dictionary strings $d_1 : (abc, 5)$ and $d_2 : (cde, 2)$ and synonym rules $r_1 : bc \rightarrow mn$ and $r_2 : c \rightarrow mp$, the ET generated by Algorithm 3 can be seen in *Fig. 3*.   □

We introduce a similar solution to perform top-k suggestions on ET as in Algorithm 4. Specifically, we find the deepest node in the trie that matches the search string as much as possible (Line 2 in Algorithm 4) and insert it into a priority queue. We also insert the target node of its synonym links (if any) to the queue. Then we iteratively pop the node with highest score as the current node. We add it to the result if it is a leaf node. Otherwise, if there is still any other remaining character, we find the corresponding child node and add it to the queue. When all characters are processed, we add all the children of the

---

Algorithm 3. Generation of ET

Input: Set of dictionary strings ($\mathbb{D}$), set of synonym rules ($\mathbb{R}$)

Output: Expansion trie $\mathcal{T}$

1 add all strings in $\mathbb{D}$ to $\mathcal{T}$
2 **foreach** $s \in \mathbb{D}$ **do**
3   **foreach** $r \in \mathbb{R}$ **do**
4   **if** *r can be applied onto s* **then**
5   **foreach** $f \in$ *all locus points of r on s* **do**
6                  $l \leftarrow$ deepest node of r.lhs in decendents of f
7                  $p \leftarrow f$.parent   // synonym nodes attach here
8   add each char of r.rhs with *score* = 0 as descendants of p
9                  $e \leftarrow$ select the last node just added (i.e. deepest node in synonym nodes)
10   e.links.add(l)
11 recursively set every score of every node in $\mathcal{T}$ to the maximum among its descendants
12 **return** $\mathcal{T}$

Algorithm 4. Top-k completions with ET

Input: Expansion trie $\mathcal{T}$, search string p and $k > 0$

Output: List of top-k completions C

1 $Q \leftarrow$ empty priority queue; $C \leftarrow$ empty priority list; $H \leftarrow$ empty hash table
2 $locus \leftarrow$ deepest locus point of p in $\mathcal{T}$

3 **if** *locus is a dictionary character* **then** Q.push($locus$)
4 **foreach** $l \leftarrow$ *target node of locus.links* **do** Q.push(l)
5 **while** $Q \neq \emptyset$ **do**
6      $m \leftarrow Q$.pop()
7   **if** *m is a leaf node* **then**
8   add full string of m to C with m.score
9   **if** $-C- = k$ **then return** C
10   **if** *m is the last node representing p, or there is more chars from p after m* **then**
11   Q.push(m.links.target)
12   **if** *there is more chars from p after m* **then**
13   Q.push(node holds next character of p after m)
14   **else** push all non-synonym nodes of m.children to Q
15 **return** C

---

current node to the queue (Line 12 to 14 in Algorithm 4). This procedure loops until k completions have been found or the priority queue becomes empty.

Complexity analyses. The worst-case time complexity of top-k on ET is $O(pm+klm\log(klm))$. According to the proposed algorithm, we need to examine up to pm nodes in the trie to find the locus node. After reaching one leaf node, we need to scan upward l nodes on the way to root, which is totally $O(kl)$ time corresponding to k completions. Add up to klm nodes to the binary heap contributes an additional $O(klm \log(klm))$ term. Although we use the same notation "m" here, one should notice that its value is larger compared to TT because the expansion of rules introduced more nodes, thus the maximum number of children is increased.

6   Hybrid Tries (HT)

By comparing the top-k time complexity of TT and ET, it can be seen that the latter will need more time as it needs to (i) look up the rule trie iteratively for every sub-string of p, (ii) check all synonym links in order to find the correct one. Therefore, we propose a solution that selects some synonym rules and expands them while leaving the remaining ones in rule trie, so that fewer synonym links need to be checked, which leads to a smaller running time. Note that *the more rules we expand, the more space it takes*. Therefore the problem can be defined as follows:

Problem 2 (Optimal construction of HT).   Let S_{TT} and S_{ET} be the space cost of TT and ET, given a space threshold $S \in [S_{TT}, S_{ET}]$, $\mathbb{D}$ and $\mathbb{R}$, our task is to build two tries $\langle \mathcal{T}_D, \mathcal{T}_R \rangle$ to minimize the top-k lookup time while satisfying the space constraint S.

With endeavors to make the lookup search based on HT more efficient, our approach is to solve Problem 2 with a combinatorial optimization algorithm based on the frequency of rules in applicable strings. Therefore, the policy of selecting rules for expansion turns into *maximizing the total number of applicable rules on dictionary strings within space constraints*.

Let r_i be the *item* (synonym rule) at index i in $\mathbb{R}$, $\{v_1, v_2, ..., v_{|\mathbb{R}|}\}$ be the frequency (time-of-use) of items, $\{w_1, w_2, ..., v_{|\mathbb{R}|}\}$ be its weight, i.e. space cost when expanding it to the dictionary trie, $\{x_1, x_2, ..., x_{|\mathbb{R}|}\}$ be a set of integers either 0 or 1, it can be seen that the problem is similar with a 0/1 knapsack problem, which is known NP-hard:

$$\text{maximize} \sum_{i=1}^{|\mathbb{R}|} v_i x_i \text{ subject to} \sum_{i=1}^{|\mathbb{R}|} w_i x_i \leq S$$

However, our problem is not such straightforward because the space cost of a synonym rule may be smaller depends on the presence of other rules in the trie. Consider dictionary string $abcde$ and two rules $r_1 : abc \rightarrow mn$ and $r_2 : abc \rightarrow mnp$. Expanding r_1 adds two synonym nodes $\underline{m}$ and $\underline{mn}$. Then when expanding r_2, it uses existing nodes $\underline{m}$ and $\underline{mn}$ which are generated by r_1 before. Thus only one new node $\underline{mnp}$ is created. By considering such interactions, we are able to formalize Problem 2 more precisely as follows:

Problem 3 (0/1 Knapsack Problem with Item Interactions).

$$\text{maximize} \sum_{i=1}^{|\mathbb{R}|} v_i x_i \text{ subject to} \sum_{i=1}^{|\mathbb{R}|} f_i(x_i, x_j | j \in P_i) \leq S$$

$f_i(\cdot)$ is the weight function that returns the weight of item r_i with knowledges of x_i, current items in knapsack (their indexes are stored in C), and P_i as indexes of all items which have interactions with r_i. Specifically, the weight function can have three types of return values: (i) $f_i(\cdot) = 0$ when $x_i = 0$, i.e. item r_i is not selected. (ii) $f_i(\cdot) = w_i$ when $x_i = 1$ and $\nexists x_j = 1 | j \in (P_i \cap C)$. (iii) $f_i(\cdot) \in (0, w_i)$, otherwise.

It is possible to adapt the dynamic programming (DP) method to Problem 3, by sorting the items so that all items which r_i depends on are located before r_i. This ensures all interacted items are processed before r_i itself. However, in our problem the cyclical cannot be ignored [1]: we can say that the weight of r_1 depends on the presence or absence of r_2, but it is also true to say r_2 depends on r_1, since r_2 can also provide the two synonym nodes which reused by r_1. Due to the hardness of the problem, some approximate methods are proposed, by grouping interacted items as a single knapsack item [8] or cutting weak interactions [11]. However, all such solutions are not able to give a bound of the estimation error. In this paper, we present a new solution following a branch and bound (B&B) fashion by tightening the upper- and lower-bound with considerations of item interactions, which gives an exact solution subject to total value.

We now introduce three terms used in our algorithm. All items can be classified into one of three categories at any specific stage of B&B algorithm [7]: (i) *Included:* the item is explicitly included in the solution. According to our definition, item r_i is an included item when $x_i = 1$. (ii) *Excluded:* the item is explicitly excluded in the solution, i.e. $x_i = 0$. (iii) *Unassigned:* the item is not processed yet. At any given stage, this type of items should only contains further items that has not been tested in any branch.

Tight upper-bounds. For knapsack with independent items, the method for obtaining an upper-bound is based on the solution of *fractional knapsack problem*, where a part of an item can be take into the knapsack when the space does not fit it as a whole. A greedy algorithm by Dantzig et al. [4] can be employed to obtain an optimal result. In our case, we sort items by assuming all interactions already existed. That is, for item r_i, we assume every item $\forall j \in P_i, x_j = 1$. We use $w_{min,i}$ to indicate this minimum weight. This can guarantee that the greedy algorithm returns a solution which is the largest one among all feasible upper-bounds.

Tight lower-bounds. A classic method to obtain the lower-bound is to look forward down the unassigned items in current solution, and greedy take (in order) items into knapsack until the weight budget left cannot fit the next item. We extend this method by assuming every interacted item r_j is either excluded or unassigned, i.e. $\forall j \in P_i, x_j = 0$.

Measuring exact weight. We add the exact space consumption for expanding r_i to knapsack in each branch operation. One straightforward solution can be "scan all items to accumulate any possible savings". Unfortunately, this method ignores that fact that most items are not interacted and will be very slow when

$|\mathbb{R}|$ is large because each scan requires $O(|\mathbb{R}|)$ time. As in our solution, we perform a *partition* prior to B&B by grouping all items to several parts: r_i has interactions with all other items in the same part, but not with items in other parts. As the result, saving can only be obtained when r_i is included together with items from the same part, otherwise it has its original weight. The partition allows us to use a heuristic in each branch operation by scanning though only items in the same part with r_i, instead of all items.

Construction of HT and top-k completions. Our algorithm for solving Problem 2 is given in Algorithm 5. Specifically, we highlight our three extensions to B&B algorithm as separate functions.

Algorithm 5. Construction of HT

Input: Set of dictionary strings ($\mathbb{D}$), set of synonym rules ($\mathbb{R}$) and space threshold (S)

Output: Hybrid tries $\langle \mathcal{T}_D, \mathcal{T}_R \rangle$

1 $P \leftarrow$ partition rules in $\mathbb{R}$
2 sort $\mathbb{R}$ by items' minimum weight (i.e. assume all interactions exist)
3 $\langle \mathbb{R}_{in}, \mathbb{R}_{ex} \rangle \leftarrow$ solve knapsack problem with branch and bound, with bound functions $upper\_bound(r_i)$ and $lower\_bound(r_i)$, r_i is the current item in the decision tree. in each *branch*, the exact weight of r_i is obtained by $exact\_weight$ (r_i, P_{r_i}, X_{inc}), where P_{r_i} is the part r_i belongs to, X_{inc} is the set of included items at current step
4 $\mathcal{T}_D \leftarrow$ build dictionary trie with $\mathbb{D}$ and expand rules in $\mathbb{R}_{in}$ following Algorithm 3
5 $\langle \mathcal{T}_D, \mathcal{T}_R \rangle \leftarrow$ build rules trie with $\mathbb{D}$ and $\mathbb{R}_{ex}$ following Algorithm 1, while let $\mathcal{T}_D$ be the ready-to-use dictionary trie
6 **return** $\langle \mathcal{T}_D, \mathcal{T}_R \rangle$

7 **Function** upper_bound (r_i)
8      $ub_i \leftarrow r_i$.weight   // take current weight
9   **while** $ub_i < S$ **do**
10   take r_i, add its *minimum* weight to ub_i; $i \leftarrow i + 1$
11          $ub_i \leftarrow ub_i + \frac{v_i}{w_{min,i}} \times (S - ub_i)$ // take a fraction of next item using its *minimum* weight
12   **return** ub_i

13 **Function** lower_bound (r_i)
14      $lb_i \leftarrow r_i$.weight
15   **while** $lb_i < S$ **do**
16   take r_i, add its *original* weight to lb_i; $i \leftarrow i + 1$
17   **return** lb_i

18 **Function** exact_weight (r_i, P_{r_i}, X_{inc})
19      $w_{real} \leftarrow w_i$
20   **foreach** $r | r \neq r_i, r \in P_{r_i}$ **do**
21   **if** $\exists r \in X_{inc}$ **then** $w_{real} \leftarrow \min(w_{real}, f_i(x_i, r))$
22   **return** w_{real}

Example 4. Given dictionary strings $d_1 : (abc, 5)$ and $d_2 : (cde, 2)$ and synonym rules $r_1 : bc \to mn$ and $r_2 : c \to mp$, the HT generated by expanding r_1 according to Algorithm 5 is illustrated in *Fig. 4.*

We can preform top-k completions queries on HT by extending Algorithm 2: every time when checking possible synonym rules in p_r (before Line 12), we push the target of *l.links* with `foreach t← target of l.links do` `Q.push(t)`.

Because the top-k completions algorithm on HT is similar with TT, their worst-time complexity is also the same. However, the value of s (i.e. maximum

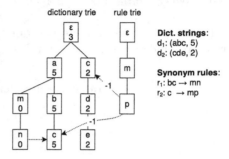

Fig. 4. HT example

number of links per synonym rule) is smaller since we reduced the number of synonym links per rule by moving some rules to the dictionary tire.

7 Experimental Analysis

To evaluate the effectiveness of the proposed top-k completion techniques, Twin Tries (**TT**), Expansion Trie (**ET**) and Hybrid Tries (**HT**), we compare their effectiveness on the following datasets from different application scenarios on a Intel i7-4770 3.4 GHz processor with 8 GB of RAM, complied with OpenJDK 1.8.0 on Ubuntu 14.04 LTS.

7.1 Datasets

We use three datasets: conference publications and book titles (**DBLP**), US addresses (**USPS**), and gene/protein data (**SPROT**). These datasets differ from each other in terms of rule-number, rule-complexity, data-size and string-length. Our goal in choosing these diverse sources is to understand the usefulness of algorithms in different real world environments.

DBLP: We collected 24,810 conference publications and book titles from DBLP website (http://dblp.uni-trier.de/). We obtained 214 synonym pairs between the common words and their abbreviations used in computer science field listed on IEEE website.

USPS: We downloaded common person names, street names, city names and states from the United States Postal Service website (http://www.usps.com). We then generated 1,000,000 records as dictionary strings, each of which contains a person name, a street name, a city name and a state. We also gathered extensive information about the common nicknames and format of addresses, from which we obtained 341 synonym pairs. The synonym pairs covers a wide range of alternate representations of common strings, e.g. `Texas → TX`.

SPROT: We obtained 1,000,000 gene/protein records from the UniProt website (http://www.uniprot.org/). Each record contains an entry name, a protein name, a gene name and its organism. In this dataset, each protein name has 5−22 synonyms. We generated 1,000 synonym rules describing relations between different names.

Table 1 gives the characteristics of the three datasets. The scores of each string are randomly generated in this experiment.

Table 1. Characteristics of datasets.

Name of dataset	# of strings	String len (avg/max)	# of synonym rules	Rules per string (avg/max)
DBLP	24,810	60/295	368	2.51/11
USPS	1,000,000	25/43	341	2.15/12
SPROT	1,000,000	20/28	1,000	2.11/12

7.2   Data Structure Construction

Space. We evaluate the compactness of the generated data structures by reporting in Table 2 the average number of bytes per string (including score and relations e.g. *node.parent*). For comparison, we also report the size of the data structure generated by the baseline method (**BL**) described in the expansion trie (see Sect. 5). Across the datasets, the baseline method produce the largest trie structure, about 4KB per dictionary string for DBLP dataset. For larger dataset like USPS and SPROT, it crashes because of exponentially increasing number of new strings. The ET structure takes the second largest space consumption, while TT consistently takes the smallest size, about 58% smaller than ET on SPROT dataset. Finally, the HT structure (we set the space threshold to $0.5 \times (S_{ET} - S_{TT})$) takes a larger size than TT but smaller than ET.

Table 2. Data structure sizes in bytes per string.

Name of dataset	BL	TT	ET	HT
DBLP	4,250.98	528.71	638.76	578.12
USPS	Failed	160.49	200.03	172.64
SPROT	Failed	128.82	217.55	161.25

To better understand how the space is used, we present in Fig. 5 the storage breakdown of each of the techniques on SPROT dataset. We break the total space down to (i) Space taken by dictionary nodes, including labels, scores and relations like *node.parent* and *node.children*, (ii) Expanded synonym nodes: size of synonym nodes in the dictionary trie and (iii) Unexpanded synonym nodes:

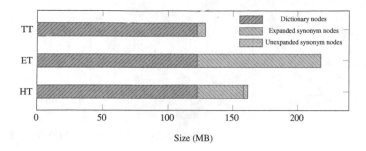

Fig. 5. Data structure size breakdown on SPROT dataset

size of synonym nodes in rule trie $\mathcal{T}_R$. For ET, the number of synonym nodes in the trie is about 15 times more than in rule trie (TT) due to the numerous different locus points. The latter eliminates multiple copies of nodes, but will incur some sacrifice in top-k speed. For HT, the most frequent rules are expanded like ET, while half size of less-frequent rules are left in the rule trie. This results a moderate space consumption between TT and ET.

Time. In addition to the space cost, we also measure their running time on three dataset and report them in Fig. 6. For small dataset like DBLP, all four methods finish within a reasonable time, however, the baseline method is nearly 30 times slower than the other three. It also failed to finish within 300 s on large dataset USPS and SPROT. For the other three methods, TT is always the fastest on all datasets, because it does not need to create synonym nodes for each application, but use the existing ones and add a new synonym link. The HT runs the slowest due to the additional computation in the B&B method.

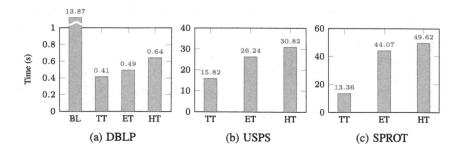

Fig. 6. Construction time

7.3   Top-k Efficiency

This set of experiments studies the overall efficiency of our auto-completion. We generate test queries by randomly applying synonym rules onto the dictionary

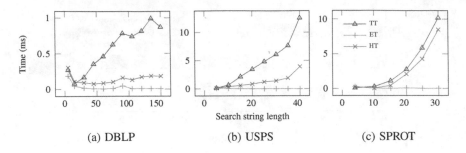

Fig. 7. Top-10 auto-completion lookup time

strings, then we randomly pick a substring of each new string, formed 50,000 query strings for each dataset. We ran every query string based on the TT, ET and HT structures and plotted the running time in Fig. 7. We observed that for shorter queries (length 2 to 10), all three algorithms runs very fast, less than 0.5ms for small dataset and 1 ms for large ones. However, the running time of TT and HT grows as the length of query becomes longer. The primary reason for this is that they need to lookup every substring of query in the rule trie, which consumes more time (Lines 4 to 7 in Algorithm 2). Determining the correct link further slows down the speed. Besides, as HT expanded some synonym rules, its speed is faster for the reason that less synonym links being checked. In contrast, ET runs the fastest in all experiments, whose running time is not affected by the length of search strings.

We observe that running time of HT is more like TT especially on SPROT dataset. As the space threshold is the key parameter to control the construction of HT, we preform one more set of experiments to deeply study the effect of this parameter on the lookup running time. We define a ratio $\alpha \in [0, 1]$ where $\alpha = \frac{S}{S_{ET} - S_{TT}}$. We select several values for α, build HT and then perform top-10 lookup. The speed of top-10 operations corresponding to different αs is illustrated in Fig. 8. The result shows that when α becomes larger, i.e. larger space threshold for HT, the top-k lookup becomes faster. When $\alpha = 0$ and $\alpha = 1$, the time is exactly the same with TT and ET, respectively. This experiment shows that if we select a space threshold between 75% and 100% of $S_{ET} - S_{TT}$, we can expect to have more than 50% performance boost compared with TT while performing lookup.

7.4   Scalability

To assess the scalability of the data structures, we compare the structure size and top-10 speed on different subsets of the USPS dataset. We generate these subsets by taking the top-N items in decreasing score order. Figure 9a shows that the sizes of all three structures increase linearly, where TT and TT are the smallest and largest, respectively. In Fig. 9b, the average time per completion for ET does not increase as the dataset grows, while TT and HT become slower as number

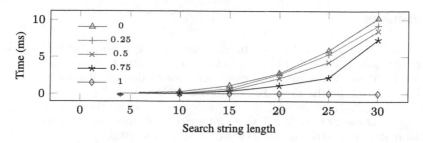

Fig. 8. Top-10 auto-completion lookup time of HT on SPROT dataset, in respect of different space ratios α

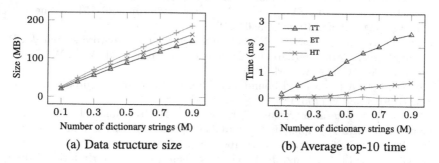

(a) Data structure size   (b) Average top-10 time

Fig. 9. Data structure size and average top-10 time related to number of dictionary strings on USPS dataset

of dictionary string becomes larger. This is because the increasing number of strings brings more synonym links need to be checked. However, compared with TT, who has a sharp increasing trend (about 3 ms per million strings), the time of HT grows slowly, only from 0.18 to 0.6 ms while data grows from 0.5 M to 0.9 M.

8 Conclusion and Future Work

In this paper, we have presented three data structures, i.e. TT, ET and HT, to address the problem of top-k completion with synonyms, each with different space and time complexity trade-offs. Experiments on large-scale datasets show that our algorithms can support synonym-based retrieval of completions of strings at about a microsecond per-completion for more than 1 million dictionary strings and thousands of synonym rule while taking small memory space. As our future work, it would be interesting to work on the problem called "*synonym ambiguity*". For instance, "DB" can be either "Database" or "Development Bank" depending on different contexts. We will explore the context of words to select appropriate synonym rules for auto-completion.

References

1. Burg, J.J., Ainsworth, J.D., Casto, B., Lang, S.: Experiments with the "oregon trail knapsack problem". Electron. Notes Discrete Math. **1**, 26–35 (1999)
2. Cai, F., Rijke, M.: A survey of query auto completion in information retrieval. Found. Trends Inf. Retrieval **10**(4), 273–363 (2016)
3. Chaudhuri, S., Kaushik, R.: Extending autocompletion to tolerate errors. In: Proceedings of the 2009 ACM SIGMOD International Conference on Management of Data, SIGMOD 2009, pp. 707–718. ACM, New York (2009)
4. Dantzig, G.B.: Discrete-variable extremum problems. Oper. Res. **5**(2), 266–277 (1957)
5. Hyvönen, E., Mäkelä, E.: Semantic autocompletion. In: Mizoguchi, R., Shi, Z., Giunchiglia, F. (eds.) ASWC 2006. LNCS, vol. 4185, pp. 739–751. Springer, Heidelberg (2006). doi:10.1007/11836025_72
6. Ji, S., Li, G., Li, C., Feng, J.: Efficient interactive fuzzy keyword search. In: Proceedings of the 18th International Conference on World Wide Web, WWW 2009, pp. 371–380. ACM, New York (2009)
7. Kolesar, P.J.: A branch and bound algorithm for the knapsack problem. Manage. Sci. **13**(9), 723–735 (1967)
8. LeFevre, J., Sankaranarayanan, J., Hacigümüs, H., Tatemura, J., Polyzotis, N., Carey, M.J.: MISO: souping up big data query processing with a multistore system. In: SIGMOD Conference, pp. 1591–1602. ACM (2014)
9. Li, G., Ji, S., Li, C., Feng, J.: Efficient type-ahead search on relational data: a TASTIER approach. In: ACM SIGMOD, pp. 695–706 (2009)
10. Lu, J., Lin, C., Wang, W., Li, C., Xiao, X.: Boosting the quality of approximate string matching by synonyms. ACM Trans. Database Syst. **40**(3), 15 (2015)
11. Schnaitter, K., Polyzotis, N., Getoor, L.: Index interactions in physical design tuning: modeling, analysis, and applications. PVLDB **2**(1), 1234–1245 (2009)
12. Singh, R., Gulwani, S.: Learning semantic string transformations from examples. PVLDB **5**(8), 740–751 (2012)
13. Tsuruoka, Y., McNaught, J., Tsujii, J., Ananiadou, S.: Learning string similarity measures for gene/protein name dictionary look-up using logistic regression. Bioinformatics **23**(20), 2768–2774 (2007)
14. Xiao, C., Qin, J., Wang, W., Ishikawa, Y., Tsuda, K., Sadakane, K.: Efficient error-tolerant query autocompletion. Proc. VLDB Endow. **6**(6), 373–384 (2013)

Efficient Regular Expression Matching on Compressed Strings

Yutong Han, Bin Wang$^{(\boxtimes)}$, Xiaochun Yang, and Huaijie Zhu

School of Computer Science and Engineering, Northeastern University,
Shenyang 110169, Liaoning, China
hanytneu@gmail.com, {binwang,yangxc}@mail.neu.edu.cn, zhuhjneu@gmail.com

Abstract. Existing methods for regular expression matching on LZ78 compressed strings do not perform efficiently. Moreover, LZ78 compression has some shortcomings, such as high compression ratio and slower decompression speed than LZ77 (a variant of LZ78). In this paper, we study regular expression matching on LZ77 compressed strings. To address this problem, we propose an efficient algorithm, namely, *RELZ*, utilizing the *positive factors*, i.e., a prefix and a suffix, and *negative factors* (Negative factors are substrings that cannot appear in an answer.) of the regular expression to prune the candidates. For the sake of quickly locating these two kinds of factors on the compressed string without decompression, we design a variant suffix trie index, called *SSLZ*. In addition, we construct bitmaps for factors of regular expression to detect potential region and propose block filtering to reduce candidates. At last, we conduct a comprehensive performance evaluation using five real datasets to validate our ideas and the proposed algorithms. The experimental result shows that our RELZ algorithm outperforms the existing algorithms significantly.

Keywords: Regular expression · LZ77 · String matching · Self-index

1 Introduction

Finding matches of a regular expression (RE) on a string are emerged in many applications such as text editing, biosequence search, shell commands, and data repair [5]. In recent years, several solutions are proposed for regular expression matching on a string (without compression). However, for a very huge and long string, these solutions are not efficient due to requiring huge storing space. And a particularly way of saving space when storing the long string in practice is using compression techniques. Meanwhile, a lot of compression techniques have been studied over the years, e.g. LZ77 [2], LZ78 [17] and compressed genome [14]. The main idea of LZ compression series is to compress the string using a static

---

This work is partially supported by the NSF of China for Outstanding Young Scholars under grant No. 61322208, the NSF of China under grant Nos. 61272178 and 61572122, and the NSF of China for Key Program under grant No. 61532021.

© Springer International Publishing AG 2017
S. Candan et al. (Eds.): DASFAA 2017, Part II, LNCS 10178, pp. 219–234, 2017.
DOI: 10.1007/978-3-319-55699-4_14

dictionary. When some substrings appeared repetitively in the preceding string, they are replaced by a portion of the dictionary. Though papers [1,7] studied regular expression matching on the LZ78 compressed string, they have two main issues: (1) LZ78 is with high compression ratio and slower decompression speed than LZ77; (2) The algorithms in [1,7] do not perform efficiently. So in this paper, we study regular expression matching on the LZ77 compressed string.

To address this problem, we propose an efficient algorithm, namely, *RELZ*, utilizing the positive and negative factors of the regular expression to prune the candidates. Negative factors [15] are proposed for improving regular expression matching on strings by pruning false negatives. The basic idea of RELZ is to obtain these high quality factors using the algorithms [15] first. Then we need to locate these factors on the compressed string. For the sake of quickly locating these two kinds of factors on the compressed string without decompression, we design a variant suffix trie index, called *SSLZ*, motivated by self-index structures on LZ77 for exact pattern matching [4]. Once we have located these factors on the compressed string, we merge several factors to confirm the candidates. For speeding up factors merging, we construct bitmaps for this. In addition, we propose blocking filtering to prune candidates. At last, we conduct a comprehensive performance evaluation using five real datasets to validate our ideas and the proposed algorithms.

The contributions made in this paper are four-fold:

- To the best of my knowledge, we are the first to study regular expression matching on the LZ77 compressed string.
- We design a variant suffix trie *SSLZ* in self-index to locate positive and negative factors of regular expression on the compressed string efficiently.
- We utilize bitmaps to speed up factor merging and propose block filtering to reduce a lot of candidates.
- We conduct extensive experiments to evaluate the proposed algorithm. The experimental result shows that our algorithm significantly outperforms the existing algorithms on the LZ78 compressed string.

The rest of our paper is organized as follows. Section 2 reviews the related works of regular expression matching on strings and compressed strings. Section 3 gives the basic concepts of regular expression, positive and negative factors and LZ77 self-index. Section 4 presents the improved self-index structure in detail. Section 5 describes our algorithm RELZ for regular expression matching on the compressed string. Section 6 reports the experiments about our algorithm in practice.

2   Related Work

There are two categories concentrating on regular expression matching, including searching on original strings and compressed strings.

2.1   Regular Expression Matching on Original String

The classical way of regular expression matching is to convert the expression into automaton. For each position in the original string the automaton checks all the characters until the end of string reached. Thompson [11] defines the concept of NFA for searching a regular expression of length m in a string of length n. They transform the regular expression into NFA with at most $2m$ states, and the searching algorithm need $O(mn)$ in the worst case. BPThompson [12] uses bit-parallel ingeniously to simulate the NFA, which represents states by a computer word. They provide bit operators instead of inspecting the active states in the NFA. BPGlushkov [8,9,16] takes the advantage of Glushkov NFA that all the arrows reach the same state marked by the same character. When searching regular expression on the large sequence, it is a hard work to inspect every position in the string. Gnu Grep picks up a collection of substrings as necessary factors from the regular expression, which must be contained by the occurrence. Negative factors [13,15] shed a different light on regular expression matching. Yang carries out the strategy to achieve the core negative factors with excellent pruning power.

2.2   Regular Expression Matching on the Compressed String

Another alternative way is trying to skip on the compressed string with the help of the filtering extracted from regular expression. Navarro gives a solution to answering regular expression on LZ78 compression in [7]. They propose a variant DFA based on bit-parallelism to search the end position of occurrences accepted by the regular expression. With the compression scheme of LZ78, the algorithm processes all the phrases (also called blocks) one by one, update the state in DFA and report the last position of occurrences in each phrase.

In [1] Philip Bille describes a solution that improves the space by a factor $\Theta(m)$ matching the same time bound of Navarro's method. They select a subset of the phrases from LZ78 compression. After computation of the corresponding active states in the Thompson NFA for every phrase in the subset, these state sets combine the last math phrase pointers recursively to report the end position of the occurrences.

3   Preliminaries

In this section, we introduce some background of our research. We give the basic concepts of regular expression, positive and negative factors, LZ77 compression and self-index.

3.1   Regular Expression

Regular expression is a language over symbols $\Sigma \cup \{\epsilon, |, \cdot, *, (,)\}$. E is defined as regular expression consist of Σ^*, $E_1 \cdot E_2$, $E_1 \mid E_2$, and E_1^* recursively. The operations in the regular expression are defined as follows:

- ϵ denotes an empty string which does not contain any character.
- E_1 denotes an expression themselves.
- $E_1 \cdot E_2$ denotes that strings in E_1 concatenate strings in E_2.
- $E_1 \mid E_2$ denotes a set of strings which be obtained by union of E_1 and E_2.
- E_1^* is the Kleene closure of expression E_1, which means the expression E_1 concatenate for any finite times including zero.

We use $|RE|$ to represent the length of regular expression RE, which is the total number of characters in RE. *lmin* denotes the length of shortest string which can be accepted by RE. In this paper, we illustrate an automaton as ThompsonNFA to recognize strings.

Example 1. As shown in Fig. 1, consider $RE = (b|e)d^*a(c|d)$ and a string $T = \mathtt{abdacadbedabbedacbdacadc}$. The first occurrence $T[1,4]$ passes through ThompsonNFA starting from position 1. The other two occurrences of RE are $T[13,16]$ and $T[17,20]$.

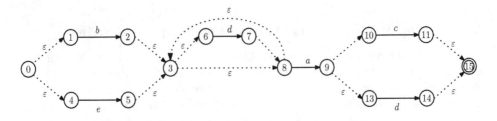

Fig. 1. ThompsonNFA of $RE = (b|e)d^*a(c|d)$

3.2 Positive and Negative Factors

Regular expression matching is aimed at recognizing multiple substrings in a long string. A lot of works focus on the common properties of these multiple strings. Based on the representation of regular expression, Yang et al. in [13] introduce *positive factor* and *negative factor*.

Definition 1 (*Positive factor*). *Positive factors are certain substrings in regular expression and must appear in the occurrence. The positive factor contains prefix and suffix factors. Given a regular expression RE, P is a collection of strings which are prefixes of RE with the length lmin. $fp \in P$ is called a prefix factor with respect to RE abbreviated as* prefix. *The definition of suffix factor is similar to the prefix.*

Definition 2 (*Negative factor*). *Given a regular expression RE, N is a collection of strings which can not appear in a string recognized by RE. $fn \in N$ is called a negative factor of RE abbreviated as* n-factor *such no string $\Sigma^* fn \Sigma^*$ can be accepted by.*

Example 2. Given a regular expression $RE = (\text{b}|\text{e})\text{d}^*\text{a}(\text{c}|\text{d})$, the prefix factors of length $lmin = 3$ are {bda, bdd, eda, edd} and the suffix factors include {dac, dad}. We see that a set of strings {aa, ab, bc, de, cc, ccc, ...} must not happen in the occurrence of RE. Let F be a factor and $F = T[i, j]$, and $T[i, j]$ be a *matching* of factor F.

3.3 LZ77 Compression and Self-index

For a finite alphabet set Σ, each character $T[i]$ in the sequence T belongs to Σ. Let $T[i, j]$ be the substring of T ranging from the i-th character to the j-th character.

The LZ77 [4] compression scheme parses a string $T[1, n]$ into a sequence $Z[1, n']$ of phrases such that $T = Z[1]Z[2]...Z[n']$. Given $T[1, i - 1]$ produced the sequence $Z[1, p - 1]$, we extract the longest prefix $T[i, i' - 1](i < i' \leqslant n)$ appeared in $T[1, n]$, then set $Z[p] = T[i, i']$ and continue with $i = i' + 1$. If $T[j, j'] = T[i, i' - 1]$ where $1 \leqslant j, j' \leqslant i - 1$, we call $T[j, j']$ is the *source* of phrase $Z[p]$. $T[i']$ is the trailing character. Absolutely, the dictionary of LZ77 is a collection of substrings appeared before.

Example 3. An example of LZ77 parsing is shown in Fig. 2. Assume we have a sequence $T = \text{abdacadbedabbedacbdacadc}$. The identifier is on the top of every phrase. The prefix of $T[2]$ does not appear in $T[1]$, so we assume ε as the source of $T[2]$ and parse $Z[2] = T[2]$. $T[4]$ is the longest prefix appeared in $T[1]$ so that the 4th phrase is $Z[4] = T[4, 5]$.

1	2	3	4	5	6	7	8	9
a	b	d	ac	ad	be	dab	bedac	bdacadc

Fig. 2. An example for LZ77 parsing

Navarro in [4] designs *self-index* based on LZ77 compression, which consists of two tries and a range structure. On the left of Fig. 3 is the reverse trie, which indexes all the reverse phrases in LZ77. The leaf nodes *revid* in rectangle denote the identifier of phrases. Suffix trie is constructed with all the phrases of LZ77 parsing and leaf nodes are denoted as *id*. The 2-D mesh on the right of Fig. 3 shows the point $id = revid + 1$ in the range structure which concatenates the ranks of all the connected phrases in both tries.

3.4 Problem Definition

Given a regular expression RE and a compressed string T of length n, the *occurrence* represents the substring in T which can be recognized by RE. The regular expression matching problem is to find all the occurrences of RE in T.

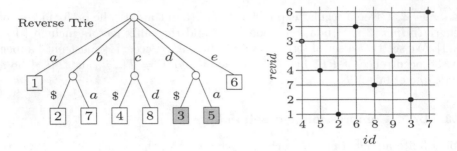

Fig. 3. LZ77 self-index for a string $T = \texttt{abdacadbedabbedacbdacadc}$

4 Data Structure: SSLZ

In Navarro's self-index [4], suffix trie indexes phrases connected with the prior phrase. However, they have to check all the sources one by one to infer all the matchings contained in the phrase. Here we put suffixes of the phrase associate with suffix trie [4] to locate factors in the compressed string. The improved suffix trie integrates all the suffixes in the phrases. The number of node *ssid* denotes the identifier of suffixes in the phrase. The gray node in the trie marks the *phrase node* which demonstrates the entire phrase taken place as a suffix. For simplicity of expression, we omit some of the duplication points. The structure of improved suffix trie in SSLZ self-index is shown in Fig. 4. In the bit vector B_z, we have symbol '1' to mark the end of the phrase. A bitmap SB_s marks the phrase node in the suffix trie by symbol '1'. For instance, the suffix of 7-th phrase ab

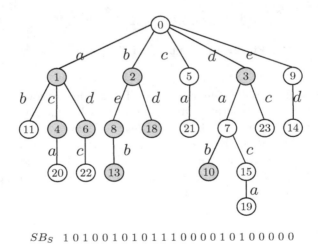

SB_S 1 0 1 0 0 1 0 1 0 1 1 1 0 0 0 0 1 0 1 0 0 0 0 0

B_z 1 1 1 0 1 0 1 0 1 0 0 1 0 0 0 1 0 0 0 0 0 0 1

Fig. 4. The suffix trie of SSLZ for a string $T = \texttt{abdacadbedabbedacbdacadc}$

is indexed in the leaf node 11. Since it is not a phrase node, $SB_s[2] = 0$ and $B_z[11] = 0$. The 4-th phrase **ab** is also the suffix of itself, therefore the leaf node 4 is a phrase node with $SB_s[3] = 1$ and $B_z[4] = 0$.

5   Regular Expression Matching on the LZ77 Compressed String

Considering the regular expression matching algorithm on compressed string, in this paper, we elaborate positive and negative factors into compressed string. To locate factors on compressed string quickly, we utilize SSLZ self-index introduced in previous section. According to the factors definitions in Sect. 3, we merge positive factors between two n-factors to detect candidates and propose a fast merging strategy with bitmaps. To prune false negative candidates, we design block filtering strategy employing active states of ThompsonNFA.

5.1   Locating Factors on SSLZ

We show how to locate factors on the SSLZ. The matching form of factors is represented as a two-tuple (s, o) for s denotes identifier of phrase that factors are started from and o denotes the offset of factors in the phrase. Two types of matchings exist exactly in T. For the matchings across the phrases, we divide the factor F into two parts Fl and Fr. For the right part, we traverse Fr on the suffix trie from the root and check the characters on the path. When we reach the branch node, only the leaf nodes of subtree rooted at the branch node indicate the phrases with prefix Fr, which must satisfy Fomula 1. Meanwhile, we also match the reversed Fl on the reverse trie. The range structure splices ranks of the nodes from two tries if they come from the adjacent phrases in the LZ77 parsing, where $revid = id + 1$. So we get $s = id$ and $o = |id| - |Fl|$, where $|id|$ denotes the length of id-th phrase and $|Fl|$ denotes the length of left part of factors.

$$id = \begin{cases} rank_1\,(SB_s,\ ssid_i) & access\,(SB_s,\ ssid_i) = 0, \\ rank_1\,(SB_s,\ ssid_i) - 1 & access\,(SB_s,\ ssid_i) = 1. \end{cases} \tag{1}$$

$access\,(B, i)$ returns the symbol at position i in B, and $rank_b\,(B, i)$ is the number of matchings of bit b in B.

We traverse the suffix trie directly to obtain matchings in the phrase. There is no need trailing matchings from the previous matchings one by one. We implement reversed suffix trie by binary search [4] to constrain the space consuming of the improved suffix trie.

Note that the offset of matchings contained in the phrase is computed by $o = select_1\,(B_z,\ id_i) - ssid_i{}^1$.

---

[1] $select_b(B,\ i)$ is the position of i-th bit$b$matchings in B.

Example 4. Considering the same regular expression $RE = (b|e)d^*a(c|d)$, dac is one of the suffix factors. For the first type matchings, S is divided into two parts $F_l = $ d and $F_r = $ ac. On the left part of Fig. 3, gray nodes in the reverse trie imply the phrases with suffix d. $F_r = $ presents as the prefix of phrases in interval $[3, 4]$ with $ssid = 4$ that $id = 4$ in Fig. 4. We traverse crossing area between interval $[3, 4]$ and $[6, 7]$, so we get a matching started from the 3rd phrase for $s = 3$ and $o = 0$.

5.2   Merging Positive Factors

After we find all the matchings of factors in the phrases based on SSLZ, the matchings of factors are integrated in a list L in ascending order by their (s, o). We can not decide which one appears first based on the phrase identifier when factors gather in the same phrase. For this purpose, we use offsets o of factors in the phrase to distinguish order of factors in the same phrase. According to the properties of positive factors, we verify all the unions of prefix and suffix factors to get all the occurrences of RE.

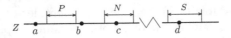

Fig. 5. N-factor in phrase

N-factors [15] can help us accelerate merging process and terminate early. Figure 5 shows the impact of n-factor on candidates in the phrases. Since n-factors in the context could not be accepted by the regular expression, the matching list is delimited into the safety margins. We only need inspect the suffixes which are dominated by the prefix between two continuous n-factors.

Assume the n-factor N, prefix P, suffix S, and N' denoting sequent factors in L. The prefix P dominates the suffix S $(P \Rightarrow S)$, iff they satisfy the following conditions:

- $N_s < P_s < S_s < N'_s$;
- If $N_s = P_s = S_s$, then $P_o > N_o$ and $P_o < S_o + lmin - 1 < N'_o + |N|$.

Figure 6 illustrates factors of Example 2 in a two-dimensional grid. We utilize the vertical axis to express the end phrase of the factors. Horizontal axis is the start phrases of factors. Disc points in the figure indicate prefixes. The prefix dominates the suffix $(P \Rightarrow S)$ marked by circle in a triangular area limited by n-factors. Intuitively, the $P \Rightarrow S$ in the phrase provides a potential region started from prefix to suffix. All the $P \Rightarrow S$ pass through the automaton to achieve the final positions of the occurrences of RE. Inspired by PNS pattern [15], we propose an efficient merging strategy based on phrases without decompression.

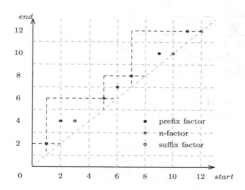

Fig. 6. Prefix dominating the suffix

5.3 Accelerating Merging Factor Using Bitmaps

Once we achieve all the matchings of factors, our next step is to merge these factors efficiently. We use bitmaps to track available $P \Rightarrow S$. If L_i is a phrase where the prefix started from, then bitmap B_p satisfies $access\,(B_p,\ i) = 1$. Similarly, bitmaps B_s and B_n are corresponding to suffixes and n-factors in L.

Algorithm 1 shows the pseudocode of merging positive factors using bitmaps. Before processing factors in L, we append a n-factor at the end of list L in lines 2–3 so as to make the last potential region consistent with the previous ones. Firstly, we figure out the next n-factor N_n in line 4, we look for the last suffix S_l in L which meet $P \Rightarrow S$ appeared before S_n in lines 5–8. Secondly, we traverse the prefixes between n-factor N_{n-1} and N_n to check out whether they dominate the suffix S_l in lines 9–12. Note that several methods address these operation on bitmaps in constant time. We execute the solution of Rodrigo Gonzälez in [3].

Example 5. Reexamine the example in Fig. 6. Table 1 shows bitmaps of the factors lying in the list L. We start at $i = 4$ in the list, the next n-factor appears at $select_1\,(B_n, i+1) = 12$. With $rank_1\,(B_s, 12) = 3$, three suffixes appear ahead of L_{12}. We get the position of 3rd suffix in L by $select_1\,(B_s, 3) = 11$. Then we detect all the prefix satisfying $P \Rightarrow S_n$. Because $select_1\,(B_n, i) = 7$ and $rank_1\,(B_p, 12) - rank_1\,(B_p, 7) = 4 - 2 = 2$, before L_{12}, there are two prefixes which dominate L_{11}. The prefixes fall in $select_1\,(B_p, 4) = 10$ and $select_1\,(B_p, 3) = 8$. We achieve a candidate set $C_{re} = (10, 12, 8, 12)$.

The positive and negative factors are constructed on regular expression RE and alphabet set Σ. However, these techniques are out of operation when the n-factors are non-uniform distribution in the string. If n-factors gather in a small area, there are still many positive factors to be verified.

Algorithm 1. MergingPositiveFactors

Input: The alignment of list L, Bitmap of prefix B_p, Bitmap of suffix B_s,
　　　　Bitmap of n-factor B_n;
Output: The candidate set $C_{re} = (P, S)$;

1　int $i = 1, k = 1$;
2　**if** L_i *is not a n-factor* **then**
3　　　Add a n-factor at the end of L

4　**while** $select_1(B_n, i+1) < L.size()$ **do**
5　　　$nc = select_1(B_n, i)$; $nn = select_1(B_n, i+1)$
6　　　$k = rank_1(B_s, nn)$; $sn = select_1(B_s, k)$
7　　　$pcount = rank_1(B_p, nn) - rank_1(B_p, nc)$;
8　　　//Check the previous prefix in L
9　　　**while** $pcount! = 0$ **do**
10　　　　int $j = rank_1(B_p, nn)$ $pn = select_1(B_p, j)$ **if** $pn > nc$ **then**
11　　　　　$C_{re}.push(L_{pn}, L_{sn})$;

12　return C_{re} ;

Table 1. Bitmaps of factors in phrases

L	1	2	3	4	5	6	7	8	9	10	11	12
id	1	2	3	4	5	6	7	8	8	9	9	9
o	0	0	0	1	1	1	1	1	1	2	0	1	3
B_p	0	1	0	0	0	1	0	1	0	1	0	0
B_s	0	0	1	0	0	0	0	0	1	0	1	0
B_n	1	0	0	1	1	0	1	0	0	0	0	1

5.4　Block Filtering

In this section, we describe how to reduce the scale of candidates with *block filtering*. Figure 7 show two ultra types of $P \Rightarrow S$ in the phrases. Figure 7(i) presents that $P \Rightarrow S$ crosses through several phrases. $P \Rightarrow S$ just takes place in the same phrase in Fig. 7(ii) denote as $P \mapsto S$.

For the first type, we utilize block filtering to detect the $P \Rightarrow S$ which are false positives. Consider regular expression RE, we maintain a state set π_λ for phrase nodes in the suffix trie and π_μ for leaf nodes in the reverse trie. π_λ describes the active states in automaton after reading phrases from the root and π_μ denotes the active states in reverse automaton [6] which accept reverse phrases.

Recall that LZ77 compression scheme parses depending on duplicate substrings. Let $Z[p]$ be the source of phrase $Z[p']$ in the parsing. Since $Z[p']$ makes one character different from $Z[p]$. If the set of active states $Z[p]$ denote as π_μ^p, and the trailing character is α, the active states of $Z[p']$ is $\pi_\mu^{p'} = \delta(\pi_\mu^p, \alpha)$ where δ is defined as the transition map.

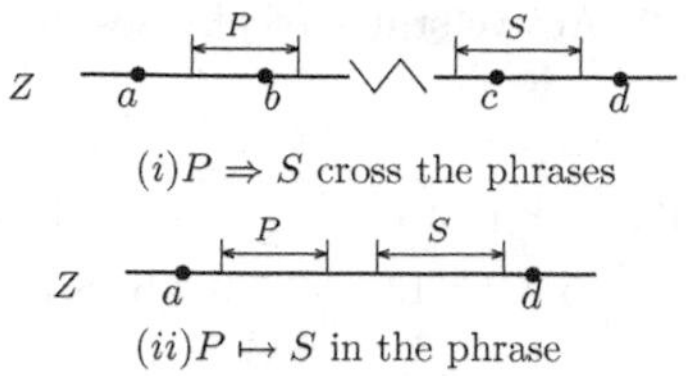

$$
\begin{array}{c}
\text{(i)}P \Rightarrow S \text{ cross the phrases} \\[4pt]
\text{(ii)}P \mapsto S \text{ in the phrase}
\end{array}
$$

Fig. 7. Two types of candidates in LZ77

Lemma 1 *(Block Filtering). Given a sequential list L of factors matchings in the phrases, the active states of r-th phrase π_μ^r and reverse active state of $(r+1)$-th phrase π_λ^{r+1}. Supposing $\pi_\mu^r \cap \pi_\lambda^{r+1} = \emptyset$, a candidate $P \Rightarrow S$ meet the following inequations that could not be a final occurrence.*

$$
P_s + P_o < Z_s^r < S_s + S_o \tag{2}
$$
$$
P_e < Z_s^{r+1} Z_e^{r+1} < S_e \tag{3}
$$

Z_s^r and Z_e^r denote the start and end position of phrase Z^r. P_e and S_e denote the end position of prefix and suffix factors.

As shown in Fig. 8, P_i is the i-th prefix in L and S_j is the j-th suffix in L. Z^r denotes the r-th phrase in LZ77 parsing. If $\pi_\mu^r \cap \pi_\lambda^{r+1} = \emptyset$, we called that there is a *break point* between phrases Z^r and Z^{r+1}. We skip (P_{i+2}, S_j) and check the other candidates (P_{i+2}, S_j) which satisfy $P_s^{i+2} + P_o^{i+2} < Z_s^{r+1} < S_s^j + S_o^j$.

Note that for the prefix crosses the break point we still have to detect the consecutive phrases. Because we have no idea the active states of suffixes of phrases which is actually the prefix of prefix factor. We need to decompress the gap between prefix and suffix factors for second type $P \mapsto S$ in the phrase.

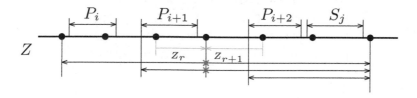

Fig. 8. Block filtering on LZ77 compression

Reexamine in Fig. 6, Table 2 presents the active states of ThompsonNFA π_μ and π_λ for every phrase in LZ77 parsing. According to Lemma 1, a break point is between $Z[1], Z[2]$ and $Z[4], Z[5]$, because $\pi_\mu^1 \cap \pi_\lambda^2 = \emptyset$.

Table 2. Active states of phrases in LZ77

id	1	2	3	4	5	6	7	8	9
π_μ	9,10,13	2,3,6,8	7,8,14,15	11,15	14,15	ϕ	ϕ	ϕ	ϕ
π_λ	8,7,3,2,5	1	6,3,2,5,13,9	8,7,3,2,5	8,7,3,2,5	ϕ	ϕ	ϕ	ϕ

6   Experiments

The experiments of our algorithm are put into effect over Intel Core CPU running at 3.40 GHz with 8 GB memory and 1 TB disk. We execute on GCC version 4.8.2 in Ubuntu 64-bit operating system. Our self-index are constructed offline.

Our experiments are evaluated on real repetitive sources DNA, Chrome, Einstein, Kernel and Leaders which are wildy used in performance measurement of compressed indexes. Einstein is an English version article provided by Wikipedia. Kernel is consist of 36 versions of Linux Kernel from 1.0.x to 1.1.x. Leaders covers the files of CIA World Leaders. The last three datasets can be downloaded from http://pizzachili.dcc.uchile.cl/repcorpus.html.

Table 3 shows the compression ratio of several methods. We study our index based on five datasets. Kreft describe two variant Ziv-Lemple compression LZ77 and LZEnd in [4]. RLCSA is another compression scheme proposed by [10]. We maintain the improved suffix trie by binary search in SSLZ-B. As shown in Table 3, due to appending all the suffixes of phrases into our suffix trie. Our index SSLZ-B almost costs 3–4 times space more than LZ77 self-index and nearly twice over LZEnd index size.

Table 3. Ratio between index and the original string

Dataset	LZ77	LZEnd	SSLZ	SSLZ-B	RLCSA
DNA	1.602	2.099	10.654	4.021	0.0632
Chrome	1.404	1.8885	11.942	3.885	0.0187
Einstein	0.045	0.065	9.505	2.786	0.928
Kernel	1.857	2.175	10.229	3.975	0.932
Leader	0.438	0.612	10.235	3.099	0.987

6.1   Performance of Locating Factor

Locating factor is the exact matching of multiple patterns on compression. We give the performance of locating factor on Kernel, Leaders, and Einstein in Fig. 9. In SSLZ-B we take the advantage of binary search for saving space. However the locating time of our two indexes very nearly. As shown in Fig. 9(a), there is a peak in LZEnd and LZ77 respectively when the pattern length is 6. In our method the lines are relatively smooth. This is because LZ77 and LZEnd are sensitive to the frequency of queries. They search the string relying on the matching of pattern in previous. Our improved self-index SSLZ blunts the effect of frequency. For pattern length of 4 in Leaders Fig. 9(b) it takes 142 ms to get 134760 matchings and LZ77 cost 874 ms on average. We also see the same trend in Fig. 9(c).

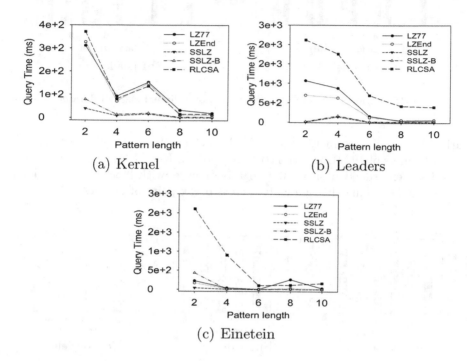

(a) Kernel   (b) Leaders

(c) Einetein

Fig. 9. Performance of factor extraction

Figure 10 shows the implementation of our algorithm with growing size of datasets. As dataset Einstein, Leaders and Kernel from 2 MB to 10 MB, the relationship between the extraction time and dataset scale is proportional. In Fig. 10(a), we choose six patterns with length from 1 to 6. The shorter the pattern is, the greater probability it appears in the dataset. Therefore the query times fall down when query length with $m = 5$ and $m = 6$ locates 3740 matchings of the pattern. SSLZ index spends 44 ms to locate pattern $m = 3$ in Kernel 6 MB and 15 ms for $m = 3$ in Leaders 6 MB as shown in Fig. 10(b).

6.2 Performance of Regular Expression Matching

Pruning Power. We study the pruning power of our algorithm in Fig. 11. Pruning power is the ratio between false positive answers which are separated out in merging processing and number of candidates. Figure 11(a) carries out that with length of regular expression $|RE|$ ranging from 5 to 10, our algorithm skips

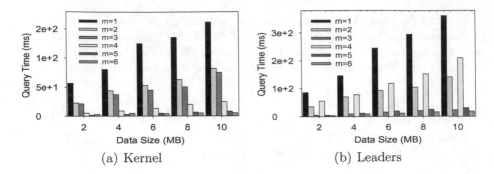

Fig. 10. Performance of factor extraction on different sizes of dataset

nearly 81% candidates in Einstein. We also achieve 85% on average when we process Leaders of 10 MB in Fig. 11(b) and 83% on Kernel. This is because our method are embedded positive and negative factors with block filtering which help us avoid checking the candidates across break point of phrases.

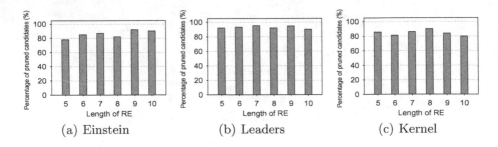

Fig. 11. Pruning power on different datasets

Figure 12 shows the pruning power of growing data size. In Fig. 12(a), we search regular expression for $|RE| = 8$, our algorithm prunes 83% candidates in Einstein. The pruning power for Chrome reaches 90% on average in Fig. 12(c). With the growth of data size in Fig. 12(b), we get the similar trend in Leaders. The characters are evenly distributed in datasets and the max length of the phrases in compressed Chrome is 14, therefore block filtering has a good performance.

Running Time. Fig. 13 shows the running time of our algorithm compared with the exiting regular expression method. Figure 13(a) presents the regular expression matching on the original string NR-Grep[2] and RE2[3]. We implement these methods on Einstein for length of regular expression $|RE|$ from 2 to 6. Our algorithm RELZ achieves a better performance than the others. Figure 13(b) shows

---

[2] http://www.dcc.uchile.cl/~gnavarro/pubcode/.
[3] https://github.com/google/re2/.

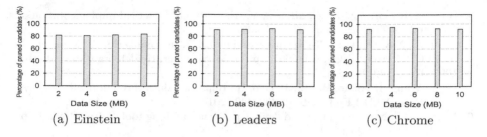

(a) Einstein   (b) Leaders   (c) Chrome

Fig. 12. Pruning power on different sizes of dataset

the comparison with regular expression matching method RELZ78 on LZ78 compression [1]. We also perform RELZ versus RELZ78 with varying datasize in Fig. 13(c).

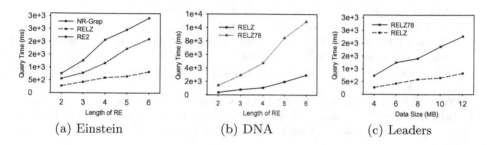

(a) Einstein   (b) DNA   (c) Leaders

Fig. 13. Comparison of running time on different datasets

7   Conclusion

In this paper, we propose an efficient solution for regular expression matching on LZ77 compressed string. The positives and negative factors are implanted into our methods. We develop a variant self-index SSLZ to extract matchings of factors on compressed string. We propose a new strategy to speed up merging factors with bitmaps and propose block filtering to reduce the scale of false positive candidates to be verified. In practise we implement our algorithm compared to existing works and show the good performance on several real datasets.

References

1. Bille, P., Fagerberg, R., Gortz, I.L.: Improved approximate string matching and regular expression matching on Ziv-Lempel compressed texts. In: Proceedings of the 18th Annual Conference on Combinatorial Pattern Matching, pp. 52–62 (2007)
2. Gagie, T., Gawrychowski, P., Kärkkäinen, J., Nekrich, Y., Puglisi, S.J.: LZ77-based self-indexing with faster pattern matching. In: Pardo, A., Viola, A. (eds.) LATIN 2014. LNCS, vol. 8392, pp. 731–742. Springer, Heidelberg (2014). doi:10.1007/978-3-642-54423-1_63

3. Gonzlez, R., Grabowski, S., Mkinen, V., Navarro, G.: Practical implementation of rank and select queries, pp. 27–38 (2005)
4. Kreft, S., Navarro, G.: Self-indexing based on LZ77. In: Giancarlo, R., Manzini, G. (eds.) CPM 2011. LNCS, vol. 6661, pp. 41–54. Springer, Heidelberg (2011). doi:10. 1007/978-3-642-21458-5_6
5. Li, Z., Wang, H., Shao, W., Li, J., Gao, H.: Repairing data through regular expressions. Proc. VLDB Endow. **9**(5), 432–443 (2016)
6. Navarro, G.: NR-grep: a fast and flexible pattern-matching tool. Softw. Pract. Exp. **31**(13), 1265–1312 (2001)
7. Navarro, G.: Regular expression searching on compressed text. J. Discrete Algorithms **1**(5–6), 423–443 (2003)
8. Navarro, G., Raffinot, M.: Fast regular expression search. In: Vitter, J.S., Zaroliagis, C.D. (eds.) WAE 1999. LNCS, vol. 1668, pp. 198–212. Springer, Heidelberg (1999). doi:10.1007/3-540-48318-7_17
9. Navarro, G., Raffinot, M.: Compact DFA representation for fast regular expression search. In: Brodal, G.S., Frigioni, D., Marchetti-Spaccamela, A. (eds.) WAE 2001. LNCS, vol. 2141, pp. 1–13. Springer, Heidelberg (2001). doi:10.1007/ 3-540-44688-5_1
10. Schneeberger, K., Hagmann, J., Ossowski, S., Warthmann, N., Gesing, S., Kohlbacher, O., Weigel, D.: Simultaneous alignment of short reads against multiple genomes. Genome Biol. **10**(9), R98 (2009)
11. Thormpson, K.: Regular expression search algorithm. Commun. ACM **11**(6), 419–422 (1968)
12. Wu, S.: Fast text searching: allowing errors. Commun. ACM **35**(10), 83–91 (1992)
13. Yang, X., Qiu, T., Wang, B., Zheng, B., Wang, Y., Li, C.: Negative factor: improving regular-expression matching in strings. ACM Trans. Database Syst. **40**(4), 1–46 (2016)
14. Yang, X., Wang, B., Li, C., Wang, J.: Efficient direct search on compressed genomic data. In: 2013 IEEE 29th International Conference on Data Engineering (ICDE), pp. 961–972 (2013)
15. Yang, X., Wang, B., Qiu, T., Wang, Y., Li, C.: Improving regular-expression matching on strings using negative factors. In: ACM SIGMOD International Conference on Management of Data, pp. 361–372 (2013)
16. Zhang, M., Zhang, Y., Hou, C.: Compact representations of automata for regular expression matching. Inf. Process. Lett. **116**(12), 750–756 (2016)
17. Ziv, J., Lempel, A.: Compression of individual sequences via variable-rate coding. IEEE Trans. Inf. Theor. **24**(5), 530–536 (1978)

Mining Top-k Distinguishing Temporal Sequential Patterns from Event Sequences

Lei Duan[1,2]([⊠]), Li Yan[1], Guozhu Dong[3], Jyrki Nummenmaa[4,5], and Hao Yang[1]

[1] School of Computer Science, Sichuan University, Chengdu, China
leiduan@scu.edu.cn, roy18@126.com, hyang.cn@outlook.com
[2] West China School of Public Health, Sichuan University, Chengdu, China
[3] Department of Computer Science and Engineering,
Wright State University, Dayton, USA
guozhu.dong@wright.edu
[4] School of Information Sciences, University of Tampere, Tampere, Finland
jyrki.nummenmaa@uta.fi
[5] Sino-Finnish Centre, Tongji University, Shanghai, China

Abstract. Sequential patterns are useful in many areas such as biomedical sequence analysis, web browsing log analysis, and historical banking transaction log analysis. Distinguishing sequential patterns can help characterize the differences between two or more sets/classes of sequences, and can be used to understand those sequence sets/classes and to identify informative features for classification and so on. However, previous studies have not considered how to mine distinguishing sequential patterns from event sequences, where each event in a sequence has an associated timestamp. To fill that gap, this paper considers the mining of distinguishing temporal event patterns (DTEP) from event sequences. After discussing the challenges on DTEP mining, we present DTEP-Miner, a mining method with various pruning techniques, for mining DTEPs with top-k contrast scores. Our empirical study using both real data and synthetic data demonstrates that DTEP-Miner is effective and efficient.

Keywords: Contrast mining · Sequential pattern · Temporal event

1 Introduction

Sequential data plays an important role in many aspects of our lives and sequence data mining has impacted those aspects in a significant way. Mining distinguishing temporal patterns from event sequences can give useful information, which can help improve the quality of our daily lives and help predict the occurrence of important events. For example, consider the issue of reducing the risk of

This work was supported in part by NSFC 61572332, the Fundamental Research Funds for the Central Universities 2016SCU04A22, and the China Postdoctoral Science Foundation 2016T90850, 2014M552371.

© Springer International Publishing AG 2017
S. Candan et al. (Eds.): DASFAA 2017, Part II, LNCS 10178, pp. 235–250, 2017.
DOI: 10.1007/978-3-319-55699-4_15

premature delivery of babies. Although the types of daily activities of most pregnant women are similar, the way, such as the order, when, and how long, they perform them vary. By comparing the daily activity sequences of pregnant women whose infants are healthy and delivered at the expected time, against those sequences of women whose infants suffer premature birth and low birth weight, harmful daily activity patterns, such as staying up late, too little physical exercises, or too much physical exercises in short periods of time, can be identified. Such patterns can guide pregnant women to reduce the risk of premature delivery.

The above need cannot be adequately addressed using existing sequential pattern mining [1] or contrast data mining [2] methods, and thus suggests a novel data mining problem. In two sets of event sequences for this scenario, we want to find the subsequences that occur frequently within certain time intervals in one data set but occur infrequently in the other data set. We refer to such a pair of a subsequence and a time interval as a *distinguishing temporal event pattern (DTEP)*. DTEP mining is an interesting problem which has many useful applications. For a second example, we may wish to mine daily activity patterns of seniors in order to give better care for seniors who need urgent attention.

While there are many existing studies on sequential pattern mining [3–5], they focus on mining frequent patterns that frequently occur in one given set of sequences. Moreover, to the best of our knowledge, all previous studies on distinguishing sequential pattern mining [6–9] do not take the temporal factor into consideration. The DTEP mining problem addressed here is different from conventional sequence data mining. Indeed, the gap between two consecutive positions for non-temporal sequences is a constant (namely 1), whereas the gap between two consecutive temporal sequence positions can vary a lot. Moreover, the two kinds of sequences are also different since timestamps offer an extra means to align temporal sequences. The above imply that the potential application and the mining methods of this problem differ significantly from the methods for sequential pattern mining and distinguishing sequential pattern mining. We will review the related work and explain the differences systematically in Sect. 3.

To tackle the DTEP mining problem, we need to address several technical challenges. First, the brute-force method, which enumerates all combinations of event type sequences and timestamp sequences, is too costly on sequence sets with a large number of event types and a large number of timestamps. We need an efficient method to avoid generating clearly useless candidates. Second, for each candidate event segment, we need to have a concise yet complete way to investigate all possible time intervals (specified by two timestamps) so that the number of candidate patterns can be reduced, and the algorithm can be efficient. Third, we need to devise techniques to efficiently find DTEPs that are ranked high with respect to an interestingness measure. This further complicates the issue since we also need to consider time lag constraints and so on.

Contributions: Besides introducing the novel problem of mining top-k DTEPs, we make several contributions in this paper. (1) Investigation of the differences among the DTEP and previously studied distinguishing sequential patterns (including contrast patterns). (2) Development of algorithmic techniques, including several pruning rules and a time interval tree, for mining top-k DTEPs efficiently. (3) Related experimentation on both synthetic and real data sets that demonstrates the effectiveness, stability, and efficiency of our method.

Road Map: The rest of the paper is organized as follows. We formulate the problem of mining top-k distinguishing temporal event patterns in Sect. 2, and review related work in Sect. 3. In Sect. 4, we present the framework of our DTEP-Miner, and discuss the critical techniques in DTEP-Miner. We report a systematic empirical study in Sect. 5, and conclude the paper in Sect. 6.

2   Problem Formulation

We start with some preliminaries. Let $\mathcal{E}$ be the set of all possible *event types*. Examples of event types include "reading" or "sleeping" etc. We use the symbol e, possibly with subscripts, to denote event types. We use a series of continuous non-negative integers starting from 0 to denote the time points; we use $\mathcal{T}$ to denote the maximum time point. Without loss of generality, we assume that the smaller the value, the earlier the time point, and that the interval between any two consecutive time points is a constant. We note that not all time points need to be associated with events.

An *event* is a pair (e, t) where $e \in \mathcal{E}$ is an event type and t is the timestamp of e. An *event sequence* S is a list of events, ordered by their timestamps, of the form $S = <(e_1, t_1), (e_2, t_2), ..., (e_n, t_n)>$, where $e_i \in \mathcal{E}$, $t_i \in [0, \mathcal{T}]$, and $t_i \leq t_j$ $(1 \leq i < j \leq n)$. The *length* of S is the number of events in S, denoted by $|S|$. We denote by $S[i]$ the i-th element in S $(1 \leq i \leq |S|)$. For $S[i]$, we use $S[i].e$ to denote the event type, and use $S[i].t$ to denote the timestamp. Taking S_1 in Table 1 as an instance, $|S_1| = 4$, $S_1[2] = (e_3, 32)$, $S_1[2].e = e_3$ and $S_1[2].t = 32$.

Table 1. A toy set of event sequences

ID	Event sequence	Set
S_1	$<(e_2, 10), (e_3, 32), (e_2, 36), (e_3, 89)>$	D_+
S_2	$<(e_2, 24), (e_1, 46), (e_2, 56), (e_3, 64), (e_1, 88)>$	
S_3	$<(e_4, 10), (e_1, 34), (e_2, 36), (e_3, 38), (e_3, 89)>$	
S_4	$<(e_3, 32), (e_2, 36), (e_1, 46), (e_3, 89)>$	
S_5	$<(e_1, 31), (e_1, 46), (e_2, 54), (e_3, 64), (e_1, 88)>$	
S_6	$<(e_1, 13), (e_2, 34), (e_1, 65), (e_3, 88), (e_4, 95)>$	D_-
S_7	$<(e_2, 19), (e_2, 36), (e_4, 46), (e_2, 56), (e_1, 99)>$	
S_8	$<(e_1, 22), (e_4, 46), (e_2, 54), (e_3, 72)>$	
S_9	$<(e_4, 10), (e_1, 34), (e_2, 43), (e_1, 46), (e_2, 56)>$	
S_{10}	$<(e_2, 19), (e_2, 36), (e_3, 45), (e_1, 99)>$	

A *time interval* w is a sub-interval of $[0, T]$ of the form $w = [w.t_s, w.t_e]$ satisfying $0 \le w.t_s < w.t_e \le T$. The *time span* of w, denoted by $||w||$, is the number of time points in w, i.e., $||w|| = w.t_e - w.t_s + 1$. For time intervals w and w', w is a *sub-interval* of w', denoted by $w \subset w'$, if $w.t_s \ge w'.t_s$ and $w.t_e \le w'.t_e$.

An *event segment* E is an ordered list of event types of the form $E = <e_1, e_2, ..., e_n>$, where $e_i \in \mathcal{E}$ $(1 \le i \le n)$. Similarly, we denote by $E[i]$ the i-th element in E $(1 \le i \le |E|)$. We say E is a *super-sequence* of E', denoted by $E' \sqsubset E$, if there exist integers $1 \le k_1 < k_2 < ... < k_{|E'|} \le |E|$ such that $E' = <E[k_1], E[k_2], ..., E[k_{|E'|}]>$. For example, $<e_2, e_4> \sqsubset <e_1, e_2, e_3, e_4>$.

A *time lag constraint* ℓ is specified by two nonnegative integers of the form $\ell = [\ell_{min}, \ell_{max}]$ satisfying $0 \le \ell_{min} \le \ell_{max} \le T$. (Time lag constraints are similar to, but essentially different from, gap constraints for standard sequences.) Given an event sequence S, we say that event segment E *matches* S within time interval w satisfying time lag constraint ℓ, denoted by $(E, w) \sqsubseteq_\ell S$, if there exist integers $1 \le k_1 < k_2 < ... < k_{|E|} \le |S|$, such that

(i) $S[k_i].e = E[i]$, $w.t_s \le S[k_i].t \le w.t_e$ for all $1 \le i \le |E|$, and
(ii) $S[k_j + 1].t - S[k_j].t \in [\ell_{min}, \ell_{max}]$ for all $1 \le j \le |E| - 1$.

We also say that $<S[k_1], S[k_2], ..., S[k_{|E|}]>$ is an instance of E in S within w. For a given time lag constraint, E may match S within different time intervals.

Example 1. Let $E = <e_1, e_2>$, $w = [46, 56]$, and $\ell = [0, 10]$. Consider S_2 in Table 1. As $S_2[2].e = E[1] = e_1$, $S_2[3].e = E[2] = e_2$, $S_2[2].t \ge 46$, $S_2[3].t \le 56$, and $S_2[3].t - S_2[2].t \in \ell$, it follows that $(E, w) \sqsubseteq_\ell S_2$.

The *support* of an event segment E within time interval w satisfying time lag constraint ℓ in an event sequence set D, denoted by $Sup(D, (E, w))$, is

$$Sup(D, (E, w)) = \frac{|\{S \in D \mid (E, w) \sqsubseteq_\ell S\}|}{|D|}. \tag{1}$$

Definition 1. *Given two sets of temporal sequences, D_+ and D_-, the* contrast score *of (E, w) targeting D_+ against D_-, denoted by $cScore(E, w)$, is*

$$cScore(E, w) = Sup(D_+, (E, w)) - Sup(D_-, (E, w)) \tag{2}$$

Example 2. Consider Table 1 again. Let $E = <e_3, e_2>$, $w = [32, 36]$, $\ell = [0, 10]$. Then $Sup(D_+, (E, w)) = 0.4$ ((E, w) matches S_1 and S_4) and $Sup(D_-, (E, w)) = 0.0$. So, $cScore(E, w) = 0.4 - 0.0 = 0.4$.

Definition 2. *Given two sets of event sequences, D_+ and D_-, time lag constraint ℓ, and event segment E, a tuple (E, w) $(w \subseteq [0, T])$ is a DTEP targeting D_+, if the following conditions are true:*

(i) *(positive contrast) $cScore(E, w) > 0$;*
(ii) *(distinguishing temporal occurrence) There does not exist w' such that (E, w') satisfies Condition (i), and*
    - *$cScore(E, w') > cScore(E, w)$, or*
    - *$cScore(E, w') = cScore(E, w)$ and $||w'|| < ||w||$.*

To select top-k DTEPs, we first define a total order on all discovered DTEPs.

Definition 3. *Given two DTEPs (E, w) and (E', w'), $(E, w) \succ (E', w')$ (called (E, w) precedes (E', w') or (E, w) has a higher precedence than (E', w')) if:*

1. $cScore(E, w) > cScore(E', w')$, *or*
2. $cScore(E, w) = cScore(E', w')$, *but* $|E| > |E'|$, *or*
3. $cScore(E, w) = cScore(E', w')$ *and* $|E| = |E'|$, *but* $||w|| < ||w'||$
4. *all of the three parameters are the same, but $E[i]$ is lexically smaller than $E'[i]$, where $i = \min\{j \mid E[j] \neq E'[j], 1 \leq j \leq |E|\}$.*

Given k and ℓ, the problem of mining top-k distinguishing temporal event patterns is to find the DTEPs with top-k precedence targeting D_+ against D_-. Table 2 lists the top-5 DTEPs discovered from Table 1.

Table 2. List of top-5 DTEPs in Table 1 ($\ell = [0, 10]$)

Rank	DTEP (E, w)	$cScore(E, w)$
1	$(<e_3>, [32, 64])$	$1.0 - 0.2 = 0.8$
2	$(<e_1, e_2, e_3>, [34, 64])$	$0.6 - 0.0 = 0.6$
3	$(<e_1>, [34, 46])$	$0.8 - 0.2 = 0.6$
4	$(<e_3, e_2>, [32, 36])$	$0.4 - 0.0 = 0.4$
5	$(<e_2, e_3>, [54, 64])$	$0.4 - 0.0 = 0.4$

3   Related Work

Our study is related to the existing work on sequential pattern mining and contrast data mining. We review the related work briefly here.

Sequential pattern mining is a well studied subject in data mining. A comprehensive review of the abundant literature on sequential pattern mining is clearly beyond the capacity of this paper. Previous work of sequential pattern mining can be categorized by the type of sequences being studied. The first category is for sequences whose elements do not have explicit timestamps. Most studies of this category mainly focus on exploring an approach to discover patterns that frequently occur in a given set of sequences. Typical algorithms include GSP [3], PrefixSpan [4], and SPAM [5]. The second category is for the sequence consisting of time-point based events. Several studies were devoted to find temporal dependencies between two events within a specific interval (time lag) [10,11]. Moreover, Tang *et al.* [11] pointed out that many temporal patterns, such as mutually dependent pattern [12], partially periodic pattern [13] and frequent episode pattern [14], can be considered as special cases of temporal dependencies with different lag intervals. The third category is for sequences that consist of interval-based events. Allen's temporal logics [15] are widely used to describe the complex relations among intervals [16,17].

Our work belongs to the second category, since we focus on mining DTEPs from time-point based event sequences. However, we consider two sets of event sequences, and we discover temporal patterns with significant discriminativeness between the two sets, i.e., the patterns occur frequently in one set but infrequently in the other. Thus, the problem of DTEP mining is fundamentally different from previous work on sequential pattern mining.

Contrast data mining discovers patterns and models that manifest significant differences between data sets [2]. One of the best known types of contrast patterns is *Distinguishing sequential pattern* (DSP) which describes the differences between two sets of sequences. DSP has many interesting applications, such as prediction in bioinformatics and computational biology [18].

Several methods have been proposed for DSP mining. Ji *et al.* [6] designed an approach to find the minimal subsequences that satisfy gap constraints and occur frequently in sequences of one class but infrequently in sequences of another class. Deng *et al.* [7] proposed a suffix tree-based method for DSP mining, and built a classifier for sequence data. Wang *et al.* [8] introduced the density concept into DSP mining, and proposed a method to find patterns satisfying both frequency and density thresholds. Yang *et al.* [9] considered a more general case of DSP mining where each element in a sequence is a set of items.

Although both DTEP mining and DSP mining try to find patterns whose supports differ significantly between two sets of sequences, there are two essential differences between DTEP mining and DSP mining. First, in DTEP mining, each element in a sequence is associated with a timestamp. Thus, the expression of a DTEP includes a temporal factor, which is absent in a DSP. Second, the match relationship between a pattern and a sequence is defined based on the position in DSP mining, while it is defined based on the time interval in DTEP mining.

4   Design of DTEP-Miner

4.1   Framework

As defined in Definition 2, a DTEP consists of an event segment and a time interval. In brief, the DTEP-Miner algorithm uses the following main steps in an iterative manner: (i) generating a candidate event segment E, (ii) for E, finding the time interval w that maximizes $cScore(E, w)$. In each iteration, DTEP-Miner keeps the collection of top-k DTEPs discovered so far.

For the sake of efficiency, there are three critical points in the design of DTEP-Miner. First, how to avoid generating clearly useless candidate event segments (Sect. 4.2). Second, for each candidate event segment, how to investigate all possible time intervals efficiently (Sect. 4.3). Third, for a candidate DTEP, how to calculate its contrast score efficiently (Sect. 4.4).

4.2   Candidate Event Segment Generation

To ensure that DTEP-Miner can find the DTEPs with largest contrast scores, we will enumerate all possible event segments in a systematic way (and we will

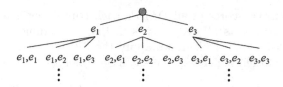

Fig. 1. An example of an event segment enumeration tree

develop pruning techniques for efficiency). Here, we adopt the set enumeration tree approach [19], which has been used in many sequential pattern mining methods. Conceptually, a set enumeration tree takes a set with a total order, the event types in the context of our problem, and then enumerates all possible combinations systematically. For example, Fig. 1 shows an example of an enumeration tree that enumerates all event segments over $\mathcal{E} = \{e_1, e_2, e_3\}$.

DTEP-Miner starts by generating $\mathcal{E} = \{S[i].e \mid S \in D_+, 1 \leq i \leq |S|\}$, i.e., the set of all event types that occur in D_+. Then, DTEP-Miner generates candidate event segments by traversing the enumeration tree in a depth-first manner.

It is time-consuming to traverse all nodes in the event segment enumeration tree. Fortunately, Theorems 1 and 2 demonstrate the monotonicity of $Sup(D, (E, w))$ with respect to E and w, respectively.

Theorem 1. *Given a set of temporal sequences D and a time lag constraint ℓ, we have $Sup(D, (E, w)) \leq Sup(D, (E', w))$ for all event segments E and E' and time interval w, provided that E is a super-sequence of E' ($E' \sqsubseteq E$).*

Theorem 2. *Given a set of temporal sequences D and a time lag constraint ℓ, then we have $Sup(D, (E, w)) \leq Sup(D, (E, w'))$ for event segment E and time intervals w and w' provided that $w \subset w'$.*

Let R denote the top-k list of DTEPs discovered targeting D_+ against D_- at a given time of the computation. Let $cScore_k$ denote the k-th largest contrast score we found so far, i.e., $cScore_k = min\{cScore(E, w)|(E, w) \in R\}$. Then, we have following corollary.

Corollary 1. *Suppose $Sup(D_+, (E', w')) < cScore_k$. Then (E, w) cannot be a top-k DTEP for all event segments E and time intervals w satisfying $E' \sqsubseteq E$ and $w \subset w'$.*

Corollary 1 leads us to a useful pruning rule, which allows us to terminate the depth-first traversal of an entire branch at the current node. Please recall that $[0, \mathcal{T}]$ is the maximal time interval.

Pruning Rule 1. *For event segment E, any super-sequence of E can be pruned, if $Sup(D_+, (E, [0, \mathcal{T}])) < cScore_k$.*

It makes sense to find DTEPs whose contrast scores are large early, so that the pruning methods give bigger impact. By this observation, DTEP-Miner first computes the contrast scores of all DTEPs each containing exactly one single event type in $\mathcal{E}$, and sorts all event types in the descending order of contrast scores. Similar to Pruning Rule 1, we apply the following rule to prune event types when none of their super-sequences can be a member of the top-k DTEPs.

Pruning Rule 2. *An event type e can be removed from $\mathcal{E}$ without loosing any valid top-k DTEPs, if $Sup(D_+, (e, [0, T])) < cScore_k$.*

Example 3. Let $k = 5$. For Table 1, $cScore_5 = 0.4$. As $Sup(D_+, (e_4, [10, 89])) = 0.1 < cScore_5$, event type e_4 can be removed from $\mathcal{E}$ in the computation.

4.3 Candidate Time Interval Generation

For each node in the event segment enumeration tree traversed by DTEP-Miner, an event segment E is generated. To generate all candidate time intervals for E, a naïve way is to enumerate all sub-intervals of $[0, T]$; this generates $O(T^2)$ candidate time intervals for E. Clearly, the time cost is high.

Observation 1. *In real datasets the number of distinct timestamps of an event type is typically much smaller than T.*

Based on Observation 1, we design a method to generate candidate time intervals based on the timestamps of elements of E that occur in the dataset under consideration. For clarity, we define the following notations.

For an event segment E, we denote by $\mathcal{W}_E$ the set of candidate time intervals of E. In addition, we use $\mathcal{W}_E^s$ and $\mathcal{W}_E^e$ to denote the sets of timestamps that can be start time points and end time points of a time interval, respectively.

To find the time intervals from $\mathcal{W}_E$ that can contribute to the top-k DTEPs efficiently, we face two challenges: (i) how to efficiently find $\mathcal{W}_E^s$ and $\mathcal{W}_E^e$, and use them to generate $\mathcal{W}_E$ (Sect. 4.3); (ii) how to efficiently find the time interval $w \in \mathcal{W}_E$ such that $cScore(E, w)$ is the maximum (Sect. 4.3).

Please note that $\mathcal{W}_E$ is computed by algorithms and different $\mathcal{W}_E$'s can be computed. Our goal is to compute smaller $\mathcal{W}_E$'s (to minimize the amount of computation) that can guarantee that we can find the correct top-k DTEPs.

Bounds of Candidate Time Intervals

Definition 4. *The* occurrence *of an event type e in a set D of event sequences, denoted by $T(D, e)$, is the set of timestamps on which e occurs in some sequence belonging to D. That is, $T(D, e) = \bigcup_{S \in D} \{S[i].t \mid S[i].e = e\}$.*

For Challenge (i), given an event segment E, one naïve way is to set $\mathcal{W}_E^s = T(D_+, E[1])$, $\mathcal{W}_E^e = T(D_+, E[|E|])$, and then get $\mathcal{W}_E$ by the cartesian product of $\mathcal{W}_E^s$ and $\mathcal{W}_E^e$

$$\mathcal{W}_E = \{[t_i, t_j] \mid t_i \in \mathcal{W}_E^s, t_j \in \mathcal{W}_E^e, t_i \leq t_j\}. \tag{3}$$

Example 4. Consider Table 1 again. For event segment $E = <e_1, e_2, e_3>$, $\mathcal{W}_E^s = T(D_+, E[1]) = \{31, 34, 46, 88\}$, and $\mathcal{W}_E^e = T(D_+, E[3]) = \{32, 38, 64, 89\}$. By Eq. 3, 10 candidate time intervals (Fig. 2) are in $\mathcal{W}_E$.

We note that the way of generating $\mathcal{W}_E$ by Eq. 3 ignores the timestamps of $E[i]$ ($1 < i < |E|$), which may reduce the size of $\mathcal{W}_E$. Thus, for the sake of efficiency, DTEP-Miner uses Eq. 3 to generate $\mathcal{W}_E$ in the case of $|E| = 1$, i.e., E is a single event type. Then, it iteratively expands the event sequences and computes the sets of candidate time intervals for longer event sequences from that for shorter event sequences.

As stated in Sect. 4.2, DTEP-Miner generates each candidate event segment by traversing the event segment enumeration tree in a depth-first manner. Let E' be the parent node of E in the tree. Then, E is composed by concatenating E' with an event type e, denoted by $E' \oplus e$: $E = E' \oplus e = < \underbrace{E'[1], E'[2], ..., E'[|E'|]}_{E'}, e>$.

Corollary 2. *Suppose $|E| > 1$ and $Sup(D, (E, w)) \geq cScore_k$. Then there exist $E' \subset E$ and $w' \subseteq w$ such that $Sup(D, (E', w')) \geq cScore_k$.*

The above corollary allows us to safely use $\hat{\mathcal{W}}_{E'} = \{w' \mid Sup(D_+, (E', w')) \geq cScore_k\}$ as the set of candidate time intervals for E'. We note this is a fairly tight approximation of the optimal $\mathcal{W}_{E'}$ although better approximation may exist. Now, we define $R(t) = \min\{w'.t_e \mid w' \in \hat{\mathcal{W}}_{E'}, w'.t_s = t\}$, i.e., $[t, R(t)]$ is the one with minimum time span among all time intervals in $\hat{\mathcal{W}}_{E'}$ starting from t. Then, for $E = E' \oplus e$, we define:

$$\mathcal{W}_E = \{[t_i, t_j] \mid t_i \in \mathcal{W}_E^s, t_j \in \mathcal{W}_E^e, R(t_i) \leq t_j\} \tag{4}$$

where, $\mathcal{W}_E^s = \{w'.t_s \mid w' \in \hat{\mathcal{W}}_{E'}\}$, $\mathcal{W}_E^e = T(D_+, e)$.

DTEP-Miner uses Eq. 4 to generate $\mathcal{W}_E$ in the case of $|E| > 1$, i.e., E is composed by more than one event type. In summary, for an event segment E, a candidate time interval w is *valid* if w satisfies the conditions specified by Eq. 3 ($|E| = 1$) or Eq. 4 ($|E| > 1$).

Evaluation with Time Interval Tree. Next, for Challenge (ii), we present our method to find the best time interval w for an event segment E such that $w = \underset{w' \in \mathcal{W}_E}{\operatorname{argmax}} cScore(E, w')$.

To evaluate all possible time intervals in a systematic and efficient way, we design a novel data structure, called *time interval tree*. The first step to build a time interval tree is sorting timestamps in $\mathcal{W}_E^s$ and $\mathcal{W}_E^e$, respectively, in ascending order. Here, we use superscripts to refer to the orders. For example, we denote by $\mathcal{W}_E^{s[i]}$ the i-th ($1 \leq i \leq |\mathcal{W}_E^s|$) smallest timestamp in $\mathcal{W}_E^s$.

Let $\mathcal{C}$ be $\mathcal{W}_E^s$ or $\mathcal{W}_E^e$. We say that $\mathcal{C}^{[i]}$ is the *predecessor* of $\mathcal{C}^{[i+1]}$, denoted by $Prec(\mathcal{C}^{[i+1]}, \mathcal{C}) = \mathcal{C}^{[i]}$; $\mathcal{C}^{[i+1]}$ is the *successor* of $\mathcal{C}^{[i]}$, denoted by $Succ(\mathcal{C}^{[i]}, \mathcal{C}) = \mathcal{C}^{[i+1]}$. For example, with $\mathcal{C} = \{31, 34, 46, 88\}$, we have $Prec(\mathcal{C}^{[3]}, \mathcal{C}) = \mathcal{C}^{[2]} = 34$ and $Succ(\mathcal{C}^{[2]}, \mathcal{C}) = \mathcal{C}^{[3]} = 46$.

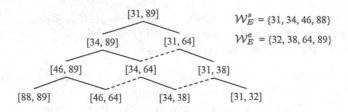

Fig. 2. A time interval tree

A time interval tree is a binary tree, in which each node is a valid candidate time interval. The rules for building a time interval tree are as follows:

1. The root is $[\mathcal{W}_E^s{}^{[1]}, \mathcal{W}_E^e{}^{[|\mathcal{W}_E^e|]}]$;
2. For each node $p = [w.t_s, w.t_e]$, p's left child is $[Succ(w.t_s, \mathcal{W}_E^s), w.t_e]$, and p's right child is $[w.t_s, Prec(w.t_e, \mathcal{W}_E^e)]$.
3. Every node is a valid time interval.

We note that, for any right child node, its left child is the same as another node's right child. For the sake of efficiency, all left child nodes whose parents are right child nodes are ignored by DTEP-Miner. Figure 2 shows an example of a time interval tree. We use dash lines to represent links to the shared nodes.

For an event segment E, DTEP-Miner evaluates each candidate time interval of E by traversing the time interval tree in a breadth-first manner. By Theorem 2, we have the following pruning rule.

Pruning Rule 3. *For event segment E and time interval w, if $Sup(D_+, (E, w)) < cScore_k$, all sub-intervals of w can be pruned.*

Although DTEP-Miner traverses all nodes in the time interval tree in the worst case, our experiments show that Pruning Rule 3 can improve the efficiency of DTEP-Miner for most cases. Algorithm 1 shows the pseudo-code of evaluating candidate time intervals using time interval tree. The time complexity of Algorithm 1 is $O(n^2)$, where n is the number of distinct timestamps.

4.4 Contrast Score Calculation

Given a DTEP (E, w), to get the contrast score of (E, w), by Eq. 2, we have to calculate $Sup(D_+, (E, w))$ and $Sup(D_-, (E, w))$ first.

Theorem 3. *Given a set of event sequences D and a DTEP (E, w), $\{S \in D \mid (E, w) \sqsubseteq_\ell S\} \subseteq \bigcap_{1 \le i \le |E|} \{S \in D \mid (E[i], w) \sqsubseteq_\ell S\}$.*

By Theorem 3, to calculate $Sup(D, (E, w))$, DTEP-Miner only checks whether $(E, w) \sqsubseteq_\ell S$ holds for $S \in \bigcap_{1 \le i \le |E|} \{S \in D \mid (E[i], w) \sqsubseteq_\ell S\}$ (instead of $S \in D$).

Finally, we present the pseudo-code of DTEP-Miner in Algorithm 2.

Algorithm 1. TimeInterval($E, \mathcal{W}_E^s, \mathcal{W}_E^e, cScore_k$)

Input: E: event segment, $\mathcal{W}_E^s$: the start time points of candidate time intervals, $\mathcal{W}_E^e$: the end time points of candidate time intervals, $cScore_k$: the k-th largest contrast score of DTEPs searched so far

Output: w: the best time interval for E.

1: $w_E \leftarrow null; max \leftarrow 0; \mathcal{W} \leftarrow \{[\mathcal{W}_E^{s[1]}, \mathcal{W}_E^{e[|\mathcal{W}_E^e|]}]\};$ // root
2: **repeat**
3:    $w \leftarrow$ the first element in $\mathcal{W}$;
4:    **if** $Sup(D_+, (E, w)) < cScore_k$ **then**
5:       prune all sub-intervals of w; // Pruning Rule 3
6:    **else**
7:       **if** $cScore(E, w) > max$ **then**
8:          $w_E \leftarrow w; max \leftarrow cScore(E, w);$
9:       **end if**
10:     **if** w is a left child node **then**
11:        $\mathcal{W} \leftarrow \mathcal{W} \cup \{[Succ(w.t_s, \mathcal{W}_E^s), w.t_e]\};$ // left child of w
12:     **end if**
13:     $\mathcal{W} \leftarrow \mathcal{W} \cup \{[w.t_s, Prec(w.t_e, \mathcal{W}_E^e)]\};$ // right child of w
14:    **end if**
15:    $\mathcal{W} \leftarrow \mathcal{W} \setminus \{w\};$
16: **until** $\mathcal{W}$ is null
17: **return** w_E;

Algorithm 2. DTEP-Miner(D_+, D_-, k, ℓ)

Input: D_+ and D_-: two sets of event sequences, k: an integer, ℓ: time lag constraint
Output: R: the set of top-k DTEPs

1: initialize $cScore_k \leftarrow 0$ and $R \leftarrow \emptyset$;
2: **for** each event type $e \in \mathcal{E}$ **do**
3:    get $\mathcal{W}_e^s, \mathcal{W}_e^e; w \leftarrow TimeInterval(e, \mathcal{W}_e^s, \mathcal{W}_e^e, cScore_k)$;
4:    **if** w is $null$ **then**
5:       remove e from $\mathcal{E}$; // Pruning Rule 2
6:    **else**
7:       update R and $cScore_k$ if $|\{P \succ (e, w) \mid P \in R\}| < k$;
8:    **end if**
9: **end for**
10: sort all event types in their contrast scores descending order;
11: **for** each event segment E searched by traversing the event segment enumeration tree in a depth-first way **do**
12:    get $\mathcal{W}_E^s, \mathcal{W}_E^e; w \leftarrow TimeInterval(E, \mathcal{W}_E^s, \mathcal{W}_E^e, cScore_k)$;
13:    **if** w is $null$ **then**
14:       prune all super-sequences of E; // Pruning Rule 1
15:    **else**
16:       update R and $cScore_k$ if $|\{P \succ (E, w) \mid P \in R\}| < k$;
17:    **end if**
18: **end for**
19: **return** R;

---

5   Empirical Evaluation

In this section, we report an empirical study using both real-world and synthetic data to verify the effectiveness and efficiency of DTEP-Miner. All experiments were conducted on a PC with an Intel Core i7-4790 3.60 GHz CPU and 16 GB main memory, running the Windows 7. All algorithms were implemented in Java and compiled by JDK 7. The unit of a timestamp is minute in our experiments.

5.1   Mining DTEPs on Activities Data Set

We apply DTEP-Miner to the ADLs data set from the UCI repository [20]. This data set records the daily living activities by two users (A and B) in their own homes ranging over 14 and 21 days, respectively. Table 3 lists all activities and their abbreviations. As each activity lasts for a certain time period, we use subscripts, 'S' and 'E', to refer to the start and the end of each activity. For example, BR_S represents the start of having breakfast.

First, we take user A and user B as the targets in turn, and apply DTEP-Miner to mine DTEPs. Table 4 lists the top-10 DTEPs targeting user A and user B, respectively. By examining the discovered DTEPs, some interesting daily activity patterns characterizing user A and user B can be identified. For example, user A frequently eats breakfast within 20 min after showering from 9:47 to 12:59 (the top-1 DTEP targeting user A), while user B frequently eats breakfast from about 9 o'clock to half past 10 (the top-10 DTEP targeting user B). Also, we can see that user A frequently does toileting before showering from about half past 9 to the noon (the top-2 DTEP targeting user A). Moreover, the top-1 DTEP targeting user B shows that user B does grooming twice within 20 min during most of the days.

Table 3. Activities and corresponding abbreviations

Breakfast (BR)	Dinner (DI)	Grooming (GR)	Leaving (LE)	Lunch (LU)
Showering (SH)	Sleeping (SL)	Snack (SN)	Spare_Time/TV (ST)	Toileting (TO)

Table 4. Top-10 DTEPs discovered from ADLs ($k = 10, \ell = [0, 20]$)

Rank	Targeting user A		Targeting user B	
	DTEP	cScore	DTEP	cScore
1	($<SH_S,SH_E,BR_S,BR_E>$, [09:47,12:59])	0.929	($<GR_S,GR_E,GR_S,GR_E>$, [00:51,17:45])	0.833
2	($<TO_S,TO_E,SH_S,SH_E>$, [09:28,12:54])	0.929	($<GR_E,GR_S,GR_E>$, [00:54,17:45])	0.833
3	($<SH_E,BR_S,BR_E>$, [09:52,12:59])	0.929	($<GR_S,GR_E,GR_S>$, [00:51,17:44])	0.833
4	($<SH_S,SH_E,BR_S>$, [09:47,12:56])	0.929	($<GR_S,GR_S,GR_E>$, [00:51,17:45])	0.833
5	($<SH_S,BR_S,BR_E>$, [09:47,12:59])	0.929	($<GR_E,GR_S>$, [00:54,17:44])	0.833
6	($<TO_E,SH_S,SH_E>$, [09:35,12:54])	0.929	($<GR_S,GR_S>$, [00:51,17:44])	0.833
7	($<TO_S,TO_E,SH_S>$, [09:28,12:50])	0.929	($<TO_S,TO_E>$, [01:40,10:21])	0.810
8	($<TO_S,SH_S,SH_E>$, [09:28,12:54])	0.929	($<TO_S>$, [01:40,10:19])	0.810
9	($<SH_E,BR_S>$, [09:52,12:56])	0.929	($<TO_E>$, [01:41,10:21])	0.810
10	($<SH_S,BR_S>$, [09:47,12:56])	0.929	($<BR_S, BR_E>$, [08:56,10:34])	0.786

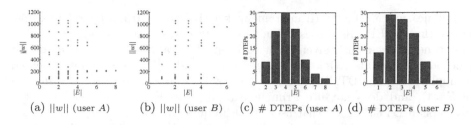

(a) $||w||$ (user A)   (b) $||w||$ (user B)   (c) # DTEPs (user A) (d) # DTEPs (user B)

Fig. 3. $||w||$ and the number of DTEPs w.r.t $|E|$ ($k = 100, \ell = [0, 20]$)

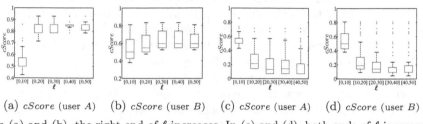

(a) $cScore$ (user A)   (b) $cScore$ (user B)   (c) $cScore$ (user A)   (d) $cScore$ (user B)

In (a) and (b), the right end of ℓ increases. In (c) and (d), both ends of ℓ increase.

Fig. 4. Contrast score w.r.t. time lag constraint ℓ ($k = 100$)

We note that many patterns for one given user in Table 4 are similar to each other. Take DTEPs targeting user A for instance. Intuitively, the top-1 DTEP is similar to the patterns whose ranks are 3, 4, 5, 9 and 10. The reason is that, as the initial work addressing DTEP mining, we did not consider the diversity of discovered DTEPs. It is an open question about how to evaluate the similarity among DTEPs; this is a future research problem.

Figure 3(a) and (b) present the time span of time interval ($||w||$) with respect to the length of event segment ($|E|$) among the top-100 DTEPs. Figure 3(c) and (d) present the event segment length distributions of the top-100 DTEPs. We see that for most DTEPs $||w||$ ranges from 100 to 200, and $|E|$ is in the range of $[2, 5]$. We also note that there is no clear correlation between $||w||$ and $|E|$.

Figure 4 presents statistics on the contrast scores of top-100 DTEPs with respect to time lag constraint. In Fig. 4(a) and (b), we increase *the right end* of the time lag constraint. Intuitively, more patterns can be found when we use "wider" time lag constraints. We can see that the contrast scores of DTEPs indeed tend to increase with wider lag constraints. In Fig. 4(c) and (d), we increase *both the left end and right end* of the time lag constraint. In this case, the time span between two adjacent event types in a pattern is increased. We can see that the contrast scores of DTEPs tend to decrease.

5.2   Efficiency

Here, we evaluate the efficiency of DTEP-Miner and its variations. (To the best of our knowledge, there were no previous methods tackling exactly the same mining problem as the one studied in this paper.) To evaluate the effectiveness of our techniques, we call the algorithm framework with Pruning Rules 1 and 2 Baseline. Recall that DTEP-Miner uses a time interval tree to speed up the finding of the best time interval for a given event segment. To evaluate the efficiency of out methods for candidate time interval generation and evaluation, we implemented a simplified version DTEP-Miner, denoted by Baseline*, that uses Eq. 3 (instead of Eq. 4), and does not use Pruning Rule 3. Moreover, we implemented the full version DTEP-Miner that uses all techniques. In our efficiency test, we set $k = 10$, and $\ell = [0, 20]$ as the default parameter values for DTEP-Miner.

We used randomly generated synthetic event sequence sets for efficiency test. There are several parameters for synthetic data generation: the size of the target set ($|D_+|$), the size of the other set ($|D_-|$), the average number of events in event sequences (denoted by NE), the number of event types ($|\mathcal{E}|$), and the maximal time point ($\mathcal{T}$). We set $|D_+| = 100$, $|D_-| = 100$, $NE = 200$, $|\mathcal{E}| = 40$, and $\mathcal{T} = 2000$ as defaults for the synthetic data generation.

Figure 5 shows the running time of DTEP-Miner with respect to k, ℓ, $|D_+|$, $|D_-|$, NE, $|\mathcal{E}|$, and $\mathcal{T}$. Logarithmic scale has been used for the runtime to better demonstrate the difference in the behavior between DTEP-Miner and the baseline methods. We see that the runtime of DTEP-Miner and the two baseline methods increases with the increase of k, ℓ, $|D_+|$, $|D_-|$ and NE. Please note that the average number of events in an event sequence is pre-fixed by parameter NE in our experiments. Thus, the larger $|\mathcal{E}|$, the smaller the average number of the occurrences of an individual event type in a sequence. Therefore, the runtime of DTEP-Miner decreases with larger $|\mathcal{E}|$. As we stated in Sect. 4.3, the complexity of DTEP-Miner for time interval evaluation depends on the number of timestamps instead of $\mathcal{T}$. The runtime of DTEP-Miner is insensitive to $\mathcal{T}$.

Clearly, DTEP-Miner runs faster than both Baseline and Baseline*, since DTEP-Miner employs a heuristic strategy to sort all event types in the descending order of their contrast scores at the beginning, and evaluates candidate time intervals using time interval tree to minimize useless computation. Baseline* is faster than Baseline because it uses the heuristic strategy of event type sorting.

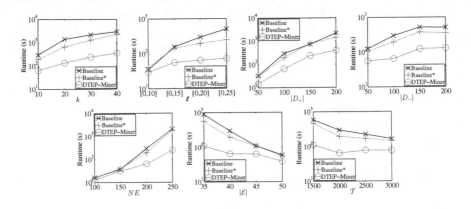

Fig. 5. Efficiency test: runtime w.r.t. k, ℓ, $|D_+|$, $|D_-|$, NE, $|\mathcal{E}|$, $\mathcal{T}$

6   Conclusions

In this paper, we studied the novel and interesting problem of mining DTEPs from event sequences. We systematically developed a method with various pruning techniques. Using both real and synthetic data sets, we verified that mining DTEPs is interesting and useful. Moreover, our experiments show that our DTEP mining method is effective and efficient.

References

1. Dong, G., Pei, J.: Sequence Data Mining. Springer, Heidelberg (2007)
2. Dong, G., Bailey, J. (eds.): Contrast Data Mining: Concepts, Algorithms, and Applications. CRC Press, Boca Raton (2013)
3. Srikant, R., Agrawal, R.: Mining sequential patterns: generalizations and performance improvements. In: Apers, P., Bouzeghoub, M., Gardarin, G. (eds.) EDBT 1996. LNCS, vol. 1057, pp. 1–17. Springer, Heidelberg (1996). doi:10.1007/BFb0014140
4. Pei, J., Han, J., Mortazavi-Asl, B., Pinto, H., Chen, Q., Dayal, U., Hsu, M.: Prefixspan: mining sequential patterns by prefix-projected growth. In: Proceedings of the 17th IEEE International Conference on Data Engineering, ICDE, pp. 215–224 (2001)
5. Ayres, J., Flannick, J., Gehrke, J., Yiu, T.: Sequential pattern mining using a bitmap representation. In: Proceedings of the 8th ACM International Conference on Knowledge Discovery and Data Mining, KDD, pp. 429–435 (2002)
6. Ji, X., Bailey, J., Dong, G.: Mining minimal distinguishing subsequence patterns with gap constraints. Knowl. Inf. Syst. **11**(3), 259–286 (2007)
7. Deng, K., Zaïane, O.R.: Contrasting sequence groups by emerging sequences. In: Gama, J., Costa, V.S., Jorge, A.M., Brazdil, P.B. (eds.) DS 2009. LNCS (LNAI), vol. 5808, pp. 377–384. Springer, Heidelberg (2009). doi:10.1007/978-3-642-04747-3_29
8. Wang, X., Duan, L., Dong, G., Yu, Z., Tang, C.: Efficient mining of density-aware distinguishing sequential patterns with gap constraints. In: Bhowmick, S.S., Dyreson, C.E., Jensen, C.S., Lee, M.L., Muliantara, A., Thalheim, B. (eds.) DASFAA 2014. LNCS, vol. 8421, pp. 372–387. Springer, Heidelberg (2014). doi:10.1007/978-3-319-05810-8_25
9. Yang, H., Duan, L., Dong, G., Nummenmaa, J., Tang, C., Li, X.: Mining itemset-based distinguishing sequential patterns with gap constraint. In: Renz, M., Shahabi, C., Zhou, X., Cheema, M.A. (eds.) DASFAA 2015. LNCS, vol. 9049, pp. 39–54. Springer, Heidelberg (2015). doi:10.1007/978-3-319-18120-2_3
10. Li, T., Ma, S.: Mining temporal patterns without predefined time windows. In: Proceedings of the 4th IEEE International Conference on Data Mining, ICDM. 451–454 (2004)
11. Tang, L., Li, T., Shwartz, L.: Discovering lag intervals for temporal dependencies. In: Proceedings of the 18th ACM International Conference on Knowledge Discovery and Data Mining, KDD, pp. 633–641 (2012)
12. Ma, S., Hellerstein, J.L.: Mining mutually dependent patterns. In: Proceedings of the 1st IEEE International Conference on Data Mining, ICDM, pp. 409–416 (2001)

13. Ma, S., Hellerstein, J.L.: Mining partially periodic event patterns with unknown periods. In: Proceedings of the 17th IEEE International Conference on Data Engineering, ICDE, 205–214 (2001)
14. Mannila, H., Toivonen, H., Verkamo, A.I.: Discovery of frequent episodes in event sequences. Data Min. Knowl. Discov. 1(3), 259–289 (1997)
15. Allen, J.F.: Maintaining knowledge about temporal intervals. Commun. ACM 26(11), 832–843 (1983)
16. Wu, S., Chen, Y.: Discovering hybrid temporal patterns from sequences consisting of point- and interval-based events. Data Knowl. Eng. 68(11), 1309–1330 (2009)
17. Mörchen, F., Ultsch, A.: Efficient mining of understandable patterns from multivariate interval time series. Data Min. Knowl. Discov. 15(2), 181–215 (2007)
18. Shah, C.C., Zhu, X., Khoshgoftaar, T.M., Beyer, J.: Contrast pattern mining with gap constraints for peptide folding prediction. In: Proceedings of the 21st International Florida Artificial Intelligence Research Society Conference, FLAIRS. pp. 95–100 (2008)
19. Rymon, R.: Search through systematic set enumeration. In: Proceedings of the 3rd International Conference on Principles of Knowledge Representation and Reasoning, KR, pp. 539–550 (1992)
20. Lichman, M.: UCI machine learning repository (2013)

Stream Data Processing

Soft Quorums: A High Availability Solution for Service Oriented Stream Systems

Chunyao Song[1,2]([✉]), Tingjian Ge[2], Cindy Chen[2], and Jie Wang[2]

[1] Nankai University, Tianjin, China
`chunyao.song@nankai.edu.cn`
[2] University of Massachusetts, Lowell, USA
`{csong,ge,cchen,wang}@cs.uml.edu`

Abstract. Large-scale information gathering becomes more and more common with the increasing popularity of smartphones, GPS, social networks, and sensor networks. Services based on this real-time data are the logical next step. Service Oriented Stream Systems (SOSS) have a focus on one-time ad hoc queries as opposed to continuous queries. High availability is crucial in these services. However, data replication has inherent costs, which are particularly burdensome for high rate, often overloaded, SOSS. To provide high availability and to cope with the problem of overloading the system, we propose a mechanism called soft quorums. Soft quorums provide high availability of data, a tradeoff between query result accuracy and performance, and adaptation to dynamic data/query stream rates. Finally, we conduct a comprehensive experimental study using real-world and synthetic datasets.

1 Introduction

With the increasing popularity of smartphones, GPS, WiFi, Web 2.0, social networks, RFID, and sensor networks, large-scale information gathering becomes more and more common. As a result, *services* based on this real-time information will be the logical next step. There have been many systems that use real-time data for services, including bus tracking, dynamic traffic routing [5], and smartphone related social network apps. Let us look at an example in detail.

Example 1. The real-world Cab dataset [7] can be used for online services. It contains GPS coordinate traces of taxis in San Francisco, USA, and is used by dispatchers to efficiently reach customers. The data is transmitted from each cab to a central station in real time, and can answer queries such as: (1) "retrieve counts of taxis grouped by their areas", (2) "retrieve the fraction of taxis that are occupied", and (3) "retrieve the free taxis that are within 5 miles away from a customer at location (x, y)". Mobile app software based on such taxi

Tingjian Ge was supported in part by the NSF, under the grants IIS-1149417, IIS-1319600, and IIS-1633271.

Jie Wang was supported in part by the NSF under grant CNS-1331632 and by Eola Solutions Inc. under a research grant.

© Springer International Publishing AG 2017
S. Candan et al. (Eds.): DASFAA 2017, Part II, LNCS 10178, pp. 253–268, 2017.
DOI: 10.1007/978-3-319-55699-4_16

data includes Uber [4], Flywheel [2], and Didi [3]. For example, Didi is a mobile service that collects GPS coordinate traces of all taxi drivers who use this app, and there are over 150 million users as of 2015. The highest number of requests per day is over 12.17 million. The data stream rates and query rates fluctuate significantly throughout a day or on different days [3].

In the big data era, there are increasingly more *service oriented stream systems* (SOSS) like this. When financial interests and real-time businesses are involved, high availability is critical. In SOSS, there is an emphasis on *one-time ad hoc* queries, in contrast to the *continuous* queries studied in most, if not all, previous work on data streams [9,15]. For instance, all three queries in Example 1 are one-time ad hoc queries issued by various users of the system on the real time data in some most recent time interval. The system does not need to be burdened with continuously answering these queries, especially given that a large number of service queries may arrive concurrently in Example 1; the users are in general content with one time answers.

1.1 Related Work

One-time ad hoc queries on data streams are mentioned in previous work (e.g., [9,15]). In particular, previous work on fault tolerance of data streams (e.g., [17]; see [11] for a survey) replicates *the states of each query operator* in the query execution graph of a continuous query. During recovery, query operators in a node resume their states before the failure. While this approach is powerful and ideal for continuous queries, it is too expensive and unnecessary for one-time ad hoc queries, which can simply be restarted as long as the data is protected and available. Moreover, the techniques in previous work [17] typically deal with intermediate data tuple queues at various operators, which does not apply to one-time queries. By contrast, we have a simple and efficient mechanism to replicate and protect the *input stream data only*, which is common to however many concurrent one-time queries that are issued to the system. A failed query processing will simply need to be re-executed over the protected data. This is clearly suitable and scalable for SOSS.

Background on quorums. In distributed computing, quorums are used for either commit protocols (voting) or **data replication control**. In this paper, we extend the quorum concept for the latter only. A *quorum system* is a collection of subsets (*quorums*) of servers, every two of which intersect [20]. This ensures that if a *write* operation is performed on a quorum, and later a *read* is performed on another quorum, then there is at least one node (i.e., server) that observes both operations and provides the up-to-date value to the reader. Strict quorums have been studied (e.g., [20]). Non-strict quorums, called *partial quorums* in [10], have been proposed too. In a partial quorum system, two quorums may not overlap. There are two types of partial quorums, k-quorums [8] and probabilistic quorums [19]. A k-quorum provides guarantees that it will return values that are within k versions of the most recent write. This does not apply to data streams as typically data stream tuples are insert-only and no updates. A probabilistic quorum system provides probabilistic guarantees on the intersection of any two quorums. The partial quorums used in Amazon Dynamo [13], Cassandra [18], Riak [12], and Voldemort [14] are termed *expanding partial quorums* [10].

At any time instant t, a node preserves a time interval τ of the most recent data, i.e., from time t back to time $t - \tau$. Consider all tuples between times $t - \tau$ and t in the stream as a relation R_t, which we call a *fresh batch*. Clearly, R_t dynamically changes with t, as old tuples leave and new tuples enter the time interval. The length τ is application dependent and is limited by memory size. An ad hoc query can use *any portion* (or all) of data inside the dynamic relation R_t; the exact semantics and how the data is used are up to the specific queries, and are not addressed in this paper. **The goal of our work is to protect the data in R_t, and ensure its high availability and non-interruption even if a number of servers in the system are down.**

Soft quorums. We propose a *soft quorum* scheme for data streams. The *logical* architecture of a soft quorum system is shown in Fig. 1. At a high level, a write quorum (of size w) is chosen randomly from the n *data nodes*, and the *coordinator* inserts a new data item into these w data nodes. Similarly, a read quorum (of size r) is chosen randomly from the n data nodes, and the coordinator reads data from these r nodes to answer a query. A key idea of the soft quorum scheme is that **the choice of w and r is dynamic and adaptive to the stream data incoming rate and users' ad hoc query rate**. By randomness, we provide a quantifiable tradeoff between performance and query result accuracy, while achieving the ultimate goal of *high availability* of data (in the event of node crashes). Our adaptive choice of parameters w and r minimizes the overall system load, and our random scheme achieves load balance. Soft quorums target SOSS, and can be regarded as a further relaxation of the probabilistic quorums in that they do not require that any two quorums have a high intersection probability. Thus, we can use smaller quorum sizes, resulting in higher availability and better performance. Moreover, we use radically different techniques, and set quorum sizes w and r adaptively with stream/query rate. Finally, soft quorums have load balancing and optimize the load.

Our contributions. We propose the soft quorum scheme along with the write and read quorum algorithms which trade query result accuracy for performance in Sect. 2. In Sect. 3, we discuss the dynamic selection of write and read quorum sizes, adaptive to the high data incoming rate and query rate while guaranteeing query result accuracy. We finally set w and r to minimize system load in Sect. 4. To provide high availability of data, in Sect. 5, we propose a recovery process and show the accuracy guarantees. Finally, we perform a systematic evaluation on two real-world and some synthetic datasets in Sect. 6.

2   Soft Quorum Scheme

2.1   Preliminaries and Notations

We illustrate the system architecture in Fig. 1. At the base level are *data nodes*—computers connected by a network, forming a distributed in-memory store. We abbreviate the data nodes as *nodes*, and denote the number of nodes as n. Above the nodes level is one or more *coordinators*—computers which take new stream tuples and insert them into the nodes or take queries and process them after reading data from nodes. The coordinators are simply the *query processing nodes* in data stream systems. We only have *nodes* and *coordinators*. Typically coordinators process queries using data (from data nodes) in a pipeline manner, and a number of coordinators distribute the workload. The number of coordinators or their work assignment is up to the system, and is beyond the scope of this work. We focus on the soft quorum system S that is at the data nodes. Importantly, Fig. 1 is a *logical* architecture because there is no strict physical boundary between nodes and coordinators. It is possible that the same computer can be both a node (for storing data) and a coordinator (for processing queries).

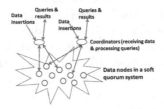

Fig. 1. Logical architecture of a soft quorum system

A *write* (*read*, resp.) *quorum of size* w (r, resp.) is a subset of w (r, resp.) nodes ($w,r \leq n$) where a write (read, resp.) operation is performed. Each node has a finite buffer size b for a data stream; we denote the buffer space of a stream at each node as a queue Q, reflecting the fact that a stream is a time series. We say that each element in Q is an *item* or an *observation* of a sample, where the whole Q is a sample of the original stream. Each item has a timestamp and a weight α (which is initially the same value as the write quorum size used when writing this item - see Sect. 2.2). The actual length of Q is denoted as $|Q|$, and each item as $Q[i]$, ($1 \leq i \leq |Q|$). We use τ to denote the length of the most recent time interval preserved from stream, i.e., the *fresh batch*. When τ approaches infinity, we essentially replicate and sample the whole stream; but typically SOSS applications are interested in data in some recent history. Finally, we denote the data incoming rate of a stream (over all coordinators and nodes) as λ, and the query rate as μ. Table 1 is a summary of notations for easy reference.

Table 1. Terminologies and notations used in the paper

Terms, symbols	Meaning	Symbols	Meaning
Node	Data node, part of a quorum system	w	Write quorum size
Coordinator	A computer that does query processing, write (insertion), and read of data	r	Read quorum size
Item/observation	A stream tuple in a quorum	Q	Buffer at a node for a stream
Sample	A set of observations	$|Q|$	Number of items in Q (i.e., length of Q)
τ	Length of the most recent time interval	b	Maximum $|Q|$ (buffer space limit)
λ, μ	Stream incoming rate and query rate	$Q[i]$	The i'th item in Q ($1 \leq i \leq b$)
n	Total number of nodes	α	Weight of an item in Q

2.2 The Quorum Scheme

The basic scheme is as follows: a write (read, resp.) quorum consists of any w (r, resp.) nodes chosen uniformly at random from the n nodes. Complexity arises from **choosing the suitable w and r parameters** to guarantee that any read quorum contains a sample that is arbitrarily close to the original stream, rendering a trade-off among accuracy, performance, and availability. We discuss the dynamic choices of w and r minimizing the system load in this and the next two sections. We now present the write and read quorum algorithms and quantify the effective sampling rate.

As this is essentially a protocol between a coordinator and a set of nodes, we present the WRITEQUORUM algorithm from both the *coordinator* side and the *node* side. We have an *adaptive* flag as input to enable the estimation of current stream rate and to select new quorum sizes (w, r). This can be done periodically. Line 2 of the Coordinator side is selection without replacement. In line 4, each acknowledgement message contains the stream incoming rate estimate from a node, and the coordinator simply averages the estimates $(\widehat{\lambda})$ and use $\widehat{\lambda}$ to get new quorum sizes (Sect. 3). Lines 4–8 are done asynchronously as the new parameter setting can be at any time without waiting. The *Node side* algorithm maintains a recent interval of size τ (fresh batch). In line 2, the item is added to Q (with an initial weight w_0; the timestamp is in v). In lines 3–7, if Q overflows, it chooses an item to evict and assigns a new weight to all items. Line 6 is such that $Q[i]$ has probability p_i to be removed. The $\widehat{\lambda}$ estimate (line 9) is discussed in Sect. 3.

We next look at READQUORUM, which has a Boolean flag indicating if a uniform sample is required. If query processing requires quantifiable estimations, then this flag is set true. If this flag is false, then more observations may be returned (but not uniformly random). Line 3 of the Coordinator side in READQUORUM sends a read request to the chosen node. In line 4, it receives (α_i, R_i) from each node in the read quorum, as described by the Node side. Each of the r nodes may have a different α_{min}; hence the α_i's are sorted in decreasing order as $\beta_1, ..., \beta_r$ (line 7). While one can again make them uniform by taking the smallest (β_r) and removing more observations, we show in Theorem 1 that they are already a uniform sample (making the returned sample size as large as possible).

Algorithm 1. WRITEQUORUM $(v, w_0, adaptive)$

Input: v: the observation to be written;
w_0: the current write quorum size;
adaptive: whether to adapt w to stream rate.
Output: new w, r setting if *adaptive* is true
/* Coordinator side: */
1 **for** *each* $i \leftarrow 1$ *to* w_0 **do**
2 | pick one node uniformly at random without replacement
3 |_ send $(v, w_0, adaptive)$ to that node

4 receive ACK$(\widehat{\lambda}_i)$ from the w_0 nodes in lines 2–3
5 **if** *adaptive* **then**
6 | $\widehat{\lambda} = \frac{\sum_{i=1}^{w_0} \widehat{\lambda}_i}{w_0}$
7 | get (w, r) with $\widehat{\lambda}$ (Sect. 3)
8 |_ **return** (w, r)

/* Node side upon receiving $(v, w_0, adaptive)$:   */
1 remove observations with age τ or older from Q
2 add (v, w_0) to Q //*Observation v has initial weight w_0*
3 **if** $|Q| = b + 1$ **then** //*if the queue exceeds bound*
4 | $\alpha \leftarrow \frac{b}{\sum_{i=1}^{b+1} \frac{1}{\alpha_i}}$, where α_i is the weight of $Q[i]$
5 | **for** *each* $i \leftarrow 1$ *to* $b+1$ **do** $p_i \leftarrow 1 - \frac{\alpha}{\alpha_i}$ randomly remove one item from Q
   | based on p_i's
6 |_ assign weight α to each remaining item
7 **if** *adaptive* **then**
8 |_ $\widehat{\lambda} \leftarrow \frac{n}{\tau} \sum_{i=1}^{|Q|} \frac{1}{\alpha_i}$

9 send $ACK(\widehat{\lambda})$ back to coordinator

Theorem 1. *When uniform is true, the sample R returned by* READQUORUM *is a uniformly random sample with rate* $\eta = 1 - \prod_{i=1}^{r}(1 - \frac{\beta_i}{n-i+1})$, *as also returned by the algorithm.*

Proof. The proof is omitted due to space constraint and can be found in [1].

Note that the weight α of an observation o in Q is initially w_0, the write quorum size (which is dynamic) at the time o enters Q. When Q is full and WRITEQUORUM performs an eviction in line 7 of its Node algorithm, all observations in Q have the same weight (i.e., all α_i's are equal), in which case $\frac{\alpha_{min}}{\alpha_i} = 1$ in line 5 of the Node side of READQUORUM (thus all $Q[i]$'s are added to R). However, observations constantly expire too and are removed from Q (line 1 of Node side of WRITEQUORUM) and new observations will be mixed in with different initial weights α_i (i.e., the write quorum size w_0 at the time of joining Q). Hence, in general, observations in Q have different weights α_i's. Since the sampling rate at each node is determined by the minimum α value α_{min} at the node, we certainly wish to maximize α_{min} at each node in order to maximize the uniform sample size.

Theorem 2. *The eviction policy in lines 4–7 of Node algorithm of* WRITE-QUORUM *maximizes the minimum sampling rate of the items in Q.*

Algorithm 2. READQUORUM $(r,\ uniform)$

Input: r: read quorum size;
uniform: true if a *strictly* uniform sample is required
Output: R: the sample read from a quorum
η: the effective sample rate (if *uniform* is true)
/* *Coordinator side:* */
1 **for** *each* $i \leftarrow 1$ *to* r **do**
2 | pick one node uniformly at random without replacement
3 | send (*uniform*) to that node
4 receive (α_i, R_i) from r nodes in line 2
5 $R \leftarrow \bigcup_{i=1}^{r} R_i$
6 **if** *uniform* **then**
7 | $(\beta_1, ..., \beta_r) \leftarrow$ sort $\alpha_1, ..., \alpha_r$ in *decreasing* order
8 | $\eta \leftarrow 1 - \prod_{i=1}^{r}(1 - \frac{\beta_i}{n-i+1})$
9 **return** (R, η)
/* *Node side upon receiving* (*uniform*): */
1 $\alpha_{min} \leftarrow minimum\ \alpha$ *value in* Q
2 **if** *uniform* **then**
3 | $R \leftarrow \emptyset$
4 | **for** *each* $i \leftarrow 1$ *to* $|Q|$ **do**
5 | | $R \leftarrow R \cup Q[i]$ with probability $\frac{\alpha_{min}}{\alpha_i}$
6 **else** //*uniform is false (i.e., not required)*
7 | $R \leftarrow Q$
8 send (α_{min}, R) back to the coordinator

Proof. The proof is omitted due to space constraint and can be found in [1].

The ultimate goal of soft quorums is data protection via replication, while sampling is just a by-product. Note that when the system load is light, strict quorums can be used; soft quorums may kick in only when the system is overloaded. Even then, if there is a high priority query that must require all data, it can have a read quorum size $r = n$. Alternatively, we may have a separate primary node that possesses all data in the fresh batch, which the special query resorts to. If the primary node fails, it is recovered from the soft quorum system.

3 Adaptation and Read Accuracy

3.1 Adaptation to Dynamic Stream Rates

Consider a node s_i that receives a write message from one of the coordinators. The message contains (v, w_0), where v is the new tuple and w_0 is the current w value used by the coordinator. Node s_i use the *maximum likelihood estimate* (MLE) [16] to estimate the stream incoming rate from s_i's own perspective.

Theorem 3. *Line 9 of the Node side of* WRITEQUORUM, *i.e.,* $\widehat{\lambda} \leftarrow \frac{n}{\tau} \sum_{i=1}^{|Q|} \frac{1}{\alpha_i}$ *is an MLE of the stream rate.*

Proof. The proof is omitted due to space constraint and can be found in [1].

Then line 6 of the coordinator side of WRITEQUORUM is equivalent to the MLE over items from all w_0 queues. Note that λ does not have to be estimated for every write message; the system can do it periodically. We will discuss how to determine w and r based on this estimate.

3.2 Read Quorum Accuracy

As in strict quorums, read consistency is a function of quorum sizes (w and r). We study the choice of w and r based on data consistency requirement, in addition to the stream rate estimate $\widehat{\lambda}$ above. We consider **three consistency modes** in this section (the detailed algorithm is omitted due to space constraint and can be found in [1]). Consistency metrics proposed in previous work [10] are designed for stored data and are not suitable for our setting. For example, k-staleness [10] requires that the value read is no more than k versions older than the latest committed version, while we are concerned with tuples in the most recent interval and stream tuples are continually appended. We explore soft data consistency through ensuring that the *statistical distance* between the sample from a read quorum and the original stream (which we also call sample) is small enough. Our definition below is based on the *Jaccard distance* between two sets [21].

Definition 1 (statistical distance between samples). *The statistical distance between two samples R and S is defined as $\mathcal{D}(R, S) := 1 - \frac{|R \cap S|}{|R \cup S|}$, where R and S are deemed sets of items. Here, an item can be multidimensional, containing the set of attributes required by the application.*

Intuitively, the statistical distance measures how different the two samples are. It is a *metric*, and is always *between 0 and 1*. It is easy to verify that when R and S are identical, their statistical distance is 0; when they have no overlap, the distance is 1. This metric informs us how close the values from a read quorum are to the original ones. In addition, it provides a means to compare two soft quorum schemes.

Definition 2 ((Δ, ϵ)-consistency). *We say that a soft quorum system is (Δ, ϵ)-consistent if $Pr[\mathcal{D}(R, S) \leq \Delta] \geq 1 - \epsilon$, where S is the original stream sample and R is the sample from a read quorum chosen uniformly at random.*

Based on the above definitions, we show Lemma 1 below which is used in the proofs of Theorems 4 and 6.

Lemma 1. *Consider a sample S that is a set of observations and a subsample R that is a subset of S. Then $\mathcal{D}(R, S) = \frac{|S - R|}{|S|}$.*

Proof. This is straightforward based on Definition 1, since $R \cap S = R$ and $R \cup S = S$.

Given the stream rate estimate $\widehat{\lambda}$ and a time interval size τ, there are $m = \widehat{\lambda}\tau$ observations in the original sample. We show the following result which will help us to select r and w when the consistency mode is 1.

Theorem 4. *We achieve (Δ, ϵ)-consistency if the quorum sizes satisfy $rw \geq$ $n \ln \frac{2}{a - \sqrt{a^2 - 4\Delta^2}}$ where $a = 2\Delta + \frac{3}{m} \ln \frac{1}{\epsilon}$.*

Proof. The proof is omitted due to space constraint and can be found in [1].

For example, if we require $(0.15, 0.1)$-consistency with $n = 20$ and $m = 500$, then from Theorem 4 we have $r \cdot w \geq 44$ (e.g., $r = 4$ and $w = 11$) is sufficient. Note that (Δ, ϵ)-consistency can be a very strong guarantee, which says that, with probability at least 1-ϵ, the statistical distance between a read quorum value and the original data is no more than Δ. We also provide weaker guarantees. We have two other modes: ($mode = 2$) expected statistical distance is no more than Δ, and ($mode = 3$) the approximate sampling rate is at least some threshold value ϵ. For $mode = 2$, from Lemma 1, and the linearity of expectation, we have: $\Delta < e^{-rw/n}$, which gives us a lower bound of $r \cdot w$ as $n \ln \frac{1}{\Delta}$. Finally, $mode = 3$ especially applies to the case when stream rates are high and $rw < n$. Then the approximate sampling rate is $w \cdot \frac{r}{n}$ (i.e., $\frac{r}{n}$ is the probability that one of the w copies of a tuple is in the read quorum); solving $w \cdot \frac{r}{n} \geq \epsilon$ gives us a lower bound of $r \cdot w$ as ϵn. We discuss how to determine r and w based on an $r \cdot w$ budget c in Sect. 4.

4   Load

In the literature, *load* is defined as the probability of accessing the *busiest* node [20]. Intuitively, one certainly wishes to balance the workloads (including network traffic) for writing and reading quorums in order to maximize parallel access of nodes. In soft quorums, however, we have asymmetry between writes and reads. When we write a quorum of w nodes, we only write one value to each node. When reading a quorum, by contrast, each of the r nodes contains a set of values. Thus, we need to generalize the classical load definition to account for the number of values accessed.

Definition 3 (load). *Let $\mathcal{T}_i (\varphi)$ denote the expected number of values written to or read from node i for a random quorum access with probability φ (1-φ, resp.) being a write (read, resp.), where the expectation is taken over the random choice of write and read quorums. Then the system load is defined as $\mathcal{L} := max_{i \in \{1, \dots, n\}} \mathcal{T}_i(\varphi)$.*

Thus, load $\mathcal{L}$ is a function of φ, the fraction of quorum access that is a write operation. $\mathcal{T}_i(\varphi)$ is the expected number of values accessed at a node, while $\mathcal{L}$ is the maximum over all nodes. Recall that the accuracy requirements in Sect. 3 only requires a lower bound of $r \cdot w$. We are now ready to show the following result that finally sets r and w. Note that the query rate μ can be learned from query statistics or workload information as in the standard database design literature.

Theorem 5. *Given a stream incoming rate λ, a read rate μ, and a quorum size constraint $rw \geq c$, the setting $w = \sqrt{\frac{cb\mu}{2\lambda}}$ and $r = \sqrt{\frac{2c\lambda}{b\mu}}$ minimizes the expected load $E[\mathcal{L}]$, where the expectation is taken over the random configuration of node queues, each with a maximum size b.*

Proof. The proof is omitted due to space constraint and can be found in [1].

Theorem 5 indicates that when the ratio between read and write increases (i.e., μ/λ increases), we should increase w and decrease r, which is intuitive as we read more. The same is true when the buffer size b for each queue increases.

5 Availability

5.1 Failure Probability and Read Accuracy

System failure probability $F_p(\mathcal{S})$, for a quorum system $\mathcal{S}$, is used to measure the availability under the fail-stop model [20]. $F_p(\mathcal{S})$ is defined as *the probability that at least one node in every quorum has failed*, where p is the probability that a node may fail independently. As such, the smaller $F_p(\mathcal{S})$ is, the more available $\mathcal{S}$ is. In soft quorums, we generalize $F_p(\mathcal{S})$ as the *probability that either read or write cannot proceed due to unavailability of a quorum*. We first note that failed nodes can only affect reads. This is because if f out of n nodes fail, a write just needs to work with the remaining $n - f$ nodes (effectively a smaller n). For reading data that exist before the nodes fail, however, we must make sure there is a quorum that is not hit by the failed nodes. As each stream in the system may have a different read quorum size, we let the maximum read quorum size be γ. Then $F_p(\mathcal{S})$ is the probability that any set of γ nodes (a read quorum) has at least one failed node, which is equal to that at least $n - \gamma + 1$ nodes fail. Hence: $F_p(\mathcal{S}) = \sum_{i=n-\gamma+1}^{n} \binom{n}{i} p^i (1-p)^{n-i}$. If we increase the maximum read quorum size γ, $F_p(\mathcal{S})$ increases, indicating a lower availability.

When f out of n nodes fail ($f \leq n - \gamma$), a read works by choosing r nodes from $n - f$ nodes (rather than n nodes) uniformly at random. At first glance it may seem that this would reduce read accuracy. However, read accuracy is *not affected* due to the random distribution of observations. In other words, a restricted subset of sample is still random within the whole sample. In particular, the equation on $F_p(\mathcal{S})$ and the whole analysis of Sect. 3.2 still hold.

5.2 Reconstructing Failed Nodes

We can recover the lost data (in the fresh batch time interval) in the f failed nodes based on the remaining $n - f$ nodes. As discussed in Sect. 3, a coordinator keeps track of an estimate of the current stream rate $\widehat{\lambda}$ and the w value that it uses. To recover data in the f failed nodes, we do a **RecoveryProcess** (the detailed algorithm is omitted due to space constraint and can be found in [1]) by first counting how many times each observation appears in the surviving nodes, then adding a number of each observation randomly to the f failed nodes so that the total count is w. Thus, what cannot be recovered are only those observations that originally only appear in the f failed nodes. Note $\frac{\lambda \tau w f}{n}$ is the expected total count of items that are in the f failed nodes. So if the number of items sent to the f nodes to reach the w total count is less than this expectation, then we divide the difference evenly to all observations, and add them to the

f nodes. The RECOVERYPROCESS is executed by the coordinator, coordinating with all nodes through messages. Let us now quantify the data quality after the RECOVERYPROCESS. It is possible that some more nodes fail after a recovery, we show that the error essentially accumulates.

Theorem 6. *The expected statistical distance between the sample in the n nodes after the* RECOVERYPROCESS *and the original sample, which we call error ϵ, satisfies $\epsilon \leq (\frac{f}{n})^w$. After such a recovery, if some nodes fail again, including f_2 new nodes that have not failed before, then after the* RECOVERYPROCESS, *we have $\epsilon \leq \epsilon_1 + (\frac{f_2}{n})^w$, where ϵ_1 is the error before this recovery.*

Proof. The proof is omitted due to space constraint and can be found in [1].

Theorem 6 indicates that a greater w results in a better data quality after recovery (for a fixed f), giving a smaller error. Thus, it provides a trade-off between availability (or recoverability) and performance.

6 Experiments

6.1 Datasets and Experiment Setup

We perform a systematic evaluation using two real-world and some synthetic datasets: **(1) Cab data.** This dataset contains mobility traces of taxi cabs in San Francisco Bay Area, USA [7]. **(2) RFID data.** RFID tracking data [6] was collected from the seventh HOPE (Hackers On Planet Earth) conference held in July 18–20, 2008. Conference attendees received RFID badges that uniquely identify and track them across the conference space. This can be used to drive social networking features which completely change the conference experience. **(3) Synthetic data.** Synthetic datasets are generated based on the real datasets, but vary parameters such as timestamps that determine the stream rates.

We implement all the algorithms presented in the paper, and run the experiments over a cluster of 16 computers. Each computer has an Intel Core2 Duo CPU (2.66 GHz), with a 6 MB cache for each core, and a 1.7 GB memory. The computers are connected by a 100 MB/sec bandwidth local area network. By default we use one computer as the *Coordinator* and 15 computers as *Nodes* (i.e., $n = 15$). All presented results are averages of at least three runs.

6.2 Experimental Results

Data schema and queries. The Cab data has a simple schema [*latitude, longitude, occupancy, time*], where *occupancy* shows if a cab has a fare (1 = occupied, 0 = free). The three queries we run are described in Example 1. The main data stream file in the RFID dataset has a schema [*time, tag_id, area_id, x, y, z*], where *tag_id* identifies a conference participant, *area_id* identifies a particular conference area. The dataset also indicates a particular person (*tag_id*)'s *interests*. We run these two queries which a conference participant may be interested

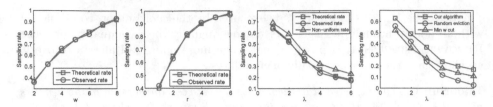

Fig. 2. (a) w vs. sampling rate, (b) r vs. sampling rate in a read quorum

Fig. 3. Stream rate vs. sampling rate

Fig. 4. Eviction policy and alternatives

in: *"get the distribution of the locations of people who are interested in network security"*, and *"get the talk room that has the most number of people"*.

Sampling rates. In the first experiment, we measure the actual sampling rate in a sample read by the Coordinator from a read quorum under various parameter settings. We calculate the theoretic sampling rate based on Theorem 1, and compare it with what we actually observe in a read quorum. The results are shown in Fig. 2. In Fig. 2(a) we fix $r = 3$ but vary w, while in Fig. 2(b) we fix $w = 6$ but vary r. The y axis is the sampling rate in a read quorum. The observed uniform sampling rate closely matches the result from Theorem 1, verifying the high accuracy of our analysis. In addition, as w or r increases, so does the sampling rate, which approaches 1. Next, fixing $w = 6$ and $r = 2$, we evaluate how well the system works with different data rates λ, especially when the buffers at each node often reach capacity and the eviction algorithm takes effect. We use $b = 100$ and $\tau = 180$ s. When $\lambda = 2$, the expected number of observations within a time interval at each Node $\frac{\lambda \tau w}{n} = 144 > b$, the eviction algorithm will need to take effect. Based on the Cab data, we generate random synthetic datasets with different (expected) stream rates λ from 1 to 6. The result is shown in Fig. 3. Recall that in READQUORUM by setting the input parameter *uniform* to be false, we get a non-uniform (but larger) sample, whose size ratio is also shown in Fig. 3. It has a slightly larger size as w values do not vary significantly within Q.

Buffer eviction policy. In Fig. 4, we compare our buffer eviction policy (WRITEQUORUM) with an intuitive alternative method ("random eviction") which picks an item uniformly at random in Q to evict and multiplies the remaining items' w values by $\frac{b}{b+1}$. We also compare with a variant of the READQUORUM algorithm ("Min w cut") which simply uses the minimum w value among all r nodes in the read quorum to further filter the sample. Figure 4 shows that our algorithm achieves the highest uniform sampling rate. Random eviction gets worse as λ increases, because Q tends to be more dynamic with various w values within it as λ increases, for which a judicious eviction algorithm shows more advantage. The result of this experiment is consistent with Theorem 2.

Adaptation and accuracy guarantees. Figure 5 indicates that our algorithm's estimates of $\widehat{\lambda}$ are very accurate. Next we study the connection between quorum sizes and read accuracy, as measured by statistical distance. The result

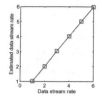 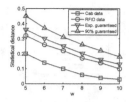 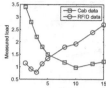 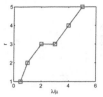

Fig. 5. Data stream rate estimates

Fig. 6. Statistical distance and guarantees

Fig. 7. w vs. load for fixed $r \cdot w$

Fig. 8. λ/μ vs. optimal r

is shown in Fig. 6 under various w values while fixing $r = 3$. We can see from Cab data and RFID data that the distance decreases as w increases, entailing more accuracy. Recall that we provide an expected statistical distance or a (Δ, ϵ)-consistency guarantee in Sect. 3. We calculate the maximal statistical distances (in both expectation and $1 - \epsilon$ confidence where $\epsilon = 0.1$) that can be guaranteed. Figure 6 shows that (Δ, ϵ)-consistency is a stronger guarantee (verifying Theorem 4), while both follow the same trend as w increases.

Load. We now examine the load $\mathcal{L}$, with the result in Fig. 7. The Cab data has $\lambda \cong 1$, and we set the query rate $\mu = 1$ and $b = 10$. By fixing $rw = 20$, we study the relationship between different values of w and the load $\mathcal{L}$. Similarly, using the same parameters, we repeat the experiment for the RFID data which has much greater average stream rate ($\lambda \cong 15$). Figure 7 shows that $\mathcal{L}$ is minimal when $w = 10$ for the Cab dataset and when $w = 3$ for the RFID dataset, which is consistent with Theorem 5. Then using synthetic datasets with various stable stream rates λ and fixed query rate μ, we observe the relationship between the write/read ratio λ/μ and the r selected using Theorem 5 based on an $rw = 20$ budget. The result is shown in Fig. 8. The selected read quorum size r gradually increases as the write/read ratio does. This is because as λ/μ increases, our algorithm will decrease the write quorum size w and increase the read quorum size r to minimize the load $\mathcal{L}$ while maintaining the same level of accuracy guarantee. In Fig. 9, we show $\mathcal{L}$ as λ increases. Perhaps surprisingly, as λ increases, $\mathcal{L}$ slightly decreases. This is because the load definition is about the number of values accessed "per quorum access". When λ increases, a quorum access is more likely to be a write, which tends to access a smaller number of values than a read.

Access delays and adaptations. Next we put all the pieces together, and evaluate the performance when queries access read quorums in real time as the real-world data streams flow in. We also compare the performance with *partial quorums* in previous work [10] for stored data (without adaptations) and simple replications. Periodically (once every five seconds), the soft quorum system adaptively learns a new w, r setting based on the current stream rate. We use $n = 11$, and thus can have up to 5 Coordinator machines. We fix the query rate $\mu = 1$, but vary the number of coordinators, and distribute the stream write and query workload evenly among the Coordinators. The result is shown in Fig. 10. We first run the experiment with Cab data. The query read quorum delay

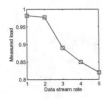
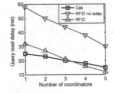
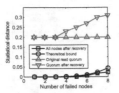
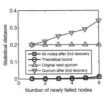

Fig. 9. Stream rate vs. load

Fig. 10. Read delay and adaption

Fig. 11. Accuracy after recovery

Fig. 12. Accuracy after 2nd recovery

is about 20 ms, and slightly decreases as the number of Coordinators increases. This is because the network traffic becomes more distributed and less congested as we distribute queries and stream input handling over multiple Coordinators.

We then immediately switch to the RFID data stream. We test the performance with and without adaptively setting w and r values based on the new stream rate of the RFID dataset. We can see from Fig. 10 that optimal quorum sizes can dramatically bring down the query's read-quorum delays. Likewise, other quorums such as probabilistic quorums [19] cannot select optimal quorum sizes based on stream rate, and result in longer delays. Expanding partial quorums [10] or other simple replication schemes that send all tuples to all n nodes have about 17 times longer read delays than the soft quorum with optimal w and r; they are omitted from the figure for clarity. Moreover, as discussed earlier, other quorum schemes cannot provide uniformly random sampling and a tuning knob to trade accuracy for performance and feasibility.

Recovery and data quality. Starting from $n = 15$ nodes and using $w = 5$ and $r = 3$, we first programmatically fail any f nodes (a variable) by clearing their Q's. We then do the RECOVERYPROCESS, and measure the quality of recovered data. The results of the Cab data are in Fig. 11 (the RFID data gives us a similar graph which we omit). We first measure the sample statistical distance between the one collected over all n nodes after recovery and the one before recovery. We compare it with the theoretical upper bound given by Theorem 6. From Fig. 11, the statistical distance is below 0.03 for f up to 8 (over *half* of the total nodes), and obeys the theoretical bound. We then test the statistical distance between the sample collected from a read quorum after recovery and the original stream sample before any failure and recovery. We compare this curve with the corresponding one without any node failure (which is a constant in Fig. 11). The difference in statistical distance is no more than 0.1 for $f \leq 8$, and no more than 0.03 for $f \leq 4$. Finally, we evaluate the data quality after additional node failures and recoveries. We start with an initial failure $f = 4$ and a recovery. We then programmatically fail an additional f_2 new nodes, followed by another recovery. We conduct the same set of comparisons as in the previous experiment, and show the results in Fig. 12, where the comparisons are with the corresponding data before the first failure. Again, the sample statistical distance over all n nodes is very small and is close to the values calculated from Theorem 6.

Summary of results. Our experiments verify that soft quorums are suitable for providing high availability for high-rate dynamic data streams, offering a tradeoff among availability, accuracy, and performance. Our sampling rate estimate is highly accurate, and the buffer eviction policy achieves optimal accuracy. Soft quorums adapt to dynamic stream rates by accurately estimating stream rates, adjusting r, w parameters, and providing rigorous accuracy guarantees. The choice of parameters and adaptivity to dynamic streams minimize system load and achieve superior performance in terms of access delays than previous quorum and replication schemes not designed for streams. Finally, recovery after node failures maintains good data quality.

7 Conclusions

In light of the increasing importance of providing high availability for SOSS, we propose soft quorums for data streams. Soft quorums address the overhead associated with replication and cope with overloads. They adapt to dynamic streams and guarantee query result accuracy. We propose algorithms to choose quorum sizes based on accuracy requirement and load minimization. Finally, we devise recovery algorithms for failed nodes and study data accuracy. Our comprehensive experiments on real and synthetic datasets show that soft quorums advance the state of the art in providing high availability for SOSS and render a delicate tradeoff among availability, accuracy, and performance.

References

1. https://drive.google.com/file/d/0B9Umiq2eYGoVdWZRMjBKMEpWSDQ/view?usp=sharing
2. http://flywheel.com/
3. http://net.chinabyte.com/238/13231738.shtml
4. https://www.uber.com/
5. http://www.inrix.com/
6. http://crawdad.cs.dartmouth.edu/hope/amd (2008)
7. http://crawdad.cs.dartmouth.edu/epfl/mobility (2009)
8. Aiyer, A., Alvisi, L., Bazzi, R.A.: On the availability of non-strict quorum systems. In: Fraigniaud, P. (ed.) DISC 2005. LNCS, vol. 3724, pp. 48–62. Springer, Heidelberg (2005). doi:10.1007/11561927_6
9. Babcock, B., Babu, S., Datar, M., Motwani, R., Widom, J.: Models and issues in data stream systems. In: PODS (2002)
10. Bailis, P., Venkataraman, S., Frnaklin, J.M., Joseph, H.M., Stoica, I.: Probabilistically bounded staleness for practical partial quorums. In: VLDB (2012)
11. Balazinska, M., Hwang, J.H., Shah, A.M.: Fault-tolerance and high availability in data stream management systems. In: Liu, L., Özsu, M.T. (eds.) Encyclopedia of Database Systems, pp. 1109–1115. Springer, Heidelberg (2009)
12. Basho, R.: http://basho.com/products/riak-overview/ (2012)
13. Decandia, G. et al.: Dynamo: Amazon's highly available key-value store. In: SOSP (2007)

14. Feinberg, A.: Project voldemort: reliable distributed storage. In: ICDE (2011)
15. Golab, L., Ozsu, M.T.: Issues in data stream management. In: SIGMOD (2003)
16. Harris, J.W., Stocker, H.: Maximum likelihood method. In: Handbook of Mathematics and Computational Science. Springer, New York (1998)
17. Hwang, J.-H., Xing, Y., Cetintemel, U., Zdonik, S.: A cooperative, self-configuring high-availability solution for stream processing. In: ICDE (2007)
18. Lakshman, A., Malik, P.: Cassandra - a decentralized structured storage system. In: LADIS (2008)
19. Malkhi, D., Reiter, M.K., Wool, A., Wright, R.N.: Probabilistic quorum systems. Inf. Commun. **170**(2), 184–206 (2001)
20. Naor, M., Wool, A.: The load, capacity, and availability of quorum systems. SIAM J. Comput. **27**(2), 423–447 (1998)
21. Tan, P.-N., Steinbach, M., Kumar, V.: Introduction to Data Mining. Pearson, Upper Saddle River (2005)

StroMAX: Partitioning-Based Scheduler for Real-Time Stream Processing System

Jiawei Jiang[1]([✉]), Zhipeng Zhang[1], Bin Cui[2], Yunhai Tong[1], and Ning Xu[1]

[1] Key Laboratory of High Confidence Software Technologies (MOE),
School of EECS, Peking University, Beijing, China
{blue.jwjiang,zhipengzhang,yhtong,ning.xu}@pku.edu.cn
[2] The Shenzhen Key Lab for Cloud Computing Technology and Applications
(SPCCTA), School of Electronics and Computer Engineering (SECE),
Peking University, Shenzhen, China
bin.cui@pku.edu.cn

Abstract. With the increasing availability and scale of data from Web 2.0, the ability to efficiently and timely analyze huge amounts of data is important for industry success. A number of real-time stream processing platforms have been developed, such as Storm, S4, and Flume. A fundamental problem of these large scale decentralized stream processing systems is how to deploy the workload to each node so as to fully utilize the available resources and optimize the overall system performance. In this paper, we present StroMAX, a graph-partitioning based approach of workload scheduling for real-time stream processing systems. StroMAX uses two advanced generic schedulers to improve the performance of stream processing systems by reducing the inter-node communication cost while keeping the workload of nodes below a certain computational load threshold. The first scheduler analyzes the workload structure when a job is committed and uses the graph-partitioning result to determine the deployment of tasks. The second scheduler analyzes the statistical information of physical nodes, and dynamically reassigns the tasks during runtime to improve the overall performance. Besides, StroMAXcan be deployed to many other state-of-the-art real-time stream processing systems easily. We implemented StroMAX on Storm, a representative real-time stream processing system. Extensive experiments conducted with real-world workloads and datasets demonstrate the superiority of our approaches against the existing solutions.

Keywords: Real-time stream processing · Task allocation · Workload scheduling · Graph partition

1 Introduction

With the unprecedented proliferation of data from web, it is natural to extend the scope to efficient processing mechanisms and methods that can handle real-time workloads [19]. For example, Twitter, the popular online social network,

© Springer International Publishing AG 2017
S. Candan et al. (Eds.): DASFAA 2017, Part II, LNCS 10178, pp. 269–288, 2017.
DOI: 10.1007/978-3-319-55699-4_17

processes over 500 million tweets every day. It is challenging to process and analyze such a big data stream in real-time. Traditional distributed processing frameworks, such as MapReduce, are designed for offline batch processing. They are ill-suited to process real-time workloads. Real-time stream computing is an effective way to process big data with low-latency. It is becoming one of the fastest and most efficient ways to obtain useful knowledge from various kinds of real-time data. Thus, many real-time stream processing frameworks have been proposed, such as Storm [21], S4 [14], and Flume [1].

Compared to the batch processing systems, resource allocation and scheduling in real-time stream processing systems are much more difficult and important due to the dynamic nature of the input data streams. An application or workload in these systems consists of several processing components. A component can produce the input of stream or execute the processing logic to generate results. In this paper, we use **Spout** and **Bolt**, borrowed from Storm, to represent the input component and processing component, respectively. Tuples emitted by a spout constitute a stream that can be transformed by passing through one or more bolts that implement the user-defined logic. Therefore, we can use a directed acyclic graph, called a topology, to denote the stream transformations. When a topology is submitted, the system schedules the tasks of each spout and each bolt to a certain physical node of the cluster. Similar to the batch data processing systems such as Hadoop, the allocation strategy impacts the performance of a real-time stream processing system. However, most of the above systems apply a round-robin method as their default scheduler which evenly distributes the components of a topology to the physical nodes. This basic scheduler is easy to implement, however, it does not take into account the cost of tuple transmission between components. Furthermore, the communication cost of tuple transmission heavily increases the average processing latency and deteriorates the overall performance of the system.

In this paper, we design and implement StroMAX, which provides two novel schedulers for the real-time stream processing systems to improve their performance. Different from the default round-robin scheduler, StroMAX aims at reducing the average processing latency of tuple by minimizing the total inter-node communication cost and keeping computational load balanced on each node. These two schedulers use graph-partitioning based algorithms to partition the topology. The first scheduler, named *Bootstrap Scheduler*, analyzes the topology graph and partitions the topology when it is submitted to the system. This strategy is simple and is executed before the topology is started, so neither the cluster workload nor the network traffic is taken into account. The second scheduler, named *Rebalance Scheduler*, goes one step further by monitoring the runtime statistics of all the topologies and the workload of cluster, then it rebalances the topologies for overall performance optimization when necessary. Besides, Rebalance Scheduler provides a heuristic to dynamically move components from the bottleneck nodes to the idle ones based on the statistical information of the cluster and the topologies.

To evaluate our schedulers, we implemented StroMAX on Storm. The performance of StroMAX is validated with several real-world workloads. The experimental results show that the proposed graph-partitioning based approaches significantly outperform the original scheduler and demonstrate superior scalability on both synthetic benchmarks and real-world scenarios.

Our contributions in this paper can be summarized as follows:

1. We propose Bootstrap Scheduler which analyzes the graph structure of the input topology and partitions the topology when it is committed to the system.
2. We design Rebalance Scheduler that generates a global-topology-graph and partitions all the topologies to the nodes so as to improve the overall performance of the system. In addition, Rebalance Scheduler provides a novel mechanism to dynamically reassign the components when necessary.
3. We implement StroMAX on Storm, a prevailing open-source real-time stream processing system. We conduct extensive experimental studies to exhibit the advantages of our approach.

The remaining of this paper is organized as follows. In Sect. 2, we review the representative real-time systems and relevant performance issues. In Sect. 3, we present the Bootstrap Scheduler and Rebalance Scheduler, followed by the architecture of StroMAX in Sect. 4. Section 5 reports the results of extensive experimental studies. Finally, we introduce the related work and conclude this paper in Sects. 6 and 7.

2   Background

In this section, we first introduce Storm on which our prototype system is built. We next introduce the weakness of the default scheduler and analyze the cost of inter-node and inner-node communication.

Architecture of Storm. Apache Storm [21] is an open-source distributed real-time stream computation system. For parallelism, Storm uses two levels of abstractions: physical and logical.

- *Physical*: Storm consists of a master node (Nimbus), a number of zookeeper nodes that serve as a control unit, and a set of slave physical nodes (Supervisors) which process stream workload as shown in Fig. 1a.
- *Logical*: As shown in Fig. 1b, a Storm workload, called a *topology*, is a *directed acyclic graph* (DAG). Each vertex represents a processing component and each edge represents data transferred between two components. As mentioned in Sect. 1, there are two types of components: spout and bolt. The spouts provide a general mechanism to emit tuples into a topology. The bolts consume tuples from spouts or other bolts, and process them in the way defined by the user. Each component consists of a group of *tasks* communicating with other groups of *tasks* connected to it. A task can be considered as an instance of a spout or bolt.

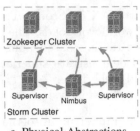

a. Physical Abstractions.

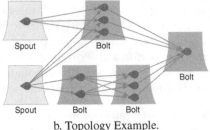

b. Topology Example.

Fig. 1. Architecture of storm.

When a topology is committed to a Storm cluster, the tasks are assigned to the physical nodes. Consequently, we need a scheduler to determine the assignment.

Scheduler in Real-time Stream Processing Systems. The default scheduler used in state-of-the-art systems is even scheduler. It enforces a round-robin strategy to balance the computation cost of each physical node, however, it lacks the consideration of communication cost. There are two types of communication among tasks. If two connected tasks are assigned to the same physical node, they use inner-node communication mechanism; otherwise, they use inter-node communication. Generally speaking, inter-node communication is much slower than inner-node communication. Therefore, we need to minimize inter-node communication while keeping computation load balanced on each node.

3   Graph-Partitioning Based Schedulers

In this section, we first introduce the notations and our graph partitioning models of the scheduling problems. Then we present the graph-partitioning based schedulers. Table 1 lists the symbols used in this paper.

3.1   Problem Definition

Graph Partitioning. Given a graph $G(V, E)$ where V denotes the set of vertices and E denotes the set of edges, we let $P = \{P_1,...,P_k\}$ be k subsets of V. P is defined to be a partition of G if: $P_i \neq \emptyset$, $P_i \cap P_j = \emptyset$, and $\cup P_i = V$ ($i, j = 1,...,k; i \neq j$). The number k is called the cardinality of the partition. Graph partitioning problem is to find an optimal partition P based on an objective function. Here we give a formal definition:

Definition 1. *The graph partitioning problem can be defined by a triplet (S, p, f). S is a discrete set of all the partitions of G. p is a predicate on S which creates a admissible solution set $S_p \in S$. All the partitions in S_p is admissible for p. The aim is to find a partition $\bar{P} \in S_p$ that minimizes the objective function $f(P)$:*

$$\bar{P} = arg \min_{P \in S_p} f(P) \tag{1}$$

Table 1. Notations.

Symbols	Description
$G_t(V_t, E_t)$	Task graph
$G_g(V_g, E_g)$	Global task graph
n_i	i-th physical node
$\varpi(n_i)$	Maximum processing capability of node n_i
$\omega(n_i)$	Capacity used in node n_i
v_i, P_i	i-th task and i-th set of tasks
$|v_i|, |P_i|$	Computation cost of task v_i and the set P_i
$N(v)$	Neighbors of vertex v
$Edgecut(P_i, P_j)$	Number of cross edges between two sets P_i, P_j
$Comm(P_i, P_j)$	Communication cost between two sets P_i, P_j
$rc(v_i, v_j)$	Bandwidth cost between two tasks v_i, v_j
$\Gamma(v_i)$	Total inter-node communication cost of v_i

Graph Partitioning in Real-time Stream Processing System. For real-time stream processing systems, we can use $G(V, E)$, a directed acyclic graph, to represent the topology T. The vertex $v_i \in V$ is the i-th component in the topology which can be a spout or a bolt. As mentioned above, each processing component consists of several tasks. Thus, we have $v_i = \{t_i^1, t_i^2, ..., t_i^m\}$, where t_i^n is the n-th task for processing component v_i and m is the number of parallelized tasks of the i-th component. The edge $(i, j) \in E$ denotes each task in v_i is connected to each task in v_j. Then, we can use a directed acyclic graph, $G_t(V_t, E_t)$, to represent the graph of tasks. A vertex v_i in V_t represents a task and an edge (i, j) in E_t represents the connection from task v_i to v_j. The data flow of the topology is organized as a graph of tasks. Here we give a formal definition of the scheduling methods based on graph-partitioning in real-time systems.

Definition 2. *Given a task graph $G_t(V_t, E_t)$, where each vertex represents a task and each edge denotes the data flows between them, the goal of a graph-partitioning based scheduling method is to partition G_t into k parts, so that each part has the same number of tasks and the number of edges between different parts is minimized. We assume that each part P_i is allocated to the i-th physical node.*

3.2 Bootstrap Scheduler

<u>Motivation.</u> In real-time stream processing systems, the key of the scheduling algorithm is to balance the computation cost and minimize the communication cost. The even scheduler achieves balanced computation, while overlooks the importance of communication cost. Since the processing latency is dominated by inter-node transfer time, reducing the tuples sent through the network can help

to improve the performance. In this section, we propose **Bootstrap Scheduler** that considers both computation cost and communication cost.

Modeling the Node Capability and Tuple Cost. We first formally model the capability of physical nodes and the cost of tuples. Given a cluster consisting of m physical nodes — $N = \{n_1, ..., n_m\}$, we define that the maximum processing capability of node n_i is $\varpi(n_i)$, and the current computation capacity of n_i is denoted as $\omega(n_i)$. For Bootstrap Scheduler, which is executed before the topology is actually executed, we cannot measure the computation cost of a task to process a tuple and the communication cost to transfer a tuple between two tasks. Therefore, we assume that the communication cost to transfer a tuple is equal to one and the computation cost to process a tuple is equal to one for all the tasks. In other words, $\varpi(n_i)$ denotes the number of tasks each node can handle, while $\omega(n_i)$ denotes that already handled.

Graph Partitioning in Bootstrap Scheduler. The goal for Bootstrap Scheduler is to partition $G_t(V_t, E_t)$ into m parts — $P = \{P_1, ..., P_m\}$, and then assign each part P_i to the physical node n_i, so that the total inter-node communication cost is minimized and the processing cost does not exceed each node's maximum capacity $\varpi(n_i)$. Therefore, we can formalize the objective function for Bootstrap Scheduler:

$$f(P) = \sum_{i,j \in [1,m], i \neq j} (Edgecut(P_i, P_j)) \tag{2}$$

$$S_p = \{P \in S, |P_i| \leq \varpi(n_i), i \in [1, m]\} \tag{3}$$

where $Edgecut(P_i, P_j)$ denotes the number of cross edges between P_i and P_j, and $|P_i|$ denotes the computation cost of P_i. Then $f(P)$ measures the total communication cost of graph $G_t(V_t, E_t)$ and S_p is the set of admissible solutions. Based on Eqs. 2 and 3, the aim is to find the partition $\bar{P} \in S_p$ that minimizes f:

$$\bar{P} = arg \min_{P \in S_p} \sum_{i,j \in [1,m], i \neq j} (Edgecut(P_i, P_j)) \tag{4}$$

This graph partitioning problem is NP-hard by reducing it to Task Allocation Problem [4]. Bootstrap Scheduler leverages a linear streaming method to solve the above graph partitioning problem.

If the vertices of the task graph arrive in some order with the set of their neighbors, and we partition the graph based on the vertex stream, it is called a streaming graph partitioning, which is fast and easy to implement. Streaming graph partitioning decides which part to assign the incoming vertex to. Once the vertex is placed, it will not be removed. This algorithm makes decisions based on incomplete information; therefore, the order of vertex stream will significantly affect the performance [20]. In this paper, we use a novel linearization approach to get the stream order.

Topology Linearization. We linearize the topology based on the property of DAG using topological sorting. Given $G_t(V_t, E_t)$, if a task v_i emits tuples into a

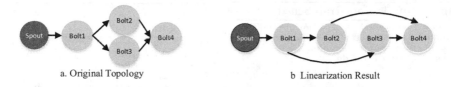

a. Original Topology   b Linearization Result

Fig. 2. An example of topology linearization.

stream that is consumed by another task v_j, then we have $v_i < v_j$, where the $<$ denotes the partial order between v_i and v_j. If $v_i < v_j$ and $v_j < v_k$, we have $v_i < v_k$ by transitivity of partial order. Since we deal with acyclic graphs, we can determine a linearization $\mathcal{L}$ of the components according to the partial order: ① If $v_i < v_j$ holds, then v_i appears in $\mathcal{L}$ before v_j. ② If neither $v_i < v_j$ nor $v_i > v_j$ holds, v_i and v_j can appear in $\mathcal{L}$ in any order. ③The first element of $\mathcal{L}$ is a random spout task v_k of the topology. Figure 2 showcases an example of a linearization of a topology with 5 components.

The linearization approach generates a linear order of tasks for the input stream. Then we study a one-pass method to partition the graph with this order. There is a streaming loader to successively read vertices (tasks), and send them to the partition program. Afterwards, the program determines the assignment of each incoming vertex (task) according to the current partition state and vertex information.

Intuition for Task Assignment. There are two intuitions the task (vertex) assignment should consider.

1. The first intuition is that we need to assign a task to the physical node that has less running tasks, in order to balance the computation cost and prevent too much computational load on one node.
2. The second intuition is that we need to assign a task to the physical node that has more neighbors of the task, in order to minimize the inter-node communication cost.

A Heuristic Solution. Motivated by these two intuitions, we use a novel streaming heuristic to solve the graph partitioning problem, i.e., to decide which part to assign the incoming vertex (task) v to.

$$index = arg \max_{i \in [1,m]} \{|P_i \cap N(v)| \left(1 - \frac{|P_i|}{\varpi(n_i)}\right)\} \tag{5}$$

In the above equation, m is the number of the partitions, $\varpi(n_i)$ is the total capacity of physical node n_i, and $N(v)$ is the set of neighbors of vertex v. For each node n_i, the first part $|P_i \cap N(v)|$ measures the number of neighbors of the incoming vertex, and the second part $(1 - |P_i|/\varpi(n_i))$ measures the computation idleness. In other words, we make a combinatorial decision considering both balancing the computation load and minimizing the inter-node communication.

Algorithm 1. Bootstrap Scheduler

Require: # of physical node: m, DAG: $G_t(V_t, E_t)$.
Ensure: Partition $P = \{P_1, P_2, ..., P_m\}$ for $G_t(V_t, E_t)$.
1: $\mathcal{L} \leftarrow \varnothing$, $S \leftarrow$ all vertices v with $in(v) = 0$, $P_i \leftarrow \varnothing$
2: **for** each vertex v in S **do**
3:    $S = S - v$
4:    $\mathcal{L} = \mathcal{L} \cup v$
5:    **for** each vertex u that has $edge(v, u) \in E_t$ **do**
6:       $E_t = E_t - edge(v, u)$
7:       **if** $in(u) = 0$ **then**
8:          $S = S \cup u$
9:       **end if**
10:    **end for**
11: **end for**
12: **if** $E_t \neq \varnothing$ **then**
13:    **return** Error: the graph is not DAG.
14: **end if**
15: $SL = streamingloader(\mathcal{L})$
16: **for** each vertex v in SL **do**
17:    $index = arg\max\{|P_i \cap N(v)| \left(1 - \frac{|P_i|}{\varpi(n_i)}\right)\};$
18:    Insert vertex v into P_{index};
19: **end for**
20: **return** P

Algorithm 2. Dynamic-Task-Reassignment

Input: Partition result: $P = \{P_1, ..., P_m\}$; θ; $G_g = (V_g, E_g)$ and $rc(v_i, v_j)\ v_i, v_j \in V_g$.
1: **for** each P_i in P **do**
2:    **if** $\sum_{v \in P_i} |v| > \theta$ **then**
3:       $List \leftarrow$ Sort $\{\Gamma(v), v \in P_i\}$ in non-descending order
4:       **while** $\sum_{v \in P_i} |v| > \theta$ **do**
5:          $v_t = pop(List)$
6:          $j = arg\max_{v_j \in P_j, i \neq j}\{\sum rc(v_t, v_j)\}$
7:          Reassign v_t to the node j
8:          $V_i = V_i - v_t$
9:       **end while**
10:    **end if**
11: **end for**

Let $in(v)$ denote the incoming degree of vertex v, we summarize Bootstrap Scheduler in Algorithm 1. The topology linearization is executed in line 1–14. With the streaming graph-partitioning heuristic, we partition the task graph $G_t(V_t, E_t)$ into m parts (line 15–19). Finally we assign the tasks to m physical nodes according to the partition result P.

3.3 Rebalance Scheduler

In this section, we propose **Rebalance Scheduler** which leverages the runtime statistics to assign all the topologies to improve the overall performance. Rebalance Scheduler uses two techniques for task reallocation, i.e., *global-topology-graph-partitioning* to repartitions all the topologies and *dynamic-task-reassignment* to move tasks from skew nodes to idle ones automatically. Following, we first discuss the motivation, and then describe these two techniques in detail.

Motivation. Bootstrap Scheduler produces an initial assignment of the tasks when it is submitted. The goal of Bootstrap Scheduler is to allocate tasks to physical nodes so as to satisfy the constraint on the number of running tasks on each physical node and minimize the inter-node communication cost. Bootstrap Scheduler is executed before the topology actually runs and only considers the newly committed topology. In practice, however, there are other topologies running on the system before the new topology is committed. Therefore, the scheduler should allocate tasks based on all the running topologies. Besides, for Bootstrap Scheduler, we assume that the communication cost to transfer a tuple equals to one and the computation cost to process a tuple equals to one for all the tasks. However, in practice, the computation cost to process a tuple and the communication cost to transfer a tuple are significantly different for different tasks.

Traditional database systems use collected historical statistics to estimate the running time for query optimization. For real-time stream processing systems, we also study the strategy that collects historical information to estimate the computation cost and communication cost of each task during the execution.

Metrics of Computation Cost and Communication Cost. To measure the runtime statistics in terms of the computation cost and communication cost of each task, we use two metrics as described below.

1. *Computation cost of a task* is measured by the average computation time for processing a tuple for a certain task. To measure this metric, we use the running logs to estimate the computation cost of a certain task during the execution. In order to deal with the heterogeneity of nodes in the clusters (different computational abilities such as different CPU frequencies), we need to consider the CPU speed of the nodes. For example, if a task takes 10 ms on a 1 GHz CPU, then migrating the task to a node with 2 GHz CPU would generate about a time cost of 5 ms. For this reason, we measure the computation cost unit as the multiplication of CPU frequency(GHz) and time(millisecond). Specially, in the above example, the computation cost for that task to process one tuple is 10.
2. *Communication cost between two tasks* is measured by the average size of data transferred between them. Similar as the measurement of the computation cost, we log the size of package for the tuple transferred from one task to the other during the execution. Then we use the average size of package for transferring one tuple as the average communication cost. We use 1024 Bytes as the unit. For example, if the average size of a tuple transferred from task i to task j is 5 KB, then the communication cost between i and j is 5.

Global-Topology-Graph Partitioning. In order to better partition the tasks of the running topologies, we model the global-topology-graph which contains all the topologies running on the cluster. The global-topology-graph is represented as a weighted directed acyclic graph $G_g(V_g, E_g)$, where V_g is the set of all the tasks in the cluster and E_g is the set of connections between tasks. The weight of each vertex v_i represents the computation cost of task v_i, denoted as $comp(v_i)$. The weight of each edge (v_i, v_j) represents the communication cost between task v_i and v_j, denoted as $comm(v_i, v_j)$. Let $G_t^i(V_t^i, E_t^i), i \in [1, m]$ be a single topology running on the system, we can generate the global-topology-graph G_g:

$$G_g = (\bigcup_{i \in [1,m]} V_t^i, \bigcup_{i \in [1,m]} E_t^i)$$

The global-topology-graph is the combination of all the topologies run on the system with the weight of the computation cost and communication cost. With the global-topology-graph, we can allocate the topologies based on the global information of all the topologies. Besides, the vertex weight and edge weight of global-topology-graph provide us the information of the heterogeneity of the tasks which further improves the accuracy of the partition result.

Given a global-topology-graph $G_g(V_g, E_g)$, our goal is to partition the graph $G_g(V_g, E_g)$ into m parts — $P = \{P_1, ..., P_m\}$, and assign each part P_i to a physical node n_i, so that the inter-node communication cost is minimized and the processing cost on each node does not exceed the maximum capacity $\varpi(n_i)$. We define our objective function $f(P)$ and the admissible solution set S_p as follows:

$$f(P) = \sum_{i,j \in [1,m], i \neq j} (Comm(P_i, P_j)) \tag{6}$$

$$S_p = \{P \in S \ and \ |P_i| < \varpi(n_i), i \in [1, m]\} \tag{7}$$

where P_i denotes the vertex set of the i-th part, and $Comm(P_i, P_j)$ denotes the sum of communication cost between P_i and P_j which can be computed as:

$$Comm(P_i, P_j) = \sum_{v_i \in P_i} \sum_{v_j \in P_j} comm(v_i, v_j) \tag{8}$$

This graph partitioning problem aims to find the partition $\bar{P} \in S_p$ that minimizes:

$$\bar{P} = arg \min_{P \in S_p} \sum_{i,j \in [1,m], i \neq j} Comm(P_i, P_j) \tag{9}$$

This graph partitioning problem is known as the k-balanced graph partitioning problem and has been proved NP-hard [3]. Similar to Bootstrap Scheduler, we use a streaming graph partitioning heuristic to solve the k-balanced graph partitioning problem. We first use topological sorting to linearize the global-topology-graph into a linearized vertex stream $\mathcal{L}$. For a vertex v from $\mathcal{L}$, we use the following partitioning heuristic to determine which node to assign it to.

$$index = arg \max_{i \in [1,m]} \left\{ \sum_{x \in P_i \cap N(v)} |x| \left(1 - \frac{|P_i|}{\varpi(n_i)}\right) \right\} \tag{10}$$

where $\varpi(n_i)$ is the maximum computation ability of physical node i, $N(v)$ is the set of neighbors of vertex v, $|x|$ is the computation cost of task x, and $|P_i|$ is the sum of computation cost of tasks on physical node P_i. Different from Bootstrap Scheduler, we take into consideration the real-time computation cost of all the tasks on each physical node. The basic intuition is quite similar, we choose a physical node most relevant to task v with most capacity remained when assigning task v.

Dynamic-Task-Reassignment. During the execution of the system, we further use the statistics of log to monitor the running status of each node. If some nodes become bottlenecks of the whole system, we dynamically reassign the tasks from these nodes to other nodes with less workload to improve the overall performance. We use a threshold θ to judge whether a node is skew enough and needed to reassign its tasks to other nodes. The threshold is defined as follows:

$$\theta = \vartheta * \frac{\sum_{i \in [1,m]} |P_i|}{m} \tag{11}$$

Here ϑ is the percentage we will discuss in Sect. 5.3, $|P_i|$ is the total computation cost of node i and m is the number of the physical nodes. The basic intuition is that, if the computation cost $|P_i|$ of physical node i is higher than the average cost among the cluster by θ, then we move some tasks from that node to others.

Here we propose a novel heuristic to move the tasks to an appropriate node. The movement tries to minimize the inter-node communication cost and rebalance the computation cost of each node in the cluster. For a task v_i in the system, StroMAX logs the bandwidth cost of the communication between two tasks, denoted by $rc(v_i, v_j)$. If v_i and v_j are on the same node, we have $rc(v_i, v_j) = 0$. Let $N(v)$ be the neighbor set of task v_i, the total inter-node bandwidth of v_i is denoted as $\Gamma(v_i)$.

$$\Gamma(v_i) = \sum_{v_j \in N(v)} rc(v_i, v_j)$$

The algorithm of reallocating tasks is illustrated in Algorithm 2. If node i needs a reassignment, we first choose the tasks with more inter-node communication according to $\Gamma(v_i)$. Specially, we sort the remote communication bandwidth of each task in a non-descending order (line 3), and greedily reassign the task with the max $\Gamma(v_i)$ to the node which communicates with v_i most frequently, i.e., maximal communication cost (line 5–7). When a task is removed, the estimated communication cost of the node will be decreased by $\Gamma(v_i)$. The reassignment stops when the total computation cost is below the threshold.

4 The StroMAX Architecture

System Architecture. Figure 3 shows how StroMAX is integrated into Storm. Note that our system can be migrated to other real-time stream processing systems as well. For the Storm cluster, there are three types of nodes — one master node called the nimbus node, zookeeper nodes, and worker nodes. When a new topology is submitted, the nimbus allocates the tasks to the workers and monitors failures. The zookeeper maintains the coordination state of nimbus and worker nodes. The topologies are executed on the worker nodes where a supervisor daemon is run on each worker for communication with zookeeper.

The components of StroMAX run on the nimbus node and worker nodes. There is a schedule manager running on the nimbus node that provides meta data for partitioning. It stores the meta data of the cluster and log statistics submitted from StroMAX monitors. When a new topology is committed, the schedule manager analyzes and partitions the topology by Bootstrap Scheduler. Then, it triggers the global-topology-graph-partitioning and dynamic-task-reassignment to rebalance the tasks running on the cluster. A schedule monitor runs on each worker node to record, collect, and report log information to the schedule manager. It also calculates the computation and communication measurements. For example, we use the Java API to log the CPU time for 1000 tuples and then calculate the average computation cost.

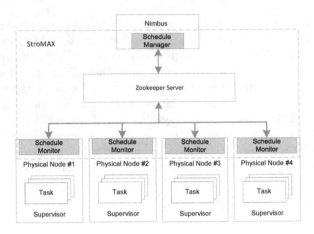

Fig. 3. Architecture of StroMAX.

Implementation of Task Reassignment. To reassign the tasks, we use the Storm infrastructure, which supports suspending and resuming tasks during runtime. Storm blocks the spouts, and thus prevents new stream from being propagated to the bolts and forwarded through the topology. Then all of the in-flight data is propagated through the bolts until all communication queues among these bolts are empty. Our scheduler then reconfigures the cluster by reassigning tasks to proper physical nodes.

5   Evaluation

In this section, we conduct extensive experiments to evaluate StroMAX. We first describe the experimental setup, then present and discuss the performance with different workload settings.

5.1   Experimental Settings

In this section, we briefly introduce the experimental settings for the evaluation, including the cluster, workload, and evaluation metrics. All the evaluation results are measured by average of five executions.

Cluster. All the experiments were conducted on a cluster of 42 nodes. Each node was equipped with two 2.80 GHz Intel Xeon E5-2680 CPU, 2 GB memory, and 48 GB SSD disk. All the nodes were connected by 1 GB bandwidth routers. We used one node for the nimbus, one for the zookeeper, and 40 for the workers.

Workload. Experiments are conducted with six data processing topologies as illustrated in Fig. 4 and described below — word count (WC), throughput test (TT), twitter trending topics (TWTT), log processing (LP), twitter stream sentiment analysis (TSSA), and synthetic communication (SC). We compared our

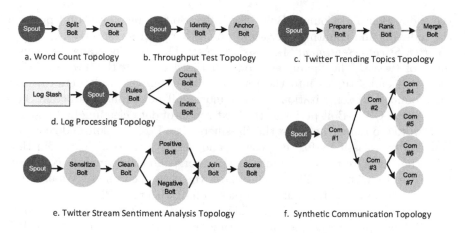

Fig. 4. Evaluated topologies.

proposed schedulers against the default scheduler in Storm. To find a proper tuple input rate for each topology, we first used a low initial rate and increased it gradually, until the average CPU usage of the cluster is above 50%. Then we used this rate in the whole evaluation.

1. *Word Count (WC)*: *WC* is a basic topology shown in Fig. 4a. It has one spout and two bolts. The Spout reads English words one line at a time from a local file which is made from 10 thousand random pages crawled from Wikipedia. The Split Bolt splits each line into words and passes them to the Count Bolt. The Count Bolt increases the counters based on distinct input word tuples and produces the results.
2. *Throughput Test (TT)*: *TT* has one spout and two bolts as shown in Fig. 4b. The Spout repeatedly generates random 10 KB strings as input tuples. The Identity Bolt emits the received tuples to the Anchor Bolt without any change. The Anchor Bolt increases a counter by one and records the processing time.
3. *Twitter Trending Topics (TWTT)*: This topology computes the top-k trending twitter topics as shown in Fig. 4c. The topology is a pipeline of one spout and three bolts. The Spout pushes tweets into the topology. Then the Prepare Bolt updates the counter of each topic, partitions topics alphabetically, and propagates the topic/count pairs. The Rank Bolt receives the topic/count pairs and maintains a list of top-k topics. Finally, the Merge Bolt merges all the lists to produce a single list of the current top-k topics. We used a dataset with one million English tweets from Twitter4j API.
4. *Log Processing (LP)*: *LP* presents a real-world case of log processing which is shown in Fig. 4d. The Spout reads tuples from an open-source log agent, Log Stash, as the input for the topology. The Log Stash reads log information from local file which is the kernel logs of Ubuntu server of our lab. The Rules Bolt performs a rule-based analysis on the log and emits log entry tuples to

the Index Bolt and Counter Bolt. The Index Bolt and Counter Bolt perform the indexing and counting operations on the log entries respectively.

5. *Twitter Stream Sentiment Analysis (TSSA)*: This topology analyzes sentiment of tweets. The Spout, as shown in Fig. 4e, parses the Twitter JSON data and emits tuples into the topology. The Sensitize Bolt performs the first-round data sensitization which removes all non-alpha characters. Following, the Clean Bolt performs the next round of data cleaning by removing stop words to reduce noise for the classifiers. The Positive Bolt and Negative Bolt are two classifiers for the positive and negative classes. Next, Join Bolt joins the scores from the two previous classifiers, and the Score Bolt compares the scores from the classifiers and declares the class accordingly. We used the same dataset for twitter trending topics from Twitter4j API.

6. *Synthetic Communication (SC)*: This is a synthetic topology as shown in Fig. 4f. The Spout reads one line at a time from the local file used in the *WC* topology. Each Communication Bolt doubles the received words and passes them to the next bolt. This is a typical communication intensive workload.

Evaluation Metrics. We use two metrics to systematically evaluate the result of our experiments.

- *Tuple Processing Time (TPT)*: It presents the average elapsed time for a tuple emitted from spout till its completion. We leveraged the timing mechanism in Storm to track each tuple's processing time. We calculated the average TPT every 30 s for performance evaluation.
- *Inter-Node Bandwidth (INB)*: It indicates the inter-node bandwidth of the topology during the execution. INB can well characterize the effect of our graph-partitioning based schedulers because they are designed to reduce the cost of inter-node communication among physical nodes.

5.2 Effect of Bootstrap Scheduler

Results and Analysis. We first evaluate the performance of Bootstrap Scheduler. We used 40 worker nodes, with 10 task slots on each node. Here we present the results on LP topology and TSSA topology as representatives. The experimental results on the other workloads yield similar improvements.

1. *Log Processing (LP)*: For this topology, we used 10 tasks for the spout, 50 tasks for the log rule bolt, 30 tasks for the index bolt, and 30 tasks for the counter bolt. Figure 5a shows the TPT of the default scheduler and Bootstrap Scheduler on this topology. The TPT of Bootstrap Scheduler is reduced by 37% on average compared to the default scheduler during the first 10 min. This is because Bootstrap Scheduler reduces inter-node communication cost by considering the structure of the committed topology while the default scheduler evenly distributes the task to the physical nodes.

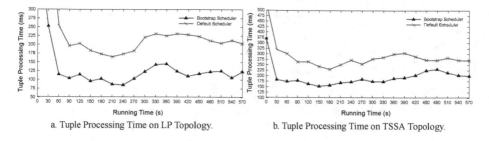

a. Tuple Processing Time on LP Topology.   b. Tuple Processing Time on TSSA Topology.

Fig. 5. Tuple processing time on two workloads.

2. *Twitter Stream Sentiment Analysis (TSSA)*: For this topology, we used 8 tasks for tuple input, and 20 tasks for each of the other bolts. As shown in Fig. 5b, the TPT of Bootstrap Scheduler is shorter than that of the default schduler by 39% on average after the system reaches a stable state. The TPT of TSSA is higher than that of LP because the graph structure of TSSA is more complicated, which incurs more inter-node communication, as shown in Fig. 4d and e.

Summary. When a new topology is committed to the system, Bootstrap Scheduler can analyze and assign its tasks to the proper physical nodes. This assignment takes inter-node communication cost into account, thus, Bootstrap Scheduler outperforms the default Storm scheduler. The above experiments prove that, with the help of graph partitioning, Bootstrap Scheduler can significantly reduce the average tuple processing time on various workloads.

5.3   Effect of Rebalance Scheduler

We next present the performance of Rebalance Scheduler. Rebalance Scheduler uses two techniques for task reallocation, i.e., *global-topology-graph-partitioning* which repartitions all the topologies on the cluster and *dynamic-task-reassignment* which moves tasks from overloaded nodes to idle ones automatically.

Effect of Global-Topology-Graph-Partitioning. In this experiment, we initially committed the WC topology to the cluster, then we committed the TT topology after 30 s and the TWTT topology after 60 s. These topologies were allocated by Bootstrap Scheduler when committed. As same as the setting in Bootstrap Scheduler, we used 40 worker nodes, with 10 task slots on each node. In addition, we used 5 tasks for each spout and 20 tasks for each bolt. We started global-topology-graph-partitioning to reassign the tasks of these topologies at 300 s.

Figure 6 summarizes the TPT of these three topologies in 10 min. Due to the space constraint, we present the results of WC and TWTT topologies, and the result of TT topology is similar. As we can see, once global-topology-graph-partitioning was triggered, it calculated a new assignment for all the topologies

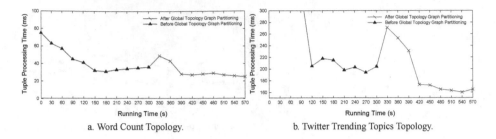

a. Word Count Topology.   b. Twitter Trending Topics Topology.

Fig. 6. Evaluation of global-topology-graph-partitioning.

a. INB before Dynamic-Task-Reassignment.   b. INB after Dynamic-Task-Reassignment.

Fig. 7. INB between components of twitter stream sentiment analysis topology.

in the cluster, which briefly increased the tuple processing time of these topologies. Afterwards, the tuple processing time dropped sharply to a normal value and outperformed the previous result, with a 10.9%–26.5% reduction of TPT on these workloads. This is because global-topology-graph-partitioning utilizes the collected runtime statistics to estimate the computation cost and communication cost of each task during the execution, and then optimizes the task assignment for all the topologies. In contrast, Bootstrap Scheduler allocates the newly committed topology via the graph partitioning result of a single topology. Therefore, when we commit three different topologies to the system, global-topology-graph-partitioning improves the tuple processing time based on the runtime result of Bootstrap Scheduler.

Effect of Dynamic-Task-Reassignment. We proceed to evaluate the effect of the dynamic-task-reassignment. We used 40 worker nodes, with 10 task slots on each node. We committed all the 6 topologies to the system with Bootstrap Scheduler. We used 3 tasks for each spout and 10 tasks for each bolt. We reassigned the tasks by global-topology-graph-partitioning after the system reached a stable state. Then we used the dynamic-task-reassignment to monitor and reallocate the tasks when they were skew enough. We recorded the inter-node bandwidth (INB) between components of the TSSA topology before and after executing the dynamic-task-reassignment.

As shown in Fig. 7, most of the inter-node bandwidths were reduced after the dynamic-task-reassignment. Specially, the communication cost between the Spout and Sensitive Bolt decreased from 116.3 Mbps to 83.5 Mbps, and the cost between the Negative Bolt and Join Bolt decreased from 9.2 Mbps to 2.6 Mbps.

Parameter Tuning of Dynamic-Task-Reassignment. As we mentioned in Sect. 3.3, we use a parameter ϑ to judge whether a node is skew enough and needed to reassign its tasks to other nodes. We have conducted a series of experiments to investigate the selection of ϑ. Due to the space limit, we do not show the details of parameter tunning.

We find that when ϑ is chosen between 10% and 15%, the dynamic-task-reassignment achieves the best performance for most of the workloads. The reason is that, when ϑ is small ($\vartheta < 10\%$), there are too many tasks reallocated during the execution. When ϑ becomes larger ($\vartheta > 15\%$), the reassignment is hard to be triggered. Thus we use $\vartheta = 12\%$ for the dynamic-task-reassignment in our experiments.

Summary. The above experiments indicate that Rebalance Scheduler can efficiently reduce the communication cost among the components of the topologies by reassigning the tasks based on the global-topology-graph. It leverages the statistics of log to monitor the running status of each node and dynamically reallocates bottleneck tasks. Therefore, Rebalance Scheduler can significantly improve the performance of real-time processing systems.

5.4   Overall Performance and Scalability of StroMAX

Overall Performance. In this part, we investigate the overall performance of StroMAX. In this experiment, we executed each of the five topologies on the cluster separately. The process was as follows: we added the topologies onto the cluster one by one. When a topology finished, we restarted the cluster and deployed a new topology onto the cluster with the help of Bootstrap Scheduler and Rebalance Scheduler.

Figure 8a illustrates the TPT of StroMAX and the default Storm scheduler on five topologies. As we can observe, compared to the default scheduler, the TPT of most workloads significantly decreases. For instance, the TPT on TT topology decreases by 87.3%. These results confirm that StroMAX can significantly reduce the tuple processing time and inter-node communication cost with the graph partitioning. Besides, the results also reveal the generality of the proposed approach that it can be applied to various workloads.

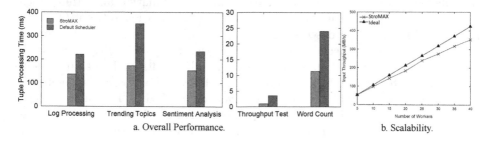

Fig. 8. Overall performance and scalability of StroMAX.

Scalability Study. Scalability is an important issue for real-time stream processing systems. We further evaluate the scalability of StroMAX by increasing the number of worker nodes. We increase the number of worker nodes from 5 to 40, and present the input throughput of the TSSA topology in Fig. 8b. We use both Bootstrap Scheduler and Rebalance Scheduler to schedule the topology. We use an ideal curve to represent the ideal execution time which assumes that the performance is linear to the number of the workers. As expected, as the number of workers increases, the throughput performance of StroMAX is close to the ideal case. This result confirms that StroMAX has a graceful scalability.

6   Related Work

The graph-partitioning based scheduling problem in real-time stream processing systems discussed in this paper is related to several fields. We briefly review the most relevant works.

Real-time Stream Processing Systems. System S [2] is a stream processing system developed by IBM. A query in System S is modeled as an event processing network which consists of a set of event processing agents. S4 [14] is another stream processing system, developed by Yahoo, where queries are designed as graphs of processing elements. Recently, Storm [10,11,21], an open-source, distributed, reliable, and fault-tolerant processing system, was proposed by Twitter for real-time stream processing. Some works [5] tried to bridge the gap between stream workload and MapReduce abstraction by proposing a stream version of the MapReduce approach. In these systems, events flow among the map and reduce stages without incurring. Wang [23] studied the problem of efficient load distribution in D-DSMS to minimize end-to-end latency. Besides, Xing [24] studied operators moving to dynamically change loads in high-performance computing clusters such as blade computers.

Graph Partitioning. Graph partitioning is a optimization problem which has been studied for decades [18,25]. The widely used k-balanced graph partitioning aims to minimize the number of edge-cut between partitions while balancing the number of vertices. Though the k-balanced graph partitioning problem is an NP-Complete problem [8], several solutions have been proposed to tackle this challenge. Andreev et al. [3] presented an approximation algorithm which guarantees polynomial running time with an approximation ratio of $\mathcal{O}(logn)$. Another solution was proposed by Even et al. [7] who gave an LP method based on spreading metrics which also gets an $\mathcal{O}(logn)$ approximation. Besides approximated solution, Karypis et al. [13] proposed a parallel multi-level graph partitioning algorithm to minimize the bisection on each level. There are some heuristic implementations like METIS [12], parallel version of METIS [16], and Chaco [9] which are widely used in many existing systems. Although these heuristics cannot provide a precise performance guarantee, they are effective.

The aforementioned methods are offline and generally require long processing time. Recently, Stanton and Kliot [20] proposed a series of online streaming partitioning method using heuristics. Fennel [22] extended the Stanton's work by proposing a streaming partitioning framework which combines some other heuristic methods. However, these methods are designed for generally graph partitioning and lack the consideration of the characteristics of DAG.

Beyond these static graph partitioning technologies, Nicosia [15] theoretically studied how to adapt to the graph structure changing without the overhead of reloading or repartitioning the graph. Some of the recent works [6,26] can cope with the changes in graph structure. However, the cost of these approaches to handle the changes is quite high. Shang et al. [17] investigated several graph algorithms and proposed simple, yet effective, policies that can achieve dynamic workload balance, while this approach uses hashing partitioning as the initial input.

7 Conclusion

In this paper, we systematically investigated the performance issues of real-time stream processing systems. We designed a novel system, StroMAX, to allocate the topology based on two graph partitioning based schedulers. The first scheduler, named Bootstrap Scheduler, analyzes the topological graph and partitions the topology when it is committed to the system. The second scheduler, named Rebalance Scheduler, goes one step further by monitoring the effectiveness of all the topologies and the load of cluster during runtime. Rebalance Scheduler then rebalances the topologies for a performance improvement when necessary. The experimental results confirmed the improvements of our proposed approaches.

Acknowledgment. This research is supported by the National Natural Science Foundation of China under Grant No. 61572039, Shenzhen Government Research Project JCYJ20151014093505032, 973 program under No. 2014CB340405, and Tecent Research Grant (PKU).

References

1. Flume. http://flume.apache.org/
2. Amini, L., Andrade, H., et al.: SPC: a distributed, scalable platform for data mining. In: DM-SSP, pp. 27–37 (2006)
3. Andreev, K., Racke, H.: Balanced graph partitioning. Theor. Comput. Syst. **39**(6), 929–939 (2006)
4. Billionnet, A., Costa, M.C., Sutter, A.: An efficient algorithm for a task allocation problem. JACM **39**(3), 502–518 (1992)
5. Brito, A., Martin, A., Knauth, T., Creutz, S., Becker, D., Weigert, S., Fetzer, C.: Scalable and low-latency data processing with stream mapreduce. In: CloudCom, pp. 48–58 (2011)
6. Cheng, R., Hong, J., Kyrola, A., Miao, Y., Weng, X., Wu, M., Yang, F., Zhou, L., Zhao, F., Chen, E.: Kineograph: taking the pulse of a fast-changing and connected world. In: EuroSys, pp. 85–98 (2012)
7. Even, G., Naor, J., Rao, S., Schieber, B.: Fast approximate graph partitioning algorithms. In: SODA, pp. 639–648 (1997)

8. Garey, M.R., Johnson, D.S., Stockmeyer, L.: Some simplified NP-complete problems. In: STOC, pp. 47–63 (1974)
9. Hendrickson, B., Leland, R.W.: A multi-level algorithm for partitioning graphs. SC **95**, 28 (1995)
10. Huang, Y., Cui, B., Jiang, J., Hong, K., Zhang, W., Xie, Y.: Real-time video recommendation exploration. In: SIGMOD, pp. 35–46 (2016)
11. Huang, Y., Cui, B., Zhang, W., Jiang, J., Xu, Y.: Tencentrec: real-time stream recommendation in practice. In: SIGMOD, pp. 227–238 (2015)
12. Karypis, G., Kumar, V.: Multilevel graph partitioning schemes. In: ICPP, pp. 113–122 (1995)
13. Karypis, G., Kumar, V.: Parallel multilevel k-way partitioning scheme for irregular graphs. In: SC (1996)
14. Neumeyer, L., Robbins, B., Nair, A., Kesari, A.: S4: distributed stream computing platform. In: ICDM, pp. 170–177 (2010)
15. Nicosia, V., Tang, J., Musolesi, M., Russo, G., Mascolo, C., Latora, V.: Components in time-varying graphs. Chaos Interdisc. J. Nonlinear Sci. **22**(2), 023101 (2012)
16. Schloegel, K., Karypis, G., Kumar, V.: Parallel static and dynamic multi-constraint graph partitioning. Concurrency Comput. Pract. Exp. **14**(3), 219–240 (2002)
17. Shang, Z., Yu, J.X.: Catch the wind: graph workload balancing on cloud. In: ICDE, pp. 553–564 (2013)
18. Shao, Y., Cui, B., Ma, L.: Page: a partition aware engine for parallel graph computation. TKDE **27**(2), 518–530 (2015)
19. Shi, X., Cui, B., Shao, Y., Tong, Y.: Tornado: a system for real-time iterative analysis over evolving data. In: SIGMOD, pp. 417–430 (2016)
20. Stanton, I., Kliot, G.: Streaming graph partitioning for large distributed graphs. In: KDD, pp. 1222–1230 (2012)
21. Toshniwal, A., Taneja, S., et al.: Storm@ twitter. In: SIGMOD, pp. 147–156 (2014)
22. Tsourakakis, C.E., Gkantsidis, C., Radunović, B., Vojnović, M.: Fennel: streaming graph partitioning for massive scale graphs. Technical report, Microsoft (2012)
23. Wang, W., Sharaf, M.A., Guo, S., Özsu, M.T.: Potential-driven load distribution for distributed data stream processing. In: SSPS, pp. 13–22 (2008)
24. Xing, Y., Zdonik, S., Hwang, J.H.: Dynamic load distribution in the borealis stream processor. In: ICDE, pp. 791–802 (2005)
25. Xu, N., Cui, B., Chen, L., Huang, Z., Shao, Y.: Heterogeneous environment aware streaming graph partitioning. TKDE **27**(6), 1560–1572 (2015)
26. Yang, S., Yan, X., Zong, B., Khan, A.: Towards effective partition management for large graphs. In: SIGMOD, pp. 517–528 (2012)

Partition-Based Clustering with Sliding Windows for Data Streams

Jonghem Youn[1]($\boxtimes$), Jihun Choi[1], Junho Shim[2], and Sang-goo Lee[1]

[1] Seoul National University, Seoul, Republic of Korea
{jonghm,jhchoi,sglee}@europa.snu.ac.kr
[2] Sookmyung Womens University, Seoul, Republic of Korea
jshim@sookmyung.ac.kr

Abstract. Data stream clustering with sliding windows generates clusters for every window movement. Because repeated clustering on all changed windows is highly inefficient in terms of memory and computation time, a clustering algorithm should be designed with considering only inserted and deleted tuples of windows. In this paper, we address this problem by sliding window aggregation technique and cluster modification strategy. We propose a novel data structure for construction and maintenance of 2-level synopses. This data structure enables to update synopses efficiently and support precise sliding window operations. We also suggest a modification strategy to decide whether to append new synopses to pre-existing clusters or perform clustering on whole synopses according to the difference between probability distributions of the original and updated clusters. Experimental results show that proposed method outperforms state-of-the-art methods.

Keywords: Data streams · Clustering · Sliding windows · Approximation algorithms

1 Introduction

Large scale data streams are generated from a variety of applications such as sensor networks, transportation monitoring, smart devices, search engines, social media, and news portals. The data streams are massive, rapid evolving, and infinitely created. Therefore, clustering algorithm for data streams are designed with low computation cost and memory limitation.

Early studies assumed that clustering is to be performed over an entire data stream, and directly applied one-pass clustering algorithms to those data streams [16]. However, data streams evolve continuously over time. In most data stream applications, the most recent tuples are considered to be more decisive and influential. This characteristic makes clustering algorithms with window models to be developed. Window models widely used by the algorithms include the landmark window [1,9,14,17] and damped window [6,10,15]. The landmark window model splits data streams into fixed-size, non-overlapping chunks and contains tuples

© Springer International Publishing AG 2017
S. Candan et al. (Eds.): DASFAA 2017, Part II, LNCS 10178, pp. 289–303, 2017.
DOI: 10.1007/978-3-319-55699-4_18

that arrived after landmark and usually used when periodic results are needed (e.g., on a daily or weekly basis). In the damped window model, also known as the fading window model, tuples are associated with weights that decrease over time. Algorithms with these window models are based on the insertion-only model, which assumes tuples received only once are not removed from the window at a later time, and give newer tuples higher weight values than older ones.

While these two window models are effective in some data stream applications, they are insufficient for domains requiring the sliding window model. In this model, the window contains only tuples with timestamps different from the current timestamp back to a certain timestamp in the past. As time passes, the window removes tuples whose timestamps have expired. Applications with the sliding window model regard the exact number of recent tuples as critical and appropriate, such as topic extraction in news portal and traffic monitoring.

Extensive research on clustering with the landmark or damped window model has been done, but only a small number of studies exist on clustering with sliding windows [4,5,7,18]. Clustering with sliding windows should produce results for every window movement, and a seemingly straightforward approach would be to perform repeated clustering. However, this is impractical and carries significant computational costs. Therefore, a clustering algorithm is needed that considers both deletion and insertion in both synopses and cluster results.

Although data stream clustering algorithms were originally designed with error tolerance and results are approximate, it is important to keep a precise range of target tuples for tracking and clustering evolving data streams. We consider general and expressive sliding window specifications in continuous query language [3] of data stream management systems (DSMS) for clustering with accurate sliding window operations. The sliding window is specified by RANGE for the length of the window and SLIDE for the movement intervals.

In this paper, we present an efficient data stream clustering algorithm with sliding windows. Main contributions of the algorithm are as follows: (1) Our algorithm supports general and precise sliding window operations [3] for data stream clustering. Our algorithm constructs and maintains 2-level synopses for the cluster features through the use of a sliding window aggregation technique that reduces space and computation time [11]. For sliding window aggregation, a window is divided into disjoint chunks, and a synopsis of the window is computed by merging the synopses of chunks. (2) We adopt an index structure for cluster features based on Locality-Sensitive Hashing (LSH) [8] to search the nearest cluster feature in synopses efficiently. Previous studies use tree-based index structure, but insertion and deletion operations occurs with average $O(\log n)$ time complexity whenever data is input. On the other hand, the hash-based index structure is suitable for the sliding windows because it has average $O(1)$ time complexity. (3) We propose a modification strategy of pre-exist clusters to avoid clustering every time data arrive. Clustering operation accesses all data objects iteratively, making it a very expensive operation. We allow appending new input tuples to pre-existing clustering if the quality of modified clustering results is acceptable. We also suggest quality measure of difference between two sets of clusters.

This paper is organized as follows. In Sect. 2, we explain tasks related to clustering algorithms for data streams. Section 3 presents background information. We show how synopses for clustering are maintained in Sect. 4 and how clustering is performed in Sect. 5. An analysis of our experimental results is described in Sect. 6. Finally, we present our concluding remarks in Sect. 7.

2   Related Work

Clustering algorithms for data streams have been extensively studied, and a detailed survey of these algorithms is presented in [13]. Clustering algorithms are categorized by grouping concepts which include partition-based [1,2], density-based [6,10], and message passing-based [14,17] clustering, and are developed under landmark window [1,14,17], damped window [6,10], and sliding window [4,5,7,18] models. The majority of these algorithms adopt the landmark window model, while density-based algorithms are designed according to the damped window model.

In contrast to landmark or damped window, only a small number of studies focus on clustering algorithms with sliding windows [4,5,7,18]. Dang et al. propose a Gaussian mixture models based clustering algorithm for sliding window [7]. They exploit the Expectation Maximization technique, and develop splitting and merging operations to remove expired tuple. Babcock et al. present a technique of maintaining variance and k-median based on *exponential histogram(EH)* for sliding window [4]. Zhou et al. focus on the problem of tracking the evolution of clusters in sliding window, developing SWClustering, a k-means clustering algorithm based on an extension of EH, *exponential histogram of cluster features(EHCF)* which combines temporal cluster features with EH [18].

In the theory community, Braverman et al. propose a merge-and-reduce based technique to transform coreset construction in the insertion-only streaming model to the sliding window model [5].

Specifically, algorithms which exploit EH as synopsis data structure support insertion and deletion [4,18]. EH is defined as a collection of buckets on a set of tuples, and generates $(\frac{k}{2}+1)((\log\frac{2N}{k}+1)+1)$ for $k = \lceil\frac{1}{\epsilon}\rceil$. Only the synopsis of the tuples in each bucket is stored by the appropriate bucket, with the synopsis containing both cluster features and the most recent timestamp of the tuples in the bucket. Because of the memory limitations, if the number of buckets exceeds the user defined number, the buckets are merged, with each merged bucket holding a number of tuples equal to or double that held in the previous unmerged buckets. For example, we assume that input tuples are $x_1, x_2, \ldots$ (x_2 newer than x_1), and the state of the buckets is $B_1 = \{x_1, x_2\}$, $B_2 = \{x_3\}$, $B3 = \{x_4\}$. As new tuples arrive, the old buckets are merged and a new bucket is created with the new tuples, i.e., $B_1 = \{x_1, x_2\}$, $B_2 = \{x_3, x_4\}$, $B3 = \{x_5\}$. When the sliding window moves, buckets whose timestamp have expired are removed. However, a small deviation occurs in the timestamp. If the size of the sliding window is 4 in the example, it should drop x_1 from the window. However, the bucket B_1 also contains x_2 which is valid for the window, so it cannot be removed. This case occurs more frequently as the window size increases. Therefore, one of objectives of our algorithm is clustering on accurate ranges of tuples.

In this paper, we propose an efficient algorithm for partition-based clustering with sliding windows. The algorithm aims to produce high-quality clustering results quickly. Unlike other algorithms, novel data structure and procedures of our algorithm enable to perform clustering on the tuples in exact ranges, and reduce the computation cost of operations such as insertion, deletion, searching and clustering.

3   Preliminaries

3.1   Data Streams

A data stream is defined as an infinite sequence of tuples.

$$S = \langle s_1, t_1 \rangle, \langle s_2, t_2 \rangle, ..., \langle s_n, t_n \rangle, ... \tag{1}$$

where s_i is a tuple, and t_i is a timestamp. A tuple s_i is represented by multi-dimensional attribute vector. A tuple of d dimensions is denoted by $s_i = (x_i^1, ..., x_i^d)$. A timestamp t_i is non-negative integer value, and t indicates current time. For simplicity, we assume that tuples arrive in chronological order, i.e. for any $i < j$, a tuple $S_i = \langle s_i, t_i \rangle$ arrives earlier than $S_j = \langle s_j, t_j \rangle$. Timestamp value denotes a sequence number in tuple-based window, and a particular time instance in time-based window.

3.2   k-means Clustering

For any two tuples s_1, s_2, we denote the Euclidean distance between s_1 and s_2 by $dist(s_1, s_2) = \sqrt{\sum_{i=1}^{d}(x_1^i - x_2^i)^2} = \|s_1 - s_2\|$, and the squared Euclidean distance by $dist^2(s_1, s_2) = \|s_1 - s_2\|^2$

For a set of tuples $\{s_1, s_2, ..., s_n\}$, k-means clustering problem is to partition n tuples into k clusters $C = \{C_1, C_2, ..., C_k\}$ such that clustering criterion is optimized. The most widely used clustering criterion is to minimize sum of squared Euclidean distance between each tuple and center of the assigned cluster. Specifically, the Euclidean k-means clustering problem is to find $\arg\min_C \sum_{i=1}^{k} \sum_{s_j \in C_i} \|s_j - c_i\|^2$ where c_i is the center of cluster C_i.

Because the problem is NP-hard even for $k = 2$, heuristic algorithms are proposed. One of the classical heuristic algorithms is *Lloyd*'s algorithm [12]. The *Lloyd*'s algorithm converges to a local optimum, and does not guarantee to converge to the global optimum.

3.3   Sliding Window

The sliding window contains only tuples whose timestamp is within the range of the current timestamp and the start timestamp of the window. In continuous queries in DSMS, sliding window is specified by RANGE for the length of the window and SLIDE for the movement intervals of the window.

Windows are categorized into tuple-based and time-based sliding windows in accordance with sliding condition and time unit. The tuple-based sliding window slides when new tuples arrive. The window contains a fixed number of tuples and also slides by fixed number of tuples. The example, S [RANGE 1000 TUPLES SLIDE 100 TUPLES] is a tuple-based window, and the time unit is the number of tuples. The time-based sliding window slides as time progresses. For example, S [RANGE 20 MINUTES SLIDE 5 MINUTES] contains tuples from the most recent 20 min and slides every 5 min. Granularity of the window depends on time units which include HOURS, MINUTES, or SECONDS. In the time-based sliding window, the number of tuples within the window is not bounded.

For ease of explanation, we only consider tuple-based windows, but the methodologies can be applied to time-based windows as well.

4   Synopses for Sliding Windows

Clustering consists of two steps in general: (1) Construction and maintenance of synopses over sliding window (2) Decision on whether to append new synopses to pre-existing clusters or perform clustering on whole synopses according to the difference between probability distributions of the original and updated clusters. In this section, we describe a method for constructing and maintaining synopses with sliding windows.

4.1   Cluster Features

Cluster Feature(CF) is a data structure for storing statistic summaries of data streams in Euclidean space. CF consists of linear sum of the tuples LS, square sum of the tuples SS, the number of the tuples N, and the most recent timestamp of the tuples T. The tuples are in the range of sliding window. LS and SS are generated by pairwise summation of tuples, i.e. for d-dimensional n tuples, $LS = \sum_{i=1}^{n} s_i = \sum_{i=1}^{n}(x_i^1, ..., x_i^d)$ and $SS = \sum_{i=1}^{n} s_i^2 = \sum_{i=1}^{n}((x_i^1)^2, ..., (x_i^d)^2)$. LS and SS are d-dimensional vectors, and the N and T are numeric values. The basic components LS, SS, and N are proposed in [16], and the timestamp component T is appended in [18].

The CFs have incrementality and additivity properties. Incrementality means that the CF is updated by adding a new tuple s_j, while additivity means that two disjoint CFs can be merged into a new CF by adding their components [16]. These properties enable to modify the synopses in a constant time.

Incrementality	Additivity

$$
\begin{array}{ll}
LS = LS_1 + s_j & LS = LS_1 + LS_2 \\
SS = SS_1 + (s_j)^2 & SS = SS_1 + SS_2 \\
N = N_1 + 1 \quad (2) & N = N_1 + N_2 \quad (3) \\
T = t_j & T = max(T_1, T_2)
\end{array}
$$

Values for clustering such as centroid can be calculated easily by using components of the CF, i.e., $Centroid = LS/N$. In our algorithm, we use only LS, N, and T.

4.2 Synopses Construction and Maintenance

The CF contains summaries of tuples in sliding window. If synopses maintain only CF of tuples in sliding window, it is possible to update CF with new arrival tuples because of incrementality. However, it is impossible to update CF with expired tuples because synopses have not preserve values of expired tuples to subtract from the CF. To retain the CF for sliding window, the values to subtract from the CF should be kept. Because it is inefficient to keep all tuples in sliding window, we propose a data structure for synopses, pane-based CF (PCF) and window-based CF (WCF).

Figure 1 shows an overview of the synopses. A window is decomposed into panes which are non-overlapping sets of tuples. Assume that RANGE is R, SLIDE is L. The number of panes is $\lceil R/L \rceil$, and each pane represents at most L tuples. For example, sliding windows, as defined by S [RANGE 1000 TUPLES SLIDE 100 TUPLES] have ten panes, with each pane containing a summary of 100 tuples. In S [RANGE 1000 TUPLES SLIDE 99 TUPLES], the window consists of 11 panes, where ten panes each contain 99 tuples, and one pane contains 10 tuples. For ease of presentation, we only discuss the case that R is divisible by L.

When the window slides, $\Delta W_{expired}$ is removed and ΔW_{new} is appended. New PCFs are generated based on tuples in ΔW_{new}. Components of the PCF are the same as for CF, which includes LS, SS, N, and T. The detailed process of creating PCF is described in Algorithm 1. Given threshold θ, tuples whose distances are below θ are grouped into the same PCF. If the distance between

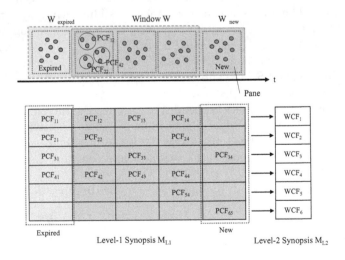

Fig. 1. Synopses for clustering

Algorithm 1. CREATEPANECF

Input: A set of tuples B, and threshold θ
Output: A set of PCFs
1: create empty set P
2: **for** each $b \in B$ **do**
3:      **if** P is empty **then**
4:          create new PCF p based on b
5:          $P \leftarrow P \cup \{p\}$
6:      **else**
7:          $p \leftarrow$ nearest PCF in P to b
8:          **if** $dist(b, p) < \theta$ or $dist(b, p) <$ radius of p **then**
9:              update p by adding b
10:         **else**
11:             create new PCF p based on b
12:             $P \leftarrow P \cup \{p\}$
13:         **end if**
14:     **end if**
15: **end for**
16: **return** P

---

a tuple and centroid of PCF, $dist(b, p)$ is below θ or radius of PCF, the PCF absorbs the tuple. Generating operation has $O(L \times m)$ time complexity, where m is the number of generated PCFs.

As shown in Fig. 1, synopses consist of level-1 synopsis and level-2 synopsis. Level-1 synopsis is a 2-dimensional array of $\lceil R/L \rceil$ width. Each row in level-1 synopsis contains PCFs whose distances are close. Generated PCFs by Algorithm 1 are inserted into the last column of level-1 synopsis. Timestamps T of the inserted PCFs are in $(R - L, R]$. When the expired tuples are removed, the first column of level-1 synopsis whose timestamps are in $[t_1, t_1 + L]$ is truncated, where t_1 is the earliest timestamp. Removing operation for level-1 synopsis has $O(1)$ time complexity if an adequate data structure is adopted such as a linked list queue. The exact number of tuples is R in the level-1 synopsis, and remains constant.

Window-based CFs WCFs in the level-2 synopsis are built by summing up PCFs which are in the same row. WCF is equal to the CF of tuples in the row due to the additivity property in Eq. (3). WCF is represented as

$$WCF_i = \sum_{j=1}^{\lceil R/L \rceil} PCF_{ij} \tag{2}$$

The algorithm updates WCFs by adding new PCFs and subtracting expired PCFs when the window slides. Because level-1 synopsis contains PCFs based on panes, we specify PCFs to be removed and quantify the values of expired tuples. The additivity property also guarantees correct WCFs for subtraction.

Algorithm 2 describes a procedure for updating synopses. The algorithm is performed through batch processing for performance. CREATEPANECF in line

---

Algorithm 2. UPDATESYNOPSES

Input: Stream S, threshold θ, range R, slide L, level-1 synopses M_{L1}, and level-2 synopses M_{L2}

Output: updated M_{L1} and M_{L2}

1: **if** exist expired tuples for R in M_{L1} **then**
2:　　$E \leftarrow$ expired PCFs in M_{L1}
3:　　subtract E from M_{L2}
4:　　truncate the column of E in M_{L1}
5:　　append new empty column in M_{L1}
6: **end if**
7: $B \leftarrow$ recent tuples of $(R - L, R]$ in S
8: $P \leftarrow$ CREATEPANECF(B, θ)
9: **for** each $p \in P$ **do**
10:　　$WCF_i \leftarrow$ nearest WCF to p found by LSH of M_{L2}
11:　　**if** $dist(p, WCF_i) < \theta$ or $dist(p, WCF_i) <$ radius of WCF_i **then**
12:　　　　$WCF_i \leftarrow WCF_i + p$
13:　　　　append p at i^{th} row and last column in M_{L1}
14:　　**else**
15:　　　　create new WCF based on p
16:　　　　append new WCF to M_{L2}
17:　　　　append p at new row and last column in M_{L1}
18:　　**end if**
19:　　update LSH of M_{L2}
20: **end for**
21: **return** M_{L1}, M_{L2}

---

8 creates PCFs of recent tuples whose timestamps are in $(R - L, R]$ from data streams. The process for removing expired tuples is presented in line 1–6. $dist(p, WCF_i)$ computes distance between centroids of PCF and WCF. $WCF_i + p$ in line 12 means that it updates components in WCF_i by adding components in PCF p, and timestamp T is replaced by timestamp t of p because p is newer than WCF_i.

4.3 Finding Nearest WCF

Updating M_{L1} and M_{L2} involves linear time complexity. The most time-consuming parts in the Algorithm 2 are CREATE-PANECF and finding the nearest WCF to a PCF. CREATEPANECF can be executed within a reasonable time by adjusting the size of the SLIDE L. However, finding the nearest WCF is a computationally heavy operation since it scans all M_{L2} and computes all distances for each PCF. The operation is well known as the nearest neighbor search problem. The searching operation is executed $N_{L2} \times N_P$ times per the window slides, where N_{L2} is the number of WCFs in M_{L2}, and N_P is the number of PCFs in P. To avoid unnecessary computation, we utilize a data structure based on Locality-Sensitive Hashing (LSH) [8] for indexing WCFs.

The basic concept of LSH is to map similar vectors to hash values which have higher probability of collision than hash values of dissimilar vectors. In other words, if two vectors are close to each other, after projection the vectors remain close. Hash function $h_{a,b}(x) : \mathcal{R}^d \rightarrow \mathcal{N}$ is a scalar projection which maps a vector x to an integer. The hash function is given by $h_{a,b}(x) = \lfloor (a \cdot x + b)/w \rfloor$, where a is a randomly drawn d-dimensional vector, w is the width of the quantization bin, and b is a random variable in the interval $[0, w)$.

General LSH generates a hash table whose hash keys are computed from hash functions to decrease the probability that dissimilar vectors fall into the same quantization bin. A hash key is obtained by concatenating values from the hash functions, e.g., when we have 2 hash functions, key $g(x)$ is $(h_{a_1,b}(x), h_{a_2,b}(x))$.

However, in data streams, the hash table need to be updated continuously as tuples are inserted and deleted. Updating the hash table and computing hash key carry with it high computational costs. Therefore, we maintain an adequate number of hash functions to update and compute distances from a target tuple to neighbors which are found in the hash table.

Figure 2 shows an example of the hash table for finding the nearest *WCF*. We utilize the assumption that if distance x between two vectors A and B is less than θ, the distance after projection is also less than θ. Therefore, by setting the width of the quantization bin w of the hash function to θ, each bucket of hash tables contains vectors whose distances are within θ. To find close vectors of target vector A, the operation first takes a bucket with the same hash value $g(A)$, and if there is no other elements except itself, it searches for buckets with adjacent hash key values from $g(A) - (1, ..., 1)$ to $g(A) + (1, ..., 1)$. Then the operation computes the real distances to the found vectors. The operation need to takes the elements in $g(A) \pm (1, ..., 1)$ adjacent buckets because the target vector A can be located near the border of the $g(A)$ bucket.

As the number of hash functions increases, the number of adjacent hash keys that need to be searched increases exponentially. For m hash functions, the operation searches for 3^m keys. To prevent this, we use a heuristic that the operation access only buckets with key values that differ by 1 in each component

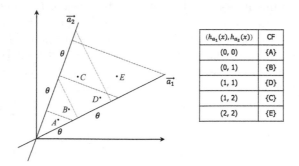

Fig. 2. LSH for searching nearest WCF

Algorithm 3. CLUSTERING

Input: Level-2 synopses M_{L2}, the number of clusters k, and error bound e
Output: Clusters C
1: $C_p \leftarrow$ pre-exist clusters
2: **if** C_p is empty **then**
3:     $C \leftarrow$ clusters which are created by k-means clustering using M_{L2}
4: **else**
5:     $W \leftarrow$ new and changed $WCFs$ in M_{L2}
6:     $C \leftarrow$ clusters which are modified by assigning each $w \in W$ to its nearest cluster
    of C_p
7:     $\alpha \leftarrow KL\text{-}divergence$ between C_p and C
8:     **if** $\alpha > e$ **then**
9:         $C \leftarrow$ clusters which are created by k-means clustering using M_{L2}
10:     **end if**
11: **end if**
12: **return** C

---

of $g(A)$, which are $2m$. For example, in Fig. 2, for $g(B) = (0, 1)$, the operation takes elements of $(-1, 1), (0, 0), (1, 1), (0, 2)$ keys.

In Algorithm 2, target vector is the centroid of PCF, and vectors in hash tables are centroids of $WCFs$.

5   Clustering with Sliding Windows

In this section, we present the clustering algorithm with sliding windows for data streams, which performs clustering based on synopses. In order to reduce the total computation cost of clustering, we add a modification step to the algorithm, which appends new synopses to pre-existing clusters based on the probability distributions of those clusters. The detailed process is presented in Algorithm 3. The algorithm is executed as the window slides.

First, the algorithm uses k-means clustering to produce the clusters based on their features in M_{L2}. Clustering based on CF has been widely studied. The basic and most commonly used methodology is to consider the centroid of WCF as a tuple, and perform clustering on them. Clustering on centroid with weight is also commonly used, where weight is the number of tuples N in WCF.

If clusters already exist, the algorithm detects new and changed $WCFs$, and assigns each of them to its nearest cluster. This produces approximate clusters. However, it is much faster than performing clustering again.

To preserve clustering quality and decide to perform clustering again, we measure quality degeneration of original and modified clusters by *Kullback-Leibler divergence* (*KL-divergence*). *KL-divergence* of probability distributions $p(x)$ and $q(x)$ is a measure of information gain achieved if $p(x)$ is used instead of $q(x)$. *KL-divergence* is defined as

$$D_{KL}(p(x)||q(x)) = \sum_x p(x) \log \frac{p(x)}{q(x)} \quad (3)$$

When $p(x)$ and $q(x)$ follow the Gaussian distribution, probability density functions are $p(x) = \mathcal{N}(\sigma_p, \mu_p^2)$ and $q(x) = \mathcal{N}(\sigma_q, \mu_q^2)$. KL-divergence is calculated from mean and deviation.

$$D_{KL}(p(x)\|q(x)) = \log \frac{\sigma_q}{\sigma_p} + \frac{\sigma_p^2 + (\mu_p - \mu_q)^2}{2\sigma_q^2} - \frac{1}{2} \tag{4}$$

When the data distribution does not follow the Gaussian distribution, the algorithm needs to select the probability distribution that best fits to a dataset first. Selecting the distribution determines how well the candidate distributions fit to the dataset using the specific goodness of fit tests such as Kolmogorov-Smirnov test.

We assume that distances between a centroid of WCF and a center of the assigned cluster follow the Gaussian distribution. k-Means clustering is designed to works well and generates high-quality clusters for the data which follow Gaussian distribution. Therefore, it is a reasonable assumption that data streams and clusters follow the Gaussian distribution. While performing clustering, cluster statistics information is easy to generate and store. We maintain sum of distances, sum of squared distances, and the number of WCFs with the cluster, and averages and deviations can be calculated from these. When a cluster is modified, the statistics information can be updated in a constant time because it has additivity property.

In Algorithm 3, the error bound e adjusts how much the algorithm tolerates the error. If the e is small, the algorithm performs clustering frequently. The appropriate value of e depends on datasets, and is tested experimentally. According to our experimental results, this step does not seriously decrease clustering quality.

6 Experiment

6.1 Experimental Setup

We evaluated efficiency and scalability of our clustering algorithm on different datasets, which included synthetic datasets and real-world datasets. We compared our algorithm with recent and frequently used clustering algorithms for data streams, *SWClustering* [18], *StreamKM++* [1] and *ClusTree* [10]. We also measured the performance of basic k-means clustering, which is implemented by *Lloyd's* algorithm [12].

All algorithms are implemented by Java. We executed all experiments with 64-Bit OpenJDK 1.8.0_91 on Intel i7-3820 3.60GHz CPU and 32GB main memory using Linux 4.4.0-43 kernel. Maximum Java heap size (-Xmx option) is set to 8GB.

Table 1 shows the overview of datasets for experiments. Syn1k30d40 and syn1k30d80 are synthetic data, and others are real-world data. We generate the data which follow the Gaussian distribution and have 30 clusters with dimensions of 40 and 80. Kddcup99 is network data streams to detect network intrusion.

Table 1. Datasets

	Size	Dim.
syn1k30d40	2,000,000	40
syn1k30d80	2,000,000	80
kddcup99	4,898,431	34
covtype	581,012	54
spambase	4,601	57
census1990	2,458,285	68

Kddcup99 contains logs of TCP connection of network at MIT Lincoln Labs of 2 weeks. This dataset is used to evaluate clustering algorithms in [2]. Covtype contains cartographic data from the Roosevelt National Forest of northern Colorado. Spambase contains email statistics to predict spam emails. Census1990 contains personal records sampled from the 1990 U.S. census data. Covtype and census1990 are used in *StreamKM++*, and a detailed description is presented in [1].

To evaluate efficiency and scalability, we measure total running time and the running time of each sliding. To evaluate the quality of clusters, we measure the sum of squared distance (SSQ) of the clusters. SSQ is defined as $\sum \|s_i - c_i\|^2$, which means the sum of squared distance between each tuple and the center of their cluster. The lower SSQ value indicates better quality of clusters. In sliding windows, the algorithm produces multiple results as the window moves. Therefore, the quality is evaluated by average SSQ of the results.

6.2 Experimental Results

In each experiment, we set threshold θ as $\theta_{syn1k30d40} = 14.1$, $\theta_{syn1k30d80} = 19$, $\theta_{kddcup99} = 60$, $\theta_{covtype} = 75$, and $\theta_{census1990} = 9$, and error bound $e = 0.2$ for our algorithm and *SWClustering*. Clustering quality and speed are in trade-off relationship with the values. We tested several thresholds and error bounds, and select the values to generate best clustering quality. We use only two hash functions, and that number is enough to reduce the computation time for the experiment datasets. Because other algorithms except *SWClustering* do not support sliding operation, we ran the clustering algorithms repeatedly on tuples which are within the range of the window.

Figure 3 shows the clustering quality of the algorithm for different values of k. We fix RANGE = 100,000 and SLIDE = 10,000. The average SSQ is transformed to log scale for presentation. We observe that our algorithm is usually better than *StreamKM++* and *ClusTree* in most datasets. Basic k-means in the experiment shows the best quality because it performs clustering on whole tuples in the window without any summarization. The results reveal that *StreamKM++* and *ClusTree* are inappropriate for sliding window, which removes tuples from the window continuously. The performance *SWClustering* is slightly better than our

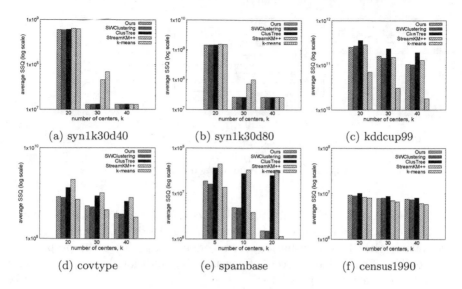

Fig. 3. Clustering quality comparison

algorithm. As described in Sect. 2, this happened because *SWClustering* contained more tuples which are invalid and expired in synopses. *SWClustering* contains synopses of 155,485 tuples, not 100,000 at 200,000 timestamp for census1990 at $k = 40$.

In terms of running time, our algorithm shows better scalability than the others. In Fig. 4, we measure the total running time of the algorithms to process 200,000 tuples for different values of RANGE. We set $k = 30$, and SLIDE = 10,000. As the size of the window increases, the running time of other algorithms also increases. However, the running time of our algorithm is almost the same, although RANGE increases.

Figure 5 shows the running times at each timestamp when the window slides. We set $k = 30$, RANGE = 100,000, SLIDE = 10,000, and the number of tuples = 200,000. Our algorithm is stable and fastest in terms of running time. *ClusTree*

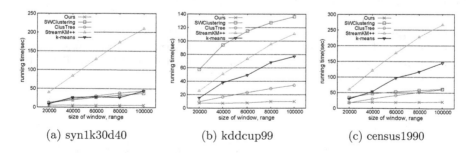

Fig. 4. Total running time comparison

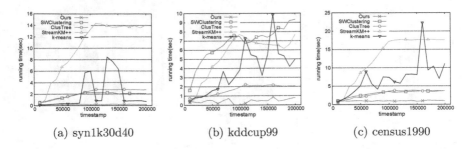

(a) syn1k30d40   (b) kddcup99   (c) census1990

Fig. 5. Running time at each timestamp

shows the second best performance, and *SWClustering* shows lowest performance in the kddcup99. Because k-means is a randomized algorithm, the running time is fluctuating with data distribution. However, k-means shows the proper performance. This means that the cost of constructing additional synopses is quite high for other streaming algorithms.

7   Conclusions

In this paper, we show our development of an efficient partition-based clustering algorithm with sliding windows for data streams. In the abstraction step, we exploit the sliding window model, which is divided into disjoint panes, and a pane-based aggregation technique for maintaining synopses. Synopses consist of pane-based CFs and window-based CFs to update cluster features efficiently. The synopses also maintain LSH of window-based CFs to reduce search time of the nearest neighbors. In the clustering step, the algorithm performs clustering on window-based CFs. A strategy for modification of clusters is proposed to avoid unnecessary clustering.

Our approach has an advantage over recent algorithms which perform clustering on whole data streams, as it provides the functionality of tracking changes in data streams by producing snapshots of every clustering, using less computational power. In future research efforts, we will extend our techniques to other clustering algorithms such as density-based clustering.

Acknowledgments. This research was supported by the MSIP (Ministry of Science, ICT and Future Planning), Korea, under the ITRC (Information Technology Research Center) support program (IITP-2016-R0992-16-1023) supervised by the IITP(Institute for Information & communications Technology Promotion).

References

1. Ackermann, M.R., Märtens, M., Raupach, C., Swierkot, K., Lammersen, C., Sohler, C.: Streamkm++: a clustering algorithm for data streams. J. Exp. Algorithmics **17**, 2.4:2.1–2.4:2.30 (2012)

2. Aggarwal, C.C., Han, J., Wang, J., Yu, P.S.: A framework for clustering evolving data streams. In: Proceedings of the 29th International Conference on Very Large Data Bases, vol. 29, pp. 81–92 (2003). VLDB Endowment

3. Arasu, A., Babu, S., Widom, J.: The CQL continuous query language: semantic foundations and query execution. VLDB J. 15(2), 121–142 (2006)

4. Babcock, B., Datar, M., Motwani, R., O'Callaghan, L.: Maintaining variance and k-medians over data stream windows. In: Proceedings of the Twenty-Second ACM SIGMOD-SIGACT-SIGART Symposium on Principles of Database Systems, pp. 234–243. ACM (2003)

5. Braverman, V., Lang, H., Levin, K., Monemizadeh, M.: Clustering problems on sliding windows. In: Proceedings of the Twenty-Seventh Annual ACM-SIAM Symposium on Discrete Algorithms, pp. 1374–1390. Society for Industrial and Applied Mathematics (2016)

6. Cao, F., Ester, M., Qian, W., Zhou, A.: Density-based clustering over an evolving data stream with noise. In: 2006 SIAM Conference on Data Mining, pp. 328–339. SIAM (2006)

7. Dang, X.H., Lee, V., Ng, W.K., Ciptadi, A., Ong, K.L.: An EM-based algorithm for clustering data streams in sliding windows. In: Zhou, X., Yokota, H., Deng, K., Liu, Q. (eds.) DASFAA 2009. LNCS, vol. 5463, pp. 230–235. Springer, Heidelberg (2009). doi:10.1007/978-3-642-00887-0_18

8. Datar, M., Immorlica, N., Indyk, P., Mirrokni, V.S.: Locality-sensitive hashing scheme based on p-stable distributions. In: Proceedings of the Twentieth Annual Symposium on Computational Geometry, pp. 253–262. ACM (2004)

9. Guha, S., Meyerson, A., Mishra, N., Motwani, R., O'Callaghan, L.: Clustering data streams: theory and practice. IEEE Trans. Knowl. Data Eng. 15(3), 515–528 (2003)

10. Kranen, P., Assent, I., Baldauf, C., Seidl, T.: The clustree: indexing micro-clusters for anytime stream mining. Knowl. Inf. Syst. 29(2), 249–272 (2011)

11. Li, J., Maier, D., Tufte, K., Papadimos, V., Tucker, P.A.: No pane, no gain: efficient evaluation of sliding-window aggregates over data streams. SIGMOD Rec. 34(1), 39–44 (2005)

12. Lloyd, S.: Least squares quantization in PCM. IEEE Trans. Inf. Theory 28(2), 129–137 (1982)

13. Silva, J.A., Faria, E.R., Barros, R.C., Hruschka, E.R., de Carvalho André, C.P.L.F., Gama, J.: Data stream clustering: a survey. ACM Comput. Surv. 46(1), 13:1–13:31 (2013)

14. Sun, L., Guo, C.: Incremental affinity propagation clustering based on message passing. IEEE Trans. Knowl. Data Eng. 26(11), 2731–2744 (2014)

15. Wan, L., Ng, W.K., Dang, X.H., Yu, P.S., Zhang, K.: Density-based clustering of data streams at multiple resolutions. ACM Trans. Knowl. Discov. Data 3(3), 14:1–14:28 (2009)

16. Zhang, T., Ramakrishnan, R., Livny, M.: Birch: an efficient data clustering method for very large databases. SIGMOD Rec. 25(2), 103–114 (1996)

17. Zhang, X., Furtlehner, C., Germain-Renaud, C., Sebag, M.: Data stream clustering with affinity propagation. IEEE Trans. Knowl. Data Eng. 26(7), 1644–1656 (2014)

18. Zhou, A., Cao, F., Qian, W., Jin, C.: Tracking clusters in evolving data streams over sliding windows. Knowl. Inf. Syst. 15(2), 181–214 (2008)

CBP: A New Parallelization Paradigm for Massively Distributed Stream Processing

Qingsong Guo[1(✉)] and Yongluan Zhou[2]

[1] North University of China, Taiyuan, China
qingsongg@gmail.com
[2] University of Southern Denmark, Odense, Denmark
zhou@imada.sdu.dk

Abstract. Resource efficiency is essential for distributed stream processing engines (DSPEs), in which a streaming application is modeled as an operator graph where each operator is parallelized into a number of instances to meet the low-latency and high-throughput requirements. The major objectives of optimizing resource efficiency in DSPEs include minimizing the communication cost by collocating the tasks that transfer a lot of data between each other, and by dynamically configuring the systems according to the load variations at runtime. In the current literature, most proposals handle these two optimizations separately, and a shallow integration of these techniques, such as performing the two optimizations one after another, would result in a suboptimal solution. In this paper, we present component-based parallelization (CBP), a new paradigm for optimizing the resource efficiency of DSPEs, which provides a framework for a deeper integration of the two optimizations. In the CBP paradigm, the operators are encapsulated into a set of non-overlapping components, in which operators are parallelized consistently, i.e., using the same partitioning key, and hence the intra-component communication is eliminated. According to the changes of workload, each component can be adaptively partitioned into multiple instances, each of which is deployed on a computing node. We build a cost model to capture both the communication cost and adaptation cost of a CBP plan, and then propose several optimization algorithms. We implement the CBP scheme and the optimization algorithms on top of Apache Storm, and verify its efficiency by an extensive experiment study.

1 Introduction

Real-time big data analysis requires processing of *continuous queries* (CQ) over fast streaming data with low latency. Usually, distributed stream processing engines (DSPEs) [1,18,22] organize CQs as an operator graph as shown in Fig. 1(a). To handle the deluge of data, one can resort to massive parallelization that each operator is cloned with a number of instances and its inputs are

The author from North University of China is supported by National Natural Science Foundation of China (61602427) and Natural Science Foundation of Shanxi(201601D202037).

S. Candan et al. (Eds.): DASFAA 2017, Part II, LNCS 10178, pp. 304–320, 2017.
DOI: 10.1007/978-3-319-55699-4_19

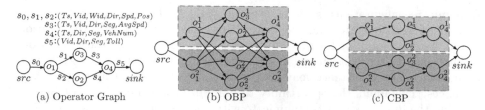

s_0, s_1, s_2:($Ts, Vid, Wid, Dir, Spd, Pos$)
s_3:($Ts, Vid, Dir, Seg, AvgSpd$)
s_4:($Ts, Dir, Seg, VehNum$)
s_5:($Vid, Dir, Seg, Toll$)

(a) Operator Graph   (b) OBP   (c) CBP

Fig. 1. Paradigms for parallelizing operator graph. This is a query that calculates the tolls of vehicles based on the source stream s_0 containing data of vehicles' speeds and positions. It consists of four operators: (1) a stateless operator o_1 that filters and partitions input data to the next operators; (2) two operators o_2 and o_3 calculate the average speed *AvgSpd* and the traffic volume respectively; and (3) o_4 calculates the toll of each vehicle, which is a function of *AvgSpd* and *SegNum*. The format of streams are specified in figure (a) and each operator o_i is designated with key k_i for partitioning the streams, where $k_1 = \{Ts, Vid, Spd, Dir, Seg, Pos\}$, $k_2 = \{Vid, Dir, Seg\}$, $k_3 = \{Dir, Seg\}$, and $k_4 = \{Dir, Seg\}$. We use two ways to parallelize the query: (1) in figure (b), the input streams of the operators are partitioned with different keys; and (2) in figure (c), the input streams of the four operators are partitioned consistently with the same key $\{Dir, Seg\}$.

partitioned into disjoint substreams. For the sake of resource efficiency, there are in general two critical optimizations to be considered:

1. **Runtime resource reconfiguration.** Load variations caused by the changes of data distribution and input rate are ubiquitous in the streaming context [22,24,25]. It is essential to provide adaptive data partitioning to achieve load balancing and to scale the number of parallel instances of each operator to avoid over-provisioning or under-provisioning.

2. **Communication cost minimization.** A large amount of data has to be continuously transmitted among the neighboring operators. Data transfer not only consumes bandwidth but also incurs significant computation overhead, including serializing and de-serializing the transmitted data. Optimizing the allocation of operator instances can to minimize cross-node communication can significantly reduce the resource consumption in a DSPE.

In existing solutions, the two problems are addressed separately. For example, M.A. Shah et al. [22] studied how to dynamically partition the input data at runtime to balance the workload across the parallel instances of an operator, while Y. Ahmad et al. [4] and P. Pietzuch et al. [19] investigated the operator placement to minimize the bandwidth usage by implicitly assumed assumption that operators do not need to be parallelized.

One can simply combine these methods to provide a complete solution. For example, we can first determine the parallelism for each operator [1], and transform the operator graph into a graph of operator instances. Thereafter, we can optimize the deployment by applying an operator placement algorithm, such as [4,19]. Suppose we have two nodes in the cluster, Fig. 1(b) shows a possible parallelization and task allocation plan for the operator graph in Fig. 1(a).

Dynamic reconfigurations, such as re-scaling and load balancing, can be performed on each operator independently. However, such a shallow integration would provide suboptimal performance. As shown in Fig. 1(b), if the 4 operators are not parallelized consistently, e.g., partitioning the input on the same key, then each operator instance may have to transfer data to all its downstream instances. This limits the opportunity to minimize communication cost by collocating the instances that communicate with each other.

On the contrary, if we can parallelize the operators consistently using a common partitioning key, then we could have a plan as shown in Fig. 1(c), which minimizes cross-node communication. Although this idea may sound simple, it is nevertheless challenging to implement in a DSPE supporting runtime reconfiguration. First of all, dynamic data repartitioning makes it difficult or even impossible to achieve consistent parallelization of multiple operators given that the operators could be reconfigured at runtime independently. Secondly, dynamic scaling and data repartitioning involve a lot of state movements [22,23]. The overhead of moving the states around operator instances in order to maintain the consistency of data partitioning may offset the benefits of collocating their communicating instances. Therefore we need a new parallelization framework that can optimize the parallelization of operators such that the total cost is minimized, including the communication and reconfiguration cost.

To address the challenges, we present **component-based parallelization (CBP)**—a new operator parallelization paradigm that considers both dynamic reconfiguration and resource optimization. In CBP, an operator graph is first decomposed into non-overlapping *components*, each being a connected subgraph. The operators in a component should have partitioning keys "compatible" with each other, i.e., sharing common attributes, and thus they can be parallelized using the same key. Each component acts as a singleton that is parallelized into a set of instances and the parallelism can be adapted at runtime in accordance with the load variations. This strategy simplifies the optimization of parallel stream processing and localizes the side-effect of reconfiguration within each component. In general, in the CBP paradigm, the more operators are grouped into a component, the less communication cost there would be, with a probable increase of the component's reconfiguration cost. This is because every time we have to re-scale or re-balance one operator within a component, we have to trigger repartitioning of all the operators within the component. Therefore, a good trade-off should be found to minimize the total cost of a CBP plan.

We propose a cost-based optimizer to compute an optimized CBP plan for a given query graph. We develop a novel cost model that integrates the reconfiguration overhead into the optimization. We formally define the optimization problem as a MINIMUM-COST-COMPONENT-BASED-PARALLELIZATION problem (MCCBP). We prove that MCCBP is NP-hard, and then present two heuristic algorithms to solve it. All the techniques have been implemented on top of Apache Storm [1]. We compare our solutions with the operator placement algorithm by using both synthetic workload and an extension of the Linear Road Benchmark [6]. The experiments show that our methods can save the network communication by up to 40%. Furthermore, our solutions can reduce the average end-to-end data latency by about 10% to 30%.

2 Background.

2.1 Parallel Stream Processing

Continuous queries(CQs) [17] over streaming data are usually organized as an operator graph [1,13,18] in a *distributed stream processing engine* (DSPE). DSPEs like Flux [22] and StreamCloud [14] exploit data parallelism [11] to cope with the deluge of data, in which an operator is cloned into a set of independent instances each working on a partition of the input data. The number of partitioning can be determined according the input rates to achieve high throughput.

Operators can be categorized as stateless and stateful. For a stateless operator, the input tuples can be processed independently by any instance of it. While the stateful operators, such as **join** and **group-by aggregate**, are "context-sensitive", so tuples with the same keys should be processed by the same instance to guarantee correctness. *Stream grouping* specifies the way how a stream of tuples is grouped and dispatched to the consumer operator instances. We consider two primitives: (1) *shuffle grouping*, where the input tuples are randomly routed to the operator instances; (2) *key grouping*, in which tuples are partitioned into a number of substreams based on a specified set of keys. Shuffle grouping is often the optimal choice for stateless operators since the load can be easily balanced, while key grouping is necessary for stateful operators.

Challenges of load variation. Usually, one can easily observe two kinds of variations over streaming data: (1) the fluctuation of input rates [22,24], and (2) change of data value distributions [22,24,25]. If an SPE does not react to the variations, applications can run into problems:

- **Unmatched provision:** the over-provisioning or under-provisioning caused by the fluctuation of the input rates can result in low system utilization, high operational cost (e.g., using pay-as-you-go cloud services), and system failures.
- **Load imbalance:** the load distribution is skewed due to the change of data value distribution. For example some stream grouping keys become more popular than the others so that some operator instances are over-loaded while the others are under-loaded. Load imbalance can harm the processing latency and system throughput if the skewness is not resolved soon.

To handle the above problems, we resort to adaptation techniques including *dynamic scaling* [18] and *load balancing* [25]. CQs use the concept of *sliding windows* of tuples over a stream to specify the operational context of an operator. For instance, to perform a windowed join, we need to buffer the tuples within the current window(s) as the context of the join operation on the newly incoming tuples. This kind of context is called as *processing state* [8]. While processing an adaptation, the substreams should be reassigned around operator instances, and the processing states needs to be reallocated accordingly. This process is called *state movement*. Note that both scaling and load balancing involve state movements, which consume both significant CPU and network bandwidth and thus cannot be ignored [22, 23].

2.2   System Model

Data model. A data stream s is an unbounded and append-only sequence of tuples $(\ldots, t_{i-1}, t_i, t_{i+1}, \ldots)$. Each tuple $t = (\tau, \alpha)$ has a timestamp $\tau \in \mathbb{T}$ and a set of attributes $\alpha = (a_1, \ldots, a_k)$. We assume that the attribute set α of every stream conforms to a relational schema. For simplicity, τ is assumed to be unique. In practice, if τ is not unique, existing systems usually use a unique sequence number to identify each tuple.

Operator model. A CQ is composed of a number of *operators*, each implementing a certain computation logic, such as *join, aggregate, filter*, or *user-defined functions*. An operator o is a 6-tuple, $(\text{IN}_o, \text{OUT}_o, \text{K}_o, \text{F}_o, \text{W}_o, \text{PS}_o)$, where IN_o and OUT_o are the input and output streams respectively. K_o is the *key*, a subset of attributes of the input streams IN_o, which used for partitioning IN_o. F_o defines the processing logic, where its operating context, i.e. the *processing state* PS_o, is specified by the sliding window W_o. For stateless operators like *map* and *filter*, $\text{PS} = \emptyset$.

We organize CQs as an operator graph $\text{G} = (\text{O}, \text{S})$, which is a directed acyclic graph of the operator set O and the stream set S. A stream $s \in \text{S}$ is represented as a directed arc (u_s, d_s), $u_s, d_s \in O$, where u_s and d_s are its producer and consumer respectively. Two special operators, *Src* and *Sink*, are responsible for spouting source streams and collecting the final results respectively. An operator graph is also referred to as a *topology* and these two terms are interchangeable throughout this paper.

Physical execution. The operator graph is executed on a cluster of identical nodes. The *execution graph* is a physical realization of the query in which each operator o is parallelized into multiple instances $\mathcal{I} = \{o^1, \ldots, o^\pi\}$, where $\pi \in \mathbb{N}^+$ is the *parallelism*. For an input stream s of o, each tuple is a key-value pair <k,v>, where v is the tuple and k $= t.\text{K}_o$. A partitioning function split the domain of K_o into p groups, where $p \gg \pi$. Then, the tuples of s, according their key values, form a number of substreams $\mathcal{S} = \{s^1, \ldots, s^p\}$. An assignment $\mathcal{F} : \mathcal{S} \to \mathcal{I}$ allocate the processing of each substream to a unique operator instance. The degree of parallelism π and the assignment $\mathcal{F}$ are adapted at runtime to handle load variations.

3   Component-Based Parallelization

3.1   CBP Abstraction

In essence, CBP decomposes an operator graph into a set of non-overlapping *components*, which act as the parallelization unit. In particular, CBP relies on two essential properties: *compatibility* and *connectivity*. Compatibility concerns if some operators can be parallelized consistently. A set of operators $\{o_1, \ldots, o_k\}$ is compatible iff the intersection of their keys is not empty, i.e., $K_{o_1} \cap \cdots \cap K_{o_k} \neq \emptyset$. Note that the compatibility property is not *transitive*. For example, suppose we have three operators o_1, o_2, and o_3 with keys $K_1 = \{a_1, a_2\}$, $K_2 = \{a_2, a_3\}$, and $K_3 = \{a_1, a_3\}$ respectively. Even though any pair of them are compatible, they as a whole are incompatible because $K_1 \cap K_2 \cap K_3 = \emptyset$. The rationale of assembling the topology into components is to reduce the communication cost. One can benefit from placing compatible operators into a node only if they are connected by streams.

Formally, we can define a component as follow.

Definition 1 (Component). *A component* $C = (O_C, S_C)$ *is an induced subgraph of the operator graph* $G = (O, S)$, *where* C *is connected and the operators in* O_C *are compatible.*

Let $IN(C)$ be the set of all input streams of the operators in component C, then $IN(C) = \cup_{o \in O_C} IN_o$. Assuming $O_C = \{o_1, \ldots, o_{|C|}\}$. The streams of $IN(C)$ can be grouped by a partition function over the key $K = K_{o_1} \cap \cdots \cap K_{o_{|C|}}$, which is the intersection of the keys of all the operators in C. Since $K \neq \emptyset$, all the streams of $IN(C)$ can be partitioned uniformly into p substreams. For the convenience of discussion, we regard the streams in $IN(C)$ as a *composite stream* cs, which is partitioned into a set of substreams $\mathcal{CS} = \{cs^1, \ldots, cs^p\}$. In addition, each component C is parallelized into a number of instances $\mathcal{CI} = \{ci^1, \ldots, ci^\pi\}$, where π is the *parallelism* of C and each instance has a clone of the computation logic of each operator in C. The parallel processing of the composite stream $\mathcal{CS}$ is specified by an assignment $\mathcal{F}_C : \mathcal{CS} \rightarrow \mathcal{CI}$, which is adapted at runtime to handle load variations.

4 MCCBP

4.1 Metrics

The cost of a CBP plan can be put into three parts: (1) *Processing cost* $\mathcal{PC}$, which is the CPU usage of the computation, (2) *Communication cost* $\mathcal{CC}$, which is the CPU and network usages of data transmission, and (3) *Adaptation cost* $\mathcal{AC}$, which is the CPU and network usages of carrying out adaptations.

In particular, we assume that $\mathcal{PC}$ keeps the same regardless of the physical execution, and thus it can be disregarded in our cost model. In addition, we categorize data communication into *inter-component communication* and *intra-component communication*. The first one involves three sequential steps: (1) data serialization, (2) network propagation, and (3) de-serialization. Steps (1) and (3) consume CPU cycles and step (2) occupies network bandwidth. In contrast, the intra-component communication is realized via local memory access, whose overhead is negligible. Therefore, we only take the overhead of inter-component communication into account.

Statistics measurements. The cost calculation relies on the statistics of execution of the operator graph. In our implementation, the statistics are measured periodically over a sequence of time intervals of length Δ, which are called as *statistics windows*. Suppose the historical data spans m statistics windows that start at the time instance $\tau = 0$, then the timespan of historical data is $[0, m\Delta]$. The following discussions are confined within the timespan $[0, m\Delta]$.

For the input stream $s \in \mathsf{S}$ of a component that is split into p partitions, the statistics are represented as a sequence of histograms $\boldsymbol{Y}(s) = (Y_1, \ldots, Y_m)$, where the histogram $Y_r = (y_{1,r}, \ldots, y_{p,r})^T$, $r = 1 \ldots m$, is a vector recording the data rate of the p partitions over the r-th statistics window. In other words, the data distribution of s at the r-th window can be approximated with Y_r. With $\boldsymbol{Y}$, we can derive other statistics on demands. For instance, denote $s = (o_i, o_j)$, then the load l_{ij} of s during $[0, m\Delta]$ is $l_{ij} = \sum_{r=1}^{m} \sum_{k=1}^{p} y_{kr}$.

The adaptation cost is closely related to the adaptation frequency f, where $\Delta = 1/f$. For simplicity, we assume that SPE performs an adaptation at each window. Let ψ_i^r be the number of state movements in the r-th adaptation of component C_i, then $\mathcal{AC} = \sum_{i=1}^{|\mathcal{C}|} \psi_i$, where $\psi_i = \sum_{r=1}^{m} \psi_i^r$ is the adaptation cost of C_i.

4.2   Problem Formulation

Consider an operator graph that is grouped into a set of disjoint components $\mathcal{C} = \{\mathsf{C}_1, \mathsf{C}_2, \ldots\}$, it is called a CBP plan if $\cup_{i=1}^{|\mathcal{C}|} \mathsf{O}_{\mathsf{C}_i} = \mathsf{O}$ and $\mathsf{O}_{\mathsf{C}_i} \cap \mathsf{O}_{\mathsf{C}_j} = \emptyset$ for any two components of $\mathcal{C}$. Let X be the streams interconnecting components in $\mathcal{C}$. Let $w(\mathsf{C}_i)$ be the adaptation cost of C_i and $c(s)$ be the communication cost incurred by stream s. Since $\mathcal{PC}$ is independent on the CBP plan, the cost of a CBP plan $\mathcal{C}$, denoted as $cost(\mathcal{C})$, is measured by the sum of the communication cost $\mathcal{CC}$ and adaptation cost $\mathcal{AC}$. That is,

$$cost(\mathcal{C}) = \mathcal{CC} + \mathcal{AC} = \sum_{s \in \mathsf{X}} c(s) + \sum_{\mathsf{C}_i \in \mathcal{C}} w(\mathsf{C}_i) \tag{1}$$

We introduce a constraint on the adaptation cost, $w(\mathsf{C}_i) \leq \beta$, to prevent any component from being the bottleneck. Consequently, the objective of optimizing a CBP plan is to minimize $cost(\mathcal{C})$. We denote this problem as MINIMUM COST COMPONENT-BASED PARALLELIZATION (MCCBP), which is a variant of graph partitioning problem under constraints of connectivity and compatibility. Formally, it is stated as follow.

Definition 2 (MCCBP). *Given an operator graph* $\mathsf{G} = (\mathsf{O}, \mathsf{S})$ *and a positive constant* β*, the MCCBP problem is to find a CBP plan, which is a partition of* G *into a set of disjoint components* $\mathcal{C} = \{\mathsf{C}_1, \mathsf{C}_2 \ldots\}$*, to achieve the following objective:*

$$minimize \ cost(\mathcal{C})$$
$$subject \ to \ \cup_{i=1}^{|\mathcal{C}|} \mathsf{O}_{\mathsf{C}_i} = \mathsf{O}$$
$$w(\mathsf{C}_i) \leq \beta$$

MCCBP can be proved to be NP-hard by simplifying it to a *Minimum-Capacity-Graph-Partitioning* (MCGP) problem, which has been shown to be NP-hard.

5   Computing CBP Plans

5.1   Greedy Algorithm

A straightforward idea is to obtain an initial CBP plan $\mathcal{C}_0$ in advance, and then make improvement incrementally. The algorithm, as shown in Algorithm 1, begins with the initial plan $\mathcal{C}_0$ (Line 2) and makes improvement step by step (Line 9–21). The initial plan $\mathcal{C}_0$ is generated by a depth-first search (DFS) of the operator graph. The traversal is tracked by an operator stack OS. In each iteration, we peek an operator from OS. Let o be the current operator being visited and $C(o)$ be the component containing o. Then o will be popped out from OS if it has no unvisited child or is a leaf node. Otherwise we choose an unvisited child v of o and then check the compatibility between v and $C(o)$. If they are compatible, v will be added into component $C(o)$; Otherwise, a new component C_i containing operator v is created.

The essence of Algorithm 1 is to reduce the cost by moving operators around components. Let $move(C_i,C_j,o_k)$ be the *potential movement* that attempts to move o_k from C_i to C_j. It is *admissible* if $o_k \in C_i$ and the new operator set $C_j \cup \{o_k\}$ can form a component. Given a CBP plan $\mathcal{C}$, the execution of the potential movement $move(C_1,C_2,o_k)$ gives rise to a new plan $\mathcal{C}'$ if it is admissible. The admissibility of it is checked in Line 8.

The movement results in the following change of costs: (1) the change of communication cost between C_1 and C_2, and (2) the change of adaptation costs of C_1 and C_2. Hence the profit $\delta_{12}(o_k)$ of $move(C_1,C_2,o_k)$ consists of two parts: the changes of the communication cost and adaptation cost, denoted as $\delta_{12}^1(o_k)$ and $\delta_{12}^2(o_k)$ respectively. Let $\varphi_1(o_k)$ be the data rate between o_k and C_1. Then, $\varphi_1(o_k) = \sum_{\substack{(o_k,o_t)\in S \vee (o_t,o_k)\in S \\ o_t \in C_1}} l_{kt}$. $\varphi_1(o_k)$ does not contribute to $\mathcal{CC}$ if $o_k \in C_1$, otherwise it does. After the movement, $\varphi_i(o_k)$ contributes to $\mathcal{CC}$, but $\varphi_j(o_k)$ does not contribute to $\mathcal{CC}$. Therefore, the gain on communication cost is $\delta_{12}^1(o_k) = \varphi_2(o_k) - \varphi_1(o_k)$. Let $\psi(o_k)$ be the new adaptation cost of a component, then we have $\delta_{12}^2(o_k) = (\psi_1 + \psi_2) - (\psi_1(o_k) + \psi_2(o_k))$.

Summing all together, we get the overall profit of the movement, $\delta_{12}(o_k) = \delta_{12}^1(o_k) + \delta_{12}^2(o_k)$. In each run, we choose an admissible movement with the maximum positive profit to execute. Suppose that $\delta_{12}(o_k)$ is the best movement in the current run, then the load and state statistics of C_1 and C_2 should be changed after the execution of $move(C_1,C_2,o_k)$ (Line 14). The movement also causes changes of the profits of any admissible movement involving C_1 or C_2. To prepare the next iteration, we should recompute the profits of these admissible movements (Line 15).

Algorithm 1. Greedy Algorithm

Input: Operator graph $G = (O, S)$, load statistics $\{Y(s_1), Y(s_2), \dots\}$, state
statistics $\{Z(o_1), Z(o_2), \dots\}$
Output: CBP plan $\mathcal{C}$

1 $\mathcal{C} \leftarrow \texttt{InitialPartition(G)}$;
2 compute load statistics $\mathcal{Y}(C_i)$, state statistics $\mathcal{Z}(C_i)$, and adaptation cost ψ_i for
   each component $C_i \in \mathcal{C}$;
3 $\delta \leftarrow 1.0$;
4 **while** $\delta > 0$ **do**
5   **foreach** $o_k \in O$ **do**
6      $C_i \leftarrow$ get the component containing o_k ;
7   **foreach** $C_j \in |\mathcal{C}|$ and $j \neq i$ **do**
8   **if** $a_{jk} \neq -1$ *and* $C_j \cup \{o_k\}$ *is compatible* **then**
9          $\delta_{ij}^1(o_k) \leftarrow \ell_{jk} - \ell_{ik}$;
10         $\delta_{ij}^2(o_j) \leftarrow (\psi_i + \psi_j) - (\psi_i(o_k) + \psi_j(o_k))$;
11         $\delta_{ij}(o_k) \leftarrow \delta_{ij}^1(o_k) + \delta_{ij}^2(o_k)$;

12   $\delta \leftarrow \max\{\delta_{ij}(o_k)\}$;
13   move o_k from C_i to C_j;
14   update the load and state statistics of C_1 and C_2;
15   recompute the profits for any admissible movement involves C_i or C_j ;

16 **return** $\mathcal{C}$;

5.2   MWSC

We proceed to consider an alternative that transforms MCCBP into the *minimum weighted set cover problem* (MWSC). Let $\Omega = \{C_1, C_2, \dots\}$ be a set containing all the possible components of O. Let N be the cardinality of Ω, i.e., $N = |\Omega|$. A CBP plan $\mathcal{C} = \{C_i | C_i \in \Omega\}$ is a subset of Ω. It is apparent that the plan $\mathcal{C}$ is a set cover of O, since $\bigcup_{i=1}^{|\mathcal{C}|} C_i = O$ and $C_i \cap C_j = \emptyset$ for $\forall C_i, C_j \in \mathcal{C}$. Therefore, MCCBP is equivalent to find a subset $\mathcal{C}$ of Ω such that $\mathcal{C}$ is a partition of O. We attempt to optimize this problem by enumerating all the feasible components and finding the optimal CBP plan from them.

Each component associates with adaptation cost ψ_i and intra-component communication cost ϕ_i, where $\phi_i = \sum_{o_i, o_j \in C} l_{ij}$. For each component $C_i \in \Omega$, we assign a weight w_i to it such that $w_i = \psi_i + l - \phi_i$, where l is the overall load, $l = \sum_{i=1}^n \sum_{j=1}^n l_{ij}$. It is obvious that $\psi_i > 0$ and $l - \phi_i \geq 0$.

Let x_i be a decision variable that indicates whether component C_i is chosen in the set cover $\mathcal{S}$, where $x_i = 1$ if C_i is picked, or $x_i = 0$ otherwise. Then the MCCBP is transformed to the weighted set cover problem. A set cover $\mathcal{S}$ of O has some redundant operators, for example $C_i \cap C_j = o_k$. Denote $\mathcal{S}'$ as the new set cover by discarding o_k. Since $\psi_i > 0$ and $l - \phi_i \geq 0$, the cost of $\mathcal{S}'$ is definitely smaller than that of the former one, i.e., $w(\mathcal{S}') < w(\mathcal{S})$. Finally, we can get the minimum set cover of O by removing all the redundant operators.

Algorithm 2. MWSC

Input: Operator graph $\mathsf{G} = (\mathsf{O}, \mathsf{S})$, load statistics $\{\mathbf{Y}(s_1), \mathbf{Y}(s_2), \dots\}$, state
    statistics $\{\mathbf{Z}(o_1), \mathbf{Z}(o_2), \dots\}$
Output: CBP plan $\mathcal{C}$

1   $l \leftarrow \sum_{i=1}^{n} \sum_{j=1}^{n} l_{ij}$;   `/* overall loads */`
2   $\Omega \leftarrow$ Enumerate(G, k) ;
3 **foreach** *component* C_i *in* Ω **do**
4 | compute the adaptation cost ψ_i ;
5 | $\phi_i \leftarrow \sum_{o_i, o_j \in \mathsf{C}} l_{ij}$;
6 | $w_i \leftarrow \psi_i + l - \phi_i$;   `/* weight of Cᵢ */`
7   $\mathcal{S} \leftarrow$ compute the MWSC of O over Ω ;
8   $\mathcal{C} \leftarrow \mathcal{S}$;
9 **return** $\mathcal{C}$;

Definition 3 (MWSC). *Given a universe* O *and a family* Ω *of subsets of* O, *the minimum weighted set cover of* O *can be expressed as an integer linear programming:*

$$minimize \ \ w(\mathcal{S}) = \mathbf{w}^T \mathbf{x} \tag{2}$$

$$subject \ to \ \ \sum_{\mathsf{C}_i : o \in \mathsf{C}_i}^{N} x_i \geq 1 \ \ \ for \ each \ operator \ o \in \mathsf{O},$$

$$x_i \in \{0, 1\}$$

where $\mathbf{w} = (w_1, \dots, w_N)$ *is the weight vector and* $\mathbf{x} = (x_1, \dots, x_N)$ *is the decision vector for* Ω *respectively.*

Apparently, a MWSC is a partition of O. Thus,

$$w(\mathcal{S}) = \sum_{i=1}^{N} x_i \psi_i + |\mathsf{S}| l - \sum_{i=1}^{N} x_i \phi_i = \underbrace{\sum_{i=1}^{N} x_i \psi_i}_{} + \underbrace{\left[l - \sum_{i=1}^{N} x_i \phi_i \right]}_{} + \underbrace{(|\mathsf{S}| - 1) l}_{} \tag{3}$$

where $|\mathsf{S}|$ is the number of edges of $\mathsf{G} = (\mathsf{O}, \mathsf{S})$.

Comparing to the cost model Eq. (1), we have the first component $\sum_{i=1}^{N} x_i \psi_i$ and the second $l - \sum_{i=1}^{N} x_i \phi_i$ of Eq. (3) equal to the adaptation cost $\mathcal{AC}$ and communication cost $\mathcal{CC}$ respectively. As the third component $(|\mathsf{S}| - 1) l$ is a constant, the best solution of MWSC is equivalent to the optimal CBP plan.

The idea is depicted in Algorithm 2. We first enumerate all the possible components of G (Line 2). Then we compute the adaptation cost ψ_i and the load ϕ_i of each component C_i, and assign a weight to each component (Line 3–6). Finally, we compute a solution $\mathcal{S}$ of MWSC and take it as a CBP plan by discarding all the redundant operators (Line 7–8). MWSC can be solved exactly with a MIP solver like Gurobi [2] when N is not too large. But we also implement a greedy routine to solve MWSC (Line 7) according to the description in [9, Chap. 35]. The greedy routine is a useful option when N is large. Since the set cover $\mathcal{S}$ obtained through the greedy routine might not be a CBP plan, we have to remove the redundant operators to get the final solution $\mathcal{C}$.

6   Evaluation

6.1   Experimental Setup

Evaluation metrics—We use the following metrics in the evaluation:

- **Communication cost** counts the number of tuples transmitted through inter-component communication.
- **Adaptation cost** counts the number of state movements in an adaptation process.
- **End-to-end latency** indicates the time completing the processing of a source tuple. It includes the time spent on processing, adaptation, and communication, and thus it is a overall metric to reflect the effectiveness of CBP.

Tested solutions—We implement the sparse-cut algorithm, a graph partition algorithm used in COLA [15], to compare with our solutions. Note that the objective of baseline is merely to minimize communication cost. In general, we evaluated the following three algorithms: (1) *greedy algorithm*, (2) the *MWSC algorithm*, and (3) the *baseline algorithm* which implements an OBP-based operator placement algorithm in [15].

We implement our algorithms in Java and integrate them with Apache Storm [1] by extending it with runtime adaptation. Part of the experiments are conducted via simulation, while the rest are conducted on Amazon's EC2 with medium VM instances (*m1.medium*), where each has 1.7 GB of RAM, moderate IO performance and one EC2 compute unit (approximately equivalent to a 1.2 GHz 2007 Xeon CPU). While these VMs have low processing capabilities, they are representatives of public cloud VMs.

6.2   Simulation Result

In the test, we used a randomized topology $\mathsf{G} = (\mathsf{O}, \mathsf{S})$. In the topology G, each operator o, except src, maintains computing states and randomly forwards the received data to downstream operators according to the selectivity $\delta(o)$. The specific setting of G is summarized in Fig. 2. Operator src generates two synthetic streams s_1 and s_2 to simulate two types of variations, where the key values of s_1 and s_2 follow the uniform distribution and Zipf respectively. Therefore, s_1 only results in scaling. In contrast, data distribution of s_2 is skewed and thus the adaptations involve both scaling and load balancing. Each operator of G randomly chooses two attribute of sch as the partition key. The data arrivals of s_1 and s_2 follow a Poisson process $X(t) : P[N(t + \tau) - N(t) = k] = (k!)^{-1}e^{-\lambda\tau}(\lambda\tau)^k$, where τ is set to 1 s and $\lambda = 10,000$. Both s_1 and s_2 conform to

(a) stream s_1 (b) stream s_2

Parameters	Settings
Random graph	$G = (O, S, d)$
Number of operators	$|O| = 100$
Average degree d	$d = \{3, 5, 10\}$
Selectivity $\delta(o)$	$\delta(o) \sim N(0.5, 1.0)$
Size of states $|PS_o|$	$|PS_o| = X(t)$

Fig. 2. Setting of parameters   **Fig. 3.** Comparison of communication costs

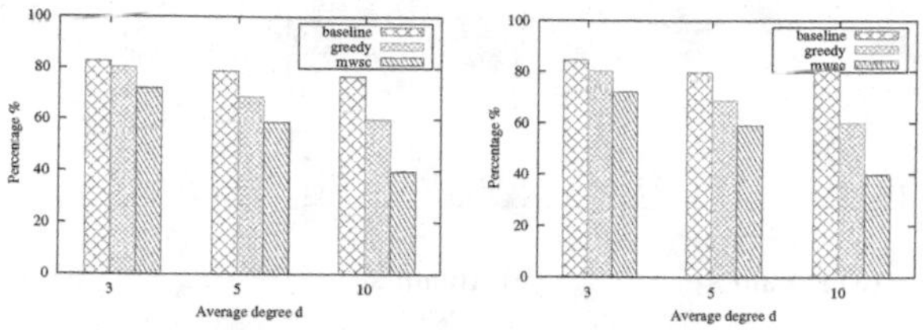

the schema: SynStream(ts:Unix timestamp, a_1:int, a_2:int, a_3:int, a_4:int), in which each attribute has 4 Bytes.

We measured the communication cost and state movements by varying the average degree d and the adaptation frequency f. Let N_1 be the number of tuples processed by all the operators and N_2 be the number of tuples in the states of all the operators at every adaptation. We calculated the percentages, $\frac{100n_c}{N_1}$ and $\frac{100n_a}{N_2}$, of tuples involved in the communication and state movement, where n_c and n_a are the communication cost and adaptation cost respectively.

Comparison of communication costs—Figure 3a and b show the percentages achieved by three algorithms. We can observe that the baseline algorithm can save the cost by at most 20%, but the CBP solutions can reduce the cost by at least 20%. In particular, the greedy algorithm saves about 20% when $d = 3$, and it increases to 40% when $d = 10$. MWSC outperforms the greedy algorithm. It significantly reduces the communication cost by about 27.8% when $d = 3$ and by nearly 60% when $d = 10$. The baseline algorithm deploys the operator graph based on a placement plan, which is generated in advance by graph partitioning. Since the operators are incompatible, the physical topology of the query changes as adaptation process. The parallelization plan is no longer optimal when the physical topology has been changed. Therefore, we cannot optimize the communication cost efficiently with operator placement.

The intra-component communications of a CBP plan are eliminated completely regardless of the change of physical topology. This is confirmed by Fig. 3. By comparing Fig. 3a and b, the communication costs of CBP solutions keep the same regardless of the difference of s_1 and s_2. However, the costs of baseline algorithm is slightly different over s_1 and s_2, where the cost is about 3% higher over s_2 than that over s_1. The frequency f shows similar impact to the algorithms.

Comparison of adaptation costs—Figures 4 and 5 shows the impact of load variation and the adaptation frequency. In this experiment, the frequency f is varied by changing the length of adaptation window from 1 min to 10 min, i.e., $1/f = \{1, 2, 5, 10\}$. Figure 4 plots the adaptation cost of each algorithm when $1/f = 1$. It is clear that CBP has larger adaptation costs than the baseline algorithm. Moreover, the adaptation cost over a skewed stream, s_2 in Fig. 4b, is higher than the uniformly distributed stream, s_1 in Fig. 4a. We observe similar

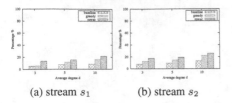

(a) stream s_1 (b) stream s_2

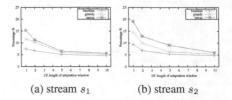

(a) stream s_1 (b) stream s_2

Fig. 4. Comparison of adaptation costs

Fig. 5. Adaptation costs with respect to $1/f$

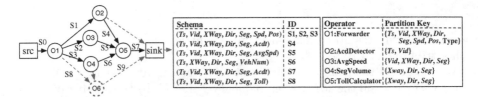

Schema	ID
$(Ts, Vid, XWay, Dir, Seg, Spd, Pos)$	S1, S2, S3
$(Ts, Vid, XWay, Dir, Seg, Acdt)$	S4
$(Ts, Vid, XWay, Dir, Seg, AvgSpd)$	S5
$(Ts, XWay, Dir, Seg, VehNum)$	S6
$(Ts, Vid, XWay, Dir, Seg, Acdt)$	S7
$(Ts, Vid, XWay, Dir, Seg, Toll)$	S8

Operator	Partition Key
O1:Forwarder	$\{Ts, Vid, XWay, Dir,$ $Seg, Spd, Pos, Type\}$
O2:AcdDetector	$\{Ts, Vid\}$
O3:AvgSpeed	$\{Vid, XWay, Dir, Seg\}$
O4:SegVolume	$\{XWay, Dir, Seg\}$
O5:TollCalculator	$\{XWay, Dir, Seg\}$

Fig. 6. Operator graph for LRB

results when $f = \{2, 5, 10\}$. The results also justify an implicit assumption in this paper that the adaptation cost is normally higher when we assemble operators into components.

Figure 5 shows the impact of adaptation frequency f. As we can see from the figure, the cost drops greatly at the beginning when we increase $1/f$. The number of state movements is determined by two factors: (1) the adaptation frequency f, and (2) the skewness of the data. The skewness usually goes serious if we increase $1/f$, i.e., it always involves more state movements in a single adaptation. As we expected, the decline of adaptation cost is much gentle when $1/f$ is larger.

6.3 End-to-End Latency

We proceed to evaluate the end-to-end latency of the tested solutions. In this experiment, we use the *Linear Road Benchmark* (LRB) [6]. LRB models a road toll network, in which tolls depend on the level of congestion. The primitive LRB gadget only has 7 operators, which is too small to represent a large-scale computation. So we extend it by connecting a number of LRB gadgets together with a road network. The road network $G = (V, E)$ is a graph where an edge $e \in E$ stands for an expressway of LRB and a vertex $v \in V$ represents the joint of expressways. This extension has a wide range of applications. If we want to measure the traffic between two locations or track the route of a vehicle, then an LRB gadget must dispatch result to its downstream LRB gadgets. Consequently, we introduce a new operator o_6 to calculate the traffic between every pairs of vertices every 1 min. Figure 6 shows the topology of the extended LRB, where some new streams (blue dashed arcs), have been added into a LRB gadget to fulfill the requirement.

Table 1. Statistics about end-to-end latency (ms)

	Mean				Median				95%				Maximum			
1/f	1	2	5	10	1	2	5	10	1	2	5	10	1	2	5	10
Greedy	677	610	566	617	141	121	109	116	1501	1236	1095	1130	2825	2223	1736	2117
MWSC	583	517	534	602	131	120	97	118	1532	1333	1054	1171	3103	2703	1853	1853
Baseline	775	710	673	681	153	137	114	127	1017	928	856	889	2109	1809	1673	1681

$G = (V, E)$ is generated with the random graph presented in Sect. 6.2. In particular, $|V| = 10$, $|E| = 30$, and the average degree $d = 3$. Therefore, we have 30 LRB gadgets and 180 operators in total excluding *srcs* and *sinks*. For each LRB gadget, the data rate of the source stream is controlled with the Poisson process used in the previous section. The experiments are conducted on EC2 with 30 VMs and accomplished in two phases: (1) We first deploy it over EC2, and keep it running for two hours to collect statistics. The length of statistic window is set to 1 min. (2) With the statistics, we partition the topology into components or subgraphs with the tested algorithms. Thereafter, we deploy the partitioned topology on EC2 and run the experiments.

We measure the end-to-end latency at 4 scales of the adaptation frequency f, i.e., $1/f = \{1, 2, 5, 10\}$. The latency values are given in Table 1, where "95%" is the 95th percentile of latency. In general, the results follow what we expected. By comparing the mean, we observe that our algorithms reduce the latencies by about 10%–25%. It shows that the CBP algorithm can indeed improve the performance and thus confirms the effectiveness of CBP. We can further identify the impacts of adaptation process and load imbalance in these values. For example, the CBP algorithms are more sensitive to adaptation process and load imbalance comparing to the greedy algorithm. The maximum latency is 3103 ms for MWSC when $1/f = 1$, which is higher than the maximum latency of Baseline.

Tuples with a latency smaller than the median are less affected by the adaptation process and load imbalance. In contrast, tuples with latencies larger than 95-th percentiles are greatly affected by the adaptation process and load imbalance. We take the latency when $1/f = 1$ as an example, the medians of MWSC and Greedy are about 75% and 87% of that of Baseline. So the results confirm that CBP can save communication cost efficiently. In contrast, the 95-th percentiles for MWSC and Greedy are about 29% and 26% greater than the baseline algorithm.

During an adaptation, input tuples are buffered by the upstream operators. The tuples will be replayed to downstream after the completion of adaptation. Therefore, adaptation process increases the end-to-end latencies for a portion of tuples. As we can see from the table, the maximum latency peaks up to about 3 s.

For each algorithm, each numeral characteristic drops with the increase of $1/f$ at first and then grow with the increase of $1/f$ on the contrary. This behavior is obvious for the 95-th percentile. In terms of the 95-th percentile, it is obvious MWSC is higher than Greedy and Baseline. This phenomena confirms the impact of adaptation process and load imbalance. The adaptation cost drops with the increase of $1/f$, but load imbalance get worse on the contrary. Thus we observe that all lines are concave. It means that the adaptation frequency is very important as it can trade off between impact of adaptation cost and load imbalance. In this experiment, $f = 1/2$ is the best choice for MWSC and $f = 1/5$ is the best choice for Greedy and Baseline.

7   Related Work

Parallel stream processing. Much work has been focused on exploiting parallelism in stream processing. The early SPEs aim at providing transparent parallelization for distributed stream processing in a shared-nothing environment. Aurora [7] and Borealis [3] supports intra-query parallelism by organizing a topology into a set of boxes and conducting parallelization via *box-splitting*.

Many SPE proposals, e.g., System S [5] and Flux [22], leverage partitioned parallelism [11] to improve scalability. They propose new "Exchange" operators between stream producers and consumers to encapsulate the adaptive state partitioning and stream routing. In recent years, many efforts have been made to improve the scalability of parallelization [12,20,21]. The MapReduce model [10] enables programmer to think in a *data-centric* fashion and hence provides a practical implementation for partitioned parallelism. Distributed SPEs like Apache Storm [1], Yahoo! S4 [18], and StreamCloud [14] are inspired by such a model.

Operator placement. If an application is geographically distributed, the transmission latency is sensitive to the communication channels. The SAND project [4] exploits the knowledge of the underlying network characteristics such as topology and link bandwidths to make intelligent in-network placement of query graph. In contrast, [19] develops a *stream-based overlay network* (SBON) over Borealis, which is a network-aware optimization framework that manages operator placement within a pool of wide-area overlay nodes in order to make efficient use of network bandwidth. The placement decisions are made based on the cost space that encodes multidimensional metrics such as latency and load.

COLA [15] employs graph-partitioning algorithms to compute an optimal allocation of operators with regard to a cost model that captures the communication and CPU costs. The operator graph is partitioned into processing elements (PE) at compile-time, which acts as a deployable unit. COLA aims at balancing load across the processing nodes and minimizing the communication cost of the PEs. It only measures the CPU cost incurred by processing and communicating, but ignores the network bandwidth usage. In addition, COLA does not consider how to parallel the operators. Moreover a partition plan obtained at compile-time is incapable to handle the load variations at runtime.

The essence of operator placement is to optimize an assignment of operators to the computing nodes based on an objective function. Unfortunately, the existing solutions are static and the cost of the state migration cannot be ignored in the presence of load variations. For more detailed comparisons of the placement strategies, please refer to a survey paper [16].

8 Conclusion

We present CBP, a succinct parallelization paradigm for DSPEs that leverages both the connectivity and compatibility of operators. CBP seamlessly integrates operator placement with parallelization and thereby provides a framework to integrate the optimizations of runtime resource reconfiguration and communication cost minimization. Furthermore, we introduce a cost model that captures the cost of communication and adaptation. Two algorithms are proposed to optimize the CBP plans for a given computation. The extensive experiments confirm that an optimized CBP plan can improve the resource efficiency of DSPEs significantly.

References

1. Apache Storm. http://storm.apache.org/
2. Gurobi Parallel MIP solver. http://www.gurobi.com/resources/getting-started/mip-basics
3. Abadi, D.J., Ahmad, Y., Balazinska, M., Cetintemel, U., Cherniack, M., Hwang, J.-H., Lindner, W., Maskey, A.S., Rasin, A., Ryvkina, E., Tatbul, N., Xing, Y., Zdonik, S.: The design of the borealis stream processing engine. In: CIDR 2005, Asilomar, CA, January 2005
4. Ahmad, Y., Çetintemel, U.: Network-aware query processing for stream-based applications. In: VLDB 2004, vol. 30, pp. 456–467 (2004)
5. Andrade, H., Gedik, B., Wu, K., Yu, P.S.: Scale-up strategies for processing high-rate data streams in system S. In: ICDE 2009
6. Arasu, A., Cherniack, M., Galvez, E., Maier, D., Maskey, A., Ryvkina, E., Stonebraker, M., Tibbetts, R.: Linear road: a stream data management benchmark. In VLDB 2004
7. Carney, D., Çetintemel, U., Cherniack, M., Convey, C., Lee, S., Seidman, G., Stonebraker, M., Tatbul, N., Zdonik, S.: Monitoring streams: a new class of data management applications. In: VLDB 2002, pp. 215–226 (2002)
8. Castro Fernandez, R., Migliavacca, M., Kalyvianaki, E., Pietzuch, P.: Integrating scale out and fault tolerance in stream processing using operator state management. In: SIGMOD 2013, pp. 725–736. ACM, New York (2013)
9. Cormen, T.H., Stein, C., Rivest, R.L., Leiserson, C.E.: Introduction to Algorithms, 3rd edn. The MIT Press, Cambridge (2009)
10. Dean, J., Ghemawat, S.: Mapreduce: simplified data processing on large clusters. In: OSDI 2004, vol. 6. USENIX Association, Berkeley (2004)
11. DeWitt, D., Gray, J.: Parallel database systems: the future of high performance database systems. Commun. ACM 35(6), 85–98 (1992)
12. Gedik, B., Schneider, S., Hirzel, M., Wu, K.-L.: Elastic scaling for data stream processing. IEEE Trans. Parallel Distrib. Syst. 25, 1447–1463 (2010)
13. Graefe, G.: Encapsulation of parallelism in the volcano query processing system. In: SIGMOD 1990, pp. 102–111. ACM (1990)
14. Gulisano, V., Jimenez-Peris, R., Patino-Martinez, M., Valduriez, P.: StreamCloud: a large scale data streaming system. In: ICDCS 2010, pp. 126–137 (2010)
15. Khandekar, R., Hildrum, K., Parekh, S., Rajan, D., Wolf, J., Wu, K.-L., Andrade, H., Gedik, B.: COLA: optimizing stream processing applications via graph partitioning. In: Bacon, J.M., Cooper, B.F. (eds.) Middleware 2009. LNCS, vol. 5896, pp. 308–327. Springer, Heidelberg (2009). doi:10.1007/978-3-642-10445-9_16

16. Lakshmanan, G.T., Li, Y., Strom, R.: Placement strategies for internet-scale data stream systems. IEEE Internet Comput. **12**(6), 50–60 (2008)
17. Motwani, R., Widom, J., et al.: Query processing, resource management, and approximation in a data stream management system. In: CIDR 2003, pp. 245–256, January 2003
18. Neumeyer, L., Robbins, B., Nair, A., Kesari, A.: S4: distributed stream computing platform. In: ICDMW 2010, pp. 170–177. IEEE Computer Society, Washington, DC (2010)
19. Pietzuch, P., Ledlie, J., Shneidman, J., Roussopoulos, M., Welsh, M., Seltzer, M.: Network-aware operator placement for stream-processing systems. In: ICDE 2006. IEEE (2006)
20. Schneider, S., Andrade, H., Gedik, B., Biem, A., Wu, K.-L.: Elastic scaling of data parallel operators in stream processing. In: IPDPS, pp. 1–12 (2009)
21. Schneider, S., Hirzel, M., Gedik, B., Wu, K.-L.: Auto-parallelizing stateful distributed streaming applications. In: PACT 2012, pp. 53–64. ACM, New York (2012)
22. Shah, M.A., Chandrasekaran, S., Hellerstein, J.M., Franklin, M.J.:. Flux: an adaptive partitioning operator for continuous query systems. In: ICDE, pp. 25–36 (2002)
23. Wu, S., Kumar, V., Wu, K.-L., Ooi, B.C.: Parallelizing stateful operators in a distributed stream processing system: how, should you and how much? In: DEBS 2012, pp. 278–289 (2012)
24. Xing, Y., Hwang, J.-H., Çetintemel, U., Zdonik, S.: Providing resiliency to load variations in distributed stream processing. In: VLDB 2006, pp. 775–786. VLDB Endowment (2006)
25. Xing, Y., Zdonik, S., Hwang, J.-H.: Dynamic load distribution in the borealis stream processor. In: ICDE 2005, pp. 791–802. IEEE Computer Society (2005)

Social Network Analytics (II)

Measuring and Maximizing Influence via Random Walk in Social Activity Networks

Pengpeng Zhao[1], Yongkun Li[1,2(✉)], Hong Xie[3], Zhiyong Wu[1],
Yinlong Xu[1,4], and John C.S. Lui[5]

[1] University of Science and Technology of China, Hefei, China
{roczhau,wzylucky}@mail.ustc.edu.cn, {ykli,ylxu}@ustc.edu.cn
[2] Collaborative Innovation Center of High Performance Computing,
National University of Defense Technology, Changsha, China
[3] National University of Singapore, Singapore, Singapore
hongx87@gmail.com
[4] AnHui Province Key Laboratory of High Performance Computing, Hefei, China
[5] The Chinese University of Hong Kong, Hong Kong, China
cslui@cse.cuhk.edu.hk

Abstract. With the popularity of OSNs, finding a set of most influential users (or nodes) so as to trigger the largest influence cascade is of significance. For example, companies may take advantage of the "word-of-mouth" effect to trigger a large cascade of purchases by offering free samples/discounts to those most influential users. This task is usually modeled as an influence maximization problem, and it has been widely studied in the past decade. However, considering that users in OSNs may participate in various kinds of online activities, e.g., giving ratings to products, joining discussion groups, etc., influence diffusion through online activities becomes even more significant.

In this paper, we study the impact of online activities by formulating the influence maximization problem for social-activity networks (SANs) containing both users and online activities. To address the computation challenge, we define an influence centrality via random walks to measure influence, then use the Monte Carlo framework to efficiently estimate the centrality in SANs. Furthermore, we develop a greedy-based algorithm with two novel optimization techniques to find the most influential users. By conducting extensive experiments with real-world datasets, we show our approach is more efficient than the state-of-the-art algorithm IMM [17] when we needs to handle large amount of online activities.

Keywords: OSN · Influence maximization · Random walk

1 Introduction

Due to the popularity of online social networks (OSNs), viral marketing which exploits the "word-of-mouth" effect is of significance to companies which want to promote product sales. Therefore, it is of interest to find the best initial set of

© Springer International Publishing AG 2017
S. Candan et al. (Eds.): DASFAA 2017, Part II, LNCS 10178, pp. 323–338, 2017.
DOI: 10.1007/978-3-319-55699-4_20

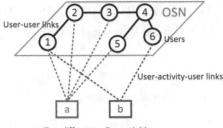

Fig. 1. An example of social-activity network (SAN).

users so as to trigger the largest influence spread. This viral marking problem can be modeled as an influence maximization problem, which was first formulated by Kempe et al. [12]. That is, given an OSN and an information diffusion model, how to select a set of k users, which is called the seed set, so as to trigger the largest influence spread. This problem is proved to be an NP-hard problem [4,6], and it has been studied extensively in the past decade [4–6,17,18].

Note that users in today's OSNs may participate in various kinds of online activities, e.g., joining a discussion group, and clicking `like` on Facebook etc. Hence, users not only can create friendship relationships, which we call *user-user* links, but can also form relationships by participating in online activities, which we call *user-activity-user* links. For example, if two users in Facebook express like to the same public page, then they form a user-activity-user link no matter they are friends or not. We call this kind of networks which contain both user-user relationships and user-activity-user relationships as *social-activity networks* (SANs).

With the consideration of online activities in SANs, influence may also spread through the user-activity-user links as well as the user-user links. In this paper, we focus on the online activities which generate positive influence, e.g., clicking `like` on the same public page in Facebook, giving high rating to the same product in online rating systems, and joining in a community sharing the same interest in online social networks. Due to the large amount of online activities, e.g., each pair of users may participate in multiple online activities, influence diffusion through the user-activity-user links becomes even more significant, and so only considering OSNs alone may not trigger the largest influence spread. Existing works on influence maximization usually focus on OSNs only and do not take the impact of online activities into consideration. *This motivates us to formulate the influence maximization problem for SANs, and to determine the most influential nodes by taking online activities into consideration.*

However, solving the influence maximization problem in SANs with online activities is challenging. First, influence maximization in OSNs without online activities was already proved to be NP-hard, and considering online activities makes this problem even more complicated. Second, the amount of online activities in a SAN is very large even for small OSNs, this is because online activities happen more frequently than friendship formation in OSNs. As a result, the underlying graph which characterizes users and their relationships may become extremely dense if we transform the user-activity-user links to user-user links, so it requires highly efficient algorithms for finding the most influential nodes. To address the above challenges, in this paper, we make the following contributions.

- We generalize the influence diffusion models for SANs by modeling SANs as hypergraphs, and approximate the influence of nodes in SANs by defining an influence centrality based on random walk.
- We employ the Monte Carlo framework to estimate the influence centrality in SANs, and also develop a greedy-based algorithm with two novel optimization techniques to solve the influence maximization problem for SANs.
- We conduct experiments with real-world datasets, and results show that our approach is more efficient while keep almost the same accuracy compared to the state-of-the-art algorithm.

This paper is organized as follows. In Sect. 2, we formulate the influence maximization problem for SANs. In Sect. 3, we present our random walk based methodology. In Sect. 4, we present the Monte Carlo method to estimate the influence centrality in SANs. In Sect. 5, we present our greedy-based algorithm and optimization techniques to solve the influence maximization problem. In Sect. 6, we present the experimental results. Related work is given in Sect. 7 and Sect. 8 concludes.

2 Problem Formulation

In this section, we first model the SAN with a hypergraph, and then formulate the influence maximization problem for SANs.

2.1 Model for SANs

We use a hypergraph $G(V, E, \mathcal{E}_1, ..., \mathcal{E}_l)$ to characterize a SAN, where V denotes the set of users, E denotes the *user-user* links, and $\mathcal{E}_i$ $(i = 1, 2, ..., l)$ denotes the set of type i hyperedges in which each hyperedge is a set of users who participated in the same online activity, and represented as a tuple. Considering Fig. 1, only activity a is of the first type, so $\mathcal{E}_1 = \{(1, 2, 3, 5)\}$. For ease of presentation, we denote $N(j)$ as the set of neighbors of user j, i.e., $N(j) = \{i | (i, j) \in E\}$, $M_e(j)$ as the set of users except for user j who connected to the hyperedge e, i.e., $M_e(j) = \{i | i \in e, \& i \neq j\}$, and denote $\mathcal{E}_t(j)$ as the set of type t hyperedges that are connected to user j, i.e., $\mathcal{E}_t(j) = \{e | e \in \mathcal{E}_t \& j \in e\}$. Considering Fig. 1, $N(1) = \{2\}$, $M_e(1) = \{2, 3, 5\}$ when $e = (1, 2, 3, 5)$ and $\mathcal{E}_1(1) = \{(1, 2, 3, 5)\}$.

2.2 Influence Maximization in SANs

Before describe the influence diffusion process for SANs, we first recall the independent cascade model(IC) which was proposed by Kempe et al. in [12]. Suppose that each user has two states, either active or inactive. At first, we initialize a set of users as active. For an active user i, she will activate each of her inactive neighbor j ($j \in N(i)$ where $N(i)$ denotes the neighbor set of user i) with probability q_{ij} ($0 \le q_{ij} \le 1$). One common setting of q_{ij} is $q_{ij} = \frac{1}{d_j}$, e.g., [4,5,12,17,18], where d_j denotes the degree of user j, i.e., $d_j = |N(j)|$. After a neighbor j being activated, then she will further activate her inactive neighbors in the set $N(j)$, and this diffusion process continues until no user can change her state. We call the expected size of the final set of active users the *influence spread*, and denote it as $\sigma(S(k))$ if the set of k initial active users is $S(k)$.

Now we describe the influence diffusion process for SANs. The key issue is to define the influence between user i and user j (i.e., g_{ij}) after taking online activities into consideration. Our definition is based on three criteria:

- A user may make a purchase due to her own interest or being influenced by others through user-user or user-activity-user links, so we define the total influence probability by one-hop neighbors as c ($0 < c < 1$), and call it the decay parameter. As we have l types of online activities, we define α_{jt} (where $0 \le \alpha_{jt} \le 1$ and $0 \le \sum_{t=1}^{l} \alpha_{jt} \le 1$) as the proportion of influence to user j through type t online activities, and call it *weight of activities*. Clearly, $1 - \sum_{t=1}^{l} \alpha_{jt}$ indicates the proportion of influence from direct neighbors.
- For the influence to user j from direct neighbors, we define the weight of each neighbor i ($i \in N(j)$) as u_{ij}, and assume that $0 \le u_{ij} \le 1$ and $\sum_{i \in N(j)} u_{ij} = 1$.
- For the influence to user j through the type t online activities, we define the weight of each online activity a as v_{aj}, where $0 \le v_{aj} \le 1$ and $\sum_{a \in N_t(j)} v_{aj} = 1$. Besides, considering that maybe multiple users participated in the same online activity a, we define the weight of each user i who participated in a as u_{ij}^a ($i \in N(a) \backslash \{j\}$), and assume that $0 \le u_{ij}^a \le 1$ and $\sum_{i \in N(a) \backslash \{j\}} u_{ij}^a = 1$.

For simplicity, we let $u_{ij} = 1/|N(j)|$ in this paper. Note that this uniform setting is exactly the same as the IC model in OSNs, which has been widely studied in [4,5,12,17,18]. Similarly, we also let $v_{aj} = 1/|\mathcal{E}_t(j)|$ and $u_{ij}^a = 1/|M_e(j)|$ by following the uniform setting. We would like to point out that our random walk approach in this paper also applies to general settings. Now we can define the influence of user i to user j, which we denote as g_{ij}:

$$g_{ij} = c \times \left(\frac{1 - \sum_{t=1}^{l} \alpha_{jt}}{|N(j)|} \times \mathbf{1}_{\{i \in N(j)\}} + \sum_{t \in [1,l]} \sum_{e \in \mathcal{E}_t(j)} \frac{\alpha_{jt}}{|\mathcal{E}_t(j)|} \times \frac{1}{|M_e(j)|} \times \mathbf{1}_{\{i \in M_e(j)\}} \right). \quad (1)$$

The first part in the right hand side of Eq. (1) denotes the influence diffusion through user-user links, and the second part represents the influence diffusion through user-activity-user links.

Now we formulate the influence maximization problem for SANs, which we denote as **IMP(SAN)**, as follows.

Definition 1. IMP(SAN): *Given a SAN $G(V, E, \mathcal{E}_1, ..., \mathcal{E}_l)$, an influence diffusion model with parameters α_{jt}, find a set of k nodes $S(k)$, where k is an integer, so as to make the influence spread $\sigma(S(k))$ maximized.*

3   Methodology

In this section, we present our methodology to address the (**IMP(SAN)**) problem. To reduce the large computation cost, we first develop a random walk framework on hypergraphs to estimate the influence diffusion process. Then, we define a centrality measure based on random walk to approximate the influence of a node set. With this centrality measure, we can approximate the influence maximization problem by solving a centrality maximization problem.

3.1   Random Walk on Hypergraph

Here, we present our random walk based framework, which is extended from the classical random walk on a simple unweighted graph $G(V, E)$, which can be stated as follows. For a random walk at vertex $i \in V$, it uniformly selects at random a neighbor j ($j \in N(i)$), and then moves to j in the next step. Mathematically, if we denote $Y(t)$ as the position of the walker at step t, then $\{Y(t)\}$ constitutes a Markov chain where the one-step transition probability p_{ij} is defined as $p_{ij} = 1/|N(i)|$ if $(i, j) \in E$, and 0 otherwise.

We now define the one-step transition probability p_{ij} when performing a random walk on the hypergraph $G(V, E, \mathcal{E}_1, ..., \mathcal{E}_l)$. Note that each hyperedge may contain more than two vertices, so we take the one-step random walk from user i to user j as a two-step process.

• **Step one:** Choose a hyperedge associated to user i. Precisely, according to the influence diffusion models in Sect. 2.2, we set the probability of selecting type t hyperedges as α_{it}, and choose hyperedges of the same type uniformly at random. Mathematically, if the walker is currently at user i, then it chooses a hyperedge e of type t with probability $\frac{\alpha_{it}}{|\mathcal{E}_t(i)|}$.

• **Step two:** Choose a user associated to the hyperedge e selected in step one as the next stop of the random walk. We consider random walks without backtrace. In particular, if a walker is currently at node i, then we select the next stop uniformly from the vertices that are connected to the same hyperedge with user i. We define the probability of choosing user j as $1/|M_e(i)|$.

By combing the two steps defined above, we can derive the transition probability from user i to j as follows, and we can find $g_{ji} = c \times p_{ij}$.

$$p_{ij} = \frac{1 - \sum_{t=1}^{l} \alpha_{it}}{|N(i)|} \times \mathbf{1}_{\{j \in N(i)\}} + \sum_{t \in [1,l]} \sum_{e \in \mathcal{E}_t(i)} \frac{\alpha_{it}}{|\mathcal{E}_t(i)|} \times \frac{1}{|M_e(i)|} \times \mathbf{1}_{\{j \in M_e(i)\}}. \quad (2)$$

3.2   Influence Centrality Measure

To address the **IMP(SAN)** problem, one key issue is to measure the influence of a node set. To achieve this, we define a centrality measure based on random walks on hypergraphs to approximate the influence of a node set S. We call it *influence centrality*, and denote it as $I(S)$, which is defined as follows.

$$I(S) = \sum_{j \in V} h(j, S), \quad (3)$$

where $h(j, S)$ aims to approximate the influence of S to j, which is called decayed hitting probability. It is defined as

$$h(j, S) = \begin{cases} \sum_{i \in V} c p_{ji} h(i, S), j \notin S, \\ 1, j \in S, \end{cases} \tag{4}$$

where c is the decay parameter defined in Sect. 2.2, and p_{ji} is the one-step transition probability defined in Eq. (2).

To solve the influence maximization problem of **IMP(SAN)**, we use the influence centrality measure $I(S)$ to approximate the influence of the node set S, and our goal is to find a set S of k users so that $I(S)$ is maximized. In other words, we approximate the influence maximization problem **IMP(SAN)** by solving the centrality maximization problem **CMP** defined as follows.

Definition 2. CMP: *Given a hypergraph $G(V, E, \mathcal{E}_1, ..., \mathcal{E}_l)$ and the corresponding parameters α_{jt}, find a set S of k nodes, where k is an integer, so as to make the influence centrality of the set S of k nodes $I(S)$ maximized.*

4 Centrality Computation

We note that the key challenge of solving the centrality maximization problem **CMP** is how to efficiently estimate the influence centrality of a node set $I(S)$, or the decayed hitting probability $h(j, S)$. We give an efficient framework to estimate $h(j, S)$ as follows. We first rewrite $h(j, S)$ in a linear expression which is an infinite converging series, and then truncate the converging series to save computation time (see Sect. 4.1). To further estimate the truncated series, we first explain the expression with a random walk approach, and then use a Monte Carlo framework via random walks to estimate it efficiently (see Sect. 4.2).

4.1 Linear Expression

We first transform $h(j, S)$ defined in Eq. (4) to a linear expression.

Theorem 1. *The decayed hitting probability $h(j, S)$ can be rewritten as*

$$h(j, S) = c e_j^T \boldsymbol{Q}' e + c^2 e_j^T \boldsymbol{Q} \boldsymbol{Q}' e + c^3 e_j^T \boldsymbol{Q}^2 \boldsymbol{Q}' e + \cdots . \tag{5}$$

where $\boldsymbol{Q}$ is a $(|V| - |S|) \times (|V| - |S|)$ dimensional matrix which describes the transition probabilities between two nodes in the set $V - S$, $\boldsymbol{Q}'$ is a $(|V| - |S|) \times |S|$ dimensional matrix which describes the transition probabilities from a node in $V - S$ to a node in S, $\boldsymbol{I}$ is an identity matrix, e is a column vector with all elements being 1, and finally e_j is a column vector with only the element corresponding to node j being 1 and 0 for all other elements.

Proof: Please refer to the technical report [21]. ■

We only keep the L leading terms of the infinite series, and denote the truncated result as $h^L(j, S)$, so we have

$$h^L(j, S) = ce_j^T Q'e + c^2 e_j^T QQ'e + \ldots + c^L e_j^T Q^{L-1} Q'e. \tag{6}$$

Since c is defined as $0 < c < 1$, the series truncation error is bounded as follows.

$$0 \le h(j, S) - h^L(j, S) \le c^{L+1}/(1 - c). \tag{7}$$

Based on the above error bound, we can see that $h^L(j, S)$ converges to $h(j, S)$ with rate c^{L+1}. This implies that if we want to compute $h(j, S)$ with a maximum error ϵ ($0 \le \epsilon \le 1$), we only need to compute $h^L(j, S)$ by taking a sufficiently large enough L, or $L \ge \lceil \frac{\log(\epsilon - \epsilon c)}{\log c} \rceil - 1$.

4.2   Monte Carlo Algorithm

In this subsection, we present a Monte Carlo algorithm to efficiently approximate $h^L(j, S)$. Our algorithm is inspired from the random walk interpretation of Eq. (6), and it can achieve a high accuracy with a small number of walks.

Consider the random walk interpretation of a particular term $e_j^T Q^{t-1} Q'e$ ($t = 1, \ldots, L$) in Eq. (6). Let us consider a L-step random walk starting from $j \notin S$ on the hypergraph. At each step, if the walker is currently at node k ($k \notin S$), then it selects a node i and transits to i with probability p_{ki}, which is defined in Eq. (2). As long as the walker hits a node in S, then it stops. Let $j^{(t)}$ be the t-th step position, and define an indicator $X(t)$ as

$$X(t) = \begin{cases} 1, & j^{(t)} \in S, \\ 0, & j^{(t)} \notin S. \end{cases}$$

We can see that $e_j^T Q^{t-1} Q'e$ is the probability that a random walk starting from j hits a node in S at the t-th step. We have

$$e_j^T Q^{t-1} Q'e = E[X(t)]. \tag{8}$$

By substituting $e_j^T Q^{t-1} Q'e$ with Eq. (8), we can rewrite $h^L(j, S)$ as

$$h^L(j, S) = cE[X(1)] + c^2 E[X(2)] + \cdots + c^L E[X(L)]. \tag{9}$$

Now we estimate $h^L(j, S)$ by using a Monte Carlo method with random walks on the hypergraph based on Eq. (9). Specifically, for each node j where $j \notin S$, we set R independent L-step random walks starting from j. We denote the t-th step position of the R random walks as $j_1^{(t)}, j_2^{(t)}, \ldots, j_R^{(t)}$, respectively, and use $X_r(t)$ to indicate whether $j_r^{(t)}$ belongs to set S or not. Precisely, we set $X_r(t) = 1$ if $j_r^{(t)} \in S$, and 0 otherwise, so $c^t E[X(t)]$ can be estimated as

$$c^t E[X(t)] \approx \frac{c^t}{R} \sum\nolimits_{r=1}^{R} X_r(t).$$

By substituting $c^t E[X(t)]$ in Eq. (9), we can approximate $h^L(j, S)$, which we denote as $\hat{h}^L(j, S)$, as follows.

$$\hat{h}^L(j, S) = \frac{c}{R} \sum_{r=1}^R X_r(1) + \cdots + \frac{c^L}{R} \sum_{r=1}^R X_r(L). \tag{10}$$

Algorithm 1 presents the process of the Monte Carlo method described above. We can see that its time complexity is $O(RL)$ as the number of types of online activities l is usually a small number. In other words, we can estimate $h^L(j, S)$ in $O(RL)$ time and compute $I(S)$ in $O(nRL)$ time as we need to estimate $h^L(j, S)$ for all nodes. The main benefit of this Monte Carlo algorithm is that its running time is independent of the graph size, so it scales well to large graphs.

Algorithm 1. Monte Carlo Estimation for $h^L(j, S)$

1: **function** $h^L(j, S)$
2:     $\sigma \leftarrow 0$;
3:     **for** $r = 1$ to R **do**
4:         $i \leftarrow j$;
5:         **for** $t = 1$ to L **do**
6:             Generate a random number $x \in [0, 1]$;
7:             **for** $T = 0$ to l **do**
8:                 **if** $x \le \alpha_{iT}$ **then** $\triangleright \alpha_{0T} = 1 - \sum_{T=1}^l \alpha_{iT}$;
9:                     $E \leftarrow \mathcal{E}_T(i)$;
10:                    break;
11:                $x \leftarrow x - \alpha_{iT}$;
12:            Select a hyperedge e from E randomly;
13:            $i \leftarrow$ select a user from $\{k | k \in e, k \ne i\}$ randomly;
14:            **if** $i \in S$ **then**
15:                $\sigma \leftarrow \sigma + c^t / R$;
16:                break;
17:     **return** σ;
18: **end function**

Note that $\hat{h}^L(j, S)$ computed with Algorithm 1 is an approximation of $h^L(j, S)$, and the approximation error depends on the sample size R. To estimate the number of samples required to compute $h^L(j, S)$ accurately, we derive the error bound by applying Hoeffding inequality [10], and the results are as follows.

Theorem 2. *Let the output of Algorithm 1 be* $\hat{h}^L(j, S)$, *then we have*

$$P\{|\hat{h}^L(j, S) - h^L(j, S)| > \epsilon\} \le 2L \exp(-2(1 - c)^2 \epsilon^2 R). \tag{11}$$

Proof: Please refer to the technical report [21].   ∎

Based on Theorem 2, we see that Algorithm 1 can estimate $h^L(j, S)$ with a maximum error ϵ with least probability $1 - \delta$ ($0 < \delta, \epsilon < 1$) by setting $R \ge \log(2L/\delta)/(2(1 - c)^2\epsilon^2)$.

Algorithm 2. Baseline Greedy Alg. for Maximizing $I(S)$

Input: A hypergraph, and a parameter k;
Output: A set S of k nodes for maximizing $I(S)$;
1: $S \leftarrow \emptyset$, $I(S) \leftarrow 0$;
2: **for** $s = 1$ to k **do**
3:    **for** $u \in (V - S)$ **do**
4:        $I(S \cup \{u\}) \leftarrow 0$;
5:        **for** $j \in (V - S \cup \{u\})$ **do**
6:            $I(S \cup \{u\}) \leftarrow I(S \cup \{u\}) + h(j, S \cup \{u\})$;
7:    $v \leftarrow \arg\max_{u \in (V-S)} I(S \cup \{u\}) - I(S)$;
8:    $S \leftarrow S \cup \{v\}$;

---

5 Centrality Maximization

In this section, we develop efficient algorithms to address the centrality maximization problem **CMP** defined in Sect. 3.2. Noted that even though we can efficiently estimate the decayed hitting probability $h(j, S)$ by using random walks (see Sect. 4), finding a set S of k nodes in a SAN to maximize its influence centrality $I(S)$ is still computationally difficult as it requires to estimate the influence centrality of all combinations of k nodes. In particular, **CMP** is NP-hard.

Theorem 3. *The centrality maximization problem* **CMP** *is NP-hard.*

Proof: Please refer to the technical report [21].   ∎

To solve the centrality maximization problem **CMP**, we develop greedy-based approximation algorithms by exploiting the submodularity property of $I(S)$. Specifically, we first show the submodularity property and present a baseline greedy algorithm to maximize $I(S)$, and then develop two novel optimization techniques to accelerate the greedy algorithm.

5.1 Baseline Greedy Algorithm

Before presenting the greedy-based approximation algorithm for maximizing $I(S)$, we first show that $I(S)$ is a non-decreasing submodular function, and the result is stated in the following theorem.

Theorem 4. *The centrality $I(S)$ is a non-decreasing submodular function.*

Proof: Please refer to the technical report [21].   ∎

Based on the submodularity property, we develop a greedy algorithm for approximation when maximizing $I(S)$, and we call it *the baseline greedy algorithm*. Algorithm 2 describes this procedure. To find a set of k nodes to maximize $I(S)$, the algorithm works for k iterations. In each iteration, it selects the node which maximizes the increment of $I(S)$.

Recall that the time complexity for estimating the influence of a set S to a particular node $j \notin S$, i.e., $h(j, S)$, is $O(RL)$ (see Sect. 4.2). Thus, the total time complexity for the baseline greedy algorithm is $O(kn^2RL)$ where n denotes the total number of users in the SAN, because estimating the influence of a set $S \cup \{u\}$ requires us to sum up its influence to all nodes, and we need to check every node u so as to select the one which maximizes the increment of $I(S)$. Although the baseline greedy algorithm gives a polynomial time complexity, it is inefficient when the number of users becomes large. To further speed up the computation, we present two novel optimization techniques in the next subsection.

5.2 Optimizations

• **Parallel Computation**: The key component in the greedy algorithm is to measure the marginal increment of the influence after adding node u, i.e., $\Delta(u) = I(S \cup \{u\}) - I(S)$, which can be derived as follows.

$$\Delta(u) = \left[1 - \sum_{h=1}^{\infty} c^h P(u, S, h)\right] \times \left[1 + \sum_{j \in (V - S \cup \{u\})} \sum_{h=1}^{\infty} c^h P^S(j, \{u\}, h)\right].$$

In the baseline greedy algorithm, $\Delta(u)$'s are computed sequentially, which as a result incurs a large time overhead. Our main idea to speed up the computation is to estimate the marginal increment of all nodes, i.e., $\Delta(u)$ for every u, *in parallel*. Specifically, when performing R random walks from a particular node j, we measure the contribution of j to the marginal increment of every node. In other words, we obtain $P^S(j, u, h)$ for every u by using only the R random walks starting from j. As a result, we need only $O(nR)$ random walks to derive the marginal increment of all nodes, i.e., $\Delta(u)$ for every u, instead of $O(n^2R)$ random walks as in the baseline greedy algorithm.

• **Walk Reuse**: The core idea is that in each iteration of choosing one node to maximize the marginal increment, we record the total $O(nR)$ random walks in memory, and apply the updates accordingly after one node is added into the result set. By doing this, we can reuse the $O(nR)$ random walks to derive the marginal increment in the next iteration instead of starting new random walks from each node again.

By incorporating the above optimization techniques, we can reduce the time complexity to $O(nRL)$, where L denotes the maximum walk length. In other words, we can use the L leading terms to estimate $\sum_{h=1}^{\infty} c^h P^S(j, \{u\}, h)$ and $\sum_{h=1}^{\infty} c^h P(u, S, h)$ as described in Sect. 4. Thus, we let each walk runs for L steps at most. Algorithm 3 states the procedure. We use $score[u]$ and $P[u]$ to record $\sum_{j \in V - S \cup \{u\}} \sum_{h=1}^{\infty} c^h P^S(j, \{u\}, h)$ and $\sum_{h=1}^{\infty} c^h P(u, S, h)$ for computing $\Delta(u)$, respectively. Algorithm 3 runs in two phases. The first phase (line 1–13) is to select the first seed node by running random walks and also record all the walking information for reuse. The second phase (line 14–18) is to select the remaining $k - 1$ nodes based on the stored information which requires to be updated after selecting each node. We give the update function in Algorithm 4.

The update function is to update the walk information stored in $score$ and P. Every time after we selecting a node v, the random walk in the following iterations should stop when it encounters v, and the values stored in $score$ and P should change accordingly. To achieve this, for each random walk that hits v (line 2),

Algorithm 3. Optimized Greedy Algorithm

Input: A hypergraph and a parameter k;
Output: A set S of k nodes for maximizing $I(S)$;
1: $S \leftarrow \emptyset$, $score[1...n] \leftarrow 0$, $P[1...n] \leftarrow 0$;
2: **for** $j \in V$ **do**
3:     **for** $r = 1$ to R **do**
4:        $i \leftarrow j$, $visited \leftarrow \emptyset$;
5:        **for** $t = 1$ to L **do**
6:           $visited \leftarrow visited \cup \{i\}$;
7:           $i \leftarrow$ Select a user according to the transition prob.;
8:           $RW[j][r][t] \leftarrow i$;
9:           **if** $i \notin visited$ **then**
10:              $index[i].add(item(j, r, t))$;
11:              $score[i] \leftarrow score[i] + \frac{c^t}{R}$;
12: $v \leftarrow \arg\max_{u \in V} score[u]$;
13: **for** $s = 2$ to k **do**
14:     Update $(RW, index, P, score, S, v, L)$, $S \leftarrow S \cup \{v\}$;
15:     $v \leftarrow \arg\max_{u \in (V-S)}(1 - P[u])(1 + score[u])$;
16: $S \leftarrow S \cup \{v\}$;

Algorithm 4. Update Function

1: **function** UPDATE $(RW, index, P, score, S, v, L)$
2:     **for** $w \in index[v]$ **do**
3:        $k \leftarrow L$;
4:        **for** $t = 1$ to L **do**
5:           **if** $RW[w.j][w.r][t] \in S$ **then**
6:              $k \leftarrow t$;
7:              break;
8:        **if** $k == L$ **then**
9:           $P[w.j] \leftarrow P[w.j] + c^t/R$;
10:        **for** $i = w.t + 1$ to k **do**
11:           $u \leftarrow RW[w.j][w.r][i]$, $score[u] \leftarrow score[u] - c^t/R$;
12: **end function**

---

we first check if it has visited any node in S (line 4–7). If not, we increase $P[w.j]$ after adding v in S (line 8,9). Since the following walks should stop when hitting v, we update $score[u]$ if node u is visited after v (line 10–12).

6 Experiments

To show the efficiency and effectiveness of our approach, we conduct experiments on real-world datasets. In particular, we first show that incorporating online activities in seed selection can lead to a significant improvement on the influence spread, i.e., influence more users with the same seed size. Then we show that our IM-RW algorithm takes much less running time than the state-of-the-art influence maximization algorithm, while achieves almost the same influence spread.

6.1 Datasets

We consider three datasets from social rating systems: Ciao [16], Yelp [1] and Flixster [11]. Such social rating networks are composed of a social network, where the links can be interpreted as either friendships (undirected link) or a following relationship (directed link), and a rating network, where a link represents that a user assigns a rating (or writes a review) to a product. Assigning a rating corresponds to an online activity, and multiple users assigning ratings to the same product means that they participate in the same online activity. In the rating network, we remove rating edges if the associated rating is less than 3 so as to filter out the users who dislike a product. Through this we guarantee that all the remaining users who give ratings to the same product have similar interests, e.g., they all like the product. Since the original Flixster dataset is too large to run the state-of-the-art influence maximization algorithms, we extract only a subset of the Flixster dataset for comparison studies. In particular, since the OSN of Flixster is almost a connected component, we randomly select a user, and run the breadth-first search algorithm until we get 300,000 users. We state the statistics of the three datasets in Table 1. All algorithms are run on a server with two Intel Xeon E5-2650 2.60 GHz CPU and 64 GB memory.

Table 1. Datasets Statistics.

Dataset	Users	Links in OSN	Products	Ratings	OSN type
Ciao	2,342	57,544	15,783	32,783	directed
Yelp	174,100	2,576,179	56,951	958,415	directed
Flixtser	300,000	6,394,798	28,262	2,195,134	undirected

6.2 The Benefit of Incorporating Activities

We first show that incorporating online activities in seed selection can lead to a significant improvement on the influence spread. We fix the seed size k as 50. To show the impact of activities, we use the state-of-the-art influence maximization algorithm IMM [17] to select the seed set on OSNs and use our IM-RW algorithm to select the seed set on SANs which take online activities into account. Then we use simulations to estimate the expected influence spread of the selected k users on SANs and denote the results as $\sigma(OSN)$ and $\sigma(SAN)$, respectively. Finally, we define the improvement ratio on the expected influence spread as $[\sigma(SAN) - \sigma(OSN)]/\sigma(OSN)$.

To present the key insights, we consider the simple case in which there is only one type of users and online activities. Namely, all users have a same value of α which indicates the weight of activities. We emphasize that our model also works in the general case of multiple types of users and online activities.

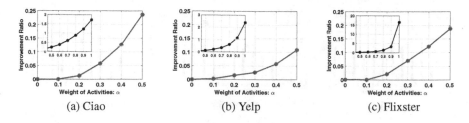

(a) Ciao   (b) Yelp   (c) Flixster

Fig. 2. Impact of online activities on influence spread.

Figure 2 depicts the improvement of incorporating online activities by varying the weight of activities α from 0 to 1. The horizontal axis shows the value of α, and the vertical axis presents the corresponding improvement ratio. From Fig. 2, one can observe that the improvement ratio is 0 when $\alpha = 0$. This is because users are not affected by other users through online activities when $\alpha = 0$. As α increases, the improvement ratio also increases. This shows that as users are more prone to be affected by other users through online activities, incorporating online activities bring larger benefit. When $\alpha = 0.5$, the improvement ratio is around 25% for Ciao dataset. That is, we can influence 25% more users when incorporating online activities in the seed selection. Similar conclusions can also be observed for the datasets of Yelp and Flixster. It is interesting to observe that as α approaches to one, the improvement ratio reaches up to 16 for Flixster, which implies a more than an order of magnitude improvement. In summary, incorporating online activities in the seed selection by using IM-RW significantly improves the selection accuracy.

6.3   Performance Evaluation of IM-RW

In this subsection, we validate the efficiency and effectiveness of IM-RW by comparing it with IMM, which is the state-of-the-art algorithm for solving influence maximization problem in OSNs, from two aspects, the running time and the influence spread. Note that IMM was originally developed for OSNs without online activities being considered, so for fair comparison, we transform SANs to a weighted graph by also taking online activities into account, and then apply IMM on the weighted graph to derive the most influential nodes.

We first compare the running time of IM-RW and IMM by varying the weight of activities α and the seed size k, and the results are presented in Figs. 3 and 4. Specifically, Fig. 3 shows that IMM takes much longer time than IM-RW, especially when the network is large and online activities become more important (i.e., with larger α). This is because as α increases, the time cost of IMM depends more on user-activity-user links than user-user links. Thus, as the amount of user-activity-user links is much more than that of user-user links in SANs, the time cost of IMM will increase. On the other hand, when we fix α as 0.8 and vary the seed size k, Fig. 4 also shows that IMM takes much longer time than IM-RW under all settings. Therefore, we can conclude that our IM-RW algorithm really improves

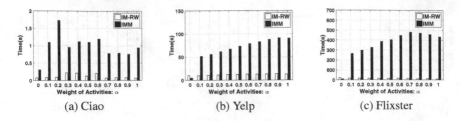

Fig. 3. Running time of IM-RW and IMM with different activity weights.

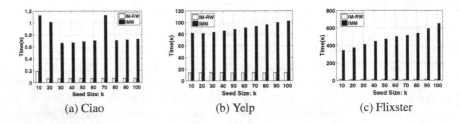

Fig. 4. Running time of IM-RW and IMM with different seed sizes.

the efficiency of solving the influence maximization problem in SANs with online activities being considered.

We further show the influence spread of the most influential users selected by the two algorithms in Fig. 5. The horizontal axis shows the values of α, and the vertical axis represents the corresponding influence spread. We see that by taking online activities into consideration, both IMM and IM-RW can achieve almost the same performance. Because IMM is an influence maximization algorithm with theoretical performance guarantees, we can conclude that our IM-RW approach also has a good performance to maximize the influence spread.

Summary: Our IM-RW algorithm achieves a good performance in both the running time and the influence spread by taking online activities into account in SANs. In particular, comparing to the state-of-the-art algorithm IMM, our IM-RW algorithm achieves almost the same performance in seed selection, while it only requires much less running time.

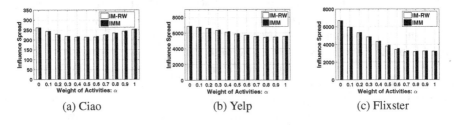

Fig. 5. Influence spread of IM-RW and IMM.

7   Related Work

Influence maximization problem in OSNs was first formulated by Kempe et al. [12], and in this seminal work, the authors proposed the IC model and the LT model. Since then, this problem receives a lot of interests in academia in the past decade [4–6]. Because of the NP-hardness under both the IC model [4] and the LT model [6], many of the previous studies focus on how to reduce the time complexity. Recently, Borgs et al. [3] developed an algorithm which maintains the performance guarantee while reduces the time complexity significantly, and Tang et al. [17,18] further improved the method and proposed the IMM algorithm, which is the state-of-the-art solution for influence maximization in OSNs. Besides reducing the computation overhead, several works improved the influence models, for example, topic-aware influence model [2], competitive influence model [14], opinion-based influence model [9] etc.

Centrality measure based approach was also studied, for example, the studies [5,7,8,19] find the most influential nodes based on degree centrality and closeness centrality. In terms of random walk, it is widely used to analyze big graphs, e.g., PageRank computation [15], graph sampling [20], and SimRank [13] etc.

We would like to emphasize that our work differs from existing studies which address the traditional influence maximization problem, while we take online activities into consideration. When we consider these online activities, only considering user-user links alone may not trigger the largest influence spread. Although we can also transform the user-activity-user links to user-user links, the underlying graph may become extremely dense so that traditional methods may not be efficient.

8   Conclusions

In this paper, we address the influence maximization problem in SANs with a random walk approach. Specifically, we propose a general framework to measure the influence of nodes in SANs via random walks on hypergraphs, and develop a greedy-based algorithm with two novel optimization techniques to find the top k most influential nodes in SANs by using random walks. Experiments with real-world datasets show that our approach greatly improves the computation efficiency, while keeps almost the same performance in seed selection accuracy compared to IMM, the state-of-the-art algorithm.

Acknowledgements. This work was supported by National Nature Science Foundation of China (61303048 and 61379038), and Anhui Provincial Natural Science Foundation (1508085SQF214).

References

1. Yelp Dataset. https://www.yelp.com/dataset_challenge/dataset
2. Barbieri, N., Bonchi, F., Manco, G.: Topic-aware social influence propagation models. In: Proceedings of ICDM (2012)

3. Borgs, C., Brautbar, M., Chayes, J., Lucier, B.: Maximizing social influence in nearly optimal time. In: Proceedings of SODA (2014)
4. Chen, W., Wang, C., Wang, Y.: Scalable influence maximization for prevalent viral marketing in large-scale social networks. In: Proceedings of ACM KDD (2010)
5. Chen, W., Wang, Y., Yang, S.: efficient influence maximization in social networks. In: Proceedings of ACM KDD (2009)
6. Chen, W., Yuan, Y., Zhang, L.: Scalable influence maximization in social networks under the linear threshold model. In: Proceedings of IEEE ICDM (2010)
7. Everett, M.G., Borgatti, S.P.: The centrality of groups and classes. J. Math. Sociol. **23**(3), 181–201 (1999)
8. Freeman, L.C.: Centrality in social networks conceptual clarification. Soc. Netw. **1**(3), 215–239 (1979)
9. Galhotra, S., Arora, A., Roy, S.: Holistic influence maximization: Combining scalability and efficiency with opinion-aware models. arXiv preprint arXiv:1602.03110 (2016)
10. Hoeffding, W.: Probability inequalities for sums of bounded random variables. J. Am. Stat. Assoc. **58**(301), 13–30 (1963)
11. Jamali, M., Ester, M.: A matrix factorization technique with trust propagation for recommendation in social networks. In: Proceedings of ACM RecSys (2010)
12. Kempe, D., Kleinberg, J., Tardos, É.: Maximizing the spread of influence through a social network. In: Proceedings of ACM KDD (2003)
13. Kusumoto, M., Maehara, T., Kawarabayashi, K.-I.: Scalable similarity search for SimRank. In: Proceedings of ACM SIGMOD (2014)
14. Lin, Y., Lui, J.C.: Analyzing competitive influence maximization problems with partial information: an approximation algorithmic framework. Perform. Eval. **91**, 187–204 (2015)
15. Page, L., Brin, S., Motwani, R., Winograd, T.: The PageRank Citation Ranking: Bringing Order to The Web. Technical report (1999)
16. Tang, J., Gao, H., Liu, H.: mTrust: discerning multi-faceted trust in a connected world. In: Proceedings of ACM WSDM (2012)
17. Tang, Y., Shi, Y., Xiao, X.: Influence maximization in near-linear time: a martingale approach. In: Proceedings of ACM SIGMOD (2015)
18. Tang, Y., Xiao, X., Shi, Y.: Influence maximization: near-optimal time complexity meets practical efficiency. In: Proceedings of ACM SIGMOD (2014)
19. Zhao, J., Lui, J., Towsley, D., Guan, X.: Measuring and maximizing group closeness centrality over disk-resident graphs. In: Proceedings of SIMPLEX (2014)
20. Zhao, J., Lui, J., Towsley, D., Wang, P., Guan, X.: A tale of three graphs: sampling design on hybrid social-affiliation networks. In: Proceedings of IEEE ICDE (2015)
21. Zhao, P., Li, Y., Xie, H., Wu, Z., Xu, Y., Lui, J.C.S.: Measuring and Maximizing Influence via Random Walk in Social Activity Networks, Technical report. https://arxiv.org/abs/1602.03966

Adaptive Overlapping Community Detection with Bayesian NonNegative Matrix Factorization

Xiaohua Shi[1,2]($\boxtimes$), Hongtao Lu[1], and Guanbo Jia[3]

[1] MOE-Microsoft Laboratory for Intelligent Computing and Intelligent Systems,
Department of Computer Science and Engineering,
Shanghai JiaoTong University, Shanghai, China
xhshi@sjtu.edu.cn
[2] Library, Shanghai Jiaotong University, Shanghai, China
[3] University of Birmingham, Birmingham, UK

Abstract. Overlapping Community Detection from a real network is unsupervised, and it is hard to know the exact community number or quantized strength of every node related to each community. Using Non-negative Matrix Factorization (NMF) for Community Detection, we can find two non-negative matrices from whole network adjacent matrix, and the product of two matrices approximates the original matrix well. With Bayesian explanation in factorizing process, we can not only catch most appropriate count of communities in a large network with Shrinkage method, but also verify good threshold how a node should be assigned to a community in fuzzy situation.

We apply our approach in some real networks and a synthetic network with benchmark. Experimental results for overlapping community detection show that our method is effective to find the communities number and overlapping degree, and achieve better performance than other existing overlapping community detection methods.

Keywords: Overlapping community detection · Non-negative matrix factorization · Bayesian inference · Automatic relevance determination

1 Introduction

Overlapping Community Detection is an important approach in complex networks to understand and analysis large network character [3,50], such as social network [30,49], collaborative network [39], and biological network [1]. We can find most correlated overlapping sub-communities to simplify global structure to understand the network topology, and keep original network with overlapping structure especially in density network.

It is a recognition with community detection that nodes in same community are densely connected, and nodes in different communities are sparsely connected. A node can be allocated into different communities in overlapping situation [55]. We can find overlapping communities with methods as clique percolation techniques [23], random walk [18], label propagation [12,51], seed

© Springer International Publishing AG 2017
S. Candan et al. (Eds.): DASFAA 2017, Part II, LNCS 10178, pp. 339–353, 2017.
DOI: 10.1007/978-3-319-55699-4_21

expansion [47], objective function optimization (modularity or other function) [35], or statistical inference [11,40,48]. Overlapping communities can also be detected based on the graph partitioning approach, which tries to find underling clusters from minimize the number of edges between communities [8,43].

Macropol *et al.* [29] propose a biologically sensitive algorithm based on repeated random walks (RRW) for discovering functional modules, e.g., complexes or pathways, within large-scale protein networks. RRW considers the element of network topology, edge weights, and long range interactions between proteins. Zhang *et al.* [53] propose a learning algorithm which can learn a node-community membership matrix via stochastic gradient descent with bootstrap sampling. Lee *et al.* [25] introduce a community assignment algorithm named Greedy Clique Expansion (GCE). GCE algorithm identifies distinct cliques as seeds and expands these seeds by greedily optimizing a local fitness function.

In many clustering applications, object data is nonnegative due to their physical nature, e.g., images are described by pixel intensities and texts are represented by vectors of word counts. As to a graph-based network, the adjacency matrix (or weighted adjacency matrix) $\mathbf{A}$ as well as the Laplacian matrix completely represents the structure of network, and $\mathbf{A}$ is non-negative naturally. Meanwhile, Nonnegative Matrix Factorization (NMF) was originally proposed as a method for dimension reduction and finding matrix factors with parts-of-whole interpretations [15,27]. Based on the consideration that there is no any physical meaning to reconstruct a network with negative adjacency matrix, using NMF to obtain new representations of network with non-negativity constraints can achieve much productive effect in overlapping community analysis [52,53]. It is likely an efficient network partition tool to find the communities because of its powerful interpretability and close relationship with other clustering methods. Overlapping community detection with NMF can capture the underlying structure of network in the low dimensional data space with its community-based representations [41]. Zhang *et al.* [54] propose a method called bounded nonnegative matrix tri-factorization (BNMTF) with three factors in the factorization, and explicitly model and learn overlapping community membership of each node as well as the interaction among communities.

NMF decomposes a given nonnegative data matrix $\mathbf{X}$ as $\mathbf{X} \approx \mathbf{UV^T}$ where $\mathbf{U} \geq \mathbf{0}$ and $\mathbf{V} \geq \mathbf{0}$ (meaning that U and V are *component-wise nonnegative*). Tan *et al.* [45] addresses the estimation of the latent dimensionality in nonnegative matrix factorization (NMF) with the β-divergence, and proposes for maximum a posteriori (MAP) estimation with majorization-minimization (MM) algorithms. Psorakis *et al.* [40] presents a novel approach to community detection that utilizes the Bayesian non-negative matrix factorization model to extract overlapping modules from a network.

In this paper, we propose an adaptive Bayesian non-negative matrix factorization (ABNMF) method for overlapping community detection. In a Bayesian framework, ABNMF assumpts that original matrix $\mathbf{X}$ with object matrix $\mathbf{U}$ and $\mathbf{V}$ follow a certain probability distribution. In this way, we expect that ABNMF can obtain a relevant count of communities and quantized strength of each node related to every community from original network data. To achieve this, we design a new non-negative matrix factorization objective function by incorporating Bayesian Detection, and suggest an adaptive node-based threshold for different communities. Our experiments show that the proposed approach can validly estimate relevant dimension in lower space, find suitable overlapping communities, and also achieve better performance than the state-of-arts overlapping methods.

2   Related Works

Let $\mathbf{X}$ be a $m \times n$ non-negative matrix, and NMF consists in finding an approximation:

$$\mathbf{X} \approx \mathbf{U}\mathbf{V}^T \tag{1}$$

where $\mathbf{U}$ and $\mathbf{V}$ are $m \times k$ and $n \times k$ non-negative matrices. The factorization rank $\mathbf{k}$ is often chosen such that $k \ll \min(m, n)$. The objective behind this choice is to summarize and split the information contained in $\mathbf{U}$ into $\mathbf{k}$ factors (the columns of $\mathbf{U}$). Depending on the application field, these factors are given different names: basis images, metagenes or source signals. In community detection, we equivalently and alternatively use the terms primary communities to refer to matrix $\mathbf{U}$, and mixture coefficient matrix or communities assignment profiles to refer with matrix $\mathbf{V}$. We examine each row of $\mathbf{V}$, and assign node $\mathbf{x}_j$ to community c if $c = \arg\max_c v_{jc}$ [44] in non-overlapping community detection like crisp clustering. If we define a proper threshold set δ, a node j can be assigned into community c if $v_{jc} \geq \delta_c$ in overlapping situation like fuzzy clustering [37].

The main approach of NMF is to estimate matrices $\mathbf{U}$ and $\mathbf{V}$ as a local minimum with a cost function in some distance metric. Generally we use β-Divergence $\mathbf{D}_\beta(\mathbf{X}; \mathbf{U}\mathbf{V}^\mathbf{T})$ [7]. When $\beta = 0, 1, 2, \mathbf{D}_\beta(\mathbf{X}; \mathbf{U}\mathbf{V}^\mathbf{T})$ is proportional to the (negative) log-likelihood of the Itakara-Saito (IS), KL and Euclidean noise models up to a constant.

Recently, Bayesian inference has been introduced into NMF with a noise E between $\mathbf{X}$ and $\mathbf{U}\mathbf{V}^\mathbf{T}$.

$$\mathbf{X} = \mathbf{U}\mathbf{V}^\mathbf{T} + \mathbf{E} \tag{2}$$

Morten et al. [31] demonstrate how a Bayesian framework for model selection based on Automatic Relevance Determination (ARD) can be adapted to the Tucker and CandeComp/PARAFAC (CP) models. By assigning priors for the model parameters and learning the hyperparameters of these priors the method is able to turn off excess components and simplify the core structure at a computational cost of fitting the conventional Tucker/CP model. Morten et al. [32] also formulate a non-parametric Bayesian model for community detection consistent with an intuitive definition of communities, and present a Markov chain Monte Carlo procedure for inferring the community structure.

Automatic Relevance Determination is a hierarchical Bayesian approach that widely used for model selection. In ARD, hyperparameters explicitly represent the relevance of different features by defining the range of variation for these features, and are usually by modeling the width of a zero-mean prior imposed on the model parameters. If the width becomes zero, the corresponding feature cannot have any effect on the prediction. Hence, ARD optimizes these hyperparameters to discover which features are relevant. While ARD based on Gaussian or Poisson priors, we can prune excess components by admitting sparse representation and retain active components. Applying ARD in some real network community detection process, we can effectively find the relevant communities number without knowing in advance.

Jin et al. [17] extend the stochastic model method to detection of overlapping communities with the virtue of autonomous determination of the number of communities. Their approach hinges upon the idea of ranking node popularities within communities and using a Bayesian method to shrink communities to optimize an objective function based on the stochastic generative model. Wang et al. [46] propose a probabilistic model, Dynamic Bayesian Nonnegative Matrix Factorization, for automatic detection of overlapping communities in temporal networks. Their model can not only give the overlapping community structure based on the probabilistic memberships of nodes in each snapshot network but also automatically determines the number of communities in each snapshot network based on automatic relevance determination.

Schmidt et al. [42] present a Bayesian treatment of NMF based on a Gaussian likelihood and exponential priors, and approximate the posterior density of the NMF factors. This model equals to minimize the squares Euclidean distance $\mathbf{D_2(X; UV^T)}$ for NMF. Cemgil [5] proposes NMF models with a KL-divergence error measure in a statistical framework with a hierarchical generative model consisting of an observation and a prior component. We can see that this models of $\mathbf{D_1(X; UV^T)}$ is equals to NMF model with Poisson noise likelihood:

$$P(n; \lambda) = \frac{\lambda^n}{n!} exp(-\lambda) \tag{3}$$

$$P(X|U, V) = \prod_i \prod_j \frac{[UV^T]_{ij}^{X_{ij}} exp(-[UV^T]_{ij})}{X_{ij}!} \tag{4}$$

We further assume that all entries of X are independent of each other (the dependency structure is later induced by the matrix product), we can write:

$$ln(P(X|U, V)) = \sum_i \sum_j X_{ij} ln[UV^T]_{ij} - [UV^T]_{ij} - ln(X_{ij}!) \tag{5}$$

We use Stirling's formula $ln(n!) \approx nln(n) - n$ for $n >> 1$ to get approximated expression:

$$ln(P(X|U, V)) \approx \sum_i \sum_j X_{ij} ln \frac{[UV^T]_{ij}}{X_{ij}} - [UV^T]_{ij} + X_{ij} \tag{6}$$

3 Overlapping Community Detection with Bayesian NMF

3.1 Bayesian NMF Model

In this section, we introduce Bayesian inference process of our Adaptive Bayesian NMF (ABNMF) method. Given a network $\mathbf{G}$ consisting of n nodes $\mathbf{a}_1, \mathbf{a}_2, \cdots, \mathbf{a}_n$, we can represent the network as matrix $\mathbf{X}$ transformed from adjacency matrix. In our ABNMF processing, the diagonal elements are defined to be 1 rather than 0 as in usual clustering cases, and $\mathbf{X}$ is $n \times n$ square matrix and non-negative.

We consider there lies a relation between original network matrix $\mathbf{X}$ and combination of factorized matrix $\mathbf{UV^T}$. The distribution of this relation can be Gaussian [42] or Poisson [26] model. As Poisson noise model algorithm have much better performance than Gaussian noise models [19,21] to achieve better sparse estimation effect, we select Poisson likelihood in our ABNMF method. In maximum-likelihood solution to find $\mathbf{U}$ and $\mathbf{V}$, $\mathbf{P(X|UV^T)}$ is maximized, or its energy function $-\mathbf{logP(X|UV^T)}$ is minimized.

To simplify likelihood with positive error [10,36], we chose the relation of $\mathbf{U}, \mathbf{V}$ and $\mathbf{X}$ as $X_{ij} \sim Poisson(\sum_k U_{ik} * V_{kj})$. In this Poisson model, the log-likelihood of $\mathbf{X}$ and $\mathbf{UV^T}$ is:

$$- ln(P(X|UV^T)) = - \sum_i \sum_j \left\{ X_{ij} ln \frac{[UV^T]_{ij}}{X_{ij}} - [UV^T]_{ij} + X_{ij} \right\} \qquad (7)$$
$$= \quad -Xln(UV^T) + 1UV^T 1^T + const(X)$$

where $\mathbf{1}$ is an n $\times$ n matrix with every elements equal to 1. We use independent half-normal prior over every column of $\mathbf{U}$ and $\mathbf{V}$, where the mean is zero and precisian is β_j:

$$p(u_{ij}|\beta_j) = \mathcal{HN}(x|0, \beta_j^{-1}) \qquad (8)$$
$$p(v_{jk}|\beta_j) = \mathcal{HN}(x|0, \beta_j^{-1})$$

when

$$\mathcal{HN}(x|0, \beta^{-1}) = \sqrt{\frac{2}{\pi}} \beta^{\frac{1}{2}} exp(-\frac{1}{2}\beta x^2) \qquad (9)$$

We define the diagonal matrix $\mathbf{B}$ with $[\beta_1, ..., \beta_K]$ and zeros elsewhere, and the negative log priors of $\mathbf{U}$ and $\mathbf{V}$ are:

$$- ln(p(U|\beta)) = \sum_i \sum_j \frac{1}{2}\beta_j u_{ij}^2 - \sum_j \frac{N}{2}log\beta_j + const \qquad (10)$$

$$-ln(p(V|\beta)) = \sum_j \sum_k \frac{1}{2}\beta_j v_{jk}^2 - \sum_j \frac{N}{2}log\beta_j + const$$

At last, we set the independent prior distribution of β_j as a Gamma distribution with parameters a_j and b_j:

$$p(\beta_j|a_j, b_j) = \frac{b_j^{a_j}}{\Gamma(a_j)} \beta_j^{a_j-1} exp(-\beta_j b_j) \tag{11}$$

The negative log of β_j is:

$$- ln(p(\beta)) = \sum_j [\beta_j b_j k - (a_j - 1) ln\beta_j] + const \tag{12}$$

The MAP(Maximum a Posteriori) of ABNMF is:

$$\mathcal{U} = -lnP(X|UV^T)) - lnP(U|\beta)) - lnP(V|\beta)) - lnP(\beta)) \tag{13}$$

3.2   Iteration Rules of ABNMF

From Eq. (13), we can derive the multiplicative update rules of ABNMF with Poisson likelihood. Let ϕ_{ij}, ψ_{jk} be the Lagrange multiplier for constraint $u_{ij} \geq 0$ and $v_{jk} \geq 0$, respectively, and $\mathbf{\Phi} = [\phi_{ij}]$, $\mathbf{\Psi} = [\psi_{jk}]$. The Lagrange function $\mathcal{L}$ is

$$\mathcal{L} = \mathcal{U} + tr(\mathbf{\Phi U^T}) + tr(\mathbf{\Psi V^T}) \tag{14}$$

Let the derivatives of $\mathcal{L}$ with respect to $\mathbf{U}$ or $\mathbf{V}$ vanish, we have:

$$\frac{\partial \mathcal{L}}{\partial U} = -2 * \frac{X}{UV^T} U + 2 * 1U + 2 * BU + \mathbf{\Phi} = 0 \tag{15}$$

$$\frac{\partial \mathcal{L}}{\partial V} = -2 * \frac{X}{UV^T} V + 2 * 1V + 2 * BV + \mathbf{\Psi} = 0 \tag{16}$$

Using the KKT conditions $\phi_{ij} u_{ij} = 0$ and $\psi_{jk} v_{jk} = 0$, we get the following equations for u_{ij}, v_{jk}:

$$u_{ij} \longleftarrow u_{ij} \left(\frac{X}{UV^T}\right)_{ij} \left(\frac{U}{1U + UB}\right)_{ij} \tag{17}$$

$$v_{jk} \longleftarrow v_{jk} \left(\frac{X}{UV^T}\right)_{jk} \left(\frac{V}{1V + VB}\right)_{jk} \tag{18}$$

and the β_j will be updated below:

$$\beta_j \longleftarrow \frac{n + a_j - 1}{\frac{1}{2}(\sum_i u_{ij}^2 + \sum_k v_{jk}^2) + b_j} \tag{19}$$

We can get an approximate fixed value in convergence for iteration. Suppose the multiplicative updates stop after t iterations with parameters from Table 1, the overall computational complexity for ABNMF will be $O(tn^2c + n^2)$. A relatively small initial c will save running time of the algorithm.

Table 1. Parameters used in complexity analysis

Parameters	Description
n	Number of network nodes
c	Number of initial communities count
β_j	Paraments of communities number
a_j	Hyper-hyperparaments a
b_j	Hyper-hyperparaments b

3.3 Determination of Overlapping Community Number K and Threshold δ

In regular NMF methods for clustering, the object factorized dimension K should be given. But in community detection situation, we just know the relation of nodes without prior information of community number K, and it's hard to count out the suitable number. If K is too small, some communities will be very large and the model can not be fitted well. On contrary, If K is too large, we can not catch the group character effectively from an entire network and occur into overfitting. We need to find K with a appropriate solution between network fineness and overfitting.

To solve this problem, we propose a statistical *shrinkage* method in a Bayesian framework to find the number of communities and build a model selection method based on Automatic Relevance Determination [31,45]. In ABNMF, we principally iterate out v_{jk} with gradual change, and the prior will try to promote a *shrinkage* to zero of v_{jk} with a rate constant proportional to β_j. A large β_j represents a belief that the half-normal distribution over v_{jk} has small variance, and hence v_{jk} is expected to get close to zero. We can see the priors and the likelihood function (quantifying how well we explain the data) are combined with the effect that columns of V which have little effect in changing how well we explain the observed data will shrink close to zero. We can effectively estimate the communities number K by computer the non-zero column number from V with initial rank c.

In overlapping fuzzy detection, a sparse or dense network may have different overlapping degree. A dense network may contain more communities overlapped. Network density p describes the portion of the potential connections in a network that are actual connections. Every node potentially has a basic probability p to connect with rest nodes in a network, regardless of whether or not they actually connect:

$$p = \frac{2 * |E|}{n * (n - 1)} \tag{20}$$

There is a fact that nodes shared multiple community memberships receive multiple chances to create a link in overlapping assumption. We may assume each overlapping sub-community is larger than a potential network, that refer every v_{ij} will large than one fixed threshold in each network. Yang *et al.* [52] suggest that the threshold value can be $\delta = \sqrt{-log(1-p)}$ to achieve good performance. Note that this process adaptively generates an increasing relationship between edge probability and the number of shared communities.

3.4   Performance Comparisons in Different Networks

We compare our algorithm with other four popular overlapping community detection methods. Five algorithms are listed below:

1. CFinder tries to find overlapping dense groups of nodes in networks, and is based on method Clique Percolation Method (CPM) [38].
2. COPRA (Community Overlap PRopagation Algorithm) is based on the label propagation technique for finding overlapping community structure in large networks [12].
3. OSLOM (Order Statistics Local Optimization Method) locally optimizes the statistical significance information of a cluster with respect to random fluctuation with Extreme and Order Statistics [24].
4. LCM (Link Communities Method) organizes community structures spanning inner-city to regional scales while maintaining pervasive overlap, and builds blocks that reveal overlap and hierarchical organization in networks [2].
5. ABNMF (Adaptive Bayesian Non-negative Matrix Factorization) with Poisson likelihood. Its overlapping threshold is related with network density.

We run OSLOM, LFM and ABNMF in different six network datasets without groundtruth to evaluate its communities number, overlap fraction and modularity. Then we generate a synthetic network with 5000 nodes in different overlap fraction [22]. The details of experiments are stated below:

(1). In ABNMF methods, we select 10 different initial communities count c and apply 10 independent experiments. Every experiment iterates for 500 times.
(2). We test the ABNMF method in Email network [14] to evaluate the performance with different initial dimension number of c in Table 1.
(3). We use six different size and different character networks to compare communities number, overlap fraction and Modularity [35]. **Football** (American College football), **Email** (Email network of University at Rovira i Virgili in Tarragona, Spain), and **PGP** (Pretty Good Privacy communication network) [9,13,14] are social networks. **Erdos** (Collaboration network with famous mathematician Erdos) and **Cmat** (Condensed matter collaborations 2003) [4,34] are collaborative networks. **Metabolic** (Metabolic Network) [16] is biological network.
(4). We use Omega Index [6,33] to evaluate overlapping communities detecting performance with benchmark in a synthetic network.

Modularity has widely used to measure the strength of non-overlapping or overlapping community structure found by community detection methods. In Eq. (21), A_{ij} is the adjacency matrix, and k_i, k_j are node degree of i, j. $\delta(c_i, c_j)$ is probability of having a link between i and j in the null model are weighted by the belonging of i and j to the same community, since $\delta(c_i, c_j)$ is equal to 1 only when i and j belong to the same community, and it is 0 otherwise.

$$Q_{ov} = \frac{1}{n} \sum_{i,j \in V} \left[A_{ij}\delta(c_i, c_j) - \frac{k_i k_j}{2n}\delta(c_i, c_j) \right] \tag{21}$$

The Omega Index can evaluate the extent of two different solutions for overlapping communities in which each pair of nodes is estimated to share same community:

$$Omega(C_1, C_2) = \frac{\sum_{j=0}^{min(J,K)} \frac{A_j}{N} - \sum_{j=0}^{min(J,K)} \frac{N_{j1}N_{j2}}{N^2}}{1 - \sum_{j=0}^{min(J,K)} \frac{N_{j1}N_{j2}}{N^2}} \tag{22}$$

where J and K represent the maximum number of communities in which any pair of nodes appears together in solution C_1 and C_2, respectively, A_j is the number of the pairs agreed by both solutions to be assigned to number of community j, and N is the number of pairs of nodes. N_{j1} is the total number of pairs assigned to number of communities j in solution C_1, and N_{j2} is the total number of pairs assigned to number of communities j in solution C_2.

Table 2. Overlapping community number K with different initial c in ABNMF

	c	K	O	Q_{ov}
1	23	22	0.3600	0.6876
2	30	26	0.3668	0.6887
3	52	30	0.3772	0.6975
4	76	34	0.3768	0.6960
5	114	35	0.3862	0.7058
6	227	36	0.3845	0.7063
7	378	35	0.3846	0.7064
8	567	37	0.3900	0.7133

We run ABNMF method to test the impact of different initial communities numbers c in Email network. Table 2 lists the different result K, relevant overlap fraction(O) and Modularity(Q_{ov}). ARD is effective well on features extraction with large initial c, and contractive communities count K is around 30 in **Email** network. We can find that different c has weak influence for O and Q_{ov} results when c is set from 567 to 52. In ABNMF, we choose the initial count c from 1/5 to 1/10 of total nodes n to keep the performance of algorithm and keep the operational efficiency.

3.5   Overlapping Community Detection in Different Network

On American Football Game real network with 115 nodes and 613 edges, we run ABNMF for case study and the visualization of our found overlapping community structure is shown in Fig. 1, where same color nodes are allocated into same overlapping community. From Fig. 1, we can see that our proposed method ABNMF automatically finds 10 strong sense communities which are gathered by crisp clustering, and most of the football teams are correctly assigned into their corresponding communities in our found overlapping community structure. Moreover, it is very interesting to note that our proposed method ABNMF detects 32 overlapping nodes in different communities in total, in which each overlapping node has two different colors indicating different communities the node belongs to. This is because, besides against other football teams in the same conference, these football teams corresponding to the overlapping nodes also frequently play many games against football teams in other conferences. Therefore, we can see that our proposed method ABNMF has a good performance in detecting overlapping community structures in this real world social network.

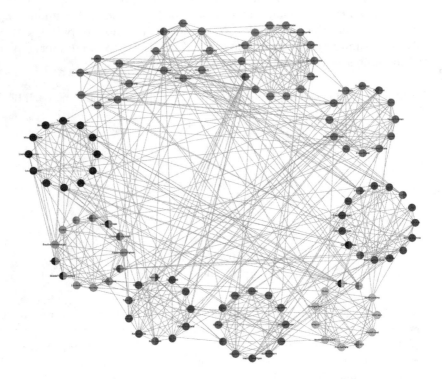

Fig. 1. Overlapping communities of Football network obtained by ABNMF.

We select 6 popular networks with different size, and compare community number(K), overlap fraction(O) and overlap modularity(Q_{ov}) in OSLOM, LCM and ABNMF. We can find from Table 3 that, our method ABNMF can effectively find overlapping community number and is highly close to results of OSLOM and LFM. ABNMF detects much dense communities in overlap fraction and achieves high overlapping modularity than other two methods. In ABNMF with

Multiplicative Update Rules [27], we achieve good performance in large **PGP** and **Cmat** networks both of which have more than ten thousands of nodes. We may also combine with Projected Gradient [28] method or Block Gradient Descent method [20] to solve Eq. (13) in larger datasets with millions of nodes.

Table 3. Overlapping community detection comparison on different networks

Network	Nodes	OSLOM			LCM			ABNMF		
		K	O	Q_{ov}	K	O	Q_{ov}	K	O	Q_{ov}
Football	115	9	0.1956	0.6032	11	0.2134	0.5992	10	0.2444	**0.6609**
Metabolic	453	32	0.4347	0.4212	31	0.4525	0.4678	27	0.5055	**0.6919**
Email	1133	28	0.2567	0.5796	27	0.2754	0.5821	26	0.3766	**0.6982**
Erdos	6927	77	0.2765	0.7187	81	0.2897	0.6837	73	0.3098	**0.8203**
PGP	10680	233	0.4688	0.8782	230	0.478	0.8843	227	0.4944	**1**
Cmat	27519	486	0.356	0.7216	483	0.4121	0.7255	475	0.534	**0.7340**

We evaluate the performance of our proposed algorithm on the LFR synthetic networks with benchmark, and compare with other four overlapping community detection algorithms. The LFR (Lancichinetti-Fortunato-Radicchi) benchmark [22] provides a class of artificial networks in which both the degrees of the nodes and the sizes of the communities follows power laws the same as many real-world networks. Here, we adopt a LFR benchmark with 5000 nodes respectively from the benchmark generator source code[1] in our experiment:

benchmark -N 5000 -k 10 -maxk 30 -mu 0.1 -minc 10 -maxc 50 -on 50 -om 2

In this LFR benchmark, we set the average degree of nodes $davg = 10$, the maximum degree of nodes $dmax = 30$, the minimum community size $minc = 10$, the maximum community size $maxc = 50$, the exponents of the power law of the community size distribution $t_1 = 1$, the exponents of the power law of the community size distribution $t_2 = 2$, the overlapping nodes in the entire network $on = 50$, and the number of communities that each overlapping node belongs to $om = 2$. Moreover, we define the mixing parameter μ as the average percentage of edges that connect a node to those in other communities which indicates that every node shares a fraction $(1 - \mu)$ edges with other nodes in its community and a fraction μ edges with nodes outside its community. The network community structure will be weakened by increasing μ.

Five algorithms are executed on the LFR benchmark network, and the average Omega Index is used to measure the similarities between the known community structure and the obtained resultant community structure by these algorithms. The results of different algorithms in the LFR networks are shown in Table 4.

It can be seen that all these five algorithms perform well and our proposed ABNMF algorithm has slightly better performance comparing with the other

---

[1] https://sites.google.com/site/santofortunato/inthepress2.

Table 4. Omega index comparison on LFR 5000 network

Overlap fraction	CPM	COPRA	OSLOM	LPM	ABNMF
0.05	0.86	0.86	0.86	0.86	**0.89**
0.1	0.83	0.855	0.855	0.855	**0.88**
0.15	0.81	0.85	0.85	0.83	**0.86**
0.2	0.75	0.84	0.84	0.81	**0.86**
0.25	0.6	0.82	0.82	0.8	**0.85**
0.3	0.47	0.82	0.83	0.8	**0.84**
0.35	0.45	0.79	0.82	0.8	**0.83**
0.4	0.4	0.77	0.81	0.8	**0.83**
0.45	0.38	0.74	0.8	0.79	**0.82**
0.5	0.32	0.71	0.79	0.75	**0.81**
0.55	0.3	0.64	0.6	0.74	**0.8**
0.6	0.22	0.62	0.62	0.73	**0.77**
0.65	0.2	0.1	0.14	0.58	**0.7**

four algorithms when the value of overlap fraction μ is small on the LFR network. Moreover, as the value of μ increasing, the performance of our proposed algorithm does not degrade rapidly as shown in Table 4. Therefore, our proposed algorithm has a good ability to detect overlapping community structures in complex networks no matter whether they have dense or sparse overlapping structure.

4   Conclusions

In this paper, we solve an overlapping community detection problem using Adaptive Bayesian NMF. We propose a model that considerate Bayesian inference process with Poisson model into NMF, and derive the updating rules and conduct experiments to valid our model. We also apply Automatic Relevance Determination method with sparse constrain to learn the community count of a network, and compare the detection impact of different initial community rank. At last, we adaptively select a most proper value related to network density as overlapping threshold for mixture coefficient matrix. Our method can be applied in real network data without any given information, and achieves good performance than other overlapping community detection methods.

Acknowledgments. This work was supported by NSFC (No. 61272247, 61533012, 61472075), the Basic Research Project of "Innovation Action Plan" (16JC1402800), the Major Basic Research Program (15JC1400103) of Shanghai Science and Technology Committee, and the Arts and Science Cross Special Fund (13JCY14) of Shanghai JiaoTong University.

References

1. Adamcsek, B., Palla, G., Farkas, I.J., Derényi, I., Vicsek, T.: Cfinder: locating cliques and overlapping modules in biological networks. Bioinformatics **22**(8), 1021–1023 (2006)
2. Ahn, Y.Y., Bagrow, J.P., Lehmann, S.: Link communities reveal multiscale complexity in networks. Nature **466**(7307), 761–764 (2010)
3. Amelio, A., Pizzuti, C.: Overlapping community discovery methods: a survey. In: Gündüz-Öğüdücü, Ş., Etaner-Uyar, A.Ş. (eds.) Social Networks: Analysis and Case Studies. LNSN, pp. 105–125. Springer, Vienna (2014). doi:10.1007/978-3-7091-1797-2_6
4. Batagelj, V., Mrvar, A.: Some analyses of Erdos collaboration graph. Soc. Netw. **22**(2), 173–186 (2000)
5. Cemgil, A.T.: Bayesian inference for nonnegative matrix factorisation models. Comput. Intell. Neurosci. **2009**, 1–17 (2009)
6. Collins, L.M., Dent, C.W.: Omega: a general formulation of the rand index of cluster recovery suitable for non-disjoint solutions. Multivar. Behav. Res. **23**(2), 231–242 (1988)
7. Fevotte, C., Idier, J.: Algorithms for nonnegative matrix factorization with the beta-divergence. Neural Comput. **23**(9), 2421–2456 (2011)
8. Gama, F., Segarra, S., Ribeiro, A.: Overlapping clustering of network data using cut metrics, pp. 6415–6419. IEEE (2016)
9. Girvan, M., Newman, M.E.: Community structure in social and biological networks. Proc. Natl. Acad. Sci. **99**(12), 7821–7826 (2002)
10. Gopalan, P., Ruiz, F.J., Ranganath, R., Blei, D.M.: Bayesian nonparametric Poisson factorization for recommendation systems. In: AISTATS, pp. 275–283 (2014)
11. Gopalan, P.K., Gerrish, S., Freedman, M., Blei, D.M., Mimno, D.M.: Scalable inference of overlapping communities. In: Advances in Neural Information Processing Systems, pp. 2249–2257 (2012)
12. Gregory, S.: Finding overlapping communities in networks by label propagation. New J. Phys. **12**(10), 103018 (2010)
13. Guardiola, X., Guimera, R., Arenas, A., Diaz-Guilera, A., Streib, D., Amaral, L.: Macro-and micro-structure of trust networks. arXiv preprint arXiv:cond-mat/0206240 (2002)
14. Guimera, R., Danon, L., Diaz-Guilera, A., Giralt, F., Arenas, A.: Self-similar community structure in a network of human interactions. Phys. Rev. E **68**(6), 065103 (2003)
15. He, Y.C., Lu, H.T., Huang, L., Shi, X.H.: Non-negative matrix factorization with pairwise constraints and graph Laplacian. Neural Process. Lett. **42**(1), 167–185 (2015)
16. Jeong, H., Tombor, B., Albert, R., Oltvai, Z.N., Barabási, A.L.: The large-scale organization of metabolic networks. Nature **407**(6804), 651–654 (2000)
17. Jin, D., Wang, H., Dang, J., He, D., Zhang, W.: Detect overlapping communities via ranking node popularities. In: Thirtieth AAAI Conference on Artificial Intelligence (2016)
18. Jin, D., Yang, B., Baquero, C., Liu, D., He, D., Liu, J.: A Markov random walk under constraint for discovering overlapping communities in complex networks. J. Stat. Mech: Theor. Exp. **2011**(05), P05031 (2011)
19. Kaganovsky, Y., Han, S., Degirmenci, S., Politte, D.G., Brady, D.J., O'Sullivan, J.A., Carin, L.: Alternating minimization algorithm with automatic relevance determination for transmission tomography under poisson noise. SIAM J. Imaging Sci. **8**(3), 2087–2132 (2015)

20. Kim, J., He, Y., Park, H.: Algorithms for nonnegative matrix and tensor factorizations: a unified view based on block coordinate descent framework. J. Global Optim. **58**(2), 285–319 (2014)

21. Kucukelbir, A., Ranganath, R., Gelman, A., Blei, D.: Automatic variational inference in stan. In: Advances in Neural Information Processing Systems, pp. 568–576 (2015)

22. Lancichinetti, A., Fortunato, S.: Benchmarks for testing community detection algorithms on directed and weighted graphs with overlapping communities. Phys. Rev. E **80**(1), 016118 (2009)

23. Lancichinetti, A., Fortunato, S., Kertész, J.: Detecting the overlapping and hierarchical community structure in complex networks. New J. Phys. **11**(3), 033015 (2009)

24. Lancichinetti, A., Radicchi, F., Ramasco, J.J., Fortunato, S.: Finding statistically significant communities in networks. PloS one **6**(4), e18961 (2011)

25. Lee, C., Reid, F., McDaid, A., Hurley, N.: Detecting highly overlapping community structure by greedy clique expansion. arXiv preprint arXiv:1002.1827 (2010)

26. Lee, D., Seung, H.: Algorithms for non-negative matrix factorization. In: Advances in Neural Information Processing Systems, vol. 13 (2001)

27. Lee, D., Seung, H., et al.: Learning the parts of objects by non-negative matrix factorization. Nature **401**(6755), 788–791 (1999)

28. Lin, C.J.: Projected gradient methods for nonnegative matrix factorization. Neural Comput. **19**(10), 2756–2779 (2007)

29. Macropol, K., Can, T., Singh, A.K.: Rrw: repeated random walks on genome-scale protein networks for local cluster discovery. BMC Bioinf. **10**(1), 1 (2009)

30. Meena, J., Devi, V.S.: Overlapping community detection in social network using disjoint community detection. In: 2015 IEEE Symposium Series on Computational Intelligence, pp. 764–771. IEEE (2015)

31. Mørup, M., Hansen, L.K.: Automatic relevance determination for multi-way models. J. Chemometr. **23**(7–8), 352–363 (2009)

32. Mørup, M., Schmidt, M.N.: Bayesian community detection. Neural Comput. **24**(9), 2434–2456 (2012)

33. Murray, G., Carenini, G., Ng, R.: Using the omega index for evaluating abstractive community detection. In: Association for Computational Linguistics, pp. 10–18 (2012)

34. Newman, M.E.: Scientific collaboration networks. i. network construction and fundamental results. Phys. Rev. E **64**(1) (2001). 016131

35. Nicosia, V., Mangioni, G., Carchiolo, V., Malgeri, M.: Extending the definition of modularity to directed graphs with overlapping communities. J. Stat. Mech: Theor. Exp. **2009**(03) (2009). P03024

36. Paisley, J., Blei, D., Jordan, M.I.: Bayesian nonnegative matrix factorization with stochastic variational inference. In: Handbook of Mixed Membership Models and Their Applications. Chapman and Hall/CRC, Boca Raton (2014)

37. Pakhira, M.K., Bandyopadhyay, S., Maulik, U.: Validity index for crisp and fuzzy clusters. Pattern Recogn. **37**(3), 487–501 (2004)

38. Palla, G., Barabási, A.L., Vicsek, T.: Quantifying social group evolution. Nature **446**(7136), 664–667 (2007)

39. Palla, G., Derényi, I., Farkas, I., Vicsek, T.: Uncovering the overlapping community structure of complex networks in nature and society. Nature **435**(7043), 814–818 (2005)

40. Psorakis, I., Roberts, S., Ebden, M., Sheldon, B.: Overlapping community detection using bayesian non-negative matrix factorization. Phys. Rev. E **83**(6). 066114 (2011)
41. Rabbany, R., Zaïane, O.R.: Generalization of clustering agreements and distances for overlapping clusters and network communities. Data Min. Knowl. Disc. **29**(5), 1458–1485 (2015)
42. Schmidt, M.N., Laurberg, H.: Nonnegative matrix factorization with Gaussian process priors. Comput. Intell. Neurosci. **2008**, 3 (2008)
43. Shankar, D.S., Bhavani, S.D.: Consensus clustering approach for discovering overlapping nodes in social networks. In: Proceedings of the 3rd IKDD Conference on Data Science, p. 21. ACM (2016)
44. Shi, X., Lu, H., He, Y., He, S.: Community detection in social network with pairwisely constrained symmetric non-negative matrix factorization. In: Proceedings of the 2015 IEEE/ACM International Conference on Advances in Social Networks Analysis and Mining 2015, ASONAM 2015, pp. 541–546. ACM, New York (2015)
45. Tan, V.Y.F., Fevotte, C.: Automatic relevance determination in nonnegative matrix factorization with the beta-divergence. IEEE Trans. Pattern Anal. Mach. Intell. **35**(7), 1592–1605 (2013)
46. Wang, W., Jiao, P., He, D., Jin, D., Pan, L., Gabrys, B.: Autonomous overlapping community detection in temporal networks: a dynamic bayesian nonnegative matrix factorization approach. Knowl.-Based Syst. **110**, 121–134 (2016)
47. Whang, J.J., Gleich, D.F., Dhillon, I.S.: Overlapping community detection using seed set expansion. In: Proceedings of the 22nd ACM International Conference on Conference on Information Knowledge Management - CIKM 2013. Association for Computing Machinery (ACM) (2013)
48. Wu, P., Fu, Q., Tang, F.: Social community detection from photo collections using Bayesian overlapping subspace clustering. In: Lee, K.-T., Tsai, W.-H., Liao, H.-Y.M., Chen, T., Hsieh, J.-W., Tseng, C.-C. (eds.) MMM 2011. LNCS, vol. 6524, pp. 57–64. Springer, Heidelberg (2011). doi:10.1007/978-3-642-17829-0_6
49. Wu, Z.H., Lin, Y.F., Gregory, S., Wan, H.Y., Tian, S.F.: Balanced multi-label propagation for overlapping community detection in social networks. J. Comput. Sci. Technol. **27**(3), 468–479 (2012)
50. Xie, J., Kelley, S., Szymanski, B.K.: Overlapping community detection in networks: the state-of-the-art and comparative study. ACM Comput. Surv. **45**(4), 43:1–43:35 (2013)
51. Xie, J., Szymanski, B.K.: Towards linear time overlapping community detection in social networks. In: Tan, P.-N., Chawla, S., Ho, C.K., Bailey, J. (eds.) PAKDD 2012. LNCS (LNAI), vol. 7302, pp. 25–36. Springer, Heidelberg (2012). doi:10.1007/978-3-642-30220-6_3
52. Yang, J., Leskovec, J.: Overlapping community detection at scale: a nonnegative matrix factorization approach. In: Proceedings of the Sixth ACM International Conference on Web Search and Data Mining, pp. 587–596. ACM (2013)
53. Zhang, H., King, I., Lyu, M.R.: Incorporating implicit link preference into overlapping community detection. In: AAAI, pp. 396–402 (2015)
54. Zhang, Y., Yeung, D.Y.: Overlapping community detection via bounded nonnegative matrix tri-factorization. In: Proceedings of the 18th ACM SIGKDD International Conference on Knowledge Discovery and Data Mining - KDD 2012. Association for Computing Machinery (ACM) (2012)
55. Zhubing, L., Jian, W., Yuzhou, L.: An overview on overlapping community detection. In: 2012 7th International Conference on Computer Science and Education (ICCSE), pp. 486–490. IEEE (2012)

A Unified Approach for Learning Expertise and Authority in Digital Libraries

B. de La Robertie[1]($\boxtimes$), L. Ermakova[2,3], Y. Pitarch[1],
A. Takasu[4], and O. Teste[1]

[1] Université de Toulouse, IRIT UMR5505, 31071 Toulouse, France
`baptiste.robertie@gmail.com`, {`o.teste,y.pitarch`}`@irit.fr`
[2] Université de Lorraine, Nancy, France
`liana.ermakova.@irit.fr`
[3] LISIS, Université de Paris-Est Marne-la-Vallée, Champs-sur-Marne, France
[4] National Institute of Informatics, 2-1-2 Hitotsunashi, Chiyoda, Tokyo, Japan
`takasu@nii.ac.jp`

Abstract. Managing individual expertise is a major concern within any industrial-wide organization. If previous works have extensively studied the related expertise and authority profiling issues, they assume a semantic independence of these two key concepts. In digital libraries, state-of-the-art models generally summarize the researchers' profile by using solely textual information. Consequently, authors with a large amount of publications are mechanically fostered to the detriment of less prolific ones with probably higher expertise. To overcome this drawback we propose to merge the two representations of expertise and authority and balance the results by capturing a mutual reinforcement principle between these two notions. Based on a graph representation of the library, the expert profiling task is formulated as an optimization problem where latent expertise and authority representations are learned simultaneously, unbiasing the expertise scores of individuals with a large amount of publications. The proposal is instanciated on a public scientific bibliographic dataset where researchers' publications are considered as a source of evidence of individuals' expertise and citation relations as a source of authoritative signals. Results from our experiments conducted over the Microsoft Academic Search database demonstrate significant efficiency improvement in comparison with state-of-the-art models for the expert retrieval task.

Keywords: Expert finding · Link analysis · Optimization · Digital libraries

1 Introduction

Keeping track and managing individuals' expertise in industrial-wide organizations or public scientific repositories is a major concern. Motivated by expertise capitalization, skill mining, or knowledge sharing purposes, strong interests on

S. Candan et al. (Eds.): DASFAA 2017, Part II, LNCS 10178, pp. 354–368, 2017.
DOI: 10.1007/978-3-319-55699-4_22

the expert finding task rapidly spawned both private and public researches [25]. For example, the Experscape platform[1], by mining the US National Library of Medicine and the National Institutes of Health databases[2], provides search functionalities to seek experts according to 26,000 topics (e.g., Alzheimer Disease, Arthritis, Brain Tumor) and geographic features (country, region, city, and institution). The system AMiner[3], resting on DBLP[4] and ACM[5], also provides search functionalities for the Computer Science field and capitalizes more than 100 million researchers and 200 million publications. Microsoft Academic Search[6] and more recently ResearchGate[7] also constitute popular examples exploiting digital libraries for profiling and discovering goals.

While expert profiling and retrieval attract significant interest by the scientific community, unified approaches that consider both expertise and quality models receive too little attention. Indeed, state-of-the-art models generally formulate the expert finding problem as a summarization task where text data, essentially associated to individuals, are used to model knowledge and expertise [1,6,20,22]. Intranet documents, reports, project descriptions, mails, or publications are used as a source of information whereas tags, key words, or flat topics are extracted to link knowledge and experts [6]. *In fine*, candidates are then ranked according to the probability of being an expert given a particular topic. The underlying matching process, generally based on standard information retrieval techniques, ignores quality or authoritative criteria. Therefore, authors with larger amounts of productions are promoted, biasing the final ranking over the candidates.

To illustrate this downside, let us consider the following example. Let $\mathcal{R} = \{r_1, r_2\}$ and $\mathcal{A} = \{a_1, a_2, a_3, a_4, a_5\}$ be 2 sets of 2 researchers and 5 articles respectively. The authoring relation between researchers and articles is given in Fig. 1. Let θ_1 and θ_2 be the profiles associated to the researchers r_1 and r_2 respectively. We consider a language model formalism for summarization. Given a query q, researchers are ranked according to the probability $p(q|\theta_i) = \prod_{w \in q} p(w|\theta_i)$. Using Bayes' rules, it holds $p(w|\theta_i) = \sum_{a_j \in \mathcal{A}} p(w|a_j)p(r_i|a_j)$, making the value of $p(w|\theta_i)$ increasing with the number articles authored by a researcher r_i. For example, given a topic query $q = \{w_1\}$ where the term probability $p(w_1|a_j)$ is the same for all articles a_j (see Fig. 1), $p(q|\theta_1) = 0.4 < p(q|\theta_2) = 0.6$. Thus, the researcher r_2 is promoted with regard to topic w_1. However, if the articles authored by r_1 are much more cited than those authored by r_2, one will probably rank researcher r_1 higher than r_2. This example motivates the need of considering quality or authority signal in a profile summarization task.

---

[1] http://expertscape.com.
[2] https://www.ncbi.nlm.nih.gov.
[3] https://aminer.org/.
[4] http://dblp.uni-trier.de/.
[5] http://dl.acm.org/.
[6] http://academic.research.microsoft.com/.
[7] https://www.researchgate.net.

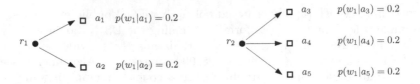

Fig. 1. Illustration of the drawback of state-of-the-art expert profiling methods. Without considering any quality or authority signal, researcher r_2 will be ranked higher than researcher r_1 for similar topic queries.

In this work, we tackle this drawback by assuming that expertise and authority influence each other. We assume that (1) experts are sources of knowledge (associated publications contain proofs of expertise), (2) experts are authoritative (associated relations contain proofs of authority) and (3) these two components, being two sides of the same coin, should have a common representation. To capture this mutual reinforcement principle, we formulate the expert profiling task as an optimization problem where both authority and expertise vectors are unified and simultaneously learned. As confirmed by the experiments, such a representation for expert profiling significantly improves the expert finding phase. To summarize, our contributions are as follows:

1. We provide a unified model capturing both individuals' expertise and authority based on an heterogeneous graph representation of digital libraries;
2. We formulate the expert profiling task as an optimization problem learning both latent topics and authoritative signals in a single process;
3. We conduct experiments over a representative subset of the Microsoft Academic Search (MAS) database and show a significant improvement as compared to state-of-the-art methods.

The rest of the paper is organized as follows: Sect. 2 discusses the related work. Section 3 formally describes our model. Section 4 discusses the experiments. Finally, concluding remarks are drawn in Sect. 5.

2   Related Work

Our model relates to both expertise and authority fields. We first provide an overview of these two research topics and then motivate the need of a unified approach for modelling authority and expertise using a single representation.

Expertise profiling and retrieval models. Historical approaches related to expert finding manually store individuals' skills in knowledge bases [7]. The distinction between the representation of knowledge and data is manually made on the basis of reports, scientific articles or employee pages but presents considerable maintenance costs. Craswell et al. [6] first propose an automatic solution, assimilating an employe's profile to the concatenate list of his/her related documents. Thus, given a topic query, standard information retrieval techniques are used to retrieve the top-n experts. State-of-the-art models generally make use of language or topic models to summarize an individual profile [3]. In this category, extensive works have been done by Balog et al. [1,2,19] by proposing a generative probabilistic modeling framework for expert profiling. Standard Information Retrieval techniques are adapted for that task, estimating a probability of a candidate being an expert in a particular topic. For a given topic query q, candidates are ranked according to the probability $p(q|\theta_{ca})$ where the representation of a candidate θ_{ca} is generally performed using a multinomial probability distribution over a vocabulary (i.e., $p(q|\theta_{ca}) = \prod_{t \in q} p(t|\theta_{ca})^{n(t,q)}$). In [20], the expert profiling task is formulated as a tagging problem where features extracted from various sources are used to model an employee. In particular, authored enterprise documents, discussion lists, and enterprise search click-through data are used to learn a tag probability of being a good descriptor for a particular employee. In [21], the web user profiling problem is tackled on the basis of topic modelling, without considering authority signal. In all these previous works, only the textual content is used for expert profiling which constitutes the introduced major drawback. Yang et al. [24] integrate authoritative features using the PageRank scores of researchers. Nevertheless pre-computed scores and some other language model features are aggregated *a postiori*, feeding a feature vector for training. The proposition cannot capture any cyclic relation between the two concepts. Deng et al. [9] construct a weighted language model to take into consideration not only the relevance between a query and documents but also the importance of the documents. Only the number of citations of an article is integrated as a prior probability. Thus, the notions of authority and expertise in the literature are generally separated and do not influence each other.

Graph-based authoritative models. In organizational networks, graph-based models, largely based on random walk [23], are widely used to estimate individual authority. In this field, extensive researches have demonstrated strong correlations between centrality and authority [10,16,26,27]. The famous PageRank algorithm proposed by L. Page et al. [18] and later the Topic-sensitive Pagerank [11] have proven the value of the citation graph for web pages. Campbell et al. [5] exploit network patterns in email communication graphs to discover experts and show that a HITS-like algorithm [14] performs better than content-based approaches for the expert finding task. The co-author graph on Wikipedia has demonstrated to carry out authority signals and help in identifying authoritative users producing high quality content [8]. Jurczyk et al. [13] also make used of a HITS-based algorithm to estimate the authority of Question and Answering platforms' members and confirm the robustness of such approach. Finally, Takasu et al. [12] employ both co-author and citation graphs to discriminate researchers' importance rating. State-of-the-art approaches demonstrate the efficiency of graph-based authority models but also the lack of unified approaches considering expertise.

Discussion. Propositions considering both expertise and authority signals have received too little attention. Unified approaches widely compute two representations then aggregate them *a posteriori* preventing from capturing a mutual reinforcement principle. Unlike previous well-established methods, we propose to formulate the expert profiling problem as a summarization task where both expertise and authority concepts are merged into a single representation. An individual is considered as an expert not only if he/she authors some articles in a particular field but also if the authored articles are credited by the community. Moreover, the unified representation enables us to strengthen the scores of poorly represented dimensions of a researcher's profile who would have authored few but highly cited articles. Conversely, this unified representation enables to balance the scores of over-represented dimensions of a researcher's profile who would have abundantly written poor quality articles. To the best of our knowledge, our proposition is the first approach connecting the two key concepts for the expertise retrieval task.

3   Model

A digital library can naturally be represented by an heterogeneous directed graph, denoted by G, where the sets of nodes U correspond to the different entities in the library and the sets of edges V to the different relations defined by the platform. In this work, G encodes the sets of articles, researchers and words in addition to the authoring and citing relations. Unlike state-of-the-art models, we assume that individuals' expertise and authority share a common representation in $\mathbb{R}^K$, encoding to what extent a researcher is an expert *and* he/she is authoritative in a particular field. We suppose that the content of the articles contains proof of expertise and the relative locations of the articles in the citation graph constitute proof of quality of the articles. Thus, we propose to compute the profile of a researcher as an aggregation of the estimated expertise and quality of the authored articles. Section 3.1 introduces the general notations for representing a digital library. Section 3.2 details the proposed unified representation for capturing the cyclical relation between expertise and authority. Finally, the objective function to learn expertise and authority simultaneously is detailed in Sect. 3.3.

3.1   Platform Representation

Let $\mathcal{R} = \{r_i\}_{1 \leq i \leq N}$, $\mathcal{A} = \{a_j\}_{1 \leq j \leq M}$ and $\mathcal{W} = \{w_s\}_{1 \leq s \leq W}$ be the sets of researchers, articles and words respectively. We define the heterogeneous graph $G = (U, V)$ over the set of nodes $U = \mathcal{R} \cup \mathcal{A} \cup \mathcal{W}$ and relations $V = V_{RA} \cup V_{AA} \cup V_{AW}$. In particular, V_{RA} is an authoring relation associating each researcher to the articles he/she authored. V_{AA} is a citing relation. Finally, V_{AW} associates each article to the set of words it contains. Corresponding adjacency matrices are denoted by X_{RA}, X_{AA} and X_{AW} respectively. Note that X_{RA} and X_{AA} are binary matrices (i.e., $X_{RA}(i, j) = 1$ if researcher r_i has authored article a_j, 0 otherwise). In this work, X_{AW} contains TF-IDF weights. Notations are summarized in Fig. 2. Latent representations of expertise and authority are detailed in the next section.

$$x_{AA}$$

$$x_{RA} \qquad \qquad x_{AW}$$

$$\mathcal{R} \longrightarrow \mathcal{A} \longrightarrow \mathcal{W}$$

$$z_R \qquad \qquad z_A \qquad \qquad z_W$$

Fig. 2. Graphical representation of a digital library.

3.2 Encoding Expertise and Authority

We propose to represent both expertise and authority in a single vector in $\mathbb{R}^K$ where the k-th dimension is aimed to estimate both expertise and authority of a particular entity in G for a latent topic k. Figure 3 illustrates the proposed formulation for a particular researcher and 4 topics. The mutual reinforcement principle between the expertise of the researcher and his/her authority in topic 3 figures out by dotted arrows. In order to unbias expertise scores associated to over-represented or under-represented dimensions, we propose to merge these two vectors. The proposed unified representation estimates without distinction both expertise and authority, balancing poor levels of expertise when the corresponding level of authority is high. In the following, we denote by $z_R \in \mathcal{M}_{N,K}$ the latent unified representation encoding both the expertise and the authority of the researchers. In particular, $z_R(i) \in \mathbb{R}^K$ is the expertise vector associated to researcher i and $z_R^k(i) \in \mathbb{R}$ reflects to what extent the researcher have expertise and is authoritative in topic k. Similarly, $z_A \in \mathcal{M}_{M,K}$ is the latent representation encoding the expertise and the authority of the articles.

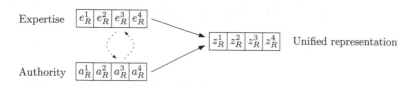

Expertise $\boxed{e_R^1\,|\,e_R^2\,|\,e_R^3\,|\,e_R^4}$

$\boxed{z_R^1\,|\,z_R^2\,|\,z_R^3\,|\,z_R^4}$ Unified representation

Authority $\boxed{a_R^1\,|\,a_R^2\,|\,a_R^3\,|\,a_R^4}$

Fig. 3. Illustration of the mutual reinforcement principle between the notions of expertise and authority using a toy example with 4 topics (left) and the proposed unified representation (right).

3.3 Problem Formulation

We capture the cyclic relation between expertise and authority and learn the introduced unified representation by minimizing the objective function $\mathcal{L}_\lambda(z_A, z_W)$ formulated by Eq. (1):

$$\mathcal{L}_\lambda(z_A, z_W) = \lambda\|X_{AW} - z_A z_W^T\|_F^2 + (1-\lambda)\|X_{AA}z_A - z_A\|_F^2 \qquad (1)$$
$$\text{s.t. } z_W > 0, z_A > 0$$

where $z_W \in \mathcal{M}_{W,K}$ is a latent matrix associating to each word a topic distribution, $||.||_F$ is the Frobenius norm, and $\lambda \in [0,1]$ is a user-parameter that controls the sensitivity of both criteria.

The first part of the objective function $||X_{AW} - z_A z_W^T||_F^2$ corresponds to a standard matrix factorization loss [15] aimed at learning latent topics from the articles content while the second part of the function $||X_{AA} z_A - z_A||_F^2$ is a slight variation of the PageRank formulation [18] applied on the citation matrix. Note that both parts share the proposed common unified representation z_A. In particular, for $\lambda = 1$, a standard non-negative matrix factorization problem is tackled over the article/vocabulary matrix. This standard expertise model, hereafter denoted as NMF, summarizes the articles' content ignoring quality signals. Conversely, for $\lambda = 0$, a PageRank-like algorithm, noted PR, is performed over the citing matrix and only the relative importance of the articles is estimated.

By gathering both objectives around the common variable z_A, we force the estimated authority (learned with PR) and expertise (learned with NMF) to influence each other during the optimization. As empirically shown in Sect. 4, authoritative features can help to improve the expertise retrieval phase, and conversely, expertise features can help to identify authoritative researchers. This mutual reinforcement principle between the notions of expertise and authority is the core of our proposed unified approach.

Finding the latent variables associated to the articles is equivalent to solve Eq. (2):

$$(z_A^*, z_W^*) = \underset{z_A, z_W}{\arg\min} \; \mathcal{L}_\lambda(z_A, z_W) \qquad (2)$$

Since the Frobenius norm is a convex function, standard gradient descent approaches can be used. In particular, we have:

$$\frac{\partial \mathcal{L}_\lambda}{\partial z_A} = 2\left((1-\lambda)D_{AA}(X_{AA} z_A) - \lambda(X_{AW} - z_A z_W^T)z_W\right)$$

$$\frac{\partial \mathcal{L}_\lambda}{\partial z_W} = -2\lambda(X_{AW} - z_A z_W^T)^T z_A$$

where $D_{AA} = \mathrm{diag}(X_{AA}\mathbb{1} - 1)$, or equivalently, $D_{AA}(i,i) = \sum_{1 \le j \le M} X_{AA}(i,j)$. It should be noted that since the parameters z_A and z_W have KM and KW decision variables respectively, the model $\theta = (z_A, z_W)$ defines a metric space in $\mathbb{R}^{K(M+W)}$. In practice, we solve Eq. (2) using the Limited-Memory BFGS [17] algorithm (L-BFGS), a quasi-Newton method for non-linear optimizations when the number of variables is high (more than 100 million in our case).

Finally, we naturally assimilate a researcher's profile to an aggregation of the obtained latent representation of his/her articles. By summing over the associated articles, we have:

$$z_R^* = X_{RA} z_A^* \qquad (3)$$

Therefore, we consider that the researcher r_i is more likely to be an expert in the topic k than the researcher r_j iff $z_R^{*k}(i) > z_R^{*k}(j)$.

4   Experiments

This section is dedicated to the presentation of our results. We evaluate the proposition along two main lines:

1. How well the proposed algorithm can be used to identify authoritative researchers in a digital library. In other words, to what extent expertise features can bring authoritative information.
2. How well the proposed solution can answer to the expert finding task by identifying experts in response to a particular topic query. In other words, to what extent authoritative features can help the expert profiling phase.

We first describe the dataset used for the experiments in Sect. 4.1. Then, Sect. 4.2 presents the protocol for evaluation. Competitors and evaluation metric are introduced in Sects. 4.3 and 4.4 respectively. Finally, quantitative and qualitative results are discussed in Sects. 4.5 and 4.6.

4.1   Data

For the evaluation, 3 data sources were merged to construct several labeled expertise graphs. The Microsoft Academic Search database[8] (MAS), the AMiner platform[9], the Core.edu portal[10], and the induced graphs are detailed thereunder.

Raw data. We made use of the digital library Microsoft Academic Search (MAS) for evaluation. The MAS portal is a semantic network providing a variety of metrics for the research community in addition to literature search. The portal has not been updated since 2013 but is still available online and contains valuable information about roughly 40 million articles and 9 million authors. For the evaluation, all articles and corresponding authors associated to the Computer Science community were crawled. Raw data, including articles titles and abstracts, stored in a relational database represents 4.1 Gb.

Quantitative evaluation. For quantitative evaluation, the AMiner portal was used. The platform provides a public list of $1,270$ experts in the computer science field according to 10 expertise domains from Boosting to Support Vector Machine. From this expert list, roughly 900 experts were retrieved in the MAS dataset to constitute a ground truth. For automatic evaluation purpose, a set of label vectors $\{\mathbf{y}_i\}_{i \leq N}$ with $\mathbf{y}_i \in \mathbb{B}^{10}$ is constructed. In particular, y_i^k is a binary label indicating if the researcher r_i is an expert in the field k ($y_i^k = 1$) or not ($y_i^k = 0$). Note that some researchers are considered as experts in different topics. The 10 considered topics are listed in Table 1.

---

[8] http://academic.research.microsoft.com/.
[9] https://aminer.org.
[10] http://www.core.edu.au/.

Qualitative evaluation. For qualitative evaluation, we made use of the Core.edu portal. The service provides assessments of major conferences in the Computer Science discipline. Standard labels, from A* for leading venues to C for conferences meeting minimum standards, are used to label the conferences. Specifically, 2,158 conferences published by the Core.edu portal were found in the MAS dataset. We used the associated labels to indirectly measure the articles quality.

Expertise graphs. To evaluate the capacity of the proposal to identify authoritative researchers, 10 expertise graphs were constructed using both previous sources of information. Given a topic k and the associated set of experts, an expertise graph G_k is constructed by iteratively adding in the set of nodes (a) the experts, (b) their co-authors, (c) the associated papers, (d) every citing and cited paper, and (e) each corresponding author. Moreover, to evaluate the capacity of the proposal to identify experts in a particular topic, a complete graph G merging the 10 previously defined expertise graphs is also constructed. Statistics of the different graphs used for the evaluation are summarized in Table 1.

Table 1. Statistics of the 11 graphs used for the experiments.

Expertise graph	Experts	Researchers	Articles
G_0 - Boosting	43	52 228	94 172
G_1 - Data Mining	221	86 786	243 071
G_2 - Information Extraction	72	36 880	80 983
G_3 - Intelligent Agents	28	36 323	60 246
G_4 - Machine Learning	52	37 277	69 025
G_5 - Language Processing	36	20 175	36 684
G_6 - Ontology Alignments	42	30 216	48 601
G_7 - Planning	13	22 809	32 710
G_8 - Semantic Web	274	81 039	244 855
G_9 - Support Vector Machine	70	33 448	60 319
G - All	851	131 303	1 427 317

4.2   Protocol

Preprocessing. The articles' content was processed using the Natural Language Toolkit[11] library for Python. Nouns were extracted from the abstracts and the titles of the articles and those appearing in more than 70% of the articles or in less than 20 articles were removed. From this preprocessing step, a vocabulary of roughly 5 000 words was obtained. Note that we voluntary restrained the size of the vocabulary for efficiency considerations and related sparseness problems.

---

[11] http://www.nltk.org/.

The remaining words were stemmed using the Lancaster Stemmer. We used TF-IDF weights to model the strength of the relations between words and articles. Thus, $X_{AV}(j, w)$ is the TF-IDF weight of the word w in the article a_j. Finally, to avoid full zero columns in the adjency matrix of the evaluations graphs, every researcher without any authored article and all articles that do not cite any paper were removed.

Optimization. The proposed objective function $\mathcal{L}_\lambda(z_A, z_V)$ was minimized using standard optimization packages for Python[12]. We made used of the Limited-Memory BFGS [17] algorithm (L-BFGS), a quasi-Newton method for non-linear optimizations handling many variables. In practice, the optimization spent roughly three days over the complete graph G. Since L-BFGS approximates the objective function locally and might return local optimums, several optimizations were performed in parallel for each value of λ. Moreover, we made the number k of latent topics vary for each run (from 5 to 100). Only the best runs according to the evaluation metric are presented.

Evaluation. We conducted two series of evaluations. The first one was associated to the authority evaluation while the second one focused on the expertise assessment.

1. We studied the capacity of the proposal to identify authoritative researchers. We wanted to show that considering textual features from articles content may help in identifying authoritative researchers. It should be noted that no reconciliation process between topic query and researchers' profile was performed. To this end, we operated as follows. For each latent topic k, the researchers were ranked by decreasing order of predicted scores z_R^{*k} and the model was evaluated, using the set of labels $\{\mathbf{y}_i^k\}_{i \leq N}$. We report here the best performances over the different discovered latent topics.
2. Secondly, we evaluated the capacity of the models to retrieve experts in response to a particular topic query. The 10 topic queries presented in Table 1 were used for evaluations over the graph G and the set of labels $\{\mathbf{y}_i^k\}_{i \leq N}$ was used as groundtruth. For each topic query q, researchers were ranked according to the vector of scores z_R^{*k} where k corresponds to the latent topic maximizing:

$$k = \arg\max_{k \leq K} \prod_{w \in q} z_W^{*k}(w)$$

where we assumed, for simplicity, that $z_W(w)$ is the entry line in the matrix z_W of the word $w \in \mathcal{W}$.

4.3   Competitors

The proposition, denoted below by **UA** (Unified Approach), was compared to the following state-of-the-art models:

---

[12] https://www.scipy.org/.

- **PR.** The proposal when $\lambda = 0$. It corresponds to a PageRank-like algorithm capturing the authority of the researchers through the quality of the articles they authored.
- **NMF.** The proposal when $\lambda = 1$. It is a standard non-negative matrix factorization approach capturing latent topics from the article/vocabulary matrix.
- **COS.** A standard Information Retrieval model assuming that the expertise score of a researcher for a topic query q is the cosine similarity between q and a researcher profile. To align with state-of-the-art approaches, a researcher profile was built from a concatenation of the authored articles. Both queries and authors were modeled using bag of words representations and TF-IDF weights.
- **LM.** The model proposed by Balog et al. [1] based on language model formalism. Given a query q, researchers were ranked according to the probability $p(q|\theta_{r_i}) = \prod_{w \in q} p(w|\theta_{r_i})$, where θ_{r_i} encodes the profile of the researcher r_i. In particular, $p(w|\theta_{r_i}) = \sum_{a_j \in \mathcal{A}} p(w|a_j) X_{RA}(i,j)$.
- **LMS.** A smoothed version of the former, also proposed by Balog et al. [1]. Probabilities were smoothed by the frequencies of the corresponding terms in the collection. Formally $\tilde{p}(w|a_j) = \alpha p(w|a_j) + (1 - \alpha)p(w|\mathcal{A})$. In our experiments, we set $\alpha = 0.5$.

4.4 Evaluation Metric

The standard classification metric AUC (Area Under the Curve) [4] was used to report the performance of the different classifiers. The metric estimates the probability of ranking a randomly chosen expert higher than a randomly chosen researcher in the final ranking by reporting the area under the ROC curve. Therefore, the closer to 1 the AUC, the better the classifier.

4.5 Quantitative Results

Results for the first set of experiments, associating to the evaluation of the authority, are summarized in Table 2. Results for the expertise assessment are given in Table 3.

Authority evaluation. We discuss here the results associated to the evaluation of the proposed method for identifying authoritative researchers in the different expertise graphs. Interestingly, we observe from Table 2 that for most of the expertise graphs there exists at least one configuration of the proposal that outperforms the PR method. Over the graph G, PR achieves 0.647 while the proposal reaches 0.661 for $\lambda = 0.1$. In general, values of λ around 0.2 improve the baseline of roughly 2%. Intuitively, these results confirm that experts constitute hubs in the different expertise graphs, relatively to the articles (they may write more articles than others) but they also form hubs regarding to the nodes associated to the vocabulary. In other words, the TF-IDF edge weights between nodes associated to articles and words, summarized in the discovered latent topics, indirectly bring authoritative information. This first important result suggests that representing the textual content of the articles in an expertise graph, in particular by considering words as nodes, can reinforce the discriminative process. It is not surprising that for $\lambda = 1$, although some results are not essentially deceptive, most of them are only slightly better than a random classifier. A single NMF approach, at least over the article/vocabulary matrix, does not suit well for authority modelling.

Table 2. Authority evaluation of the proposal using the AUC metric.

λ	PR	0.1	0.2	0.3	0.4	0.5	0.6	0.7	0.8	0.9	NMF
G_0	0.683	0.682	**0.693**	0.681	0.683	0.683	0.678	0.672	0.671	0.663	0.621
G_1	0.666	0.671	0.664	0.666	0.660	**0.672**	0.665	0.671	0.669	0.668	0.659
G_2	0.644	0.643	0.642	**0.653**	0.653	0.641	0.642	0.636	0.639	0.631	0.558
G_3	0.671	**0.684**	0.676	0.675	0.672	0.669	0.681	0.682	0.674	0.662	0.565
G_4	0.674	**0.680**	0.667	0.673	0.675	0.672	0.671	0.677	0.670	0.667	0.582
G_5	0.635	0.636	**0.644**	0.634	0.641	0.629	0.643	0.644	0.637	0.628	0.548
G_6	0.642	0.641	**0.651**	0.639	0.634	0.643	0.648	0.638	0.631	0.622	0.586
G_7	0.688	0.688	**0.694**	0.692	0.691	0.688	0.690	0.688	0.672	0.664	0.612
G_8	**0.667**	0.648	0.655	0.662	0.656	0.647	0.658	0.641	0.654	0.649	0.593
G_9	0.671	**0.674**	0.673	0.672	0.670	0.663	0.665	0.667	0.663	0.666	0.559
G	0.647	**0.661**	0.649	0.653	0.659	0.658	0.656	0.649	0.649	0.651	0.553

Expertise evaluation. Here are discussed the results associated to the expert finding task for the 10 topic queries presented in Table 1 over the graph G. Results associated to the PR method are not available since textual content is not taken into account by this approach and, therefore, matching between query and researchers' profile is not possible. It should be noted that results of the UA method were obtained by minimizing the proposed objective function for $\lambda = 0.2$, $K = 20$ and $|\mathcal{V}| = 5\,300$. Table 3, by reporting the AUC of the five competitors for each topic query, shows that on average, the proposal (UA) outperforms all the competitors (AUC ≈ 0.7), especially the strong baseline LMS (AUC ≈ 0.65). It means that the proposal is more likely to rank the experts higher than the competitors. Considering the queries individually, we observe quite important differences between the performances of LMS and UA. For example, UA is very efficient for retrieving the experts in the Boosting and Planning topic but is outperformed by LMS for the Intelligent Agents or Information Extraction fields. Such irregularities in the results might be explained by the quality of the latent topics and more particularly by the way we have performed the preprocessing step. Topics Boosting and Planning are relatively more specific than others and the associated clusters are easier to learn.

Table 3. Expertise evaluation of the competitors using the AUC metric.

Topic query	NMF	COS	LM	LMS	UA
Boosting	0.829	0.703	0.703	0.703	**0.842**
Data Mining	0.671	0.664	0.635	**0.682**	0.681
Information Extraction	0.607	0.601	0.676	**0.696**	0.623
Intelligent Agents	0.628	0.766	0.676	**0.771**	0.717
Machine Learning	0.745	0.622	0.553	0.635	**0.781**
Language Processing	0.464	0.488	0.492	0.487	**0.567**
Ontology Alignments	0.386	0.492	0.499	0.492	**0.512**
Planning	0.837	0.607	0.617	0.617	**0.904**
Semantic Web	0.541	0.648	0.550	**0.651**	0.622
Support Vector Machine	0.723	0.712	**0.786**	0.779	0.743
Mean	0.643	0.646	0.619	0.651	**0.699**

4.6   Qualitative Results

In this section, we study the publications of the top-5 researchers returned by the different models. The repartition of the conferences classes associated to the publications authored by the different top-5 experts is summarized in Table 4. Results are straightforward. The top-5 experts retrieved by UA publish more than 40% of their articles in A* conferences while this number for the researchers retrieved by other competitors is around 20%. More importantly, only 8% of the articles authored by the experts retrieved by UA are published in C conferences. In general, we see that all competitors that do not integrate any authority feature (i.e., NMF, COS, LM and LMS) lead to similar results in term of class repartition while the proposal is more sensitive to the two extremes. This important result puts forward the interest of considering quality signals for profiling since experts seem to be more concerned by the quality of the productions.

Table 4. Percentage of the publications of the top-5 researchers per conference class.

Model	A*	A	B	C
NMF	20.83	41.66	22.91	14.58
COS	21.90	34.28	28.25	15.55
LM/LMS	21.54	31.64	27.60	19.19
UA	**40.85**	**35.10**	15.74	**8.29**

5   Conclusions

Expert profiling and retrieval constitute challenging problematics for the scientific community. Although authority and expertise are widely studied in literature, these concepts are assumed to be independent biasing expert retrieval to authors with a large amount of publications. To overcome this issue, we defined a unified model based on an heterogeneous graph representation of digital libraries where authority and expertise vectors are learned simultaneously to capture a mutual reinforcement principle. The evaluation conducted on the Microsoft Academic Search data collection showed that capturing both individuals' expertise and authority significantly outperforms strong baselines. In perspective we will study how to integrate new authoritative criteria such as the co-authoring relation. Temporal and cold-start aspects constitute also challenging questions to refine the results.

References

1. Balog, K., Azzopardi, L., de Rijke, M.: Formal models for expert finding in enterprise Corpora. In: Proceedings of the 29th Annual International ACM SIGIR Conference on Research and Development in Information Retrieval. SIGIR 2006, pp. 43–50. ACM, New York (2006)
2. Balog, K., de Rijke, M.: Determining expert profiles (with an application to expert finding). In: IJCAI 2007, Proceedings of the 20th International Joint Conference on Artifical Intelligence, pp. 2657–2662. Morgan Kaufmann Publishers Inc., San Francisco (2007)
3. Blei, D.M., Ng, A.Y., Jordan, M.I.: Latent Dirichlet allocation. J. Mach. Learn. Res. **3**, 993–1022 (2003)
4. Bradley, A.P.: The use of the area under the ROC curve in the evaluation of machine learning algorithms. Pattern Recogn. **30**(7), 1145–1159 (1997)
5. Campbell, C.S., Maglio, P.P., Cozzi, A., Dom, B.: Expertise identification using email communications. In: Proceedings of the Twelfth International Conference on Information and Knowledge Management, CIKM 2003, pp. 528–531. ACM, New York (2003)
6. Craswell, N., Hawking, D., Vercoustre, A.-M., Wilkins, P.: P@noptic expert: searching for experts not just for documents. In: Ausweb, pp. 21–25 (2001)
7. Davenport, T.H., Prusak, L., Prusak, L.: Working Knowledge: How Organizations Manage What They Know. Harvard Business School Press, Boston (1997)
8. de La Robertie, B., Pitarch, Y., Teste, O.: Measuring article quality in Wikipedia using the collaboration network. In: Proceedings of the 2015 IEEE/ACM International Conference on Advances in Social Networks Analysis and Mining 2015, ASONAM 2015, pp. 464–471. ACM, New York (2015)
9. Deng, H., King, I., Lyu, M.R.: Formal models for expert finding on DBLP bibliography data. In: Proceedings of the 2008 Eighth IEEE International Conference on Data Mining, ICDM 2008, pp. 163–172. IEEE Computer Society, Washington, D.C. (2008)
10. Gollapalli, S.D., Mitra, P., Giles, C.L.: Ranking experts using author-document-topic graphs. In: Proceedings of the 13th ACM/IEEE-CS Joint Conference on Digital Libraries. JCDL 2013, pp. 87–96. ACM, New York (2013)
11. Haveliwala, T.H.: Topic-sensitive pagerank. In: Proceedings of the 11th International Conference on World Wide Web, WWW 2002, pp. 517–526. ACM, New York (2002)
12. Huynh, T., Takasu, A., Masada, T., Hoang, K.: Collaborator recommendation for isolated researchers. In: Proceedings of the 2014 28th International Conference on Advanced Information Networking and Applications Workshops, WAINA 2014, pp. 639–644. IEEE Computer Society, Washington, D.C. (2014)

13. Jurczyk, P., Agichtein, E.: Discovering authorities in question answer communities by using link analysis. In: Proceedings of the Sixteenth ACM Conference on Conference on Information and Knowledge Management, CIKM 2007, pp. 919–922. ACM, New York (2007)

14. Kleinberg, J.M.: Authoritative sources in a hyperlinked environment. J. ACM 46(5), 604–632 (1999)

15. Lee, D., Seung, H.: Algorithms for non-negative matrix factorization. Adv. Neural Inf. Process. Syst. 1, 556–562 (2001)

16. Li, C.-L., Su, Y.-C., Lin, T.-W., Tsai, C.-H., Chang, W.-C., Huang, K.-H., Kuo, T.-M., Lin, S.-W., Lin, Y.-S., Lu, Y.-C., Yang, C.-P., Chang, C.-X., Chin, W.-S., Juan, Y.-C., Tung, H.-Y., Wang, J.-P., Wei, C.-K., Wu, F., Yin, T.-C., Yu, T., Zhuang, Y., Lin, S.-D., Lin, H.-T., Lin, C.-J.: Combination of feature engineering and ranking models for paper-author identification in KDD cup 2013. In: Proceedings of the 2013 KDD Cup 2013 Workshop, KDD Cup 2013, pp. 2:1–2:7. ACM, New York (2013)

17. Nocedal, J.: Updating quasi-Newton matrices with limited storage. Math. Comput. 35(151), 773–782 (1980)

18. Page, L., Brin, S., Motwani, R., Winograd, T.: The pagerank citation ranking: bringing order to the web. In: Proceedings of the 7th International World Wide Web Conference, pp. 161–172 (1998)

19. Rybak, J., Balog, K., Nørvåg, K.: Temporal expertise profiling. In: Proceedings of the 36th European Conference on Advances in Information Retrieval, ECIR 2014, pp. 540–546 (2014)

20. Serdyukov, P., Taylor, M., Vinay, V., Richardson, M., White, R.W.: Automatic people tagging for expertise profiling in the enterprise. In: Proceedings of the 33rd European Conference on Advances in Information Retrieval, ECIR 2011 (2011)

21. Tang, J., Yao, L., Zhang, D., Zhang, J.: A combination approach to web user profiling. ACM Trans. Knowl. Discov. Data 5(1), 2:1–2:44 (2010)

22. Tang, J., Zhang, J., Jin, R., Yang, Z., Cai, K., Zhang, L., Su, Z.: Topic level expertise search over heterogeneous networks. Mach. Learn. 82(2), 211–237 (2011)

23. White, S., Smyth, P.: Algorithms for estimating relative importance in networks. In: Proceedings of the Ninth ACM SIGKDD International Conference on Knowledge Discovery and Data Mining, KDD 2003, pp. 266–275. ACM, New York (2003)

24. Yang, Z., Tang, J., Wang, B., Guo, J., Li, J., Chen, S.: Expert2bole: from expert finding to bole search. In: Knowledge Discovery and Data Mining (2009)

25. Yimam-Seid, D., Kobsa, A.: Expert-finding systems for organizations: problem and domain analysis and the DEMOIR approach. J. Org. Comput. Electron. Commer. 13(1), 1–24 (2003)

26. Zhang, J., Ackerman, M.S., Adamic, L.: Expertise networks in online communities: structure and algorithms. In: Proceedings of the 16th International Conference on World Wide Web, WWW 2007, pp. 221–230. ACM, New York (2007)

27. Zhou, D., Zhu, S., Yu, K., Song, X., Tseng, B.L., Zha, H., Giles, C.L.: Learning multiple graphs for document recommendations. In: Proceedings of the 17th International Conference on World Wide Web, WWW 2008, pp. 141–150. ACM, New York (2008)

Graph and Network Data Processing

Efficient Local Clustering Coefficient Estimation in Massive Graphs

Hao Zhang[1], Yuanyuan Zhu[1(✉)], Lu Qin[2], Hong Cheng[3], and Jeffrey Xu Yu[3]

[1] State Key Laboratory of Software Engineering, Computer School,
Wuhan University, Wuhan, China
zhanghaowuda12@gmail.com, yyzhu@whu.edu.cn
[2] Centre for Quantum Computation and Intelligent Systems,
University of Technology Sydney, Ultimo, Australia
lu.qin@uts.edu.au
[3] The Chinese University of Hong Kong, Hong Kong, China
{hcheng,yu}@se.cuhk.edu.hk

Abstract. Graph is a powerful tool to model interactions in disparate applications, and how to assess the structure of a graph is an essential task across all the domains. As a classic measure to characterize the connectivity of graphs, clustering coefficient and its variants are of particular interest in graph structural analysis. However, the largest of today's graphs may have nodes and edges in billion scale, which makes the simple task of computing clustering coefficients quite complicated and expensive. Thus, approximate solutions have attracted much attention from researchers recently. However, they only target global and binned degree-wise clustering coefficient estimation, and their techniques are not suitable for local clustering coefficient estimation that is of great importance for individual nodes. In this paper, we propose a new sampling scheme to estimate the local clustering coefficient with error bounded, where global and binned degree-wise clustering coefficients can be considered as special cases. Meanwhile, based on our sampling scheme, we propose a new framework to estimate all the three clustering coefficients in a unified way. To make it scalable on massive graphs, we further design an efficient MapReduce algorithm under this framework. Extensive experiments validate the efficiency and effectiveness of our algorithms, which significantly outperform state-of-the-art exact and approximate algorithms on many real graph datasets.

Keywords: Clustering coefficient · Massive graph · Sampling · MapReduce

1 Introduction

Graph is a powerful tool to model the interactions in a variety of contexts, such as social network, web graph, co-author network, citation network, and so on. This popularity has increased interest in analyzing the properties of graphs across all

© Springer International Publishing AG 2017
S. Candan et al. (Eds.): DASFAA 2017, Part II, LNCS 10178, pp. 371–386, 2017.
DOI: 10.1007/978-3-319-55699-4_23

the domains. Among the classic indexes for measuring graph structural properties, clustering coefficient and its variants have received much attention in recent studies. The global clustering coefficient of a graph is the ratio of the number of closed wedges (triangles) to the number of wedges (2-hop paths), which is an important metric to indicate how tightly the communities are connected in a graph. The binned degree-wise clustering coefficient, which is the average of the clustering coefficients for nodes of a specified degree group, measures how tightly the neighbors are connected for nodes with certain degrees. It can be used to find the relation between degree distribution and clustering coefficients [27] and assess the quality of generative models [26]. Another important variant is local clustering coefficient of a node, which is defined as the fraction of wedges centered at this node that participate in triangles to measure how tightly the neighbors of a node are connected among themselves. It has a great impact on network dynamics, such as game theory [17], cascading [16], synchronization [36] and spreading [31]. Meanwhile, it can also be used to identify fake accounts [37] and influential nodes [3] in social networks, and detect spam pages [1] and hidden thematic layers [6] in web mining. An overview of these and other applications can be referred to [35].

To compute clustering coefficients, we usually need to involve triangle enumeration, followed by several post-processing steps. In the literature, a large number of triangle enumeration algorithms have been proposed, including in-memory algorithms [13], external algorithms [4,8,19], and parallel algorithms [5,22,23,30]. However, explicitly enumerating all the triangles can be quite expensive due to the extremely large number of triangles, especially on large graphs. Therefore, approximate methods are further investigated to compute clustering coefficients without triangle enumeration. Most of them adopt sampling mechanisms, including edge sampling [33–35] and wedge sampling [10,27,28]. However, edge sampling suffers from high variances. Although existing wedge sampling methods [10,27,28] can bound the error of global and binned degree-wise clustering coefficients with high confidence by sampling a constant number of wedges, they are not suitable for local clustering coefficient estimation, because sampling a constant number of wedges for each node will result in enormous total number of sampled wedges.

In this paper, we study the problem of local clustering coefficient estimation and make the contributions as follows: (1) We propose a new wedge sampling scheme for local clustering coefficient estimation, where global and binned degree-wise clustering coefficients can be considered as special cases. Compared to existing wedge sampling methods, our method can reduce a large number of sampled wedges while making the error bounded with high probability. (2) We develop a unified clustering coefficient estimation (UCCE) framework to estimate all the three variants of clustering coefficient based on our proposed sampling scheme. (3) We also devise an efficient MapReduce algorithm under this framework with several optimization techniques to make it scalable on large graphs. (4) Finally, we conduct extensive experiments to compare our algorithm with the state-of-the-art exact and approximate algorithms and validate the efficiency and effectiveness of our approach.

The rest of the paper is organized as follows. Section 2 reviews related works. Section 3 presents problem definition and previous approaches on wedge sampling. In Sect. 4, we introduce our new sampling scheme and unified computation framework. Section 5 gives our MapReduce algorithm under this framework. Section 6 gives the experimental analysis and Sect. 7 concludes this paper.

2   Related Work

In this section, we will first review studies on triangle enumeration since it is involved in the exact computation of clustering coefficients. Then we discuss approximate methods for estimating triangle number and clustering coefficients.

Existing triangle enumeration algorithms usually adopt the node-iterator or edge-iterator framework, which examines every triplet to check whether a triangle exists. A large number of in-memory triangle listing algorithms can be found in an excellent survey [13]. To handle large graphs which cannot fit into memory, researchers have been dedicated to devising external algorithms [4,8,19]. Recently, parallel algorithms have been also widely investigated to accelerate the computation [5,22,23,30]. Existing parallel triangle listing algorithms in MapReduce mainly fall into two categories. One is graph partition based approaches such as GP [30], TTP [22] and CTTP [23], which partition a graph into a series of subgraphs and enumerate triangles in parallel. The other is NodeIterator algorithms [5,30], which first enumerate all the wedges in a MapReduce round and then check whether an wedge can form a triangle in another round.

To avoid explicitly enumerating triangles in massive graphs, approximate methods based on sampling are further proposed to estimate triangle number and clustering coefficients. Tsourakakis et al. [32–35] proposed several approximate approaches based on graph sparsification, among which the most representative approach is Doulion [34]. It sparsifies a graph by sampling each edge with a certain probability and then conducts estimation on the sparsified graph. They also developed another sampling approach which invokes both edge and triple-node sampling [11] and gave a MapReduce implementation in [21]. However, these sampling methods suffer from high variance. Schank and Wagner [24] proposed wedge sampling to estimate the total number of triangles with a hard bound on the variance. Seshadhri et al. [27,28] extended this idea to compute the triangle-based metrics including global and binned degree-wise clustering coefficients, and also gave a MapReduce implementation in [10]. However, their wedge sampling scheme is not suitable for local clustering coefficient estimation, as it requires to sample a large constant number of wedges for each node. Alternative sampling mechanisms for triangle counting have been proposed for streaming and semi-streaming algorithms, and interested readers can refer to most recent works in [9,15,29] and earlier works surveyed in [18]. Note that although [15,29] can estimate the local triangle numbers based on edge sampling in streaming model, they also suffer from high variance.

3   Preliminaries

In this section, we give the problem definition, and briefly introduce previous approaches based on wedge sampling closely related to our work.

3.1   Problem Definition

In this paper we target an *undirected simple graph* $G = (V, E)$ with no self loops and parallel edges, where V is the vertex set and $E \subseteq V \times V$ is the edge set. We use $|V|$ and $|E|$ to denote the number of vertices and the number of edges respectively. An edge between nodes u and v is represented as $(u, v) \in E$. For a node $v \in V$, we denote its neighbor set by $N(v) = \{u \in V | (u, v) \in E\}$ and its degree by $d_v = |N(v)|$.

A *triangle* $\triangle(u, v, w)$ in G, is a set of three nodes $\{u, v, w\} \subseteq V$ such that $(u, v), (v, w), (u, w) \in E$. We define the set of local triangles incident to node v as $\triangle(v) = \{\triangle(u, v, w) | u, w \in N(v)\}$. A *wedge* $\vee(u, v, w)$ is a triplet such that $(u, v), (v, w) \in E$. We define the set of local wedges centered at v as $\vee(v) = \{\vee(u, v, w) | u, w \in N(v)\}$. Obviously, we have $|\vee(v)| = \binom{d_v}{2} = \frac{d_v \times (d_v - 1)}{2}$. We use $T_v = |\triangle(v)|$ to denote the number of triangles incident to v, and use $W_v = |\vee(v)|$ to denote the number of wedges centered at v. Accordingly, we use T and W to denote the triangle number and wedge number in the whole graph respectively.

Definition 1 (Local Clustering Coefficient). *The local clustering coefficient (LCC) of node v is the ratio of the number of incident triangles to the number of wedges centered at v, which is formally defined as*

$$C_v = \frac{T_v}{\binom{d_v}{2}}. \tag{1}$$

The local clustering coefficient measures how tightly the neighbors of a node are connected among themselves. At the global level, this property is an indicator of how tightly the communities are connected in a graph.

Definition 2 (Global Clustering Coefficient). *The global clustering coefficient (GCC) of a graph G is the ratio of number of wedges closed (forming a triangle) to wedge number in the graph, which can be formally defined as*

$$C_g = \frac{\sum_{v \in V} T_v}{\sum_{v \in V} \binom{d_v}{2}} = \frac{3T}{W}. \tag{2}$$

Note that, a triangle $\triangle(u, v, w)$ are counted three times, one is in $\triangle(v)$, a second in $\triangle(u)$, and a third in $\triangle(w)$.

Another variant of clustering coefficient, targeting the nodes of a specified degree group, are defined below.

Definition 3 (Binned Degree-wise Clustering Coefficient). *Let $D_G = \{i | \exists v \in V \text{ such that } d_v = i\}$ be the degree set of graph G, and $D \subset D_G$ be a* *subset of degrees (we ignore degree-zero nodes). We define $V_D = \{v|d_v \in D\}$. The binned degree-wise clustering coefficient (DCC) for degree set D is the average local clustering coefficient on all nodes in V_D, which is*

$$C_D = \frac{\sum_{v \in V_D} T_v}{\sum_{v \in V_D} \binom{d_v}{2}}. \tag{3}$$

3.2 Previous Work Based on Wedge Sampling

In this subsection, we will introduce the method in [27], including theoretical basis of wedge sampling and algorithms for estimating global and binned degree-wise clustering coefficients. We also discuss its limitations and explain why it cannot be directly applied to local clustering coefficient estimation.

The main idea of wedge sampling is to sample a number of wedges from the whole wedge set and calculate the percentage of sampled wedges closed (forming a triangle) to estimate the clustering coefficient. The estimation error can be bounded by the Hoeffding's inequality given as follows.

Theorem 1 (Hoeffding's Inequality [7]). *Let $X_1, X_2, ..., X_n$ be independent random variables in range [0, 1]. We define their empirical mean by $\bar{X} = \frac{1}{n} \sum_{i=1}^{n} X_i$ and the expected value of $\bar{X}$ as $u = E[\bar{X}]$. For any positive t and δ, setting $n \geq \lceil \frac{1}{2} t^{-2} \ln(2/\delta) \rceil$ yields*

$$Pr\{|\bar{X} - u| \geq t\} \leq \delta \tag{4}$$

If we randomly sample at least $\lceil \frac{1}{2} t^{-2} \ln(2/\delta) \rceil$ wedges uniformly with replacement from the whole wedge set of a graph, and let $X_i = 1$ if the i-th selected wedge form a triangle and $X_i = 0$ otherwise, we have $Pr\{|\bar{X} - C_g| \geq t\} \leq \delta$. By doing so, the estimation error can be bounded by t with confidence at least $1 - \delta$. Note that k, which is determined by t and δ, is independent of graph size.

GCC Estimation Algorithm. The basic premise for GCC estimation is to select a number of wedges uniformly at random and check whether or not each is closed. Their strategy is to select vertex v with probability $p_v = \frac{\binom{d_v}{2}}{W}$, and then select two of its neighbors uniformly at random without replacement. Thus, the overall probability of selecting a particular wedge is $p_v \times \binom{d_v}{2} = 1/W$.

DCC Estimation Algorithm. The strategy to make above procedure work for binned degree-wise situation is straightforward. Instead of uniformly selecting a number of wedges from the whole wedge set, they uniformly choose a number of wedges from the wedge set in a specific degree bin. Specifically, they select vertex v with probability p_v only from the set of nodes in a specific degree bin and then select two of its neighbors uniformly at random without replacement.

Discussion. The above wedge sampling method is very effective for global clustering coefficient as only a constant number of wedges need to be sampled to achieve high accuracy with high probability. However, for binned degree-wise clustering coefficient, sampling the same constant number of wedges for each degree bin will lead to a large total number of sampled wedges if the degrees are binned per degree. More seriously, it become a curse when estimating node clustering coefficient, because we have to sample a constant number of wedges for each node, which might be even larger than the number of its local wedges. Take $t = 0.01$ and $\delta = 0.001$ for example, we need to sample 38005 wedges for nodes whose wedge number exceeds such value, and sample all local wedges for the other nodes. Such enormous sampled wedges will cause costly computation, and generate huge shuffled data in the MapReduce implementation [10].

4 Our Sampling Scheme and Computation Framework

In this section, we will first present our new sampling scheme for local clustering coefficient estimation, where global and binned degree-wise clustering coefficients can be considered as special cases. Then we propose a novel clustering coefficient estimation framework based on our wedge sampling scheme, in which different variants of clustering coefficient are computed in a unified way.

4.1 A New Wedge Sampling Scheme

Recall that in Hoeffding's Inequality, X is assumed to be a random variable following arbitrary distribution. In fact, we observe that in the case of clustering coefficient, X follows hypergeometric distribution. It describes the probability of k successes in n draws without replacement, from a finite population of size N that contains K successes, wherein each draw is either a success or a failure.

For a node v, suppose we sample n wedges from its local wedge set $\vee(v)$, and check whether each wedge is closed (forming a triangle) or open. The probability that k wedges are closed is $P(X = k) = \binom{K}{k}\binom{N-K}{n-k}/\binom{N}{n}$, where $N = W_v$ and $K = T_v$. Obviously, such probability mass function satisfies the condition of hypergeometric distribution, i.e., $X \sim Hypergeometric(K, N, n)$. Let $p = K/N = T_v/W_v$ be the local clustering coefficient of node v. Then the estimated value is X/n, whose error can be bounded as follows [7].

Theorem 2. Let $X \sim Hypergeometric(K, N, n)$ and $p = K/N$. Then we can derive the following bound.

$$\Pr[|\frac{X}{n} - p| \geq t] \leq 2\exp(-2t^2 n). \tag{5}$$

Note that setting $n \geq \lceil \frac{1}{2}t^{-2}\ln\frac{2}{\delta} \rceil$ yields $\Pr[|\frac{X}{n} - p| \geq t] \leq \delta$. This means that we need to sample at least $\lceil \frac{1}{2}t^{-2}\ln\frac{2}{\delta} \rceil$ wedges to make the error is at most t with confidence at least $1 - \delta$, which is the same as that in Hoeffding's Inequality.

Next, we introduce how to improve above bound. As we know, the degrees of most nodes in real life graphs are not high because real life graphs usually follow power-law degree distribution. Thus, for a node v, to achieve small error with high confidence, the number of required sampled wedges n in Eq. 5 can be even larger than $W_v/2$. In this case, we can get a tighter bound by inverting the bound according to the symmetry of hypergeometric distribution as follows [25].

Theorem 3. *Let* $X \sim Hypergeometric(K, N, n)$ *and* $p = K/N$. *For* $n \geq N/2$, *we can derive the following bound.*

$$\Pr[|\frac{X}{n} - p| \geq t] \leq 2\exp(-2t^2 n \frac{n}{N-n}). \tag{6}$$

Setting $n \geq \lceil \frac{1}{4}t^{-2}(\sqrt{\ln\frac{2}{\delta}(\ln\frac{2}{\delta} + 8Nt^2)} - \ln\frac{2}{\delta}) \rceil$ yeilds $\Pr[|\frac{X}{n} - p| \geq t] \leq \delta$.

Now the question is which bound should be chosen for a specific node v? We answer this by giving the following theorem.

Theorem 4. *Let* $X \sim Hypergeometric(K, N, n)$ *and* $p = K/N$. *To achieve the error bound* t *with probability* $1 - \delta$, *the least possible value of* n *is:*

$$\begin{aligned} n &= \lceil \frac{1}{2}t^{-2}\ln\frac{2}{\delta} \rceil && if \ N \geq \lceil t^{-2}\ln\frac{2}{\delta} \rceil, \\ n &= \lceil \frac{1}{4}t^{-2}(\sqrt{\ln\frac{2}{\delta}(\ln\frac{2}{\delta} + 8Nt^2)} - \ln\frac{2}{\delta}) \rceil && otherwise. \end{aligned} \tag{7}$$

Proof: From Theorems 2 and 3, we can know that they achieve the same error and confidence when they have the same number of sampled wedges, i.e., $\frac{1}{2}t^{-2}\ln\frac{2}{\delta} = \frac{1}{4}t^{-2}(\sqrt{\ln\frac{2}{\delta}(\ln\frac{2}{\delta} + 8Nt^2)} - \ln\frac{2}{\delta})$. By solving this equation, we derive $N = t^{-2}\ln\frac{2}{\delta}$, which completes the proof.

Based on this theorem, we can know that, given t and δ, for a node v with $W_v \geq t^{-2}\ln\frac{2}{\delta}$, we can sample a constant number of wedges $\lceil \frac{1}{2}t^{-2}\ln\frac{2}{\delta} \rceil$; otherwise, we sample $\lceil \frac{1}{4}t^{-2}(\sqrt{\ln\frac{2}{\delta}(\ln\frac{2}{\delta} + 8W_vt^2)} - \ln\frac{2}{\delta}) \rceil$ wedges, which is related to W_v. For example, given $t = 0.01$ and $\delta = 0.001$, for nodes with $W_v \geq 76010$, we need to sample 38005 wedges, otherwise, the number of sampled wedges is smaller than 38005. We also give the theoretical numbers of sampled wedges for different degrees in Fig. 1, where UCCE represents our sampling scheme and WS represents wedge sampling scheme in [27]. It shows that WS needs to sample all local wedges for node with degree less than 277, and UCCE only needs to sample part of local wedges. Thus, we can significantly reduce the number of sampled wedges than that in [27].

Moreover, our sampling scheme can also be applied to GCC and DCC estimation. For GCC, since the number of total wedges in a real life graph is usually very large (far larger than $t^{-2}\ln\frac{2}{\delta}$), we can sample a constant number of wedges $\frac{1}{2}t^{-2}\ln\frac{2}{\delta}$. For DCC, we can also determine the number of sampled

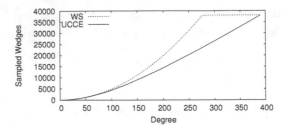

Fig. 1. Theoretical number of sampled wedges

wedges based on the total number of wedges in a specific degree bin as local clustering coefficient does. After we obtain the required number of sampled wedges n for global or binned degree-wise clustering coefficients, we divide it to each node by $n_v = nW_v/W$ or $n_v = nW_v/W_D$.

4.2   A Unified Clustering Coefficient Estimation Framework

To integrate the computation of local, global, and binned degree-wise clustering coefficients into a unified framework, we first design a general mapping function F to map graph nodes into different groups. The group numbers are different for each variant of clustering coefficient. For LCC, we divide the nodes into $|V|$ groups, where each group contains only one node. For degree set $D_1, D_2, ..., D_l \subseteq D$, we divide nodes into groups $V_{D_1}, V_{D_2}, ..., V_{D_l}$, where group V_{D_i} contains nodes with degree in D_i. For GCC, we can simply consider the nodes as one single group V.

Equipped with above mapping function, we present our unified framework UCCE in Algorithm 1. The whole process of clustering coefficient estimation is divided into three phases: Gathering Information, Sampling and Verification, and Grouping Results. In the first phase, we will collect the statistical information about each node and each group, and calculate the number of sampled wedges for each node. Specifically, we collect the neighbors for each node v to compute the total number of wedges for each group. Then we obtain the number of sampled wedges for each group based on Theorem 4, and divide it to obtain the number of sampled wedges W_v^s for each node v. In the second phase, we sample W_v^s wedges for each node v, and check whether each of them can form a triangle. The number of triangles formed by sampled wedges is denoted as T_v^s. Finally, in the last phase, we group the number of sampled wedges and the number of sampled triangles for each group to obtain the final result.

Algorithm 1. UCCE Framework

---

1 Gathering Information
2 Sampling and Verification
3 Grouping Results

---

5 Our MapReduce Algorithm

In this section, we present a new MapReduce algorithm UCCE under our framework and give its optimization techniques.

5.1 UCCE Algorithm

We will first introduce the algorithms for gathering information and grouping results, and give the optimized algorithm for sampling and verification later.

Gathering Information. The process of gathering information is shown in Algorithm 2, which can be completed in one MapReduce round. In map function, we partition the edges into different groups by applying mapping function F to node u for each edge (u, v). In reduce function, we use $List(u, v)$ to denote the list of edges which shares the same key value $F(u)$, and use $V_{F(u)}$ to denote the set of nodes in the group that u belong to. $S(u)$ is the sample rate of node u calculated based on $List(u, v)$ by the method discussed in Sect. 4.1.

Grouping Results. We present the process of grouping results in Algorithm 3, which can also be completed in one MapReduce job. The map function takes the number of sampled wedges W_u^s and number of sampled triangles T_u^s for each node u as input, and maps nodes to group $F(u)$. The reduce function aggregates sampled wedges and triangles to calculate the clustering coefficient for each group.

Algorithm 2. Gathering Information

1   map $(\langle (u, v); \emptyset \rangle)$
2   **begin**
3 | Emit $\langle F(u), (u, v) \rangle$

4   reduce $(\langle F(u); List(u, v) \rangle)$
5   **begin**
6 | **foreach** $u \in V_{F(u)}$ **do**
7 | | Emit $\langle u, S(u) \rangle$

Algorithm 3. Grouping Results

1   map $(\langle u; (W_u^s, T_u^s) \rangle)$
2   **begin**
3 | Emit $\langle F(u), (W_u^s, T_u^s) \rangle$

4   reduce $(\langle F(u); (W_u^s, T_u^s) \rangle)$
5   **begin**
6 | **foreach** $(W_u^s, T_u^s) \in List(W_u^s, T_u^s)$ **do**
7 | | $T_s + = T_v^s; \ W_s + = W_v^s$
8 | Emit $\langle F(u), T_s / W_s \rangle$

5.2   Optimized Algorithm for Sampling and Verification

The task of sampling and verification in MapReduce can be completed by slightly modifying triangle enumeration algorithms such as NodeIterator++ [5]. It contains two MapReduce rounds. The first is to sample and output wedges, and the second is to check whether each wedge is closed or not. Although it is logically simple, a large amount of intermediate data in the form of wedges are shuffled, which severely hinder the scalability. Therefore, we devise an optimized algorithm to significantly reduce the intermediate data to speed up.

Reduction of Intermediate Data. We reduce the intermediate data by utilizing a light weight data structure Bloom filter. It can help check whether a wedge is closed with high accuracy without outputting sampled wedges. Bloom filter [2] is a space-efficient probabilistic data structure to test whether an element is a member of a set. Given the edge set E, we can build a bloom filter B of m bits by mapping each edge in E to B with k' hash functions, and query an element in B with false positives rate $p_f = (1 - e^{k'|E|/m})^{k'}$ [20]. Although Bloom filter is already space-efficient, it can still cost a lot of memory if we want to achieve a very small false positive rate for extremely large graphs. To utilize Bloom filter

---

Algorithm 4. Sampling and Verification

---
1 Distributed Bloom Filter
2 **foreach** $i < L$ **do**
3 | map $(\langle u; v \rangle)$
4 | **begin**
5 | |__ Emit $\langle u, v \rangle$
6 | reduce-setup
7 | **foreach** $j = 1$ *to* s **do**
8 | | **if** $j\%L = i$ **then**
9 | |__ |__ load $bloomfilter[j]$
10 | reduce $(\langle u; N(u) \rangle)$
11 | **begin**
12 | | $(T_u^s, W_u^s) = \texttt{SampleWedges}(u, S(u), i)$
13 | |__ Emit $\langle u, (T_u^s - p_f W_u^s, W_u^s) \rangle$
14 |__ $i++$

15 **Function** SampleWedges$(u, S(u), i)$
16 randomly sample wedges into I_u with sample rate $S(u)$
17 **foreach** $\vee(v, u, w) \in I_u$ **do**
18 | **if** $hash(v)\%L = i$ **then**
19 | | $++W_u^s$
20 | | **if** Check(v, w) **then**
21 | |__ |__ $++T_u^s$

22 return (T_u^s, W_u^s);

---

with limited memory space, we can distributedly generate the Bloom filter and partially load them in multiple rounds.

Optimized Sampling and Verification. We present our optimized sampling and verification process in Algorithm 4, which includes a preliminary step of generating distributed Bloom filter. We partially load distributed Bloom filter into memory in L rounds (line 2), and a small L is usually enough for real life graphs. Suppose that the Bloom filter are partitioned into s parts. In round i, we only load $bloomfilter[j]$ such that $hash(j)\%L = i$ into the memory for each reducer (lines 7–9). In the reduce function, we sample the wedges for each node by function `SampleWedges`. Note that not all sampled wedges can be verified by Bloom filter as it is only partially loaded. Therefore, for each sampled wedge $\vee(v, u, w)$, we first obtain the hash value $hash(v)$ and then examine whether $bloomfilter[hash(v)]$ is loaded (line 18). If yes, the number of sampled wedges are increased, otherwise we abandon this wedge (lines 17–21). Note that before we output the numbers of sampled wedges and triangles, we amend the number of sampled triangles by subtracting the number of false positive triangles caused by the false positive rate of Bloom filter (line 13).

6 Experiments

In this section, we evaluate the performance of UCCE and compare it with competing methods. Especially, we will answer the following two questions.

Q1 How accurate is UCCE for local, global, and binned-wise degree clustering coefficient estimation?
Q2 How efficient is UCCE compared with the state-of-the-art exact and approximate competitors?

6.1 Experimental Setup

Datasets. We use five publicly available real graph datasets which have been intensively investigated in previous works [10,15,23,30]. These datasets are from diverse domains such as social networks, hyperlinks in webpages, collaboration networks, etc. Their meta information is displayed in Table 1. TW is from [12],

Table 1. Summary of the testing graph datasets

Datasets	Name	$|V|$	$|E|$	$|T|$	$|W|$	C_g
as-Skitter	AS	1.7×10^6	1.1×10^7	2.9×10^7	1.6×10^{10}	0.0054
com-livejournal	LJ	4.0×10^6	3.5×10^7	1.8×10^8	4.2×10^9	0.1253
orkut	OK	3.0×10^6	1.2×10^8	6.3×10^9	4.6×10^{10}	0.0413
twitter-2010	TW	4.2×10^7	1.2×10^9	3.5×10^{10}	1.2×10^{14}	0.0008
com-friendster	FD	6.5×10^7	1.8×10^9	4.2×10^9	7.2×10^{11}	0.0174

and the rest are from SNAP [14]. In our experiments, each dataset is preprocessed to be a simple undirected graph.

Implementation. We compare our algorithm with state-of-the-art exact and approximate algorithms. The exact algorithms are two recent triangle enumeration methods NodeIterator++ (NI++ for short) [30] and CTTP [23]. We add several post-processing steps to compute the clustering coefficients based on the outputted triangles. The approximate approaches we compare with are the recent wedge sampling methods for clustering coefficient estimation in MapReduce [10,27] discussed in Sect. 3. Specifically, we denote their method for estimating GCC as WSG, and denote their method for estimating DCC as WSD. We also further modified their MapReduce algorithm in [10] to adapt the computation of LCC, denoted as WSL. All the methods are implemented in Java and running on a cluster of 20 nodes. Each node is equipped with 2 Intel Xeon E5-2630-v2 cpu (6 core, each core running at 2.6 GHZ) and 4×2 TB hard drives, and connected with each other by 2×1 GB Ethernet. We use 240 mappers/reducers at the same time, and each one is allocated with 2 GB memory.

6.2   Performance of UCCE

In this subsection, we will answer $Q1$. To answer it, we fix δ to 0.001 and test UCCE under different values of t for LCC and GCC. To display the experimental results for DCC, we fix δ to 0.001 and t to 0.01, and show the exact and estimated clustering coefficients for each degree bin. For GCC, we measure the accuracy of algorithms by absolute error between C_g and C'_g, where C_g is exact clustering coefficient and C'_g is the estimated clustering coefficient. For LCC, we measure the accuracy by average absolute error defined as $\frac{\sum_{v \in V} |C_v - C'_v|}{|V|}$, where C_v and C'_v are exact and estimated clustering coefficients for node v respectively.

Figure 2 shows the curves of estimation error of UCCE under different t for LCC and GCC. It can be seen in Fig. 2(a) that as the value of t decreases, the average error is also decreasing. It is worth noticing that the average error is always smaller than the theoretical bound t. For example, when t is 0.05, the

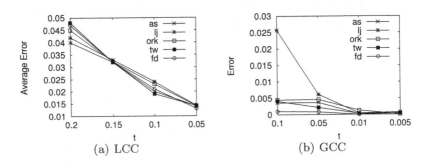

Fig. 2. Accuracy for LCC and GCC

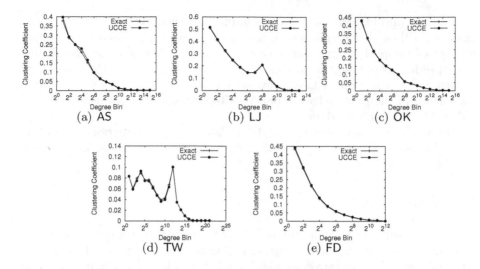

Fig. 3. Accuracy for DCC

average error is less than 0.015 for all datasets we tested. The case for GCC in Fig. 2(b) is similar. Although there is a small fluctuation for $t = 0.05$ for dataset AS, the estimation error is still far smaller than the theoretical bound.

Next we show the accuracy of our methods for DCC by plotting the curves of exact and estimated clustering coefficient for each graph dataset. The results are shown in Fig. 3. It can be seen that the curves for exact and estimation results are very close for all datasets.

6.3 Comparison with Competitors

In this subsection, we will answer $Q2$. We compare the running time between our method with competitors and show the results in Fig. 4.

For LCC, from Fig. 4(a), we can see that our algorithm is faster than all the competing algorithms on all testing dataset ($t = 0.05$ and $\delta = 0.001$). It can be over an order of magnitude faster than exact algorithms on large graph datasets

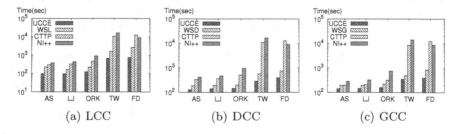

Fig. 4. Running time comparison

such as TW and FD. Previous wedge sampling method WSL is also faster than exact algorithms, but it consumes more running time than ours due to the large number of sampled wedges and large size of intermediate data in MapReduce. The speedup of our algorithm comes from two sides. One is that we sample smaller number of wedges than previous methods; the other is that we avoid the output of sampled wedges in our MapReduce algorithm, which can largely reduce the intermediate data.

For DCC and GCC, Fig. 4(b) and (c) also show the outperformance of our algorithm over competitors ($t = 0.01$ and $\delta = 0.001$). The curves are similar as that in Fig. 4(a).

7   Conclusion

Clustering coefficient is an important metric to characterize graph properties. In this paper, we propose a new wedge sampling scheme to estimate local clustering coefficient where the error are well bounded with high probability. Moreover, we propose a new framework which can estimate the local, global, and binned degree-wise clustering coefficients in a unified way. Under this framework, we further devise a MapReduce algorithm for clustering coefficient estimation, which is also the first efficient MapReduce algorithm that can estimates local clustering coefficient. We also give the optimization techniques for reducing the intermediate data to further accelerate the computation. Extensive experiments validate the efficiency and effectiveness of our algorithms.

Acknowledgements. This work was partially supported by the grants from the National Science Foundation of China (61502349), Hubei Provincial Natural Science Foundation of China (2015CFB339), the Scientific and Technologic Development Program of SuZhou (SYG201442), Research Grants Council of the Hong Kong SAR, China (14209314 and 14221716), Australian Research Council (DE140100999 and DP160101513), Microsoft Research Asia Collaborative Research Grant and Chinese University of Hong Kong Direct Grant (4055048). Yuanyuan Zhu is a corresponding author.

References

1. Becchetti, L., Boldi, P., Castillo, C., Gionis, A.: Efficient algorithms for large-scale local triangle counting. TKDD **4**(3) (2010). Article no. 13
2. Bloom, B.H.: Space/time trade-offs in hash coding with allowable errors. Commun. ACM **13**(7), 422–426 (1970)
3. Chen, D.-B., Gao, H., Lü, L., Zhou, T.: Identifying influential nodes in large-scale directed networks: the role of clustering. PloS one **8**(10), e77455 (2013)
4. Chu, S., Cheng, J.: Triangle listing in massive networks and its applications. In: KDD, pp. 672–680. ACM (2011)
5. Cohen, J.: Graph twiddling in a mapreduce world. Comput. Sci. Eng. **11**(4), 29–41 (2009)
6. Eckmann, J.-P., Moses, E.: Curvature of co-links uncovers hidden thematic layers in the world wide web. PNAS **99**(9), 5825–5829 (2002)

7. Hoeffding, W.: Probability inequalities for sums of bounded random variables. J. Am. Stat. Assoc. **58**(301), 13–30 (1963)
8. Hu, X., Tao, Y., Chung, C.-W.: Massive graph triangulation. In: SIGMOD, pp. 325–336. ACM (2013)
9. Jha, M., Seshadhri, C., Pinar, A.: A space-efficient streaming algorithm for estimating transitivity and triangle counts using the birthday paradox. TKDD **9**(3), 15:1–15:21 (2015)
10. Kolda, T.G., Pinar, A., Plantenga, T., Seshadhri, C., Task, C.: Counting triangles in massive graphs with mapreduce. SISC **36**(5), S48–S77 (2014)
11. Kolountzakis, M.N., Miller, G.L., Peng, R., Tsourakakis, C.E.: Efficient triangle counting in large graphs via degree-based vertex partitioning. Internet Math. **8**(1–2), 161–185 (2012)
12. Kwak, H., Lee, C., Park, H., Moon, S.: What is Twitter, a social network or a news media? In: WWW 2010: Proceedings of the 19th International Conference on World wide web, pp. 591–600. ACM, New York (2010)
13. Latapy, M.: Main-memory triangle computations for very large (sparse (power-law)) graphs. Theoret. Comput. Sci. **407**(1), 458–473 (2008)
14. Leskovec, J., Krevl, A.: SNAP Datasets: Stanford large network dataset collection, June 2014. http://snap.stanford.edu/data
15. Lim, Y., Kang, U.: MASCOT: memory-efficient and accurate sampling for counting local triangles in graph streams. In: KDD, pp. 685–694 (2015)
16. Lin, Y., Xiong, H., Chen, M., Ding, L., Cao, Y., Wang, G., Liu, M.: Dynamical model and analysis of cascading failures on the complex power grids. Kybernetes **40**(5/6), 814–823 (2011)
17. Masuda, N.: Clustering in large networks does not promote upstream reciprocity. PloS one **6**(10), e25190 (2011)
18. McGregor, A.: Graph stream algorithms: a survey. SIGMOD Rec. **43**(1), 9–20 (2014)
19. Menegola, B.: An external memory algorithm for listing triangles (2010)
20. Mitzenmacher, M., Upfal, E.: Probability and Computing: Randomized Algorithms and Probabilistic Analysis. Cambridge University Press, New York (2005)
21. Pagh, R., Tsourakakis, C.E.: Colorful triangle counting and a mapreduce implementation. Inf. Process. Lett. **112**(7), 277–281 (2012)
22. Park, H.-M., Chung, C.-W.: An efficient mapreduce algorithm for counting triangles in a very large graph. In: CIKM, pp. 539–548. ACM (2013)
23. Park, H.-M., Silvestri, F., Kang, U., Pagh, R.: Mapreduce triangle enumeration with guarantees. In: CIKM, pp. 1739–1748. ACM (2014)
24. Schank, T., Wagner, D.: Approximating clustering coefficient and transitivity. J. Graph Algorithms Appl. **9**(2), 265–275 (2005)
25. Serfling, R.J.: Probability inequalities for the sum in sampling without replacement. Ann. Stat. **2**(1), 39–48 (1974)
26. Seshadhri, C., Kolda, T.G., Pinar, A.: Community structure and scale-free collections of erdős-rényi graphs. Phys. Rev. E **85**(5), 056109 (2012)
27. Seshadhri, C., Pinar, A., Kolda, T.G.: Fast triangle counting through wedge sampling. In: SDM, vol. 4, p. 5. Citeseer (2013)
28. Seshadhri, C., Pinar, A., Kolda, T.G.: Triadic measures on graphs: the power of wedge sampling. In: SDM, pp. 10–18. SIAM (2013)
29. Stefani, L.D., Epasto, A., Riondato, M., Upfal, E.: Trièst: counting local and global triangles in fully-dynamic streams with fixed memory size. In: KDD, pp. 825–834 (2016)

30. Suri, S., Vassilvitskii, S.: Counting triangles and the curse of the last reducer. In: WWW, pp. 607–614. ACM (2011)
31. Trpevski, D., Tang, W.K., Kocarev, L.: Model for rumor spreading over networks. Phys. Rev. E **81**(5), 056102 (2010)
32. Tsourakakis, C.E.: Fast counting of triangles in large real networks without counting: algorithms and laws. In: ICDM, pp 608–617 (2008)
33. Tsourakakis, C.E., Drineas, P., Michelakis, E., Koutis, I., Faloutsos, C.: Spectral counting of triangles via element-wise sparsification and triangle-based link recommendation. Soc. Netw. Anal. Mining **1**(2), 75–81 (2011)
34. Tsourakakis, C.E., Kang, U., Miller, G.L., Faloutsos, C.: Doulion: counting triangles in massive graphs with a coin. In: KDD, pp. 837–846. ACM (2009)
35. Tsourakakis, C.E., Kolountzakis, M.N., Miller, G.L.: Triangle sparsifiers. J. Graph Algorithms Appl. **15**(6), 703–726 (2011)
36. Wu, X., Lu, H.: Cluster synchronization in the adaptive complex dynamical networks via a novel approach. Phys. Lett. A **375**(14), 1559–1565 (2011)
37. Yang, Z., Wilson, C., Wang, X., Gao, T., Zhao, B.Y., Dai, Y.: Uncovering social network sybils in the wild. TKDD **8**(1), 2 (2014)

Efficient Processing of Growing Temporal Graphs

Huanhuan Wu[✉], Yunjian Zhao, James Cheng, and Da Yan

Department of Computer Science and Engineering,
The Chinese University of Hong Kong, Hong Kong, China
{hhwu,yjzhao,jcheng,yanda}@cse.cuhk.edu.hk

Abstract. Temporal graphs are useful in modeling real-world networks. For example, in a phone call network, people may communicate with each other in multiple time periods, which can be modeled as multiple temporal edges. However, the size of real-world temporal graphs keeps increasing rapidly (e.g., considering the number of phone calls recorded each day), which makes it difficult to efficiently store and analyze the complete temporal graphs. We propose a new model, called *equal-weight damped time window model*, to efficiently manage temporal graphs. In this model, each time window is assigned a unified weight. This model is flexible as it allows users to control the tradeoff between the required storage space and the information loss. It also supports efficient maintenance of the windows as new data come in. We then discuss applications that use the model for analyzing temporal graphs. Our experiments demonstrated that we can handle massive temporal graphs efficiently with limited space.

1 Introduction

A temporal graph is a graph in which the relationship between vertices is not just modeled by an edge between them, but the time period when the relationship happens is also recorded. For example, two persons A and B talked on the phone in time periods $[t_1, t_2]$ and $[t_3, t_4]$ are modeled as two temporal edges, $(A, B, [t_1, t_2])$ and $(A, B, [t_3, t_4])$. An example of a temporal graph is shown in Fig. 1(a).

Graphs are used ubiquitously to model relationships between objects in real world. However, the graph data in many applications are actually better to be modeled as temporal graphs. For example, in communication networks, including online social networks, messaging networks, phone call networks, etc., people communicate with each other in different time periods. Temporal graphs collected from these applications carry rich time information, and have been shown to possess many important time-related patterns that cannot be found from non-temporal graphs [8–10,13,17,19,25].

However, existing work overlooks one serious problem presented by temporal graphs in real world applications, that is, the number of temporal edges (or temporal records) can be extremely huge so that it becomes overly expensive

© Springer International Publishing AG 2017
S. Candan et al. (Eds.): DASFAA 2017, Part II, LNCS 10178, pp. 387–403, 2017.
DOI: 10.1007/978-3-319-55699-4_24

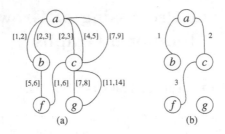

Fig. 1. (a) A **temporal graph** $\mathbb{G}$, and (b) the **weighted graph** $G_{[2,5]}$

to store and process a temporal graph. For example, in a temporal graph that models phone-call records, a person may talk on the phone many times in different time periods in a day, where each phone call is represented by a temporal edge with the corresponding time period. The total number of temporal edges accumulated over time for all persons can easily become overwhelming. Note that while the number of temporal edges usually increases at a steady rate over time, the number of vertices, on the other hand, does not increase too much over time after passing the growth stage.

The problem in the above example is actually a real problem presented to us by a telecommunications operator, who collects phone-call and messaging records represented as a temporal graph that becomes too large over time for them to manage (millions to tens of millions of new temporal edges added each day). While analyzing only a short recent window of the data is useful, the telecom operator is also very keen in storing and analyzing the temporal graph over a long period of time (e.g., in recent years), and possibly the entire history, in an efficient way. Motivated by this, we propose a new model to efficiently manage a temporal graph.

Our new model considers the input temporal graph as a continuous stream, which captures how the temporal graph is collected in real-life applications (e.g., new call/message records are accumulated in the order of the calling/sending time). However, the sheer size of the stream over the entire time history renders analysis (and even storage) of the original temporal graph too costly. To address this problem, we consider a *damped time window model* (also called *tilted time window*) [5], where a decay function is applied to depreciate the importance of records in an older window. However, the windows defined by existing damped time window models do not have a unified weight and hence the importance of records in different windows cannot be easily compared. For example, which of the following patterns is more important: a pattern that A and B communicated 10 times in a recent window (e.g., last week), or a pattern that A and B communicated 10,000 times in an older window (e.g., last year)?

We design a new damped time window model that gives a unified weight to each time window, called *equal-weight damped time window*, and represents the temporal graph falling into each window (i.e., a time period) as a weighted graph. The weighted graphs from different time windows can then be compared and analyzed.

The main contributions of our work are as follows:

- Our equal-weight damped time window distributes a unified weight to each time window, which makes it easy to compare different time windows.
- Our model can handle massive temporal graphs with limited space requirement, and support efficient graph analysis with little information loss.
- The equal-weight design in our model also leads to natural and efficient update maintenance of the entire window (within a bounded storage space).
- We present an application that analyzes the connectivity of a temporal graph with our new model. More applications such as community finding can be found in [23]. We verified the effectiveness and efficiency of our method by extensive experiments on large temporal graph datasets.

Outline. Section 2 presents the equal-weight time window model. Section 3 discusses one application based on our model. Section 4 reports experimental results. Section 5 discusses related work. Section 6 concludes the paper.

2   Equal-Weight Damped Time Window

Different window models have been proposed for processing a data stream. Among which, the *landmark window model* [14] considers the entire history of a stream without distinguishing the importance of recent and old records, while the *sliding window model* [6] focuses on the most recent window only. Our work is motivated by application needs from a telecom operator that requires to analyze historical data while giving more importance to recent data. For this purpose, the *damped time window model* [5] seems to suit the requirement. We introduce our damped time window model in this section, and discuss its difference with existing ones.

We first define the notations related to a temporal graph. Let $\mathbb{G} = (\mathbb{V}, \mathbb{E})$ be a temporal graph, where $\mathbb{V}$ is the set of vertices and $\mathbb{E}$ is the set of edges in $\mathbb{G}$. An edge $e \in \mathbb{E}$ is a quadruple $(u, v, [t_i, t_j])$, where $u, v \in \mathbb{V}$, and $[t_i, t_j]$ is the time that e is *active*. We focus our discussion on undirected temporal graphs, while we note that it is not difficult to extend our method to directed temporal graphs.

2.1   The Weight Function

In a damped time window model, a decaying weight function is used to depreciate the importance of a record over time. In the setting of a temporal graph, we use such a function to assign weight to temporal edges in the graph. We first present the weight density function as follows.

Definition 1 (Weight density function). *Let t_τ be the current time. The weight density of a record at time t (with respect to t_τ) is defined as*

$$f(t) = e^{\lambda(t - t_\tau)},$$

where $\lambda \geq 0$ is a decaying constant.

Note that $t \leq t_\tau$, and t is a time in the past if $t < t_\tau$.

In Definition 1, $f(t)$ is an exponentially decaying function, which is used throughout the paper as it has been widely adopted [11,24]. But $f(t)$ can also be defined differently (e.g., as a linear decaying function) depending on the application. Based on $f(t)$, we define our weight function as follows.

Definition 2 (Weight function). *The weight of a temporal edge* $(u, v, [t_1, t_2])$ *is given as the integral*

$$F(t_1, t_2) = \int_{t_1}^{t_2} f(t)dt.$$

Let $W = [t_x, t_y]$ be a given time window. With the weight function, we represent the part of a temporal graph $\mathbb{G} = (\mathbb{V}, \mathbb{E})$ that falls into W as a weighted graph G_W defined as follows.

Definition 3 (Weighted graph). *The weighted graph of a temporal graph* $\mathbb{G} = (\mathbb{V}, \mathbb{E})$ *within a time window* $W = [t_x, t_y]$ *is given by* $G_W = (V_W, E_W, \Pi_W)$, *where:*

– $V_W = \mathbb{V}$,
– $E_W = \{(u, v) : (u, v, [t_i, t_j]) \in \mathbb{E}_W\}$, *where* $\mathbb{E}_W = \{(u, v, [t_i, t_j]) \in \mathbb{E} : [t_i, t_j] \cap [t_x, t_y] \neq \emptyset\}$,
– Π_W *is a function that assigns each edge* $e = (u, v) \in E_W$ *a weight* $\Pi_W(e) = \sum_{(u,v,[t_i,t_j]) \in \mathbb{E}_W} F(\max(t_i, t_x), \min(t_j, t_y))$.

Example 1. Figure 1(a) shows a temporal graph and Fig. 1(b) shows the corresponding weighted graph $G_{[2,5]}$ within the time window $[2, 5]$. For simplicity, we assume that $\lambda = 0$ and hence $f(t) = 1$. Thus, the weight of edge (a, b) is $F(2, 3) = 1$, the weight of edge (a, c) is $F(2, 3) + F(4, 5) = 2$, and the weight of edge (c, f) is $F(2, 5) = 3$. The weight of all other edges is 0.

2.2   The Equal-Weight Window Model

Next, we determine the size of each window in a data stream given the weight function.

Existing damped time window model [5] usually sets the sizes of the windows in a stream by an exponentially increasing function (e.g., $2^0 T, 2^1 T, 2^2 T, 2^3 T, \ldots$, where the windows are disjoint and the most recent window has a size $2^0 T$), or by the lengths of conventional time units (e.g., hour, day, month, year, ...). These window size settings may seem to be intuitive, but they are primarily designed for mining frequent itemsets from a stream and are not suitable for our problem of handling a temporal graph stream (see more discussion in "Advantages of the new model" at the end of this subsection). We introduce an equal-weight scheme as follows.

Let $[t_0, t_\tau]$ be the time period of the entire stream up to the current time t_τ. To limit the space requirement for handling a large temporal graph, we divide the stream into θ windows for a given constant number θ. We first define the equal-weight window condition as follows.

Definition 4 (Equal-weight window condition). *Consider that the probability distribution of any edge being active at any time follows a uniform distribution. Under this distribution, the equal-weight condition is satisfied if the stream is divided into θ windows such that the weighted graph of each window is the same in expectation.*

Intuitively, Definition 4 states that *if the probability of any edge being active does not change over time, then the weight of the edge should not change in any of the θ windows.*

Let the time periods of the θ windows be $[t_0, t_1], [t_1, t_2], \ldots, [t_{\theta-1}, t_\tau]$. Then, applying Definition 4, we have

$$\int_{t_0}^{t_1} f(t)dt = \int_{t_1}^{t_2} f(t)dt = \ldots = \int_{t_{\theta-1}}^{t_\tau} f(t)dt = \frac{1}{\theta}\int_{t_0}^{t_\tau} f(t)dt.$$

Take $f(t) = e^{\lambda(t-t_\tau)}$. We first determine t_1 as follows:

$$\int_{t_0}^{t_1} f(t)dt = \frac{1}{\theta}\int_{t_0}^{t_\tau} f(t)dt$$

$$\Rightarrow \frac{1}{\lambda}(e^{\lambda(t_1-t_\tau)} - e^{\lambda(t_0-t_\tau)}) = \frac{1}{\theta\lambda}(e^{\lambda(t_\tau-t_\tau)} - e^{\lambda(t_0-t_\tau)})$$

$$\Rightarrow \theta(e^{\lambda t_1} - e^{\lambda t_0}) = e^{\lambda t_\tau} - e^{\lambda t_0}$$

$$\Rightarrow e^{\lambda t_1} = \frac{e^{\lambda t_\tau} + (\theta-1)e^{\lambda t_0}}{\theta}$$

$$\Rightarrow t_1 = \frac{1}{\lambda}\ln\frac{e^{\lambda t_\tau} + (\theta-1)e^{\lambda t_0}}{\theta}.$$

Similarly, we obtain t_i, where $1 \le i \le \theta - 1$, as follows:

$$t_i = \frac{1}{\lambda}\ln\frac{i \times e^{\lambda t_\tau} + (\theta-i)e^{\lambda t_0}}{\theta}.$$

Based on the above analysis, we define *equal-weight damped time window model* as follows.

Definition 5 (Equal-weight damped window model). *Given a stream that spans the time period $[t_0, t_\tau]$, and an integer θ, equal-weight damped time window model divides the stream into θ windows spanning time periods $[t_0, t_1]$, $[t_1, t_2], \ldots, [t_{\theta-1}, t_\tau]$, where $t_i = \frac{1}{\lambda}\ln\frac{i \times e^{\lambda t_\tau} + (\theta-i)e^{\lambda t_0}}{\theta}$, for $1 \le i \le \theta - 1$.*

Based on Definition 5, we obtain θ weighted graphs derived from the temporal graph that falls into each of the θ windows in the stream. The value of θ is determined by users, which controls the space requirement and the efficiency of graph analysis, as well as the degree of information loss (from the original temporal graph to the θ weighted graphs). The larger the value of θ, the finer is the granularity of the windows and the less is the information loss, but also the more is the memory space needed.

Advantages of the new model. The equal-weight damped time window model has the following advantages: (A1) it is a generalization of existing damped time window models; (A2) it gives equal importance to each window, which makes it easy to compare the graphs from different windows; and (A3) it provides a systematical way for update maintenance of the windows.

For (A1), by defining an appropriate weight density function, we can apply our proposed equal-weight scheme to compute the size of each window for existing damped time window models. Take the logarithmic tilted-time window model as an example, where an exponentially increasing function (e.g., 2^0T, 2^1T, 2^2T, 2^3T, ...) is used. Assume that the time span of the entire stream is $[0, 2^\theta - 1]$, then the weight density function is defined as follows:

$$f(t) = \begin{cases} 1, & 0 \le t < 2^{\theta-1} \\ 2, & 2^{\theta-1} \le t < 2^{\theta-1} + 2^{\theta-2} \\ ..., & ... \\ 2^{\theta-1}, & 2^\theta - 2 \le t \le 2^\theta - 1 \end{cases}$$

For (A2), if the weights of an edge (u, v) in two different windows W_1 and W_2 are the same, then it implies that the probability of (u, v) being active remains the same in W_1 and W_2. Now if the probability of (u, v) being active is higher in W_1, then apparently the weight of (u, v) in W_1 is also higher than that in W_2. This may not be true if existing damped time windows are used unless we define an appropriate $f(t)$ function for them, and apply our scheme proposed in this section to determine the window sizes.

For (A3), we show that our model provides a systematical way for update maintenance of the windows in Sect. 2.3.

2.3 Window Maintenance

As time goes on, new temporal edges are created and the windows need to be updated. We devise an update scheme for our window model as follows.

Let $[t_0, t_1]$, $[t_1, t_2]$, ..., $[t_{\theta-1}, t_\theta]$ denote the θ existing windows, and $[t_\theta, t_{\theta+1}]$ denote the new window. As the current time changes from $t_\tau = t_\theta$ to $t'_\tau = t_{\theta+1}$, the weight density function $f(t)$ changes from $f(t) = e^{\lambda(t-t_\theta)}$ to $f(t) = e^{\lambda(t-t_{\theta+1})}$. Lemmas 1 and 2 state the change needed.

Due to the space limitation, the proofs for all the lemmas and theorems are given in the full version of this paper [23].

Lemma 1. *If the current time changes from t_τ to t'_τ, for any temporal edge whose weight w is last updated at time t_τ, the weight should be updated as follows:*

$$w \leftarrow w \times e^{\lambda(t_\tau - t'_\tau)}.$$

Lemma 2. *Given a weighted graph $G = (V, E, \Pi)$ of any window, if the current time changes from t_τ to t'_τ, the weight w of each edge in E which is computed at time t_τ should be updated as follows*

$$w \leftarrow w \times e^{\lambda(t_\tau - t'_\tau)}.$$

Lemma 2 shows that it is simple to update the weighted graphs of the existing windows as new windows are created in the stream. However, we still need to determine at what point a new window, i.e., $[t_\theta, t_{\theta+1}]$, should be created in the stream, which is to determine $t_{\theta+1}$. Following our discussion in Sect. 2.2, we have

$$\int_{t_{\theta-1}}^{t_\theta} f(t)dt = \int_{t_\theta}^{t_{\theta+1}} f(t)dt$$

$$\Rightarrow e^{\lambda t_\theta} - e^{\lambda t_{\theta-1}} = e^{\lambda t_{\theta+1}} - e^{\lambda t_\theta}$$

$$\Rightarrow t_{\theta+1} = \frac{1}{\lambda} \ln(2e^{\lambda t_\theta} - e^{\lambda t_{\theta-1}}).$$

Similarly, we can also compute the windows that are to follow in the stream, i.e., $[t_{\theta+1}, t_{\theta+2}], \ldots$. However, in this way, the number of windows keeps increasing, and the size of a new window (i.e., the time span of the window) becomes smaller and smaller. To solve these issues, we propose to merge windows to bound the number of windows in the stream within the range $[\theta, 2\theta]$. Specifically, when the number of windows reaches 2θ, we merge every two consecutive windows into one window. In this way, every window in the stream after merging still satisfies the equal-weight window condition. In fact, we can also merge more than two windows into a single window if necessary.

3   Window-Based Network Analysis

We now discuss network analysis based on the equal-weight damped time window model, which we illustrate using an application of connectivity analysis in this section. We also discuss other applications (e.g., community finding) and give a list of open problems about analyzing large temporal graphs using our model, but due to the space limitation we present the details in the full version of this paper [23].

Let $G_1, G_2, \ldots, G_\theta$ denote the θ weighted graphs derived from the θ windows in the stream.

3.1   Connectivity Analysis

Given a weighted graph $G = (V, E, \Pi)$ of a window in the stream (defined in Definition 3, and here the window W is omitted for simplicity), we define a measure of connectivity between two vertices u and v in G as follows.

Algorithm 1. Compute $\gamma(u, v)$

Input : A weighted graph $G = (V, E, \Pi)$, two query vertices u and v
Output: $\gamma(u, v)$

1 Initialize $c[u] \leftarrow \infty$, $c[x] \leftarrow 0$ for every vertex $x \in V \setminus \{u\}$;
2 Let Q be a maximum priority queue, where an element of Q is a pair $(x, c[x])$
   and $c[x]$ being the key;
3 Initialize Q by inserting a single element $(u, c[u])$;
4 **while** Q *is not empty* **do**
5      $(x, c[x]) \leftarrow$ Extract-Max(Q);
6   **if** $x = v$ **then**
7   Goto Line 12;
8   **foreach** *neighbor vertex, y, of x* **do**
9   **if** $c[y] < \min(c[x], \Pi(x, y))$ **then**
10             $c[y] \leftarrow \min(c[x], \Pi(x, y))$;
11   If y is not in Q, push $(y, c[y])$ into Q; otherwise, update $c[y]$ in Q;

12 **return** $\gamma(u, v) = c[v]$;

Definition 6 (Connectivity). *Let* $\mathbb{P}(u, v) = \{P(u, v) : P(u, v)$ *is a path from* u *to* v *in* $G\}$. *The* connectivity *of a path* $P(u, v)$, *denoted by* $\gamma(P(u, v))$, *is defined as the minimum edge weight among the edges on* $P(u, v)$. *The* connectivity *between* u *and* v, *denoted by* $\gamma(u, v)$, *is defined as* $\gamma(u, v) = \max\{\gamma(P(u, v)) : P(u, v) \in \mathbb{P}(u, v)\}$.

Since the weight of each edge in a weighted graph indicates the strength of relationship (or interaction, communication, etc.) between the two end points in the corresponding temporal graph within the time span of the window, the value of $\gamma(u, v)$ reflects the connectivity between u and v within the window, for example, the amount of information that can be passed between u and v via any path within the time span.

Given a *connectivity query* $\gamma(u, v)$, we can answer it using an algorithm similar to Dijkstra's algorithm, as shown in Algorithm 1. Algorithm 1 uses a maximum priority queue Q to keep the current largest connectivity value, $c[x]$, of a path from u to a visited vertex $x \in V$. The algorithm starts from one of the query vertices, u, greedily grows the paths by extending to u's neighbors, and then further grows to the neighbors' neighbors until reaching the other query vertex v. During the greedy process, the $c[x]$ value of a vertex x is updated whenever a larger connectivity value from u to x is found. At each iteration, the vertex with maximum $c[.]$ is extracted from Q to update the $c[.]$ values of its neighbors.

We now show the correctness and complexity (the proof is given in [23]).

Theorem 1. *Algorithm 1 correctly computes the connectivity value* $\gamma(u, v)$ *in* $O((|E| + |V|) \log |V|)$ *time.*

The complexity of Algorithm 1, even if Fibonacci heap is used, is too high to process a connectivity query online. One may pre-compute the connectivity values for all pairs of vertices. However, the space complexity of this method is $O(|V|^2)$, and the pre-computation requires $O((|E| + |V|)|V| \log |V|)$ time, both of which are impractical for handling a large graph. We propose a more efficient way to process connectivity queries, with linear index space.

First, we compute a *maximum spanning tree, MaxST*, of the weighted graph G. Without loss of generality, we assume G is connected. If not, we can consider each connected component of G separately. A MaxST has the cut property. A *cut* is a partition of the vertex set of a graph into two disjoint subsets. We say that *an edge crosses the cut* if it has one endpoint in each subset of the partition. The *cut property* states that for any cut C in the graph, if the weight of an edge e crossing C is larger than the weights of all the other edges crossing C, then e must be contained in every MaxST.

Given a MaxST, T, there is a unique path connecting any two vertices in T. Let $\gamma_T(u, v)$ denote the connectivity value between u and v in the MaxST T. Based on the cut property, we have the following lemma (the proof is given in [23]).

Lemma 3. *Given a MaxST T of a weighted graph G, $\gamma(u, v) = \gamma_T(u, v)$, for any pair of vertices u and v in G.*

Based on Lemma 3, a connectivity query $\gamma(u, v)$ can be answered by first finding the unique path between u and v in the MaxST T, and then returning the minimum edge weight on the path. The query time complexity is $O(|V|)$, which is much better than that of Algorithm 1. Next, we show that we can further reduce the querying time complexity to $O(1)$ time.

We first introduce the concept of *Cartesian tree* [18], which is a binary tree derived from a sequence of numbers. Given an array A of n numbers ($A[0]$ to $A[n-1]$), the root of the Cartesian tree is the minimum number among all the n numbers. Let $A[i]$ be the minimum number, i.e., the root. Then, its left subtree is computed recursively on the numbers $A[0]$ to $A[i-1]$, while its right subtree is computed recursively on the numbers $A[i+1]$ to $A[n-1]$.

We construct a Cartesian tree, C_T, based on a MaxST T. The root node of C_T is the edge with the minimum weight among all the edges in T. Then, by removing this edge, T will be partitioned into two subtrees. Following a similar procedure, we can recursively construct the left and right subtrees of the root node. When removing an edge (u, v), if u (and/or v) is not an end point of any remaining edges in T, then we also create a leaf node u (and/or v) as the child of the node (u, v) in C_T. Thus, the set of leaf nodes in the tree C_T corresponds to the set of vertices in T.

Based on the Cartesian tree C_T, given a connectivity query $\gamma(u, v)$, we first find the lowest common ancestor (LCA) of the two leaves u and v in C_T, and then return the weight of the edge in T that corresponds to the LCA. The following example demonstrates the concepts of MaxST, Cartesian tree, and how to answer a connectivity query.

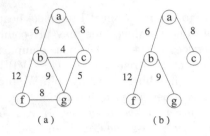

Fig. 2. (a) A **weighted graph** G, and (b) a **MaxST** T

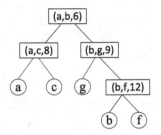

Fig. 3. The **Cartesian tree** C_T of T in Fig. 2(b)

Example 2. Figure 2(a) shows a weighted graph G and Fig. 2(b) shows a MaxST T of G. It is easy to verify $\gamma(u,v) = \gamma_T(u,v)$. For example, $\gamma(c,f) = 6$ in G and $\gamma_T(c,f) = 6$ in T. Figure 3 shows the Cartesian tree C_T of T. The root node of C_T is the edge $(a,b,6)$, since this edge is the one with the minimum weight in T. Removing $(a,b,6)$ partitions T into two components $\{a,c\}$ and $\{b,f,g\}$. Following a similar procedure recursively, we obtain C_T. The leaves of C_T are the vertices in T. Then, to find the connectivity value between any two vertices, we find the LCA of these two vertices in C_T. For example, given a connectivity query $\gamma(f,g)$, we find that the edge $(b,g,9)$ is the LCA of the leaves f and g in C_T. Thus, we return 9 as the answer for $\gamma(f,g)$. It is easy to verify that the answer is correct.

Now, we give the complexity of processing a connectivity query and of constructing the index (the proof is given in [23]).

Theorem 2. *A connectivity query $\gamma(u,v)$ can be answered in $O(1)$ time with an index using $O(|V|)$ space, and the index construction time is $O((|E| + |V|) \log |V|)$.*

Given the θ weighted graphs $G_1, G_2, \ldots, G_\theta$ from the θ windows in the stream, we define the connectivity between u and v in the entire θ windows as $\Gamma(u,v) = \min\{\gamma_1(u,v), \ldots, \gamma_\theta(u,v)\}$, where $\gamma_i(u,v)$ is the connectivity value $\gamma(u,v)$ in the weighted graph G_i, for $1 \le i \le \theta$. Since θ is a constant, the query $\Gamma(u,v)$ can be answered in constant time with indexes of size linear to the number of vertices.

3.2   Queries on a Random Window

Besides the need of analysis on a temporal graph in the whole time window, users may also be interested in analyzing the graph in any time period. For example, user A is interested in time window $[1, 20]$, while user B is interested in time range $[10, 40]$. To satisfy each user's need, the naive way is to store the complete temporal graph and extract the temporal subgraph from the required time range, which is not practical due to the massive size of the complete graph. We discuss how to efficiently obtain a weighted graph of any time period based on the equal-weight damped time window model.

Given a random window $W = [t_x, t_y]$, we are required to return $G_W = (V, E_W, \Pi_W)$. Given θ weighted graphs, $G_1 = (V, E_1, \Pi_1)$, $G_2 = (V, E_2, \Pi_2)$, ..., $G_\theta = (V, E_\theta, \Pi_\theta)$, we return an approximate weighted graph G'_W of G_W as follows.

Let $t_i < t_x \leq t_{i+1}$ and $t_j \leq t_y < t_{j+1}$. First, we return an approximate weighted graph $G'_{[t_x, t_{i+1}]}$ of $G_{[t_x, t_{i+1}]}$. $G'_{[t_x, t_{i+1}]} = (V, E'_{[t_x, t_{i+1}]}, \Pi'_{[t_x, t_{i+1}]})$ is computed as follows:

- $E'_{[t_x, t_{i+1}]} = E_{i+1}$,
- $\Pi'_{[t_x, t_{i+1}]}(e) = \Pi_{i+1}(e) \times \dfrac{\int_{t_x}^{t_{i+1}} f(t)dt}{\int_{t_i}^{t_{i+1}} f(t)dt}$, for each $e \in E'_{[t_x, t_{i+1}]}$.

In other words, $G'_{[t_x, t_{i+1}]}$ is computed based on $G_{i+1} = (V, E_{i+1}, \Pi_{i+1})$ in expectation. Similarly, we compute an approximate weighted graph $G'_{[t_j, t_y]}$ of $G_{[t_j, t_y]}$. Then, we have $G'_W = (V, E'_W, \Pi'_W)$ as follows:

- $E'_W = E'_{[t_x, t_{i+1}]} \cup E_{i+2} \cup \ldots \cup E'_{[t_j, t_y]}$,
- $\Pi'_W(e) = \Pi'_{[t_x, t_{i+1}]}(e) + \Pi_{i+2}(e) + \ldots + \Pi'_{[t_j, t_y]}(e)$, for each $e \in E'_W$.

4   Experimental Results

We evaluated the usefulness of our equal-weight window model by showing the quality of the θ weighted graphs obtained based on the model, and the efficiency and quality of graph analysis based on these weighted graphs. We also verified the efficiency of dynamic update maintenance and the scalability of our method, where the results of them are reported in [23] due to the space limitation. All the experiments were run on a Linux machine with an Intel 3.3 GHz CPU and 16 GB RAM. All the programs were implemented in C++ and complied using G++ 4.8.2.

We used 8 real temporal graphs for our experiments, as shown in Table 1, where we list the number of vertices and edges in each graph $\mathbb{G}$, the average degree in $\mathbb{G}$ (denoted by $d_{avg}(v, \mathbb{G})$), and the number of distinct time instances in $\mathbb{G}$ (denoted by $|T_\mathbb{G}|$). The **phone** graph consists of call records in Ivory Coast [1], where the call records were collected over a span of 150 days. The other 7 graphs were obtained from the Koblenz Large Network Collection (http://konect.uni-koblenz.de/), where one large temporal graph was selected from each

Table 1. Real temporal graphs ($K = 10^3$)

Dataset	$|\mathbb{V}|$	$|\mathbb{E}|$	$d_{avg}(v, \mathbb{G})$	$|T_{\mathbb{G}}|$
phone	1,237	338,008,540	273,248.62	3,369
arxiv	28,094	9,193,606	327.24	2,337
elec	8,298	214,028	25.79	101,063
enron	87,274	2,282,904	26.16	220,364
facebook	46,953	1,730,624	36.86	867,939
lastfm	174,078	38,254,660	219.76	17,498,009
email	168	164,613	979.84	57,842
conflict	118,101	5,903,522	49.99	312,457

of the following 7 categories: arxiv-HepPh (arxiv) from the arxiv networks; elec from the network of English Wikipedia; enron from the email networks; facebook-links (facebook) from the facebook network; lastfm-band (lastfm) from the music website last.fm; radoslaw-email (email) from the internal email communication network between employees of a mid-sized manufacturing company; wikiconflict (conflict) indicating conflicts between users of Wikipedia.

4.1   Results on Weighted Graph Construction

In this experiment, we evaluated the space requirement and the construction time of the θ weighted graphs for each of the temporal graphs, and then we measured the quality of the weighted graphs. We tested θ from 10 to 50. We set the value of $\lambda = 10^{-x}$ for the weight density function given in Definition 1, where $10^x \leq |T_{\mathbb{G}}| < 10^{x+1}$, that is, $\lambda = 10^{-\lfloor \log_{10} |T_{\mathbb{G}}| \rfloor}$. For example, for the phone graph, $\lambda = 10^{-3}$.

Space requirement. We first report the space requirement for the θ weighted graphs, as a percentage of the original temporal graph shown in Fig. 4. As the value of θ increases, the total size of the θ weighted graphs also increases. But the rate of increase is slow. For graphs with high average degree, the total size of the θ weighted graphs is only a small percentage of the original temporal

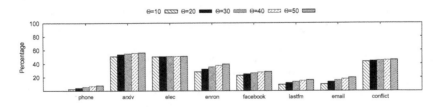

Fig. 4. The total size of the θ weighted graphs compared with the original temporal graph $\mathbb{G}$

Table 2. Construction time of θ weighted graphs (in seconds)

Dataset	$\theta = 10$	$\theta = 20$	$\theta = 30$	$\theta = 40$	$\theta = 50$
phone	130.3067	137.6559	143.6432	148.6535	153.1762
arxiv	4.0591	4.2070	4.3718	4.4772	4.5788
elec	0.1110	0.1168	0.1229	0.1266	0.1292
enron	0.8419	0.9031	0.9600	1.0041	1.0473
facebook	0.6245	0.6743	0.6996	0.7325	0.7581
lastfm	12.6525	13.2842	14.0147	14.6400	15.5061
email	0.0511	0.0548	0.0575	0.0607	0.0617
conflict	2.8762	2.9693	3.0447	3.1341	3.2189

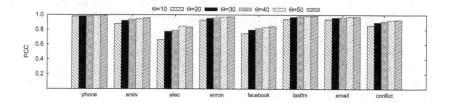

Fig. 5. PCC between G'_W and G_W for different θ

graph. For example, for the phone graph, even the total size of 50 weighted graphs is less than 10% of the original temporal graph. We emphasize that for temporal graphs, the set of vertices remains relatively stable while the number of temporal edges grows linearly over time, and thus the result verifies that our method can handle large temporal graphs as they grow over time, with small space requirement.

Construction time. Table 2 reports the time taken to read the temporal graphs from disk and construct the corresponding θ weighted graphs, for different values of θ. The construction is fast for all graphs as we only need to scan the graphs once, regardless of the value of θ. The construction time increases as θ increases because more weighted graphs need to be constructed, but the rate of increase is slow as scanning the original temporal graph dominates the cost.

Quality of results. Next, we examine the quality of the weighted graphs. To do this, we constructed a weighted graph, G_W, directly from the original temporal graph within a time window W, as defined in Definition 3. We also constructed an approximate weighted graph G'_W of G_W from the θ weighted graphs as discussed in Sect. 3.2. Then, we compared G_W and G'_W.

We computed G_W and G'_W for 100 randomly generated windows, $W = [t_x, t_y]$, where we ensured that W is a valid window by ensuring $t_x < t_y$. We use *Pearson correlation coefficient (PCC)* to measure the degree of linear correlation between G'_W and G_W, and report the results in Fig. 5.

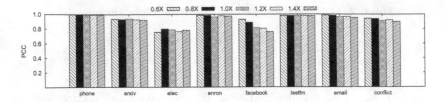

Fig. 6. PCC between G'_W and G_W for different λ ($\theta = 30$)

The result shows that we obtain high PCC values in most of the cases, which implies that analysis conducted on the approximate graph G'_W shares similar patterns/trends with that conducted on the exact graph G_W (we further verify this point by applying the application in Sect. 3. The results can be found in [23] (e.g., Fig. 8 in [23]) which lead to a similar conclusion as Fig. 5. As θ increases from 10 to 50, the PCC values also increase, verifying that a larger θ leads to less information loss and hence higher correlation between G'_W and G_W. For a number of graphs, the PCC values are close to 1. The results are particularly impressive for the **phone** graph, for which the space requirement is also very small as shown in Fig. 4.

Next, we tested the effect of different values of λ. In all the other experiments, we set $\lambda = 10^{-\lfloor \log_{10} |T_G| \rfloor}$ as default. In this experiment, we tested λ at 0.6, 0.8, 1.0, 1.2, and 1.4 of its default value, and fixed $\theta = 30$. The result, as reported in Fig. 6, shows that the PCC values are not much affected by the change in λ, and in all cases the PCC values are high.

Efficiency of graph analysis. We also evaluated the efficiency of using the θ weighted graphs for connectivity analysis. We varied θ from 10 to 50, and tested 1000 randomly generated connectivity queries. We used the index presented in Sect. 3.1 to answer the queries, and compared with the online algorithm given in Algorithm 1. We denote these two methods by *Index* and *Online*, respectively. Table 3 reports the average processing time per query. The result shows that *Index* is more than 3 orders of magnitude faster than *Online*, verifying the efficiency of our method. The index construction time and the index size are also small, which are linear to the number of vertices (as shown Table 1).

Due to the space limitation, we report more results in [23], which show that our method is efficient and effective for core community analysis in temporal graphs, is fast in dynamic update maintenance, and has good scalability.

5   Related Work

Much of the work on temporal graphs, also called time-varying graphs or timetable graphs, was related to temporal paths. Temporal paths have been applied to study the connectivity of a temporal graph [9], the information latency in a temporal network [10], small-world behavior [17], and to find temporal connected components [16]. Temporal paths have also been used to define metrics for

Table 3. Average query processing time of connectivity queries (in milliseconds)

	$\theta = 10$		$\theta = 20$		$\theta = 30$		$\theta = 40$		$\theta = 50$	
	Index	Online	Index	Online	Index	Online	Index	Online	Index	Online
phone	0.0041	24.0973	0.0059	42.0291	0.0082	56.7214	0.0095	71.6784	0.0116	85.7897
arxiv	0.0045	21.6051	0.0081	19.8686	0.0127	18.9631	0.0160	18.3361	0.0188	17.8703
elec	0.0029	0.5434	0.0057	0.5460	0.0083	0.6127	0.0108	0.6866	0.0140	0.7909
enron	0.0049	5.6191	0.0103	6.2276	0.0132	6.8081	0.0186	7.7741	0.0230	8.7257
facebook	0.0052	5.7762	0.0095	5.5650	0.0141	5.6486	0.0185	5.9857	0.0231	6.4586
lastfm	0.0062	31.5103	0.0122	34.3018	0.0201	40.0901	0.0249	43.6900	0.0331	46.0317
email	0.0004	0.1239	0.0008	0.1856	0.0016	0.2295	0.0022	0.2649	0.0030	0.3012
conflict	0.0052	15.5628	0.0098	14.3276	0.0149	13.6799	0.0195	14.3760	0.0255	15.2124

temporal network analysis, such as temporal efficiency and temporal clustering coefficient [15,16], and temporal closeness [13]. Algorithms for computing temporal paths were discussed in [19,20,25]. Indexing method for answering reachability and time-based path queries in a temporal graph was proposed in [22]. Diversified subgraph pattern mining in a temporal graph was introduced in [30]. Core decomposition in a large temporal graph was studied in [21]. Readers can also refer to more comprehensive surveys on temporal graphs [4,8,12], and more related work can be found in the full version of this paper [23].

There are also works on storing temporal graphs in a compact way [2,3,7]. In [2], a compressed suffix array strategy was proposed to store temporal graphs. In [3], two data structures, compact adjacency sequence and compact events ordered by time, were proposed to represent temporal graphs. However, all these methods need to store each temporal edge, and their performance is not better than the *gzip* compression.

6 Conclusions

We proposed the *equal-weight damped time window* model for processing massive growing temporal graphs. Our model allows users to set the number of windows to trade off between the required space and the information loss. Based on this model, we presented an application of connectivity analysis to analyze the temporal graph. We conducted comprehensive experiments to verify the usefulness and efficiency of our method for analyzing large temporal graphs. As for future work, we plan to explore how to integrate the proposed time window model into our prior work on distributed graph processing systems [26–29] to analyze massive dynamic temporal graphs.

Acknowledgements. We thank the reviewers for their valuable comments. The authors are supported by the Hong Kong GRF 2150851 and 2150895, ITF 6904079, MSRA grant 6904224, and CUHK Grants 3132964 and 3132821.

References

1. Blondel, V.D., Esch, M., Chan, C., Clérot, F., Deville, P., Huens, E., Morlot, F., Smoreda, Z., Ziemlicki, C.: Data for development: the D4D challenge on mobile phone data. CoRR, abs/1210.0137 (2012)
2. Brisaboa, N.R., Caro, D., Fariña, A., Rodríguez, M.A.: A compressed suffix-array strategy for temporal-graph indexing. In: Moura, E., Crochemore, M. (eds.) SPIRE 2014. LNCS, vol. 8799, pp. 77–88. Springer, Heidelberg (2014). doi:10.1007/978-3-319-11918-2_8
3. Caro, D., Rodríguez, M.A., Brisaboa, N.R.: Data structures for temporal graphs based on compact sequence representations. Inf. Syst. **51**, 1–26 (2015)
4. Casteigts, A., Flocchini, P., Quattrociocchi, W., Santoro, N.: Time-varying graphs and dynamic networks. Int. J. Parallel Emergent Distrib. Syst. **27**(5), 387–408 (2012)
5. Chen, Y., Dong, G., Han, J., Wah, B.W., Wang, J.: Multi-dimensional regression analysis of time-series data streams. In: VLDB, pp. 323–334 (2002)
6. Datar, M., Gionis, A., Indyk, P., Motwani, R.: Maintaining stream statistics over sliding windows. SIAM J. Comput. **31**(6), 1794–1813 (2002)
7. Bernardo, G., Brisaboa, N.R., Caro, D., Rodríguez, M.A.: Compact data structures for temporal graphs. In: DCC, p. 477 (2013)
8. Holme, P., Saramäki, J.: Temporal networks. CoRR, abs/1108.1780 (2011)
9. Kempe, D., Kleinberg, J.M., Kumar, A.: Connectivity and inference problems for temporal networks. J. Comput. Syst. Sci. **64**(4), 820–842 (2002)
10. Kossinets, G., Kleinberg, J.M., Watts, D.J.: The structure of information pathways in a social communication network. In: KDD, pp. 435–443 (2008)
11. Lai, J., Wang, C., Yu, P.S.: Dynamic community detection in weighted graph streams. In: SDM, pp. 151–161 (2013)
12. Müller-Hannemann, M., Schulz, F., Wagner, D., Zaroliagis, C.: Timetable information: models and algorithms. In: Geraets, F., Kroon, L., Schoebel, A., Wagner, D., Zaroliagis, C.D. (eds.) Algorithmic Methods for Railway Optimization. LNCS, vol. 4359, pp. 67–90. Springer, Heidelberg (2007). doi:10.1007/978-3-540-74247-0_3
13. Pan, R.K., Saramäki, J.: Path lengths, correlations, and centrality in temporal networks. Phys. Rev. E **84**, 016105 (2011)
14. Perng, C., Wang, H., Zhang, S.R., Jr., D.S.P.: Landmarks: a new model for similarity-based pattern querying in time series databases. In: ICDE, pp. 33–42 (2000)
15. Tang, J., Musolesi, M., Mascolo, C., Latora, V.: Temporal distance metrics for social network analysis. In: WOSN, pp. 31–36 (2009)
16. Tang, J., Musolesi, M., Mascolo, C., Latora, V.: Characterising temporal distance and reachability in mobile and online social networks. Comput. Commun. Rev. **40**(1), 118–124 (2010)
17. Tang, J., Scellato, S., Musolesi, M., Mascolo, C., Latora, V.: Small-world behavior in time-varying graphs. Phys. Rev. E **81**(5), 055101 (2010)
18. Vuillemin, J.: A unifying look at data structures. Commun. ACM **23**(4), 229–239 (1980)
19. Wu, H., Cheng, J., Huang, S., Ke, Y., Lu, Y., Xu, Y.: Path problems in temporal graphs. PVLDB **7**(9), 721–732 (2014)
20. Wu, H., Cheng, J., Ke, Y., Huang, S., Huang, Y., Wu, H.: Efficient algorithms for temporal path computation. IEEE Trans. Knowl. Data Eng. **28**(11), 2927–2942 (2016)

21. Wu, H., Cheng, J., Lu, Y., Ke, Y., Huang, Y., Yan, D., Wu, H.: Core decomposition in large temporal graphs. In: IEEE International Conference on Big Data, pp. 649–658 (2015)
22. Wu, H., Huang, Y., Cheng, J., Li, J., Ke, Y.: Reachability and time-based path queries in temporal graphs. In: ICDE, pp. 145–156 (2016)
23. Wu, H., Zhao, Y., Cheng, J., Yan, D.: Efficient processing of growing temporal graphs (2016). http://www.cse.cuhk.edu.hk/%7ejcheng/papers/tm_tr.pdf
24. Xie, W., Tian, Y., Sismanis, Y., Balmin, A., Haas, P.J.: Dynamic interaction graphs with probabilistic edge decay. In: ICDE, pp. 1143–1154 (2015)
25. Xuan, B.-M.B., Ferreira, A., Jarry, A.: Computing shortest, fastest, and foremost journeys in dynamic networks. Int. J. Found. Comput. Sci. **14**(2), 267–285 (2003)
26. Yan, D., Cheng, J., Lu, Y., Ng, W.: Blogel: a block-centric framework for distributed computation on real-world graphs. PVLDB **7**(14), 1981–1992 (2014)
27. Yan, D., Cheng, J., Lu, Y., Ng, W.: Effective techniques for message reduction and load balancing in distributed graph computation. In: WWW, pp. 1307–1317 (2015)
28. Yan, D., Cheng, J., Özsu, M.T., Yang, F., Lu, Y., Lui, J.C.S., Zhang, Q., Ng, W.: A general-purpose query-centric framework for querying big graphs. PVLDB **9**(7), 564–575 (2016)
29. Yang, F., Li, J., Cheng, J.: Husky: towards a more efficient and expressive distributed computing framework. PVLDB **9**(5), 420–431 (2016)
30. Yang, Y., Yan, D., Wu, H., Cheng, J., Zhou, S., Lui, J.C.S.: Diversified temporal subgraph pattern mining. In: SIGKDD, pp. 1965–1974 (2016)

Effective k-Vertex Connected Component Detection in Large-Scale Networks

Yuan Li[1], Yuhai Zhao[1(✉)], Guoren Wang[1], Feida Zhu[2],
Yubao Wu[3], and Shengle Shi[1]

[1] Northeastern University, Shenyang, China
zhaoyuhai@ise.neu.edu.cn
[2] Singapore Management University, Singapore, Singapore
[3] Georgia State University, Atlanta, USA

Abstract. Finding components with high connectivity is an important problem in component detection with a wide range of applications, e.g., social network analysis, web-page research and bioinformatics. In particular, k-edge connected component (k-ECC) has recently been extensively studied to discover disjoint components. Yet many real applications present needs and challenges for overlapping components. In this paper, we propose a k-vertex connected component (k-VCC) model, which is much more cohesive and therefore allows overlapping between components. To find k-VCCs, a top-down framework is first developed to find the exact k-VCCs. To further reduce the high computational cost for input networks of large sizes, a bottom-up framework is then proposed. Instead of using the structure of the entire network, it locally identifies the seed subgraphs, and obtains the heuristic k-VCCs by expanding and merging these seed subgraphs. Comprehensive experimental results on large real and synthetic networks demonstrate the efficiency and effectiveness of our approaches.

1 Introduction

Component detection is a fundamental problem [11,19] in the analysis of large-scale networks. Many real applications can benefit from finding highly connected components. For example, groups of intimate entities discovered in social networks can be exploited to analyze their social behaviors [11]; a set of servers with common contents in web server networks can be used to construct the network index [4]; clusters of interactive genetic markers discovered in genetic interaction networks can be utilized to infer the corresponding cause of diseases [25].

The existing methods of component detection can be roughly divided into two main categories, i.e. clique-based methods and clique-relaxed methods. According to differently relaxed constraints, clique-relaxed methods can be further divided into degree-relaxed [3,6,27], distance-relaxed [14,16] and triangulation-relaxed [22,23] methods. Although succeeding to some extent, these methods still have respective drawbacks.

© Springer International Publishing AG 2017
S. Candan et al. (Eds.): DASFAA 2017, Part II, LNCS 10178, pp. 404–421, 2017.
DOI: 10.1007/978-3-319-55699-4_25

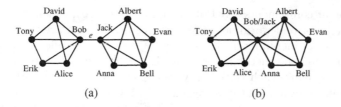

Fig. 1. A toy co-friendship network.

A toy co-friendship network is considered in Fig. 1(a) as an example. Degree-relaxed and distance-relaxed methods often consider the network as an indivisible whole, saying a k-core with $k = 3$ and a n-club with $n = 3$, since the degree of every vertex is no less than 3 and the maximum length of shortest pathes of all pairs of vertices is no larger than 3, respectively. This contradicts with the intuition that Fig. 1(a) should be disconnected into two components by deleting edge e. The reason is that these two methods only concern about the degree and distance but ignore the connectivity of any pair of vertices. Although triangulation-relaxed methods can detect these two components, they do not work when there are no triangles in graph, e.g. bipartite graphs.

Connectivity is measured by the number of disjoint paths between vertices. Intuitively, high connectivity would contribute to the steadiness and robustness of the component. A component with high connectivity could still be connected, even if losing a few relations or entities. Therefore, recently, connectivity-based methods such as k-edge connected components (k-ECC) have drawn great attention [1,5,28]. A k-ECC refers to a maximal subgraph, the remaining subgraph of which is still connected after any $k - 1$ edges are removed from it. The network in Fig. 1(a) is a 1-ECC. Since the low edge connectivity, it is naturally divided into two separate 3-ECCs, {*Bob, David, Tony, Erik, Alice*} and {*Jack, Anna, Bell, Evan, Albert*}.

However, k-ECC still has its own limitation. For example, if *Bob* is an alias of *Jack*, Fig. 1(a) is equivalent to Fig. 1(b). In this case, Fig. 1(b) is identified as a whole 3-ECC, although there are practically two separate components, since once we remove the vertex *Bob/Jack*, the network becomes disconnected. Thus, high edge connectivity does not necessarily indicate a component of strong connectivity.

In this paper, we study the *k-vertex connected component (k-VCC) detection problem*, which focuses on vertex connectivity instead of edge connectivity, of networks. Given a graph G, the goal is to find all such induced subgraphs, g', of G that g' is still connected after removing any $k - 1$ vertices from it and no supergraph of g' has the same property. According to this informal definition, the network in Fig. 1(b) can be identified as two 3-VCCs, {*Bob/Jack, David, Tony, Erik, Alice*} and {*Bob/Jack, Anna, Bell, Evan, Albert*}. Unlike k-ECC, k-VCCs has three unique advantages: (1) k-VCC captures more connectivity of networks than k-ECC. According to [8], a component of high vertex connectivity must be of high edge connectivity, but not vice versa; (2) k-VCC allows overlap among different components, say Bob/Jack in Figure 1(b), which is more natural and reasonable for real-world networks [18]; (3) k-VCC can prevent the detected communities from including irrelevant subgraphs, i.e. free rider effect [24].

Unfortunately, the methods for k-ECC [1,5] cannot be directly utilized to find k-VCCs. Because each vertex only belongs to at most one k-ECC [5], these methods could obtain k-ECCs by vertex contraction [21]. In our case, however, each vertex can be in more than one k-VCCs, which makes the former trick not work. In this paper, we devise two novel frameworks to tackle the k-VCC detection problem, namely top-down and bottom-up frameworks for k-VCC detection, respectively. The top-down iteratively divides the networks by finding minimum vertex cut set, which could find all exact k-VCCs. The bottom-up framework, instead of using the entire network structure, locally identifies the seed subgraphs, and obtains the heuristic k-VCCs by expanding and merging these seed subgraphs.

Our contributions are summarized as below: (1) a novel k-VCC detection problem is proposed from the perspective of vertex connectivity; (2) the top-down and bottom-up frameworks are developed to solve the problem. Specifically, in the bottom-up framework, a concept of local k-vertex connected subgraph is proposed to accelerate k-VCC detection, which enables identifying seed subgraphs locally instead of globally. In addition, several optimization techniques are proposed to further reduce the search space; (3) extensive experiments on both real and synthetic datasets demonstrate the efficiency and effectiveness of our frameworks.

The rest of our paper is organized as follows. We give the notions and problem statement in Sect. 2. In Sects. 3 and 4, we present the top-down and bottom-up k-VCC detection frameworks, respectively. Extensive experiments are reported in Sect. 5. The related work is discussed in Sect. 6. Section 7 concludes this work.

2   Notions and Problem Statement

In this paper, we focus on an undirected and unweighted graph $G(V, E)$, where V is the set of vertices and E is the set of edges. We denote the number of vertices and the number of edges by $n = |V|$ and $m = |E|$, respectively. A graph $G[S]$ is called an *induced subgraph* of G when $S \subseteq V$, and $E(S) = \{(u, v) \in E | u, v \in S\}$. We use $nb_G(v)$ to denote the set of neighbors of a vertex v in G, that is, $nb_G(v) = \{u | (u, v) \in E\}$. We define the degree of v in G as $deg_G(v) = |nb(v)|$. If there is no ambiguity, we denote them as $nb(v)$ and $deg(v)$. In addition, d_{max} denotes the maximum vertex degree of G.

Notions. We first give some formal definitions used in this work.

Definition 1 *(Vertex connectivity of two vertices). Let u and v be two vertices in graph G. If $(u, v) \notin E$, we define the vertex connectivity between u and v, $\kappa(u, v)$ as the least number of vertices chosen from $V - \{u, v\}$, whose deletion from G will disconnect u and v (destroy every vertex disjoint path between u and v), and if $(u, v) \in E$, then set $\kappa(u, v) = n - 1$.*

Definition 2 *(Vertex connectivity of a graph). The vertex connectivity of a graph G denoted as $\kappa(G)$ is the least cardinality $|S|$ of a vertex set $S \subseteq V$ such that $G[V \backslash S]$ is either disconnected or trivial (graph with only one vertex). Such a vertex set S is called a* minimum vertex cut set.

Obviously, $\kappa(G)$ can be expressed in terms of $\kappa(v, w)$ as follows: $\kappa(G) = \min\{\kappa(v, w)|unordered\ pair\ v, w\ in\ G\}$.

Definition 3 *(k-vertex connected graph). A graph $G(V, E)$ is k-vertex connected if the remaining graph is still connected after the removal of any $(k-1)$ vertices from G, in other words, $\kappa(G) \geq k$.*

Specially, we define the graph with only one vertex is trivial and the vertex connectivity of a complete graph K_n is $(n-1)$. In other words, if a graph G is k-vertex connected, there are at least $(k+1)$ vertices in it.

Definition 4 *(k-vertex connected component). Given a graph $G(V, E)$, a subgraph $G[S]$ ($S \subseteq V$) of G is a k-vertex connected component (k-VCC) if (1) $G[S]$ is k-vertex connected, and (2) any supergraph $G[S']$ ($S \subset S' \subseteq V$) is not k-vertex connected.*

For example, in Fig. 2, graph G_1, G_2, G_3 and G_4 are all 3-vertex connected subgraphs, while only G_3 and G_4 are 3-VCCs, because G_1 and G_2 are contained in G_4.

Problem statement. Here, we give the formal problem statement.

Problem 1 (k-VCC detection problem). Given a graph $G(V, E)$ and an integer k, we study the problem of efficiently computing all k-VCCs of G.

In theory, the value of k in k-VCC ranges from 1 to $n-1$, however, it is unlikely to reach $n-1$ in practice, because if k is large enough, the final result of k-VCCs is probably an empty set. Here, we give the upper bound of parameter k. It is highly related with k-core, which is the maximal subgraph $G[C_k]$ of G such that $\forall v \in C_k, deg_{G[C_k]}(v) \geq k$. The core number of a vertex $v \in V$, denoted as $\psi(v)$, is the largest k such that v is in $G[C_k]$. In other words, $\psi(v) = k$ means that $v \in C_k$ and $v \notin C_{k+1}$.

Lemma 1. *All the k-VCCs in graph G are included in the k-core subgraph of G.*

Proof. Based on Definition 3, in each k-VCC $G[S], \forall u \in S, deg_{G[S]}(u) \geq k$. And, k-core is the maximal subgraph $G[V_k]$ of G such that $\forall v \in V_k, deg_{G[V_k]}(v) \geq k$. Thus, for any vertex in the k-VCC, it must be contained in the corresponding k-core subgraph. Also, k-VCC and k-core are induced subgraphs, hence all the edges in k-VCC are contained in the k-core.

Definition 5. *(Degeneracy of G). The degeneracy $\mathcal{D}$ of $G(V, E)$ is the largest k for which G has a non-empty k-core, i.e., $\mathcal{D} = \max\limits_{v \in V} \psi(v)$.*

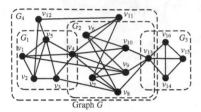

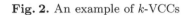

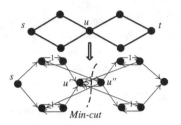

Fig. 2. An example of k-VCCs

Fig. 3. An example of computing minimum vertex cut

Theorem 1. *The value of k in k-VCC is upper bounded by the degeneracy $\mathcal{D}$ of G.*

Theorem 1 could be directly induced from Lemma 1. The degeneracy $\mathcal{D}$ of graph G could be calculated efficiently by the algorithm proposed by Batagelj and Zaversnik [2], which repeatedly remove vertices from G whose degree is less than k until no more vertices can be removed and use bin-sort to order the vertices to achieve $O(m+n)$ time complexity. Therefore, once finding out the given k value is larger than $\mathcal{D}$, we will make sure that the result for k-VCCs in G is $\emptyset$.

3   Top-Down Framework for k-VCCs Detection

In this section, we detail the top-down framework for k-VCCs detection. The main idea of this framework is to iteratively compute the minimum vertex cut set V_{cut} of the current graph $G[C]$, if $|V_{cut}| \geq k$, then $G[C]$ is a k-VCC; otherwise V_{cut} and their incident edges are copied to each remaining connected subgraph after deleting V_{cut} and their induced edges from $G[C]$ and these newly constructed subgraphs are saved in a queue structure Q for further consideration. Algorithm 1 summarizes this process. First, Lemma 1 enable us to exploit k-core to shrink the scale of G, which is possible to divide the original big graph G into several subgraphs of much smaller scale (line 1). In this way, the computation cost of the minimum vertex cut set V_{cut} of G is largely reduced.

Another important problem is how to find the minimum vertex cut set (line 9). Unlike the min-cut [21] and GomoryHu tree [12] methods, which can efficiently find the minimum edge cut set in an undirected graph, we have to reduce the problem of computing $\kappa(G)$ into a maximum flow problem in directed graph.

For each input $G(V,E)$ and vertices s,t, we construct the directed flow network $G'(V',E')$ as follows.

1. For each $v \in V$ ($v \neq s$ and $v \neq t$), add two vertices v', v'' into V', and the directed edge (v', v'') and (v'', v') into E'. The edge (v', v'') has weight 1 and (v'', v') has weight ∞.

Algorithm 1. Top-down Framework

    Input: $G(V, E)$, k
    Output: The set of k-VCCs $\mathcal{G}[\mathcal{R}]$
1 Find the k-core $G[V_k]$ of G;
2 Initialize queue $Q \leftarrow \emptyset$;
3 $Q.enqueue$(all the connected subgraph of $G[V_k]$);
4 **while** $Q \neq \emptyset$ **do**
5 | $G[C] \leftarrow Q.dequeue()$;
6 | **if** $|G[C]| > k$ **then**
7 | | Find the minimum vertex cut set V_{cut} of $G[C]$;
8 | | **if** $|V_{cut}| < k$ **then**
9 | | | Find all the connected subgraphs of $G[C \backslash V_{cut}]$, denoted as $G[C_i]$;
10 | | | Add V_{cut} and the induced edges into each $G[C_i]$;
11 | | | $Q.enqueue$(all $G[C_i]$);
12 | | **else**
13 | | $\lfloor$ Put $G[C]$ into $\mathcal{G}[\mathcal{R}]$;

14 Return $\mathcal{G}[\mathcal{R}]$;

---

2. For each edge $(s, v) \in E$, add edge (s, v') to E'; for each edge $(v, t) \in E$, add edge (v'', t) to E'; for each other edge $(u, v) \in E$, add two edges (u'', v') and (v'', u') to E'. Each edge has capacity ∞.
3. Assign s as the source vertex and t as the sink vertex.

Even and Tarjan [10] prove that $\kappa(s, t)$ in an undirected graph G is equivalent to the maximal value of flow from s to t in the corresponding constructed directed graph G'. Figure 3 shows an example of the above process. Also, $\kappa(G)$ can be calculated in $O(n - \delta - 1 + \delta(\delta - 1)/2)$ calls to maximum flow algorithm where $\delta = d_{max}$ [9]. Actually, we do not need to find the minimum vertex cut set every time. For the current subgraph $G[C]$, once we discover a $\kappa(u, v) < k$ where $u, v \in G[C]K$, $G[C]$ can not be a k-VCC, we can safely terminate this process and use the vertex cut set corresponding to u, v to separate $G[C]$. Theorem 2 guarantees the top-down framework is correct.

Complexity analysis. Computing $\kappa(v, w)$ on graph $G[C]$ needs $O(m'n'^{2/3})$ time where n' is the average size of C and m' is the average size of $E(G[C])$. In the worst case, it needs to be invoked $O(n' - \delta - 1 + \delta(\delta - 1)/2)$ times. Let L represent the total number of $G[C]$ detected in the algorithm. The overall running time of the top-down framework is $O((n' - \delta - 1 + \delta(\delta - 1)/2) \cdot m'n'^{2/3} \cdot L)$.

Theorem 2. *Given a graph G and a value k, the top-down framework for k-VCC detection can correctly find all the k-VCCs.*

Proof. Suppose a graph $G[V_0] \in \mathcal{G}[\mathcal{R}]$ is k-vertex connected subgraph, but not a k-VCC, which indicates it is not maximal, then there must be a k-VCC $G[V_{max}]$ such that $V_0 \subset V_{max}$, and there must also exist a vertex cut set in some loop which separates a vertex or some vertices in V_{max} away from V_0. However, this cannot happen because $G[V_{max}]$ is supposed to be k-vertex connected. Therefore, $G[V_0]$ is a k-VCC. In addition, Algorithm 1 operates until Q is empty, which means all the subgraphs have been processed. Thus, the theorem is correct.

Algorithm 2. Bottom-up Framework for k-VCC Detection

Input: $G(V, E)$, k
Output: The set of k-VCCs, $\mathcal{G}[\mathcal{R}]$
1   $\mathcal{G}[\mathcal{R}] \leftarrow \emptyset$; $\mathcal{G}[\mathcal{S}] \leftarrow \emptyset$; $\mathcal{G}[\mathcal{S}'] \leftarrow \emptyset$;
2   Find the k-core $G[V_k]$ of G;
3   $\mathcal{G}[\mathcal{S}] \leftarrow \text{Seeding}(G[V_k], k)$; //detailed in Subsection 4.1;
4   **while** $\mathcal{G}[\mathcal{S}'] \neq \mathcal{G}[\mathcal{S}]$ **do**
5      $\mathcal{G}[\mathcal{S}'] \leftarrow \mathcal{G}[\mathcal{S}]$;
6      $\mathcal{G}[\mathcal{S}] \leftarrow \text{Expanding}(G[V_k], k, \mathcal{G}[\mathcal{S}])$; //detailed in Subsection 4.2;
7      $\mathcal{G}[\mathcal{S}] \leftarrow \text{Merging}(G[V_k], k, \mathcal{G}[\mathcal{S}])$; //detailed in Subsection 4.2;
8   $\mathcal{G}[\mathcal{R}] \leftarrow \mathcal{G}[\mathcal{R}] \cup \mathcal{G}[\mathcal{S}]$;
9   **return** $\mathcal{G}[\mathcal{R}]$;

---

4   Bottom-Up Framework for k-VCCs Detection

The top-down framework highly depends on global structure of graph, which may not be efficient and practical when graph scale becomes huge. In this section, we develop a bottom-up framework for k-VCCs detection.

The Bottom-up framework focuses on the microscopic structure when dealing with large networks and thus is able to find target components with computational cost proportional to their size. The idea is to locally find seed subgraphs around the neighborhood of vertices and obtain the k-VCCs heuristically by expanding and merging these subgraphs. The overall framework is summarized in Algorithm 2. First, we utilize Seeding() to find local k-vertex connected subgraphs around the neighborhood of vertices as seed subgraphs (line 3). Then, we exploit Expanding() and Merging() to expand and merge these seed subgraphs (lines 4–7). Although this framework is heuristic, the experiment in Sect. 5.3 shows that the result is comparable to the real.

4.1   Identifying Seed Subgraphs

We propose the local k-vertex connected subgraph as seed subgraph, which only considers the neighborhood structure of a vertex. Moreover, unlike maximal clique, which is adopted as seed subgraph in [15], the local k-vertex connected subgraph is more generalized as seed than clique, whose structure is too strict [19].

In this section, seeding() is developed to identify such graphs. And there exists two important problems: (1) how to find the seed subgraph for a given vertex; (2) how to efficiently identify seed subgraphs for the whole network. Correspondingly, we first give the formal definition of seed subgraph and propose the LkVCS method for its discovery, and then we devise two optimization strategies to accelerate the process of identifying seed subgraphs, in which we do not have to identify seed subgraphs for all the vertices.

Identifying seed subgraph for a given vertex. Here, we study how to define and find local k-vertex connected subgraph for a given vertex u. We first give the definition of 2-ego neighborhood. We then exploit 2-ego neighborhood to add additional constraint on the path length between vertices in the defined local k-vertex connected subgraph, which makes it possible to find the seed subgraph within the neighborhood of a vertex.

Definition 6 *(2-ego neighborhood). Given a graph $G(V,E)$ and a vertex u in G, the 2-ego neighborhood of u, $N_2(u)$, denotes the set of vertices in G whose distance to u is no more than 2, i.e., $\{v|dist(u,v)leq 2\}$. Specially, u also belongs to $N_2(u)$.*

Definition 7 *(Local k-vertex connected subgraph). Given a graph $G(V,E)$ and a vertex $u \in V$, an induced subgraph $G[S]$ is a local k-vertex connected subgraph if and only if (1) $\forall v, w \in S$, if $(v,w) \notin E(S)$, $|nb_{G[S]}(v) \cap nb_{G[S]}(w)| \geq k$; (2) $|S| > k$; (3) $u \in S$.*

Based on [7], in real networks, community is usually existed in the neighborhood of vertices, hence it is rational to define the local k-vertex connected subgraph as seed subgraph. Furthermore, a k-VCC is usually composed of many adjacent local k-vertex connected subgraphs. Thus, if we can obtain these local subgraphs in advance, we probably retrieve the original k-VCC from them.

Further, we propose the LkVCS method, which can find one of the local k-vertex connected subgraphs from the induced subgraph $G[N_2(u)]$ as the seed subgraph for u. The main idea of LkVCS is to start with every different subset of vertices of size k from the neighborhood of u, denoted as R and then continue bring vertices from $P'\backslash R$ into R until $G[R]$ is a local k-vertex connected subgraph or there exists $x, y \in R$ and $(x,y) \notin E$ such that $|nb_{G[P']}(x) \cap nb_{G[P']}(y)| < k$ where P' is the vertex set of k-core of $G[N_2(u)]$. When the combination number $\binom{|nb_{G[P']}(u)|}{k}$ is larger than a threshold α, γ subsets are randomly tested. Example 1 illustrates how the LkVCS method works.

Example 1. We apply LkVCS on the graph G in Fig. 2 for $u = v_1$. We set $k = 3$. First, we obtain the 3-core of $G[N_2(v_1)]$ is G_4. We arbitrarily select 3 vertices from $nb_{G_4}(v_1)$, that is $\{v_2, v_3, v_4\}$ and $R = \{v_2, v_3, v_4\} \cup \{v_1\}$. Because $G[R]$ is not a 3-vertex connected subgraph, we continue to add the common neighbor v_5 of v_1, v_3 and v_2, v_4 into R. Now, $G[R]$ is a 3-vertex connected subgraph. We output $G[R]$ as the local 3-vertex connected subgraph of v_1.

Identifying seed subgraphs for the whole network. A naive way is to compute the local k-vertex connected subgraph for every vertex in the network by LkVCS method. However, it is not efficient enough. Here, we devise two optimization strategies to further reduce the computational cost.

Optimization 1: Vertex order priority based strategy. In vertex order priority strategy, we assign the vertices with smaller degree have higher priority than that with larger degree. We observe that the value of $\binom{deg(u)}{k}$ is not large for a vertex u with small degree. Hence, if a vertex u having smaller degree, we can take less time to detect whether there exists a local k-vertex connected subgraph for u.

Theorem 3. *Given a vertex u in graph $G(V, E)$, it can be contained in at most 0 for $deg(u) < k$ and $1 + \lceil \frac{deg(u)-k}{k-1} \rceil$ for $deg(u) \geq k$ k-VCCs, simultaneously.*

Theorem 3 gives the upper bound of number of k-VCCs, that a vertex could be contained in at the same time. We can see that the vertices with larger degree can be contained in more different k-VCCs and even larger amount of k-vertex connected subgraphs. In particular, for a specific vertex u with $deg(u) = k$, it can only be contained in at most 1 k-VCC. Recall that the LkVCS method will take more computational time for the input vertex u with larger vertex degree, because it will enumerate much more combinations of vertices than that of small degree to find the local k-vertex connected subgraphs. To avoid visiting the vertices with larger degree first, we set the vertices with larger degree having lower priority.

Optimization 2: Non-redundancy based strategy. We observe that if we find the seed subgraph for every vertex in the network, we will acquire many duplicate subgraphs or highly overlapping subgraphs. In order to reduce redundance, we design the non-redundancy based strategy such that for a given vertex, if it has already been contained in the discovered seed subgraphs of other vertices, there is no need to find its own seed subgraph. Note that a vertex can be included in different seed subgraphs, even if it is not processed.

Together with the vertex order priority strategy, we can find seed subgraphs for the uncovered vertices with smaller degree as soon as possible. Besides, based on the long-tail theory, the vertices with larger vertex degree are probably included in the discovered seed subgraphs, which means we do not need to detect seed subgraphs for these vertices. Thus, making use of these two strategies, we can find constraint number of seed subgraphs with higher efficiency.

Example 2. We apply the above two strategies in seeding() procedure on the graph shown in Fig. 2 to identify the seed subgraphs. We rank all the vertices in G according to their non-decreasing order of their vertex degree denoted as, $v_{12}(3) \preceq v_{14}(3) \preceq \ldots \preceq v_9(5) \preceq v_{13}(6) \preceq v_4(7)$. The numbers in the brackets are their vertex degree. We first find the seed subgraph for v_{12}, which is $\emptyset$. Then, we successively visit the vertex according to the vertex priority. For example, we find the seed subgraph G_3 for v_{14}. As G_3 contains $\{v_{13}, v_{15}, v_{16}\}$, we do not need to detect the seed subgraphs for these vertices. At last, we identify three seed subgraphs including G_3 for v_{14}, G_1 for v_1 and G_2 for v_6.

4.2   Expanding and Merging Seed Subgraphs

In this section, we focus on solving the problem of detecting k-VCCs from the discovered seed subgraphs. We observe that the k-VCCs do not satisfy the property of *downward closeness*. That is for a k-VCC denoted as $G[S], \exists S' \subseteq S$, the induced subgraph $G[S']$ is not a k-vertex connected subgraph. Thus, we cannot simply expand the discovered seed subgraphs by adding a series of vertices adjacent to their neighborhood to obtain the target k-VCCs.

Menger's Theorem [8] indicates that a graph is k-vertex connected if and only if it contains k vertex independent paths between any two vertices. Based on this relationship between independent path and vertex connectivity, we devise two algorithmic approaches, Expanding() and Merging(), in which, we could safely add vertices into the current subgraph and combine different subgraphs to form a bigger k-vertex connected subgraph, respectively. Next, we detail these two approaches.

Expanding. We first give some explanations of the notions used here. Given a graph $G(V, E)$ and a vertex set $S \subseteq V, \overline{S}$ represents the complement of S in G, i.e., $\overline{S} = V \backslash S$, and δS denotes the boundary of the induced subgraph $G[S]$, which means for any vertex $v \in \delta S$, there exists a vertex $u \in nb(v)$ and $u \notin S$.

In Expanding(), we add vertices that are connected to at least k vertices in $G[S]$ into the current subgraph as shown in Fig. 4. The specific process is as follows. For each k-vertex connected subgraph $G[S]$ now we have, if there exists vertex u in $\delta \overline{S}$ that is adjacent to at least k vertices in δS, we add every vertex like this into the current S and update $S \cup u$ as the new S. We iteratively conduct this procedure until there is no such vertex in $\delta \overline{S}$ that can be added into S. Expanding() ensures that for each generated subgraph $G[S]$ in the result set $\mathcal{G}[\mathcal{S}]$, $\forall u \in V \backslash S, G[S \cup u]$ is not a k-vertex connected subgraph. Theorem 4 guarantees the correctness of the Expanding() approach.

Theorem 4. *Suppose $G[S]$ is a k-vertex connected subgraph. If vertex u is adjacent to at least k vertices in $\delta S, G[S \cup u]$ is also a k-vertex connected subgraph.*

Proof. Theorem 4 can be induced from Menger's Theorem [8].

Example 3. We use the graph in Fig. 2 as an example to illustrate Expanding(). Assume that we have already obtained the 3-vertex connected subgraph $G[S]$ where $G[S] = G_2$. We find one of the boundary vertices v_{11} is adjacent to v_4, v_6 and v_7 in $G[S]$. Thus, we add v_{11} into $G[S]$ and $G[S \cup v_{11}]$ is also a 3-vertex connected subgraph.

Merging. When the obtained k-vertex connected subgraphs cannot be further expanded by Expanding(), we expect to combine some of the adjacent subgraphs together to acquire larger subgraphs with k-vertex connectivity shown in Fig. 5. Here, we develop Merging() to integrate different k-vertex connected subgraphs with at least k direct independent paths into a new one.

The process of Merging() is described as bellow. We first detect whether the input subgraphs in $\mathcal{G}[\mathcal{S}]$ are k-VCCs. If so, we put these subgraphs into the result set $\mathcal{G}[\mathcal{R}]$. Then, we iteratively merge the subgraphs satisfying the condition in Theorem 5 that will be detailed later until no subgraphs can be merged. If the subgraph $G[S \cup S']$ after combined meets the conditions in Corollary 1, we directly put it into result set $\mathcal{G}[\mathcal{R}]$, otherwise we put it back the candidate set $\mathcal{G}[\mathcal{S}]$ for further processing. In the implementation, we use the disjoint sets structure to accelerate the merging operation.

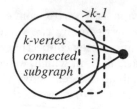
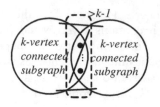

Fig. 4. The process of Expanding   **Fig. 5.** The process of Merging

Formally, Theorem 5 provides the sufficient condition to guarantee Merging() is correct. If the sum of the number of overlapping vertices and length-1 independent paths between two k-vertex connected subgraphs is more than or equal to k, they can be combined together.

Theorem 5. *Let $G(V, E)$ be a graph, $S \subseteq V, S' \subseteq V$ and $G[S], G[S']$ be two k-vertex connected subgraphs. If the following condition is satisfied, we say $G[S \cup S']$ is a k-vertex connected subgraph: $|S \cap S'| + \min\{|\delta S \cap \delta \overline{S'}|, |\delta \overline{S} \cap \delta S'|\} \geq k$.*

Proof. The idea of this proof is similar to that of Theorem 4.

Furthermore, based on Theorems 4 and 5, we obtain Corollary 1. In Merging(), Corollary 1 can be exploited as an *early termination condition*. That is once finding some subgraphs which satisfy the conditions in Corollary 1, we can put these subgraphs into the result set, because it is impossible to be combined with any other subgraphs. This can significantly reduce the number of subgraph combining operations.

Corollary 1. *Let $G[S]$ be a k-vertex connected subgraph. If the following two conditions are satisfied, we say $G[S]$ is a k-VCC:*
(1) $\nexists v \in \delta(\overline{S}), |nb(v) \cap \delta S| > k$;
(2) $\min\{|\delta S|, |\delta \overline{S}|\} < k$.

Example 4. A running example of Merging() is given using G in Fig. 2. Suppose we already have three 3-vertex connected subgraphs G_1, G_2 and G_3. We observe that G_3 satisfies the conditions in Corollary 1. That is, v_8, v_9 and v_{10} are only adjacent to one boundary vertex v_{13} of G_3, and $\min\{|\delta V(G_3)|, |\delta \overline{V(G_3)}|\} = \min\{1, 3\} = 1 < 3$. Thus, G_3 is a 3-VCC. For G_1 and G_2, we have that $|V(G_1) \cap V(G_2)| + \min\{|\delta V(G_1) \cap \delta \overline{V(G_2)}|, |\delta \overline{V(G_1)} \cap \delta V(G_2)|\} = 1 + \min\{2, 2\} = 3 \geq 3$. Thus, we can merge G_1 and G_2 together based on Theorem 5.

5   Experiments

We conduct extensive experiments to evaluate the effectiveness and efficiency of the proposed methods by using a variety of real and synthetic datasets. All algorithms are implemented in C++. All the experiments are conducted on a Linux Server with Intel Xeon 3.2 GHz CPU and 64 GB main memory.

5.1   Datasets and Compared Methods

The statistics of real networks used in the experiments are shown in Table 1. d_{max} denotes the maximum vertex degree of G. $\mathcal{D}$ is the degeneracy of G in Definition 5. $\#C$ is the number of ground-truth communities. The first Yeast dataset is a protein-protein interaction network downloaded from BioGRID[1]. The other four datasets are networks with ground-truth communities[2]. We abbreviate these datasets as YA, AZ, DP, YT and LJ.

Table 1. Statistics of real networks ($K = 10^3$ and $M = 10^6$)

Network	Abbr.	$|V(G)|$	$|E(G)|$	d_{max}	$\mathcal{D}$	$\#C$
Yeast	YA	$6.5K$	$229K$	2587	86	–
Amazon	AZ	$335K$	$926K$	549	6	$151K$
DBLP	DP	$317K$	$1M$	343	113	$13K$
Youtube	YT	$1.1M$	$3M$	28754	51	$8K$
LiveJournal	LJ	$4M$	$35M$	14815	360	$287K$

We compare our k-VCC with k-CC [2] and k-ECC [5] for effectiveness evaluation. Further, we evaluate the following algorithms for efficiency comparison:

- TkVCC: the top-down framework for k-VCC detection shown in Algorithm 1, discussed in Sect. 3. This is also used as the baseline method.
- BkVCC-Ran: the bottom-up framework for k-VCC detection shown in Algorithm 2 with random vertex order priority strategy in Sect. 4.1.
- BkVCC-NI: the bottom-up framework for k-VCC detection shown in Algorithm 2 with non-increasing vertex order priority strategy in Sect. 4.1.
- BkVCC-ND: the bottom-up framework for k-VCC detection shown in Algorithm 2 with non-decreasing vertex order priority strategy in Sect. 4.1.

5.2   Evaluation on Real Networks

Effectiveness Evaluation. To evaluate the effectiveness of different community models, we compare the proposed k-VCC with k-CC [2] and k-ECC [5] on 4 real datasets including AZ, DP, YT and LJ with ground-truth communities [26] under different types of criteria.

First, we use F-score to measure the accuracy of the detected communities with regard to the ground-truth communities. Given the discovered community $G[S]$ and the ground-truth community $G[T]$, F-score is defined as $F(S,T){=}2*\frac{prec(S,T)*rec(S,T)}{prec(S,T)+rec(S,T)}$ where $prec(S,T) = \frac{|S \cap T|}{|S|}$ represents the precision

---

[1] thebiogrid.org.
[2] http://snap.stanford.edu.

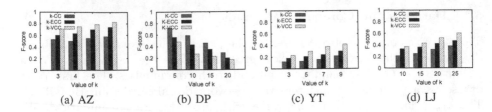

Fig. 6. F-score on different real networks

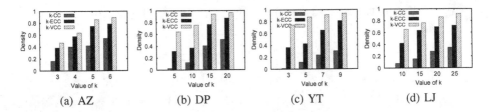

Fig. 7. Density on different real networks

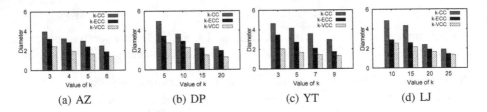

Fig. 8. Diameter on different real networks

and $rec(S,T) = \frac{|S \cap T|}{|T|}$ represents the recall. We can see that higher F-score value means the detected community is more similar with the ground-truth.

In the experiments, for different input k, we detect the k-VCCs by BkVCC-ND, the k-CC by the method in [2] and the k-ECCs by the method in [5] as communities, respectively. For each discovered community S_i, we compute the F-score with every ground-truth community T_j of the dataset and choose the largest $F(S_i, T_j)$ as the final F-score, F_i of S_i. Further, we use the average value of all F_i, denote as $\overline{F}$ to represent the F-score corresponding to a given dataset. Figure 6 shows the F-scores of the compared methods for different value of k. We find that the k-VCCs have the highest F-score on AZ, YT and LJ datasets. This is because in these datasets, they defined the ground-truth communities based on common interest or function, which is very cohesive. However, the ground-truth community in DP is defined based on publication venues. The authors publishing papers in the same conference or journal may be not densely connected [26]. Thus, we see k-VCCs have the lowest F-score on DP.

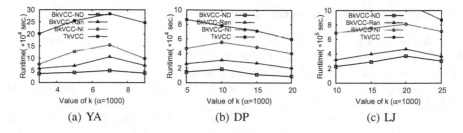

Fig. 9. Runtime on different real networks

Then, we use *density* and *diameter* to measure the goodness of the detected communities. *Density* is defined as the fraction of the edges that appear between the vertices to that of all possible edges and *diameter* is defined as the longest distance among all shortest paths between vertices in G. Given a community $G[S]$, the density and diameter of $G[S]$ is denoted $dens(G[S]) = \frac{2|E(S)|}{|S||S-1|}$ and $diam(G[S])$, respectively. The communities are more cohesive when they have larger *density* and smaller *diameter*. Figures 7 and 8 show the density and diameter of the detected communities on different networks. It can be seen that with the increasing of k, the density becomes larger and the diameter become smaller for all the methods. Morcover, for the same k value, k-VCC has the best performance. It has the highest density and lowest diameter, that is, the results of k-VCC are more cohesive than that of k-ECC and k-CC.

Efficiency Evaluation. In this section, we conduct experiments to study the efficiency of different methods to detect k-VCCs on different real networks. Figure 9 shows the comparison on overall running time of TkVCC, BkVCC-Ran, BkVCC-NI and BkVCC-ND for varying parameter k. We can see that the TkVCC method always runs slowest on all the datasets. This is because that it exploits the structure of the entire graph to find the minimum vertex cut set. When the scale of the graph getting larger, it will be very time-consuming. Thus, for large real network such as LJ in Fig. 9(c), it even cannot finish within the required time.

On the contrary, BkVCC-ND method runs much faster than BkVCC-Ran and BkVCC-NI over all the datasets. Recall that in BkVCC-ND, we assign vertices with smaller vertex degree have higher priority, which reduce the combination number of the neighbors for a given vertex. When we visit vertices with large vertex degree, they are very probably having been included in the vertices with small degree, which reduces the running time a lot. On the other hand, along with the increasing of parameter k, the running time of these methods first increase. This is because it needs more time to compute minimum vertex cut for TkVCC and seed subgraphs for the bottom-up based methods. Then, when the k value reaches a turning point, the running time begin to decrease. The reason is that with k becoming larger, more and more vertices are pruned by the k-core component.

5.3   Evaluation on Synthetic Networks

We generate a set of synthetic bipartite networks to evaluate the performance of the selected methods. The number of vertices are balanced in each part of these bipartite networks. The degree of both parts follow the power-law distribution with exponent γ and $d_{max} = n/2$. We set $\gamma = 2$. The vertices in the networks are linked according to [17].

We evaluate the efficiency and effectiveness of the TkVCC and BkVCC-ND methods. We set $k = 4$ for all the situations. Figure 10(a) shows the running time when varying the number of vertex in the network. We can see that BkVCC-ND method is much more efficient than TkVCC method, which is consistent with the results on real datasets. Figure 10(b) shows the F-score of the result of BkVCC-ND corresponding to that of TkVCC. Since the result of TkVCC are exact solution, the relative high values of F-score indicates that although the BkVCC-ND method is heuristic, it could generate results with high quality and hence proves its effectiveness.

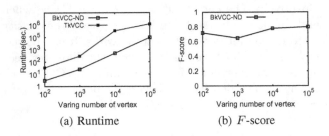

(a) Runtime   (b) F-score

Fig. 10. Results on synthetic network

5.4   Case Study

We construct an author collaboration network on KDD conference extracted from the raw DBLP dataset[3] for case study. A vertex represents an author, and an edge between two authors indicates they have co-authored. Figure 11 presents three 4-VCCs containing professor Jiawei Han. Based on the background knowledge, Fig. 11(a) shows the research group when he worked at SFU. Figure 11(b) shows his cooperation with the group of his colleague Chengxiang Zhai, when he began to work at UIUC. And, Fig. 11(c) shows his research group at UIUC and some very famous professors. In particular, we can find that professor Jian Pei often cooperates with Jiawei Han. However, if we use 4-ECC, we can only acquire one community containing Jiawei Han. Thus, we say that the communities detected by k-VCCs are more reasonable and interpretable, which effectively reduces the free rider effect.

---

[3] http://dblp.uni-trier.de/xml/.

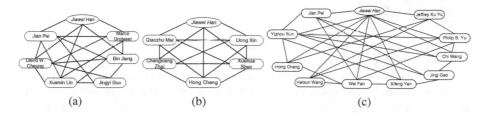

Fig. 11. Real examples of 4-VCCs containing Jiawei Han

6   Related Works

Our work relates to two main streams of research, concerning graph connectivity and component detection, respectively.

Graph connectivity. Graph connectivity has an extricably bound with minimum cut, since the minimum cut contributes to the graph connectivity. A large number of algorithms have been designed for computing the global minimum connectivity of the whole graphs [9,10,12,21]. Recently, there exists several research on finding k-edge connected subgraphs, which concern about the local edge connectivity in the subgraphs [1,5,28]. Whereas, in this paper, we focus on the k-vertex connectivity of subgraphs.

Component detection. The existing component detection methods can be roughly devided into clique-based [19] and clique-relaxed-based methods. Since the definition of clique is too strict, the clique-relaxed based methods have recently drawn a great deal of attentions. It can be classified into the following several categories: (1) **Distance-based relaxed methods.** n-clique is a maximal subgraph such that the distance of each pair of its vertices is not larger than n in the whole network [14]. n-clan is an n-clique whose diameter is no larger than n [16]. n-club is a maximal subgraph whose diameter is no larger than n [16]. (2) **Degree-based relaxed methods.** k-plex is defined as a maximal subgraph in which each vertex is adjacent to all other vertices except at most k of them [3]. Similarly, k-core is a maximal subgraph in which each vertex is adjacent to at least k other vertices of the subgraph. Efficient global search methods [6,20] and local search method [7] have been developed to discover the k-core communities or the community k-core containing given entities. *Quasi-clique* with a parameter γ is a subgraph with n vertices and $\gamma*\binom{n}{2}$ edges [27]. (3) **Triangulation-based relaxed methods.** DN-graph [23] is a connected subgraph in which the lower bound of shared neighborhood between any connected vertices is locally maximized. k-truss [13,22] is the largest subgraph in which every edge is contained in at least $(k-2)$ triangles within the subgraph. Based on the limitation of the above methods detailed in Sect. 1, we propose the k-VCC model, which has high vertex connectivity.

7   Conclusion

Component detection is a fundamental problem in network analysis and has attracted intensive interests. Most existing component detection methods suffer from the low connectivity issue. In this paper, we propose the k-vertex connected component model, which focuses on the vertex connectivity of networks. We study the k-VCC detection problem and develop the top-down and bottom-up frameworks for k-VCC detection. Extensive experimental results on large real and synthetic networks demonstrate the effectiveness and efficiency of our proposed approaches.

Acknowledgments. This research is partially supported by the National NSFC (No. 61272182, 61100028, 61332014, U1401256, 61672144), the Fundamental Research Funds for the Central Universities (N150402002, N150404008), the National Research Foundation, Prime Ministers Office, Singapore under its International Research Centres in Singapore Funding Initiative and the Pinnacle lab for Analytics at SMU.

References

1. Akiba, T., Iwata, Y., Yoshida, Y.: Linear-time enumeration of maximal k-edge-connected subgraphs in large networks by random contraction. In: CIKM, pp. 909–918 (2013)
2. Batagelj, V., Zaversnik, M.: An o(m) algorithm for cores decomposition of networks. arXiv preprint cs/0310049 (2003)
3. Berlowitz, D., Cohen, S., Kimelfeld, B.: Efficient enumeration of maximal k-plexes. In: SIGMOD, pp. 431–444 (2015)
4. Broder, A., Kumar, R., Maghoul, F., Raghavan, P., Rajagopalan, S., Stata, R., Tomkins, A., Wiener, J.: Graph structure in the web. Comput. Netw. **33**(1), 309–320 (2000)
5. Chang, L., Yu, J.X., Qin, L., Lin, X., Liu, C., Liang, W.: Efficiently computing k-edge connected components via graph decomposition. In: SIGMOD, pp. 205–216 (2013)
6. Cheng, J., Ke, Y., Chu, S., Özsu, M.T.: Efficient core decomposition in massive networks. In: ICDE, pp. 51–62 (2011)
7. Cui, W., Xiao, Y., Wang, H., Wang, W.: Local search of communities in large graphs. In: SIGMOD, pp. 991–1002 (2014)
8. Diestel, R.: Graph Theory. Graduate Texts in Mathematics. Springer, Heidelberg (2005)
9. Esfahanian, A.H., Louis Hakimi, S.: On computing the connectivities of graphs and digraphs. Networks **14**(2), 355–366 (1984)
10. Even, S., Tarjan, R.E.: Network flow and testing graph connectivity. SIAM J. Comput. **4**(4), 507–518 (1975)
11. Fortunato, S.: Community detection in graphs. Phys. Rep. **486**(3), 75–174 (2010)
12. Hariharan, R., Kavitha, T., Panigrahi, D., Bhalgat, A.: An o(mn) gomory-hu tree construction algorithm for unweighted graphs. In: ACM Symposium on Theory of Computing, pp. 605–614 (2007)
13. Huang, X., Cheng, H., Qin, L., Tian, W., Yu, J.X.: Querying k-truss community in large and dynamic graphs. In: SIGMOD, pp. 1311–1322 (2014)

14. Kargar, M., An, A.: Keyword search in graphs: finding r-cliques. PVLDB **4**(10), 681–692 (2011)
15. Lee, C., Reid, F., McDaid, A., Hurley, N.: Detecting highly overlapping community structure by greedy clique expansion. arXiv preprint arXiv:1002.1827 (2010)
16. Mokken, R.J.: Cliques, clubs and clans. Qual. Quant. **13**(2), 161–173 (1979)
17. Molloy, M., Reed, B.: The size of the giant component of a random graph with a given degree sequence. Comb. Probab. Comput. **7**(3), 295–305 (1998)
18. Palla, G., Derényi, I., Farkas, I., Vicsek, T.: Uncovering the overlapping community structure of complex networks in nature and society. Nature **435**(7043), 814–818 (2005)
19. Pattillo, J., Youssef, N., Butenko, S.: On clique relaxation models in network analysis. Eur. J. Oper. Res. **226**(1), 9–18 (2013)
20. Sozio, M., Gionis, A.: The community-search problem and how to plan a successful cocktail party. In: SIGKDD, pp. 939–948 (2010)
21. Stoer, M., Wagner, F.: A simple min-cut algorithm. J. ACM (JACM) **44**(4), 585–591 (1997)
22. Wang, J., Cheng, J.: Truss decomposition in massive networks. PVLDB **5**(9), 812–823 (2012)
23. Wang, N., Zhang, J., Tan, K.L., Tung, A.K.: On triangulation-based dense neighborhood graph discovery. PVLDB **4**(2), 58–68 (2010)
24. Wu, Y., Jin, R., Li, J., Zhang, X.: Robust local community detection: on free rider effect and its elimination. PVLDB **8**(7), 798–809 (2015)
25. Wu, Y., Jin, R., Zhu, X., Zhang, X.: Finding dense and connected subgraphs in dual networks. In: ICDE, pp. 915–926 (2015)
26. Yang, J., Leskovec, J.: Defining and evaluating network communities based on ground-truth. In: ICDM, pp. 745–754 (2012)
27. Zeng, Z., Wang, J., Zhou, L., Karypis, G.: Coherent closed quasi-clique discovery from large dense graph databases. In: KDD, pp. 797–802 (2006)
28. Zhou, R., Liu, C., Yu, J.X., Liang, W., Chen, B., Li, J.: Finding maximal k-edge-connected subgraphs from a large graph. In: EDBT, pp. 480–491 (2012)

Spatial Databases

Efficient Landmark-Based Candidate Generation for kNN Queries on Road Networks

Tenindra Abeywickrama$^{(\boxtimes)}$ and Muhammad Aamir Cheema

Monash University, Melbourne, Australia
{tenindra.abeywickrama,aamir.cheema}@monash.edu

Abstract. The k nearest neighbor (kNN) query on road networks finds the k closest points of interest (POIs) by network distance from a query point. A past study showed that a kNN technique using a simple Euclidean distance heuristic to generate candidate POIs significantly outperforms more complex techniques. While Euclidean distance is an effective lower bound when network distances represent physical distance, its accuracy degrades greatly for metrics such as travel time. Landmarks have been used to compute tighter lower bounds in such cases, however past attempts to use them in kNN querying failed to retrieve candidates efficiently. We present two techniques to address this problem, one using ordered Object Lists for each landmark and another using a combination of landmarks and Network Voronoi Diagrams (NVDs) to only compute lower bounds to a small subset of objects that may be kNNs. Our extensive experimental study shows these techniques (particularly NVDs) significantly improve on the previous best techniques in terms of both heuristic and query time performance.

Keywords: Road networks · Nearest neighbor · Landmark Lower Bounds

1 Introduction

The k nearest neighbor (kNN) query on road networks finds the k closest points of interest (POIs) by their shortest path distances in the road network from a query point. Incremental Euclidean Restriction (IER) [10] is a kNN method that uses a simple Euclidean distance heuristic. IER retrieves Euclidean kNNs as *candidates* and computes network distances to each one using a shortest path algorithm (e.g., Dijkstra). The kth furthest candidate implies an upper bound network distance to the kth NN. IER then iteratively retrieves further Euclidean NNs, computes network distances and updates the k candidates if closer POIs are found. Since Euclidean distance is a lower bound on network distance, IER terminates when the distance to the next Euclidean NN is larger than the network distance to the kth candidate. In a recent PVLDB experimental study [1], when IER was combined with a fast shortest path technique (instead of Dijkstra) it was found to be significantly faster than the state-of-the-art kNN methods.

© Springer International Publishing AG 2017
S. Candan et al. (Eds.): DASFAA 2017, Part II, LNCS 10178, pp. 425–440, 2017.
DOI: 10.1007/978-3-319-55699-4_26

Inspired by the observation of a simple heuristic being so effective, our study seeks to improve on this by employing better heuristics.

Euclidean distance is a lower bound on the network distance between vertices in road network graphs with travel distance edge weights. It can also be adapted for use when edge weights represent other metrics. For example when they represent travel time, we can divide the Euclidean distance by the maximum speed for any edge to obtain the minimum possible travel time. However a lower bound obtained in this way is looser and IER is likely to retrieve far more candidates that are not real kNNs (false hits) on travel times. This was evident as IER's advantage was smaller in several travel time experiments in [1]. This identifies the need for improvement by using better heuristics.

A popular alternative lower bounding technique is based on using *landmarks*, which Goldberg et al. [4] employed to improve the A* shortest path algorithm. By using distances to *landmark* vertices and the triangle inequality, they were able to compute more accurate lower bounds leading to a significant speed-up of A* search. Naturally, as the shortest path problem is closely related to the kNN problem, this raises the question whether these *Landmark Lower Bounds* (LLBs) can be used to similarly improve IER's kNN query performance. Until our study, the answer to this question has been "no".

Past attempts to use landmarks [7,8] computed lower bounds for *all* POIs and stored them in a sorted list. This is necessary for *every* query as LLBs depend on the query vertex. Candidates with the next smallest lower bound are retrieved iteratively from the list. This approach may be reasonable for small numbers of POIs as computing lower bounds is a relatively fast operation. However it will not scale well with increasing numbers of POIs. Some POI sets number in the tens of thousands e.g., the 25,000 fast food outlets in the US [1]. In such scenarios, it is desirable to incrementally retrieve POIs in order of their LLBs without computing LLBs for all POIs. For example, Euclidean NNs can be incrementally retrieved quite efficiently using an R-tree or similar structure.

Figure 1 demonstrates this problem using the US travel time road network. In Fig. 1(a) the LLB-based method is orders of magnitude worse on query time (left y-axis) despite producing fewer false hits (right y-axis). This is due to there being more POIs in total with increasing density (i.e., ratio of POIs to vertices), requiring more LLBs to be computed. The query time in Fig. 1(b) is constant as it is dominated by the computation of LLBs irrespective of k. In both figures we clearly see that (1) LLB-based methods provide better lower bounds as evident from the fewer false hits and (2) LLB-based methods perform very poorly without the ability to retrieve candidates incrementally.

Inspired by the performance of Euclidean heuristics in the PVLDB study and using the observation that landmark lower bounds appreciably reduce false hits over Euclidean distance, we investigate how to efficiently employ landmarks to improve kNN query performance. To summarise our contributions:

- We present two techniques to generate kNN candidates using landmarks. The first, Object Lists, demonstrates the difficulties in using LLBs and is efficient in several scenarios. We further improve on this and present another

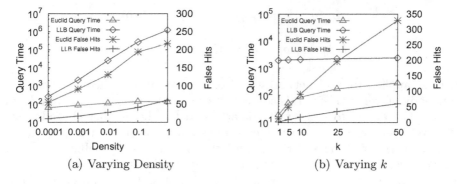

(a) Varying Density   (b) Varying k

Fig. 1. Euclidean kNN vs. Landmark kNN (US, $d = 0.001$, $k = 10$, uniform)

technique which, using a novel combination of Network Voronoi Diagrams and landmarks, provides significant improvement over existing techniques on both heuristic and running time performance.

- LLBs are expected to be more accurate than Euclidean distance, especially for edge weights such as travel times. However, to the best of our knowledge, they have not been empirically compared for kNN queries. In addition to other experimental results, we present a detailed study into the number of "false hits" (i.e., candidates which are not real kNNs). This machine independent metric is applicable to any experimental setting or shortest path technique, allowing a better understanding of the usefulness of LLBs.

2   Preliminaries

We consider a road network graph $G = (V, E)$ where V is a vertex set and E is an edge set. Edge $(u, v) \in E$ connects two adjacent vertices with weight $w(u, v) \in \mathbb{R}_{>0}$ representing any real positive metric such as distance or travel time. The shortest path $P(u, v)$ with network distance $d(u, v)$ represents the minimum sum of weights connecting any two vertices u and v. Similar to almost all existing studies we consider POIs (objects) and query points located on vertices in V. So, given a query vertex q and a set of object vertices O, a kNN query retrieves the k closest objects in O based on their network distances from q in G.

2.1   Landmark Lower Bounds

To compute a Landmark Lower Bound (LLB), firstly a set of m *landmark* vertices $L = \{l_1, \ldots, l_m\} \subseteq V$ is selected. From each landmark $l_i \in L$ we compute the distances to all vertices in V. Now given a source vertex q and destination vertex o, we can compute a lower bound LB_{l_i} on the network distance $d(q, o)$ using the distances to landmark l_i and the triangle inequality as defined in (1). We obtain the "tightest" lower bound (i.e., closest to $d(q, o)$) by choosing the maximum lower bound LB_{max} over all m landmarks as defined in (2).

$$LB_{l_i}(q, o) = |d(l_i, q) - d(l_i, o)| \leq d(q, o) \tag{1}$$

$$LB_{max}(q, o) = \max_{l_i \in L}(|d(l_i, q) - d(l_i, o)|) \tag{2}$$

First applied to road networks by Goldberg et al. [4], there now exists a large body of work utilising this concept. Two considerations arising from LLBs are (a) the vertices to select as landmarks and (b) the number of landmarks. Intuitively vertices whose shortest path trees cover longer shortest paths (e.g., those appearing at fringes of the graph [4]) have a higher probability of giving tighter lower bounds. A larger number of landmarks similarly increases this probability, but at the expense of higher space cost and computing more lower bounds to find the tightest overall. Since our study is concerned with using LLBs efficiently for kNNs rather than improving them, we refer the reader to past studies [4,5] for discussion on these choices. Note that (1) and (2) are only applicable on undirected graphs, but the idea can easily be extended to directed graphs by computing distances to and from landmarks.

3   Techniques

As detailed in Sect. 1, Incremental Euclidean Restriction (IER) is a kNN technique that computes network distances to candidate objects retrieved by their Euclidean distance until the k candidates cannot be improved. This technique can be generalized to consider *any* lower bounding technique, such as Landmark Lower Bounds (LLBs) discussed earlier. Let us refer to the equivalent kNN algorithm to IER using LLBs as Incremental Lower Bound Restriction (ILBR). ILBR works in exactly the same way as IER except we retrieve the candidate with the smallest LLB. IER can incrementally retrieve candidates by Euclidean NN search on an R-tree, avoiding computing Euclidean distances to all objects. But there is no efficient analogous method for LLBs and past studies [7,8] resort to computing LB_{max} for all objects which is not practicable. Here we present two techniques that incrementally retrieve candidates for ILBR.

3.1   ILBR by Landmark Object Lists

By computing LLBs for all objects as in past studies more LLBs are computed than necessary. We introduce the pre-computed Object List (OL) index that enables LLBs to be computed more optimistically. The OL approach solves the same underlying problem, i.e., to find the object $p \in O$ with the smallest $LB_{max}(q, o)$ as defined by (3). p is then returned to ILBR as a candidate.

$$p = \min_{p \in O}(\max_{l_i \in L}(|d(l_i, q) - d(l_i, p)|)) \tag{3}$$

Pre-processing: Given object set O, we pre-compute an *Object List* OL_i for each landmark $l_i \in L$ as shown in Fig. 2. The list OL_i contains an element for

$$OL_1 \quad \boxed{o_1, d(o_1, l_1) \mid o_2, d(o_2, l_1) \mid \ldots \mid o_{|O|}, d(o_{|O|}, l_1)}$$

$$\vdots$$

$$OL_m \quad \boxed{o_1, d(o_1, l_m) \mid o_2, d(o_2, l_m) \mid \ldots \mid o_{|O|}, d(o_{|O|}, l_m)}$$

Fig. 2. Unsorted object lists for m landmarks

$$OL_q \quad \boxed{o_6, 1 \mid o_2, 2 \mid o_4, 3 \mid o_5, 4 \mid o_7, 7 \mid o_1, 8 \mid o_3, 9}$$

Index   0   1   2   3   4   5   6

Fig. 3. Sample query object list OL_q

every object $o \in O$, with each element consisting of o and its distance from the landmark $d(l_i, o)$. Finally each list is sorted on $d(l_i, o)$. Since the Object List index only depends on the object set O, which is known beforehand, it is created and sorted entirely in the pre-processing stage.

Query Algorithm: Given a query vertex q and its nearest landmark l_q, we use object list OL_q of l_q to populate a set of *potential candidates*. The first potential candidate is the object that will minimise (1) for l_q. This object can be found by binary search on OL_q for the object p whose distance $d(l_q, p)$ is closest to $d(l_q, q)$. For example Fig. 3 depicts OL_q for a set of 7 objects $o_1, ..., o_7$ with distances from l_q. Let us say $d(l_q, q) = 4$, then the binary search will find the element at index 3 (shaded) as closest to 4. Therefore $p = o_5$ minimises (1) with $LB_{l_q}(q, o_5) = |4 - 4| = 0$. Finally p is inserted into a minimum priority queue Q keyed by $LB_{max}(q, p)$ computed using the ALT index [4]. For each vertex in V, ALT contains a list with its distances to each landmark. $LB_{max}(q, p)$ can be efficiently computed by iterating over the lists for q and p.

Lemma 1. *Given an object p and a landmark l_q, any object o with $LB_{l_q}(q, o) < LB_{max}(q, p)$ may also have $LB_{max}(q, o) < LB_{max}(q, p)$.*

Lemma 1 is trivially true when $LB_{max}(q, o) = LB_{l_q}(q, o)$. Now object p_n with the next smallest lower bound by l_q is immediately to the left or right of p (found above by binary search in OL_q). If $LB_{l_q}(q, p_n) < Top(Q)$ then by Lemma 1, p_n may have smaller LB_{max} than any object in Q. While $LB_{l_q}(q, p_n) < Top(Q)$, we search left or right from p. When an object satisfies the condition we compute LB_{max} and insert it into Q. When neither the next left or right object satisfies the condition, the algorithm terminates, and the top element in Q is returned as the object that minimises (3). This is correct as any object further left or right must have $LB_{l_q}(q, p_n) \geq Top(Q)$ and cannot satisfy Lemma 1.

Let us say in our example $LB_{max}(q, o_5) = 2$ so $Top(Q) = 2$ after inserting o_5. In Fig. 3, the objects to the left and right of o_5 are o_4 and o_7, with lower bounds $LB_{l_q}(q, o_4) = |4 - 3| = 1$ and $LB_{l_q}(q, o_7) = |4 - 7| = 3$ respectively. By Lemma 1,

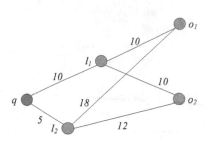

Fig. 4. Query landmark choices

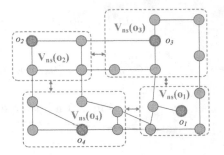

Fig. 5. Network Voronoi Diagram

o_4 may be a better candidate so we compute $LB_{max}(q, o_4)$ and insert it into Q. Let us say $Top(Q) = 2$ after inserting o_4. For the next element to the left, o_2, we have $LB_{l_q}(q, o_2) = |4 - 2| = 2$. In that case neither lower bounds for the object to the left or right is less than $Top(Q)$ and therefore cannot improve on the objects in Q and the search terminates. By saving Q and the indices in OL_q of the last left and right elements evaluated, we can continue to incrementally retrieve the object with the next smallest LB_{max}.

We choose l_q as the landmark closest to query vertex q. This heuristic increases the probability that objects are found further from l_q than q, thus producing a higher lower bound by (1). For example Fig. 4 shows the landmark distances for two landmarks, q and two objects. q, o_1 and o_2 are equally close to l_1, so $LB_{l_1}(q, o_1) = 0$ and $LB_{l_1}(q, o_2) = 0$ even though o_1 is further away. The closer landmark l_2 gives $LB_{l_2}(q, o_1) = 13$ and $LB_{l_2}(q, o_2) = 7$, correctly distinguishing the objects. Object Lists produce fewer false hits than IER on all datasets and is more efficient on low density datasets. However, this problem still occurs on datasets with higher density as landmarks are sparse, and IER is faster on these datasets. We now present a further improved technique.

3.2 ILBR by Network Voronoi Diagrams

Our second technique employs a Network Voronoi Diagram (NVD) [9] of the object set O to improve on Object Lists. Unlike its Euclidean counterpart, NVD generators are limited to the network and shortest paths represent distances.

We define $V_{ns}(o_i)$ as the *Voronoi node set* by (4), which identifies the vertices in V for which o_i is the nearest neighbor by their network distances to o_i.

$$V_{ns}(o_i) = \{v | v \in V, d(v, o_i) \leq d(v, o_j) \forall o_j \in O \setminus o_i\} \tag{4}$$

For any edge (u, v) where $u \in V_{ns}(o_i)$ and $v \in V_{ns}(o_j)$, then $V_{ns}(o_i)$ and $V_{ns}(o_j)$ are *adjacent*. The Network Voronoi Diagram for object set O is the collection of Voronoi node sets for all objects in O. Figure 5 shows an example NVD for a graph with unit edge weights and four objects. Each Voronoi node set is surrounded by a dotted container and arrows indicate adjacency.

Algorithm 1. GetNearestCandidateByNVD(q,c_l,NVD,ALT,Q,H)

Input   : q: a query vertex, c_l: candidate returned by last call (or 1NN from
              NVD if first call), NVD: Network Voronoi Diagram for object set O,
              ALT: index to compute maximum lower bounds, Q: priority queue
              with potential candidates, H: hash-table containing IDs of all
              Voronoi node sets previously evaluated
Output : c: candidate object, $LB_{max}(q,c)$: lower bound distance to c over L

1 $GenerateAdjacentCandidates(q, c_l, NVD, ALT, Q, H)$;
2 $(c, LB_{max}(q,c)) \leftarrow Dequeue(Q)$;
3 **return** $(c, LB_{max}(q,c))$;
4
5 **Function** $GenerateAdjacentCandidates(q, c_l, NVD, ALT, Q, H)$
6 $\quad V_{ns}(c_l) \leftarrow GetVoronoiNodeSet(c_l, NVD)$;
7 $\quad$ **for each** $V_{ns}(p) \in AdjacentVoronoiSets(V_{ns}(c_l))$ **do**
8 $\quad\quad$ **if** $!H.contains(V_{ns}(p))$ **then**
9 $\quad\quad\quad$ $Enqueue(Q, (p, ALT.ComputeLBMax(q,p)))$;
10 $\quad\quad\quad$ $H.insert(V_{ns}(p))$;

NVDs are not new in the context of kNNs [6,15]. VN^3 [6] utilises an NVD to retrieve candidates using the observation that the next NN is contained in a Voronoi node set adjacent to the sets of NNs found so far. VN^3 also pre-computes certain network distances. For each Voronoi node set this includes the distance from each border vertex to every other border and from each border to every contained vertex in the set. This allows VN^3 to also compute the network distance to retrieved candidates, but entails huge pre-processing and query cost. We instead relax the original observation to consider candidate NNs rather than NNs, which allows us to incrementally retrieve candidates for computing cheap LLBs. Through this novel combination of the standard NVD and landmarks we are able to consider significantly fewer candidates than Object Lists and avoid the large pre-processing overhead of VN^3.

Pre-processing: An NVD can be computed optimally in $O(|V|\log|V|)$ time with $O(|V|)$ space [3] using simultaneous Dijkstra's searches from all objects using a single priority queue. When a vertex v_d inserted by the search from o_i is dequeued and v_d is not assigned to a Voronoi node set, it is assigned to $V_{ns}(o_i)$. This is correct as v_d is the minimum element in the queue and so cannot be closer to another object. However if v_d *is* assigned to another Voronoi node set $V_{ns}(o_j)$, then $V_{ns}(o_j)$ is added to the list of adjacent sets for $V_{ns}(o_i)$ (the search from o_j creates the reciprocal entry). The search from o_i is pruned at v_d, i.e., neighbor vertices are not inserted into the queue as they cannot belong to $V_{ns}(o_i)$.

Query Algorithm: Algorithm 1 describes how to use NVDs to retrieve candidates. By their definition, an NVD can quickly return the 1NN by looking up the Voronoi node set (and hence the associated object) containing the query vertex. If $k > 1$, Algorithm 1 returns the next candidate by first retrieving the adjacent Voronoi node sets of the last candidate object. Note that in the first call to Algorithm 1 the last candidate is the 1NN. Each adjacent set generates a new potential candidate, to which we compute its LB_{max} by (2) using the ALT index and insert it into priority queue Q. We use hash-table H to avoid repeated computations for previously evaluated adjacent Voronoi node sets. Once all adjacent sets are processed in this way, we return the element in Q with the minimum LB_{max} as the next candidate.

Figure 6 shows a simplified NVD, assume the dotted containers capture the Voronoi node sets of each object and when containers share an edge they are adjacent. So for query vertex q in the figure, we can retrieve the 1NN o_1 as q is contained in $V_{ns}(o_1)$. Then the adjacent Voronoi node sets of $V_{ns}(o_1)$ (lightly shaded) are used to retrieve potential candidates, which are inserted into Q by their LB_{max} values. Let us say the candidate with the minimum key is now o_7, then o_7 would be returned by the algorithm. In the next call to the algorithm, the Voronoi node sets adjacent to $V_{ns}(o_7)$ would be retrieved and, for sets not already evaluated, new potential candidates inserted into Q.

Recall that ILBR (like IER) terminates when the network distance to the kth candidate is less than the lower bound distance to the next candidate. We propose Theorem 1 to show that Algorithm 1 is correct when this occurs.

Theorem 1. *When ILBR terminates the following are true (1) priority queue Q does not contain any objects with distance smaller than the kth candidate (2) there are no objects outside the Voronoi node sets visited so far (i.e., for objects in Q or returned as candidates) with distance smaller than the kth candidate.*

Proof. Let D_k be the network distance to the kth candidate. We now prove each case of Theorem 1 individually as follows:

Case 1: When ILBR terminates we have $D_k \leq Top(Q)$. We also have $Top(Q) \leq d(q, c)$ for any object c in Q as they are inserted using a lower bound distance and Q is a minimum priority queue. Thus we also have $D_k \leq d(q, c)$ and no object c in Q has a network distance smaller than the kth candidate.

Case 2: Let $C \subseteq O$ be the set of objects inserted into Q and let $S = \{V_{ns}(o) | o \in C\}$ be the set of associated Voronoi node sets. We prove Case 2 by contradiction in a similar but simpler manner to [6]. Let us assume there exists an object $p_k \notin C$ such that $d(q, p_k) < D_k$. Algorithm 1 inserts objects into Q from adjacent Voronoi node sets beginning with the set containing q, thus all Voronoi node sets in S are adjacent to at least one other set in S. So the shortest path $P(q, p_k)$ must pass through some Voronoi node set $V_{ns}(x) \in S$ since $p_k \notin C$. Thus $P(q, p_k)$ must contain at least one vertex $v_x \in V_{ns}(x)$, as illustrated in Fig. 6 with $x = o_2$ and $p_k = o_8$. By the definition of an NVD we have $d(v_x, x) \leq d(v_x, p_k)$ as all vertices in $V_{ns}(x)$ are closer to x than any other object. Adding $d(q, v_x)$ to both sides results in $d(q, v_x) + d(v_x, x) \leq d(q, v_x) + d(v_x, p_k)$. This simplifies to $d(q, x) \leq d(q, p_k)$ as v_x is on the shortest path $P(q, p_k)$ and $d(q, x) \leq d(q, v_x) + d(v_x, x)$. Since x is in Q, we have $Top(Q) \leq d(q, x)$, so we must have $D_k \leq d(q, x)$. This implies $D_k \leq d(q, p_k)$, contradicting our assumption.

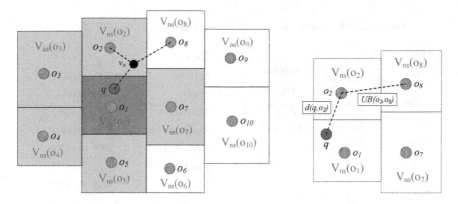

Fig. 6. Network Voronoi Diagram query   **Fig. 7.** LB optimisation

NVD Lower Bound Optimisation: When parallel Dijkstra's searches meet during NVD construction, we naturally compute an upper bound distance between objects of adjacent Voronoi node sets. This is an upper bound and not an exact distance because searches are pruned (e.g., shorter paths may exist through other adjacent Voronoi node sets). A lower bound distance to an adjacent object can be computed by applying the triangle inequality to the network distance (computed by ILBR) from q to the last candidate object and this upper bound distance. For example in Fig. 7, let o_2 be the last candidate returned with network distance $d(q, o_2)$. While evaluating the adjacent set $V_{ns}(o_8)$, we use the pre-computed upper bound distance $UB(o_2, o_8)$ between o_2 and o_8 to compute a lower bound $LB_{nvd}(q, o_8) = d(q, o_2) - UB(o_2, o_8)$. Note that it is not an absolute value due to the upper bound. In Algorithm 1, we insert o_8 into Q keyed by $LB_{nvd}(q, o_8)$ if $LB_{nvd}(q, o_8) > LB_{max}(q, o_8)$. The new lower bound may be tighter than the one computed using ALT especially when objects are further away from q and comes at a cheap pre-processing and query time cost.

4 Experiments

4.1 Experimental Setup

Environment: All experiments were run on a 3.2 GHz Intel Core i5-4570 CPU and 32 GB RAM running 64-bit Linux (kernel 4.2). Code was written in single-threaded C++ and compiled using g++ 5.2 with the O3 flag. We implemented ILBR, ALT and the candidate generation techniques ourselves. We obtained implementations of existing techniques, experimental scripts and datasets from [1]. All queries were executed in-memory for fast performance.

Datasets: We use 10 travel time road networks as in Table 1 with the largest US the default. We use a combination of synthetic and real object sets. We choose

Table 1. Road networks

Region	$|V|$	$|E|$
DE	48,812	119,004
VT	95,672	209,288
ME	187,315	412,352
CO	435,666	1,042,400
NW-US	1,089,933	2,545,844
CA	1,890,815	4,630,444
E-US	3,598,623	8,708,058
W-US	6,262,104	15,119,284
C-US	14,081,816	33,866,826
US	23,947,347	57,708,624

Table 2. Real POI sets (US)

Type	Size	Density
Schools	160,525	0.007
Parks	69,338	0.003
Fast food	25,069	0.0011
Post offices	21,319	0.0009
Hospitals	11,417	0.0005
Hotels	8,742	0.0004
Universities	3,954	0.0002
Courthouses	2,161	0.00009

synthetic objects uniformly at random based on density d where $d = |O|/|V|$. In addition we use 8 real POI sets extracted from OSM[1] as in Table 2.

Parameters: We vary object set density from 0.0001 to 1 and k from 1 to 50. We use the same default parameters as [1] with default density $d = 0.001$ and $k = 10$. We generate 25 uniform object sets and execute methods for 200 randomly selected query vertices, averaging running time over 5000 queries.

Techniques: Like IER, ILBR uses a different road network index to compute network distances. We combine ILBR with Pruned Highway Labelling (PHL) [2] as it is one of the fastest techniques. We use an ALT [4] index with 16 random landmarks to compute lower bounds and construct Object Lists. Finally we compare our techniques against the current fastest state-of-the-art technique, IER (similarly using PHL) [1]. For real-world object sets we also compare against two other techniques G-tree [16] and INE [10] for comparison with [1].

4.2   Index Pre-processing

Table 3 details the index pre-processing time and space. PHL and ALT are the road network indexes employed by ILBR and IER. While PHL is faster to construct for travel time road networks, G-tree consumes less space making it more appropriate with limited memory. The index size of ALT is small, but this is dependent on m the number of landmarks used (16 in our case). It can be reduced by using fewer landmarks at the expense of looser lower bounds. We also observe the performance of object indexes used by ILBR (Object Lists and Network Voronoi Diagrams) and IER (R-trees) for the default density $d = 0.001$. Object Lists and R-trees are fast to construct and their index sizes are small. However since the space cost is a function of object set size we expect it to

---

[1] http://www.openstreetmap.org.

Table 3. Index statistics

Road network		PHL	G-tree	ALT (m = 16)	OL (d = 0.1%)	NVD (d = 0.1%)	R-tree (d = 0.1%)
NW	Time	16 s	47 s	2 s	0.8 ms	264 ms	0.2 ms
	Space	325 MB	104 MB	67 MB	136 KB	4.2 MB	44 KB
US	Time	30 m	71 m	60 s	15 ms	12 s	4 ms
	Space	15.8 GB	2.9 GB	1.43 GB	1.8 MB	92 MB	0.9 MB

increase with density. NVDs take longer and occupy more space as the time and space complexity are functions of $|V|$. But both costs are still significantly smaller than road network indexes making it feasible to compute an NVD for each object set. NVDs may also be compressed using the geometric area of Voronoi node sets to capture vertex containment. For example it may be stored as a polygon in an R-tree [6] or as merged cells in a Quadtree [12].

4.3 Query Performance

We evaluate query performance of each technique on two metrics, namely running time and false hits per query. A *false hit* occurs when a candidate NN is not a real kNN. The greater the number of false hits, the more unnecessary network distance computations ILBR must perform. Thus false hits are an indication of a heuristic's performance irrespective of the experimental setting (disk based or main memory) or the network distance technique used. We refer to the two ILBR methods as NVD-X and OL-X as variants employing Network Voronoi Diagrams and Object Lists respectively (and X is the road network index used).

Effect of Network Size: Figure 8 shows query performance as the number of road network vertices $|V|$ increases. In Fig. 8(a), NVD-PHL is consistently the best performing method and is 2–3× faster than IER-PHL. OL-PHL is

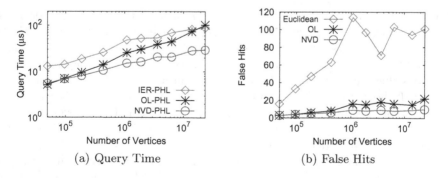

(a) Query Time   (b) False Hits

Fig. 8. Effect of Network Size $|V|$ ($d = 0.001$, $k = 10$, uniform)

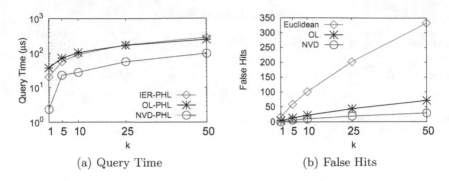

(a) Query Time   (b) False Hits

Fig. 9. Effect of k (US, $d = 0.001$, uniform)

comparable to NVD-PHL for the first few datasets after which its advantage over IER-PHL narrows until being on par with it for the largest dataset. With increasing $|V|$ the total number of objects increases for the same density causing Object Lists to become larger. E.g., we expect there to be more fast food outlets in larger regions. OL is susceptible to objects that appear close when they are similar distances from the landmark as the query vertex. When $|V|$ increases, landmarks become more distant from query vertices on average (as the number of landmarks m is constant), so the probability of such objects appearing increases. While these are only *potential candidates* and are not reflected in false hits, OL must still compute their LB_{max} values. This is evident in Fig. 8(b) as the number of false hits for OL is still lower than Euclidean distance.

Effect of k: Figure 9 shows the query performance with increasing k. For $k = 1$, NVD-based methods are essentially optimal as only a single look-up operation is needed. NVD-PHL once again outperforms all other methods, being at least 2–3× faster than IER-PHL over all k, again showing that it *is* possible to efficiently use landmarks for kNNs. Landmarks display significant improvement on false hits over Euclidean distance in Fig. 9(b). But earlier trends are also seen here and OL-PHL's query time does not improve on IER-PHL.

Effect of Density: We observe query performance with increasing object set density in Fig. 10. As density increases, the average distance between objects decreases. This makes kNNs appear closer to the query vertex and they should be easier to find. IER-PHL is an exception, as objects become closer and more numerous they become more difficult to differentiate using Euclidean distance. NVD-PHL shows this problem can be remedied using landmarks as it is an order of magnitude better than IER-PHL in Fig. 10(a). OL-PHL however degrades with increasing density to the point that its running time is an order of magnitude worse than IER-PHL. With more objects, more of them will produce inaccurate lower bounds similar to the scenario depicted in Fig. 4, making distant objects appear close to the query vertex. NVD-PHL does not suffer from this as using adjacent Voronoi node sets acts as a filter avoiding objects that "seem" close by inaccurate lower bounds. As a result NVD-PHL experiences far fewer false hits in Fig. 10(b).

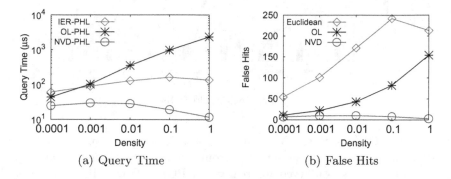

(a) Query Time

(b) False Hits

Fig. 10. Effect of Density (US, $k = 10$, uniform)

Lower Bounds Computed: Figure 10(b) showed that with increasing density, OL experiences fewer false hits than Euclidean distance even at its worst. This suggests that the poor running time of OL for high densities in Fig. 10(a) is not caused by ILBR making additional network distance computations due to false hits. It is actually due to the number of lower bounds computed by OL, which increases with density, as illustrated in Fig. 11(a). NVD computes very few lower bounds thanks to its filtering property. The final evidence of this is the behaviour of OL on the two datasets in Fig. 11. The US road network with 24 million vertices requires more lower bounds to be computed than the smaller NW dataset with 1 million vertices. The US has more objects for the same density, resulting in a larger Object List and hence a larger search space to find the best object. We note however, computing all lower bounds would require significantly more computations than OL. While OL is a substantial improvement, its utility is still dependent on the number of objects.

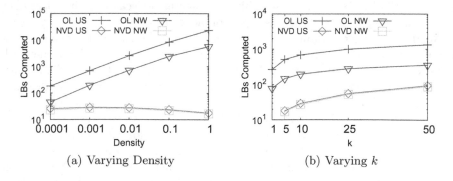

(a) Varying Density

(b) Varying k

Fig. 11. No. of lower bounds computed (US, $d = 0.001$, $k = 10$, uniform)

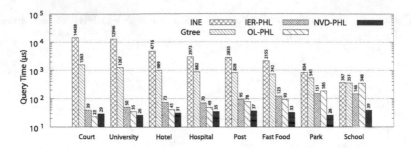

Fig. 12. Real-World POIs

Real-World Object Sets: We verify our observations on real-world POIs in
Fig. 12 with increasing object set sizes from left to right. Trends seen in previous figures are also observed for real-world POIs. NVD-PHL is the overall
best performing method, while OL-PHL is competitive except on larger object
sets like parks and schools. For small object sets like courts, IER-PHL remains
competitive as there are so few objects that Euclidean distance has a smaller
probability of making a false hit. A more typical object set such as fast food
outlets demonstrates the significant superiority of NVD-PHL.

5 Related Work

A recent experimental study [1] on the kNN problem provided an in-depth review
of the state-of-the-art. The main outcomes of this study were the surprising
performance of IER and the implications this had on heuristics used in kNN.
We refer the reader to this paper for a detailed review of kNN techniques, while
in this section we discuss the work most relevant to our study.

VN^3 [6] uses Network Voronoi Diagrams to answer kNN queries as explained
in Sect. 3.2. Landmarks have been used to answer kNN queries by Kriegal et al.
[7,8] based on the multi-step kNN paradigm [13] that we refer to as ILBR. These
studies propose interesting improvements to using landmarks. But in both cases
the kNN algorithms require creating a ranking by computing landmark lower
bounds to all objects. As discussed this approach is not scalable with object set
size and not competitive with existing approaches in practice.

Road Network Embedding (RNE) [14] involves transforming the road network
into higher dimensional space and using Minkowski metrics to estimate network
distance. However the proposed kNN method is approximate. Qiao et al. also propose an approximate technique [11] using shortest path trees to compute distance
estimates based on tree distance, but their solution applies to the k nearest *keyword* problem where objects are not split into sets.

There is a wide body of work on utilising landmarks for shortest path queries.
As previously discussed, ALT [4] was among the first. [8] proposes a hierarchical
landmark scheme to reduce index space cost. Other work [5] has focussed on
improving lower bounds for example through better landmark selection. These
compliment our work, e.g., better lower bounds reduce the number of false hits.

6 Conclusion

In this paper we present two techniques to efficiently retrieve kNN candidates by landmark-based lower bounds in an incremental manner for effective integration with the ILBR framework. We empirically compare the heuristic performance of landmark lower bounds and Euclidean distance on kNN search for the first time. We show that both methods significantly improve on the number of false hits (by up to an order of magnitude) incurred in candidate generation than the Euclidean distance heuristic used by IER. In our experimental investigation on travel time road networks, the Object List technique demonstrates the difficulties in using landmarks but outperforms IER for smaller object sets. The second technique employing a Network Voronoi Diagram outperforms IER by at least 2–3× on query time across all datasets and parameters. Thus we show that it is indeed possible to use landmark-based lower bounds to improve kNN search.

Acknowledgements. We sincerely thank the anonymous reviewers for their feedback which helped improve our work. The research of Muhammad Aamir Cheema is supported by ARC DE130101002 and DP130103405. Tenindra Abeywickrama is supported by an Australian Government RTP Scholarship.

References

1. Abeywickrama, T., Cheema, M.A., Taniar, D.: K-nearest neighbors on road networks: a journey in experimentation and in-memory implementation. PVLDB **9**(6), 492–503 (2016)
2. Akiba, T., Iwata, Y.: Kawarabayashi, K.I., Kawata, Y.: Fast shortest-path distance queries on road networks by pruned highway labeling. In: ALENEX, pp. 147–154 (2014)
3. Erwig, M., Hagen, F.: The graph voronoi diagram with applications. Networks **36**, 156–163 (2000)
4. Goldberg, A.V., Harrelson, C.: Computing the shortest path: a search meets graph theory. In: SODA, pp. 156–165 (2005)
5. Goldberg, A.V., Werneck, R.F.F.: Computing point-to-point shortest paths from external memory. In: ALENEX, pp. 26–40 (2005)
6. Kolahdouzan, M., Shahabi, C.: Voronoi-based k nearest neighbor search for spatial network databases. In: VLDB, pp. 840–851 (2004)
7. Kriegel, H.-P., Kröger, P., Kunath, P., Renz, M.: Generalizing the optimality of multi-step k-nearest neighbor query processing. In: Papadias, D., Zhang, D., Kollios, G. (eds.) SSTD 2007. LNCS, vol. 4605, pp. 75–92. Springer, Heidelberg (2007). doi:10.1007/978-3-540-73540-3_5
8. Kriegel, H.-P., Kröger, P., Renz, M., Schmidt, T.: Hierarchical graph embedding for efficient query processing in very large traffic networks. In: Ludäscher, B., Mamoulis, N. (eds.) SSDBM 2008. LNCS, vol. 5069, pp. 150–167. Springer, Heidelberg (2008). doi:10.1007/978-3-540-69497-7_12
9. Okabe, A., Boots, B., Sugihara, K.: Spatial Tessellations: Concepts and Applications of Voronoi Diagrams, 2nd edn. Wiley, Hoboken (2000)
10. Papadias, D., Zhang, J., Mamoulis, N., Tao, Y.: Query processing in spatial network databases. In: VLDB, pp. 802–813 (2003)

11. Qiao, M., Qin, L., Cheng, H., Yu, J.X., Tian, W.: Top-k nearest keyword search on large graphs. PVLDB **6**(10), 901–912 (2013)
12. Samet, H., Sankaranarayanan, J., Alborzi, H.: Scalable network distance browsing in spatial databases. In: SIGMOD, pp. 43–54 (2008)
13. Seidl, T., Kriegel, H.P.: Optimal multi-step k-nearest neighbor search. In: SIGMOD, pp. 154–165 (1998)
14. Shahabi, C., Kolahdouzan, M., Sharifzadeh, M.: A road network embedding technique for k-nearest neighbor search in moving object databases. GeoInformatica **7**(3), 255–273 (2003)
15. Zheng, B., Zheng, K., Xiao, X., Su, H., Yin, H., Zhou, X., Li, G.: Keyword-aware continuous kNN query on road networks. In: ICDE, pp. 871–882 (2016)
16. Zhong, R., Li, G., Tan, K., Zhou, L., Gong, Z.: G-tree: an efficient and scalable index for spatial search on road networks. TKDE **27**(8), 2175–2189 (2015)

MinSum Based Optimal Location Query in Road Networks

Lv Xu[1], Ganglin Mai[1], Zitong Chen[2], Yubao Liu[1(✉)], and Genan Dai[1]

[1] Sun Yat-sen University, Guangzhou, China
liuyubao@mail.sysu.edu.cn
[2] The Chinese University of Hong Kong, Sha Tin, Hong Kong

Abstract. Consider a road network G on which a set C of clients and a set S of servers are located. Optimal location query (OLQ) in road networks is to find a location such that when a new server is set up at this location, a certain objective function computed based on the clients and the servers serving the clients is optimized. In this paper, we study the OLQ with the MinSum objective function in road networks, namely the MinSum query. This problem has been studied before, but the state-of-the-art is still not efficient enough. We propose an efficient algorithm based on the two-level pruning technique. We also study the extension of the MinSum query problem, namely the optimal multiple-location MinSum query. Since this extension is shown to be NP-hard, we propose a greedy algorithm. Moreover, we give an approximate guarantee for our solution. Extensive experiments on the real road networks were conducted to show the efficiency of our proposed algorithms.

1 Introduction

Given a set C of clients and a set S of servers in a road network $G = (V, E)$ where V is a vertex set and E is an edge set, an optimal location query (OLQ) is to find a location such that when a new server is set up at this location, a certain objective function computed based on the clients and servers (including the new server) is optimized. This optimal location query is very important since it is used as a basic operation in many real applications such as location planning, location based service and profit-based marketing [1–4,10,11,16,23,26].

In OLQ, the new server can be built at any location of the road network except the locations of existing servers. There are three popular objective functions used to determine the optimal location for the new server, namely Min-Max function, MaxSum function and MinSum function [23,26]. One kind of OLQ adopting the MinSum function called the *MinSum* query was studied in [23,26]. The chain of fast food restaurants is an example to illustrate this Min-Sum query. For example, different branches of the fast food restaurants have their collaborative relationship and deliver fast food together to customers at different locations. Minimizing the average distance from customers to their nearest branches of the restaurants can determine the location of a new branch of the restaurants. The best-known algorithm for this query/problem was presented

© Springer International Publishing AG 2017
S. Candan et al. (Eds.): DASFAA 2017, Part II, LNCS 10178, pp. 441–457, 2017.
DOI: 10.1007/978-3-319-55699-4_27

in [23, 26]. The major idea is to first augment the road network by creating a vertex for each client and each server in the road network and then partition the augmented road network into sub-networks/sub-graphs for solving the problem.

However, the state-of-the-art algorithm in [23, 26] has several shortcomings as follows. First, the algorithm performance heavily depends on the pruning of the partitions of augmented road network. The pruning effect needs to be improved since the upper bounds for the partitions are quite coarse. Second, the algorithm relies on an augmented road network which could be prohibitively large. Specifically, the augmented road network has the number of its vertices as large as $|V| + |S| + |C|$ and the number of its edges as large as $|E| + |S| + |C|$, both of which become very large when there is a large number of servers or clients. Third, in practice, we sometimes want to find multiple locations and set up a server at each of these locations. The algorithm cannot support such query with multiple locations.

Motivated by the shortcomings of the existing approach in [23, 26], in this paper, we design an efficient algorithm called *MinSum-Alg* which avoids the above shortcomings. Specifically, we make the following contributions.

Firstly, we propose an efficient algorithm for the optimal location query (i.e., the single-location MinSum query) in road networks. The proposed query algorithm *MinSum-Alg* is executed on the original road network without generating any new road network where the number of the vertices to be examined is equal to $|V|$. We also present the two-level pruning technique based on the idea of *nearest location component (NLC)* of the clients in [3, 4], which can dramatically reduce the algorithm search space. Secondly, we study to find multiple locations (instead of a single location) for the MinSum query, which has not been studied in the literature. We show that this problem is NP-hard and propose a greedy algorithm for this general problem. Moreover, our solution has an approximate guarantee. Note that the multi-server version of OLQ with the MinMax and MaxSum objective functions is also shown to be NP-hard [3, 4]. However, the proposed solution has no theoretical guarantee. Thirdly, we conducted extensive experiments to verify the efficiency of our algorithm based on the real road networks *SF* (San Francisco) and *CA* (California). Our empirical studies show the greedy algorithm has good approximate ratio when compared to the optimal solutions.

The rest of this paper is organized as follows. Section 2 gives the problem definition. Section 3 introduces our algorithm *MinSum-Alg* for the single-location MinSum query. Section 4 introduces the greedy algorithm *GA(MinSum)* for the multiple-location MinSum query. Section 5 gives the empirical study and Sect. 6 reviews the related work. Section 7 concludes the paper.

2   Problem Definition

Let $G = (V, E)$ be a road network, and C (S) be a set of clients (servers) on G. For any edge $e = (v_l, v_r)$ of G, v_l (v_r) is the left (right) vertex of e. We adopt the network distance metric to define the distance between two locations on the road network, denoted by $d(\cdot, \cdot)$. Let c be a client in C. We denote c's closest server in S by $NN_S(c)$. Besides, we denote the distance between c and its closest server in S by $c.dist$, i.e., $c.dist = d(c, NN_S(c))$. Each client $c \in C$ is associated with a positive weight, denoted by $w(c)$, which denotes the importance of the client. For example, if c is a residential estate, then $w(c)$ could be the number of residents living at c. We define the *cost value* of c, denoted by $Cost(c)$, to be $w(c) \cdot c.dist$.

The purpose of MinSum query is to find a location such that once a new server is set up at this location, the sum of the cost values of all the clients in C is minimized. Formally, this problem is defined in Problem 1. For simplicity, we sometimes use a location p on the road network to represent the server located at p.

Problem 1. Given a road network $G = (V, E)$, a set C (S) of clients (servers) on G, the optimal location query problem is to find a location p which minimizes $\sum_{c \in C} w(c) \cdot d(c, NN_{S \cup \{p\}}(c))$.

Consider an example of a road network G in Fig. 1(a). In this figure, each dot corresponds to a vertex or a client or a server in the road network. In this example, there are 8 vertices, namely $v_1, v_2, ..., v_8$, 3 servers, namely s_1, s_2 and s_3, and 5 clients, namely $c_1, c_2, ..., c_5$. The number near to each line segment in the figure denotes the distance between the two end-points of the line segment. Since c_1 has the same location as vertex v_1 in the network, we write "v_1/c_1" in the figure.

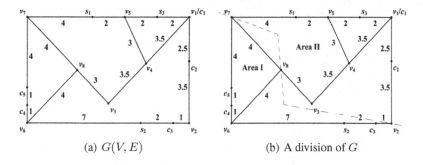

(a) $G(V, E)$ (b) A division of G

Fig. 1. A running example

Given two points p_1 and p_2 on an edge $e = (v_l, v_r)$, we define a *point interval* on e in the form of $[p_1, p_2]$ and p_1 (p_2) is said to be the *start point (end point)* of this interval. Note that a point interval is a portion (or a whole portion) of an edge. Suppose that the start point p_1 is nearer to the left vertex v_l than the end point p_2. In particular, we use $l(p_1, p_2)$ to denote the *length* of point interval $[p_1, p_2]$ along the edge e. For example, the length of point interval $[v_7, s_1]$, $l(v_7, s_1)$, is equal to 4 in Fig. 1(a). Similar to [3,4], the *nearest location component* is defined as follows.

Definition 1 (Nearest Location Component). *For each client $c \in C$, the nearest location component of c, denoted by $NLC(c)$, is defined to be a set of all points on the edges in G such that each of these points has its distance to c at most c.dist. Formally, $NLC(c) = \{p | d(c, p) \leq c.dist$ and p is a point on the edges of $G\}$.*

Given any point p on an edge in G, if $p \in NLC(c)$, then we say that the client c is *attracted* by a server to be built at the location p. For simplicity, we say that c is *attracted* by p. The solutions of [3,4] based on NLC are proposed for the OLQ with MinMax and MaxSum objective functions. They cannot be applied to the MinSum query problem.

Definition 2 (Gain). *The gain of any point p on G, is defined to be $Gain(p) = \sum_{c \in C} w(c) \cdot (c.dist - d(c, p))$ if $p \in NLC(c)$ and 0 otherwise.*

Intuitively, the gain of p corresponds to the total cost change values when the new server is build at p. We can derive that a larger gain of p will lead to a smaller cost value. So, the MinSum query problem can be changed into finding the optimal location with the largest gain. For the sake of convenience, we summarize the notations used in the paper in Table 1.

Table 1. Notations

Notation	Description
G	A road network
V/v	The set of vertices/a vertex
E/e	The set of edges/an edge
C/c	The set of clients/a client
S/s	The set of servers/a server
$w(c)$	Importance of client c
p	A location on the road network
$c.dist$	The distance between c and its nearest server
$Cost(c)$	The cost value of c
$NN_S(c)$	The server in S nearest to client c
$[p_1, p_2]$	A point interval on a single edge where p_1 and p_2 are two points on this edge
$l(p_1, p_2)$	The length of point interval $[p_1, p_2]$
$NLC(c)$	The nearest location component of c
$Gain(p)$	The gain value of p

3   Single-Location MinSum Query Algorithm

In this section, we introduce the best-known algorithm in Sect. 3.1. Then, we propose the optimization techniques and the MinSum-Alg algorithm in Sect. 3.2.

3.1   The Best-Known Algorithm

The best-known algorithm [23, 26] needs to generate the augmented network G' by creating a vertex for each client and each server in the given road network G. Then, the number of the vertices of G' is equal to $|V| + |C| + |S|$. In order to distinguish from the best-known, we call the vertex and the edge of G' as the endpoint and the road segment, respectively. For example, in Fig. 1(a), the edge (v_6, v_7) of G corresponds to three segments (v_6, c_4), (c_4, c_5), (c_5, v_7) of G' in which v_6, c_4, c_5 and v_7 are the endpoints.

In essence, the MinSum query is to find the optimal locations with the largest gain as mentioned in Sect. 2. In particular, the best-known algorithm relies on the key *findings* as follows. (1) The gain of any point on a single road segment is always maximized at one endpoint of this segment. (2) If both endpoints of a road segment have the same gain value, then there are two cases. Case 1: The gain of any point on this segment is equal to that of both endpoints. Case 2: The gain of both endpoints is larger than that of any point on this segment. So, the best-known algorithm examines at most three points for each road segment, namely two endpoints and any interior point on the segment. The set of endpoints correspond to the set $S \cup V \cup C$.

Since the number of road segments of G' is often large, the best-known algorithm divides G' into a number of subnetworks/partitions and applies the upper bounds for the partitions to reduce the number of road segments to be examined. The algorithm performance heavily depends on the pruning of the partitions. However, the existing upper bound is relaxed and the pruning effect may be improved further.

3.2   Optimization

The purpose of optimization technique is to reduce further the algorithm search space. In this section, we propose the two-level pruning method, namely area pruning and edge pruning.

Area Pruning. Given an area A. The existing upper bound is defined to be $UPP(A) = \sum_{c \in C_A} w(c) \cdot c.dist$, where C_A is the set of all possible clients attracted by the new server if it is built at any point in A. The reasons are as follows. Consider any point p in A. According to Definition 2, $Gain(p) = \sum_{c \in C} w(c) \cdot (c.dist - d(c, NN_{S \cup \{p\}})) = \sum_{c \in C - C_A} w(c) \cdot (c.dist - d(c, NN_{S \cup \{p\}})) + \sum_{c \in C_A} w(c) \cdot (c.dist - d(c, NN_{S \cup \{p\}}))$. Since the clients in $C - C_A$ cannot be attracted by the new server at p, $\sum_{c \in C - C_A} w(c) \cdot (c.dist - d(c, NN_{S \cup \{p\}})) = 0$. Then, we have $Gain(p) = \sum_{c \in C_A} w(c) \cdot (c.dist - d(c, NN_{S \cup \{p\}})) \leq UPP(A)$. Thus, the upper bound holds. We observed that the existing upper bound is relaxed. This is because C_A includes lots of clients whose cost values cannot contribute to the gain of the optimal location.

Let $d(c, A)$ be the shortest distance between the client c in C_A and any point in the area A. For each edge $e = (v_l, v_r)$ in the area A, $d(c, A) = \min\{d(c, v_l), d(c, v_r)\}$. Both C_A and $d(c, A)$ can be computed based on Dijkstra's algorithm [5].

We divide the clients in C_A into $\overline{C_A}$ and $\widetilde{C_A}$ which denote the set of clients in the area A and outside the area A, respectively. Then, we list the more tightening upper bounds for the partitions/areas of the road network in Lemma 1.

Lemma 1. $UPP(A) = \sum_{c \in \overline{C_A}} w(c) \cdot c.dist + \sum_{c \in \widetilde{C_A}} w(c) \cdot (c.dist - d(c, A))$.

Proof. For any point p in A, $Gain(p) = \sum_{c \in C_A} w(c) \cdot (c.dist - d(c, NN_{S \cup \{p\}})) = \sum_{c \in \overline{C_A}} w(c) \cdot (c.dist - d(c, NN_{S \cup \{p\}})) + \sum_{c \in \widetilde{C_A}} w(c) \cdot (c.dist - d(c, NN_{S \cup \{p\}}))$. For $c \in \widetilde{C_A}$, $d(c, NN_{S \cup \{p\}}) \geq d(c, A)$. Then, $Gain(p) \leq \sum_{c \in \overline{C_A}} w(c) \cdot c.dist + \sum_{c \in \widetilde{C_A}} w(c) \cdot (c.dist - d(c, A))$. The lemma holds.

Specifically, the area pruning works as follows. Each area is examined in descending order of their upper bounds. For each examined area, we can identify the local optimal locations by checking each edge in the area. If the gains of the local optimal locations found so far are larger than the upper bounds of all unexamined areas, we terminate the search; Otherwise we move on to the next area.

Edge Pruning. Next, we try to reduce the number of edges to be examined in an area. Consider an edge $e = (v_l, v_r)$ and any point p on e. Let C_e denote the set of clients on e. We list the upper bound for each edge in Lemma 2.

Lemma 2. $UPP(e) = \max_{v \in \{v_l, v_r\}} \{Gain(v) + \sum_{c \in C_e} w(c) \cdot \min\{l(c, v), c.dist\}\}$,
where $l(c, v)$ denotes the length of point interval on the edge e.

Proof. Let $C_1(p)$ be the set of clients on the edge e and attracted by the new server at p, namely $C_1(p) = \{c | c \in C_e \wedge p \in NLC(c)\}$. Let $C_2(p)$ be the set of clients not on the edge e and attracted by the new server at p, namely $C_2(p) = \{c | c \notin C_e \wedge p \in NLC(c)\}$. There are two cases.

Case 1: $C_e = \emptyset$. Since p is an interior point on the edge $e = (v_l, v_r)$, $Gain(p) \leq \max\{Gain(v_l), Gain(v_r)\}$ according to the key findings of the best-known algorithm. Then, this lemma holds.

Case 2: $C_e \neq \emptyset$. Consider $C_1(p) \neq \emptyset$ and $C_2(p) \neq \emptyset$. For each client c on e, if $l(v_l, c) < c.dist$, then $Gain(p) \leq Gain(v_l) + w(c) \cdot \min\{l(v_l, c), c.dist\}$. Otherwise if $l(v_l, c) \geq c.dist$, then $c.dist$ can be reduced to 0 and $Gain(p)$ may be equal to $w(c) \cdot \min\{l(v_l, c), c.dist\}$. The case for v_r can be handled similarly. Then, $Gain(p) \leq \max_{v \in \{v_l, v_r\}} \{Gain(v) + \sum_{c \in C_e} w(c) \cdot \min\{l(c, v), c.dist\}\}$. Thus, this lemma holds.

For the edge pruning, each edge is examined in descending order of their upper bounds. According to the key *findings* of the best-known algorithm, we can identify the local optimal location for each edge by computing the gains for each vertex and client on the edge. Note that we only need to check the optimal locations among the set V of vertices and the set C of clients. We need not to check the set S of servers. This is because the new server to be built cannot overlap with the existing servers.

Algorithm 1. The algorithm *MinSum-Alg*

Input: $G = (V, E)$, S, C and the number of areas m

Output: The set of all optimal locations

1 $OL \leftarrow \emptyset$, $Max \leftarrow 0$;

2 build $NLC(c)$ for each client $c \in C$;

3 divide the road network G into m areas ;

4 compute the upper bound $UPP(A_i)$ $(1 \le i \le m)$ using Lemma 1 and sort these areas by their upper bounds in descending order ;

5 **for** *each area A_i $(1 \le i \le m)$* **do**

6 $\quad$ **if** $UPP(A_i) < Max$ **then**

7 $\quad\quad$ **break** ;

8 $\quad$ compute the upper bound $UPP(e)$ for each edge e of of A_i using Lemma 2 and sort these edges by their upper bounds in descending order ;

9 $\quad$ **for** *each edge $e = (v_l, v_r)$ of A_i* **do**

10 $\quad\quad$ **if** $UPP(e) < Max$ **then**

11 $\quad\quad\quad$ **break** ;

12 $\quad\quad$ $Max' \leftarrow \max\{Gain(v_l), Gain(v_r), Gain(c_1), ..., Gain(c_i)\}$, where c_1, c_2, ..., c_i are the clients on the edge e ;

13 $\quad\quad$ **if** $Max < Max'$ **then**

14 $\quad\quad\quad$ $Max \leftarrow Max'$;

15 $\quad\quad\quad$ $OL \leftarrow$ the set of vertices and clients with the gain value of Max ;

16 **return** OL ;

Based on the above two-level pruning, we propose the *MinSum-Alg* algorithm which involves two key steps as follows.

- The first step is to divide G into a number m of subnetworks/areas such that any two areas have not any shared edges. The similar partition process of the best-known algorithm can be used. Specially, we randomly choose m points among $V \cup C$ as the center points of each area. An edge is assigned to an area if their distance is smaller than the distances between the edge and the other areas. The distance between an edge (v_l, v_r) and an area is defined to be the distance between the edge and the center point p of the area, namely $\min\{d(p, v_l), d(p, v_r)\}$.
- The second step is to apply the above area pruning and edge pruning to reduce the number of areas and edges to be examined and return the optimal locations with the largest gain value.

The description for *MinSum-Alg* is given in Algorithm 1. Next, we take an example to illustrate the algorithm process. Consider Fig. 1(a). The weight $w(c)$ for each client c is set to be 1. Let $m = 2$. We divide the road network G into two areas, A_1 and A_2. We randomly choose v_4 and v_6 as the center points of A_2 and A_1. Consider the edge $e_1 = (v_7, v_8)$. The distance between e_1 and v_6 is equal to $d_1 = \min\{d(v_6, v_7), d(v_6, v_8)\} = 4$. The distance between e_1 and v_4 is equal to $d_2 = \min\{d(v_4, v_7), d(v_4, v_8)\} = 6.5$. Since $d_1 < d_2$, the edge e_1 is assigned to A_1. Similarly, the other edges are assigned to A_1 and A_2 as shown in Fig. 1(b).

Next, by Lemma 1, the upper bounds for A_1 and A_2 are equal to $UPP(A_1) = 19$ and $UPP(A_2) = 14.5$, respectively. Then, A_1 is first examined. There are four edges, namely $e_1 = (v_7, v_8)$, $e_2 = (v_6, v_7)$, $e_3 = (v_2, v_6)$ and $e_4 = (v_6, v_8)$. Since $Gain(v_6) = 13$ and $Gain(v_7) = 7$, by Lemma 2, the upper bound for e_2 is equal to $UPP(e_2) = \max\{13 + 1 \times 1 + 1 \times 2, 7 + 1 \times 4 + 1 \times 5\} = 16$. Similarly, the upper bounds for e_1, e_3 and e_4 are equal to $UPP(e_1) = 7$, $UPP(e_3) = 15$ and $UPP(e_4) = 13$, respectively. Then, the edge e_2 is first examined. The optimal location on e_2 corresponds to the whole point interval $[c_4, c_5]$. Similarly, the optimal location on e_3 can be found. The gain of optimal location is equal to 15 which is larger than the upper bounds for e_1 and e_4. Thus, the two edges are pruned. Similarly, A_2 is pruned.

4   Multiple-Location MinSum Query Algorithm

We give the multiple-location MinSum query problem and the greedy algorithm for the problem in Sect. 4.1. Then, we propose the approximate guarantee for the algorithm in Sect. 4.2.

4.1   The Greedy Algorithm

Problem 2. Given a road network $G = (V, E)$, a set C (S) of clients (servers) on G, a positive integer $k \geq 2$, the multiple-location MinSum query problem is to find k locations for new servers in the road network, namely p_1, p_2, p_3, ..., p_k, which minimize $\sum_{c \in C} w(c) \cdot d(c, NN_{S \cup \{p_1, p_2, ..., p_k\}}(c))$.

Firstly, we introduce Lemma 3 which is similar to the key *findings* of single-location query problem.

Lemma 3. *There exist k locations in $V \cup C$ which is a solution to the multiple-location MinSum query problem.*

Proof. Consider a solution $\{p_1, p_2, ..., p_k\}$ to our problem. Suppose that p_j ($1 \leq j \leq k$) is not in $V \cup C$. Let p_j be on the edge $e = (v_l, v_r)$ in which the clients, c_1, c_2, ..., c_i are included. Suppose that we remove the server built at p_j. Then, our multi-location problem is equivalent to the single-location problem with the set of servers $S' = S \cup \{p_1, ..., p_{j-1}, p_{j+1}, ..., p_k\}$. By the key *findings* of the best-known algorithm, there exists an optimal location p which is in $\{v_l, v_r, c_1, ..., c_i\}$. Then, p_j can be replaced with $p \in V \cup C$. Similarly, each location of the solution which is not in $V \cup C$ can be replaced with a location in $V \cup C$. This lemma holds.

Theorem 1. *The multiple-location MinSum query problem is NP-hard.*

Proof. If we can solve Problem 2 in polynomial time, then we can solve maximum coverage problem in polynomial time. Suppose that we have a list of sets v_i. Each v_i is a set of clients c_j. We want to find k of v_i says v_{i_j}, $1 \leq j \leq k$ to make $|\bigcup_{j=1}^{k} v_{i_j}|$ is maximal. We assume that $|v_i| \geq 2$ (if there exists some $|v_i|$ which is equal to 1, just add a common client to each v_i). Now we can form a road network $G(V, E)$. V includes all clients, all v_i, and a server s. There is an edge between s and each clients with length equal to 3. There is an edge between c_j and v_i with length equal to 1 if $c_j \in v_i$. An example is in Fig. 2. As we know, the

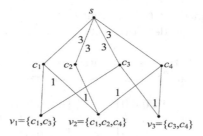

Fig. 2. A proof example

k new servers should be located at the vertices or the clients (i.e., Lemma 3). In our case, the k new servers shouldn't be located at the clients. Since if there is one server on a client c_j (only $c_j.dist$ changes from 3 to 0, the gain is 3), then it is always better to move the server to any vertex v_i where v_i contains c_j (at least two clients' $c.dist$ change from 3 to 1, the gain is 4). It is easy to see that the solution of Problem 2 is the solution of the maximum coverage problem (the maximum cover of clients, the maximum number of clients whose $c.dist$ will decrease to 1). So if we can solve Problem 2 in polynomial time, then we can solve the maximum coverage problem in polynomial time.

Since this problem is NP-hard, we propose a greedy algorithm *GA(MinSum)*, a heuristic-based method, which involves three steps. The first step is to execute *MinSum-Alg* based on the current set S of servers for finding an optimal location for a new server. The second step is to insert the new server into S. The third step is an iterative step which executes the first step and the second step iteratively until k new servers are found.

4.2   Approximate Ratio

Let $|C| = m$ and $|V \cup C| = n$. By Lemma 3, the k new servers should be located at the vertices or the clients. Consider a cost matrix M with m rows and n columns. An element M_{ij} denotes the reduction value of $Cost(c_i)$ if a new server is built at the j-th location. It is easy to know that $M_{ij} \geq 0$. According to the description of *GA(MinSum)*, in each iteration, it selects the column of cost matrix where the sum of the corresponding reduction values for each row is the largest. After k iterations, the selected k columns correspond to the approximate solution to our problem.

Consider the i-th $(1 \leq i \leq k)$ iteration. We have built $i - 1$ new servers. The cost matrix is M_i. Suppose that O_i denotes the reduction of objective value if the remaining $k - i + 1$ servers have been built at the exactly optimal locations. In particular, if $i = 1$, O_1 denotes the reduction of objective value of the exact solution to our problem. Suppose that the column selected by *GA(MinSum)* is the z-th column of M_i. Then, the gain value of the i-th location (i.e., the z-th column) is $B_i = \sum_{1 \leq x \leq m} M_{i_{xz}}$ which is the largest among all columns of M_i.

The i-th location obtained by *GA(MinSum)* may not be the exactly optimal location. Let D_i denote the deviation value, namely $D_i = O_i - (B_i + O_{i+1})$ $(D_i \geq 0)$.

Let us take an example to illustrate these notations. Let $k = 3$ and the cost matrix is with three rows and five columns as follows.

$$M_1 = \begin{bmatrix} 5\,9\,2\,3\,2 \\ 5\,3\,8\,6\,3 \\ 5\,1\,1\,5\,7 \end{bmatrix} M_2 = \begin{bmatrix} 0\,4\,0\,0\,0 \\ 0\,0\,3\,1\,0 \\ 0\,0\,0\,0\,2 \end{bmatrix} M_3 = \begin{bmatrix} 0\,0\,0\,0\,0 \\ 0\,0\,3\,1\,0 \\ 0\,0\,0\,0\,2 \end{bmatrix} M' = \begin{bmatrix} 5\,9\,0\,3\,0 \\ 5\,0\,8\,6\,0 \\ 5\,0\,0\,5\,7 \end{bmatrix}$$

Consider M_1 and the first iteration of *GA(MinSum)*. The exact solution for 3 new servers is $\{2, 3, 5\}$, namely the 2nd column, the 3rd column and the 5th column of M_1. $O_1 = 9 + 8 + 7 = 24$. The first iteration will select the 1st column of M_1 since the sum of elements of this column is the largest, namely $B_1 = 5 + 5 + 5 = 15$. After a new server has been built at the 1st column, there are 2 remaining servers. Then, M_1 is changed to M_2 by subtracting the selected column (i.e., the 1st column). If the difference value for the corresponding element is less than 0, the corresponding element of M_2 is set to be 0. The set of exactly optimal locations for 2 new servers is $\{2, 3\}$, namely the 2nd column and the 3rd column of M_2. $O_2 = 4 + 3 = 7$. Then, $D_1 = O_1 - (B_1 + O_2) = 2$. The second iteration will select the 2nd column of M_2 and $B_2 = 4 + 0 + 0 = 4$. After another server has been built at the 2nd column, there is 1 remaining server. Then, M_2 is changed to M_3 by subtracting the 2nd column. Then, $O_3 = 3$ and $D_2 = O_2 - (B_2 + O_3) = 0$. Next, we list Lemmas 4 and 5.

Lemma 4. $\frac{B_i}{O_i} \geq \frac{1}{k-i+1}$.

Proof. By the definition of B_i, $B_i = \max_{1 \leq j \leq n}\{\sum_{1 \leq x \leq m} M_{i x j}\}$. Suppose that the exactly optimal locations for the remaining $k-i+1$ new servers correspond to the columns $u_1, u_2, ..., u_{k-i+1}$ of M_i. Then, $O_i \leq \sum_{1 \leq j \leq k-i+1} \sum_{1 \leq x \leq m} M_{i x u_j} \leq \sum_{1 \leq j \leq k-i+1} B_i$. Thus, $\frac{B_i}{O_i} \geq \frac{B_i}{\sum_{1 \leq j \leq k-i+1} B_i} \geq \frac{1}{k-i+1}$.

Lemma 5. $\frac{D_i}{O_i} \leq \frac{k-i}{(k-i+1)^2}$.

Proof. Suppose that the exactly optimal locations for the remaining $k - i + 1$ new servers correspond to the columns $u_1, u_2, ..., u_{k-i+1}$ of M_i. Suppose that the column selected by *GA(MinSum)* in the i-th iteration is u' of M_i. If u' is equal to u_x $(1 \leq x \leq k - i + 1)$, then $D_i = 0$. This lemma holds.

Otherwise, we can replace one exactly optimal location with u'. Let $N(u_y) = \sum_{1 \leq j \leq m} M_{i j u_y} \cdot F(j, u_y)$ in which $F(j, u_y) = 1$ if $M_{i j u_y}$ is the largest among these exactly optimal $k - i + 1$ columns in the j-th row of M_i. Otherwise, $F(j, u_y) = 0$. Let $N'(u_y) = \sum_{1 \leq j \leq m} M_{i j u'} \cdot F(j, u_y)$.

Intuitively, $N(u_y)$ denotes the reduction value contributed by the column u_y if the new server is built at u_y. $N'(u_y)$ denotes the reduction value contributed by the column u' if u_y is replaced with u'. For the above example, M_1 is equivalent to M'.

Since O_i denotes the reduction of objective value for the remaining $k - i + 1$ new servers, we have $D_i \leq \min_{1 \leq y \leq k-i+1}\{N(u_y) - N'(u_y)\} \leq \frac{1}{k-i+1}(\sum_{1 \leq y \leq k-i+1} N(u_y) - \sum_{1 \leq y \leq k-i+1} N'(u_y)) = \frac{O_i - B_i}{k-i+1}$.

By Lemma 4, $\frac{D_i}{O_i} \leq \frac{k-i}{(k-i+1)^2}$. This lemma holds.

Theorem 2. *The approximate ratio for our solution is $R \geq \frac{k+1}{2k}$.*

Proof. By the definition of O_i, the reduction of objective value for our problem is equal to O_1. Let the reduction of objective value obtained by *GA(MinSum)* be equal to O'. Then, $R = \frac{O'}{O_1}$. If $k = 1$, $O' = O_1$. This lemma holds.

Suppose that if $k = r$, $R = \frac{O'}{O_1} \geq \frac{r+1}{2r}$ holds. For $k = r+1$, $O_2 = O_1 - B_1 - D_1$. For O', since $k = r$ holds, $O' \geq B_1 + O_2 \cdot \frac{r+1}{2r}$. Then, by Lemmas 4 and 5,

$$\frac{O'}{O_1} \geq \frac{B_1 + O_2 \cdot \frac{r+1}{2r}}{O_1} = \frac{B_1}{O_1} + \frac{r+1}{2r}(1 - \frac{B_1}{O_1} - \frac{D_1}{O_1}) = \frac{B_1}{O_1}(1 - \frac{r+1}{2r}) + \frac{r+1}{2r}(1 - \frac{D_1}{O_1}) \geq$$

$$\frac{1}{r+1}(1 - \frac{r+1}{2r}) + \frac{r+1}{2r}(1 - \frac{r}{(r+1)^2}) = \frac{(r+1)+1}{2(r+1)}. \text{ Then, this lemma also holds for}$$
$k = r + 1$.

5   Experiments

In this section, we evaluated the performance of our proposed algorithms. We ran all experiments on a machine with a 2.10 Ghz Pent. T4300 CPU and 2 GB RAM, running Windows 7 OS. All algorithms were implemented in C++. We used two real world road networks, *SF* and *CA*, for San Francisco and California, respectively. Both datasets were downloaded from http://www.cs. utah.edu/~lifeifei/SpatialDataset.htm. *SF* contains 174,955 vertices and 223,000 edges, and *CA* contains 21,047 vertices and 21,692 edges, in which most vertices involve at most 4 edges each. There is only one vertex adjacent to 8 edges. Both datasets include the real clients and servers which are generated by the way similar to [3,4,23,26]. Specially, we obtained a large number of real building locations in *SF* (*CA*) from the *OpenStreetMap* project. Each building location was projected on one of the road network edges nearest to this building. Then we randomly sampled these locations as servers and clients and they were stored in two separate lists. In both *SF* and *CA*, each client is associated with a weight generated randomly from a Zipf distribution with a skewness parameter $\alpha > 1$. By default, α is set to ∞ and this means that the weight of each client is equal to 1. In the experiments, the default value for $|S|$ for *SF* (*CA*) is 1,000 (100) and the default value for $|C|$ for *SF* (*CA*) is 300,000 (30,000). The default value of m (i.e., the number of partitions) is set to the same value as the best-known algorithm with the best performance. The default value for k is set to be 4.

5.1   Experiments for the Single-Location MinSum Query

In this set of experiments, we study the effects of $|S|$, $|C|$, α and the two-level pruning.

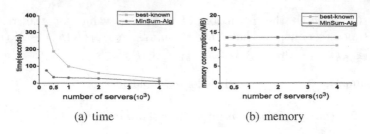

(a) time   (b) memory

Fig. 3. Effect of $|S|$ on SF for $MinSum\text{-}Alg$ and $best\text{-}known$

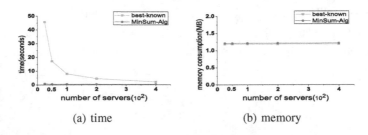

(a) time   (b) memory

Fig. 4. Effect of $|S|$ on CA for $MinSum\text{-}Alg$ and $best\text{-}known$

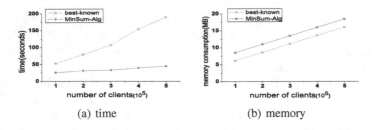

(a) time   (b) memory

Fig. 5. Effect of $|C|$ on SF for $MinSum\text{-}Alg$ and $best\text{-}known$

Effect of $|S|$. We study the effect of $|S|$ on the SF dataset in Fig. 3(a) and (b). It is obvious that the running time of $MinSum\text{-}Alg$ is smaller than that of the best-known algorithm. This is because the proposed two-level pruning method often reduces the search space of our algorithm. The running time of both algorithm is reduced with the increased number of servers. This is because the values for $c.dist$ of each client c are uniformly distributed with more servers. Then, more clients may be pruned. The memory storage of $MinSum\text{-}Alg$ is a little larger than that of the best-known algorithm. This is because $MinSum\text{-}Alg$ needs some extra space to save the related information for the two-level pruning method, including the upper bounds and $c.dist$. But the memory consumption of $MinSum\text{-}Alg$ is still small and acceptable. Similar trend can be found on the CA dataset in Fig. 4(a) and (b).

Effect of $|C|$. We study the effect of $|S|$ on the SF dataset in Fig. 5(a) and (b). Since the search space of our algorithm is smaller, the running time of

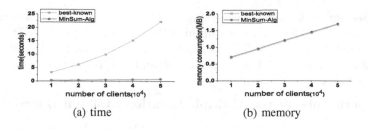

(a) time   (b) memory

Fig. 6. Effect of $|C|$ on CA for $MinSum$-Alg and $best$-$known$

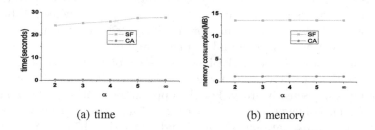

(a) time   (b) memory

Fig. 7. Effect of α on SF and CA for $MinSum$-Alg

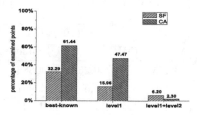

Fig. 8. Effect of two-level pruning

$MinSum$-Alg is smaller than that of the best-known algorithm. With the increased sizes of $|C|$, the running time is also increased for both algorithms. This is because more clients need to be examined for both algorithms. Besides, the memory consumption are sensitive to the number of clients and increased with the increased sizes of $|C|$. The memory consumption of $MinSum$-Alg is larger than that of the best-known algorithm. This is because $MinSum$-Alg needs extra space for the two-level pruning method. Similar trend can be found on the CA dataset in Fig. 6(a) and (b).

Effect of α. As shown in Fig. 7, the running time and the memory consumption of $MinSum$-Alg are insensitive to α.

Effect of the two-level pruning. As shown in Fig. 8, the two-level pruning method can dramatically reduce the number of points to be examined. Specially, only 15.96% and 47.47% of points in $V \cup C$ need to be examined for the area pruning (level1) on SF and CA, respectively. Since less partitions are pruned, there is a larger percentage for the area pruning on CA. Only 6.2% and 2.3% of points need to be examined for the combination of both area pruning and edge pruning (level1 + level2) on SF and CA, respectively. The best-known algorithm needs to check 32.29% and 61.44% of points on SF and CA, respectively.

5.2   Experiments for the Multiple-Location MinSum Query

In this set of experiments, we study the effects of $|S|$, $|C|$, k and the approximate ratio for $GA(MinSum)$. As shown in Table 2, the running time is decreased with the increased sizes of $|S|$. The memory consumption is not sensitive to the sizes of $|S|$. Similar reasons can be found in Figs. 3 and 4. As shown in Table 3, the running time and memory consumption are increased with the increased sizes of $|C|$. Similar reasons can be found in Figs. 5 and 6. As shown in Table 4, the running time is increased with the increased sizes of k. This is because there are more iterations to be executed when the value of k is larger. The memory consumption is not sensitive to k. For the approximate ratio, we have made 10 experiments. The value of k is varied from 2 to 5 in each experiment. As shown in Table 5, the approximate ratio is good and larger than 95% in most cases.

Table 2. Effect of $|S|$ on SF and CA for GA(MinSum)

$|S|$ (10^3)	SF		$|S|$ (10^2)	CA	
	Time (s)	Memory (MB)		Time (s)	Memory (MB)
0.25	390.642	13.475	0.25	4.883	1.211
0.5	255.062	13.479	0.5	2.667	1.211
1	159.058	13.487	1	1.342	1.212
2	92.742	13.502	2	1.045	1.214
4	52.743	13.532	4	0.656	1.217

Table 3. Effect of $|C|$ on SF and CA for GA(MinSum)

$|C|$ (10^5)	SF		$|C|$ (10^4)	CA	
	Time (s)	Memory (MB)		Time (s)	Memory (MB)
1	96.409	8.51	1	0.998	0.715
2	127.609	10.991	2	1.404	0.963
3	134.395	13.471	3	2.012	1.212
4	146.079	15.952	4	2.106	1.462
5	169.775	18.432	5	2.652	1.713

Table 4. Effect of k on SF and CA for GA(MinSum)

k	SF		CA	
	Time (s)	Memory (MB)	Time (s)	Memory (MB)
2	61.933	13.471	0.889	1.21
3	98.64	13.471	1.216	1.21
4	134.395	13.471	1.528	1.21
5	177.389	13.471	1.996	1.21
10	398.271	13.471	3.962	1.21

Table 5. Approximate ratio of GA(MinSum) (%)

k	No. 1	No. 2	No. 3	No. 4	No. 5	No. 6	No. 7	No. 8	No. 9	No. 10	Average
2	100	99.5	100	100	100	100	100	95.8	97.7	100	99.3
3	95.9	99.0	99.9	100	96.9	99.1	98.0	96.0	93.6	100	97.8
4	96.2	99.1	99.7	100	100	99.5	98.3	96.2	97.8	99.2	98.6
5	96.9	98.6	99.5	100	100	99.6	99.6	96.8	97.8	99.7	98.9

6 Related Work

The optimal location query, originating from the facility location problem, also known as location analysis [1,2,10,16], has been extensively studied in past years. Recently, researchers in the database community are paying attention to this problem because of its broad applications. The MaxBRNN problem [1] is to find an optimal region such that the total number of clients attracted by a new server to be set up is maximized. A solution with an exponential-time complexity was presented for the MaxBRNN problem in [1]. The first polynomial-time complexity algorithm for the problem was introduced in [20] and the extensions of this algorithm were studied in [21]. Moreover, an approximate algorithm was presented for the MaxBRNN problem in [25]. Recently, an improved algorithm for the MaxBRNN problem was given in [12]. The problem studied in [28] is a generalized MaxBRkNN problem in the L_p-norm space. Different from the MaxBRNN problem, our problem is to find an optimal location instead of an optimal region. Besides, our problem is based on the road network environment instead of the L_p-norm space. The algorithm in [6] was proposed to find an optimal location instead of an optimal region for the L_1-norm space. The algorithm in [27] was studied to find a location which minimizes the average distance from each client to its closest server when a new server is built at this location. The problem studied in [14,15] was to select a location from a given set of potential locations for a new server so that the average distance between a client and its nearest server is minimized. The BRNN problem [7] and the reverse top-k problem [19] are also related. The spatial matching problem is also with the non-road network setting [13,17,18,22]. Xiao et al. [23] first studied the OLQ problem with the road network setting and presented an efficient algorithm. An extension of [23] was given in [26] in which the OLQ problem with dynamic clients and servers was studied. Recently, a more efficient algorithm for the OLQ problem was given in [3,4]. The exact solution for the multi-server version of OLQ with MinMax objective function was first studied in [11]. [9] studied a static version and a dynamic version of OLQ with the MaxSum objective function. However, the MinSum query is not studied in [3,4,9,11]. There is a study about isochrone queries in a multimodal network [8], which is similar to the OLQ query. In addition, the problem of proximity queries among sets of moving objects in road networks was studied in [24].

7   Conclusion

In this paper, we study the OLQ with the MinSum objective function in road networks, namely the MinSum query. We propose an improved algorithm *MinSum-Alg* with two new pruning technologies, namely area pruning and edge pruning. We also study the optimal multiple-location MinSum query. Since this problem is NP-hard, we propose a greedy algorithm based on *MinSum-Alg*. Moreover, we give the theoretical guarantee for our solution. We verify the performance of the proposed algorithms on the real road networks. The MinSum query with dynamic settings and directed graph is our future work.

Acknowledgements. We are very thankful to the anonymous reviewers for the very useful comments. This paper is supported by the NSFC (61572537, U1401256, U1501252) and the Science and Technology Planning Project of Guangdong Province, China (2014A080802003).

References

1. Cabello, S., Diaz-Banez, J.M., Langerman, S., Seara, C., Ventura, I.: Reverse facility location problems. In: CCCG (2005)
2. Cardinal, J., Langerman, S.: Min-max-min geometric facility location problems. In: EWCG (2006)
3. Chen, Z., Liu, Y., Wong, R.C.W., Xiong, J., Mai, G., Long, C.: Efficient algorithms for optimal location queries in road networks. In: SIGMOD (2014)
4. Chen, Z., Liu, Y., Wong, R.C.W., Xiong, J., Mai, G., Long, C.: Optimal location queries in road networks. ACM Trans. Database Syst. **40**(3), 17 (2015)
5. Dijkstra, E.W.: A note on two problems in connexion with graphs. Numer. Math. **1**(1), 269–271 (1959)
6. Du, Y., Zhang, D., Xia, T.: The optimal-location query. In: Bauzer Medeiros, C., Egenhofer, M.J., Bertino, E. (eds.) SSTD 2005. LNCS, vol. 3633, pp. 163–180. Springer, Heidelberg (2005). doi:10.1007/11535331_10
7. Korn, F., Muthukrishnan, S.: Influence sets based on reverse nearest neighbor queries. In: SIGMOD (2000)
8. Gamper, J., Böhlen, M., Innerebner, M.: Scalable computation of isochrones with network expiration. In: Ailamaki, A., Bowers, S. (eds.) SSDBM 2012. LNCS, vol. 7338, pp. 526–543. Springer, Heidelberg (2012). doi:10.1007/978-3-642-31235-9_35

9. Ghaemi, P., Shahabi, K., Wilson, J.P., Kashani, F.B.: A comparative study of two approaches for supporting optimal network location queries. GeoInformatica **18**(2), 229–251 (2014)

10. Krarup, J., Pruzan, P.M.: The simple plant location problem: survey and synthesis. Eur. J. Oper. Res. **12**(1), 36–57 (1983)

11. Liu, R., Fu, A.W.C., Chen, Z., Huang, S., Liu, Y.: Finding multiple new optimal locations in a road network. In: SIGSPATIAL (2016)

12. Liu, Y., Wong, R.C.W., Wang, K., Li, Z., Chen, C., Chen, Z.: A new approach for maximizing bichromatic reverse nearest neighbor search. Knowl. Inf. Syst. **36**(1), 23–58 (2013)

13. Long, C., Wong, R.C.W., Yu, P.S., Jiang, M.: On optimal worst-case matching. In: SIGMOD (2013)

14. Qi, J., Zhang, R., Kulik, L., Lin, D., Xue, Y.: The min-dist location selection query. In: ICDE (2012)

15. Qi, J., Zhang, R., Wang, Y., Xue, A.Y., Yu, G., Kulik, L.: The min-dist location selection and facility replacement queries. World Wide Web **17**(6), 1261–1293 (2014)

16. Tansel, B.C., Francis, R.L., Lowe, T.J.: Location on networks: a survey. Manage. Sci. **29**(4), 498–511 (1983)

17. Tong, Y., She, J., Ding, B., Chen, L., Wo, T., Xu, K.: Online minimum matching in real-time spatial data: experiments and analysis. PVLDB **9**(12), 1053–1064 (2016)

18. U, L.H., Yiu, M.L., Mouratidis, K., Mamoulis, N.: Capacity constrained assignment in spatial databases. In: SIGMOD (2008)

19. Vlachou, A., Doulkeridis, C., Kotidis, Y., Norvag, K.: Monochromatic and bichromatic reverse top-k queries. IEEE Trans. Knowl. Data Eng. **23**(8), 1215–1229 (2011)

20. Wong, R.C.W., Ozsu, M.T., Fu, A.W.C., Yu, P.S., Liu, L.: Efficient method for maximizing bichromatic reverse nearest neighbor. PVLDB **2**(1), 1126–1137 (2009)

21. Wong, R.C.W., Ozsu, M.T., Fu, A.W.C., Yu, P.S., Liu, L., Liu, Y.: Maximizing bichromatic reverse nearest neighbor for lp-norm in two- and three-dimensional spaces. VLDB J. **20**, 893–919 (2011)

22. Wong, R.C.W., Tao, Y., Fu, A.W.C., Xiao, X.: On efficient spatial matching. In: VLDB (2007)

23. Xiao, X., Yao, B., Li, F.: Optimal location queries in road network databases. In: ICDE (2011)

24. Xu, Z., Jacobsen, H.A.: Processing proximity relations in road networks. In: SIGMOD (2010)

25. Yan, D., Wong, R.C.W., Ng, W.: Efficient methods for finding influential locations with adaptive grids. In: CIKM (2011)

26. Yao, B., Xiao, X., Li, F., Wu, Y.: Dynamic monitoring of optimal locations in road network databases. VLDB J. **23**(5), 697–720 (2014)

27. Zhang, D., Du, Y., Xia, T., Tao, Y.: Progressive computation of the min-dist optimal-location query. In: VLDB (2006)

28. Zhou, Z., Wu, W., Li, X., Lee, M.L., Hsu, W.: MaxFirst for MaxBRkNN. In: ICDE (2011)

Efficiently Mining High Utility Co-location Patterns from Spatial Data Sets with Instance-Specific Utilities

Lizhen Wang, Wanguo Jiang, Hongmei Chen[$\boxtimes$], and Yuan Fang

Department of Computer Science and Engineering, School of Information
Science and Engineering, Yunnan University, Kunming 650091, Yunnan, China
hmchen@ynu.edu.cn

Abstract. Traditional spatial co-location pattern mining attempts to find the subsets of spatial features whose instances are frequently located together in some regions. Most previous studies take the prevalence of co-locations as the interestingness measure. However, it is more meaningful to take the utility value of each instance into account in spatial co-location pattern mining in some cases. In this paper, we present a new interestingness measure for mining high utility co-location patterns from spatial data sets with instance-specific utilities. In the new interestingness measure, we take the intra-utility and inter-utility into consideration to capture the global influence of each feature in co-locations. We present a basic algorithm for mining high utility co-locations. In order to reduce high computational cost, some pruning strategies are given to improve the efficiency. The experiments on synthetic and real-world data sets show that the proposed method is effective and the pruning strategies are efficient.

Keywords: Spatial co-location patterns · High utility co-location patterns · Intra-utility · Inter-utility · Pruning

1 Introduction

In recent years, spatial data are rapidly generated and the size of spatial data sets is getting huger and huger. For example, NASA Earth Observing System has been generating more than 1 TB of spatial data per day. With the popularity of mobile devices, spatial data with location would increase faster and faster. The vast amounts of spatial data contain potential and valuable information which can help us make important decisions. There are a lot of researches on spatial data mining, including spatial association rule analysis, spatial clustering, spatial classification, and so on.

In spatial data, if the distance between two spatial instances is no more than a given distance threshold, the two instances satisfy the *neighbor relationship*. Traditional spatial co-location pattern mining aims at finding the subsets of spatial features whose instances are frequently located in neighborhoods. A *row instance* of a co-location c represents a subset of instances, which includes an instance of each feature in c and forms a clique under the neighbor relationship. All row instances of a co-location c make up its *table instance* denoted as $T(c)$. Similar to the support in Association

© Springer International Publishing AG 2017
S. Candan et al. (Eds.): DASFAA 2017, Part II, LNCS 10178, pp. 458–474, 2017.
DOI: 10.1007/978-3-319-55699-4_28

Rules Mining (ARM), *Participation Index* (PI) is used to evaluate the prevalence of co-locations. The *PI* of a co-location c is defined as $PI(c) = \min_{f_i \in c}\{PR(c, f_i)\}$, where $PR(c, f_i)$ is the *Participation Ratio* (PR) of feature f_i in a co-location c, that is $PR(c, f_i) = \dfrac{|\pi_{f_i} T(c)|}{\text{Number of instances of } f_i}$, where π is the relational projection operation with duplication elimination. Participation ratio is used to evaluate the prevalence of features, and participation index is used to measure the prevalence of co-locations, which is the interestingness measure in traditional co-location mining.

Mining co-locations is very significant in the real world. For example, botanists have found that there are orchids in 80% of the area where the middle-wetness green-broad-leaf forest grows. A mobile service provider may be interested in mobile service patterns frequently requested by geographical neighboring users. Other applications include Earth science, public health, biology, transportation, etc.

In most previous studies, the importance of all features and instances are treated similarly. However, there exist some difference between features and even instances belonging to the same feature. For instance, the economic value of the rosewood is much greater than that of the ordinary pine. What's more, the value of rosewoods with different sizes is also different. So, only checking the prevalence of co-locations might be insufficient for identifying real interesting patterns. Traditional co-location mining can't find some low frequency but high interesting patterns [19], and some prevalent patterns which just reflect the common sense may be worthless to users.

Here, we use an example to illustrate the problem. Figure 1 shows the locations of instances of six kinds of plants (features), and each instance is denoted by the plant type and a numeric id, e.g. A.1, and edges among instances indicate neighboring relationships. The superscript of each instance represents its utility value, which can be considered as its price. Table 1 gives the total utility value of each kind of plant which is the sum of utility value of all instances belonging to the plant type.

In Fig. 1, according to traditional co-location mining, for the co-location $\{A, B, C\}$, $PI(\{A, B, C\}) = 1/4$. And if the prevalence threshold is 0.3, $\{A, B, C\}$ would be

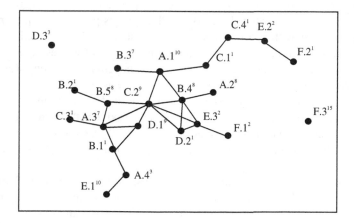

Fig. 1. An example spatial data set

Table 1. Total utility value of each plant in Fig. 1

Features	Instances	Total utility values
A	$A.1^{10}$, $A.2^{8}$, $A.3^{7}$, $A.4^{3}$	28
B	$B.1^{1}$, $B.2^{1}$, $B.3^{7}$, $B.4^{8}$, $B.5^{8}$	25
C	$C.1^{1}$, $C.2^{9}$, $C.3^{1}$, $C.4^{1}$	12
D	$D.1^{9}$, $D.2^{1}$, $D.3^{3}$	13
E	$E.1^{10}$, $E.2^{2}$, $E.3^{2}$	14
F	$F.1^{2}$, $F.2^{1}$, $F.3^{15}$	18

regarded as a non-interesting co-location. However, according to $T(\{A, B, C\}) = \{\{A.1^{10}, B.4^{8}, C.2^{9}\}, \{A.3^{7}, B.5^{8}, C.2^{9}\}\}$, the utility value of feature A's instances in $\{A, B, C\}$ is 17, which accounts for 17/28 of total utility value of A. Similarly, the proportion of B is 16/25 and C is 9/12. So the utility of each feature in $\{A, B, C\}$ account for a large proportion of its total utility. $\{A, B, C\}$ may be interesting. However, as to the pattern $\{E, F\}$, $PI(\{E, F\}) = 2/3$, $T(\{E, F\}) = \{\{E.2^{2}, F.2^{1}\}, \{E.3^{2}, F.1^{2}\}\}$. But the proportion of E is 4/14 and F is 3/18. So the utility of each feature in $\{E, F\}$ is less than 30%. $\{E, F\}$ may be non-interesting.

Therefore, the traditional measure may not find interesting co-locations because the utilities of features and instances are ignored. In this paper, we focus on high utility co-location mining from spatial data sets with instance-specific utilities.

1.1 Related Work

The problem of mining spatial association rules was first discussed in [1]. The participation index for prevalent co-location mining and join-based algorithm was presented in [2, 3]. Then a lot of existing works about co-location mining are based on the participation index which satisfies the downward closure property. Join-less algorithm was introduced in [4], using a novel model to materialize spatial neighbor relationships and an instance-lookup scheme to reduce the computational cost of identifying table instances. An efficient algorithm based on iCPI-Tree was proposed in [8]. In order to mine the co-locations with rare features, a new prevalence measure called the maximal participation ratio was proposed in [9]. A new general class of interestingness measures based on the spatial distribution of co-locations and information entropy was proposed in [5]. Probabilistic prevalent co-location mining was introduced in [6] to find co-locations in the context of uncertain data. [7] studied co-location rule mining on interval data and defined new related concepts based on the semantic proximity neighborhood. An optimal candidate generation method was proposed in [17]. Complex spatial co-location mining which can deal with complex spatial relationships was introduced in [18].

The research on high utility mining was first discussed in ARM [10]. The utility of each item consists of *internal utility* and *external utility*. The internal utility represents the quantity of items in transactions and the external utility is the unit profit values of items. But the utility of itemsets doesn't satisfy the downward closure property which can improve the mining efficiency, and a two-phase algorithm for fast mining high utility itemsets was proposed in [11]. [12] introduced a novel framework to mine the interesting high utility pattern with a strong frequency affinity. An incremental mining algorithm for efficiently mining high utility itemsets was proposed to handle the intermittent data environment in [13]. UP-Growth proposed in [14] enhances the mining performance through maintaining the information of high utility itemsets by UP-tree. A novel algorithm named GUIDE and a special data structure named TMUI-tree were proposed for mining temporal maximal utility itemsets from data stream environment in [15]. [16] introduced an efficient algorithm named USpan to mine high utility sequences from large scale data with very low minimum utility.

There are more and more studies on co-location mining and high utility itemsets mining, but there were rare literatures about high utility co-location mining [19, 20]. Similar to ARM [10], [19] divided the utility of features in a co-location into *external utility* and *internal utility*. The external utility represents the unit profile and the internal utility represents the quantity of different instances of features in a table instance. The utility of a feature in a co-location is equal to the product of external and internal utilities. And a framework for mining high utility co-locations was proposed in [19]. By following the definitions in [19], [20] discussed a problem of updating high utility co-locations on evolving spatial databases.

In some real-world data, the utilities of features are different from each other and even instances belonging to the same feature may have an obvious difference in utilities. Furthermore, in some cases, the data set can't map into the model of external and internal utility. Considering the complexity of real-world data, there exist two major challenges in high utility co-location mining from spatial data sets with instance-specific utilities. One is how to define the interestingness measure reasonably to judge high utility co-locations, and another is how to mine high utility co-locations efficiently. In this paper, we try to tackle these challenges.

1.2 Our Contributions

Different from previous researches, we make the following contributions in this paper:

First, we take the instances with utilities as study objects, and the importance of features and instances is treated differently.

Second, we propose a new interestingness measure to identify high utility co-locations in spatial data sets with instance-specific utilities.

Third, we present a basic algorithm to mine high utility co-locations. In order to reduce the computational cost, some pruning strategies are given.

Finally, the extensive experiments on synthetic and real-world data sets verify that the proposed method is effective and efficient.

The remainder of the paper is organized as follows: Sect. 2 gives the related concepts for mining high utility co-locations from spatial data sets with instance-specific utilities, and a basic algorithm is presented in Sect. 3. In Sect. 4, the pruning strategies are detailed. Experimental results and evaluation are shown in Sect. 5. The conclusion and future work are discussed in Sect. 6.

2 Related Concepts

In the real world, the importance of each instance may be different. Thus, we take the instances with utilities as study objects and the utilities reflect their importance. The related concepts for mining high utility co-locations are given in this section, and Table 2 summarizes notations frequently used throughout the paper.

Table 2. Summary of notations

Notation	Definition	Notation	Definition
F	Set of spatial features	$u(f_i)$	Utility of feature f_i
f_i	i-th spatial feature	$u(f_i, c)$	Utility of feature f_i in co-location c
S	Set of features' instances	$IntraUR$ (f_i, c)	Intra-utility ratio of f_i in c
$f_i \cdot j^v$	j-th instance with utility v of f_i	$InterUR$ (f_i, c)	Inter-utility ratio of f_i in c
c	A co-location pattern	$UPR(f_i, c)$	Utility participation ratio of f_i in c
k	Size of c	$UPI(c)$	Utility participation index of c
R	A spatial neighbor relationship	w_1	Weighted value of $IntraUR$ in computing UPR
$T(c)$	Table instance of c	w_2	Weighted value of $InterUR$ in computing UPR
$u(f_i \cdot j)$	Utility of instance $f_i \cdot j$	M	A UPI threshold

Definition 1 (spatial instance with utility value). *Given a set of spatial features F and a set of their instances S. Let spatial instance $f_i \cdot j^v \in S$ be the j-th instance of feature $f_i \in F$. The utility value of $f_i \cdot j^v$ is expressed by the superscript v. We denote the utility of spatial instance $f_i \cdot j^v$ as $u(f_i \cdot j) = v$.*

According to Definition 1, every instance may have distinct utility, even if they belong to the same feature. For example, the feature A represents the rosewood. $A.1^{1000}$ is a 100-year-old rosewood and worth \$1000, i.e., $u(A.1) = 1000$. $A.2^{25}$ is a 10-year-old rosewood, which is worth \$25, and $u(A.2) = 25$.

The *total utility* of a feature $f_i \in F$ is the sum of utilities of its instances, denoted as $u(f_i) = \sum_{j=1}^{m} u(f_i \cdot j)$, where m is the number of instances belonging to f_i. For example, the total utility of feature A in Fig. 1 is $u(A) = u(A.1) + u(A.2) + u(A.3) + u(A.4) = 10 + 8 + 7 + 3 = 28$.

Definition 2 (utility of feature in co-location). *Given a size k co-location $c = \{f_1, f_2, ..., f_k\}$, we further define the sum of utilities of instances belonging to feature $f_i \in c$ in table instance $T(c)$ as the utility of f_i in c, denoted as $u(f_i, c) = \sum_{f_i \cdot j \in \pi_{f_i}(T(c))} u(f_i \cdot j)$, where π is the relational projection operation with duplication elimination.*

For example, for $c = \{A, B, C\}$ in Fig. 1, $T(c) = \{\{A.1^{10}, B.4^8, C.2^9\}, \{A.3^7, B.5^8, C.2^9\}\}$. The utility of A in c is $u(A, c) = u(A.1) + u(A.3) = 10 + 7 = 17$.

Definition 3 (intra-utility ratio). *Given a size k co-location* $c = \{f_1, f_2, ..., f_k\}$, *the intra-utility ratio of feature* f_i *in co-location c is defined as the proportion of* f_i's *utility in c to its total utility, i.e.,* $IntraUR(f_i, c) = \frac{u(f_i,c)}{u(f_i)}$.

$IntraUR(f_i, c)$ indicates the direct utility of feature f_i in co-location c, which can be regarded as its direct influence on c.

For example, for $c = \{A, B, C\}$ in Fig. 1, $T(c) = \{\{A.1^{10}, B.4^8, C.2^9\}, \{A.3^7, B.5^8, C.2^9\}\}$. The intra-utility ratio of each feature in c is calculated as

$$IntraUR(A, c) = \frac{u(A.1)+u(A.3)}{u(A)} = 17/28, \; IntraUR(B, c) = \frac{u(B.2)+u(B.5)}{u(B)} = 16/25,$$
$$IntraUR(C, c) = \frac{u(C.2)}{u(C)} = 9/12.$$

Definition 4 (inter-utility ratio). *Given a size k co-location* $c = \{f_1, f_2, ..., f_k\}$, *the inter-utility ratio of feature* f_i *in co-location c is defined as* $InterUR(f_i, c) = \frac{\sum_{f_j \in c, j \neq i} u(f_j,c)}{\sum_{f_j \in c, j \neq i} u(f_j)}$.

The inter-utility ratio is regard as the influence of feature f_i on other features in co-location c, which is an indirect influence of f_i on c. In a co-location, some instances of features often co-occur in neighborhoods. Thus, in a co-location $c = \{f_1, f_2, ..., f_k\}$, the change of feature $f_i \in c$ probably impact on the utility of other features in c. For example, there are various services in Location-based Service. In the package service, the sales of service A might promote the sales of service B. So, we use the inter-utility ratio to indicate the effect of a feature on other features in a co-location. In Fig. 1, the effect of feature A in co-location $\{A, B, C\}$ on other features B and C is compute as

$$InterUR(A, c) = \frac{u(B, c)+u(C, c)}{u(B)+u(C)} = 25/37.$$

We divide the influence of feature f_i into two parts to evaluate a co-location c comprehensively and reasonably. One is the influence of its utility in c denoted as $IntraUR(f_i, c)$, and another is the indirect influence of f_i on c denoted as $InterUR(f_i, c)$.

Definition 5 (Utility Participation Ratio, UPR). *Given a size k co-location* $c = \{f_1, f_2, ..., f_k\}$, *the weighted sum of* $IntraUR(f_i, c)$ *and* $InterUR(f_i, c)$ *is defined as the utility participation ratio of feature* f_i *in co-location c, which is denoted as* $UPR(f_i, c) = w_1 \times IntraUR(f_i, c) + w_2 \times InterUR(f_i, c)$, *where* $0 \leq w_1, w_2 \leq 1$ *and* $w_1 + w_2 = 1$, w_1 *represents the weighted value of* $IntraUR(f_i, c)$ *and* w_2 *represents that of* $InterUR(f_i, c)$.

The w_1 and w_2 in Definition 5 can be used to adjust the effect of *IntraUR* and *InterUR*, which are assigned the specified values by users in application. For example, in sales volume promotion of supermarkets, if we are more care the promoted sale volume of different goods, $w_1 \leq w_2$ may be reasonable. Usually, w_1 and w_2 satisfy $w_1 \geq w_2$.

For example, in Fig. 1, if we suppose $w_1 = 0.7$ and $w_2 = 0.3$, then the UPR of each feature in $c = \{A, B, C\}$ is computed as

$$UPR(A, c) = 0.7 \times IntraUR(A, c) + 0.3 \times InterUR(A, c)$$
$$= 0.7 \times (17/28) + 0.3 \times (25/37) = 0.628.$$
$$UPR(B, c) = 0.7 \times IntraUR(B, c) + 0.3 \times InterUR(B, c)$$
$$= 0.7 \times (16/25) + 0.3 \times (26/40) = 0.643.$$
$$UPR(C, c) = 0.7 \times IntraUR(C, c) + 0.3 \times InterUR(C, c)$$
$$= 0.7 \times (9/12) + 0.3 \times (33/53) = 0.711.$$

Definition 6 (Utility Participation Index, UPI). *Given a size k co-location $c = \{f_1, f_2, ..., f_k\}$, We define the minimum utility participation ratio among all features in co-location c as the utility participation index of c, i.e., $UPI(c) = min\{UPR(f_i, c), f_i \in c\}$.*

A co-location pattern c is a **high utility co-location pattern** if and only if $UPI(c) \geq M$ holds, where M is a UPI threshold given by users.

The UPI measure extends the traditional PI measure only based on prevalence. If the utilities of instances and the influence between features in a co-location are ignored, UPI is equal to the traditional PI.

The prevalent patterns may not be high utility patterns and the high utility patterns may not be prevalent as well, which can be proved by patterns $\{E, F\}$ and $\{A, B, C\}$ in Fig. 1. If $w_1 = w_2 = 0.5$ and $M = 0.3$, $UPI(\{E, F\}) = 0.226$ and $PI(\{E, F\}) = 0.667$, while $UPI(\{A, B, C\}) = 0.628$ and $PI(\{A, B, C\}) = 0.25$. Because of full consideration into the difference of each instance, our interestingness measure is more reasonable. However, different from the traditional interestingness measure, UPI does not satisfy the downward closure property which is a very efficient pruning strategy for mining prevalent co-locations. Therefore, finding all high utility patterns directly is time-consuming. For example, for $c = \{A, D\}$ in Fig. 1, $T(c) = \{\{A.3^7, D.1^9\}\}$. Given $w_1 = w_2 = 0.5$, we can get $UPI(\{A, D\}) = 0.471$. But the super pattern $c' = \{A, C, D\}$ of $c, T(c') = \{\{A.3^7, C.2^9, D.1^9\}\}$, and $UPI(\{A, C, D\}) = 0.485$. So, we have the inequality $UPI(c') > UPI(c)$.

3   A Basic Algorithm

In this section, we present a basic algorithm for mining the high utility co-locations defined in Sect. 2. The basic algorithm has three phases. The first one is to materialize the spatial neighbor relationships. The spatial data set is converted into the star neighborhood partition model in [4]. The second one is to generate candidate co-locations and compute their table instances. The third one is to compute the UPI of each candidate co-location and find high utility co-locations. The second and third phases are repeated with the increment of co-locations' size. Algorithm 1 shows the pseudocode of the basic algorithm.

Algorithm 1. Basic Algorithm

Input: $F=\{f_1, f_2, ..., f_n\}$: a set of spatial features; S: a set of spatial instances; R: a spatial neighbor relationship; w_1: the weighted value of *IntraUR*; w_2: the weighted value of *InterUR*; M: the UPI threshold

Output: a set of all co-location patterns with UPI $\geq M$

Steps:

    1) $SN = gen\_star\_neighborhoods(F, S, R)$;
    2) $H_1 = F$; $k=2$;   //H_k: a set of size k high utility co-locations
    3) **While** (not empty H_{k-1} and $NonH_{k-1}$) **do**
                //$NonH_k$: a set of size k non-high utility co-locations
    4) $C_k=gen\_candi\_colocations(H_{k-1}, NonH_{k-1})$;
           //C_k: a set of size k candidate co-locations
    5) $CTI_k=gen\_table\text{-}instances(C_k, SN)$;
           //CTI_k: a set of table instances of size k candidates
    6) $Compute\_UPI(C_k, w_1, w_2)$;
    7) $H_k=select\_utility\_colocations(C_k, CTI_k, M)$;
    8) $NonH_k=C_k - H_k$;
    9) $k=k + 1$;
    10) End do
    11) Return $\cup \{H_1, H_2, \cdots, H_{k-1}\}$

Initialization (Step 1–2): Given a spatial data set and a spatial neighbor relationship, find all neighboring instance pairs using a geometric method such as mesh generation or plane sweep [4]. The star neighborhoods can be generated from the neighbor instance pairs by lexicographical order [4]. After generating the star neighborhood set (*SN*), we initialize all size 1 co-locations with utility participation index 1.0, which means all size 1 co-locations are high utility co-locations. Then, we add all size 1 co-locations into H_1.

Generating Candidate Co-locations (Step 4): A size k ($k \geq 2$) candidate co-locations in C_k is generated from a size $k-1$ co-location c in H_{k-1} or $NonH_{k-1}$ and a new feature f_s which is not included in c and greater than all features of c in lexicographical order, i.e., $C_k = \{c'|c' = c \cup \{f_s\}, \forall c \in H_{k-1} \cup NonH_{k-1}, f_s > \forall f_i \in c\}$.

Specially, the size 2 candidate co-locations in C_2 can be generated from the star neighborhood set directly.

Calculating the UPIs of Candidate Co-locations (Step 5–6): The size 2 co-locations' table instances can be gathered from the star neighborhood set directly. For size k ($k > 2$) co-locations, their table instances need to be extended by size $k-1$ co-locations' table instances. For example, the table instance of co-location $\{A, B, C\}$ can be generated from the table instance of co-location $\{A, B\}$. Then, we can compute the UPI of each candidate co-location according to the Definitions 5 and 6.

Identifying High Utility Co-locations (Step 7–8): We can filter high utility co-locations by the UPIs of candidate co-locations and the given UPI threshold M. Then, high utility co-locations are added into H_k and non-high utility co-locations are added into $NonH_k$.

Steps 3–10 are repeated with the increment of size k.

In Fig. 1, if $w_1 = w_2 = 0.5$ and $M = 0.5$, we can get the high utility co-locations {A, B}, {A, C}, {A, B, C}, {B, C}, {B, D}, {C, D}, and {C, E}. The basic algorithm tests all possible patterns and computes their UPI accurately. So, it is complete and correct, but it is inefficient. In the next section, we would give some pruning strategies to improve the efficiency of the basic algorithm.

4 Pruning Strategies

In this section, we will introduce some pruning strategies to promote the efficiency of the basic algorithm. Traditional co-location mining based on PI can efficiently find all prevalent co-locations due to the downward closure property. But there is no a similar method to find all high utility co-locations due to the non-existence of the downward closure property. Similar to traditional co-location mining, the most time-consuming component in mining high utility co-locations is to generate the table instances of candidate patterns. In order to improve the efficiency of the basic algorithm, we have to early identify some non-high utility candidate co-locations without generating their table instances. The following pruning strategies are used to prune the non-high utility candidate patterns ahead of time.

Lemma 1. *For $n_1 \geq m_1 > 0$, $n_2 \geq m_2 > 0$, there exists the following inequality:*

$$\frac{m_1 + m_2}{n_1 + n_2} \leq \max\{\frac{m_1}{n_1}, \frac{m_2}{n_2}\}$$

Proof: Given $n_1 \geq m_1 > 0$, $n_2 \geq m_2 > 0$. If $\frac{m_1}{n_1} \geq \frac{m_2}{n_2}$, then there exists $\frac{m_1 + m_2}{n_1 + n_2} - \frac{m_1}{n_1} = \frac{m_2 n_1 - m_1 n_2}{n_1(n_1 + n_2)} \leq 0$. So, $\frac{m_1 + m_2}{n_1 + n_2} \leq \frac{m_1}{n_1}$ holds. Similarly, if $\frac{m_2}{n_2} \geq \frac{m_1}{n_1}$, then $\frac{m_1 + m_2}{n_1 + n_2} \leq \frac{m_2}{n_2}$. Therefore, $\frac{m_1 + m_2}{n_1 + n_2} \leq \max\{\frac{m_1}{n_1}, \frac{m_2}{n_2}\}$ holds.   □

Corollary 1. *For $k(k > 1)$ pairs m_i and n_i ($i = 1, 2, ..., k$), if $n_i \geq m_i > 0$, there exists the following inequality:* $\frac{\sum_{i=1}^{k} m_i}{\sum_{i=1}^{k} n_i} \leq \max_{i=1}^{k}\{\frac{m_i}{n_i}\}$.

Definition 7 (non-high utility feature set). *Given a size k co-location $c = \{f_1, f_2, ..., f_k\}$, we call the set of all features in co-location c whose UPR is less than the UPI threshold M as the non-high utility feature set of c.*

For example, for $c = \{A, B, D\}$ in Fig. 1, if $M = 0.4$ and $w_1 = w_2 = 0.5$, then *UPR* (A, c) = 0.257, *UPR*(B, c) = 0.215 and *UPR*(C, c) = 0.422. The non-high utility feature set of c is {A, B}.

Theorem 1. *If c_1 and c_2 are two non-high utility co-locations, and they have and only have one common feature f_i and it is a non-high utility feature, then the pattern $c = c_1 \cup c_2$ must be a non-high utility pattern, i.e., $c = c_1 \cup c_2$ can be pruned.*

Proof: Because f_i is a non-high utility feature in c_1 and c_2, we have:

$$UPR(f_i, c_1) = w_1 \frac{u(f_i, c_1)}{u(f_i)} + w_2 \frac{m_1}{n_1} < M \tag{1}$$

where $m_1 = \sum_{f_j \in c_1, j \neq i} u(f_j, c_1)$ and $n_1 = \sum_{f_j \in c_1, j \neq i} u(f_j)$. And

$$UPR(f_i, c_2) = w_1 \frac{u(f_i, c_2)}{u(f_i)} + w_2 \frac{m_2}{n_2} < M \tag{2}$$

where $m_2 = \sum_{f_j \in c_2, j \neq i} u(f_j, c_2)$ and $n_2 = \sum_{f_j \in c_2, j \neq i} u(f_j)$.

For the co-location $c = c_1 \cup c_2$, the UPR of f_i in c satisfies:

$$UPR(f_i, c) = w_1 \frac{u(f_i, c)}{u(f_i)} + w_2 \frac{m_1 + m_2}{n_1 + n_2} \tag{3}$$

due to f_i is the unique common feature in c_1 and c_2.

According to Definition 2 and the concept of table instance, we have $u(f, c) \leq u(f, c')$ if f is the common feature in co-locations c and c', and $c' \subseteq c$.

And according to Lemma 1, $\frac{m_1 + m_2}{n_1 + n_2} \leq \max\{\frac{m_1}{n_1}, \frac{m_2}{n_2}\}$.

Therefore, we can infer $UPR(f_i, c) < M$ by (1), (2) and (3), which can judge that $c = c_1 \cup c_2$ is a non-high utility co-location. □

For example, for $c_1 = \{A, B, D\}$ and $c_2 = \{B, E\}$ in Fig. 1, if $w_1 = w_2 = 0.5$ and $M = 0.5$, $T(c_1) = \{\{A.3^7, B.1^1, D.1^9\}\}$ and $T(c_2) = \{\{B.4^8, E.3^2\}\}$. The UPRs of common feature B in c_1 and c_2 are $UPR(B, c_1) = 0.215 < M$ and $UPR(B, c_2) = 0.231 < M$ respectively, which satisfy the conditions of Theorem 2. So, $c = c_1 \cup c_2 = \{A, B, C, D\}$ must be a non-high utility co-location and can be pruned.

According to the Theorem 1 and Corollary 1, we can infer the Corollary 2.

Corollary 2. *For size 2 non-high utility co-locations c_1, c_2, ..., c_k ($k > 1$), if they have a common non-high utility feature f, then the pattern $c = c_1 \cup c_2 \cup ... \cup c_k$ must be a non-high utility pattern, i.e., c can be pruned.*

When the spatial data set is sparser or the UPI threshold M is higher, there would be large amounts of size 2 non-high utility co-locations. At that time, we could prune a large number of higher size non-high utility co-locations by combining those size 2 non-high utility co-locations.

In Fig. 1, if $w_1 = w_2 = 0.5$ and $M = 0.5$, there are size 2 non-high utility co-locations $\{A, E\}$, $\{B, E\}$, $\{D, E\}$ and $\{E, F\}$. And E is a non-high utility feature. The co-locations $\{A, B, E\}$, $\{A, D, E\}$, $\{A, E, F\}$, $\{B, D, E\}$, $\{B, E, F\}$, $\{D, E, F\}$, $\{A, B, D, E\}$, $\{A, B, E, F\}$, $\{A, D, E, F\}$, $\{B, D, E, F\}$ and $\{A, B, D, E, F\}$ can be pruned by Corollary 2.

According to Definition 2, for a size k co-location $c = \{f_1, f_2, ..., f_k\}$ and $f_i \in c$, $u(f_i, c) \leq u(f_i, c')$ holds, where c' is an arbitrary size $k-1$ sub-pattern of c including f_i. So, we call the minimum of utilities of f_i in size $k-1$ sub-patterns of c including f_i as the upper bound utility of f_i in c, donated as $upbound\_u(f_i, c)$.

For example, for $c = \{A, B, C\}$ in Fig. 1, the upper bound utility of feature A in c is $upbound\_u(A, c) = \min\{u(A, \{A, C\}), u(A, \{A, B\})\} = \min\{17, 28\} = 17$.

Lemma 2. *Given a size k co-location $c = \{f_1, f_2, ..., f_k\}$ and its size $k+1$ super-pattern $c' = c \cup \{f_{k+1}\}$, the upper bound of $UPI(c')$ is computed as follows:*

$$\min\{w_1 \frac{u(f_i, c)}{u(f_i)} + w_2 \frac{\sum_{f_j \in c, j \neq i} u(f_j, c) + upbound\_u(f_{k+1}, c')}{\sum_{f_j \in c, j \neq i} u(f_j) + u(f_{k+1})}, 1 \leq i \leq k\}$$

Proof: If $c = \{f_1, f_2, ..., f_k\}$ and $c' = c \cup \{f_{k+1}\}$. As to any feature $f_i \in c$, the inequality $u(f_i, c') \leq u(f_i, c)$ holds.

So, we have $UPR(f_i, c') \leq w_1 \frac{u(f_i, c)}{u(f_i)} + w_2 \frac{\sum_{f_j \in c, j \neq i} u(f_j, c) + upbound\_u(f_{k+1}, c')}{\sum_{f_j \in c, j \neq i} u(f_j) + u(f_{k+1})}$.

Based on Definition 6, we can infer that:

$$UPI(c') \leq \min\{w_1 \frac{u(f_i, c)}{u(f_i)} + w_2 \frac{\sum_{f_j \in c, j \neq i} u(f_j, c) + upbound\_u(f_{k+1}, c')}{\sum_{f_j \in c, j \neq i} u(f_j) + u(f_{k+1})}, 1 \leq i \leq k\}. \qquad \square$$

Theorem 2. *Given a size k non-high utility co-location $c = \{f_1, f_2, ..., f_k\}$ and its size $k+1$ super-pattern $c' = c \cup \{f_{k+1}\}$, if there is a non-high utility feature $f_i \in c$ which satisfies $\frac{\sum_{f_j \in c, j \neq i} u(f_j, c)}{\sum_{f_j \in c, j \neq i} u(f_j)} > \frac{upbound\_u(f_{k+1}, c')}{u(f_{k+1})}$, then c' is a non-high utility co-location, i.e., c can be pruned.*

Proof: For a non-high utility co-location $c = \{f_1, f_2, ..., f_k\}$ and $c' = c \cup \{f_{k+1}\}$, if f_i is a non-high utility feature in c and M is the UPI threshold, we have

$$UPR(f_i, c) = w_1 \frac{u(f_i, c)}{u(f_i)} + w_2 \frac{m}{n} < M \tag{4}$$

where $m = \sum_{f_j \in c, j \neq i} u(f_j, c)$ and $n = \sum_{f_j \in c, j \neq i} u(f_j)$.

According to Lemma 2, the UPR of f_i in c' satisfies the following inequality:

$$UPR(f_i, c') \leq w_1 \frac{u(f_i, c)}{u(f_i)} + w_2 \frac{m + upbound\_u(f_{k+1}, c')}{n + u(f_{k+1})}$$

According to Lemma 1, we have

$$\frac{m + upbound\_u(f_{k+1}, c')}{n + u(f_{k+1})} \leq \max\left\{\frac{m}{n}, \frac{upbound\_u(f_{k+1}, c')}{u(f_{k+1})}\right\}$$

If $\frac{m}{n} > \frac{upbound\_u(f_{k+1}, c')}{u(f_{k+1})}$, the following inequality holds.

$$UPR(f_i, c') \leq w_1 \frac{u(f_i, c)}{u(f_i)} + w_2 \frac{m}{n}$$

Based on the inequality (4), we can infer that $UPR(f_i, c') < M$. So, c' must be a non-high utility co-location. □

For example, for $c = \{B, C, D\}$ in Fig. 1, if $w_1 = w_2 = 0.5$ and $M = 0.5$, due to T $(c) = \{\{B.4^8, C.2^9, D.2^1\}\}$, $UPR(B, c) = 0.36$, $UPR(C, c) = 0.493$ and $UPR(D, c) = 0.268$, c is a non-high utility co-location pattern. For the supper-pattern $c' = \{B, C, D, E\}$ of c, $upbound\_u(E, c') = \min\{u(E, \{B, C, E\}) + u(E, \{B, D, E\}) + u(E, \{C, D, E\})\} = \min\{2, 2, 2\} = 2$. As to the feature B in $\{B, C, D\}$, we have:

$$\frac{u(C, c) + u(D, c)}{u(C) + u(D)} = \frac{9 + 1}{12 + 13} > \frac{upbound\_u(E, c')}{u(E)} = \frac{2}{14}$$

So, based on the computing results of size 3 co-locations, we can infer that the size 4 co-location $\{B, C, D, E\}$ must be a non-high utility co-location.

Theorem 1, Corollary 2 and Theorem 2 are regarded as three pruning strategies to identify some non-high utility co-locations ahead of time.

5 Experimental Analysis

This section verifies the effect and efficiency of the basic algorithm and the algorithm with pruning strategies on synthetic and real data sets through experiments. The algorithms are implemented in Java 1.7 and run on a windows 8 operating system with 3.10 GHz Intel Core i5 CPU and 4 GB memory.

5.1 Data Sets

We conduct the experiments on synthetic data sets and plant data sets of the "Three Parallel Rivers of Yunnan Protected Areas". Synthetic data sets are generated using a spatial data generator similar to [3, 4], and the utilities of instances are assigned randomly between 0 and 20. In the plant data sets, we compute the utilities of plant instances according the plant price associated with size and kind of plant. The efficiency of the basic algorithm and the algorithm with pruning strategies are examined on the synthetic and real data sets.

5.2 The Quality of Mining Results

We aim at finding the high utility co-locations whose instances are frequently located together in geographic space and which have high utilities. So, we take the criterion $Q(c) = \sum_{f \in c} u(f, c) / \sum_{f \in c} u(f)$ to evaluate the quality of a mined co-location c.

In order to illustrate the interestingness measure UPI proposed in this paper is more reasonable, we compare the quality of mining results identified by different interestingness measures. They are the traditional participation index measure (PI), the traditional pattern utility ratio (PUR) proposed in [19] and the UPI proposed in our paper.

In the experiments of Fig. 2, we take the number of spatial features $|F|$ is 15, the total number of instances $|S|$ is 10000, the neighboring distance threshold d is 30, and $w_1 = 0.9$, $w_2 = 0.1$. Figure 2(a) shows the sum of quality of top-k interesting co-locations identified by the measure PI, PUR and UPI respectively. The x-axis refers to the value k, while y-axis is the sum of quality of top-k interesting patterns. Figure 2(b) shows the average quality of top-20 interesting patterns identified by the three measures over different sizes. The x-axis is the sizes of co-locations, while the y-axis is the average quality of top-20 interesting patterns. The results show that our UPI measure can identify higher quality co-locations, and it can extract top co-locations with higher average utility.

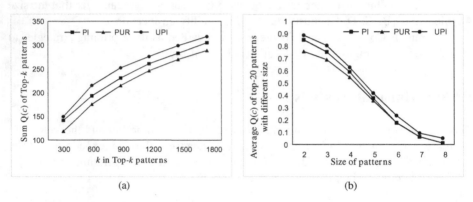

(a)   (b)

Fig. 2. Testing the quality of mining results, where (a) the sum of quality of top-k interesting patterns; (b) the average of quality of top-20 interesting patterns with different sizes.

5.3 Evaluation of Pruning Strategies

We evaluate the effect of pruning strategies with several workloads, e.g. different numbers of instances, neighbor distance thresholds, UPI thresholds and pruned rate on synthetic and real data sets.

5.3.1 Influence of the Number of Instances

We compare the running time of the basic algorithm and the algorithm with pruning strategies on synthetic and real data sets. We set $|F| = 20$, $d = 20$, $M = 0.3$, $w_1 = w_2 = 0.5$, the running time of two algorithms by increasing the number of instances is shown in Fig. 3. The x-axis represents the number of total instances and the y-axis is the running time. As a result, the performance of the algorithm with pruning strategies is better than the basic algorithm both in synthetic and real data sets. Compared with synthetic data sets, the neighbor relationships of real data sets are relatively fewer,

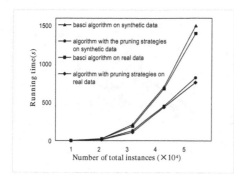

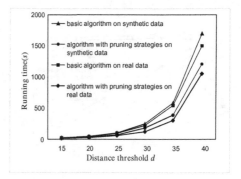

Fig. 3. The influence of the number of instances over synthetic and real data sets

Fig. 4. The influence of the distance thresholds over synthetic and real data sets

which results in less row instances to be computed. So, in our experiments, the runtime of the algorithms on real data sets is less than that on synthetic data sets.

5.3.2   Influence of the Distance Threshold d

In Fig. 4, we set $|F| = 20$, $|S| = 10000$, $M = 0.3$, $w_1 = w_2 = 0.5$. We compare the running time of two algorithms by changing the distance threshold d, where the x-axis denotes the distances threshold d and the y-axis represents the running time in different data sets. From Fig. 4, we can see that the algorithm with pruning strategies is still faster than the basic algorithm. However, both algorithms have a huge time-cost with the increase of d. This is because with increasing d, there are more cliques formed, which results in huge row instances to be computed and more time consumption.

5.3.3   Influence of the UPI Threshold M

The parameters are set $|F| = 20$, $|S| = 10000$, $d = 15$, $w_1 = w_2 = 0.5$ in this experiment. The running time of two algorithms by changing the UPI threshold M is shown in Fig. 5. The x-axis denotes the value of the UPI threshold M and the y-axis is the running time. With the increase of M, the more non-high utility co-locations are pruned ahead of time, which improve the efficiency of the algorithm with pruning strategies.

5.3.4   Pruned Rate

In order to examine efficiency of the three pruning strategies (Theorem 1, Corollary 2, and Theorem 2), we count the number of candidates pruned by each pruning strategy respectively. In the experiment, we set $|F| = 15$, $|S| = 4000$, $d = 20$, $M = 0.3$, $w_1 = w_2 = 0.5$, and we randomly generate 5 different data sets whose size is similar to each other. We independently run the algorithm with pruning strategies on 5 different data sets and compute the average proportion of the candidates pruned by each strategy. The statistic result is shown in Fig. 6.

The result shows that the pruning strategies are very efficient. However, the efficiency of pruning strategies doesn't be improved in the same degree. There are two reasons. First, the process of pruning candidates would cost some time. Second, some pruned co-locations may be used to generate the table instances of super co-locations,

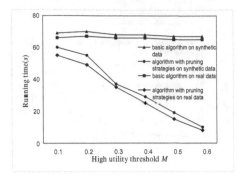

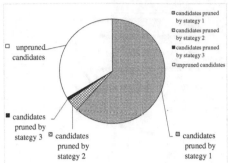

Fig. 5. The influence of the UPI threshold M over synthetic and real data sets

Fig. 6. The proportion of candidates pruned by each strategy

so we might have to generate the table instances of pruned co-locations, which has a negative effect on the algorithm. Fortunately, it rarely occurs in the experiments. Thus, the average efficiency of pruning strategies is obvious.

In addition, the basic algorithm and the algorithm with pruning strategies presented in this paper convert spatial data sets into the star neighborhood partition model in [4]. Algorithms in both papers store spatial neighbor relationships and table instances of current candidates. Therefore, the memory cost of our algorithms is similar to the join-less algorithm in [4]. Due to the non-existence of the downward closure property, the scalability of the basic algorithm requires improvement. From Figs. 3, 4, 5 and 6, we can see that the pruning strategies significantly reduce the overall runtime of the basic algorithm, while in some extreme cases less so. Further improvement of scalability is left for future work.

6 Conclusion and Future Work

Different from the previous researches, in this paper we take the instances with utilities as study objects which are near to real world and a new interesting measure is proposed. We combine the intra-utility ratio and the inter-utility ratio into the utility participation index for identifying high utility co-locations, which is comprehensive and reasonable. Because the utility participation index does not satisfy the downward closure property, we propose the effective pruning strategies to improve the efficiency of finding high utility co-locations. The experiments on synthetic and real data sets show that the pruning strategies significantly reduce the overall runtime of the basic algorithm. Although the algorithm with pruning strategies is better than the basic algorithm, it also shows less improvement in some extreme case. Our future work focuses on designing algorithms for bigger data sets and better pruning strategies.

Acknowledgments. This work is supported by the National Natural Science Foundation of China (61472346, 61662086), the Natural Science Foundation of Yunnan Province (2015FB114, 2016FA026), the Spectrum Sensing and borderlands Security Key Laboratory of Universities in Yunnan (C6165903), and the Program for Young and Middle-aged Skeleton Teachers of Yunnan University.

References

1. Koperski, K., Han, J.: Discovery of spatial association rules in geographic information databases. In: Egenhofer, M.J., Herring, J.R. (eds.) SSD 1995. LNCS, vol. 951, pp. 47–66. Springer, Heidelberg (1995). doi:10.1007/3-540-60159-7_4

2. Shekhar, S., Huang, Y.: Co-location rules mining: a summary of results. In: Spatio-Temporal Symposium on Databases (2001)

3. Huang, Y., Shekhar, S., Xiong, H.: Discovering co-location patterns from spatial data sets: a general approach. IEEE Trans. Knowl. Data Eng. **16**(12), 1472–1485 (2004)

4. Yoo, J.S., Shekhar, S.: A joinless approach for mining spatial co-location patterns. IEEE Trans. Knowl. Data Eng. **18**(10), 1323–1337 (2006)

5. Sengstock, C., Gertz, M., Tran Van, C.: Spatial interestingness measures for co-location pattern mining. In: SSTDM (ICDM Workshop 2012), pp. 821–826. IEEE Press, New York (2012)

6. Wang, L., Wu, P., Chen, H.: Finding probabilistic prevalent colocations in spatially uncertain data sets. IEEE Trans. Knowl. Data Eng. **25**(4), 790–804 (2013)

7. Wang, L., Chen, H., Zhao, L., Zhou, L.: Efficiently mining co-location rules on interval data. In: Cao, L., Feng, Y., Zhong, J. (eds.) ADMA 2010. LNCS (LNAI), vol. 6440, pp. 477–488. Springer, Heidelberg (2010). doi:10.1007/978-3-642-17316-5_45

8. Wang, L., Bao, Y., Lu, Z.: Efficient discovery of spatial co-location patterns using the iCPI-tree. Open Inf. Syst. J. **3**(2), 69–80 (2009)

9. Huang, Y., Pei, J., Xiong, H.: Mining co-location patterns with rare events from spatial data sets. Geoinformatica **10**(3), 239–260 (2006)

10. Yao, H., Hamilton, H.J., Butz, C.J.: A foundational approach to mining itemset utilities from databases. In: 4th SIAM International Conference on Data Mining, pp. 482–486 (2004)

11. Liu, Y., Liao, W.-K., Choudhary, A.: A two-phase algorithm for fast discovery of high utility itemsets. In: Ho, T.B., Cheung, D., Liu, H. (eds.) PAKDD 2005. LNCS (LNAI), vol. 3518, pp. 689–695. Springer, Heidelberg (2005). doi:10.1007/11430919_79

12. Ahmed, C.F., Tanbeer, S.K., Jeong, B.S., et al.: A framework for mining interesting high utility patterns with a strong frequency affinity. Inf. Sci. **181**(21), 4878–4894 (2011)

13. Hong, T.P., Lee, C.H., Wang, S.L.: An incremental mining algorithm for high average-utility itemsets. Expert Syst. Appl. **39**(8), 7173–7180 (2012)

14. Tseng, V.S., Wu, C.W., Shie, B.E., et al.: UP-Growth: an efficient algorithm for high utility itemset mining. In: 16th ACM SIGKDD International Conference on Knowledge Discovery and Data Mining, pp. 253–262. ACM, New York (2010)

15. Shie, B.E., Tseng, V.S., Yu, P.S.: Online mining of temporal maximal utility itemsets from data streams. In: ACM Symposium on Applied Computing, pp. 1622–1626. ACM, New York (2010)

16. Yin, J., Zheng, Z., Cao, L.: USpan: an efficient algorithm for mining high utility sequential patterns. In: 18th ACM SIGKDD International Conference on Knowledge Discovery and Data Mining, pp. 660–668. ACM, New York (2012)

17. Lin, Z., Lim, S.: Optimal candidate generation in spatial co-location mining. In: ACM Symposium on Applied Computing, pp. 1441–1445. ACM, New York (2009)
18. Verhein, F., Al-Naymat, G.: Fast mining of complex spatial co-location patterns using GLIMIT. In: SSTDM (ICDM Workshop 2007), pp. 679–984. IEEE Press, New York (2007)
19. Yang, S., Wang, L., Bao, X., Lu, J.: A framework for mining spatial high utility co-location patterns. In: 12th International Conference on Fuzzy Systems and Knowledge Discovery (FSKD 2015), pp. 595–601. IEEE Press, New York (2015)
20. Wang, X., Wang, L., Lu, J., Zhou, L.: Effectively updating high utility co-location patterns in evolving spatial databases. In: Cui, B., Zhang, N., Xu, J., Lian, X., Liu, D. (eds.) WAIM 2016. LNCS, vol. 9658, pp. 67–81. Springer, Heidelberg (2016). doi:10.1007/978-3-319-39937-9_6

Real Time Data Processing

Supporting Real-Time Analytic Queries in Big and Fast Data Environments

Guangjun Wu[1($\boxtimes$)], Xiaochun Yun[1], Chao Li[2($\boxtimes$)], Shupeng Wang[1],
Yipeng Wang[1], Xiaoyu Zhang[1], Siyu Jia[1], and Guangyan Zhang[3]

[1] Institute of Information Engineering, CAS, Beijing 100029, China
wuguangjun@gmail.com
[2] National Computer Network and Information Security Administration Center,
Beijing 100031, China
lc_lichao@126.com
[3] Department of Computer Science and Technology, Tsinghua University,
Beijing 100084, China

Abstract. Recently there has been a significant interest to perform real-time analytical queries in systems that can handle both "big data" and "fast data". In this paper, we propose an approximate answering approach, called ROSE, which can manage the big and fast data streams and support complex analytical queries against the data streams. To achieve this goal, we start with an analysis of existing query processing techniques in big data systems to understand the requirements of building a distributed analytic sketch. We then propose a sampling-based sketch that can extract multi-faced samples from asynchronous data streams, and augment its usability with accuracy-lossless distributed sketch construction operations, such as splitting, merging and union. The experimental results with real-world data sets indicate that compared with state-of-the-art approximate answering engine BlinkDB, our techniques can obtain more accurate estimates and improve 2 times of system throughput. When compared with distributed memory-computing system Spark, our system can achieve 2 orders of magnitude improvement on query response time.

Keywords: Approximate answering · Big data · Data streams · Distributed computing · Sampling

1 Introduction

In recent years, many applications produce large-volume and continuous data streams. It is necessary to perform real-time query processing in the streams to detect anomalies and explore their development trends "on the fly", e.g., data flow analysis [1,2], real-time data mining [3] and financial indicator trackers [4]. As a result, we have seen a flurry of activities in the area of building big data systems to handle not only "big data" but also "fast data" for analytics. Here "fast data" refers to high-speed real-time and near real-time data streams.

© Springer International Publishing AG 2017
S. Candan et al. (Eds.): DASFAA 2017, Part II, LNCS 10178, pp. 477–493, 2017.
DOI: 10.1007/978-3-319-55699-4_29

To meet the low-latency requirements of analytical queries, a number of data streams processing systems have been developed to handle the big and fast data, including Spark [5], StreamMapReduce [6], StreamCloud [7], Incremental Hadoop [8], and academic prototypes [9,10]. Despite various differences in implementations, these systems share some common features to boost the performance of analytics: (1) They employ complex resource planning and fine-granularity jobs scheduling in a cluster to maximize the throughput of a system. (2) They increase data parallelism, which is to partition a large dataset into smaller subsets, by scaling the number of nodes of a cluster.

However, it is not easy for common users to deploy the complex planning and scheduling techniques in their production environments or to scale their systems to hundreds or even thousands of nodes in a cluster. Meanwhile, many emergent applications tolerate some error for analytical queries. For example, in streams of stock market data, a service application may need to track the moving tendency of the price hourly over all observations efficiently. In intrusion detection and prevention systems, it is useful to obtain real-time statistics of network traffic, such as packages counting, bytes summarization, and the Top-k flows searching, to detect network-level anomalies.

Methods for *approximate query processing* (AQP) are essential for dealing with massive datasets and improving the performance of emergent queries via running the query of interest against a sketch rather than the entire dataset [11]. The challenging problem of building a general sketch of AQP is the requirements of solving high-speed asynchronous data streams in distributed environments and obtaining accurate results for different queries. Although methods for AQP have been widely studied in data warehouse and commercial databases, efficient approaches against big data have not been studied until recently [12–15]. However, those works mainly focused on OLAP queries over off-line datasets. Literatures about approximate answering algorithms have also been studied in the context of data streams, such as VAROPT [16], BSBH [17] and ECM-sketch [18]. But those algorithms assumed that data streams subject to presumed data distribution. Practical data streams abound with dynamic and irregular elements which will increase unpredictable errors for analytical queries.

In this paper, we propose an approach, called *ROSE* (accu*R*ate *O*nline *S*ketch of *E*stimation), whose aim is to build a general approach of AQP for analytics. ROSE can solve high-speed asynchronous data streams and provide the capability of answering real-time analytical queries. Moreover, it is also compatible with current distributed query processing framework in big data systems. ROSE employs three key ideas as follows.

- First, by analyzing the features of data source and properties of query processing in big data systems, we extracts multi-faced samples from data streams via a combination of sliding-window sampling and probability aggregation sampling. The selected samples are devoted to accuracy-guaranteed estimation for different analytical queries.
- Second, to solve the problem of shuffling and partition operations for synopsis in distributed systems, we propose accuracy-lossless sketch construction and maintaining methods, including such operations as sketch splitting, sketch

merging and sketch union. Those operations can natively support data shuffling and partition between nodes in a distributed query processing framework.
– Third, by leveraging the sampling-based sketch, we demonstrate that ROSE
can support a wide variety of analytical queries, such as aggregation-based
queries (e.g., sum and count) and quantile-based queries (e.g., quantiles,
median and Top-k queries), with a uniform approximation schema using million seconds over distributed production environments. The schema can provide (ξ, δ) accuracy of estimation such that for any given positive $\xi < 1$ and
$\delta < 1$, the estimate of ROSE is within a relative error ξ of the true value with
probability at least $1 - \delta$.

We implement ROSE on the Linux platform and carry out a wide variety of
experimental evaluation over synthetic and real-world data sets. When compared
with existing memory-computing system Spark, our approach can achieve 5 times
of throughput improvement and 2 orders of magnitude improvement in query
response time, while the relative error is less than 2%. Compared with existing
approximate answering engines (e.g., BlinkDB, Asy-sketch), our method can
obtain more accurate results using the same space under real-world datasets.

2   Related Work

Obtaining statistics over data streams is a fundamental work for many emerging
applications. It needs to conduct different types of queries over one-pass and
asynchronous data streams. Related problems have been studied in the research
of sliding-window model, approximate answering engines and big data analytic
stacks.

The sliding-window is a well-known streams processing model which focuses
on computing estimates for elements seen so far. Many sliding-window methods have been explored over the past decades, such as Exponential Histogram
[21], random waves [20] and asynchronous streaming sketch [19] etc. However,
traditional sliding-window models are designed as one sketch for one operator
service schema. For example, Exponential Histogram [21] and random waves [20]
are efficient on answering aggregation queries. The G-K and q-digits algorithms
can provide an ξ-approximation estimate for quantiles queries [22]. Composed
sketches are also proposed by Arasu and Manku within the context of sliding-
window to solve the approximate counts and quantiles in a same structure [23].
The preceding studies provide theory and baseline algorithms for Approximate
Query Processing. We combine the core idea of these techniques to produce
multi-faced samples and improve query accuracy for practical big data analytics.

We have also noticed that there are plentiful researches on data stream
processing [18–21, 25]. The traditional approximate engines focus on evaluating
ad-hoc queries over static data sets [12]. Several emerging applications require
answering on dynamic streaming data, which is highly distributed and constantly
updated. The techniques of sketching over dynamic data streams have been proposed in recent years [17, 24]. But these work can not solve asynchronous data
streams in distributed systems, which need to split and merge individual sketches
in an accuracy lossless manner.

Big data analytic systems, based on Hadoop, have experienced tremendous growth over past few years. Many approximate answering engines have been built on top of the Hadoop software stacks, but few of them focused on low-latency query processing requirements of high-speed data streams. For example, the Hadoop-based approximate answering engines, such as BlinkDB [13] and G-OLA tools [14], extract offline samples from HDFS and then support OLAP queries over the samples with relative complex clauses. The latency of offline sampling techniques make these approximate answering engines do not meet the strict low-latency requirement of analytical queries over data streams.

3   System Design

We consider Hadoop-based systems as our prototype to depict the features of the input dataset and properties of analytical query processing. We first present preliminaries of our system design in Sect. 3.1, including formulation of input data source and classification of query processing in big data systems. Based on the preliminaries, we describe details of our approach, which aim to meet the requirements of a practical query in a big data processing system. The details include approach overview, multi-faced sampling, sketching operations in a distributed system, and approximate answering for different queries, and they are discussed in Sects. 3.1, 3.2, 3.3 and 3.4 respectively.

3.1   Preliminaries

At a high level, a big data system is first to exact a list of tuples from inputs at the *Map* phase, then shuffles tuples between nodes, and finally processes them with a user-defined function at *Reduce* phase. The output stream from the *Reduce* side is pushed to Web UI, DFS or other downstream MR pairs.

Inputs formulation. The initial Map paradigm encodes dataset D into key-value tuples and performs user-defined functions among the tuples for analytic queries.

$$map(D) \rightarrow list(k, v)$$

In a latency-aware data stream processing system, dataset D is encoded into a list of triples [10], such as $<k, v, ts>$, where ts is the system time when the element arrives. The triple can be used for resource planning and latency-aware scheduling. Like before, a *map* function is applied to transform input elements to a list of output triples.

$$map(D) \rightarrow list(k, v, ts)$$

We consider the same triple elements as input data stream. More precisely, suppose an asynchronous stream $D = e_1, e_2, ..., e_n$, and element e_i is a triple (k_i, v_i, ts_i), where k_i is an identifier of e_i in D, v_i is a numerical data attribute,

Table 1. Operations classification in big data systems.

Name	Description	Classification (symbol)
Aggregation	Return elements counting or values summarization	(1) Aggregation-based (ab)
q-quantile	Return the element at position of $100q\%$ in a sorted sequence	(2) Quantiles-based (qb)
Median	Return the median element in a sorted sequence	(2) Quantiles-based (qb)
Top-k	Return the largest k elements in a sorted sequence	(2) Quantiles-based (qb)
Range	Search statistics within a key range or temporal range	(3) Range-based (rb)
Shuffling	Redistribute and sort elements based on their keys	(4) Splitting& Merging ($s\&m$)
Partition	Partition larger dataset into multiple smaller subset	(4) Splitting& Merging ($s\&m$)

and ts_i is the timestamp of e_i. For a real-world data stream, D is usually not an ordered set (e.g., $ts_i > ts_j$ and $i < j$), and D maybe abound with duplicated elements (e.g., $k_i = k_j$ and $i \neq j$).

Queries Classification. As for different types of query processing in big data systems, we enumerate some fundamental queries and operations in Table 1. More fundamentally, we classify these operations into four types and they are shown in the last column of Table 1: (1) ab, i.e., achieving the numerical value of summarization or counting of elements, (2) qb, i.e., returning an element or k elements from a sorted sequence, (3) rb, i.e., obtaining statistics in a specified range, such as within a key-range or a temporal-range and (4) $s\&m$, which is used to shuffle and partition tuples between nodes in a cluster. As for type (3), we mainly focus on the temporal-range query in the following discussions (a.k.a, sliding-window queries when considering only one current temporal-range). Some operations might not be called by regular applications, such as type (4), but they are prevalent and very important for distributed query processing systems. Some other operators might not be listed in Table 1, such as AVG, VARIANCE, STDDEV, and GROUP-BY etc., while they can easily be computed by current estimates or introducing specified sampling algorithms into the presented framework of ROSE.

3.2 Approach Overview

We first state formalized description of our problem. For a query Q over a stream D with operation Opr, ROSE can deliver an estimate $Est(Q)$ and the statement is shown in Eq. 1:

$$Est(Q) = ROSE_{(\xi,\delta)}(D, Opr), \tag{1}$$

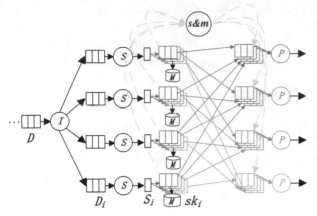

Table 2. Different sample block size.

Query types	Block size
Aggregation-based query (ab)	$O(\frac{\log(\delta^{-1})}{\xi^2})$
Quantiles-based query (qb)	$O(\frac{\log(\xi\delta)^{-1}}{\xi^2})$

Fig. 1. Distributed query processing framework of ROSE.

where Opr is an operation listed in Table 1, and $Est(Q)$ is an (ξ, δ)-approximation answer for Q.

We next explain the architectural design of ROSE, combing with skeleton of data stream processing systems. The outlines of designed framework is shown in Fig. 1. The input data stream D is dispatched by a distributor I (e.g., map function or a publish-subscribe messaging system, such as Apache Kafka). Each node in a cluster receives a subset of D (denoted as D_i in Fig. 1), and extracts multi-faced samples S_i from D_i by online sampling operator (S) to compact the data volume significantly, so we can maintain the sketch sk_i in the memory (M). To answer a query, a coordinator of the system asks each node to send local sketch to a reduce side based on a same key. The local sketches can be merged into a union one by the operator $s\&m$ to support analytical queries. Since the volume of local sketch (sk_i) is much smaller than data itself (D_i) (reducing two or three orders of magnitude relative to D_i), it is more efficient to shuffle and merge local sketches in distributed environments. The merged sketch can support analytical queries with a uniform estimation operator P to produce the final estimate. Some enhancement features in a distributed system, such as fault tolerance and load balance, can be achieved by existing techniques, and we do not discuss them in this paper for space limitation.

3.3 Multi-faced Sampling

We start introduce an online multi-faced sampling technique, which conceptually includes two key points: extended sliding-window sampling (ESS) and probability aggregation sampling (PAS). In general, for an input element $e_i = (k_i, v_i, ts_i)$, we generate a multi-faced sample $s_i = (k_i, cnt_{uni}, cnt_{exp}, v_{uni}, v_{exp}, ts_i)$, where cnt_{uni} and cnt_{exp} are attributes for element counting, $s_i.cnt_{uni} + s_i.cnt_{exp} = 1$ (resp. v_{uni} and v_{exp} are attributes for values summarization, $s_i.v_{uni} + s_i.v_{exp} = e_i.v_i$). The two-dimensional features for element counting and summarization are used to depict the division of a large element in PAS. Also, we assume that the number of elements is significantly larger than $\frac{1}{\xi^2}$. Otherwise, a simple structure can store all elements and can compute an exact return for Q. Meanwhile, $\frac{1}{\xi^2}$ is supposed to be an integer to avoid the floors and ceiling in expressions.

Extended Sliding-Window Sampling. We first consider maintaining temporal properties in our sketch to meet the needs of queries in type (3) (rb). The temporal properties are often depicted as parameters of window length and time interval by user-defined APIs in a stream-computing system. Correspondingly, our sketch maintains consecutive multiple windows, and the length of a window equals to a predefined time interval t_w. Within a window, it conceptually consists of $L + 1$ "levels", which are numbered sequentially as 0, 1, ..., L. A "level" constitutes a number of samples sorted by identifiers, and all levels keep the same number of samples. At any given point in time, a sample s_i is inserted into a temporal window $[sp_i, ep_i]$, $ts_i \in [sp_i, ep_i]$, and sketch sk_i is constructed and is responsible for types (1) and (2) queries within the window.

More precisely, a sample s_i is generated from e_i and is inserted into Level l of sk_i at probability 2^{-l} via PAS, where $0 \leq l \leq L$, $L = \log \sum e_i.v_i$. If the number of samples in a level reaches to predefined level size, we say that the level of the sketch is "full" (or the sketch is full for short), and the following inserting will lead to the oldest sample being discarded from the level. Thus, the level size reflects the sampling rate at different probability.

To support queries within any temporal range (type (3)), we arrange samples of a level into two symmetrical sample blocks to sustain bi-directional aggregation queries in a window. One block keeps samples which are close to the start-point of a window (denoted as $Block_{min}$) and the other block keeps samples which are close to the end-point of a window (denoted as $Block_{max}$). The maximum block size is limited to M. If the size of $Block^l_{min}$ is larger than M, $0 \leq l \leq L$, the maximum sample in $Block^l_{min}$ will be pushed into $Block^l_{max}$. If the size of $Block^l_{max}$ is larger than M, the minimum sample will be discarded directly. According to statistical theories, we can achieve (ξ, δ) accuracy guarantees for aggregation queries by configuring different sample block size [19]. Table 2 presents some of the configurations, and more details for analysis of M will be discussed in Sect. 3.5.

Probability Aggregation Sampling. The goal of PAS is to keep the optimal number of samples in a window while it can support (ξ, δ)-approximation for queries of types (1) and (2). It is well known that when we keep samples of size $O(\frac{1}{\xi^2} \log(\xi\delta)^{-1})$ in a stream, we can achieve an ξ-approximation for quantile queries with probability at least $1 - \delta$. However, if we keep $O(\frac{1}{\xi^2} \log(\xi\delta)^{-1})$ samples in all levels, it will cost too much space for a smaller value of ξ. An optimized Greewald-Khanna algorithm has been devised in [23] to combat this problem, which arrives at that a randomized algorithm requires only $O(\frac{1}{\xi} \log(\frac{1}{\xi} log(\xi\delta)^{-1}))$

Algorithm 1. Inserting(e, sk).

---

input : (e, sk);
 e: an input element;
 sk: a sketch for a temporal window.
output: (sk).
 1 $s.k \leftarrow e.k$;
 2 $s.(cnt_{uni}, cnt_{exp}) \leftarrow (1, 0)$;
 3 while ($e.v > 2 \times sk.avg$) do
 4 $\quad$ Let z be a random in
     $\quad$ $[\frac{1}{2}sk.avg, sk.avg]$;
 5 $\quad$ $s.(v_{uni}, v_{exp}) \leftarrow (z, 0)$;
 6 $\quad$ PAS(s, sk);
 7 $\quad$ $s.(cnt_{uni}, cnt_{exp}) \leftarrow (0, 0)$;
 8 $\quad$ $e.v \leftarrow e.v - z$;
 9 if ($e.v > 0$) then
10 $\quad$ $s.(v_{uni}, v_{exp}) \leftarrow (e.v, 0)$;
11 $\quad$ PAS(s, sk);
12 return sk.
13 Function PAS(s_i, sk_i)
14 $l \leftarrow f(s_i)$;
15 Let sk_i^l be the l-th level of sk_i;
16 Insert s_i into sk_i^l;
17 $s_i.(cnt_{exp}, v_{exp}) \leftarrow s_i.(cnt_{uni}, v_{uni})$;
18 $s_i.(cnt_{uni}, v_{uni}) \leftarrow (0, 0)$;
19 for ($j = 0$ to $l - 1$) do
20 $\quad$ Insert s_i into sk_i^j;

---

space parameterized by ξ and δ. Therefore, we keep $Max\{O(\frac{1}{\xi} \log(\frac{1}{\xi}log(\xi\delta)^{-1}))$, $O(\frac{\log(\delta^{-1})}{\xi^2})\}$ samples in a level of a sketch.

We use a hash function f to boost performance of PAS and keep samples at Level-l with probability $1/2^l$, $0 \leq l \leq L$. The hash function f is computed as follows: we choose two random numbers x and y independently from space field 2^L. The desired level number for s_i is $f(s_i) = L - \lfloor\log(x \times s_i.v_{uni} + y)\rfloor - 1$. We insert the sample s_i into all levels whose level number is no bigger than $f(s_i)$ respectively to make samples at Level-l work with probability $1/2^l$ in the sketch. Readers can refer to literature [20] for more details about probability guaranteed inserting in randomized sampling techniques.

Essentially, the PAS belongs to a type of randomized sampling technique, and it is error sensitive when large value element occurs [19]. We introduce subsampling techniques to solve the large elements discounting problem. We divide a larger value element into multiple smaller value elements on the fly and subsample them to decrease the error of the larger elements estimation. More precisely, if the value of an input element e is bigger than two times of current average of a window, we regard it as a large value element. We divide the large element $e = (k, v)$ into j smaller elements $(k, v_1), (k, v_2), ..., (k, v_j)$, $v = \sum_{i=1}^{j} v_i$, where $v_i = a_i + \Delta_i$, a_i is an estimator in the sketch before (k, v_i) inserting, and Δ_i is the value quantified by error ξ. In the current context, the a_i is current average of samples in the window, and Δ_i is a random in $[1, a_i/2]$. As shown in step 3 to step 8 of Algorithm 1, we produce smaller elements from e and insert them into the sketch sk through PAS correspondingly.

3.4 Sketch Construction

Notice that we need to solve data streams with out-of-order arrival, thus the disordered elements will lead to sketch splitting and merging between windows frequently. Moreover, when we shuffle and partition sketches between nodes, we also need to reconstruct sketch from individual sketches. These requirements can be achieved by leveraging accuracy lossless operations such as sketch splitting, sketch merging and sketch union operations.

Sketch Splitting. The sketch splitting is to move some samples from a full sketch to its neighboring sketch to balance the amount of samples between sketches. To satisfy accuracy-lossless splitting in sketch sk_i, we increase the maximum sample size of a block to $(1 + \rho)M$, $\rho < 1$. As shown in Fig. 2, the samples blocks of a sketch are divided into two parts: The white parts are used for answering queries with error-guaranteed accuracy. The gray parts are used for splitting and merging operations between sketches.

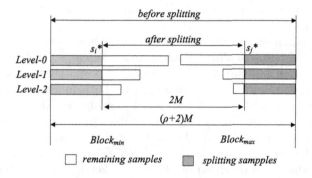

Fig. 2. Sketch splitting of a window.

In general, we insert samples into sk_i, until the length of Level-0 reaches to $(\rho + 2)M$. Within Level-0, we select a label sample s_i^* in $Block_{min}^0$ which makes M samples larger than s_i^* (resp. we select s_j^* in $Block_{max}^0$ which makes M samples smaller than s_j^*). We search at all levels to fetch samples which are smaller than s_i^* and merge them to sketch sk_{i-1}. Respectively, we search samples which are bigger than s_j^* and merge them to sketch sk_{i+1}. After splitting, the $Block_{min}^l$ and $Block_{max}^l$ of sketch sk_i, $0 \leq l \leq L$, keep sufficient samples to compute error-guaranteed estimation using the remaining samples in a level.

Sketch Merging. The sketch merging is to merge samples from individual sketches into a merged one and preserve temporal properties in the merged sketch. The native operation of sketch merging is to insert samples into the merged sketch via *Inserting()* operation intuitively. To boost the performance of sketches merging, we can conjoin samples level by level. For example, if a sample is from Level-l of $Block_{min}^l$ in sk_i, it will be inserted at Level-l of sk_{i-1} directly, when we use a same hash function in PAS.

The combination of sketch splitting and merging can be used to optimize the process of ESS (discussed in Sect. 3.3). For example, a balanced window-preserving method can be achieved when we split a large sketch (e.g., sk_i) into smaller parts and merge the split parts into its neighbouring sketch (e.g., sk_{i-1} or sk_{i+1}) to balance the number of samples between sketches. We can also use sketch merging operation to limit the space of our approach. For example, when we want to control the number of sketches smaller than B (i.e., there are B windows totally), we can periodically merge two neighbouring sketches together, which contain the minimal number of samples totally. Readers can refer to recent literature [18, 19] for details of other sketch construction techniques, such as space limited or variance optimal sketch construction methods.

Sketch Union. A big data system needs to shuffle data between nodes based on keys and then conducts specified functions at reduce side to obtain analytical answers. We introduce the operation of sketch union, which is to merge individual sketches into a union one, i.e., $SK = \bigcup_{i=1}^{n} sk_i$, and we can perform ad-hoc queries using SK directly.

To build a union sketch, we consider that individual sketches maintain same boundaries of temporal windows. We first build an empty sketch SK and then insert samples from individual sketches $\{sk_i | 1 \leq i \leq n\}$ into SK sequentially. Notice that the level number l is computed by hash function f independently among sketches (Algorithm 1), so we do not need to rehash the samples when they are merged into SK. We can conjoin samples from local sketches in a level-wise manner, and sorted the samples by their identifiers. For error-guaranteed union of sketches $sk_1, ..., sk_n$, the space of SK can be described as $\sum_{i=1}^{n} |sk_i|$, where $|sk_i|$ is the space overhead of sk_i. Since sketches are much smaller than data streams themselves, it is more resource efficient to deliver and compute a union sketch SK from individual sketches, and we can conduct the reduce side functions to obtain estimates for ad-hoc queries from SK directly.

3.5   Query Processing

In this section, we present details of query processing for queries of types (1) and (2) combining with queries of type (3) to facilitate our discussion, i.e., searching aggregates or quantiles within a temporal range.

Aggregation-Based Queries. For a query Q, which computes aggregates within temporal interval r_x, we search the sketches which are covered by r_x. If the sketches are fully-covered by r_x, we can easily obtain the aggregates from the covered windows. Therefore, we just describe details of query processing within a window w_i, which is partially covered by a queried range.

Let r_x^* be a temporal range which covers a window w_i partially, $r_x^* \in r_x$ and $r_x^* \in w_i$. If $r_x > tw$, we come to the conclusion that one of the boundaries of w_i is covered by r_x^*, i.e., for a window $w_i = [sp, ep]$, $sp \in r_x^*$ or $ep \in r_x^*$. Recall that we arrange samples in a bi-directional aggregation manner in a temporal window, thus we can support estimation for the two cases using the multi-faced samples in the corresponding samples blocks. The basic principle for aggregation-based query processing is to find the minimum level number l^*, which make samples in Level l^* contain the queried range r_x^*, and use included samples in Level l^* to compute the estimate. For details, if $w_i.sp \in r_x^*$, $r_x^* \subset Block_{min}^{l^*}$, otherwise $sk.ep \in r_x^*$, $r_x^* \subset (Block_{min}^{l^*} \cup Block_{max}^{l^*})$.

For queries of elements counting, we summarize cnt_{uni} and cnt_{exp} of the included samples in $Block_{min}^{l^*}$ or $Block_{min}^{l^*} \cup Block_{max}^{l^*}$ and enlarge 2^{l^*} times to produce the final estimate $Est(cnt)$. The formalized estimator for elements counting is shown in Eq. 2. For queries of summarization, after we obtain the minimum level number l^*, we summarize v_{exp} and v_{uni} of samples which are covered by r_x^* at Level l^* and enlarge 2^{l^*} times to get the estimate. The estimator $Est(sum)$ for queries of summarization is depicted in Eq. 3.

$$Est(cnt) \leftarrow 2^{l^*} \times \sum_{s_i \in r_x^*} (s_i.cnt_{uni} + s_i.cnt_{exp}), \tag{2}$$

$$Est(sum) \leftarrow 2^{l^*} \times \sum_{s_i \in r_x^*} (s_i.v_{exp} + s_i.v_{uni}). \tag{3}$$

Quantile-Based Queries. The quantiles queries are to obtain elements at desired positions of a sorted sequence. They are often called ϕ-quantiles searching in data stream processing systems. For example, when $\phi=0.5$, the 0.5-quantile query is to search the element of ranking median from data streams. It is well-known that it costs linear space and requires processing nearly all elements to obtain an exact answering for quantiles queries, thus we consider computing (ξ, δ)-approximation estimates for quantiles-based queries which search quantiles in a sequence sorted by frequency and value of elements.

In probability theory and statistics, it has been proved that if we keep sufficient randomized samples, we can provide error-guaranteed accuracy for quantiles queries. Recall that a sample s_i is inserted into Level-l with probability p_l, $p_l = 1/2^l$. The cnt_{uni} is attribute value for elements counting (resp. v_{uni} is attribute value for elements summarization). Thus cnt_{uni}/p_l can be considered as an estimator for the element count over the randomized sampling (resp. v_{uni}/p_l can be considered as an estimator for the element value). We use the included samples to construct an ordered set and find the approximate quantiles from the sorted sequence.

More precisely, we first search for the minimum level number l^*, which contains queried range r_x, and then use the samples in r_x to compute the approximate quantiles in the sorted sequence by samples frequency or samples values. Equation 4 shows the estimate $Est(med)$ for a median query in elements frequency, and Eq. 5 shows the estimate $Est(qf)$ for ϕ-quantiles query in elements value sequence, where $qf_{\phi=0.5}$ and $qf_{\phi=q}$ are functions to search a median or a q-quantile in a sorted sequence.

$$Est(med) \leftarrow 2^{l^*} \times \underset{\phi=0.5}{qf} (r_x \cap s_i.cnt_{uni}), \tag{4}$$

$$Est(qua) \leftarrow 2^{l^*} \times \underset{\phi=q}{qf} (r_x \cap s_i.v_{uni}). \tag{5}$$

We notice that ROSE can also be used for other approximate answering queries, which can be extended from aggregation-based and quantiles-based queries, such as AVG, VARIANCE, STDDEV, Top-k queries etc., and we do not discuss them further for space limitation.

4   Experimental Evaluation

We develop ROSE on Linux platform using JDK 1.8 packages with 64-bit addressing. The evaluation is performed under eleven nodes cluster, which are connected by 1 Gbit Ethernet switch. One node acts as master and ten nodes act as slaves nodes. Each server works with 6×2.0 GHz processors, 64 GB RAM, and two 1 TB SATA disks.

4.1   Methodology

Our elementary error metric for query accuracy is the relative error of queries which is defined for aggregation-based and quantiles-based queries respectively as follows. For an aggregation query, the estimate $\hat{E}$ obtained from ROSE compared with the actual value E, the corresponding relative error is $\xi = |E - \hat{E}|/E$. A q-quantile query is to find an element at $q|S|$ position in sorted sequence S. The relative error of approximate q-quantile query is defined as $(|rank(\hat{E}) - q|S||)/|S|$, where $rank(\hat{E})$ is the position of $\hat{E}$ in S.

We implement the research work in ASY-sketch which can support aggregation queries in a sliding-window [19], and conduct the comparisons between ASY-sketch and ROSE in a centralized environment to examine the micro features, such as query accuracy and space consumption, between the two approaches. We also evaluate performance of ROSE in production environments and compare it with big data analytic systems, such as BlinkDB and Spark. We conduct different types of queries, such as aggregation queries, median queries, quantiles queries and Top-k queries in the these big data analytical systems, and examine query response time and query accuracy against billions of real-word records.

4.2   Datasets

We use real-world page traffic from Wikipedia, nearly 1.4 billion records in a day (100 GB uncompressed data), as input stream to test system performance such as throughput and query response time. The page traffic stream is formatted as a record with four fields, including "language type", "page label", "view count" and "traffic bytes number". We build online sketch for fields of "language type", "view count" and "traffic bytes number" to support such queries as "What is the number of total page bytes and page view count for English web-pages during time period 00:00:00 May 1st, 2016 to 09:59:59 May 1st, 2016. To overcome the limitation of real-world data sets, we generate a high-speed synthetic data stream to examine the space consumption and query accuracy of our approach. We generate two types of synthetic data sets, which obey different data distributions, i.e., the Normal Form distribution $N(\mu = 1000, \sigma = 50)$ and the Zipf distribution $Z(deg = 0.5)$.

4.3 Micro Evaluation

We examine micro features of our approach by comparison with ASY-sketch in a centralized environment. ASY-sketch is a type of randomized sampling sketch, and it can solve counting and summarization problem against asynchronous data streams. We configure the same relative error ($\xi = 0.25$) during the two sketches construction. We load three types of data streams into the two sketches, (1) 10 million real-world records; (2) synthetic data streams obeying $N(\mu = 1000, \sigma = 50)$; (3) synthetic data streams obeying $Z(deg = 0.5)$. The space consumption of the two sketches within different data sets has been shown in Figs. 3(a), (b) and (c). To test queries accuracy, we generate 800 randomized temporal-range aggregation clauses and the temporal length in a clause is a random length in a window. The comparisons between query accuracy are shown in Figs. 4(a), (b) and (c) with the three types of data streams.

ROSE divides large value element into multiple equalized value elements via PAS, but the smaller elements are inserted into a sketch only once. But in ASY-sketch, any element will be divided and inserted into a level with probability 2^{-l}, where l is the level number of a sketch, so an element will be inserted repeatedly into many levels of a sketch, even if it is much smaller than the average of the current sketch. ROSE and ASY-sketch keep the same number of samples in a level, while ROSE can keep more different samples and thus it can achieve better queries accuracy for aggregation-based queries in the real-world data sets. We also notice that when data distribution changes from Normal Form or Zipf

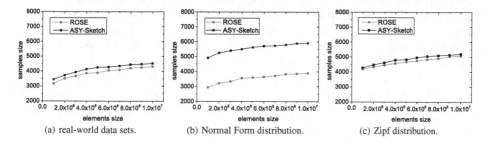

(a) real-world data sets.   (b) Normal Form distribution.   (c) Zipf distribution.

Fig. 3. Space consumption.

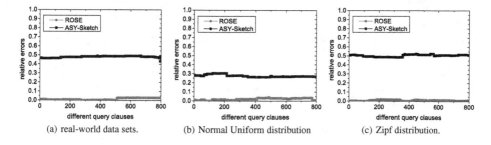

(a) real-world data sets.   (b) Normal Uniform distribution   (c) Zipf distribution.

Fig. 4. Query accuracy.

distribution to Uniform distribution, i.e., all elements work with almost same value, ASY-sketch and ROSE can achieve nearly same query accuracy.

4.4 Macro Evaluation

The macro evaluation is focused on practical performance examination of a big data system, such as system throughout and query response time, when processing real-world data sets. We select memory-computing system Spark (version 1.4.1) and Hadoop-based approximate answering engine BlinkDB (version 0.2.0) for our comparisons. Spark is a memory-computing system and it is often used as a large-scale framework for data streams system. Spark constructs different memory block, called RDDs, to support ad-hoc queries. We use Spark-SQL to conduct analytical queries. BlinkDB is a well-known approximate engine on the Hadoop platform [13]. BlinkDB can obtain estimates for analytical queries with bounded errors or bounded response times on very large data sets.

We load seven days of the page traffics from Wikipedia into the three systems. For ROSE and BlinkDB, we configure the same sampling rate (=0.01) when loading the data streams. i.e., the two approaches keep the same number of samples in their sketches. We examine capability of high-speed data stream processing for the three different systems, in Figs. 5(a), 6(a) and 8(a), when loading the real-world data sets. For BlinkDB and Spark, they first load data into HDFS, and then reload fields of interest into memory, so they produce latency for data streams processing. Rose processes an element in $O(\log(\frac{1}{\xi^2}\log\delta^{-1}))$ time, and it can achieve about 2 times improvement on system throughput than Spark and 1.5 times improvement than BlinkDB in our testing. Moreover, the throughput of ROSE is unchanged when the volume of data streams increases, while the throughput of BlinkDB and Spark decreases significantly when the volume of data streams increases. In seven days page traffic testing, the throughput decreases from 5.6 million records per second to 4.5 million records per second in BlinkDB and from 3.9 million records per second to 3.3 million records per second. Therefore, ROSE costs low latency for data streams processing in big and fast data applications.

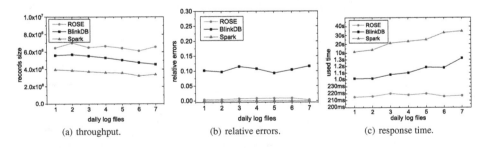

(a) throughput.   (b) relative errors.   (c) response time.

Fig. 5. Aggregation queries.

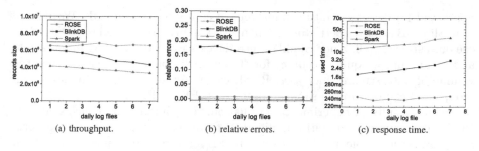

(a) throughput.   (b) relative errors.   (c) response time.

Fig. 6. Median queries.

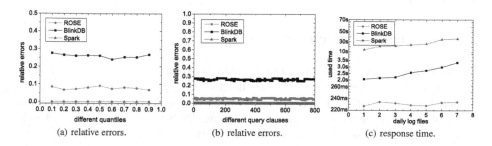

(a) relative errors.   (b) relative errors.   (c) response time.

Fig. 7. Quantiles queries.

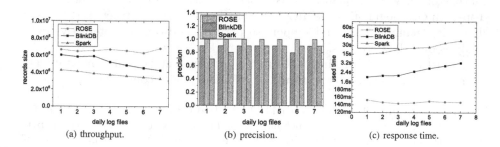

(a) throughput.   (b) precision.   (c) response time.

Fig. 8. Top-k queries.

We also compare the query accuracy and query response time among the three systems to demonstrate queries performance of our approach. We test aggregation-based queries and quantiles-based queries respectively. In aggregation-based queries, we compute the sum of bytes from page traffic within an randomized temporal range. In quantiles-based queries, we search median, q-quantiles ($q = 0.1$, 0.2, 0.4, 0.5, 0.6, 0.7, 0.8, and 0.9 respectively) and Top-k ($k = 10$) over the data streams. The results are shown in Figs. 6(b), 7(a) and (b).

As shown in Figs. 5(c), 6(c), 7(c) and 8(c), Spark and BlinkDB cost about 20 s and 2 s respectively, to compute results of the same queries. While ROSE uses about 0.2 s to compute the approximate answers for the queries. ROSE achieves about 2 orders of magnitude improvement on query response time compared to

Spark and 1 order of magnitude improvement compared to BlinkDB. We observe that BlinkDB achieves better precision in Top-k queries compared to ROSE when the data size is small. Since BlinkDB introduces stratified sampling in its sketch, and it can improve query accuracy for "Group-by" operation greatly. When the volume of data streams increases, ROSE can maintain enough samples for all groups and can obtain the same accuracy with BlinkDB.

As the results of experimental evaluation, we come to the conclusion that ROSE can support to perform different analytical queries with a uniform ξ-approximation schema. Meanwhile, ROSE can perform more efficient on throughput and query response time compared to current big data analytic systems (i.e., Spark and BlinkDB).

5   Conclusion and Future Work

Plentiful applications need strict requirement on query response time for analytics over big and fast data environments. In this paper, we propose an approach called ROSE, which can accept elements with out-of-order data arrival and obtain statistics over large volume data streams. ROSE can construct its sketch in an accuracy lossless and compact manner without pori knowledge in distributed environments. Meanwhile it can accept an input element in $O(\frac{1}{\xi^2}\log(1/\delta))$ time and obtain estimates for aggregation-based queries (such as counting, summarization, avg) and quantile-based queries (such as median, quantiles, and Top-k) in $O(\frac{1}{\xi^2}\log(1/\delta))$ time. ROSE enables systems to construct sketches over dynamic streaming data and support to perform real-time or interactive data analysis over data streams.

In further, we plan to introduce ROSE into current big data analytic platforms or systems to support real-time OLAP queries with complex query clauses.

Acknowledgment. The authors would like to thank the anonymous reviewers for their comments and suggestions which have helped to improve the quality of this paper. This work was supported by the National Key Research and Development Program of China (2016YFB0801305).

References

1. Katsipoulakis, N.R., Thoma, C., Gratta, E.A., Labrinidis, A., Lee, A.J., Chrysanthis, P.K.: CE-Storm: confidential elastic processing of data streams. In: SIGMOD, pp. 859–864 (2015)
2. Goodstein, M.L., Chen, S., Gibbons, P.B., Kozuch, M.A., Mowry, T.C.: Chrysalis analysis: incorporating synchronization arcs in dataflow-analysis-based parallel monitoring. In: PACT, pp. 201–212 (2012)
3. Zhang, Y., Chen, S., Wang, Q., Yu, G.: i2MapReduce: incremental MapReduce for mining evolving big data. In: KDD, pp. 1906–1919 (2012)
4. Preis, T., Moat, H.S., Stanley, E.H.: Quantifying trading behavior in financial markets using Google trends. Sci. Rep. **3**, 1684 (2013)
5. Zaharia, M., Das, T., Li, H., Hunter, T., Shenker, S., Stoica, I.: Discretized streams: fault-tolerant streaming computation at scale. In: SOSP, pp. 423–438 (2013)

6. Brito, A., Martin, A., Knauth, T., Creutz, S., Becker, D., Weigert, S., Fetzer, C.: Scalable and low-latency data processing with stream MapReduce. In: CloudComp, pp. 48–58 (2011)

7. Li, B., Mazur, E., Diao, Y., McGregor, A., Shenoy, P.: Scalla: a platform for scalable one-pass analytics using MapReduce. ACM Trans. Database Syst. **37**(4), 27:1–27:43 (2012)

8. Gulisano, V., Jimenez-Peris, R., Patino-Martinez, M., Soriente, C., Valduriez, P.: StreamCloud: an elastic and scalable data streaming system. Parallel Distrib. Syst. **23**(12), 2351–2365 (2012)

9. Qian, Z., He, Y., Su, C., Wu, Z., Zhu, H., Zhang, T., Zhou, L., Yu, Y., Zhang, Z.: TimeStream: reliable stream computation in the cloud. In: EuroSys, pp. 1–14 (2013)

10. Li, B., Diao, Y., Shenoy, P.: Supporting scalable analytics with latency constraints. Proc. VLDB Endow. **8**(11), 1166–1177 (2015)

11. Cormode, G., Garofalakis, M., Haas, P.J., Jermaine, C.: Synopses for massive data: samples, histograms, wavelets, sketches. Found. Trends Databases **4**(1–3), 1–294 (2012)

12. Yun, X., Wu, G., Zhang, G., Li, K., Wang, S.: FastRAQ: a fast approach to range-aggregate queries in big data environments. IEEE Trans. Cloud Comput. **3**(2), 206–218 (2014)

13. Agarwal, S., Mozafari, B., Panda, A., Milner, H., Madden, S., Stoica, I.: BlinkDB: queries with bounded errors and bounded response times on very large data. In: EuroSys, pp. 29–42 (2013)

14. Zeng, K., Agarwal, S., Dave, A., Armbrust, M., Stoica, I.: G-OLA: generalized on-line aggregation for interactive analysis on big data. In: SIGMOD, pp. 913–918 (2015)

15. Condie, T., Conway, N., Alvaro, P., Hellerstein, J.M., Gerth, J., Talbot, J., Elmeleegy, K., Sears, R.: Online aggregation and continuous query support in MapReduce. In: SIGMOD, pp. 1115–1118 (2010)

16. Chen, C., Li, F., Ooi, B.C., Wu, S.: TI: an efficient indexing mechanism for real-time search on tweets. In: SIGMOD, pp. 649–660 (2011)

17. Mousavi, H., Zaniolo, C.: Fast computation of approximate biased histograms on sliding windows over data streams. In: SSDBM, pp. 13:1–13:12 (2013)

18. Papapetrou, O., Garofalakis, M., Deligiannakis, A.: Sketching distributed sliding-window data streams. VLDB J. **24**(3), 345–368 (2015)

19. Tirthapura, S., Xu, B., Busch, C.: Sketching asynchronous streams over a sliding window. In: PODC, pp. 82–91 (2006)

20. Gibbons, P.B., Tirthapura, S.: Distributed streams algorithms for sliding windows. In: SPAA, pp. 63–72 (2002)

21. Datar, M., Gionis, A., Indyk, P., Motwani, R.: Maintaining stream statistics over sliding windows. In: SODA, pp. 635–644 (2002)

22. Wang, L., Luo, G., Yi, K., Cormode, G.: Quantiles over data streams: an experimental study. In: SIGMOD, pp. 737–748 (2013)

23. Arasu, A., Manku, G.S.: Approximate counts and quantiles over sliding windows. In: PODS, pp. 286–296 (2004)

24. Gibbons, P.B., Matias, Y., Poosala, V.: Fast incremental maintenance of approximate histograms. ACM Trans. Database Syst. **27**(3), 261–298 (2002)

25. Sharfman, I., Schuster, A., Keren, D.: A geometric approach to monitoring threshold functions over distributed data streams. ACM Trans. Database Syst. **32**(4), 23 (2007)

Boosting Moving Average Reversion Strategy for Online Portfolio Selection: A Meta-learning Approach

Xiao Lin[1]([⊠]), Min Zhang[2], Yongfeng Zhang[3], Zhaoquan Gu[4], Yiqun Liu[2], and Shaoping Ma[2]

[1] Institute of Interdisciplinary Information Sciences,
Tsinghua University, Beijing, China
jackielinxiao@gmail.com

[2] Tsinghua National Laboratory for Information Science and Technology,
Department of Computer Science and Technology,
Tsinghua University, Beijing 100084, China
{z-m,yiqunliu,msp}@tsinghua.edu.cn

[3] College of Information and Computer Science,
University of Massachusetts Amherst, Amherst, MA 01003, USA
yongfeng@cs.umass.edu

[4] Department of Computer Science, Hong Kong University, Hong Kong, China
demin456@gmail.com

Abstract. In this paper, we study the online portfolio selection problem from the perspective of meta learning for mean reversion. The online portfolio selection problem aims to maximize the final accumulated wealth by rebalancing the portfolio at each time period based on the portfolio prices announced before. Mean Reversion is a typical principle in portfolio theory and strategies that utilize this principle achieve the superior empirical performances so far. However there are some important limits of existing Mean Reversion strategies: First, the mean reversion strategies have to set a fixed window size, where the optimal window size can only be chosen in hindsight. Second, most existing mean reversion techniques ignore the temporal heterogeneity of historical price relatives from different periods. Moreover, most mean reversion methods suffer from noises and outliers in the data, which greatly affects the performances. In order to tackle the limits of previous approaches, we exploit mean reversion principle from a meta learning perspective and propose a boosting method for price relative prediction. More specifically, we generate several experts where each expert follows a specific mean reversion policy and predict the final price relatives with meta learning techniques. The sampling of multiple experts involves mean reversion strategies with various window sizes; while the meta learning technique brings temporal heterogeneity and stronger robustness for prediction. We adopt online passive-aggressive learning for portfolio optimization with the predicted price relatives. Extensive experiments have been conducted on real-world datasets and our approach outperforms the state-of-the-art approaches significantly.

© Springer International Publishing AG 2017
S. Candan et al. (Eds.): DASFAA 2017, Part II, LNCS 10178, pp. 494–510, 2017.
DOI: 10.1007/978-3-319-55699-4_30

1   Introduction

Online portfolio selection problem aims to allocate the wealth among different assets at different time periods to maximize the long-term wealth. There are two models describing the problem: the Mean-Variance model [19] and Kelly's Capital Growth model [11]. The first model uses a weighted sum of expected return (mean) and risk (variance of the return) as a trade-off between the two objectives, and it is suitable for single-period portfolio selection; the second model sees the problem as a sequential decision problem that aims to maximize the expected return at the end of multiple time periods. Kelly's Capital Growth model has a nature of online decision making, which is widely adopted by the studies from AI and Machine Learning researchers.

In online portfolio selection problem, each asset is associated with a price in each period. The ratio of prices between current period and last period is called price relative, which reflects the return of wealth invested on the assets after one period. The agent allocates the wealth among different assets based on their price relatives at different periods. Most portfolio selection strategies follow a two-phase scheme: price relative prediction phase and portfolio optimization phase. The first phase aims to predict the price relative at next period based on historical data; while the second phase aims to compute the optimal portfolio given the prediction of price relative.

One common methodology for this problem is Mean Reversion, which assumes the portfolios that perform poorly at current period will perform well next (and vice versa). The methods include PAMR (Passive Aggressive Mean Reversion) [17], CWMR (Confidence Weighting Mean Reversion) [15], OLMAR (OnLine Moving Average Reversion) [14] and RMR (Robust Mean Reversion) [9], which adopt the mean reversion idea in different ways. They achieve superior empirical results in experiments compared with other state-of-art methods. This proves the effectiveness of mean reversion policy.

Although PAMR, CWMR and OLMAR achieve good performances, they still face some difficulties. All existing mean reversion strategies do not fully consider the noisy data and outliers (RMR is proposed to alleviate the problem), which often leads to estimation error (see [20]). Furthermore, the assumption of single-period prediction [15,17] also leads to estimation error, which makes the performance poor. RMR (Robust Mean Reversion) [9] and OLMAR [14] uses multi-period prediction, but the algorithm sees each period equally, which ignores the temporal heterogeneity of historical price relatives and causes inaccuracy of predictions. We utilize meta learning to exploit the benefit of multi-period prediction and the periods are assigned with weights according to their performances. Moreover, this alleviates the impact of noisy data and outliers. The results show that our strategy outperforms RMR and OLMAR.

More specifically, in order to utilize multi-period historical data, we generate multiple experts for price relative prediction following typical MAR methods. Then we adopt the meta learning method for price prediction. Each expert is assigned a weight that is updated according to their performances and a weighted aggregation is used as the final prediction. Meanwhile, we choose the typical passive-aggressive learning method for portfolio optimization. This method captures the recent portfolio performances and the objective of enhancing the wealth return at each time period.

The contributions of our work are: first, to our knowledge, we are the first to exploit the Mean Reversion strategy with meta learning in online portfolio selection problems; second, we make a better use of multiple-period history, which is robust to the outliers and noises in historical data; third, we conduct extensive experiments on real-world datasets and achieve superior results compared with other state-of-the-art approaches.

The remainder of the paper is organized as follows: the next section gives a brief introduction to the related work; Sect. 3 formally introduces the online portfolio selection problem while Sect. 4 introduces some preliminary works about mean reversion theory and some related concepts. Section 5 proposes BMAR strategy that utilizes meta learning in online portfolio selection problem. Section 6 presents the results of experiments conducted on real-world datasets and a thorough comparison with the baselines. The conclusion and the future work are presented in Sect. 7.

2   Related Work

The study of online portfolio selection problem first concentrates on some benchmark algorithms, including Buy and Hold, Best Stock and Constant Rebalanced Portfolios. The Buy and Hold strategy means the agent invests wealth with an initial portfolio and holds it to the end without changing the portfolio. The Best Stock strategy means that one puts all the wealth on the stock whose performance is best in hindsight. Constant Rebalanced Portfolios is a strategy that rebalances the wealth to a fixed portfolio in all periods. The best CRP strategy which achieves highest accumulated wealth is called BCRP. BCRP is an optimal strategy if the market is i.i.d. [4]. Successive Constantly Rebalanced Portfolios (SCRP) [5] and Online Newton Step (ONS) [1] implicitly estimate next price relative via all historical price relatives with a uniform probability. However, both Best Stock and BCRP strategies have to be computed in hindsight.

There are two main categories of algorithms: follow-the-winner approach and follow-the-loser approach. The intuition behind first approach is to track the stock with best performance in history and raise the weights in the portfolio of these stocks. Most of the Follow-the-Winner approaches aim to imitate the BCRP strategy: including the universal portfolio selection (UP) [10], Exponential Gradient (EG) [8], follow-the-leader and follow-the-regularized-leader approaches. However, the prices of assets are unstable, even a good following of winner assets can not guarantee superior performances.

Follow-the-loser approach utilizes a typical assumption of mean reversion [18], which means that the good (poor)-performing assets will perform poor (good) in the following periods. The approaches in this category include Anti-Correlation (Anticor) [2], Passive-Aggressive Mean Reversion (PAMR) [17], Confidence-weighted Mean Reversion (CWMR) [15], Online Moving Average Reversion

(OLMAR) [14] and Robust Mean Reversion (RMR) [9]. CRP [4,5] implicitly envolves follow-the-loser approach since rebalancing the wealth means to transfer the wealth from winning stocks to losing stocks in some extent.

Another important category of the portfolio selection algorithms is pattern-matching, which estimates the portfolio price based on sampled similar historical patterns. Nonparametric kernel based moving window (BK) [7] measures the similarity by kernel method. Following the same framework, Nonparametric Nearest Neighbor (BNN) [12] locates the set of price relatives via nearest neighbor methods. [16] proposed Correlation-driven Nonparametric learning (CORN), which measures the similarity via correlation.

Since the mean-reversion technique is widely adopted in financial fields, it is useful in online learning algorithms as well. Passive Aggressive Mean Reversion (PAMR) [17] and Confidence Weighted Mean Reversion (CWMR) [15] estimate next price relative as the inverse of last price relative, which is in essence the mean reversion principle. Recently, [14] proposed On-Line Moving Average Reversion (OLMAR), which predicts the next price relative using moving averages and explores the multi-period mean reversion. Robust Mean Reversion is proposed to alleviate the impact of outliers and noises existing in the data. The empirical experiments indicate that OLMAR and RMR outperforms the other state-of-the-art algorithms. However, most of the Mean Reversion algorithms have important limits: first, the mean reversion strategies require to select a fixed time window for prediction, which can not be easily determined; second, the strategies treat each time period equally in prediction, which ignores the temporal heterogeneity; third, the strategies do not have strong robustness against noises and outliers. We utilize mean reversion strategies for its good depiction of reality and further use meta learning approach to tackle the limits. A detailed comparison between our strategy and existing Mean Reversion strategies is presented in Sect. 4.

3   Problem Setting

In this section, we formally introduce the Online Portfolio Selection problem. Assume that there exist m assets in market, and time is divided into T periods. Each asset i has a closing price $p_{t,i}$ at period t and p_t is denoted as the closing price vector (column vector for all vectors mentioned): $p_t = [p_{t,1}, p_{t,2}, ..., p_{t,m}]$. $x_{t,i}$ is the price relative that captures the ratio of closing prices between two consecutive periods: $x_{t,i} = \frac{p_{t,i}}{p_{t-1,i}}$ and $x_t = [x_{t,1}, x_{t,2}, ..., x_{t,m}]$ is the price relative vector.

In each period, the market reveals the closing prices of assets and the investor has to assign the capital with a portfolio vector: $b_t = [b_{t,1}, b_{t,2}, ..., b_{t,m}]$ where $b_{t,i}$ represents the proportion of wealth assigned to asset i at time t. We follow the typical assumption that no margin/short sale is allowed, therefore $b_{t,i} \geq 0, \forall t, i$ and $\sum_{i=1}^{m} b_{t,i} = 1, \forall t$. An investment means to select a portfolio b_t from the simplex: $\{b_{t,i} \geq 0, \sum_{i=1}^{m} b_{t,i} = 1\}$ at period t. Usually we assume the portfolio is uniformly distributed in the beginning: $b_0 = [\frac{1}{m}, ..., \frac{1}{m}]$. The sequence of

---

Algorithm 1. ONLINE PORTFOLIO SELECTION

Input: x_1^n : Historical market price relative sequence
Output: W_t : Final cumulative wealth
Procedure:

1: Initialize the portfolio $b_1 = \frac{1}{m}\mathbf{1}$, the wealth is initialized as $W_0 = 1$;
2: **for** $t = 1, 2, ..., n$ **do**
3:     Portfolio manager learns the portfolio $\mathbf{b_t}$;
4:     Market reveals the market price relative $\mathbf{x_t}$;
5:     Portfolio incurs period return $w_t = \mathbf{b_t^T x_t}$ and updates cumulative return $W_t = W_{t-1} \times (\mathbf{b_t^T x_t})$;
6:     Portfolio manager updates the online portfolio selection rules;
7: **end for**;

---

portfolios from t_1 to t_2 is denoted as $b_{t_1}^{t_2}$. Denote the wealth accumulated at t as W_t, w.l.g. the initial wealth is assumed to be 1: $W_0 = 1$. Therefore given the selected portfolio b_t at period t, the wealth becomes $W_t = W_{t-1} b_t^T x_t = \sum_{i=1}^m b_{t,i} x_{t,i} W_{t-1} = \prod_{\tau=1}^t b_\tau^T x_\tau$ (T is transpose here).

Given the notations and introduction above, the online portfolio selection problem refers to a sequential decision making problem with periods from $t = 1$ to $t = n$. In each period, the investor has to decide the portfolio based on historical closing prices of assets and the market reveals the newest closing price of assets, which leads to the change of wealth. The investor needs to strategically design portfolios b_1^n so that the accumulated wealth W_n at time $t = n$ is maximized.

We summarize the procedures from the introduction above and formulate the whole online portfolio selection process in Algorithm 1 as [13].

4   Preliminary

In this section, we briefly introduce the mean reversion principle and how former works exploit this principle.

4.1   Mean Reversion

In each period, the algorithm tries to estimate the price relatives of assets in price prediction phase and compute the portfolio with Passive-Aggressive or Confidence Weighted Learning given the predicted price relatives. The mean reversion principle is reflected in the first phase by assuming that the poor-performing assets will have good performances in next periods (and vice versa). Denote the estimated price relative as $\tilde{x}_t$ and the estimated closing price as $\tilde{p}_t$. PAMR and CWMR assumes that the assets with high/low price relatives will have low/high price relatives in next period: $\tilde{x}_{t+1} = \frac{1}{x_t}, \forall t$, which means $\frac{\tilde{p}_{t+1}}{p_t} = \frac{p_{t-1}}{p_t}$. Therefore the principle assumes that $\tilde{p}_{t+1} = p_{t-1}$. Although the two methods work well, they can not perform consistently on some datasets.

There are two reasons that lead to this inefficiency: first, the fluctuating prices may contain noises that affect the precision of mean reversion principle; second, the single period price reversion effect may not exist widely as expected.

4.2   Online Moving Average Reversion

The OLMAR (online moving average reversion) principle is proposed to model the mean reversion principle with multiple-period historical data. Denote the time window of OLMAR as w, the closing price p_t at period t is assumed to be: $\tilde{p}_t = \frac{1}{w} \sum_{\tau=t-w}^{t-1} p_\tau$. The price is therefore considered as the average of prices in a time window and the price relative becomes: $\tilde{x}_t = \frac{p_t}{p_{t-1}} = \frac{1}{w}(1 + \frac{1}{x_{t-1}} + ... + \frac{1}{\bigotimes_{\tau=t-w}^{t-1} x_\tau})$. This Moving Average Reversion strategy with time window is denoted as SMAR.

Usually, it is assumed that the price at current period is closer to the price at recent periods due to the continuity of price changes. Therefore the price can be estimated with MAR by adding a decay factor α: $\tilde{p}_t = \alpha p_{t-1} + (1-\alpha \tilde{p}_{t-1})$, which results with a price relative: $\tilde{x}_t = \alpha + (1-\alpha)\frac{\tilde{x}_{t-1}}{x_{t-1}}$. Therefore the decay factor frees the algorithm from choosing a time window and utilizes the prices from the whole history. This Moving Average Reversion strategy with time window is denoted as EMAR.

Given the price relative predictions, OLMAR method further utilizes the online passive aggressive learning policy for portfolio optimization, which is also adopted by PAMR. Notice that the choice of time window size and decay factor determines the performance of this method and can not be pre-defined in hindsight.

4.3   Robust Mean Reversion

RMR uses L1 estimator to estimate the closing prices of assets so that the resulting price has a better robustness compared to other methods: with a window size of historical periods w, RMR estimates the closing at t by minimizing this objective: $\sum_{\tau=t-1}^{t-w} \|\tilde{p}_t - p_\tau\|_2$, and the price relative is estimated as $\frac{\tilde{p}_t}{p_{t-1}}$. This estimation is named as L1 estimator and is relatively more robust to noises.

4.4   Temporal Heterogeneity

The price relatives of different periods are correlated in different extents. Usually, it is assumed that the price at current period is closer to the price at recent periods due to the continuity of price changes. Meanwhile, the changes of price relatives in some time windows are more similar than others (which is the foundation of CORN (CORrelation-driven Nonparametric learning) [16]). Therefore, when predicting the price relatives in the future, the algorithm should consider how to make use of the historical time windows differently.

Given these existing algorithms exploiting Mean Reversion principles, we make a comparison between our approach (Boosting Moving Average Reversion: BMAR) and these algorithms in Table 1. The multi-period column shows

Table 1. Comparison between our strategy (BMAR) with existing mean reversion approaches.

Approaches	Mean reversion	Multi-period	Robustness	Temporal heterogenity
PAMR	$\checkmark$	$\backslash$	$\backslash$	$\backslash$
CWMR	$\checkmark$	$\backslash$	$\backslash$	$\backslash$
OLMAR	$\checkmark$	$\checkmark$	$\backslash$	$\backslash$
RMR	$\checkmark$	$\checkmark$	$\checkmark$	$\backslash$
BMAR	$\checkmark$	$\checkmark$	$\checkmark$	$\checkmark$

whether the approaches use multiple-period historical data for prediction; the Robustness shows whether the approaches are robust to noises and outliers; the Temporal Heterogenity column shows whether the approaches can utilize the data from different periods Heterogeneously. The table shows that our strategy (BMAR) preserves all the good properties, which shows the superiority.

5   Proposed Strategy: Boosting Moving Average Reversion

Like most methods, we solve the problem with two phases: price relative prediction and portfolio optimization. We first generate a set of experts and each expert is a predictor of the price relative following a mean reversion policy. In each period, each expert first makes its predictions on the price relatives in next period; then we compute the cumulated losses induced by each expert from their historical predictions and the true past price relatives. With the cumulated losses, we compute the weights assigned to each expert following the boosting methods introduced later and make a final prediction. Then we use Online Passive-Aggressive learning method to compute an optimized portfolio with the final prediction of price relatives in next period. When the true price relatives are revealed, we can update the cumulated losses of each predictor. The process of BMAR strategy for online portfolio selection is illustrated in Fig. 1.

5.1   Boosting Moving Average Reversion for Price Relative Prediction

We generate the experts of price relative prediction with different parameters from OLMAR. Denote the set of experts as E and the SMAR expert with time window w is denoted as E_w, the EMAR expert with Decaying factor α as E_α.

Uniform Sampling: We generate the experts by sampling the parameters uniformly from the range: $w \sim U(w_{min}, w_{max})$ and $\alpha \sim U(0, 1)$. We generate $M = w_{max} - w_{min} + 1$ experts from MAR ($w = w_{min}, w_{min}+1, ..., w_{max}$) and N experts from EMAR ($\alpha = \{0.1, 0.2, ..., 0.9\}$ when $N = 9$). Based on the different ways of generating experts, we denote the strategy of generating experts with

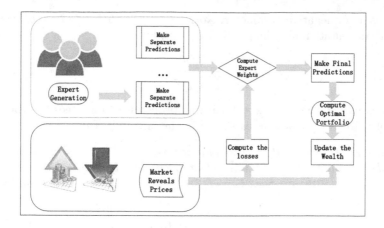

Fig. 1. BMAR strategy for online portfolio selection

time window as BMAR-1 and the strategy of generating experts by sampling α as BMAR-2. With the generated experts, we use weighted aggregation of their decisions to predict the price relatives at different periods. Each expert represents an approximation of the price relative in next period with a certain time window. By utilizing the predictions of these experts with a weighted scheme, we can induce temporal heterogeneity into our approach and the details will be introduced later.

As shown in Theorem 1 later, the regret of our strategy is closely related to the number of experts we generate. We will present the influence of expert numbers and sampling methods on the performances in the experiment.

We assume a weighted sum of the experts as the estimator of the price relative:

$$BMAR-1 : \tilde{x}(t) = \frac{\sum_{i=1}^{i=N} \theta_{i,t-1}\tilde{x}(t,w_i)}{\sum_{i=1}^{i=N} \theta_{i,t-1}}; \quad BMAR-2 : \tilde{x}(t) = \frac{\sum_{j=1}^{j=M} \theta_{j,t-1}\tilde{x}(t,\alpha_j)}{\sum_{j=1}^{j=M} \theta_{j,t-1}} \quad (1)$$

where $\tilde{x}(t,w)$ and $\tilde{x}(t,\alpha)$ are the predicted price relatives of expert E_w and E_α; $\theta_{w,t-1}$ and $\theta_{\alpha,t-1}$ are the weights assigned to these experts given their performances until period $t-1$.

Denote the loss of expert E_w by period t as $l(w,t)$ and the loss of weighted expert (BMAR) by period t as $l(t)$. The cumulated losses of expert E_w and weighted expert by period T are denoted as $L(w,T) = \sum_{t=1}^{t=T} l(w,t)$ and $L(T) = \sum_{t=1}^{t=T} l(t)$ respectively. The difference between the two losses is seen as the regret of weighted expert with respect to expert E_w: $R(w,T) = L(T) - L(w,T)$.

We introduce the weights assigned to the expert, i.e. exponential weights:

$$\theta_{i,t-1} = \frac{e^{-\eta R_{i,t-1}}}{\sum_{j=1}^{N} e^{-\eta R_{j,t-1}}} \quad (2)$$

where η is a nonnegative parameter. Notice that $R(w,T) = L(T) - L(w,T)$, the exponential weights make the predictions simpler:

$$\theta_{i,t-1} = \frac{e^{-\eta L_{i,t-1}}}{\sum_{j=1}^{N} e^{-\eta L_{j,t-1}}} \tag{3}$$

It has been proved that this expert learning procedure guarantees a proper upper bound of regret in prediction, as shown in the theorems from [3]:

Theorem 1. *Assume that the loss function is convex in its first argument and takes values from $[0,1]$, then the regret of exponentially weighted average predictor satisfies (N is the number of experts, n is the number of periods and η is the parameter in exponential weights):*

$$\hat{L}_n - \min_{i=1,\dots,N} L_{i,n} \le \frac{lnN}{\eta} + \frac{n\eta}{2} \tag{4}$$

The details of the proofs can be found in [3] and we omit the details here. Given these theorems, we can design loss functions that satisfy the requirements:

$$l(\hat{p}, y) = \frac{1}{N\varepsilon} \times \|\hat{p} - y\|_2^2 \tag{5}$$

where ε is the constant that rescales $l(\hat{p}, y)$ into $[0,1]$. And it is easy to verify that the function is convex, which satisfies the requirement of the theorems. Notice that ε actually works as coefficients of $l(\hat{p}, y)$ with η, we can simply tune the value of η to adjust the performances, therefore when using this loss function, we do not explicitly set the value of ε.

Remarks on Robustness: Notice that we do not explicitly model robustness in our prediction, however the utilization of multiple experts involves robustness: if the outliers and noises causes degradation of the experts' prediction accuracy, the weights assigned to these affected experts are lowered, which prevents the final predictions suffering from the noises and outliers.

5.2   Portfolio Optimization

Given the predicted price relatives shown in former section, we utilize the passive aggressive learning procedure to solve an optimal portfolio. The basic idea of passive aggressive learning is to keep the portfolio the same if the predefined requirement is satisfied, otherwise the portfolio is computed to satisfy the requirement with a minimal change. More specifically, we formulate the optimization problem as follows:

$$
\begin{aligned}
min. \quad & \|b_t - b_{t-1}\|^2 \\
s.t. \quad & b_t \tilde{x}_t \ge \epsilon, \text{ and } b_t \succeq \mathbf{0}
\end{aligned}
\tag{6}
$$

where ϵ is the threshold for the return at each period. Usually ϵ is a constant greater than 1 to ensure the return under predicted price relative is increasing.

The optimum to this problem is the portfolio assigned for period t. Notice that if we keep the portfolio same with that in last period and the return under predicted price relatives still exceeds the required value, we will keep the portfolios unchanged; otherwise, we will try to minimize the change between current portfolio and that in last period as long as the return can exceed the requirement. Since this optimization problem is convex, we can derive the portfolio in a closed form. The solution without considering the nonnegativity constraint is presented in the following proposition:

$$b_{t+1} = b_t - \alpha_{t+1}(\hat{x}_{t+1} - \bar{x}_{t+1} \cdot \mathbf{1}) \tag{7}$$

where $\bar{x}_{t+1} = \frac{1}{d}(\mathbf{1}\dot{\hat{x}}_{t+1})$ denotes the average predicted price relative and α_{t+1} is the Lagrangian multiplier calculated as,

$$\alpha_{t+1} = \min\{0, \frac{\hat{x}_{t+1}b_t - \epsilon}{\|\hat{x}_{t+1} - \bar{x}_{t+1} \cdot \mathbf{1}\|^2}\} \tag{8}$$

In order to ensure that the portfolio is non-negative, we project the above portfolio into the simplex domain as [14].

5.3   Transaction Costs

In this section, we will introduce the transaction cost, which is an important factor in practical scenarios. In practice, each transaction of wealth from one asset to another is charged with transaction fees. The transaction cost is imposed by markets, and a portfolios behavior cannot change the properties of transaction costs, such as commission rates or tax rates. Usually we assume the transaction fee follows a proportional model, which means rebalancing a portfolio incurs transaction costs on every buy and sell operation, based upon a transaction cost rate of $\gamma \in (0, 1)$. Therefore the transaction cost for a rebalancing from $\hat{b}_{t-1}$ to b_t is computed as:

$$\frac{\gamma}{2} \times \sum_{i=1}^{m} |b_{t,i} - \hat{b}_{t-1,i}| \tag{9}$$

Therefore the cumulated wealth after n periods becomes:

$$W_n^\gamma = W_0 \prod_{t=1}^{n} [(b_t \cdot x_t) \times (1 - \frac{\gamma}{2} \times \sum_{i=1}^{m} |b_{t,i} - \hat{b}_{t-1,i}|)] \tag{10}$$

Notice that the main intuition of Passive-Aggressive portfolio optimization is to keep the portfolio unchanged unless the requirement can not be satisfied. This avoids unnecessary rebalancing of wealth among assets and saves the transaction costs induced.

6   Experiment

We conduct extensive experiments on several real-world datasets to evaluate the performances of our strategy and make comparisons with state-of-the-art approaches.

6.1 Experiment Setting

In our experiment, we use the real-world datasets that are frequently used in related works. There are four datasets that contain price relatives of assets from US and Global markets. The time frames of these datasets range from decades to years, which reflect the performances of both long-term and short-term portfolio selections. The details of the datasets are listed in Table 2. In the experiment, we use the metrics that are adopted in the literatures for evaluation: i.e. the total wealth achieved at the final period.

Table 2. Statistics of the real-world datasets for experiment.

Dataset	Region	Time frame	# periods	# assets
NYSE(O)	US	Jul. 3rd 1962–Dec. 31st 1984	5651	36
NYSE(N)	US	Jan. 1st 1985–Jun. 30th 2010	6431	23
SP500	US	Jan. 2nd 1998–Jan. 31st 2003	1276	25
MSCI	Global	Apr. 1st 2006–Mar. 31st 2010	1043	24

6.2 Comparison Approaches

We select the state-of-art algorithms (most of them have been introduced in the related works) for comparison, including those Benchmark algorithms (Market, Best-Stock and BCRP), follow-the-winner algorithms (UP, EG, ONS), pattern-matching algorithms (B^k, B^{NN}, CORN, Anticor) and all the variants of mean reversion algorithms (PAMR, CWMR, OLMAR, RMR). For all the algorithms above, we choose the parameters with best performances as reported in related works. Notice that we select some algorithms that use information in hindsight for comparison (which are strong baseline algorithms). For the default setting of BMAR, we set $W = 8$ and $N = 9$ for BMAR-1 and BMAR-2. The other parameters are chosen as: $\eta = 1, \varepsilon = 5$ for all datasets.

6.3 Performance Evaluation

We present the cumulative wealth of our strategy and the comparative approaches in Table 3. As shown in the table, our strategy achieves the best performance on all datasets and outperforms other comparative algorithms significantly on long-term portfolio selection problems, i.e. on NYSE(O) and NYSE(N). Notice that the parameters fit for each dataset can be different, we also list the best performances of algorithms for comparison. The results show that the (including best or conventional) performances achieved by BMAR are better than other comparison algorithms. We also conduct significance test (following [6]) on the performances and the results are listed in Table 4. The significance tests shown above indicate that the performance of our strategy is significantly better on all datasets, which is not the consequence of luck. Notice that

Table 3. Cumulative Wealth on four datasets.

Categories	Approaches	NYSE(O)	NYSE(N)	SP500	MSCI
Baselines	Market	14.5	18.06	1.34	0.91
	Best Stock	54.14	83.51	3.78	1.50
	BCRP	250.60	120.32	4.07	1.51
Follow the winner	UP	26.68	31.49	1.62	0.92
	EG	27.09	31.00	1.63	0.93
	ONS	109.19	21.59	3.34	0.86
Pattern matching	B^K	1.08E+09	4.64E+03	2.24	2.64
	B^{NN}	3.35E+11	6.80E+04	3.07	13.47
	$CORN$	1.48E+13	5.37E+05	6.35	26.10
	Anticor	2.41E+08	6.21E+06	5.89	3.22
Mean reversion	PAMR	5.14E+15	1.25E+06	5.09	15.23
	CWMR	6.49E+15	1.41E+06	5.90	17.28
	OLMAR-1	3.68E+16	2.54E+08	5.83	16.39
	OLMAR-1(max)	1.62E+17	3.95E+08	20.91	25.49
	OLMAR-2	1.09E+18	5.10E+08	8.63	21.21
	OLMAR-2(max)	2.19E+18	2.84E+09	14.63	27.05
	RMR	1.64E+17	3.25E+08	8.28	16.76
	RMR(max)	2.81E+17	4.73E+08	17.05	19.07
Our Strategy	BMAR-1	**2.02E+18**	**8.95E+08**	**9.99**	23.01
	BMAR-1(max)	**4.59E+18**	**5.04E+09**	**16.02**	**30.98**
	BMAR-2	**8.11E+18**	**1.16E+09**	**10.11**	22.93
	BMAR-2(max)	**8.33E+18**	**6.64E+09**	**26.45**	**30.04**

Table 4. Significance Test of the Performances on experimental datasets. MER means Mean Excess Return; All the statistics are preferred to be higher; p-value is expected to be low.

Metrics	NYSE(O)	NYSE(N)	SP500	MSCI
Size	5651	6431	1276	1043
MER (BMAR-1)	0.0081	0.0038	0.0024	0.0033
MER (Market)	0.0005	0.0005	0.0003	0.0000
Winning ratio	0.5735	0.5400	0.5306	0.5916
α	0.0075	0.0032	0.0020	0.0033
β	1.2884	1.1166	1.2852	1.2085
t-statistics	16.7284	8.1011	2.5402	6.3251
p-value	0.0000	0.0000	0.0056	0.0000

our approach outperforms other baselines significantly, especially on datasets NYSE(O)and NYSE(N). The reason is that the NYSE datasets contain relatively long time periods, and the wealth accumulated has a "Matthew Effect" (the accumulated wealth will be increased with time passing by, the longer it goes, the more wealth will be accumulated).

6.4   Parameter Sensitivity

Notice that our strategy has several parameters: W (which is the maximum window size of experts generated from moving mean reversion) and η for strategy BMAR-1; α and η for strategy BMAR-2. The threshold for portfolio optimization ε is also a key parameter. We conduct experiments on all datasets with different values of the parameters.

Impact of η. Notice that we use P_2 norm of the difference between prediction and true price relative as loss function, the value of η therefore has two effects: first, it scales the loss function into $[0, 1]$; second, it evaluates the weights assigned to each expert considering their performances. Therefore we conduct experiments with different values of η on the datasets to show their impact in Figs. 2 and 3. Notice that the two strategies are different according to their ways of estimating the price relatives at each period. The impact of η also varies with the two strategies. Generally, choosing $\eta = 1$ guarantees relative good performances on all the datasets, which is adopted in the experiments shown in Table.3.

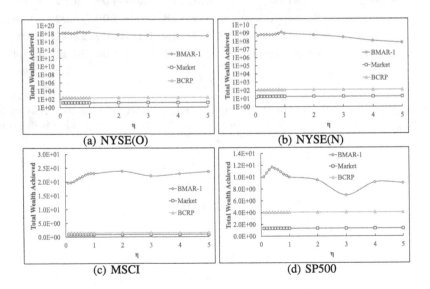

Fig. 2. Parameter sensitivity of η with $\varepsilon = 5, W = 8$ in BMAR-1

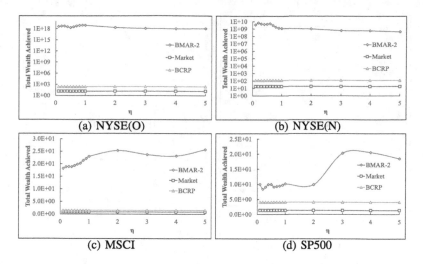

Fig. 3. Parameter sensitivity of η with $\varepsilon = 5, N = 9$ in BMAR-2

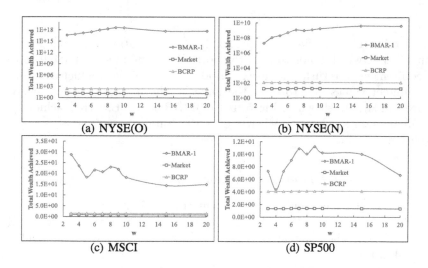

Fig. 4. Parameter sensitivity of W with $\varepsilon = 5, \eta = 1$ in BMAR-1

Impact of W. BMAR-1 generates experts with different window sizes $w \in [2, W]$, each expert estimates the asset price as the mean of prices in most recent w periods. We choose different values of W to generate experts, where each W means $W - 1$ experts with window sizes are generated. The results are shown in Fig. 4. Notice that the impacts of W are different on the datasets, this is due to the fact that the optimal window sizes for the experts to work on different datasets are also different: the optimal window size for MSCI can be relatively low compared with other datasets. We find that the strategy can achieve consistently good performances when $W \in [8, 10]$ and set it as a conventional value.

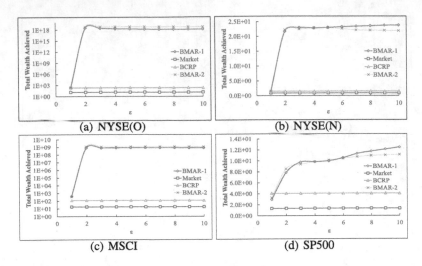

Fig. 5. Parameter sensitivity of ε with $\eta = 1, w = 8$ in BMAR-1 and $\eta = 1, N = 9$ in BMAR-2

Impact of ε. The passive-aggressive portfolio optimization technique is applied in our scheme, which tends to keep the portfolios same unless they fail to reach the requirement of return from each period. Therefore, we conduct experiments with different values of ε to show the impact. The impact of ε on our strategies are similar. As shown in Fig. 5, both BMAR-1 and BMAR-2 achieve good performances on all datasets when $\varepsilon \in [5, 10]$. Similarly, we choose $\varepsilon = 5$ as a conventional setting.

6.5   Performance Under Transaction Costs

We also conduct experiments with different transaction cost ratio since it is an unavoidable issue in practice. We alter the transaction cost ratio from 0% to 1% and compute the cumulative wealth of different strategies. The results are presented in Fig. 6.

Judging from the results, the transaction costs has a significant impact on the wealth return. When the transaction cost ratio is greater than 0.005, the wealth achieved on most datasets is close to 0. Since our algorithms can still outperform the baselines, they have good scalability for transaction costs. Notice that the real transaction cost ratio is usually below 0.005, our algorithm can work well in practice.

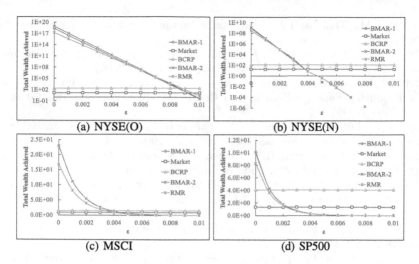

Fig. 6. Performance with different transaction cost ratios

7   Conclusion

In this paper, we consider the online portfolio selection problem from the perspective of mean reversion and meta learning. So far, mean reversion strategies have achieved best empirical results, however they face limits of unknown window size and ignores the temporal heterogeneity of different periods. Meanwhile they are easily affected by outliers and noises in the data. We utilize meta learning to tackle the limits and propose Boosting Moving Average Reversion (BMAR) strategies. The experiments on real-world datasets show that BMAR outperforms state-of-the-art strategies. We believe more accurate prediction of price relatives can further improve the performances and we will consider this as future works.

Acknowledgement. This work was supported by the Natural Science Foundation (61532011, 61672311) of China and the National Key Basic Research Program (2015CB358700). The third author was supported by the Center for Intelligent Information Retrieval and NSF grant under number IIS-1160894 and IIS-1419693.

References

1. Agarwal, A., Hazan, E., Kale, S., Schapire, R.E.: Algorithms for portfolio management based on the newton method, ICML 2006, pp. 9–16. ACM, New York (2006)
2. Borodin, A., Elyaniv, R., Gogan, V.: Can we learn to beat the best stock. J. Artif. Intell. Res. **21**(1), 579–594 (2004)
3. Cesa-Bianchi, N., Lugosi, G.: Prediction, Learning, and Games. Cambridge University Press, New York (2006)

4. Cover, T.M., Thomas, J.A.: Elements of Information Theory. Wiley, New York (2012)
5. Gaivoronski, A.A., Stella, F.: Stochastic nonstationary optimization for finding universal portfolios. Ann. Oper. Res. **100**(1), 165–188 (2000)
6. Grinold, R., Kahn, R.: Active Portfolio Management: A Quantitative Approach for Producing Superior Returns and Controlling Risk. McGraw-Hill Education, New York (1999)
7. Gyorfi, L., Lugosi, G., Udina, F.: Nonparametric kernel-based sequential investment strategies. Math. Financ. **16**(2), 337–357 (2006)
8. Helmbold, D.P., Schapire, R.E., Singer, Y., Warmuth, M.K.: On line portfolio selection using multiplicative updates. Math. Financ. **8**(4), 325–347 (1998)
9. Huang, D.J., Zhou, J., Li, B., Hoi, S., Zhou, S.: Robust median reversion strategy for online portfolio selection. IEEE Trans. Knowl. Data Eng. **28**(9), 2480–2493 (2016)
10. Kalai, A., Vempala, S.: Efficient algorithms for universal portfolios. J. Mach. Learn. Res. **3**(3), 423–440 (2003)
11. Kelly, J.L.: A new interpretation of information rate. Bell Syst. Tech. J. **35**(4), 917–926 (1956)
12. Laszlo, G., Frederic, U., Harro, W.: Nonparametric nearest neighbor based empirical portfolio selection strategies. Stat. Decis. **26**(2), 145–157 (2008)
13. Li, B., Hoi, S.C.H.: Online portfolio selection: a survey. ACM Comput. Surv. **46**(3), 1–36 (2014)
14. Li, B., Hoi, S.C.H., Sahoo, D., Liu, Z.: Moving average reversion strategy for on-line portfolio selection. Artif. Intell. **222**, 104–123 (2015)
15. Li, B., Hoi, S.C.H., Zhao, P., Gopalkrishnan, V.: Confidence weighted mean reversion strategy for online portfolio selection. ACM Trans. Knowl. Disc. Data **7**(1), 1–38 (2013)
16. Li, B., Hoi, S.C., Gopalkrishnan, V.: CORN: correlation-driven nonparametric learning approach for portfolio selection. ACM Trans. Intell. Syst. Technol. **2**(3), 1–29 (2011)
17. Li, B., Zhao, P., Hoi, S.C.H., Gopalkrishnan, V.: PAMR: passive aggressive mean reversion strategy for portfolio selection. Mach. Learn. **87**(2), 221–258 (2012)
18. Lo, A.W., Mackinlay, A.C.: When are contrarian profits due to stock market overreaction. Rev. Financ. Stud. **3**(2), 175–205 (1989)
19. Markowitz, H.: Portfolio selection. J. Financ. **7**(1), 77–91 (1952)
20. Merton, R.C.: On estimating the expected return on the market: an exploratory investigation. J. Financ. Econ. **8**(4), 323–361 (1980)

Continuous Summarization over Microblog Threads

Liangjun Song[1], Ping Zhang[3], Zhifeng Bao[1(✉)], and Timos Sellis[2]

[1] School of Science, RMIT, Melbourne, Australia
{liangjun.song,zhifeng.bao}@rmit.edu.au
[2] Swinburne University of Technology, Melbourne, Australia
tsellis@swin.edu.au
[3] Wuhan University, Wuhan, China
pingzhang@whu.edu.cn

Abstract. With the dramatic growth of social media users, microblogs are created and shared at an unprecedented rate. The high velocity and large volumes of short text posts (microblogs) bring redundancies and noise, making it hard for users and analysts to elicit useful information. In this paper, we formalize the problem from a summarization angle – Continuous Summarization over Microblog Threads (CSMT), which considers three facets: information gain of the microblog dialogue, diversity, and temporal information. This summarization problem is different from the classic ones in two aspects: (i) It is considered over a large-scale, dynamic data with high updating frequency; (ii) the context between microblogs are taken into account. We first prove that the CSMT problem is NP-hard. Then we propose a greedy algorithm with $(1 - 1/e)$ performance guarantee. Finally we extend the greedy algorithm on the sliding window to continuously summarize microblogs for threads. Our experimental results on large-scale datasets show that our method is more superior than other two baselines in terms of summary diversity and information gain, with a close time cost to the best performed baseline.

1 Introduction

Twitter is an online social networking service that enables users to send and read short 140-character messages called "microblogs". This service rapidly gained worldwide popularity, with more than 300 million users posting approximately 500 million tweets a day in 2015. People are thus overwhelmed by a large amount of data and gradually get lost in details without a clear sight of the big picture. However, the big picture is crucial in the decision making process, from marketing plans to political decisions. Therefore, how to summarize tweets and help people get a clear view of the big picture is one of the most important problems in this area, which has already absorbed much attention.

Many tweet summarization approaches have been introduced in recent years [1, 3–6, 10, 12, 14, 18]. In summary, most of them treat tweets as a set of independent documents and aim to model a topic by extracting list of relevant words or

© Springer International Publishing AG 2017
S. Candan et al. (Eds.): DASFAA 2017, Part II, LNCS 10178, pp. 511–526, 2017.
DOI: 10.1007/978-3-319-55699-4_31

sentences. However, microblogs are very short may only provide limited information. These natures actually hinder the existing approaches to extract comprehensible summarizations. Consider an example as follows:

Example 1. As shown in Fig. 1(a), NBA official account posted a tweet T "Lakers vs Celtics 100:115", user A replied to T "lets go celtics, well played team", and user B replied to T "well played team lakers". B got a lot of replies from D, "our guard is such a great player, he carried the whole team" and A got replied from C, "our guard is so good, he activated the whole team", E replied to A:"our coach is such a treasure".

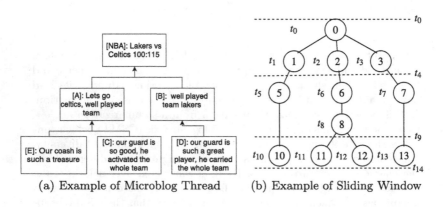

(a) Example of Microblog Thread   (b) Example of Sliding Window

Fig. 1. Examples of microblog thread and sliding window

Traditional summarization methods treat this structure (both the original and interactive tweets) as a bag of microblogs, which will cluster the posts of A and B together because both of them describes NBA teams. Descriptions about guards in different teams may also be clustered and represented as one. However, this process is counter-intuitive, because players belong to different teams and NBA covers the topic "team", and combining them together loses the "parent-child" relationships among topics. Ideally, the summary microblog should contain each individual discussion under the original NBA post, for users who wants to know more under this topic. Thus, we propose to utilize the contextual information in our processing of summarizing microblogs.

Recall the example, the sub-topics are built from the replies and forwards. Together, all the replies and forwards could form contextual links for a whole information unit, which we refer to as a *thread*. Due to the fact that one thread may contain a huge amount of replies, it may be hard for traditional methods because of the efficiency. Moreover, the high updating frequency of social media makes the problem of summarizing the microblog thread (MT) even challenging. In this paper, we focus on extracting microblogs with *broad coverage* of the microblogs and *minimum amount of repetitions* from microblog threads of the microblog stream efficiently.

We hereby formally define the problem as *continuous summarization over microblog threads* (CSMT), which aims at representing a summary of microblogs that appear in the MT from the stream of microblogs.

Although CSMT shares the same spirit of text summarization, the intrinsic characteristics of social media make it different from the classic summarization problem in several aspects: (1) Efficiency. In order to support a lot of users simultaneously requesting summary over a same MT, the efficiency is important, since a high latency may lead to a loss of users in real applications. (2) Continuousness. The MT changes over time as the new microblogs come, summary should update as well in a continuous manner. Thus, users may request a continuous summarization of MT, which brings challenges for traditional static summarization methods. (3) Contexts. The user replies indicate the conversations in a chronological order, which may help users to understand the summary and sub-topics, and thus should be retained. However, the primary social property of microblogs has been largely ignored in the literature. The works closest to this problem are Sumblr [18] and Lex-rank [8].

They have studied continuous tweet summarization problem thoroughly. Their key findings and proposed criteria can also help in solving the scenario of streaming microblogs. However, lex-rank cannot fit the problem on the continuousness aspect very well. Even it could be adopted by using sliding window, the time cost is quite high due to its calculation of the page rank algorithm. Sumblr could fit the continuousness, but it does not consider any contextual aspect of the microblogs.

Contributions of our work are as following:

(1) We first introduce the Continuous Summarization over Microblog Thread (CSMT) problem and prove it is NP-hard.
(2) After proving the submodularity of objective function of CSMT problem, we propose a greedy algorithm that significantly speeds up the calculations providing $(1 - 1/e)$-approximation.
(3) We made experiments to test the performance of the methods proposed in this paper. Experimental results show the information gain, diversity and efficiency of our methods.

The remainder of this paper is organized as follows: Sect. 2 gives a comprehensive review of the related literatures. Section 3 presents the definition of CSMT problem and shows that it is NP-hard. Section 4 first proves the submodularity of CSMT problem function, and then proposes a greedy algorithm with $(1 - 1/e)$-approximation to solve CSMT problem efficiently. Section 5 proposes update of the dialogue coverage and diversity for greedy algorithm. Section 6 presents experimental results and performance studies. At last, Sect. 7 concludes and discusses future work.

2   Related Works

There have been extensive studies on summarizing over microblogs. Broadly speaking, those works fall into three categorise based on different types of features used in the summarization process: (i) *Textual summarization*, which measures the relatedness of two microblogs based textual similarity. (ii) *Temporal summarization*, which mainly considers the temporal factor in the summarization process, and mostly aims at generating the storyline. (iii) *Contextual summarization*, which often employs external resources as a reference and gives a succinct yet highly accurate summarization. We will discuss them shortly.

2.1 Textual Summarization

Summarizing microblogs based only on the text information mostly depend on the semantic information, so the majority of approaches falling in this category use Natural Language Processing (NLP) techniques. Those NLP based approaches extract important sentences and then perform summarization based on the terms co-occurrence in microblogs [6,19]. One state-of-the-art NLP method, Lex-rank [8], also benefits from the network structure, which is constructed based on the sentence similarity. The core idea of Lex-rank is to select important sentence as the component of final summarization. In order to identify important sentences, firstly, Erkan and Radev define the sentence salience and then apply random walk in order to score those sentence based on PageRank mechanism. Finally, top ranked sentences are output and organized as the summarization of the set of considered microblogs.

Intuitively, due to the limited length of a microblog, summarizing over microblogs based only on textual information may suffer from the problem of insufficient context, and thus fails to represent the full story of considered set of microblogs. As there are rich features in microblogs, many researchers consider the problem of summarizing over microblogs beyond the textual feature only.

2.2 Temporal Summarization

One natural feature of microblogs is the temporal information. For example, a popular topic may only appear within some certain period of time. Instead of summarizing over the entire set of microblogs, it is more feasible to perform summarization within this time duration, obtaining the summary of this particular topic, and compensating existing summary established so far.

The idea of summarizing emerging microblogs and update to current summary is close to the task of Twitter Timeline Generation (TTG) [11,13,15] started in TREC 2014 [14]. The task is designed with the aim of performing real-time extraction from Twitter API. The key to accomplishing this task is to model the emerging topics when new tweets appear. There are a few works done based on this idea. For example, Ren et al. [17] proposed probabilistic models in order to describe how topic shifts as time goes by. However, the sophisticated models consume so many computational resources so that the method can hardly be applied to a large-scale size of data. Zhao et al. [20] argued that except textual and temporal information, other features such as social attention should also be considered as one of the most effective that helps in timeline generating task. Their empirical studies show that it is possible to incorporate the classic features used in traditional text mining in order to achieve a better effectiveness. To this end, more researchers started to use more features, such as web link [15], or to perform topic expansion [11] for generating the final timeline.

Although TTG task is close to the problem we are going to study in this paper, most works are done in an off-line style, regarding the set of microblogs as a set of static document collection. This static assumption makes most of the algorithms in this category difficult to be adapted to a dynamic environment. The only work that is apt to the dynamic settings is Sumblr, which is done by Shou et al. [18]. In order to give a high quality summarization, they use a multi-stage strategy, including online clustering, statistics gathering and topic modeling. Although maintaining all useful statistics of all time points may enable Sumblr effectively and efficiently identify summarization over the streamed tweets, the growing size of statistics may become a bottleneck in practice.

2.3 Contextual Summarization

Other important characteristics such as the interactive information among users also attract lots of intention. Because of the short-text problem of social media, single microblogs can only provide limited information.

That contextual information can provide extra evidence for adjusting the results output from machine learning approaches. For example, the contextual information used by Gao et al. [10] is from the news media, providing the reference for topic modeling, which gives reliable summarization. Different from directly using contextual information, Chakrabarti et al. [3] defined sub-events as context and used Hidden Markov Model (HMM) to generate the summary. Similar learning models have also been explored in [1,2,5]. They all benefit from making use of various types of contextual features and improved effectiveness to some extent, based on publicly available TREC[1] datasets.

The work closest to us is done by Wang et al. [19], which utilizes the relationships among users. Those relationships are similar to the interactions we considered in microblogs. Their summarization approaches score individual microblogs based on the linear combination of social and textual metrics. The set of microblogs can maximize the influence will be the summary as results. However, it is known that learning techniques require more time in order to achieve excellent performance for the accuracy of learning patterns, which is infeasible to be applied over a large amount of microblogs.

3   Problem Formulation

In this section, we formally define the Summarization over Microblog Thread (SMT) problem and the Continuous Summarization over Microblog Thread (CSMT) problem. Notice that, SMT is a static version of CSMT, which is very helpful for us to understand CSMT problem. For these purposes, we first give the basic concept of Microblog Thread as preliminaries.

---

[1] http://trec.nist.gov/.

3.1   Representation of the Microblogs

Definition 1 *Microblog Thread.* *A Microblog Thread (MT) is represented as a tree structure* $mt = (N, E, \text{root})$*, where* N *is the set of microblogs* $\{mb_1, mb_2, ..., mb_{|N|}\}$*,* E *is the set of edges representing the interactions between microblogs (e.g., reply, forward) and root is the original microblog. The edges connect those microblogs together as a whole information unit.*

Definition 2 *Microblog Dialogue.* *Microblog Dialogue is a path* $md = \{mb_1, ..., mb_P\}$ *in* mt*, where* mb_1 *is the root, and* mb_P *is a leaf node in* mt*.*

Thread   Dialogue

Fig. 2. Types of microblogs

Figure 2 gives two example to illustrate MT and MD respectively. A path from leaf to root corresponding to an MD, and it can be viewed as a storyline in MT. Therefore, if a set of microblogs can best represent their current dialogues, they should be a good summary for the whole microblog thread. From this point of view, we define the SMT problem as follows:

Problem 1 *Summarization over Microblog Thread (SMT).* *Given a MT which contains a set of microblogs* $N = \{m_0, ..., m_{n-1}\}$*, a* k *which is the desire size of summarization. SMT problem aims to return a set of* k *most representative microblogs.*

$$S = \arg\max{}_{S \in N} G(S), \tag{1}$$

where S *is the summary set and* G *is the objective function that used to quantify the representativeness.*

The core of SMT problem is how to evaluate the representativeness. Clearly, an ideal summary of microblog thread should cover the most information for all dialogues and meanwhile, maintain high diversity. However, quantitative analysis of the information coverage in our problem is challenged, as it is hard to integrate the contextual and textual information into one metric. To overcome this issue, we simplify each dialogue in the same thread as a cluster, and assume that the importance of any dialogue is equal. The coverage of summary can be defined as follows:

Coverage of summary. Let $D_l \in D$ be a dialogue in MT and V_S (or V_l) be the set of keywords that extract from the summary set S (or D_l), the coverage of summary in CSMT is

$$DC_l = \frac{|V_l \cap V_S| + 1}{|V_l| + 1}, V_l \in D_l \tag{2}$$

$$CC(S) = \frac{\sum_{l=1}^{|D|} DC_l}{|D|} \tag{3}$$

Diversity of summary. Diversity of summary metric is to minimize the redundancy of selected microblogs.

$$Div(S) = \frac{\sum_{j \in S} \arg \min_{i \in S, i \neq j} (1 - Jaccard(mb_i, mb_j))}{|S|} \tag{4}$$

So Diversity of any microblog in the current summary set is the dissimilarity with the most similar microblog in the set. Jaccard is the Jaccard similarity of the keywords in two microblogs.

Objective Function. We consider one possible realization of our optimization goal. Because the targeted summarization is obtained by considering both Content Coverage (CC) and Diversity (Div), so a straightforward form is to linearly combine both, which is:

$$G(S) = \alpha \times Div(S) + (1 - \alpha) \times CC(S), \qquad s.t. |S| \leq k, \tag{5}$$

in which $CC(S)$ and $Div(S)$ denote the content coverage and diversity, respectively. α is a user-defined parameter, balancing between the content coverage $CC(S)$ and diversity $Div(S)$. When a large value of α is used, our model favors more diversified results, while a smaller α tends to have higher content coverage.

Definition 3 *Microblog Stream.* *A microblog stream comprises a sequence of microblogs, each denoted by a triple $ms = \langle mt, mb, t \rangle$, where mt and mb represent microblog thread and microblog respectively, t denotes the post time of mb.*

Intuitively, given a microblog stream about a certain microblog thread, our goal is to continuously select a small number of microblogs from the stream that can best represent this thread. Before we introduce the CSMT, we first introduce the concept of sliding window.

Definition 4 *Count Based Sliding Window over Microblog Stream.* *Given W, τ, and microblog stream $M = \langle mb_0, mb_1, ... \rangle$, a sliding window is a window that capture W microblogs in this range on the stream, each time shift τ microblogs in the chronological order.*

For example, it starts at $win_0 = \{mb_0, ...mb_{W-1}\}$, and the next one is after τ, $win_1 = \{mb_\tau, ...mb_{W-1+\tau}\}$. It could also be time based, but we use count based sliding window for instance.

Problem 2 *Continuous   Summarization   over   Microblog   Thread (CSMT). Let $M = \langle mb_0, mb_1, ... \rangle$ be an infinite sequence of microblogs that belongs to an MT in order of arrival time, a sliding window win_i, CSMT aims to continuously feed the users k microblogs that ranked most representative in the sliding window win_i.*

Even though microblogs in the thread come in sequence, each one of them must have a reply or forward relationship with one of the previous microblog. These relationships form the dialogues of the microblog thread. For example, in Fig. 1(b), if $W = 8$ and $\tau = 4$, then $win_1 = \{mb_5, ..., mb_{13}\}$. Moreover, replies of m_{10} to m_5 and others are utilized in the process of our summarization.

Theorem 1. *The Optimization problem of Eq. 5 is NP-hard.*

Proof. The coverage equation is known as the NP-hard as in the dense k-subgraph problem [9]. Moreover, the diversity equation is also proved that it is NP-hard [7]. Thus, the linear combination of both of them must be NP-hard.   □

4   Optimization of Objective Function

The goal of this section is to present algorithmic treatments for CSMT Problem. Specifically, we first prove that CSMT Problem is the instances of the submodular set function maximization with cardinality constraint problem. Based on this, we propose a greedy algorithm with $1 - 1/e$ approximation factor to solve it effectively.

4.1   Submodularity and Greedy Algorithms

Before we proceed, let us give a definition of the non-decreasing submodular set function.

Definition 5 *(SubModularity). A function $f : 2^Q \rightarrow R$ is submodular if for every Set $A \subseteq B \subseteq Q$ and element $i \in Q \backslash B$ it holds that $\Delta(i\|A) \geq \Delta(i\|B)$. Equivalently, a function $f : 2^Q \rightarrow R$ is submodular if for every $A, B \subseteq Q, f(A \cap B) + f(A \cup B) \leq f(A) + f(B)$.*

By the above definition, we show that the objective functions of CSMT Problem is non-decreasing and submodular.

Theorem 2. *The objective function $G(S)$ is monotonically non-decreasing submodular set functions with $G(\varnothing) = 0$.*

Proof. We first prove (P1). For any $T_1 \subset T_2$ and any given example $x \not\subseteq T_2$, we have

$$
\begin{aligned}
E_1 &= G(T_1 \cup x) - G(T_1) \\
&= Div(T_1 \cup x) - Div(T_1) + CC(T_1 \cup x) - CC(T_1) \\
&= \arg\min_{j \in T_1}(1 - Jaccard(mb_x, mb_j)) + CC(T_1) \times \frac{\sum_{k_i \in (DC_l \backslash T_1) \cap x} k_i}{\sum_{k_j \in DC_l} k_j} \quad (6) \\
&= \Delta Div(T_1) + \Delta CC(T_1) \geqslant 0
\end{aligned}
$$

So we can prove that equation is monotonic. We have another: $E_2 = G(T_2 \cup x) - G(T_2)$

$$
\begin{aligned}
E_2 - E_1 &= \Delta Div(T_2) - \Delta Div(T_1) + \Delta CC(T_2) - \Delta CC(T_1) \\
&\leqslant \Delta CC(T_2) - \Delta CC(T_1)
\end{aligned} \quad (7)
$$

Because $T_2 \supset T_1, (DC_l \backslash T_2) \leq (DC_l \backslash T_1)$. So we have $\Delta CC(T_2) - \Delta CC(T_1) \leqslant 0$. Thus $E_2 - E_1 \leqslant 0$. This Eq. 5 is submodular.

4.2   Greedy Algorithm

As discussed above, we present our algorithm for solving this problem and analyses its performance with its complexity.

Our greedy algorithm exploits the two properties of the objective function, monotonicity and submodularity, to have a provably near-optimal solution. The algorithm iteratively expands the selections of microblogs by adding the microblog that maximize the objective function. To have k results, our algorithm requires k iterations.

Algorithm 1. Greedy Algorithm

1: **procedure** GREEDY(mt, k)
2:     //Find the microblog that has the best objective value each time.
3:     $|S| \leftarrow \emptyset,$
4:     **while** $|S| < k$ **do**
5:         $score_{max} \leftarrow 0, mb_t$
6:         **for** $t \in M$ **do**
7:             $score_t \leftarrow \alpha \times Div(S \cup t) + (1 - \alpha) \times CC(S \cup t)$
8:             **if** $score_t > score_{max}$ **then**
9:                 $score_{max} \leftarrow score_t, mb_t \leftarrow t$
10:        $S \leftarrow S.mb_t$
          return S

Greedy Algorithm is a $(1 - 1/e)$ approximation. For any monotone, submodular function f with $f(\emptyset) = 0$ it is known that an iterative procedure which selects the element e with the maximal value of $f(S \cup e) - f(S)$ with S as the elements selected so far has a performance guarantee of $(1 - 1/e) \approx 0.63$ [16].

Complexity. The content coverage could be pre-calculated with a complexity of $\mathcal{O}(|N \times V|)$ and online merging with $\mathcal{O}(|V|)$, where V is the set of keywords used in the Microblog Thread The diversity needs to be online calculated with a complexity of $\mathcal{O}(|N|^2 \times V)$.

5   Continuous Summarization

When the Microblog Threads keep updating, a continuous summarization is needed. In Sect. 4, we proposed our greedy algorithm based on the monotonicity and submodularity of $G()$. However, new microblogs mb will come to the thread and the full set will be changed into $M_1 = M_0 \cup mb$. In order to avoid re-summarizing over M_1, we adapt greedy algorithm to solve the problem of continuous summarization.

The challenges for continuous summarizations are: (1) The content of the dialogue may be changed which means the summary coverage should be re-calculated; and (2) During the calculation of the diversity, the most similar pair of the summary microblogs may be removed. Thus, the new most similar pair needs to be located. To solve the challenges, we propose the continuous calculations of these two metrics.

5.1   Continuous Calculation of Content Coverage

When new microblogs δ_n come, the earliest microblogs δ_o will be removed. Let the full set of MT be N and the summary set be S. If $\delta_o \cap S \neq \varnothing$, then $\forall mb \in S \backslash (S \cap \delta_o) \cup \delta_n$, we need to re-score and then select the top ranked microblogs as our candidates. Since the content coverage is calculated from dialogue coverage, so if $\delta_n \cup \delta_o$ are not belong to any dialogue that has no intersection with S, there is no re-ranking process needed.

Slices of Dialogues. Since the window moves in a step of τ, we can process the summary rank in the unit of τ for dialogues. We slice the microblogs that are in the same τ steps together. For example, in Fig. 2 mb_0, mb_2 are sliced into a group, so are mb_6, mb_8. During the calculations, we store the summary set with the coverage of each τ for each dialogue. When the old set of δ_o are out of the range for sliding window, we just need to reduce the keywords that are unique in δ_o during the calculation of dialogue coverage for the $S \backslash S_o$. In the algorithm, we enumerate the slices instead of the whole set of microblogs. The benefit is when we reduce or add microblogs, the calculation results could be utilized rather than re-calculate all over again. For example, we set the $\tau = 4$ in the Fig. 2. Assume that the window size $W=8$, and $k=3$. First we calculate the summary of mb_0 to mb_8, $S = \{mb_3, mb_2, mb_5\}$. When the sliding window moves to next step, we have mb_0, mb_1, mb_2, mb_3 removed and $mb_{10}, mb_{11}, mb_{12}, mb_{13}$

added. Since mb_3, mb_2 are in the removed set, we will need to generate 3 new summaries. We need to calculate the dialogue coverage of new 4 microblogs. Based on Algorithm 2, during the calculation of the dialogue coverage for mb_{10}, we can store the calculation and reuse it when we update the mb_5.

Algorithm 2. Update of the Dialogue Coverage

1: **procedure** UPDATE OF THE DIALOGUE COVERAGE(S, DC,δ,Div)
2:     //Update the results when the sliding window moves
3:     $S_o \in \delta_o, S_r \leftarrow S \backslash S_o, S_n \in \delta_n$
4:     DC is the set of dialogue coverage for each of the summary.
5:     **for** $i \in \delta_n$ **do**
6:         $DC_i = \frac{V_i}{V_{S_r \cup \delta_n}}, DC_{S_r} = \frac{(V_S \backslash V_{S_o} \cup V_{\delta_n}) \cap (V_{S_r} \cup V_{\delta_n})}{V_{S_r} \cup V_{\delta_n}}$
        return DC

5.2 Continuous Calculation of Summary Diversity

Recall Eq. 4, it is calculated by the aggregated dissimilarities for each microblog in summary with the most similar one in the set. When the δ_o are out of the window, the most similar one could be removed for the $S \backslash \delta_o$. Fortunately, we could solve this problem by storing the most similar pair by slices. For each microblog in the summary set, we store the most similar pair $p_s(i) = \{mb_{slice_0}, ..., mb_{slice_{|W/\tau|}}\}$ in each slice in arrival order. If the dissimilarity value of mb_{slice_j} is larger than $mb_{slice_{j+1}}$, which means the mb_{slice_j} will never be used. Then we can just remove this one, until the dissimilarity of mb_{slice_j} is smaller than $mb_{slice_{j+1}}$. When the δ_n microblogs come, we only update the similar pair set of each $S \backslash S_o$ with the δ_n. For example, let us have $p_s(mb_5) = mb_1, mb_7$, if the first $\tau = 4$ microblogs are out of the sliding window, then the next most similar pair for mb_5 is mb_7. However, if mb_7 is more similar to mb_5 comparing to mb_1, then the existence of mb_1 is useless because it will never be used after mb_7 is out.

6 Experiments

In this section, we evaluate and compare our algorithms against two baselines: (i) Lex-rank [8] and (ii) Sumblr [18]. All algorithms are implemented using Python 2.7. Our experiments are performed on a machine equipped with a 3876 GHz CPU, 24 GiB memory, and running Linux OpenSuse 42.1. All experiments are performed using SNAP Twitter dataset. The dataset is here[2].

---

[2] http://ow.ly/Dh5d307HVGj.

6.1   Experimental Setup

Baseline methods. As is discussed, we can model MT as a bag of microblogs, each of which is independent of each other. In this sense, existing summarization methods can be applied to solve this problem. We consider two state-of-the-art algorithms as our baseline methods: (i) Lex-rank [8] and (ii) Sumblr [18]. The lex-rank method constructs a network based on the cluster of sentence similarity. A random walk method is then applied to the graph, in order to select important sentences based on the eigenvector centrality. Compared to Lex-rank, Sumblr can generate summarization for the entire storyline. This is because this several stage algorithm maintains statistics for all tweets at any arbitrary time point. Individual summarizations at different time points are then combined using a topic model proposed by the authors.

Parameter settings. In our experiments, there are three parameters. The first one is α (in Eq. 5), which is used to balance the weight of dialogue information gain and diversity of selected microblogs. The second parameter is k, which is the number of microblogs to be identified. Usually, a larger k value leads to more iterations in the summarization steps, which may be more time consuming. The last parameter, n, is the total number of microblogs in one MT. We demonstrate the sensitivity of our methods relative to the parameter changing by varying the value listed in Table 1.

Table 1. The Parameter values used, and bolded ones are default values used.

Parameter	Values
α	**0.1**, 0.3, 0.5, 0.7, 0.9
k	**5**, 10, 20, 40, 80
n	50, 60, 70, 80, 90, **100**, 110, 120, 130, 140, 150

Effectiveness measurement. We evaluate the effectiveness of the summarization algorithms from both information gain and the summary diversity. Let v be the keyword, S be the summarization and N be MT, the information gain is calculated as:

$$IG(S) = \frac{|V_S|}{|V_N|}, \tag{8}$$

which measures the keyword coverage between the summarization and the entire MT, $freq(v)$ is the frequency of keyword v. The diversity of the summarization is calculated using Eq. 4.

6.2   Evaluating MT Summarization

Effect of MT size. We measure the effectiveness under the static settings, but the conclusion can be similarly generated under the dynamic setting. Because the two

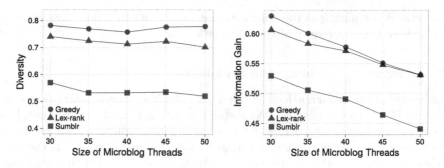

Fig. 3. Effectiveness comparison when $k = 5$. "Greedy" is our proposed method and the other two are considered baselines.

baselines merely consider content coverage, ignoring the fact of diversity, so we set $\alpha = 0.0$ for a fair comparison. Both diversity and information gain are measured relative to the changing of MT size. When $k = 5$, the results are shown in Figure 3.

As is shown in the left of Fig. 3, our model consistently outperforms the two baselines, regardless of the MT size. However, all three models show only slightly observable changes in the measured diversity. Not surprisingly, the information gains of all three methods decrease when the size of MT grows, and when it reaches 50, which is the maximum size measured in this experiments, our methods is close to Lex-rank. Method "Sumblr" performs the worst amongst the three, measured by both diversity and information gain.

Impact of α. As we discussed, the α affects the two measured effectiveness parts, therefore, we explored how they will be affected when α varies, and the results are shown in Fig. 4. The results confirm our idea that α is a knob that trade-offs between information gain and diversity. It is shown in the first of Fig. 4 that when α becomes larger, the diversity also increases, and, correspondingly, the information gain decreases. However, the diversity is more sensitive to

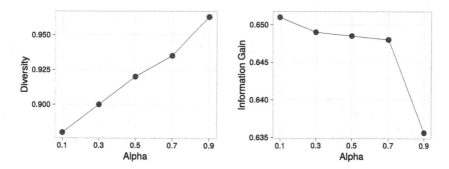

Fig. 4. The impact of α on the effectiveness, each dot is the measured effectiveness averaged across the n values listed in Table 1.

parameter α comparing to the information gain. But when α decreases from 0.7 to 0.9, there is a sharp drop in the information gain.

6.3   Comparing Efficiency and Scalability

Efficiency of summarization. We first consider the time consumption when varying MT size (N), as shown in the first of Fig. 5. The results indicate that, Lex-rank is the most expensive one, because of the random walk process is applied to the entire graph that is constructed from similar sentences. Sumblr is the most efficient algorithm, which is, not surprisingly, because this algorithm maintains microblogs statistics at arbitrary time point. Our proposed method is close to Sumblr and consistently outperforms Lex-rank. However, we don't maintain all statistical information which makes our algorithm more practical in coping with streaming data. Moreover, considering the effectiveness measured in Fig. 3, our algorithm achieves a better effectiveness with a little sacrifice in the efficiency.

Impact of k. We fix all our other settings to default values in Table 1 and then proceed to explore the time cost w.r.t. parameter k, as is shown in the RHS of Fig. 5. Because of k represents the number of desired representative microblogs, therefore, the time cost should increase as k goes larger, which is confirmed in our experiments.

Impact of parameters in the continuous setting. As we defined, the sliding window settings in CSMT problem contain two parameters: step size (τ) and window size (W). Same as previous, we fixed all of our other parameters to the default values listed in Table 1. The trend in Fig. 5 still holds in this experiments: that Lex-rank is the most expensive solution; that the Sumblr performs the best and our proposed method is close to the best performance.

In the first figure, the τ has a huge impact on Lex-rank while another two methods are relatively stable to this change. When varying the window size, all three methods show measurable changes in terms of efficiency. Still, the Lex-rank

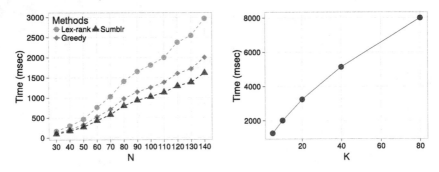

Fig. 5. Total time cost of summarization algorithms. The first figure shows the time cost (millisec) on all MTs, when varying MT sizes (labeled as N). The second figure shows the impact of parameter k on the time cost.

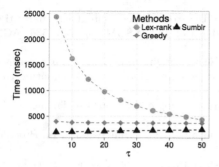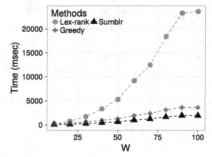

Fig. 6. Impact of τ (LHS) and W (RHS) on the efficiency of solutions to continuous summarization. The time cost is averaged by the total number of MTs.

is the most sensitive one relative to the parameter, while our proposed method is competitive to the best one, especially when a smaller window size is used (Fig. 6).

7 Conclusion

This paper proposed techniques to support a user to continuously get the summaries of a microblog thread in the microblog stream. Based on the different scenarios, we formally defined two problems: *Summarization over Microblog Thread* and *Continuous Summarization over Microblog Thread*. Both of them are related to the selection of a subset that could maximum the dialogue coverage and diversity. We showed that both problems are intractable. Hence, we developed greedy algorithm for the approximation with performance guarantees. Our evaluation showed that our technique outperforms two baselines Lex-rank and Sumblr on information gain and diversity, without the lose of much efficiency.

In future work, we plan to explore from two directions: incorporating relationships between the dialogues into more sophisticated models for higher effectiveness, and we will consider to apply stream based algorithms in order to improve the efficiency.

Acknowledgement. This work was partially supported by ARC DP170102726, DP170102231 and National Natural Science Foundation of China (NSFC) 91646204.

References

1. Bian, J., Yang, Y., Chua, T.-S.: Multimedia summarization for trending topics in microblogs. In: Proceedings of the CIKM, pp. 1807–1812 (2013)
2. Bian, J., Yang, Y., Zhang, H., Chua, T.-S.: Multimedia summarization for social events in microblog stream. IEEE Trans. Multimedia **17**(2), 216 (2015)
3. Chakrabarti, D., Punera, K.: Event summarization using tweets. In: ICWSM, vol. 11, pp. 66–73 (2011)

4. Chang, Y., Wang, X., Mei, Q., Liu, Y.: Towards twitter context summarization with user influence models. In: Proceedings of the WSDM, pp. 527–536. ACM (2013)

5. Chen, Y., Zhang, X., Li, Z., Ng, J.P.: Search engine reinforced semi-supervised classification and graph-based summarization of microblogs. Neurocomputing **152**, 274–286 (2015)

6. Chua, F., Asur, S.: Automatic summarization of events from social media. In: ICWSM (2013)

7. Drosou, M., Pitoura, E.: Dynamic diversification of continuous data. In: Proceedings of the EDBT, pp. 216–227 (2012)

8. Erkan, G., Radev, D.R.: Lexrank: graph-based lexical centrality as salience in text summarization. J. Artif. Intell. Res. **22**, 457–479 (2004)

9. Feige, U., Peleg, D., Kortsarz, G.: The dense k-subgraph problem. Algorithmica **29**(3), 410–421 (2001)

10. Gao, W., Li, P., Darwish, K.: Joint topic modeling for event summarization across news and social media streams. In: Proceedings of the CIKM, pp. 1173–1182 (2012)

11. Hasanain, M., Elsayed, T.: QU at TREC-2014: online clustering with temporal and topical expansion for tweet timeline generation. Technical report (2014)

12. Khan, M., Bollegala, D., Liu, G.: Multi-tweet summarization of real-time events. In: Proceedings of the SocialCom, pp. 128–133 (2013)

13. Li, J., Cardie, C.: Timeline generation: tracking individuals on twitter. In: Proceedings of the WWW, pp. 643–652 (2014)

14. Lin, J., Efron, M., Wang, Y., Sherman, G.: Overview of the TREC-2014 Microblog track. In: Proceedings of the TREC (2014)

15. Magdy, W., Gao, W., Elganainy, T., Wei, Z.: QCRI at TREC 2014: applying the kiss principle for the TTG task in the microblog track. Technical report (2014)

16. Nemhauser, G.L., Wolsey, L.A., Fisher, M.L.: An analysis of approximations for maximizing submodular set functions. Math. Program. **14**(1), 265–294 (1978)

17. Ren, Z., Liang, S., Meij, E., de Rijke, M.: Personalized time-aware tweets summarization. In: Proceedings of the SIGIR, pp. 513–522 (2013)

18. Shou, L., Wang, Z., Chen, K., Chen, G.: Sumblr: continuous summarization of evolving tweet streams. In: Proceedings of the SIGIR, pp. 533–542 (2013)

19. Wang, C., Yu, X., Li, Y., Zhai, C., Han, J.: Content coverage maximization on word networks for hierarchical topic summarization. In: Proceedings of the CIKM, pp. 249–258 (2013)

20. Zhao, X.W., Guo, Y., Yan, R., He, Y., Li, X.: Timeline generation with social attention. In: Proceedings of the SIGIR, pp. 1061–1064 (2013)

Drawing Density Core-Sets from Incomplete Relational Data

Yongnan Liu$^{(\boxtimes)}$, Jianzhong Li, and Hong Gao

Harbin Institute of Technology, Harbin, China
{liuyn,lijzh,honggao}@hit.edu.cn

Abstract. Incompleteness is a ubiquitous issue and brings challenges
to answer queries with completeness guaranteed. A density core-set is
a subset of an incomplete dataset, whose completeness is approximate
to the completeness of the entire dataset. Density core-sets are effective
mechanisms to estimate completeness of queries on incomplete datasets.
This paper studies the problems of drawing density core-sets on incomplete relational data. To the best of our knowledge, there is no such proposal in the past. (1) We study the problems of drawing density core-sets
in different requirements, and prove the problems are all NP-Complete
whether functional dependencies are given. (2) An efficient approximate
algorithm to draw an approximate density core-set is proposed, where an
approximate Knapsack algorithm and weighted sampling techniques are
employed to select important candidate tuples. (3) Analysis of the proposed approximate algorithm shows the relative error between completeness of the approximate density core-set and that of a density core-set
with same size is within a given relative error bound with high probability. (4) Experiments on both real-world and synthetic datasets demonstrate the effectiveness and efficiency of the algorithm.

Keywords: Data quality · Density core-sets · Incomplete data · Query
completeness estimation

1 Introduction

Incompleteness is a ubiquitous issue [24] on relational datasets, which severely
affects business [26], and brings new challenges to operations over them [4,7,11,
23] as well. Besides time complexity of processing queries on incomplete data,
completeness of query answers [23] is an emerging data quality problem. Low
completeness of query answers leads to biased answers to queries, which may
do harm to follow-up or potential applications [11]. To fix the missing values,
data imputation [25] or data repairing [12] methods are extensively studied. But
usually, such methods are continuous time-consuming tasks, and the number of
missing values can be fixed can not be estimated ahead. Therefore, queries on
incomplete data can not be answered fast with completeness guaranteed using
these methods.

© Springer International Publishing AG 2017
S. Candan et al. (Eds.): DASFAA 2017, Part II, LNCS 10178, pp. 527–542, 2017.
DOI: 10.1007/978-3-319-55699-4_32

In this paper, small subset, drawn from original dataset, is proposed to estimate the completeness of queries on incomplete data fast. Given queries, such small subset provides ratio of complete values fast, which is approximate to the ratio of complete values on the original dataset. Such small subset is called a *density core-set*. There may be many density core-sets on the original dataset, and any of them can give similar completeness information. Density core-sets are like indexes on datasets, which are drawn for frequently posed queries. With a density core-set, methods based on bootstrap [20] can estimate completeness of queries quickly.

Besides the size constraint of density core-sets, there are other challenges to draw density core-sets. To estimate accurately, a density core-set should contain various tuples. Moreover, density core-sets should consider *real* complete values, since some missing values can be determined by taking advantage of functional dependencies [5]. Usually, there is only one candidate for a missing value in some databases, such as scientific databases. And that there are multiple candidates for one missing value is not considered in this paper. Such number of real complete values can give a lower bound of number of missing values repaired by rule-based repairing methods [12]. Therefore, completeness of a density core-set can be used to estimate completeness of query answers and improvement of query answers by some repairing methods, which avoids useless queries and data repairing or imputation methods.

The idea of answering queries based on core-sets stems from [2]. But existing methods [2,15,19] to draw core-sets only consider numeric attribute values, which cannot be used in recommendation systems [10] including categoric attribute values. Moreover, these methods do not consider challenges from missing values, which cannot be used to estimate completeness of query answers on incomplete data.

Methods based on sampling [3,8,9,22] can only process aggregate queries on numeric attribute values. Therefore, these methods lose many features of different categories of tuples, which cannot answer queries like *group-by*. Congressional sampling [1] can answer group-by queries, but like other sampling-based methods, does not consider biases from missing values on the samples. Therefore, some samples cannot provide sufficient information for query answers.

To overcome the drawbacks of existing methods, we propose density core-sets to approximately show the completeness of tuples of interest. By taking advantage of functional dependencies, a scoring function is designed to measure the number of real complete values and a cost function is designed to measure the cost of fixing missing values. When missing values are missing completely at random [21], an efficient approximate algorithm is proposed to draw an approximate density core-set, where approximate equi-depth histograms [6] are constructed to estimate some real complete values efficiently, and a sampling strategy based on an approximate knapsack algorithm [16] controls the relative error of completeness between the approximate density core-set and a density core-set with the same size. The main contributions of this paper are summarized as follows:

1. Density core-sets used to show completeness of tuples of interest are firstly investigated, where functional dependencies can be employed to accurately show real completeness of a dataset. The problems of drawing a density core-set in different requirements are proved to be NP-Complete.
2. An efficient approximate algorithm to draw an approximate density core-set is proposed, where relative error of completeness between the drawn approximate density core-set and a density core-set with the same size is proved to be within given error bound with high probability.
3. Experiments over real-world and synthetic datasets demonstrate the proposed algorithm effective and efficient.

2   Problem Definition and Computational Complexity

In this section, we first introduce a definition of a density core-set, and then we propose problems of drawing a density core-set under different properties, followed by theorems giving time complexity of each problem.

To give definitions, we introduce following symbols. The density core-set of a dataset T is denoted by T_{CD}, and the completeness of the dataset T, denoted as $Cpl\,(T)$, is defined as below:

$$Cpl\,(T) = \frac{\sum\limits_{t \in T} \sum\limits_{A} C\,(A)}{mn}$$

where $C(A)$ denote the completeness of an attribute A: $C(A)$ is 1, if the value of attribute A on a tuple t is determined as not missing given a functional dependencies set F, otherwise $C(A)$ is 0. And m is the number of attributes on the dataset T, where the size of T is n, i.e. $|T| = n$. The symbol T^d denotes the dataset where another d tuples are inserted into the dataset T.

Definition 1 (Density Core-set-A). *Given a relational dataset T, the attributes $A = \{A_j\}$, $1 \leq j \leq m$, a functional dependencies set F, $\epsilon \in (0,1]$, and positive integers d and k, a density core-set T_{CD} is a subset of original dataset with minimum size satisfying:*

1. *(Property A) the completeness of the density core-set is approximate to that of the entire dataset, i.e., $|Cpl\,(T) - Cpl\,(T_{CD})| \leq \epsilon$,*
2. *(Property B) there are at least k attributes, on which there is at least one complete value,*
3. *(Property C) the completeness of current density core-set can be improved after d tuples are read, i.e., $Cpl\,(T_{CD}^d) \geq Cpl\,(T_{CD})$.*

In the definition above, the property A is the density property, which contributes to estimation of completeness of query answers on incomplete data, and property B makes a density core-set contain different categories of tuples, and property C makes a density core-set can provide more accurate answers if more time cost is permitted.

Property C seems to increase the hardness of drawing a density core-set, and sometimes no extra time is provided. Therefore, without property C in the first definition, a new density core-set can be defined, which is called *Density Core-set-B*. Though property C is removed, the hardness of drawing such a density core-set is proved to be not easier.

Property B seems to only keep one complete value, which may not contain sufficient information under some applications. Therefore, one may ask for more fruitful complete values on density core-sets, which gives rise to another new density core-set, called *Density Core-set-C*, by only changing property B in Definition 1 into a new property B' that there are at least k attributes, on which there are at least h complete values, with an extra input positive integer h.

The problem of drawing a density core-set is to draw a Density Core-set-A, Density Core-set-B, or Density Core-set-C from original dataset under different properties. The hardness of the three problems is given below. Due to space limit, proofs are omitted. The three problems can be reduced from *SetCover* problem with tuples constructed from given sets.

Theorem 1. *Whether functional dependencies are given or not, drawing a Density Core-set-A of size N is NP-Complete.*

Theorem 2. *Whether functional dependencies are given or not, drawing a Density Core-set-B of size N is NP-Complete.*

Theorem 3. *Whether functional dependencies are given or not, drawing a Density Core-set-C of size N is NP-Complete.*

3 Drawing an Approximate Density Core-Set

Since the problem of drawing a density core-set under different properties is NP-Complete, we shall propose an efficient approximate algorithm to draw an approximate density core-set. With given high probability, the relative error between completeness of drawn approximate density core-set and a density core-set with the same size is within the given relative error bound if the assumption of random missing [21] holds.

Intuitively, though some missing values can be determined by given functional dependencies (FDs for short), it takes a long time to determine all the missing values. Though there are strategies, such as blocking or grouping, can cluster tuples sharing the same values in the left-hand-side of an FD, it's hard to choose grouping keys to make the cluster small enough to determine the missing values fast, due to large size of various values on categoric attributes. To efficiently draw an approximate density core-set, only some missing values should be determined. Moreover, it's necessary to draw sufficient tuples to guarantee that the relative error is not too large.

Therefore, approximate equi-depth histograms [6] are efficiently constructed to control the number of missing values determined. And an approximate knapsack algorithm is employed to select sufficient tuples for small relative error of completeness with a scoring function and a cost function designed based on given FDs.

In this section, two functions based on given FDs are designed to measure importance of a tuple, which are used for the approximate knapsack algorithm. Based on this, an approximate algorithm ADC with approximate histograms constructed are proposed to draw an approximate density core-set efficiently. And then, the analysis of the proposed approximate algorithm is shown.

3.1   Measuring Importance of a Tuple

In this subsection, a scoring function and a cost function based on given FDs are designed to measure the completeness of a tuple, and the time cost of determining all the missing values.

A Scoring Function. Given FDs, a tuple t can be partitioned into three parts on attributes. The first part, denoted as $A(t)$, contains the values of attributes not involved in any FD, where the number of complete values is denoted as $C^A(t)$. And the missing values in this part cannot be determined easily unless more knowledge is provided. The second part, denoted as $Key(t)$, contains the values of attributes involved in the left-hand side of an FD, where the number of complete values is denoted as $C^{Key}(t)$. And if any one of such values is missing, the tuple cannot find other tuples sharing the same values of left-hand side attributes on the FD. The third part, denoted as $Val(t)$, contains the values of attributes involved in the right-hand side of an FD, where the number of complete values is denoted as $C^{Val}(t)$. And missing values can be determined by complete values in other tuples sharing the same $Val(t)$. Therefore, a score of a tuple is a weighted sum of the number of complete values in the three parts. But due to the analysis of the proposed algorithm shown below, the weight of the three should be 1 to guarantee that the property A holds. Therefore, given FDs, a score of a tuple, denoted as $V(t)$, is given by the function below:

$$V(t) = C^A(t) + C^{Key}(t) + C^{Val}(t) \qquad (1)$$

Since some missing values will be determined in the approximate histograms, the score gives an estimate of *real* completeness of a tuple, which contributes to the real completeness of the approximate density core-set.

A Cost Function. Since determining all the missing values by given FDs takes a long time, only some of them should be determined accurately by the approximate histograms based on given FDs, which makes many tuples still contain missing values. To estimate the completeness accurately, tuples with higher score and fewer missing values should be selected for the approximate density core-set.

Due to the analysis of the proposed algorithm shown below, the cost of a tuple should be number of missing attribute values in the tuple plus 1. But to invoke the approximate knapsack algorithm [16], the cost should be normalized. In [16], the authors consider an approximate knapsack problem of picking maximum sum of profits of items with given total weights no more than a constant, where total weight of input items is normalized to 1.

3.2 An Approximate Algorithm

With the importance of each tuple measured, and some extra parameters used in the construction of approximate equi-depth histograms, an approximate algorithm ADC is proposed in Algorithm 1. Since not all the missing values are determined, the real completeness of the entire dataset cannot be known, and a density core-set has to be drawn based on such incomplete information, which makes completeness of a density core-set can be below the real completeness of the entire dataset. But as is shown in the analysis in Subsect. 3.3, with sufficient tuples, relative error between completeness of the approximate density core-set and completeness of a density core-set can be small.

Algorithm 1 gives steps to draw an approximate density core-set. Step (1) the algorithm first constructs an approximate equi-depth histogram for each FD, and the missing values are contained in such a histogram; Step (2) the missing values in the first half of each bucket are determined accurately after sorting the values in the same bucket, since tuples sharing the same value of left-hand side of the same FD contain the same value of right-hand side of the same FD. The values that are not determined accurately are called *tail* values; Step (3) both the scores and costs are computed by the functions given in Subsect. 3.1. And the scores and costs are normalized for the approximate knapsack algorithm [16]; Step (4) given an error bound ϵ, probability δ, and other information from normalized costs and scores, the size of tuples for constructing a new Knapsack instance can be computed. By algorithm Knapsack in [16], the approximate Knapsack algorithm can find *better* tuples by their scores and costs, where sample size is at least $\lceil 1000\, \epsilon^{-4} \log \epsilon^{-1} \rceil$, given the constraint $WeightBound$ on the total weight. But to make the relative error small, the sample size in the approximate knapsack algorithm Knapsack should also be at least $\frac{\phi_{\delta/2}^2}{\epsilon^2} \left(\frac{1}{\inf(h_j)} \times \frac{1}{N} \times \frac{\sup(h_j)}{\inf(s_j)} - 1 \right)$ as the analysis shown in Subsect. 3.3, where $\phi_{\delta/2}$ is the $\delta/2$ fractile of the standard normal distribution, and $\inf(h_j)$ and $\sup(h_j)$ are the minimum and maximum number of the complete tail values in a tuple respectively, and $\inf(s_j)$ is the minimum cost and N is the number of total tuples in D. Then the algorithm Knapsack returns small size of tuples whose total score is within the given relative error bound. And such tuples are the approximate density core-set, whose completeness is within the given relative error bound as the analysis shows in Subsect. 3.3. The size of approximate density core-set can be very small relative to the entire dataset as the experimental results show in Sect. 4, since the algorithm Knapsack reconstructs a new knapsack problem instance.

Note that in Step (1) of Algorithm 1, the parameters ϵ_h and δ_h are used to control the differences between the heights of buckets in an approximate equi-depth histogram and the heights of buckets in an equi-depth histogram. Since

---

Algorithm 1. ADC

Input : a relational dataset D, where $|D| = N$, a functional dependencies set
F, real number $\epsilon_h, \epsilon > 0$, $\delta_h, \delta > 0$, $WeightBound > 0$, integer $k > 0$

Output: An approximate density core-set Sol

1: $Sol \leftarrow \phi$;

// Construct approximate histograms for each FD

2: **for** each tuple t in D **do**

3:     **for** each FD f in F **do**

4:         Sample enough tuples, as in Theorem 5 in [6], to find separator values between buckets;

5:         Construct an approximate equi-depth histogram H based on the value of left-hand side of f, with buckets number k, error bound ϵ_h and probability δ_h;

// Determine some missing values

6: **for** each histogram H **do**

7:     **for** each bucket b in first half of all buckets in H **do**

8:         sort the values in b, determine the missing values;

9:         for the last bucket, if the next buckets containing the same value of left-hand side as in the last bucket, determine them as well;

// Measure importance of each tuple

10: **for** each tuple t in D **do**

11:     compute score and cost of t according to functions in Subsect. 3.1;

12: Normalize the score and cost of each tuple such that both total score and total cost are 1, and the normalized set is denoted as I_{Nor};

// Draw an approximate density core-set

13: As is shown in [16], the Algorithm Knapsack takes as input
$(X = (I_{Nor}, WeightBound), \epsilon)$, where the sample size $|I|$ in the algorithm should be changed, which satisfies

$$N \geq |I| \geq \max\left(\lceil 1000\epsilon^{-4}\log\epsilon^{-1}\rceil, \frac{\phi_{\delta/2}^2}{\epsilon^2}\left(\frac{1}{\inf(h_j)} \times \frac{1}{N} \times \frac{\sup(h_j)}{\inf(s_j)} - 1\right)\right), \text{ and}$$

returns the approximate solution Sol;

14: **return** Sol;

---

the number of tuples to be determined accurately in Line 6 is also an approximate number, the differences are not necessarily very small. For example, in experiments ϵ_h is set to be 0.4 and δ_h is set to be 0.4. In Step (2) of Algorithm 1, only first half of missing values is accurately determined for efficiency. With more missing values determined, the density core-set can be better approximate to the real completeness of the entire dataset, but more time is required, which may make it impossible to fast estimate the completeness of a query. The constraint $WeightBound$ on total weight cost controls the costs of tuples, which also provides a way to control the size of the approximate density core-set. Since h_j is the number of complete tail values in a tuple, $\inf(h_j)$ is at least 1, and $\sup(h_j)$ is at most m.

3.3   Analysis of the Approximate Algorithm

In this subsection, we shall analysis the relative error between the approximate density core-set and a density core-set with the same size. To show the analysis result, an assumption of random missing and a definition of (ϵ, δ) - approximation are introduced. And based on these, two theorems will be proposed.

An Assumption of Random Missing. If values are randomly missing [21], then the tail values are randomly missing, so the average number of complete values in the tail values is approximate to the average number of complete values in the tail values in a density core-set, which is the case that there are few values in the tail values, or the tail values are almost complete. The assumption can be satisfied on some dataset from relative high-completeness database such as Wikipedia and DBLP.

Therefore, the completeness of an approximate density core-set can be defined as

$$\hat{C} = \frac{\hat{V} + |S| \times l \times \hat{L}}{|S| \times m} = \frac{\hat{V}}{|S| \times m} + \frac{l \times \hat{L}}{m} \tag{2}$$

where $\hat{V}$ is the completeness of the values that are not in the tail values as in the scoring function in Eq. (1), and $\hat{L}$ is the average number of complete tail values in a tuple, and l is the average number of tail values contained in a tuple, and S is the approximate density core-set with m attributes. And the completeness of a density core-set with the same size is defined as

$$C = \frac{V + |S| \times l \times L}{|S| \times m} \tag{3}$$

where the variables are the counterpart in a density core-set with the same size $|S|$.

Definition 2. $((\epsilon, \delta)$ - approximation) $\hat{I}$ is called as an (ϵ, δ) - approximation of I if $\Pr\left(\left|\frac{\hat{I} - I}{I}\right| \geq \epsilon\right) \leq \delta$ for any $\epsilon \geq 0$ and $0 \leq \delta \leq 1$, where $Pr(X)$ is the probability of a random event X.

Two Estimators. Since the completeness of the approximate density core-set can be partitioned into two variables: V and L, and the first variable has an $(\epsilon, \frac{1}{3})$ - approximation [16], we should design an (ϵ, δ) - approximation estimator for the second variable. With the assumption of random missing, an estimator of the second variable is given below:

$$\hat{L} = \frac{1}{|S| \times N} \times \sum_{j=1}^{|S|} \frac{h_j}{s_j}$$

where variables are defined as in Subsect. 3.2. Sufficient tuples should be sampled to make $\hat{L}$ is an (ϵ, δ) - approximation of L. And the size of sampled tuples is given in Theorem 4.

Theorem 4. *The estimator $\hat{L}$ defined as below is an (ϵ, δ) - approximation of L, $\hat{L} = \frac{1}{|S| \times N} \times \sum_{j=1}^{|S|} \frac{h_j}{s_j}$, if the size of sampled tuples, denoted as $|S|$, satisfies:*

$$|S| \geqslant \frac{\phi_{\delta/2}^2}{\epsilon^2} \left(\frac{1}{\inf(h_j)} \times \frac{1}{N} \times \frac{\sup(h_j)}{\inf(s_j)} - 1 \right)$$

where variables are defined as in Subsect. 3.2.

Combining Two Estimators. By setting proper parameters in the two estimators, an (ϵ, δ) - approximation of the completeness of a density core-set can be obtained, as is shown in Theorem 5.

Theorem 5. *If $\hat{V}$ is an $(\epsilon, \frac{1}{3})$ - approximation of V, and $\hat{L}$ is an $(\epsilon, \delta - \frac{1}{3})$ - approximation of L, then $\hat{C}$ as in Eq. (2) is an (ϵ, δ) - approximation of C as in Eq. (3). where*

$$C = \frac{V + |S| \times l \times L}{|S| \times m}$$

Theorem 6. *Time complexity of algorithm ADC is $O(N)$, where N is the number of tuples in the dataset.*

4   Experimental Results

In this section, we will show the experimental results of the proposed algorithm ADC on different datasets with different parameters. All the experiments are implemented on a Microsoft Windows 7 machine with an Intel(R) Core i5-2400 CPU 3.1 GHz and 4 GB main memory. Programs are compiled by Microsoft Visual Studio 2013 with C++ language. Each program is run 10 times on the same dataset to show stability, and the time cost given below is the average time cost. The completeness is defined as in Sect. 2.

4.1   Datasets

DBLP Proceedings. The real-world data set is extracted from DBLP[1] dataset of size 1.72 G. We built a relation *Conf* consisting of all extracted conference information and a relation *Proc* consisting of all extracted proceeding information. Based on these two relations without changing any value, we built the relation *DBLP* for the experiments. The statistics of relations used in the experiments are shown in Table 1. The DBLP consists of 8 attributes, which shows the *Author* and the *Title* of a paper, the *Conf* where the paper presented, the *Year* when the conference was held, the title of the proceedings (*Btitle*), the *Editor*

---

[1] DBLP data from http://dblp.uni-trier.de/xml/. Since DBLP is always updating, the data set was downloaded on July 30, 2016.

of the proceedings and the press where the proceedings were published(Pub). And ID is an identifier of each tuple. There are three functional dependencies: ($Conf$, $Year$) → $Btitle$, ($Conf$, $Year$) → Pub, ($Conf$, $Year$) → $Editor$. The completeness of the relation Conf is 0.948, and the completeness of the relation Proc is 0.996. The completeness of the relation DBLP is 0.938, while the real completeness given by taking advantage of FDs is 0.992.

Table 1. Statistics of relations used in the experiments

Relation name	# of tuples	# of attributes	Size (×1 MB)
Conf	30794	5	7.8
Proc	1819273	6	378.1
DBLP	1819273	8	425.1

Cars Information. The relation $Cars$ are crawled from a website[2] providing services on cars. Without changing any value, we built the relation Cars including information from dealers selling new cars. The relation Cars consists of 8 attributes of a dealer: name, stock, star-rate, ratings, location, distance from the main company, phone number, and email. There are 542850 tuples in this relation, and the size is 162 MB. To investigate performance of our algorithm without FDs, no FDs are considered in this relation. But to evaluate all parts of our algorithm, a trivial FD: $name$ → $name$ is used. The completeness of the relation Cars is 0.843, which is also the real completeness, since there are actually no FDs.

Synthetic Relations. To evaluate efficiency of our algorithm ADC, we generate a collection of synthetic relations with size varying from 1310 K to 394 M. The synthetic datasets are generated in two steps. First, there are 2×10^4 tuples with complete attribute values; Second, there are other tuples with incomplete attribute values, where a value at any attribute is missing with probability 0.2. And there are different numbers of FDs on different attributes on different dataset. Detailed information are shown in Subsect. 4.3.

4.2   Experiments on Real Datasets

In this subsection, we shall show the experimental results on the two real datasets: DBLP and Cars. We shall evaluate our algorithm by three measures: accuracy, efficiency and size of the approximate density core-set. *Accuracy* measures the relative error between the completeness of the approximate density core-set and the real completeness of the entire dataset, since the completeness of the approximate density core-set is approximate to the completeness of a density

---

[2] http://www.cars.com.

core-set with the same size. *Efficiency* measures the time cost of our algorithm ADC, and *Size* measures the size of the approximate density core-set drawn by our algorithm ADC, which also measures the possibility that the approximate density core-set contains various tuples. Since the parameter *WeightBound* borrowed from the knapsack algorithm controls the efficiency and size, experiments with different WeightBounds are conducted to observe the impacts of Weight-Bound on the three measures. Since approximate equi-depth histograms are used to control approximate number of tuples determined by given FDs, ϵ_h is set to be 0.4, which is f in [6], δ_h is set to be 0.4, and buckets number k is 10000 in all the experiments on the two real datasets, which makes the approximate equi-depth histograms contribute to the three measures equally and efficiently. To tackle the problems from the incomplete data, a simple method is to remove the tuples containing missing values, which is the first step of the baseline algorithm to prepare complete tuples for density core-sets selection. Since there are no other algorithms to select approximate density core-sets, the baseline algorithm invokes our algorithm ADC to select approximate density core-sets.

Accuracy. To investigate accuracy of our algorithm given relative error bounds, we ran a group of experiments, as is shown in Fig. 1(a), where WeightBound is set to be 1.0 both in DBLP and Cars, and the failure probability δ is set to be 0.41. The required relative error bound shown varies from 0.01 to 0.1, which is actually from 0.07 to 0.7, since algorithm Knapsack should be invoked by given $\frac{\epsilon}{7}$ to obtain a ϵ error bound with the optimal value [16].

As is depicted in Fig. 1, actual relative errors are all within the required relative error bound. As the required relative error bounds get bigger, the actual errors become bigger. Notice that the actual error on Cars is smaller than that on DBLP, since there are more missing values can be determined by given FDs on DBLP. The baseline algorithm obtain smaller actual error on both datsets, but there are two points should be noticed. One is that the actual error trends of both datasets are not obvious, which makes it hard to estimate the actual error, and the other is that the actual completeness of DBLP is very close to 1, which makes it easier to get a smaller actual error.

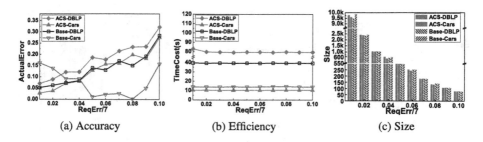

(a) Accuracy   (b) Efficiency   (c) Size

Fig. 1. Experimental results on DBLP and Cars with required relative error varying

Efficiency. To investigate efficiency of our algorithm, we ran a group of experiments with the same parameters as in Accuracy, and the results are depicted in Fig. 1(b).

As is depicted in Fig. 1(b), the values of time cost are almost the same on each dataset. Actually, the Knapsack only constructs a few items for further computation to draw an approximate density core-set, which takes a little time compared to the whole time cost. But other steps take too much time, which makes the time cost vary a little. For example, in Fig. 1(b), when required error shown is 0.04 on the DBLP dataset, only 600 new items are constructed by Knapsack, and they are all drawn as an approximate density core-set, which takes only a little time, i.e., 3.791 s, compared to the total time cost 79.166 s. Therefore, if the first three steps can be finished offline, then the approximate density core-set can be drawn online, which makes it possible to estimate the completeness of a given query fast. The baseline algorithm takes less time since there are no missing values to determine.

Size. To investigate the sizes of approximate density core-sets drawn by ADC, we ran a group of experiments with the same parameters as in Accuracy, and the results are depicted in Fig. 1(c).

As is depicted in Fig. 1(c), the size becomes smaller as the required error bound grows. Since the larger required error bound makes the items constructed by Knapsack become smaller [16], there are fewer candidate tuples, which makes the sizes of approximate density core-sets even smaller. The sizes of approximate density core-sets obtained by the baseline algorithm are almost the same as our algorithm, since the two algorithms share the same selection algorithm. Notice that sometimes the sizes of approximate density core-sets are even smaller than the size given by Theorem 4. Because the algorithm Knapsack reconstructs a new knapsack problem instance, whose completeness can be estimated by smaller tuples.

Impacts of WeightBound. As mentioned above, the parameter WeightBound controls the sizes of approximate density core-sets as well. To investigate the impacts of WeightBound on the three measures, we ran a group of experiments

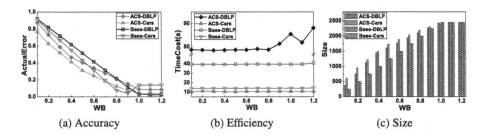

(a) Accuracy   (b) Efficiency   (c) Size

Fig. 2. Experimental results on DBLP and Cars with WeightBound varying

on the two real datasets with WeightBound varying from 0.1 to 1.2. ϵ is set to be 0.02 and δ is set to be 0.41. The results are depicted in Fig. 2.

As is depicted in Fig. 2(a), the actual error is getting smaller with Weight-Bound grows, and stays at the same value when WeightBound is too big, which begins after 1.0. Since bigger WeightBound permits more samples, the actual error become smaller. And this stops at a large WeightBound, since all the tuples reconstructed are selected. Since the baseline algorithm actually estimates the completeness of complete dataset, i.e., 1 actually, the trends of actual error are not obvious, which makes it hard to give a proper WeightBound.

Figure 2(b) depicts the values of time cost. Though WeightBound can control the size of an approximate density core-set, the first three steps take most of the time cost, which makes the time cost almost the same, as is mentioned above.

The sizes of each approximate density core-set are depicted in Fig. 2(c). As the WeightBound grows bigger, more tuples are permitted into the approximate density core-set on both datasets. Bigger WeightBound permits more samples, but this stops at a too large WeightBound, since all the tuples reconstructed are selected.

4.3   Experiments on Synthetic Relations

In this subsection, we shall show the experimental results on synthetic relations. There are two groups of datasets. One contains 8 attributes, and the other contains 10 attributes, denoted as Att-8 and Att-10 respectively. To evaluate scalability of our algorithm, two groups of dataset were generated with tuples from 2×10^4 to 4×10^6. For all experiments on synthetic datasets, ϵ is set to be 0.02, which mean the relative error bound is 0.14, δ is 0.41, WeightBound is set to be 0.95, ϵ_h is set to be 0.4, δ_h is set to be 0.4, and buckets number k is 10000. Three FDs are specified between attributes. And the experimental results of the three measures are depicted in Fig. 3, where the x-axis shows the logarithmic results of the tuples' number, and only the y-axis in Fig. 3(b) shows the logarithmic results of the TimeCost.

As is shown in Fig. 3(a), all the values of actual error are in the required error bound, i.e., 0.14. The values of actual error slightly change as the number of tuples grows, since missing values are randomly introduced.

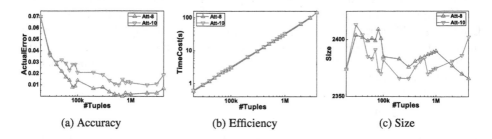

(a) Accuracy   (b) Efficiency   (c) Size

Fig. 3. Experimental results on synthetic datasets

As is shown in Fig. 3(b), the time cost grows slowly, as is observed above. Since the first three steps take most of the time, and they are all in linear complexity, the time cost grows almost linearly. Since there are two more attributes, experiments on Att-10 take more time.

The sizes of approximate density core-sets are depicted in Fig. 3(c), the size almost stays the same as the tuples' number grows. Because the number of candidates for the approximate density core-sets is computed by the required error bound ϵ, which is fixed in all the experiments, and WeightBound which controls the number of size is fixed as well. There are more attributes on the tuples in Att-10 datasets, which makes the scores and costs are smaller in normalization. Therefore slightly more tuples are permitted given a fixed WeightBound. At most 2413 tuples are drawn as the approximate density core-set, which is a very small number compared to entire tuples' number, for example 4×10^6.

5 Related Work

Numeric Core-Sets. The idea of answering queries by core-sets is from [2], and many methods of drawing various core-sets are surveyed in [19]. Merging two core-sets with coverage preserved is studied in [15]. Drawing dominating set to answer some queries are studied in [8]. There are two main drawbacks of methods above. One is that they can only draw numeric core-sets, and the other is that many values are aggregated, which makes many features described in the origin dataset removed.

Sampling Methods for Approximate Query Answers. There are many sampling methods to select small subset of entire dataset to answer queries approximately, such as [1,3,9,22]. On complete data, sample sizes can be provided by these researches in that each sample can provide effective information for approximately answering queries. But on incomplete data, not every sample contains effective information, which makes it hard to determine sample sizes for kinds of queries.

Completeness Estimation. Many researchers focus on the completeness of queries over a database, *i.e.* they decide whether a query over a given database can get a complete answer with different kinds of extra knowledge [13,14,17,18,23]. Their problems can be called *query completeness* for short. The computational complexity of query completeness is usually beyond NP-Complete, which makes it very hard to fast estimate completeness of a given query on the entire dataset. A density core-set is only a small subset of the entire dataset, which makes it easier to estimate completeness of a query fast by methods based on bootstrap.

6 Conclusion

This paper studies the problems of drawing density core-sets on incomplete data. We analyze the time complexity of the problems and propose an algorithm to draw an approximate density core-set. We prove that the relative error of completeness between the approximate density core-set and a density core-set with the same size is within given error bound with high probability. Experimental results on both real-world and synthetic datasets show effectiveness and scalability of our algorithms. With an approximate density core-set, the completeness of a given query can be estimated fast.

Acknowledgments. This work is supported in part by the Key Research and Development Plan of National Ministry of Science and Technology under grant No. 2016YFB1000703, and the Key Program of the National Natural Science Foundation of China under Grant No. 61190115, 61632010 and U1509216.

References

1. Acharya, S., Gibbons, P.B., Poosala, V.: Congressional samples for approximate answering of group-by queries. In: ACM SIGMOD Record, vol. 29, pp. 487–498. ACM (2000)
2. Agarwal, P.K., Har-Peled, S., Varadarajan, K.R.: Approximating extent measures of points. J. ACM **51**(4), 606–635 (2004)
3. Agarwal, S., Mozafari, B., Panda, A., Milner, H., Madden, S., Stoica, I.: BlinkDB: queries with bounded errors and bounded response times on very large data. In: Proceedings of the 8th ACM European Conference on Computer Systems, pp. 29–42. ACM (2013)
4. Arocena, P.C., Glavic, B., Miller, R.J.: Value invention in data exchange. In: Proceedings of the 2013 ACM SIGMOD International Conference on Management of Data, pp. 157–168. ACM (2013)
5. Beskales, G., Ilyas, I.F., Golab, L., Galiullin, A.: Sampling from repairs of conditional functional dependency violations. VLDB J. **23**(1), 103–128 (2014)
6. Chaudhuri, S., Motwani, R., Narasayya, V.: Random sampling for histogram construction: how much is enough? ACM SIGMOD Rec. **27**, 436–447 (1998). ACM
7. Chen, K., Chen, H., Conway, N., Hellerstein, J.M., Parikh, T.S.: Usher: improving data quality with dynamic forms. IEEE Trans. Knowl. Data Eng. **23**(8), 1138–1153 (2011)
8. Cheng, S., Cai, Z., Li, J., Fang, X.: Drawing dominant dataset from big sensory data in wireless sensor networks. In: 2015 IEEE Conference on Computer Communications (INFOCOM), pp. 531–539. IEEE (2015)
9. Cormode, G., Garofalakis, M., Haas, P.J., Jermaine, C.: Synopses for massive data: samples, histograms, wavelets, sketches. Found. Trends Databases **4**(1–3), 1–294 (2012)
10. Deng, T., Fan, W., Geerts, F.: On recommendation problems beyond points of interest. Inf. Syst. **48**, 64–88 (2015)
11. Dong, X.L., Gabrilovich, E., Murphy, K., Dang, V., Horn, W., Lugaresi, C., Sun, S., Zhang, W.: Knowledge-based trust: estimating the trustworthiness of web sources. Proc. VLDB Endow. **8**(9), 938–949 (2015)
12. Fan, W.: Dependencies revisited for improving data quality. In: Proceedings of the Twenty-Seventh ACM SIGMOD-SIGACT-SIGART Symposium on Principles of Database Systems, pp. 159–170. ACM (2008)

13. Fan, W., Geerts, F.: Capturing missing tuples and missing values. In: Proceedings of the Twenty-Ninth ACM SIGMOD-SIGACT-SIGART Symposium on Principles of Database Systems, pp. 169–178. ACM, June 2010

14. Fan, W., Geerts, F.: Relative information completeness. ACM Trans. Database Syst. **35**(4), 27 (2010)

15. Indyk, P., Mahabadi, S., Mahdian, M., Mirrokni, V.S.: Composable core-sets for diversity and coverage maximization. In: Proceedings of the 33rd ACM SIGMOD-SIGACT-SIGART Symposium on Principles of Database Systems, PODS 2014, pp. 100–108. ACM (2014)

16. Ito, H., Kiyoshima, S., Yoshida, Y.: Constant-time approximation algorithms for the knapsack problem. In: Theory and Applications of Models of Computation, pp. 131–142 (2012)

17. Levy, A.Y.: Obtaining complete answers from incomplete databases. In: Proceedings of the 22th International Conference on Very Large Data Bases, pp. 402–412. Morgan Kaufmann Publishers Inc. (1996)

18. Motro, A.: Integrity = validity + completeness. ACM Trans. Database Syst. **14**(4), 480–502 (1989)

19. Phillips, J.M.: Coresets and sketches. http://arxiv.org/abs/1601.00617

20. Pol, A., Jermaine, C.: Relational confidence bounds are easy with the bootstrap. In: Proceedings of the 2005 ACM SIGMOD International Conference on Management of Data, pp. 587–598. ACM (2005)

21. Poleto, F.Z., Singer, J.M., Paulino, C.D.: Missing data mechanisms and their implications on the analysis of categorical data. Stat. Comput. **21**(1), 31–43 (2011)

22. Potti, N., Patel, J.M.: DAQ: a new paradigm for approximate query processing. Proc. VLDB Endow. **8**(9), 898–909 (2015)

23. Razniewski, S., Nutt, W.: Completeness of queries over incomplete databases. Proc. VLDB Endow. **4**(11), 749–760 (2011)

24. Saha, B., Srivastava, D.: Data quality: the other face of big data. In: 2014 IEEE 30th International Conference on Data Engineering (ICDE), pp. 1294–1297. IEEE (2014)

25. Song, S., Zhang, A., Chen, L., Wang, J.: Enriching data imputation with extensive similarity neighbors. Proc. VLDB Endow. **8**(11), 1286–1297 (2015)

26. Wayne, W.: Data quality and the bottom line: achieving business success through a commitment to high quality data. The Data warehouse Institute (TDWI) report (2004). www.dw-institute.com

Big Data (Industrial)

Co-training an Improved Recurrent Neural Network with Probability Statistic Models for Named Entity Recognition

Yueqing Sun[1], Lin Li[1]([✉]), Zhongwei Xie[1], Qing Xie[1], Xin Li[2],
and Guandong Xu[3]

[1] School of Computer Science and Technlogy, Wuhan University of Technology,
Wuhan, China
{yqsuan,cathylilin,kevinsnest,felixxq}@whut.edu.cn
[2] iFLYTEK Big Data Research Institute, Hefei 230088, China
xinli2@iflytex.com
[3] School of Software, University of Technology Sydney, Ultimo 2007, Australia
Guandong.Xu@ust.edu.au

Abstract. Named Entity Recognition (NER) is a subtask of information extraction in Natural Language Processing (NLP) field and thus
being wildly studied. Currently Recurrent Neural Network (RNN) has
become a popular way to do NER task, but it needs a lot of train data.
The lack of labeled train data is one of the hard problems and traditional
co-training strategy is a way to alleviate it. In this paper, we consider
this situation and focus on doing NER with co-training using RNN and
two probability statistic models i.e. Hidden Markov Model (HMM) and
Conditional Random Field (CRF). We proposed a modified RNN model
by redefining its activation function. Compared to traditional sigmoid
function, our new function avoids saturation to some degree and makes
its output scope very close to [0, 1], thus improving recognition accuracy. Our experiments are conducted ATIS benchmark. First, supervised
learning using those models are compared when using different train data
size. The experimental results show that it is not necessary to use whole
data, even small part of train data can also get good performance. Then,
we compare the results of our modified RNN with original RNN. 0.5%
improvement is obtained. Last, we compare the co-training results. HMM
and CRF get higher improvement than RNN after co-training. Moreover,
using our modified RNN in co-training, their performances are improved
further.

Keywords: Named entity recognition · Co-training · Recurrent neural
network · Probability statistic model · Natural language processing

---

This research project is supported by the National Social Science Foundation of
China (Grant No:15BGL048), National Natural Science Foundation of China (Grant
No:61602353, 61303029), 863 Program (2015AA015403), Hubei Province Science and
Technology Support Project (2015BAA072).

S. Candan et al. (Eds.): DASFAA 2017, Part II, LNCS 10178, pp. 545–555, 2017.
DOI: 10.1007/978-3-319-55699-4_33

1   Introduction

NER is a fundamental step in Natural Language Processing (NLP) which aims to identify boundaries and type of entities in text. In big data time, plenty of valuable information lies in disordered raw texts that cannot be directly used for many tasks. By doing NER we can know which category each word belongs. This technology is useful in information extraction (IE) field. Hence NER has been an essential task in several research teams, such as the Message Understanding Conferences (MUC), the Conferences on Natural Language Learning (CoNLL), etc. [1]. Also, in industry, google brain and baidu brain are very hot now and have been used in many specific applications. For example, in college entrance examination robot plan, a kind of brain-simulation program, NER is a vital subtask. However, there are some problems that should be noted:

1. The difficulty of feature-design. Most of NER researches commonly based on traditional machine learning methods. They often rely on the construction of complex hand-designed features which are derived from various linguistic analyses and maybe only adapted to specified area [2].
2. The lack of labeled data. Many NLP tasks base on big data and need large corpus especially labeled data, so is NER. But compared to the oceans of raw data that is produced every day, data with labels is in urgently lack.

For problem 1, a modified deep learning architecture named RNN is proposed and compared with two popular probability statistical models i.e. HMM and CRF. The two statistical models can learn statistical rules from a large number of training samples, so as to make predictions about the unknown. RNN belongs to deep learning which is a branch of machine learning and a development of neural network. It shakes off the requirement of hand-designed features and frees people from complex templates design. As described in [3], for tasks that involve sequential inputs, such as speech and language, it is often better to use RNN. In this paper, we modify the RNN activation function since selection of a good activation function is an important part to design a neural network. The experimental results prove that our modification gets a better achievement.

For problem 2, we utilize a co-training strategy, a kind of semi-supervised learning for the situation when train data is much less than test data. Co-training, originally proposed by A. Blum and T. Mitchell [4], is a popular strategy in semi-supervised learning. In this paper we cotrain RNN with the above two statistical models by selecting data with high confidence level to update the train set. Experimental results show that after co-training, all the models are improved.

The rest of this paper is organized as follows: Sect. 2 introduces the related works about NER researches. Section 3 describes our improved RNN and the co-training strategy. Experiments and result analysis are shown in Sect. 4. Finally, conclusion and future work are discussed in Sect. 5.

2   Related Work

As described in [5], there are three kind of methods for named entity recognition: dictionary-based methods, rule-based methods and statistical machine learning methods which rely on different theories. NER can be solved by machine learning methods, such as CRF [6,7], Support Vector Machine (SVM) [8], HMM [9] etc. These methods are commonly used for NER these years in a way of supervised learning. In addition, semi-supervised methods are also one road to this task when labeled data is difficulty to obtain.

Recently, while the probability statistical models perform well in many fields, deep neural networks as a new wave tide in machine learning, have achieved great performances in many domains such as image classification [10], knowledge discovery [11] and translation [12] etc. Collobert et al. [13] propose a unified neural network architecture and learning algorithm to do various NLP tasks and also achieved a better result for NER task. Compared to the well-known Convolutional Neural Network (CNN) which has achieved remarkable performances in image domain, RNN can exploit the time-connection feedback thus capture dependencies beyond the input window. Therefore, RNN architecture is more suitable for NER. Song et al. [14] build a simple and efficient system for bio-NER based on Recurrent Neural Network (RNN). Jason P.C. Chiu and Eric Nichols [15] present a novel neural network architecture that can automatically detect word and character level features using a hybrid bidirectional Long Short-Term Memory (LSTM) and CNN architecture.

On the other hand, as described in [16], a deep neural network is characterized by a set of weight matrices, bias vectors, and a nonlinear activation function, which gives a deep neural network the learning ability of hierarchical nonlinear mapping. But in model parameter training, weight matrices and bias vectors are updated using an error back-propagation algorithm whereas activation function is not. So the change of activation function is important for a neural network, which can speed up model training [17], enhance stability [18]. In this paper, we adopt the RNN model and modify its activation function to do NER task.

Another problem for RNN is that it needs plenty of train data. Hence in this paper we consider a co-training method which is one of useful solutions when train data is in lack. Co-training, one of the semi-supervised learning methods, was first proposed in 1998 and also has been used in NER. Tsendsuren et al. [19] present an Active Co-Training (ACT) algorithm for biomedical named-entity recognition. Li et al. [20] propose a semi-supervised approach to extract bilingual named entity and used a bilingual co-training algorithm to improve the named entity annotation quality. But using RNN to do co-training is a few in NER researches [21] and most of them are about biomedical domain. In this paper, we aim to explore the performance when co-training an improved RNN with probability statistic models for NER task.

3   Methodology

3.1   RNN

RNN has been applied in many fields and got great achievements in recent years. In this paper, we propose an improved RNN and use it to do co-training for the NER task. As one of the most successful and well-known neural network, CNN has made the remarkable achievements in multiple cross domains so it is valuable to try deep learning method for NER. Contrast to CNN, for text and language processing, RNN is proved to be good. It is a neural network model whose architecture can exploit the time-connection feedback [22]. Generally speaking, deep convolutional nets have brought breakthroughs in processing of image, video and audio etc., whereas recurrent nets have shone light on sequential data such as text and speech. A RNN and its unfolding structure in time of the computation involved in its forward computation is shown in Fig. 1. When unfolded, RNN can be regarded as a deep feedforward network and each layer shares same weights.

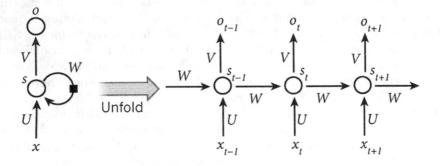

Fig. 1. A RNN and its unfold state

RNNs have many architectures and variants such as Elman-type and Jordan-type. Mesnil et al. [23] have implemented and compared the above two important RNN architectures to investigate spoken language understanding. Our RNN is based on the Elman-type described in [23] in this paper and we amend its activation function. Actually, as described in [16], many rectifier-type nonlinear functions have been proposed as activation functions, but the best nonlinear functions for any given task domain remain unknown. A same activation function performance may differ dramatically when applying it to different tasks. As for NER, compared to other well-known activation functions i.e. tanh, ReLu, PReLu, the sigmoid function, defined as Eq. 1, shows best in our experiments.

$$F(x) = \frac{1}{1 + e^{-x}} \tag{1}$$

However sigmoid function has the saturation phenomenon (derivative tends to zero when the argument x approaches infinity) both at its left and right, which may make the training process harder. Still, sigmoid is most similar to the

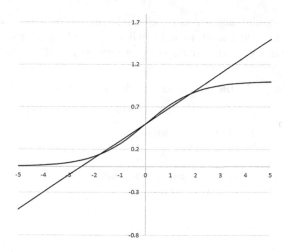

Fig. 2. Sigmoid and its linear approximation function

reflex mechanism of biological neuron and its output is always between 0 and 1, which can represent the label's prediction probability. The linear approximation function of sigmoid, expressed by Eq. 2, also performs well in our experiments. Figure 2 shows the two functions more intuitively.

$$L(x) = 0.2 \times x + 0.5 \tag{2}$$

Focusing on our NER task, we propose to combine the two above functions as our new activation function (shown in Eq. 3). Parameters a and b are coefficients which determined by experimental performances.

$$A(x) = a \times F(x) + b \times L(x) \tag{3}$$

The new activation function ameliorates the sigmoid's saturation phenomenon on the one hand and smooths the linear function on the other hand. Our co-training results using the new activation function are better than that using sigmoid or linear sigmoid function.

3.2 Co-training

In this paper, we modify activation function in deep neural network. On the other hand, we aim to explore the effect when co-training with RNN. As is known to all, co-training was proposed early years ago and wildly adopted in many tasks. To our best of knowledge, co-training using RNN is a few in NER [19,20]. Co-training is a kind of strategy in semi-supervised learning which fits for the situation when train data is limited. It uses two (or more) learners (model A and B). Wit the first same input as training data, according to different learning rules, learners produce labeled data respectively. Then the A's new labeled k data with highest confidence levels is selected and added into learner B's train set, vice versa. It will do this iteration until unlabeled data are all tagged. The co-training algorithm used in our paper is described as follows in Table 1.

Table 1. The co-training algorithm

Input:
A small set T of original labeled samples.
A big set U of unlabeled samples.
Test set V.
Classifiers C_1 and C_2 and their train set s_1 and s_2.
Number k selected data for each iteration.
Output:
bestC_1, bestC_2
Initialization:
$s_1=s_2=T$
k=300
$r_1=r_2=0$ //initialize the test results when testing V by classifiers
do:
modelA=C_1.train(s_1) //train classifiers with train set and save the model
modelB=C_2.train(s_2)
tempR_1= modelA.test(V)
tempR_2= modelB.test(V)
if temp$R_1 > r_1$ or temp$R_2 > r_2$:
$r_1=$ tempR_1
$r_2=$ tempR_2
bestC_1=modelA
bestC_2=modelB
newLabeledDataA=modelA.predict(U) //tag the unlabeled data
newLabeledDataB=modelB.predict(U)
newTrainA=newLabeledDataA.getTop(k)// select k new labeled data
newTrainB=newLabeledDataB.getTop(k)
s_1.add(newTrainA) //update train set
s_2.add(newTrainB)
U.remove(newTrainA)
U.remove(newTrainB)
until U.isEmpty()
return bestC_1, bestC_2

4   Experiments

We conduct two sets of experiments. One is supervised learning and the other is semi-supervised learning with co-training, to make comparisons and explore the situation that using RNN to do co-training. The two experiments' train data sizes were totally different since semi-supervised learning works in the situation that training data is much less than testing data.

4.1   Dataset and Evaluation

Both the two sets of experiments are based on standard Airline Travel Information Systems (ATIS) benchmark [23] which contains 127 classes and uses the in/out/begin (IOB) representation. For example, a sentence can be expressed as in Table 2.

Table 2. Sentences in ATIS

sentence	find	flight	from	memphis	to	tacoma	dinner
label	O	O	O	B-fromloc.city_name	O	B-toloc.city_name	B-meal_description
sentence	cost	of	limousine	service	at	logan	airport
label	O	O	B-transport_type	O	O	B-toloc.airport_name	I-airport_name

All the results were evaluated by precision (P), recall (R) and F1-score, defined by Eqs. 4, 5 and 6 respectively.

$$P = \frac{TP}{TP + FP} \tag{4}$$

$$R = \frac{TP}{TP + FN} \tag{5}$$

$$F1 = \frac{2 \times P \times R}{P + R} \tag{6}$$

TP means true positives, FP means false positives and FN means false negatives.

4.2   Experiment 1: Supervised Learning

In this part we train four NER models as supervised learning by using different training data size. Based on ATIS original train/test proportion, i.e., train/test sets were 3983/893 sentences, we randomly select 20%, 40%, 60%, 80% and 100% of total train data (3983 sentences) as training set to train the HMM, CRF and RNN (including our modified RNN) models. The results through K-fold cross-validations are shown in Table 3. K is different based on the different size of training set. For example, when using 20%, K is five, and when using 40% and 60%, K is three. In Table 3 we can observe that how the three models perform when training on different data size. Generally speaking, the performances of all the models in NER are gradually improved when the training data size becomes larger. However, when the training data size goes larger and larger, the percentage of improvement becomes smaller. Actually, when the training data size increases from 20% to 40%, these models generally get a highest improvement. After that, increasing data only brings little benefit. This gives the reason that semi-supervised learning is feasible to achieve good results when train data is less. From Table 3, it shows that our modified RNN performs better than original RNN when train data is less than test data.

Table 3. P, R and F1 on different training data size

Training data size	20%	40%	60%	80%	100%
Models	P, R, F	P, R, F	P, R, F	P, R, F	P, R, F
HMM	0.6575, 0.6295, 0.6432	0.6776, 0.6588, 0.6680	0.6874, 0.6700, 0.6786	0.6914, 0.6750, 0.6831	0.6959, 0.6806, 0.6882
Improvements	-	**3.05%, 4.66%, 3.86%**	1.45%, 1.7%, 1.59%	0.58%, 0.75%, 0.66%	0.65%, 0.83%, 0.75%
CRF	0.9015, 0.7870, 0.8419	0.9252, 0.8543, 0.8884	0.9264, 0.8697, 0.8972	0.9295, 0.8863, 0.9074	0.9317, 0.8950, 0.9130
Improvements	-	**2.63%, 8.55%, 5.52%**	0.13%, 1.80%, 0.99%	0.33%, 1.91%, 1.14%	0.24%, 0.98%, 0.62%
RNN	0.8965, 0.8871, 0.8917	0.9350, 0.9311, 0.9333	0.9389, 0.9363, 0.9376	0.9475, 0.9415, 0.9445	0.9517, 0.9369, 0.9442
Improvements	-	**4.30%, 4.96%, 4.67%**	0.42%, 0.56%, 0.46%	0.92%, 0.56%, 0.74%	0.44%, −0.49%, −0.03%
Our Modified RNN	0.9030, 0.8992, 0.9011	0.9436, 0.9376, 0.9406	0.9380, 0.9383, 0.9381	0.9426, 0.9376, 0.9401	0.9527, 0.9450, 0.9489
Improvements	-	**4.50%, 4.27%, 4.38%**	−0.59%, 0.07%, −0.27%	0.49%, −0.07%, 0.21%	1.07%, 0.79%, 0.94%

4.3   Experiment 2: Semi-supervised Learning

To do semi-supervised learning, we assume that the labeled data are further less than the unlabeled. Thus here we reorganize the whole training data set (4876 sentences) and randomly select 1000 sentences about 20.5% as our new training set and the left as unlabeled data set to do co-training. In each iteration 300 high confidence level samples are picked from learner A and B, respectively. Those selected 300 samples are labeled by learner A or B and will be added into B or A as training set. Here we have done two group co-training: (A = HMM, B = RNN) and (A = CRF, B = RNN).

Co-training Using Original RNN with HMM and CRF. First we use the original RNN to do the above two group co-training. The before/after co-training performances are shown in Figs. 3 and 4. Since Precision and Recall show similar changing curve with F1 scores, we only report F1 scores in Figs. 3 and 4. After co-training, generally speaking, both HMM and CRF performs better and better with increasing iterations. But RNN need less iterations to achieve highest F1 scores. For example, in Fig. 3, when iterations go to 3, RNN shows best performance and its F1 score is 0.9129. We can say that RNN, a deep learning method, is good at NER and much better than traditional HMM and CRF. In addition, RNN helps them to obtain higher F1 scores by using RNN as a learner in co-training.

Through co-training, the recognition performances of two probability statistic models (here is HMM and CRF) are improved. For example, in Fig. 3 before co-training, the F1 score of HMM is 0.6236. After co-training with RNN, its F1 score rises to 0.6833 with 9.6% improvement. For CRF in Fig. 4, the largest

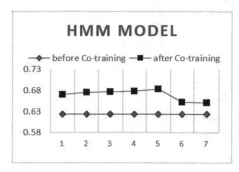

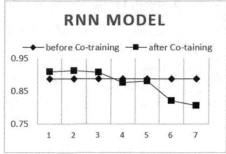

Fig. 3. Co-training results of HMM and RNN where X axis represents iteration times and Y axis represents F1 scores.

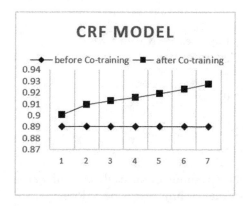

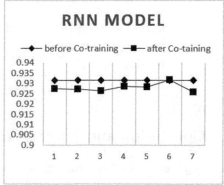

Fig. 4. Co-training results of CRF and RNN where X axis represents iteration times and Y axis represents F1 scores.

improvement is 4.14%. (0.8907 VS. 0.9276). In a word, by using co-training we can achieve better results with less training data, which give a solution when labeled data is in lack.

Co-training Using Improved RNN with HMM and CRF. In this part the improved RNN is used to redo the two group co-training. First we did several check experiments when setting a and b the different values according to their corresponding function curve trend and the (0.8, 0.2) pair is proved best in our experiments. Thus we set the new activation function coefficients a = 0.8 and b = 0.2 here. Compared to the results of co-training with original RNN, HMM and CRF get a little more improvement when co-training with improved RNN. The comparison results are reported in Table 4.

From Table 4, we can see that our improved RNN using modified activation function show better performance than HMM and CRF and the highest improvement is 5.4% compared with original RNN. Although HMM and CRF benefit a little from modified activation function, improved RNN obtains larger F1 score.

Table 4. Co-training using improved RNN with HMM and CRF

Model	HMM with RNN	HMM with Improved RNN
F1	0.6786	0.6811
Iteration	5	5
Improvement		0.4%
Model	RNN with HMM	Improved RNN with HMM
F1	0.8528	0.8987
Iteration	8	8
Improvement		5.4%
Model	CRF with RNN	CRF with Improved RNN
F1	0.8921	0.8966
Iteration	2	2
Improvement		0.5%
Model	RNN with CRF	Improved RNN with CRF
F1	0.8983	0.9008
Iteration	7	7
Improvement		0.3%

5   Conclusion and Future Work

In this paper, we consider the situation of less training data, study the influences that data sizes make on the model performance improvements. Moreover, we conduct the co-training experiments using original RNN and our modified RNN. The results of supervised learning indicate that even small train data size can get pretty good or even better achievement than that when data is bigger. The results of semi-supervised learning show that using RNN in co-training for NER task can achieve better performances when training data is less than testing data. In the future, it is worth to combine the co-training with RNN or other deep neural networks. In addition, we only change the activation function here and in the future, we are going to explore the RNN more deeply, for example improving its architecture to do NER or other related tasks.

References

1. Wahiba, B.A.K.: Named entity recognition using web document corpus. CoRR abs/1102.5728 (2011)
2. Lishuang, L., Liuke, J., Zhenchao, J., et al.: Biomedical named entity recognition based on extended Recurrent Neural Networks. In: BIBM, pp. 649–652 (2015)
3. LeCun, Y., Bengio, Y., Hinton, G.: Deep learning. Nature **521**(7553), 436–444 (2015)
4. Blum, A., Mitchell, T.: Combining labeled and unlabeled data with co-training. In: Eleventh Conference on Computational Learning Theory, pp. 92–100 (1998)

5. Li, L., Fan, W., Huang, D., et al.: Boosting performance of gene mention tagging system by hybrid methods. J. Biomed. Inform. **45**(1), 156–164 (2012)
6. Padmaja, S., Utpal, S., Jugal, K.: Named entity recognition in Assamese using CRFS and rules. In: IALP, pp. 15–18 (2014)
7. Tang, Z., Lingang, J., Yang, L., et al.: CRFs based parallel biomedical named entity recognition algorithm employing MapReduce framework. Cluster Comput. **18**(2), 493–505 (2015)
8. Ki-Joong, L., Young-Sook, H., Kim, S., et al.: Biomedical named entity recognition using two-phase model based on SVMs. J. Biomed. Inform. **37**(6), 436–447 (2004)
9. Gayen, V., Sarkar, K.: An HMM based named entity recognition system for indian languages: the JU system at ICON 2013. CoRR abs/1405.7397 (2014)
10. Sladojevic, S., Arsenovic, M., Anderia, A., et al.: Deep neural networks based recognition of plant diseases by leaf image classification. Comp. Int. Neurosc. **2016**(6), 1–11 (2016)
11. Janosek, M., Voln, E., Kotyrba, M.: Knowledge discovery in dynamic data using neural networks. Cluster Comput. **18**(4), 1411–1421 (2015)
12. Chollampatt, S., Kaveh, T., Hwee, T.N.: Neural network translation models for grammatical error correction. In: IJCAI, pp. 2768–2774 (2016)
13. Collobert, R., Weston, J., Bottou, L., et al.: Natural language processing (almost) from scratch. Mach. Learn. Res. **12**, 2493–2537 (2011)
14. Dingxin, S., Lishuang, L., Liuke, J., et al.: Biomedical named entity recognition based on recurrent neural networks with different extended methods. IJDMB **16**(1), 17–31 (2016)
15. Chiu, J.P.C., Nichols, E.: Named entity recognition with bidirectional LSTM-CNNs. TACL **4**, 357–370 (2016)
16. Hoon, C., Sung, J.L., Jeon, G.P.: Deep neural network using trainable activation functions. In: IJCNN, pp. 348–352 (2016)
17. Anhao, X., Qingwei, Z., Yonghong, Y.: Speeding up deep neural networks in speech recognition with piecewise quantized sigmoidal activation function. IEICE Trans. **99-D**(10), 2558–2561 (2016)
18. Liew, S.S., Khalil-Hani, M., Bakhteri, R.: Bounded activation functions for enhanced training stability of deep neural networks on visual pattern recognition problems. Neurocomputing **216**, 718–734 (2016)
19. Tsendsuren, M., Meijing, L., Unil, Y., et al.: An active co-training algorithm for biomedical named-entity recognition. JIPS **8**(4), 575–588 (2012)
20. Li, Y., Huang, H., Zhao, X., Shi, S.: Named entity recognition based on bilingual co-training. In: Liu, P., Su, Q. (eds.) CLSW 2013. LNCS (LNAI), vol. 8229, pp. 480–489. Springer, Heidelberg (2013). doi:10.1007/978-3-642-45185-0_50
21. Qikang, W., Tao, C., Ruifeng, X., et al.: Disease named entity recognition by combining conditional random fields and bidirectional recurrent neural networks. In: Database (2016)
22. Mikolov, T., Kara_t, M., Burget, L., et al.: Recurrent neural network based language model. In: INTERSPEECH, pp. 1045–1048 (2010)
23. Mesnil, G., He, X., Deng, L., et al.: Investigation of recurrent neural network architectures and learning methods for spoken language understanding. In: INTERSPEECH, pp. 3771–3775 (2013)

EtherQL: A Query Layer for Blockchain System

Yang Li[1], Kai Zheng[1](✉), Ying Yan[2], Qi Liu[2], and Xiaofang Zhou[1,3]

[1] School of Computer Science and Technology, Soochow University, Suzhou, China
yli6@stu.suda.edu.cn, zhengkai@suda.edu.cn
[2] Microsoft Research, Beijing, China
ying.yan@microsoft.com, v-lqii@microsoft.com
[3] The University of Queensland, Brisbane, Australia
zxf@itee.uq.edu.au

Abstract. Blockchain - the innovation behind Bitcoin - enables people to exchange digital money with complete trust, and seems to be completely transforming the way we think about trust. While blockchain is designed for secured, immutable funds transfer in trustless and decentralized environment, the underlying storage of blockchain is very simple with only limited supports for data access. Moreover, blockchain data are highly compressed before flushing to hard disk, making it harder to have an insight of these valuable data set. In this work, we develop EtherQL, an efficient query layer for Ethereum – the most representative open-source blockchain system. EtherQL provides highly efficient query primitives for analyzing blockchain data, including range queries and top-k queries, which can be integrated with other applications with much flexibility. Moreover, EtherQL is designed to provide different levels of abstraction, which are suitable for data analysts, researchers and application developers.

1 Introduction

Cryptocurrencies such as Bitcoin and its successors, serve as a new form of digital currencies, facilitating instant payment to anyone in a decentralized manner [18]. As the underlying technology of cryptocurrencies, blockchain is a distributed ledger that records all the transactions and states in the entire network. Blockchain consists of a series of blocks that are chained together by keeping a reference to the previous block. A block is mainly composed of a block header, which holds the metadata of the block such as timestamp and the hash code of the previous block, and a block body containing the corresponding list of transactions recorded in that block [25]. Each full node in the network maintains a copy of the ledger, and various consensus algorithms such as Proof of Work [6], Proof of Stake [5] and PBFT [11] are adopted to achieve data consistency.

As a concept of architecture, blockchain can be implemented in various ways. At current stage, there are two most influential blockchain platforms, Bitcoin [18] and Ethereum [25]. While Bitcoin gains its popularity for being the first decentralized asset transfer platform, Ethereum, as an extension of Bitcoin, is intended to implement a Turing-complete computation platform on top of blockchain for

© Springer International Publishing AG 2017
S. Candan et al. (Eds.): DASFAA 2017, Part II, LNCS 10178, pp. 556–567, 2017.
DOI: 10.1007/978-3-319-55699-4_34

decentralized applications. A lot of applications such as asset transfer, Internet of Thing, clearing & settlement are implemented on Ethereum [10].

Ethereum stores the block data in a simple key/value data store (in most cases, LevelDB [20] is used). LevelDB is a fast in-process database with excellent writing performance, which has the ability to process a large amount of data and eliminates the need to run another database software separately. Block data are automatically compressed before synchronized to the disk, which makes LevelDB space-efficient.

Due to the limited query primitives of LevelDB, the native implementation of the developer interfaces only supports limited queries, most of which are simply retrieving an item by its hash code. Moreover, Ethereum stores values in a special tree structure and accessing values associated with hash code requires random reading from the hundreds of thousands of files scattered all over the disk, which is time consuming. Therefore, LevelDB's limitation combined with the way Ethereum utilizes it make it unsuitable for analytical applications that normally involve complex queries on the transaction data in a blockchain.

In this paper, we address the problem of efficient querying blockchain data by adding a query layer for a public blockchain system, where a set of useful analytical queries are supported, such as the most popular ones like aggregation and top-k queries. In view of rich Ethereum applications and ecosystem, we use Ethereum to demonstrate the feasibility of our prototype in this paper, while the proposed techniques can be applied to other blockchain instances as well. Our contribution can be summarized as follows:

1. We identify existing problem of efficient querying blockchain data and propose a query layer to support efficient analytical queries in blockchain system.
2. A range of analytical queries such as range query and top-k query are supported with flexible interfaces provided so that users can issue the query with both local API or RESTful service.
3. EtherQL automatically synchronizes new block data in a timely manner and store them in a dedicated database to ensure the query is both accurate and efficient.

The remainder of this paper is organized as follows. Section 2 reviews the recent development on blockchain. Section 3 gives a brief introduction to the concepts and terminologies about Blockchain. Sections 4 and 5 present the overview and details of the system design respectively. Sections 6 and 7 provide detailed information of interfaces and report empirical evaluation results for different queries. Section 8 concludes this paper.

2   Related Work

Blockchain is a distributed ledger that is maintained by all participants. There are mainly two categories of blockchain systems. The first one is UTXO (Unspent Transaction Output) systems, such as Bitcoin [18] and Ripple [16]. Those systems records a UTXO set S. Each UTXO represents a fund that could be spent for future transactions. When a new transaction comes, referenced UTXOs are removed from S. Those systems will verify the provided signatures of the transaction. If the signatures match the owner, those funds are claimed and transactions could add new UTXOs into S with funds. Thus, the applications of UTXO systems are restricted to fund transfer. The other one is account based such as Ethereum [25]. Each user is associated with an account that records relevant information such as balance, code and storages. A series of applications have been built based on blockchain technology, such as payment facility [10], health care [26], Internet of Thing [7] and file sharing [24]. However, due to limited query supports from the underlying LevelDB, it is not easy to query the blockchain data. There are some systems that provide basic query interfaces for blockchain data. For example, the Blockchain.info [1] wraps the Bitcoin's block, transaction and address APIs to provide a RESTful service, which is convenient for developers to use. Project Toshi [12] implements the Bitcoin protocols and backed by PostgreSQL, providing RESTful service for basic query operations and limited paging functionality. Etherscan [13] is an explorer and analysis platform powered by Ethereum foundation. Etherscan has a modified version of virtual machine that could be used to extract more information from blockchain, such as internal transactions. Etherchain [2] extends the Ethereum basic APIs and provides simple statistics, such as the block time and transaction count. Those systems could have rich information and provide users with basic interfaces to explore blocks, transactions and accounts. Nevertheless, more complex queries for blockchain data are not supported. For example, graph queries and top-K queries are not supported in those systems. Besides, utilization of these public APIs is restricted. Taking Etherscan for an example, the request to these API is limited to 5 requests/sec. In a word, those explorers are helpful for blockchain participants but not for data-driven applications.

There are a number of analytical tasks carried out by developers. [14] performs graph analysis with Bitcoin transactions. It is proven that Bitcoin is not a fully anonymous system with graph analysis. Rigorous quantitative analysis of Bitcoin transactions is carried out in [19]. Authors identify the transaction patterns that users employed to hide their identities. To detect money laundering, [17] proposes a multi-variant relation model with time series dataset. Besides, [8] builds a reputation network for blockchain users to reduce transaction risks. Those systems demonstrate the needs of scalable and efficient technique that can support data analytical tasks on top of blockchain systems.

3   Preliminary Concept

In this section we briefly introduce the concept of blockchain and how it works to reach consensus among participants. Our discussion is restricted to the most popular public blockchain platform, Ethereum, though other blockchain platforms adopt similar concepts and techniques.

3.1 Blockchain Architecture

There are mainly four layers in most blockchain systems, i.e. consensus layer, data layer, execution environment and hosted applications. The consensus layer is responsible for maintaining data consistency among all participants which tackles the Byzantine Fault Tolerance problem. The data layer contains the data structures and corresponding operations. The execution layer such as Ethereum Virtual Machine enables the runtime environment of blockchain applications. At last, a rich classes of applications are implemented on top of that. The remaining discussion focuses on the data layer.

3.2 Blockchain Data Structure

Let us look at basic structure of block data. Without loss of generality, we omit minor details to avoid falling into complicated implementations of Ethereum, but readers can refer to the Ethereum yellow paper [25] for details. In the block header, there is a block hash chained with the previous block, a nonce that represents the proof of work in that block, and tree roots generated based on the information derived from the block body.

State. Accounts play a central role in Ethereum, which are referenced by the unique identifiers called addresses. In Ethereum, every account has their storage spaces, for maintaining the balances and other information. The state of all accounts is the state of Ethereum network that is updated when a new block is generated. Different from Bitcoin, Ethereum maintains the current state of all accounts in the modified version of Merkle tree [22], called Merkle Patricia tree/Trie [9], which is a mapping between addresses and account states. The Merkle tree root is stored in the block header as state root.

Transaction. The term 'transaction' is used in Ethereum to refer to the signed data package that stores a message to be sent from an account to another account in the blockchain network. A transaction mainly contains the following fields:

– Address of the sender and recipient of the transaction.
– Amount of value to transfer from the sender to the recipient.
– An optional data field, which may contain the message sent to the recipient.

The transaction root is the Merkle tree root of all transactions contained in that block.

Receipt. In order to serve as zero-knowledge proof and indexing for searches, execution environment concerning a transaction is encoded into a receipt. The transaction receipt is a tuple of four items comprising the transaction, together with the post-transaction state, the cumulative gas used in the block as of immediately after the transaction has happened and a set of logs created through execution of the transaction. Merkle tree root of all receipts in the block is added to the block header as receipt root.

3.3   Storage with Trie

Ethereum adopts LevelDB as the backend data storage to maintain blocks, transactions and accounts' states. Basically, data structures are stored in Merkle Patricia tries/Trie. A Trie originates from radix tree, and Ethereum implementation introduces a few modifications to improve its efficiency. In a normal radix tree, a key is an actual path taken through the tree to get the corresponding value. That is, beginning from the root node of the tree, each character in the key tells you which child node to follow in order to get the corresponding value. Ethereum keeps the full current state in state Trie, updating accordingly with the execution of transactions. The key is encrypted utilizing SHA3 to avoid DoS attack. Note that this actually makes it more difficult to traverse the tree and enumerate all the values.

4   System Overview

The proposed architecture of the querying layer can be demonstrated in Fig. 1. We develop a middleware that automatically synchronizes blockchain data from Ethereum public network in real-time with a built-in Ethereum client, and provides developers or data analysts with an out-of-the-box data query layer for convenient access to the whole blockchain data. The layer consists of four modules: sync manager, handler chain, persistence framework and developer interface.

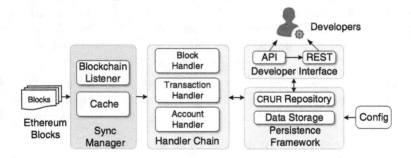

Fig. 1. System overview

In order to keep up with the latest block data from other peers, our system continuously monitors block changes by setting up a blockchain event listener. Upon receiving a blockchain data, the event listener puts it in the cache to deal with unintentional forks in blockchain network, details of which can be found in Sect. 5.1. The handler chain continuously tries to get blockchain data from the cache and extracts the data into three kinds of data structures: block, transaction and account, with different data handler. We describe details of handler chain in Sect. 5.2. With the CRUD repository (short for creation, retrieval, updating and deletion) defined in Sect. 5.3, data is persisted to a different data storage that supports SQL queries. Both the data storage and the CRUD repository can be configured through the config module. Developer interface is built on top of the CRUD repository, which hides the complexity and provides easy access to blockchain data set. We will illustrate the detailed design of developer interface in Sect. 5.4. While the API module can be used directly to interact with the underlying data, REST [23] module offers more flexibility from maintenance and deployment perspective.

5   Design Details

5.1   Sync Manager

Due to the possible delay in the distributed network, more than one miners may broadcast their version of a new block before 'hearing' from others. This is when an unintentional consensus fork takes place. That means the data synchronized from other peers has the possibility of not being on the main chain. This problem is minor for Ethereum client because it save the other branch in case it becomes longer, which make the overhead for switching to the main chain negligible. However, in relational databases, undoing the operations concerning blocks, transactions and accounts states will bring about extra overhead and is also vulnerable to potential bugs. In order to deal with this inconsistent state without lowering the performance, we maintain a cache for incoming blockchain data before entering the handler chain. As a result, we can identify a potential fork in advance and reduce the chance of falling into forks. The number of blocks to be cached can be configured in the configuration module.

5.2   Handler Chain

The module handler chain can be viewed as a plugin to the Ethereum client, transforming incoming blockchain data to fit into SQL schemas. In the data structure in Ethereum, the latest state of all accounts are stored in the Merkle Patricia trees (Trie). Actually, the Trie structure is not contained in the raw blockchain data but built on the fly through execution of the transactions in the blockchain data. Ethereum clients continuously update their local view of the latest state if the execution result obeys the pre-defined consensus rules. Handler chain updates another SQL data storage just like Ethereum updates the Trie. The difference is that handler chain has to firstly extract the corresponding information from the transactions' execution results according to the consensus rules. We divide the logical data structure of Ethereum Trie into three kinds of entities: block, transaction and account. As shown in Fig. 1, there are three components in *handler chain*: block handler that takes the incoming blockchain data as a whole and create new block entity, transaction handler that tracks the transaction list contained in current blockchain data and adds corresponding transaction entities, account handler that updates account entities to reflect the state changes described in Sect. 3.2. We chain all these handlers together [21], such that each handler undertakes its responsibility and passes the control to the next one until the blockchain data is properly processed. The main benefit of this design is that it decouples the blockchain data with the handlers so that we can dynamically add other handlers later on without affecting the client interfaces.

5.3   Persistence Framework

We intend to build a middleware for blockchain to facilitate efficient queries. To this end, it is of vital importance to provide a data persistence framework with the support of custom SQL, such as retrieving blocks with the block number in a specified range. On the one hand, raw blockchain data must be properly persisted to ensure scalability. On the other hand, custom queries must be performed efficiently on the storage to support enterprise-level applications. To support data persistence while maintaining the query flexibility, we adopt MongoDB as the external data storage to store the output result of handler chain. MongoDB [4] is an open source cross-platform NoSQL database with the support of flexible schema and can be easily configured for scalability. We have set up a default connection property for MongoDB instance but the configurations can be overridden in the configuration module.

We provide a level of abstraction above the diverse implementations of data storages so that support for other databases can be added in the following releases. In the center of data persistence framework is the CRUD repository, which is a bundle of SQL query templates. A query template encapsulates the data access interfaces exposed by the underlying data storage, and converts the function calls to real data manipulation operations. For example, the account template can be used to create an account, validate the existence of an account, update the balance, and delete an existing account.

5.4   Developer Interface

To meet different requirement of developers, we provide two types of interfaces, namely API and REST. While API is a native implementation of query interfaces, REST is a wrapper that provides RESTful services. The API module provides query interfaces for Ethereum accounts, transactions and blocks. More specifically, each module exposes 4 types of query interfaces: (1) Ethereum supported queries [3], (2) extended queries (eg. retrieving for transactions related to a specified account), (3) range queries (eg. listing transactions within a given time window, (4) top K queries (eg. top K accounts regarding the balance). Application developers can utilize these encapsulated interfaces without knowing all the details of underlying storage.

For front-end developers, it is not always feasible to know all the underlying technologies and learn how to prepare data before developing user interfaces. Therefore we wrap all the APIs above and build a high level developer interface in REST-like mode. The main purpose of integrating REST into our developer interface is that users of our system hold all the blockchain data, thus they can provide DaaS (Data as a Service) to serve applications on top of blockchain platform.

6   Implementation

Joining in the Ethereum blockchain network requires implementation of the peer discovery protocol and blockchain synchronizing logics just like the Ethereum clients. In order to avoid re-inventing wheels, our system integrates EthereumJ [15], a pure-Java implementation of the Ethereum protocols.

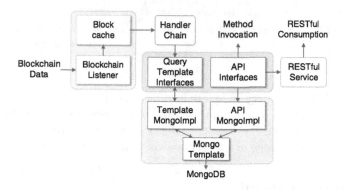

Fig. 2. Implementation details of EtherQL

As depicted in Fig. 2, **blockchain listener**, as the frontier of the system, is responsible for synchronizing blockchain data from EthereumJ in real time. The **block cache** maintains the latest N blocks and detects potential forks in the blockchain. When a fork takes place, it rebuilds the main branch. Considering blockchain's immutability feature and tolerance of time delay, N is set to 5 by default, similar to the setting of go-ethereum, which is one of the most popular Ethereum clients.

Handler chain is composed of 3 sequential handlers that deals with a block from different perspectives. For example, the first block handler stores the basic block structure in underlying storage. In each handler, there is a **query template interface** that encapsulates a bunch of operations to manipulate the blockchain data. In the current release, we have implemented MongoDB version of the query template interface that encapsulates the **MongoTemplate**, a class that actually interacts with MongoDB.

The supported APIs are classified into five categories according to their functionality, each of which can be applied to three types of data, namely account, transaction and block. Table 1 lists the major APIs implemented by our system. Once developers add our system as a dependency to their application, they are capable of utilizing these interfaces to query the blockchain data. **REST service** is built to serve web applications, for instance, blockchain explorers. More details about the supported API interfaces have to be omitted due to space limitation and can instead be found on our project site hosted by GitHub[1].

---

[1] https://github.com/LeonSpark/EtherQL.

Table 1. API interface

Functionality	Account	Transaction	Block
`Retrieve one by`	address	transaction hash	block hash or block number
`Range query by`	balance	transaction value	block number
`Aggregate by`		sender or recipient	miner
`Top K by`	balance	transaction value	
`Paging (offset & limit)`	balance	transaction value	block number or timestamp

Note that both the query template interface and the API interface provide an abstraction to decouple our system with the underlying databases. The support for MySQL is still under development. Developers can realize the two interfaces to support other databases of their choice as well.

7   Experiment Results

7.1   Experiment Setting

Data Set. The raw blockchain data we use in this experiment is publicly available over the Ethereum homestead network. The data set contains approximately 2.7 million blocks, 850k accounts and 13 million transactions [13] since 2013. An account can be viewed as a tuple of $<address, balance, nonce, code, storage>$, where nonce is a decimal value indicating the number of transactions sent from this account and storage is where accounts utilize and maintain accounts' inner states. Code stores compiled smart contracts [25], which is beyond the discussion of this paper. A transaction can be represented as $<hash, sender, receiver, value, gas, gasprice, data, nonce>$. Gas in Ethereum serves as an incentive to charge transaction senders for transaction inclusion, while the gas price is set by transaction senders to indicate that they are willing to pay at most $gas * gasprice$ for the execution of this transaction. The field data is encoded value of other arguments of this transaction. Nonce is used in order to prevent relay attacks. The block is a collection of information which can be viewed as a tuple of $<hash, parenthash, number, miner, nonce, timestamp, transactions>$ for simplicity. The field nonce is the outcome of mining procedure that indicates the amount of work of the miner. Detailed structure of the data set can be found on https://github.com/LeonSpark/EtherQL.

Environments. All the experiments are running on a PC (Intel Core I5 CPU of 8 cores, 3.2 GHz for each core and 16 GB memory on Ubuntu 14.04 LTS) with JDK 1.8 installed.

7.2   Performance Evaluation

Due to the fact that blockchain data are immutable, the process of synchronization is not repeatable unless synchronizing from scratch. Generally speaking, high-speed network bandwidth and fairly good performance of physical machine will speed up the synchronization process. Otherwise, it may take days to catch up with the other peers. Therefore, we will not include evaluation of blockchain synchronization in this paper. Our evaluation is composed of two parts. First, we compare the throughput (following the common practice of 1 s time interval) of our system against native Ethereum clients for supported queries. We adopt the latest go-ethereum (v1.5.4) clients and compare the throughput on three queries: (1) get a block by block's number, (2) get a transaction by transaction's hash and (3) get the balance of an account by its address. Please note that range queries and top-k queries are not tested since they are not supported by Ethereum client. We perform 10K queries in each thread and gradually increase the number of loading threads. The throughput in Fig. 3 is the maximum number of all iterations. Then, we evaluate the performance of our system on query's response time and concurrency.

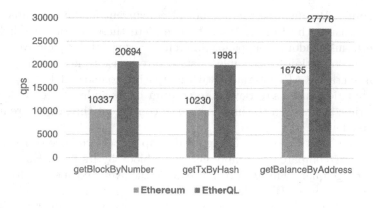

Fig. 3. Throughput comparison between go-ethereum and EtherQL

Throughput. Figure 3 shows that the throughput in EtherQL is almost 2 times as that in go-ethereum. Ethereum's lower throughput is partially due to the fact that only a single process (possibly multi-threads) can access the file at a particular time in Ethereum's underlying LevelDB storage. Moreover, Ethereum uses hash as the key to locate the entity in the Trie, which involves unpredictable access of files scattered all over the disks. Both systems are capable of processing over 15K requests/s on account states because of the relatively smaller amount of accounts.

Response time. We query for a randomly selected list of accounts, transactions and blocks, and record how long it takes to complete the query. Since a single query usually completes in milliseconds, the reported time is the average of 1000 independent queries to minimize the impact of random errors.

Table 2. Response time (in millisecond) of representative queries

Category	Typical queries	Number of concurrent threads				
		1	10	20	40	80
Block	Basic By block number	0.57	2.303	4.264	5.146	9.484
	Range Query By block number (range 100)	6.32	25.28	42.75	67.15	111.71
Account	Basic By address	0.517	1.718	3.744	4.737	7.999
	Top K By balance rank (top 100)	3.84	13.85	28.2	39.5	65.55
Transaction	Basic By hash	0.536	2.077	3.994	6.666	8.253
	Top K By value rank (top 100)	8.7	28.5	39.61	59.83	93.54

Concurrency. We increase the number of concurrent loading threads step by step and measure the time it takes to complete these queries. Each loading thread performs random lookups for different category of data and we calculate the average completion time. Table 2 lists the response time (in millisecond) of some representative API interfaces with increasing concurrent loading threads. The basic queries, such as retrieving an account by its address, are blinking fast. Response time grows with the incremental of concurrent threads but remains at a reasonable level.

There is at least an order of magnitude gap in performance of retrieving a range of blocks compared to simply getting a block by its hash. However, considering a range query will retrieve 100 blocks, the average time cost for a single block is short. MongoDB's indexes are used for retrieving top K accounts with the highest balance and transactions with the highest value, which significantly improves the performance. Note that MongoDB occupies memory space greedily if there exists sufficient amount of physical memory. However, if the data can just fit into the memory, MongoDB runs at high efficiency.

8   Conclusion

Due to the limited query supports of the underlying data storage, blockchain data have not brought all its potential into full play. In this paper, we propose EtherQL, the first query layer (to our knowledge) for Ethereum blockchain data. EtherQL has a built-in java implementation of Ethereum consensus protocols, thus can be utilized to synchronize data from any Ethereum blockchain networks. We also provide two levels of developer interfaces that can be used to retrieve data efficiently from blockchain or serve as a RESTful blockchain data provider. Experiments show the effectiveness and efficiency of EtherQL in querying blockchain data.

Acknowledgment. This work is partially supported by NSFC 61502324, 61532018.

References

1. Blockchain.info. https://blockchain.info/
2. Etherchain. https://etherchain.org/
3. Ethereum main wiki. https://github.com/ethereum/go-ethereum/wiki
4. MongoDB. https://www.mongodb.com/
5. Proof of Stake. https://en.bitcoin.it/wiki/Proof_of_Stake
6. Proof of Work. https://en.bitcoin.it/wiki/Proof_of_work
7. Bahga, A., Madisetti, V.K.: Blockchain platform for industrial Internet of Things. J. Softw. Eng. Appl. **9**(10), 533 (2016)
8. Buechler, M., Eerabathini, M., Hockenbrocht, C., Wan, D.: Decentralized reputation system for transaction networks. Technical report, University of Pennsylvania (2015)
9. Buterin, V.: Merkling in ethereum. https://blog.ethereum.org/2015/11/15/merkling-in-ethereum/
10. Buterin, V., et al.: Ethereum white paper (2013)
11. Castro, M., Liskov, B.: Practical byzantine fault tolerance. In: Symposium on Operating Systems Design and Implementation, pp. 173–186 (1999)
12. Coinbase: Toshi project. https://github.com/coinbase/toshi
13. EtherScan: EtherScan.io. https://etherscan.io/
14. Fleder, M., Kester, M.S., Pillai, S.: Bitcoin transaction graph analysis. arXiv preprint arXiv:1502.01657 (2015)
15. Ethereum Foundation: EthereumJ project. https://github.com/ethereum/ethereumj
16. R Foundation: Ripple project. https://ripple.com/
17. Krishnapriya, G., Prabakaran, M.: An multi-variant relational model for money laundering identification using time series data set. Int. J. Eng. Sci **3**, 43–47 (2014)
18. Nakamoto, S.: Bitcoin: a peer-to-peer electronic cash system. Consulted (2009)
19. Ron, D., Shamir, A.: Quantitative analysis of the full bitcoin transaction graph. In: Sadeghi, A.-R. (ed.) FC 2013. LNCS, vol. 7859, pp. 6–24. Springer, Heidelberg (2013). doi:10.1007/978-3-642-39884-1_2
20. Sanjay Ghemawat, J.D.: Leveldb github page. https://github.com/google/leveldb
21. Wikipedia: Chain of responsibility. https://en.wikipedia.org/wiki/Chain-of-responsibility_pattern
22. Wikipedia: Merkle tree. https://en.wikipedia.org/wiki/Merkle_tree
23. Wikipedia: Remote procedure call. https://en.wikipedia.org/wiki/Remote_procedure_call
24. Wilkinson, S., Boshevski, T., Brandoff, J., Buterin, V.: Storj: a peer-to-peer cloud storage network (2014)
25. Wood, G.: Ethereum: a secure decentralised generalised transaction ledger. Ethereum Project Yellow Paper (2014)
26. Yue, X., Wang, H., Jin, D., Li, M., Jiang, W.: Healthcare data gateways: found healthcare intelligence on blockchain with novel privacy risk control. J. Med. Syst. **40**(10), 218 (2016)

Optimizing Scalar User-Defined Functions in In-Memory Column-Store Database Systems

Cheol Ryu[3], Sunho Lee[3], Kihong Kim[1], Kunsoo Park[3(✉)], Yongsik Kwon[1], Sang Kyun Cha[1,3], Changbin Song[1], Emanuel Ziegler[2], and Stephan Muench[2]

[1] SAP Labs Korea, Seoul, Korea
{ki.kim,yong.sik.kwon,sang.k.cha,chang.bin.song}@sap.com
[2] SAP SE Germany, Walldorf, Germany
{emanuel.ziegler,stephan.muench}@sap.com
[3] Seoul National University, Seoul, Korea
{cheol.yoo,sunho.lee,kunsoo.park01}@sap.com

Abstract. User-defined functions such as currency conversion and factory calendar are important ingredients in many business applications. Since currency conversion and factory calendar are expensive user-defined functions, optimizing these functions is essential to high performance business applications. We optimize scalar user-defined functions by caching function call results. In this paper we investigate which method for function result caching is best in the context of in-memory column-store database systems. Experiments show that our method, which implements a function result cache as an array, combined with SAP HANA in-memory column store provides the high performance required by real-time global business applications.

Keywords: Scalar user-defined functions · Currency conversion · Function result caching · In-memory column-store database system

1 Introduction

User-defined functions and stored procedures are commonly used to execute application logics inside database servers [5,10,19]. They often lead to enormous performance improvements by executing data-intensive application logics at the place where data reside and thus minimizing the amount of data transfer between application servers and database servers [8]. In this paper we focus on scalar user-defined functions, which receive scalar arguments and return a scalar value.

Suppose that a global company stores its sales records in the SALES table, as illustrated in Table 1, where a sales amount is given in a local currency. Therefore, the total sales amount cannot be obtained by a simple aggregation query, but it requires currency conversion which applies the exchange rate on the sales date.

Assume that we have a scalar currency conversion function, CURR-CONV (source-currency, target-currency, date), which returns the exchange rate from 'source-currency' to 'target-currency' on the 'date'. If such a function is available,

© Springer International Publishing AG 2017
S. Candan et al. (Eds.): DASFAA 2017, Part II, LNCS 10178, pp. 568–580, 2017.
DOI: 10.1007/978-3-319-55699-4_35

Table 1. Sales table with sales date, amount, local currency, country, etc.

Date	Amount	Currency	Country	...
2015-01-02	1,140	USD	US	
2015-01-03	95,680	JPY	JP	
2015-01-03	1,098	EUR	DK	
2015-01-06	800	EUR	DE	
2015-01-07	647	GBP	GB	
...				

we can obtain the total revenue of year 2015 directly from an OLTP system [29] using a query shown in Eq. (1), which calculates the revenue of 2015 in euro.

$$
\begin{aligned}
&\texttt{select sum(Amount * CURR-CONV (Currency, 'EUR', Date))} \\
&\texttt{from SALES where Date between '2015-01-01' and '2015-12-31'}
\end{aligned}
\tag{1}
$$

Currency conversion in business applications is a complex user-defined function because each business has its own currency conversion scheme with its own corner-case handlings [27] (e.g., one of the bid, ask, and middle rates can be chosen as the exchange rate, and each bank or financial institute has slightly different exchange rates, etc.). Hence, it cannot be done by a simple lookup of a table (i.e., precomputing the exchange rate table for all possible application-specific options and for all possible dates is too costly). An execution plan of the query in Eq. (1) is to invoke the currency conversion function for *every* row of the SALES table. Since each function call requires queries against underlying exchange rate tables and configuration tables in the SAP ERP implementation [27], the query will take a prohibitively long time when the SALES table is big.

An alternative is to rewrite the query in Eq. (1) into the following two steps.

1. Compute exchange rates for distinct pairs of Currency and Date by running "select Currency, Date, CURR-CONV (Currency, 'EUR', Date) Rate from SALES group by Currency, Date".
2. Join the exchange rate table computed in step 1 with the SALES table.

In this way, we can perform the aggregation without calling the expensive function for every row of the SALES table. However, this alternative is still expensive. It scans the big SALES table twice, once for obtaining the distinct pairs and once more for joining and aggregation.

Factory calendar is another example of expensive user-defined functions, which is important in manufacturing planning applications. Each factory has its own work-day calendar, which is stored in a database table. A scalar function FACTORY-DAYS(date1, date2) returns the number of factory work-days between two calendar dates 'date1' and 'date2' by looking up the factory calendar table. Application-managed authorization functions are other examples. SAP ERP and many other applications do their own authorizations for business processes. Again, application users, application roles and application privileges are stored in database tables. An authorization function is invoked to check if a user has a proper privilege for a given application function/process by accessing the database tables.

Due to the high performance of SAP HANA in-memory engine, the overhead of these user-defined functions becomes conspicuous in global business applications. To meet various requirements from real-time business applications, we need to support general queries containing user-defined functions in an efficient way [8].

A natural optimization for expensive user-defined functions is to cache function call results so that subsequent calls with the same arguments are hit by the cache. Such a scheme is called *function result caching*, which has been used in commercial database systems such as Oracle [23], IBM DB2 [18], and MS SQL [4] as well as in research investigations [15,25]. The most common method for function result caching is hashing, even though many methods for caching have been proposed [15].

In this paper we propose a generic framework to apply function result caching to any expensive user-defined functions such as currency conversion and factory calendar, and we also present fast parallel implementations of user-defined functions within the framework. We explore various ways of optimizing function result caching, especially in the context of SAP HANA in-memory column store [9]. It is straightforward to implement a function result cache as a hash table, though in parallel execution we need to decide whether we will use a single hash table for all threads or multiple hash tables, one per thread. In the column store with dictionary encoding, however, a function result cache can be implemented more efficiently as an array. Dictionary encoding transforms each field value into a small integer. For instance, the local currency column stores value IDs in the range of 0–29 if there are 30 local currencies in total, and the sales date column stores value IDs in the range of 0–364 if the table has one year of sales records. Since the function call in Eq. (1) has at most 30×365 combinations of input arguments, the function result cache can be represented as an array of 30×365 entries. Since this maximum size of the array can be easily calculated from the column store of SAP HANA, the column store enables us to implement a function result cache as an array.

In our experiments we compare the performances of four methods for function result caching: array method, hash table method, Join query in the column store, and Join query in the row store. Experiments show that the array method and the Join query which are implemented in the column store outperform the hash table method and the Join query which are implemented in the row store. In particular, the array method of currency conversion implemented in the column store is the fastest, and it takes 2.1 s with 16 CPU cores for a SALES table with 151 million rows. Hence, the proposed framework combined with HANA in-memory database provides the high performance required by real-time global business applications.

2   Related Work

Optimizing user-defined functions has been studied in the context of finding an optimal execution plan of an SQL query [5,16]. That is, when a user-defined function is expensive and its cost function is given, there are many execution plans of an SQL query, and techniques to find an optimal execution plan among them have been studied [5]. These studies are orthogonal to our work, which investigates methods to speed up user-defined functions themselves.

Hellerstein and Naughton [15] describe three caching methods. The first is hashing in main memory, and the second is sorting (of all rows of SALES based on Date and Currency) and maintaining a cache of only the last function result. Since the first method suffers from lack of enough memory space and the second from the expensive cost of sorting, they propose a hybrid method that combines in-memory hashing and disk buffers. It should be noted that the machine used in the experiments of [15] had 64 MB memory. Since a typical machine for SAP HANA has enough memory (e.g., 1 TB), we consider only in-memory caching methods for best performances.

A materialized view is a kind of cache for base tables. There is a large body of work on how to select and maintain materialized views [22,26]. This is closely related to finding common subexpressions [20,28]. These works are orthogonal to ours.

Result caching has been a common technique in other areas such as web applications and programming languages. In web search engines, caching query results has been an effective way to boost performance [11,12]. In programming languages, too, function result caching is used to improve performance [17].

Parallelizing user-defined aggregate and table functions has been studied by [10,19]. We also consider parallelization but we focus on parallel execution of function result caching for scalar user-defined functions.

3   Framework

3.1   Framework for Function Result Caching

In SAP HANA, users can define scalar functions in the HANA SQLScript language and they can use such functions in SQL queries [3]. Since user-defined functions usually include operations on database tables, they are very expensive compared to built-in functions such as arithmetic functions. We implemented a generic framework for function result caching to optimize such expensive function calls inside SAP HANA.

For a scalar user-defined function $f(a_1, a_2, \ldots, a_n)$, HANA internally generates a wrapper function, $fw(sc, fid, a_1, a_2, \ldots, a_n)$, to enable the function result caching, where a_1 to a_n denote function arguments and sc denotes a statement context object, which is initialized before executing a query statement and cleared after completing the statement. The function result cache is stored in this statement context object. The value fid is an integer ID, sequentially assigned to each function when compiling a query. It is used to locate a proper cache entry for the given function when a query has multiple expensive functions.

In SAP HANA, there are multiple levels of run-time contexts. See Fig. 1. The execution context is located at the bottom. When a query is executed in parallel with multiple threads, each thread is given its own execution context, which thus can be accessed without any latching. The next level is called the statement context, which is shared by the entire threads that fulfill a statement either in one server or in multiple distributed servers. And then the transaction context and the session context follow.

Run-Time Context	Scope of Run-Time Context
session context	session
transaction context	transaction
statement context	SQL statement
execution context	execution thread

Fig. 1. Scopes of run-time contexts

3.2   Column Store of SAP HANA

Since the column store of SAP HANA is closely related to our methods for function result caching, we describe characteristics of HANA relevant to this paper. The heart of SAP HANA consists of a set of in-memory processing engines [9]. Relational data resides in tables in column or row layout in the combined column and row engine. The column store of SAP HANA stores relational tables in a column-wise manner. The column store is optimized for high performance of read operations, while providing good performance for write operations [1]. Efficient data compression is used in column tables in order to save memory and to speed up searches and calculations [1]. The row store of SAP HANA [9] is a row-based in-memory storage. It is optimized for high performance of concurrent write and read operations and for fast restart.

We now describe the representation of columns in memory and data compression used in column store tables, which are relevant to function result caching. For each column, the distinct values appearing in the column are stored in a sorted dictionary. Each value in the dictionary is identified by its value ID which is implicitly given by its position in the sorted dictionary. The actual data in the column is stored as an array of value IDs. See Fig. 2. To store the value ID array, the value IDs are bit-encoded, which means that the minimum number of bits is used to store a value ID. If the dictionary contains N values, the required number of bits is $\lceil \log_2 N \rceil$.

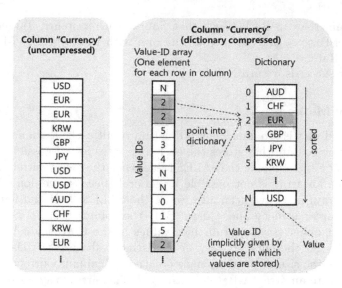

Fig. 2. Representation of a column in column store

4 Methods for Function Result Caching

We describe our methods for function result caching by using currency conversion as an example. Currency conversion needs to access an exchange rate table RATE as shown in Table 2.

Table 2. Rate table with source currency, target currency, reference date, and exchange rate.

Source-Curr	Target-Curr	Ref-Date	Rate	...
USD	EUR	2015-01-02	1.377	
AUD	EUR	2015-01-02	1.548	
GBP	EUR	2015-01-02	0.830	
USD	EUR	2015-01-03	1.367	
JPY	EUR	2015-01-03	143.200	
...				

The scalar currency conversion function CURR-CONV(scurr, tcurr, date) used in our experiments is defined as follows.

1. If there are rows in table RATE such that Source-Curr = scurr, Target-Curr = tcurr, Ref-Date ≤ date, then take the one among such rows with the largest Ref-Date, and return Rate.

2. If there are rows in RATE such that Source-Curr = tcurr, Target-Curr = scurr, Ref-Date ≤ date, then take the one with the largest Ref-Date, and return 1/Rate.
3. If neither succeeds, return NULL.

4.1   Array Method

In this method we wish to implement a function result cache as an *array*, i.e., we create an array A-Rate that stores the exchange rates for all possible pairs of a source currency and a date in the SALES table when the target currency is EUR. Creating such an array is not possible in general (because we don't know the size of the array) unless we scan all rows of the SALES table and find distinct pairs of a source currency and a date. But, the column store of SAP HANA provides the number (say, c) of distinct values in the Currency column of the SALES table and the number (say, d) of distinct values in the Date column. Note that c is the size of the dictionary created for column Currency in Fig. 2. Hence, we create an array A-Rate of size $c \times d$. This array may have pairs of a source currency and a date which do not exist in the SALES table. However, since the number of entries in A-Rate is much smaller than the number of rows of SALES in typical applications and the task of building the A-Rate array is fast by a batch computation described below, the extra computation in A-Rate is negligible when compared to the total cost of currency conversion.

Once the A-Rate array is created, a position in the array can be used as a representation of a pair because source currencies and dates in the SALES table have value IDs in the column store. For example, if a source currency has value ID 3 and a date has value ID 5, the position of the pair in the array is $3d + 5$.

It would be straightforward to compute the entries of the A-Rate array by calling the scalar function CURR-CONV for every entry. Since business applications require more efficiency, however, we compute the entries of A-Rate in a batch way as follows. This is a custom operation special to currency conversion. See Fig. 3. Suppose that the rows of the SALES table between 2015-01-02 and 2015-01-07 are as shown in Fig. 3. Then the entries of the A-Rate array with source currency s are created as in Fig. 3 (where we used actual values rather than value IDs for exposition purpose). Suppose also that the sorted entries of the RATE table with source currency s and target currency EUR are as shown in Fig. 3. Then we get the exchange rate for the pair $(s, 2015-01-02)$ from RATE and store it into the entries $(s, 2015-01-02)$, $(s, 2015-01-04)$, and $(s, 2015-01-05)$ of the A-Rate array. And we get the rate for $(s, 2015-01-06)$ from RATE and store it into $(s, 2015-01-06)$ and $(s, 2015-01-07)$ of A-Rate. This process can be done by scanning simultaneously the RATE table and the A-Rate array. The whole process is repeated one more time by reversing sources currencies and target currencies of the RATE table to perform step 2 in the definition of CURR-CONV (i.e., to apply 1/Rate). Therefore, the exchange rate computation for one pair is amortized over many entries of the A-Rate array.

Once the A-Rate array has been computed, we read the rows of the SALES table one by one. For each of the source currency and the date in a row, we get the

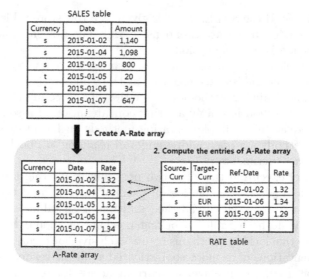

Fig. 3. Building the A-Rate array

value ID (from the value ID array in Fig. 2). From the two value IDs we compute the position of the pair in the A-Rate array, get the exchange rate corresponding to the pair, and finally multiply the exchange rate and the amount. This task can be parallelized by many threads, and a block of rows of the SALES table can be assigned to each thread. Since the A-Rate array is accessed by multiple threads, it is stored in the statement context in Fig. 1.

The array method of currency conversion was implemented in the engine level of SAP HANA [9], and the internal parallelization by many CPU cores was used in our experiments.

4.2   Join

The Join query consists of three tasks: it first finds all distinct pairs of a source currency and a date from the SALES table (which we call the *distinct pairs task*), then it builds the J-Rate table with source currency, date, and rate columns by invoking currency conversion functions for the distinct pairs (called the *build J-Rate task*), and finally it performs a join operation on the SALES table and the J-Rate table (called the *join task*).

Both the distinct pairs task and the join task are usually performed with hash-based algorithms [2]. In dictionary-encoded column tables, the same hash-based algorithms can be applied to value IDs instead of actual values. This simple difference results in a big performance boost.

4.3   Hash Table Method

The array method in Sect. 4.1 is possible because the maximum number of entries in the A-Rate array can be easily computed from the column store representation of the SALES table. If the SALES table is given as a row store table, computing the number of entries in A-Rate is expensive and thus a hash table method of currency conversion is a viable option.

In the hash table method we use a pair of a source currency and a date as the key of the hash table, and its exchange rate as the value of the hash table. Since value IDs are not available without dictionaries of the column store, the key in the hash table method is simply a string concatenation of a source currency and a date. There are many implementation methods for hash tables [6,21], and the two basic methods are open addressing and chaining [6], both of which are considered in our experiments.

In the parallel execution there are two possibilities for the hash table. One is to maintain a single hash table for all parallel threads, and the other is to have one hash table for each thread (i.e., multiple hash tables). In the case of maintaining a single hash table, a locking mechanism should be used in the hash table. For the locking mechanism, an exclusive lock can be used for a simple implementation, or a more sophisticated multiple-readers/single-writer lock [7] may be used, which allows concurrent access for read-only operations while write operations require exclusive access. We chose the best combination of these possibilities for our applications by experiments.

5   Experiments

The dataset used in our experiments is a SALES table of a global company consisting of about 151 million rows and numerous columns (including date, amount, local currency). The total size of the SALES table is 69.1 GB when it is stored as a row store table. The size of the three columns date, amount, and local currency is 556 MB when they are stored as a column store table and it is 7.1 GB when stored as a row store table. We also use a RATE table of 184,000 rows and four columns (source currency, target currency, reference date, exchange rate). The size of the RATE table is 8 MB as a column store table. The machine we used in our experiments is Intel Xeon CPU E7-8890 v3 2.5 GHz with 72 cores, L3 cache of 45 MB, and memory of 1 TB.

The three columns (date, amount, local currency) of the SALES table can be given as a column store table or a row store table in SAP HANA and we consider both cases in our experiments. If the SALES table is given in the row store, we need to convert the three columns into a column table and build dictionaries for them, which takes 44.8 s. If the SALES table is given in the column store, this task is not needed. The RATE table is stored as a column store table in the experiments.

The naïve implementation for currency conversion in Eq. (1) takes 125.8 s for 100,000 rows of the SALES table, and the execution time increases linearly to the number of rows. Hence the naïve implementation takes a prohibitively long time for all 151 million rows of the SALES table.

We will compare the performances of four methods for function result caching: array method, hash table method, Join query in the column store, and

Join query in the row store. Since the array method is an option for the column store and the hash table method for the row store, we divide the four methods into two categories: array method and Join in the column store, and hash table method and Join in the row store.

5.1 Column Store

Table 3 shows the performances of the array method and the Join query in the column store. The parallel execution of the array method is straightforward: Building the A-Rate array is executed by a single thread, and the remaining tasks are executed by multiple threads. Hence the time for building array A-Rate remains the same even though the number of cores is increasing. The remaining tasks of the array method are (1) computing the position of a pair in A-Rate from two value IDs, (2) looking up A-Rate to get the exchange rate corresponding to the pair, and (3) multiplying the exchange rate and the amount. The speedup of the array method when the number of cores is 16 is 11.0.

For the Join query, we compute the entries of the J-Rate table by calling scalar function CURR-CONV for every entry, whereas building A-Rate uses the custom operation described in Sect. 4.1. This is the reason why building J-Rate takes much more time than building A-Rate does. The join task includes multiplications of exchange rates and amounts. The speedup for the build J-Rate task is less than the speedups for the distinct pairs task and the join task because there are more variations between parallel threads in computing scalar functions. Overall, the speedup of the Join query is 11.8 when the number of cores is 16.

Table 3. Comparison of array and Join methods in column store (in seconds).

Method	Number of cores			
	1	4	8	16
Build A-Rate	0.5	0.5	0.5	0.5
Lookup+multiply	22.6	5.8	3.0	1.6
Array (total)	23.1	6.3	3.5	2.1
Distinct pairs	1.6	0.4	0.2	0.1
Build J-Rate	70.6	23.3	12.2	6.3
Join	23.3	6.0	3.1	1.7
Join (total)	95.5	29.7	15.5	8.1

5.2 Row Store

Link-hash is our own implementation of a hash table with chaining, and it is a two-layer hash table, i.e., it contains many bins (default number is 211), each of which has a hash table. We will use one lock for each bin. Google-hash is the Google dense hash [13] which implements open addressing with quadratic probing [6]. Google-hash is the best performer in hash table benchmark tests [14,24]. For the hash table method we first compared Link-hash and Google-hash.

For the parallel execution of the hash table method, we need to choose one from a single hash table and multiple hash tables, and one from exclusive lock and shared-exclusive lock (also known as multiple-readers/single-writer lock) in the case of a single hash table. Experiments showed that Link-hash with a single hash table and the shared-exclusive lock is the best combination in our applications. Therefore, we used it for the hash table method. The single hash table is stored in the statement context in Fig. 1 because it is accessed by multiple threads.

Table 4. Comparison of hash table and join methods in row store (in seconds).

Method	Number of cores			
	1	4	8	16
Hash table (total)	144.2	43.7	20.7	11.2
Distinct pairs	55.3	14.8	7.6	3.9
Build J-Rate	70.6	23.3	12.2	6.3
Join	117.0	30.6	17.3	8.9
Join (total)	242.9	68.7	37.1	19.1

Table 4 shows the performances of the hash table method and the Join query in the row store. The tasks of the hash table method are (1) generating the key of the hash table (i.e., concatenating a source currency and a date), (2) getting the exchange rate from the hash table corresponding to the key if it exists, (3) if not, inserting the rate into the hash table after invoking the scalar function CURR-CONV, and (4) multiplying the exchange rate and the amount. The speedup of the hash table method is 12.9 when the number of cores is 16.

For the Join query in the row store, the distinct pairs task and the join task take more time than those in the column store due to the reason described in Sect. 4.2. The speedup for the build J-Rate task is again less than the speedups for other tasks due to variations between parallel threads in computing scalar functions. Overall, the speedup is 12.7 when the number of cores is 16.

In summary, the array method and the hash table method are faster than the Join query in their respective contexts. The array method is the fastest, and it takes 2.1 s with 16 CPU cores for 151 million rows of the SALES table.

Acknowledgments. The work of Ryu, Lee and Park was supported in part by the National Research Foundation of Korea (NRF) funded by the Ministry of Science, ICT & Future Planning (No. 2012M3A9D1054622).

References

1. Abadi, D.J., Madden, S.R., Hachem, N.: Column-stores vs. row-stores: how different are they really? In: Proceedings of the 2008 ACM SIGMOD International Conference on Management of Data, pp. 967–980 (2008)
2. Balkesen, Ç., Teubner, J., Alonso, G., Özsu, M.T.: Main-memory hash joins on modern processor architectures. IEEE Trans. Knowl. Data Eng. **27**(7), 1754–1766 (2015)
3. Binnig, C., May, N., Mindnich, T.: SQLScript: efficiently analyzing big enterprise data in SAP HANA. In: Database Systems for Business, Technology, and Web, pp. 363–382 (2013)
4. Books online for SQL server 2016. https://msdn.microsoft.com/en-us/library/ms191007.aspx
5. Chaudhuri, S., Shim, K.: Optimization of queries with user-defined predicates. ACM Trans. Database Syst. **24**(2), 177–228 (1999)
6. Cormen, T.H., Leiserson, C.E., Rivest, R.L., Stein, C.: Introduction to Algorithms, 3rd edn. The MIT Press, Cambridge (2009)
7. Courtois, P.J., Heymans, F., Parnas, D.L.: Concurrent control with "readers" and "writers". Commun. ACM **14**(10), 667–668 (1971)
8. Färber, F., Cha, S.K., Primsch, J., Bornhövd, C., Sigg, S., Lehner, W.: SAP HANA database: data management for modern business applications. SIGMOD Rec. **40**(4), 45–51 (2012)
9. Färber, F., May, N., Lehner, W., Große, P., Müller, I., Rauhe, H., Dees, J.: The SAP HANA database an architecture overview. IEEE Data Eng. Bull. **35**(1), 423–434 (2012)
10. Friedman, E., Pawlowski, P., Cieslewicz, J.: SQL/Mapreduce: a practical approach to self-describing, polymorphic, and parallelizable user-defined functions. Proc. VLDB Endow. **2**(2), 1402–1413 (2009)
11. Gan, Q., Suel, T.: Improved techniques for result caching in web search engines. In: Proceedings of the 18th International Conference on WWW, pp. 431–440 (2009)
12. Garrod, C., Manjhi, A., Ailamaki, A., Maggs, B., Mowry, T., Olston, C., Tomasic, A.: Scalable query result caching for web applications. Proc. VLDB Endow. **1**(1), 550–561 (2008)
13. Google sparsehash. http://goog-sparsehash.sourceforge.net/
14. Hash table benchmarks. http://incise.org/hash-table-benchmarks.html
15. Hellerstein, J.M., Naughton, J.F.: Query execution techniques for caching expensive methods. SIGMOD Rec. **25**(2), 423–434 (1996)
16. Hellerstein, J.M., Stonebraker, M.: Predicate migration: optimizing queries with expensive predicates. SIGMOD Rec. **22**(2), 267–276 (1993)
17. Heydon, A., Levin, R., Yu, Y.: Caching function calls using precise dependencies. SIGPLAN Not. **35**(5), 311–320 (2000)
18. IBM i version 7.2, database SQL programming. https://www.ibm.com/support/knowledgecenter/ssw_ibm_i_72/sqlp/rbafypdf.pdf
19. Jaedicke, M., Mitschang, B.: On parallel processing of aggregate and scalar functions in object-relational DBMS. SIGMOD Rec. **27**(2), 379–389 (1998)
20. Jarke, M.: Common subexpression isolation in multiple query optimization. In: Query Processing in Database Systems, pp. 191–205 (1985)
21. Knuth, D.E.: The Art of Computer Programming, vol. 3: Sorting and Searching, 2nd edn. Addison Wesley Longman Publishing Co., Inc, Boston (1998)

22. Mistry, H., Roy, P., Sudarshan, S., Ramamritham, K.: Materialized view selection and maintenance using multi-query optimization. SIGMOD Rec. **30**(2), 307–318 (2001)
23. Oracle database performance tuning guide, 12c release 1. https://docs.oracle.com/database/121/TGDBA/toc.htm
24. Performance notes. http://goog-sparsehash.sourceforge.net/doc/performance.html
25. Richardson, S.E.: Caching function results: faster arithmetic by avoiding unnecessary computation. Technical report, Mountain View, CA, USA (1992)
26. Ross, K.A., Srivastava, D., Sudarshan, S.: Materialized view maintenance and integrity constraint checking: trading space for time. SIGMOD Rec. **25**(2), 447–458 (1996)
27. Sap, ERP 6.0 enhancement package 8. http://help.sap.com/erp2005_ehp_08/helpdata/en/59/cdc8109ce34bca896115f8ae660a69/content.htm
28. Sellis, T.K.: Multiple-query optimization. ACM Trans. Database Syst. **13**(1), 23–52 (1988)
29. Sikka, V., Färber, F., Lehner, W., Cha, S.K., Peh, T., Bornhövd, C.: Efficient transaction processing in SAP HANA database: The end of a column store myth. In: Proceedings of the 2012 ACM SIGMOD International Conference on Management of Data, pp. 731–742 (2012)

GPS-Simulated Trajectory Detection

Han Su[1], Wei Chen[1], Rong Liu[1], Min Nie[2], Bolong Zheng[3]([✉]), Zehao Huang[1], and Defu Lian[1]

[1] University of Electronic Science and Technology of China, Chengdu, China
{hansu,weichen,rongliu,dove}@uestc.edu.cn, 1078351005hzh@gmail.com
[2] Xundao Inc., Shenzhen, China
minnie@ebigdata.org
[3] University of Queensland, Brisbane, Australia
b.zheng@uq.edu.au

Abstract. Due to the prevalence of GPS-enabled devices and wireless communication technology, spatial trajectories have become the basis of many location based applications, e.g., Didi. However, trajectory data suffers low quality problems causing by sensor errors and artificial forgeries. Sensor errors are inevitable while forgeries are always constructed on bad purpose. For example, some Didi drivers use GPS simulators to generate forgery trajectories and make fake transactions. In this work we aim to distinguish whether a given trajectory is a GPS simulated trajectory. By formulating this task as the problem of traffic speed extracting and irregular measuring, we propose a simulated trajectory detection framework. In traffic speed extracting phase, we first divide time into time slots and then extract the regular speed of each road during each time slot. In irregular measuring phase, we propose three methods to measure the distance between the speed of the given trajectory and the real traffic speeds. For empirical study, we apply our solution to a real trajectory dataset and have found that the simulated trajectory detection framework can detect most forgery trajectories.

1 Introduction

Driven by major advances in sensor technology, GPS-enabled mobile devices and wireless communications, a large amount of data recording the motion history of moving objects, known as *trajectories*, are currently being generated and managed in scores of application domains. In the past few years, a lot of research works focused on the trajectory analyzing. Effective index structures [1,2,17,20,24] are built to manage trajectories and support high performance trajectory queries. Data mining methods are applied on trajectories to detect important points of interest (POI) and find the popular route from a source to a destination [10,11,13,15]. Attentions are also drawn to semantic representation or interpretation of trajectory data by associating or annotating GPS locations with semantic entities [21,25]. Despite decades of research efforts on

---

W. Chen—Equal Contribution.

© Springer International Publishing AG 2017
S. Candan et al. (Eds.): DASFAA 2017, Part II, LNCS 10178, pp. 581–593, 2017.
DOI: 10.1007/978-3-319-55699-4_36

spatial databases, people are still witnessing data quality issues widely existing in trajectory data. Specifically, trajectory has been seen to have quality issues at 2 different levels: the data level, which affects the quality of trajectory dataset itself, and the service level, which affects the qualify of trajectory-based applications. In this paper, we focus on dealing with data level quality issues.

For data level, trajectory data suffers from GPS errors and artificial forgeries quality issues. GPS-error-caused trajectory anomaly is inevitable due to the inaccuracy of GPS modules. Artificial-forgery-caused trajectory anomaly is constructed by humans for some purposes.

GPS-error-caused Anomaly. A trajectory contains two dimensions of information, i.e., spatial information (latitude and longitude) and temporal information (timestamp). The inaccuracy of GPS modules mainly appears in spatial dimension. A GPS module may record a location which is far from the real location of a moving object. This situation results to a spatial-error-based trajectory anomaly.

Artificial-forgery-caused Anomaly. Different with the inevitable GPS errors, some people use GPS simulators to construct fake trajectories on purpose. Due to the difference between hardwares and softwares of GPS simulators, there are mainly three trajectory simulation strategies, i.e., equal distance simulation, equal time simulation and direction changing simulation. A GPS simulator with equal distance simulation strategy generates trajectory sample points that every two sample points have the equal distance. A GPS simulator with equal time simulation strategy generates trajectory sample points that every two sample points have the equal time interval. A GPS simulator with direction changing simulation strategy generates trajectory sample points every time the simulated trajectory changes its direction.

Detecting these anomalies is important since trajectory data are widely used in industry and these anomaly can do a lot of harm to some businesses. Taking simulated trajectory as an example, there are some people using simulated trajectories to cheat money. Some Didi drivers use GPS simulators to generate trajectories without really passing through these roads. With the simulated trajectories, they can make fake transactions to win bonus awarded by Didi Inc. Didi Inc. losses a lot of money on these fake transactions. Several works have focusing on spatial-error-based trajectory anomaly detection, such as the maximum-range method. Since few methods can detect the simulated trajectories, we build a system which can detect the them.

The speed of roads is a feature which can detect the simulated trajectories. It is a commonsense that the traffic condition of different time of different days of a week is quite different. For example, the traffic of 5 pm should be far more slower than that of 5 am; also the traffic of weekdays has significant differences with the traffic of weekends. As for simulated trajectories, almost all GPS simulators simulate trajectories without considering the real-time traffic. Therefore, it is of high probability that the speed of a simulated trajectory has a big difference with real-time traffic. With the observation and awareness of that real speed feature can distinguish whether a trajectory is simulated or not, an *irregular measuring* process is to measure the speed gap between the given trajectory and the real traffic speeds.

It is a non-trivial task to perform simulated trajectory detection. First, the real traffic speed is not easy to get. There are mainly three sources to get real traffic speeds, i.e., sensor data, real-time trajectory data and historical trajectory data. Due to private issues, sensor data and real-time trajectory data are not easy to get. Historical trajectory data needs extraction before it can be used as real traffic speeds. Besides, historical trajectory data always suffers sparse and asymmetric distributed problems, that these problems cause the extracted real traffic speeds of low accuracy. Second, with real traffic speeds, a proper distance metric is needed to well measure the distance between the speed of the given trajectory and the real traffic speeds.

In order to address the two problems mentioned above, in this work, we propose a novel speed extraction method from historical trajectory data and three distance metrics to measure the distance between the speed of the given trajectory and the real traffic speeds. The speed extraction method divides time into fix time slots. Then calculate the average speed of each road during each time slot. For roads or time slots with little historical trajectories, we use probabilistic matrix factorization to estimate its speed. As for irregular measuring methods, we design three distance measures, i.e., voting-based detection, integral-based detection and model-based detection. Voting-based detection utilizes the rated voting system to decide whether a trajectory is a simulated trajectory. Integral-based detection measures the aggregate speed variance between the given trajectory and the real traffic speeds. Model-based detection leverages the power of SVM to classify whether a trajectory is a simulated trajectory or not.

To sum up, we make the following major contributions.

- We make a key observation that the existing of simulated trajectories, thus calls for a simulated trajectory detection to avoid the bad impact on trajectory-based applications such as Didi.
- We design a novel speed extraction method from historical trajectory data and three distance metrics to measure the distance between the speed of the given trajectory and the real traffic speeds.
- We conduct extensive experiments based on large-scale real trajectory dataset, which empirically demonstrates that the simulated trajectory detection system can detect most simulated trajectories.

The remainder of this paper is organized as follows. The preliminary concepts are shown in Sect. 2. We discuss the speed extraction and irregular measuring methods in Sects. 3 and 4 respectively. The experimental observations are presented in Sect. 5, followed by a brief review of related work in Sect. 6. Section 7 concludes the paper and outlines some future work.

2   Preliminary

In this section, we present some preliminary concepts. Table 1 summarizes the major notations used in the rest of the paper.

Table 1. Summarize of notations

Notation	Definition
T	a raw trajectory
$\overline{T}$	a calibrated trajectory
p	a sample point of a trajectory
r	a road in a road network
t	a timestamp
TS_i	a trajectory segment connecting p_i and p_{i+1} of T
S_t^r	traffic speed of road r at timestamp t
$speed_r(T)$	the moving speed of T on road r

Definition 1 (Trajectory). *A trajectory T is a finite sequence of locations sampled from the original route of a moving object and their associated timestamps, i.e., $T = [(p_1, t_1), (p_2, t_2), \cdots, (p_n, t_n)]$.*

A symbolic trajectory $\overline{T}$ is a sequence of roads and their corresponding timestamps, i.e., $\overline{T} = [(r_1, t_1), (r_2, t_2), \cdots, (r_m, t_m)]$. Utilizing the spatial-tempo map matching algorithm proposed in [23], we can map $T = [(p_1, t_1), \cdots, (p_n, t_n)]$ to the road network and get a symbolic trajectory $\overline{T} = [(r_1, t_1), \cdots, (r_m, t_m)]$.

Definition 2 (Simulated Trajectory). *A trajectory T is a GPS-simulated if T is generated by GPS simulators and no moving objects have passing through the roads of $\overline{T}$, the symbolic trajectory of T.*

Definition 3 (Trajectory Segment). *An segment $\overline{TS}_i$ of a symbolic trajectory $\overline{T}$ is a sub-trajectory which connects two consecutive landmarks l_i and l_{i+1} of $\overline{T}$.*

Problem: Given a trajectory T and the regular traffic speeds of a road network, distinguish whether T is a GPS-simulated trajectory.

3   Ground Truth Traffic

As mentioned above, speed feature can well distinguish whether a trajectory is GPS-simulated or not. In the following section we will introduce three sources of extracting traffic speeds.

Sensor Data. Speed sensors are very dense along highways in most cities. These sensors report real-time speeds from time to time. So the advantage of utilizing sensor data as real-time traffic ground truth is that the reported speed is very accurate. However, there are few sensors along several artery roads and most branch roads, so we cannot get all real-time speed of all the roads of a city from the source of sensors. Besides, the sensor data are private data of governments that it is hard to get sensor data.

Real-Time Trajectory Data. Another source of real-time speed data is real-time trajectory data. For a specific timestamp t and a specific road r, the speed of a car on r at t can roughly reflect the speed of the real-time traffic of the road. Similar with sensor data, the real-time trajectory data are hard to get due to private issues.

Historical Trajectory Data. Historical trajectory data is a rough source of real-time speed data. Since the traffic speed falls into a periodic pattern, we can estimate the speed of route at a time stamp based on previous trajectory data. For example, if the average traffic speed of moving objects on El Camino Real at 8 am on Monday is 50 km/h, then it is of high probability that the traffic speed of El Camino Real at 8 am on the up coming Monday is 50 km/h. The advantage of utilizing historical trajectory data as real-time ground truth is that several location-based services, e.g., open street map, provide historical trajectory data (easy to get). However, since the traffic speed changes from time to time, even a small accident or event may affect the traffic speed, so the estimated speed of historical trajectory data is less accurate than sensor data and real-time trajectory data.

Extracting traffic speed from sensor data and real-time trajectory data is straightforward, so we do not go into details. In the following section, we will introduce how to extract traffic speed from historical trajectory data. Since the traffic speeds vary with time and traffic speeds have periodical patterns, so we divide time into several time slots. In our experiments, we use every hour of a week as a time slot, thus there are $24 \times 7 = 168$ time slots in total. For a road r at a time slot t, we use the following method to calculate traffic speed $\mathcal{S}_t^r$ of road r at time slot t.

$$\mathcal{S}_t^r = \sum_{T \in \mathbb{T}_t^r} speed_r(T)$$

$\mathbb{T}_t^r$ are all the trajectories passing road r during time slot t and $speed_r(T)$ is the speed of T on road r. In a trajectory $T = [(p_1, t_1), \cdots, (p_n, t_n)]$, we can easily extract the speed $speed(TS_i) = \frac{d(p_{i+1}, p_i)}{t_{i+1} - t_i}$ for each trajectory segment TS_i. By utilizing the map matching algorithm [23], we can map a trajectory to roads. Thus if TS_i is mapped to r, then $speed_r(T)$ can be roughly equal to $speed(TS_i)$. However, some road in certain time slots may not have enough historical trajectories which causes the low accuracy of average speeds. So we set a threshold $thres_{support}$ to ensure that every average speed $\mathcal{S}_t^r$ has at least $thres_{support}$ historical trajectories passing r during t; otherwise $\mathcal{S}_t^r$ will be set to 0. With all the n roads and 168 time slots in our system, a $n \times 168$ matrix M with $m_{ij} = \mathcal{S}_{t_j}^{r_i}$ is built. If the historical trajectory dataset is small or trajectories are asymmetric distributed, M can be very sparse. So the next step is to estimate a proper average speed m_{ij}.

The average speed of different pairs of (road, time slot) are determined by some unweighed or even unobserved factors, which are regarded as some hidden factors. However, we do not manually specify these factors, as hard-coded factors are usually limited and biased. Instead, we assume the average speed of each road-time-slot pair is a linear combination of two groups of speeds, i.e. (1) how speeds change on different roads, and (2) how speeds change with time. Then we employ Probabilistic Matrix Factorization (PMF) [16] to factorize M into two latent feature matrices, $W \in R^{d \times n}$ and $L^{d \times 168}$, which are the latent road and time slot feature matrices, respectively. Further, we assume there exists observation uncertainty R, and the uncertain follows a normal distribution. Thus the distribution of a new road-time-slot familiarity matrix M' conditioned on W and L is defined as follows:

$$p(M'|W, L, \sigma^2) = \prod_{i=1}^{n} \prod_{j=1}^{168} [\mathcal{N}(M_{ij}|W_i^T L_j, \sigma^2)]^{I_{ij}}$$

where $\mathcal{N}(x|\mu, \theta^2)$ is the probability density function of the normal distribution with mean μ and variance θ^2, and I_{ij} is a indicator which is equal to 1 if M_{ij} is not zero, otherwise 0. The prior of W and L are defined as follows:

$$p(W|\sigma_W^2) = \prod_{i=1}^{n} \mathcal{N}(W_i|0, \sigma_W^2 I) \qquad p(L|\sigma_L^2) = \prod_{i=1}^{168} \mathcal{N}(L_i|0, \sigma_L^2 I)$$

where I is identity matrix. The following objective function maximizes the posterior of W and L with regularization terms, which minimizes the prediction difference between our model and the observed M, and also automatically detects the appropriate number of factors d through the regularization terms:

$$\sum_{i=1}^{n} \sum_{j=1}^{168} I_{ij}(M_{ij} - W_i^T L_j)^2 + \lambda_W \sum_{i=1}^{n} \|W_i\|_F^2 + \lambda_L \sum_{j=1}^{168} \|L_j\|_F^2$$

where $\lambda_W = \theta^2/\theta_W^2$, $\lambda_L = \theta^2/\theta_L^2$, and $\|\cdot\|_F^2$ denotes the Frobenius norm. A local minimum of the objective function can be found by performing gradient descent in W and L. Afterwards, more speeds between roads and time slots are inferred in M.

4   Irregular Measuring

In this section, we discuss in detail about irregular measuring based on traffic ground truths. Given a ground truth traffic set $\mathbb{S}$ and a trajectory $T = [(p_1, t_1'), \cdots, (p_m, t_m')]$, our system will tell whether T is simulated or not. We propose three methods to measure the distance between the speed of the given trajectory and the real traffic speeds. All of these methods share the same process that they need to extract two features: *ground truth speed* and *given trajectory speed*. The ground truth speed $\mathcal{S}_t^r$ of road r at time t is provided by ground truth traffic set $\mathbb{S}$. Recall Sect. 3, we extract the given trajectory speed $speed_r(T)$ of trajectory T on road r. Utilizing the spatial-tempo map matching algorithm proposed in [23], we can map T to the road network and get a symbolic trajectory $\overline{T} = [(r_1, t_1), \cdots, (r_n, t_n)]$. The speed of trajectory $\overline{T}$ on road r_i is denoted by $speed_{r_i}(\overline{T})$. $speed_{r_i}T$ can be evaluated as $\frac{d(p_b, p_a)}{t_b - t_a}$ where p_a and p_b are the starting point and end point of T on road r_i respectively.

4.1   Voting-Based Detection

In this part we present a voting-based anomaly detection method. The detection process is quite similar to rated voting system [18], of which the wining option is chosen according to the voters preferences score of options and the number of voters preferring the options. In our system, we can treat each road of passing by roads of $\overline{T}$ as a voter. Options are 'anomaly trajectory' and 'normal trajectory' indicating whether the trajectory T is simulated or not. Adopting the idea of rated voting system, we can measure the road r's preference of whether T is simulate by the following two steps:

- measure the variance $\mathcal{V}(speed_r(T), \mathcal{S}_t^r)$ between the given trajectory speed of road r and the ground truth speed of r at the associating timestamp t;
- if $\mathcal{V}(speed_r(T), \mathcal{S}_t^r)$ is bigger than a user define threshold $thres_{var}$, then r will vote $\overline{T}$ as an 'anomaly trajectory'; otherwise r will vote $\overline{T}$ as a 'normal trajectory'.

Then we sum up the preferences of each option voted by roads and road importances. The importance of each voter/road is associate with the length of the road. We choose the option with higher adding up preference scores as the result.

4.2   Integral-Based Detection

In this subsection we introduce an integral-based anomaly detection. The main idea of the integral-based anomaly detection is to measure the aggregate speed variance between the given trajectory and the ground truth traffic. If the aggregate variance is bigger than a user define threshold $thres_{int}$, then trajectory T is regarded as a simulated trajectory. Therefore the aggregate speed variance $\mathcal{V}_T$ of symbolical trajectory $\overline{T} = [(r_1, t_1), \cdots, (r_n, t_n)]$ is measured as following:

$$\mathcal{V}_T = \frac{1}{t_n - t_1} \int \frac{|speed_r(T) - \mathcal{S}_t^r|}{speed_r(T)} dt$$

where $\frac{1}{t_n - t_1}$ is a normalization factor. For ground truth traffic extracted from historical trajectory data, it cannot provide fine-grained speeds associated with timestamps. So we use the following equation to measure the aggregate variance for coarse-grained ground truth traffic.

$$\mathcal{V}_T = \sum_{r \in \overline{T}} \frac{length(r)}{length(T)} \cdot \frac{|speed_r(T) - \mathcal{S}_t^r|}{speed_r(T)}$$

where $length(r)$ and $length(T)$ are the length of road r and trajectory T respectively.

4.3 Model-Based Detection

To further improve the detection performance, we propose a more advanced model-based anomaly detection approach, which leverages the power of Support Vector Machine (SVM) to classify whether a trajectory is a simulated trajectory. The basic idea of the model-based detection algorithm is to describe a trajectory by its speed features on all the trajectory segments, and build a classifier to separate abnormal trajectories from the normal ones.

The speed on each trajectory segment could be quite different, e.g., a car can move much faster on a high way than on local street; the speeds could have much wider variance on a very busy road. Therefore, we need to normalize all the speed features. In the model-based method, we follow the standard approaches in estimating speed, and assume the ground truth speed follows an normal distribution, i.e., $f(x|\mu, \sigma^2)$. The mean μ_r of the distribution is ground truth speed S_t^r of road r at timestamp t and the standard variance σ is computed from all the trajectories on road r at timestamp t. Then we can compute the normalized speed x' of the given trajectory on road r at timestamp t is measured as following:

$$x' = \frac{speed_r(T) - \mu_r}{\sigma_r}$$

With the normalized speed of all roads of the given trajectory, a vector ns is built. Notably, the dimension size of the vector is the number of all the roads in the road network. And the vector only has values on dimensions of its passing roads. For example, if there are 1000 roads in the road network, then ns is a 1000×1 vector. And if trajectory T passes through roads r_{100}, r_{500} and r_{1000}, then the normalized vector ns_T of T only has value on dimension 100, 500 and 1000; the value of all other dimensions are 0.

The second step is to build a supervised classification model to classify whether a vector is a simulated trajectory. In our experiment, we utilize the SVM model. In order to train a best model, we need to minimize the following objective function:

$$\frac{1}{n} \sum \max(0, 1 - y_r(wx_r + b)) + \lambda ||w||^2$$

A global minimum value can be found by performing gradient descent algorithm.

5 Experiment

In this section, we conduct extensive experiments to validate the effectiveness of our GPS-simulated trajectory detection system. Our system is implemented in Java. All the experiments are run on a computer with Intel Core i7-2600 CPU (3.40 GHz) and 16 GB memory.

5.1   Experiment Setup

Commercial Map: We use the commercial map of a large city—Beijing—provided by a collaborating company. The commercial map is used to build the road network and to provide the length of a road.

Trajectory Dataset: We use a real-world trajectory dataset generated by 33,000+ taxis in Beijing over three months. This dataset has more than 100,000 trajectories.

Simulated Trajectory Dataset: We use GPS simulators to generate a simulated trajectory dataset using three sampling strategies. We set the simulated speed by varying the traffic speed to several degrees. We decelerate the traffic speed by 30%, 50%, 70% and 90%, denoted as *S30, S50, S70 and S90*. Similarly, we accelerate the traffic speed by 30%, 50%, 70% and 90%, denoted as *F30, F50, F70 and F90*.

5.2   Performance Evaluation

We study both the effectiveness and efficiency of our GPS-simulated trajectory detection system. In all our algorithms, we set the value of threshold $thres_{support}$, $thres_{var}$ and $thres_{int}$ as 5, 0.4 and 0.4 respectively. To study the effectiveness, we study the following 2 aspects of the GPS-simulated trajectory detection: (1) the accuracy rate of distinguishing whether a given trajectory is a GPS-simulated trajectory; (2) the impact of thresholds $thres_{var}$ and $thres_{int}$ to our detection system. As to the efficiency test, we record the time cost for distinguishing a single trajectory.

Accuracies. In the first set of experiments, we evaluate the accuracy rate of each irregular measuring method. We randomly select 5000 real trajectories from the trajectory dataset and 5000 $S30$ simulated trajectories from the simulated trajectory dataset to form the $S30$ test dataset. Repetitive eight times, we construct eight test datasets, i.e., $S30$, $S50$, $S70$, $S90$, $F30$, $F50$, $F70$ and $F90$ test datasets. For every trajectory of every test dataset, we use these three methods to distinguish whether it is a simulated trajectory. And then we calculate the accurate rate of each method on each test datasets. Figure 1 shows the results of accuracies of three methods on eight test datasets. Not surprisingly, accuracies of all methods gradually increase with the increase of speed variation. And all accuracy values are over 50%. The model-based method turns to be the best approach in terms of the capability of distinguishing simulated trajectory, as we can see that the accuracies of model-based method are always the highest on all the datasets. However, voting-based method and integral-based method have a significant drop on $S30$ dataset and $F30$ dataset.

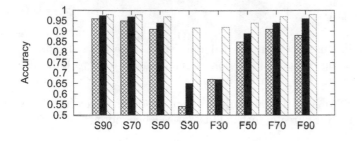

Fig. 1. User interface of STMaker

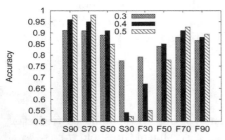

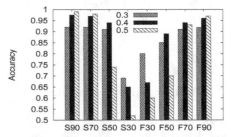

(a) Accuracy of voting-based method with different thresholds

(b) Accuracy of integral-based method with different thresholds

Fig. 2. Impact of threshold

Effect of Thresholds. Next we test how the thresholds $thres_{var}$ and $thres_{int}$ used in the voting-based method and integral-based method respectively affect the simulated trajectory detection. Recall that a higher $thres_{var}$ and a higher $thres_{int}$ result in that a trajectory with bigger speed variation will be detected as a simulated trajectory. In other words, a simulated trajectory has a lower probability to be treated as a simulated trajectory while a normal trajectory has a higher probability to be treated as a normal trajectory; In order to work out a good trade-off to get a high correctness of distinguishing both simulated trajectory and normal trajectory, we tune the threshold $thres_{var}$ and $thres_{int}$ from 0.3 to 0.5 with the step of 0.1. Meanwhile.we calculate the accuracies of voting-based method and integral-based method on eight test datasets. As shown in Fig. 2, generally all the accuracies increase with the increase of speed variation. When the threshold is low (0.3), these methods have higher accuracies comparing to those with higher thresholds on $S50$, $S30$, $F30$ and $F50$ test datasets; however, methods with high threshold (0.5) haver higher accuracies comparing to those with lower thresholds on $S90$, $S70$, $F70$ and $F90$ test datasets. Based on the observations of this experiment, we recommend the threshold with the value 0.4 to be appropriate.

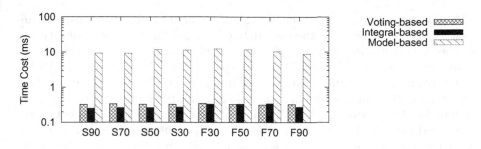

Fig. 3. Average time cost for distinguishing one trajectory

Time Cost. We also evaluate the time cost of simulated trajectory detection, which is especially important for online systems. The average time cost for distinguishing a single trajectory is shown in Fig. 3, from which we observe that all the methods can distinguish a trajectory within tens of milliseconds. Integral-based method turns out to be the most efficient approach. The voting-based approach constantly runs faster than the model-based approach, since the model-based approach involves expensive inference.

6 Related Work

In this section, we review these existing anomaly detection problems. We also review existing works on trajectory noise and map-matching problem.

Anomaly Detection. Many methods for anomaly detection have previously been proposed. Paper [3,7] provide extensive surveys of different anomaly detection methods and application. The paper [8] focuses on the similarity-based approach to anomaly detection. This approach typically involves transforming similarities between a test sample and training samples into an anomaly score. Other related methods for anomaly detection include using the sum of the distance between a sample and its kth-nearest neighbor as the anomaly score for the sample [5] and non-parametric adaptive anomaly detection methods using geometric entropy minimization [22].

Trajectory Noise. Existing approaches rely on point-by-point matching to map individual GPS points to a road segment. However, GPS data is imprecise due to noise in GPS signals. GPS coordinates can have errors of several meters so that direct mapping of individual points is error prone. Paper [14] proposes a radically different method to overcome inaccuracy in GPS data as every GPS point is potentially noisy, which considers the set of relevant GPS points in a trajectory that can be mapped together to a road segment. This algorithm outperforms state-of-the-art methods in terms of both accuracy and computational cost, even with even with highly noisy GPS measurements.

Map-matching Problem. There are a host of studies on map-matching problem. These methods used to be classified into three types: local/incremental approach, global aproach, and statistical approach. The local/incremental method try to find a local match of geometried [12]. The incremental methods usetwo similarity measures to evaluate the candidate edges, one for distance similarity and another for orientation similarity. The combined similarity measure is computed as the sum of individual scores. The time complexity is $O(n)$ once we find adjacent edges for each sample, where n is the number of GPS points to be matched. The method of "adaptive clipping" uses Dijkstra alogrithm to construct shortest path on local free space graph [4]. The aim of the global methods is matching the entire trajectory with the road network. An offline snapping method that aims to find a minimum weight path based on edit distance proposed by paper [9]. Other methods [6] are based on Frchet distance or its variants. Matching GPS observation also use statistical models. Paper [19] introduce a method for reliable matching of position and orientation measurements from a standard GPS receiver to a digital map.

7   Conclusions

In this paper we have taken an important step towards simulated trajectory detection. We propose a novel speed extraction method from historical trajectory data and three distance metrics to measure the distance between the speed of the given trajectory and the real traffic speeds. We conducted extensive experiments on a real-life trajectory dataset. The experiment results show that our simulated trajectory detection methods can detect most simulated trajectories. We expect this work will attract more attentions on improving the quality of trajectory data.

Acknowledgments. This research is supported by UESTC (Grant No: ZYGX2016KYQD135).

References

1. Cai, Y., Ng, R.: Indexing spatio-temporal trajectories with chebyshev polynomials. In: SIGMOD, pp. 599–610 (2004)
2. Chakka, V., Everspaugh, A., Patel, J.: Indexing large trajectory data sets with seti. In: CIDR (2003)
3. Chandola, V., Banerjee, A., Kumar, V.: Anomaly detection: a survey. ACM Comput. Surv. **41**(3), 75–79 (2009)
4. Wenk, C., Salas, R., Pfoser, D.: Addressing the need for map-matching speed: localizing global curvematching algorithms. In: Proceedings of the 18th International Conference on Scientific and Statistical Database Management (2006)
5. Eskin, E., Arnold, A., Prerau, M.: A geometric framework for unsupervised anomaly detection: detecting intrusions in unlabeled data. In: Barbará, D., Jajodia, S. (eds.) Applications of Data Mining in Computer Security. Springer, New York (2002)
6. Alt, H., Efrat, A., Rote, G., Wenk, C.: Matching planar maps. J. Algorithms **49**(2), 262–283 (2003)
7. Hodge, V.J., Austin, J.: A survey of outlier detection methodologies. Artif. Intell. Rev. **22**(2), 85–126 (2004)

8. Hsiao, K.-J., Xu, K.S., Calder, J., Hero III, A.O.: Multi-criteria anomaly detection using pareto depth analysis. Eprint Arxiv, vol. 25, pp. 854–862 (2011)
9. Yin, H., Wolfson, O.: A weight-based map matching method in moving objects databases. In: International Conference on Scientific and Statistical Database Management, pp. 437–438 (2004)
10. Jeung, H., Shen, H., Zhou, X.: Convoy queries in spatio-temporal databases. In: ICDE, pp. 1457–1459 (2008)
11. Jeung, H., Yiu, M., Zhou, X., Jensen, C., Shen, H.: Discovery of convoys in trajectory databases. PVLDB **1**, 1068–1080 (2008). VLDB Endowment
12. Greenfeld, J.: Matching GPS observations to locations on a digital map. In: Proceedings of 81th Annual Meeting of the Transportantion Research Board (2002)
13. Lee, J., Han, J., Whang, K.: Trajectory clustering: a partition-and-group framework. In: SIGMOD, pp. 593–604. ACM (2007)
14. Li, H., Kulik, L., Ramamohanarao, K.: Robust inferences of travel paths from GPS trajectories. Int. J. Geog. Inform. Sci. **29**(12), 2194–2222 (2015)
15. Li, Z., Ding, B., Han, J., Kays, R.: Swarm: mining relaxed temporal moving object clusters. PVLDB **3**, 723–734 (2010)
16. Mnih, A., Salakhutdinov, R.: Probabilistic matrix factorization. In: NIPS, pp. 1257–1264 (2007)
17. Ni, J., Ravishankar, C.: Indexing spatio-temporal trajectories with efficient polynomial approximations. TKDE **19**(5), 663–678 (2007)
18. Nordmann, L., Pham, H.: Weighted voting systems. IEEE Trans. Reliab. **48**(1), 42–49 (1999)
19. Pink, O., Hummel, B.: A statistical approach to map matching using road network geometry, topology and vehicular motion constraints. In: International IEEE Conference on Intelligent Transportation Systems, pp. 862–867 (2008)
20. Pfoser, D., Jensen, C., Theodoridis, Y.: Novel approaches to the indexing of moving object trajectories. In: VLDB, pp. 395–406 (2000)
21. Spaccapietra, S., Parent, C., Damiani, M.L., de Macedo, J.A., Porto, F., Vangenot, C.: A conceptual view on trajectories. Data Knowl. Eng. **65**(1), 126–146 (2008)
22. Sricharan, K., Hero III, A.O.: Efficient anomaly detection using bipartite k-NN graphs. In: Advances in Neural Information Processing Systems, vol. 24, pp. 478–486 (2011)
23. Su, H., Zheng, K., Huang, J., Wang, H., Zhou, X.: Calibrating trajectory data for spatio-temporal similarity analysis. VLDB J. **24**(1), 93–116 (2015)
24. Wang, H., Zheng, K., Xu, J., Zheng, B., Zhou, X., Sadiq, S.: SharkDB: an in-memory column-oriented trajectory storage. In Proceedings of the 23rd ACM International Conference on Conference on Information and Knowledge Management, pp. 1409–1418. ACM (2014)
25. Yan, Z., Spaccapietra, S., et al.: Towards semantic trajectory data analysis: a conceptual and computational approach. In: VLDB Ph.D Workshop (2009)

Social Networks and Graphs
(Industrial)

Predicting Academic Performance
via Semi-supervised Learning
with Constructed Campus Social Network

Huaxiu Yao, Min Nie, Han Su, Hu Xia, and Defu Lian$^{(\boxtimes)}$

Big Data Research Center, University of Electronic Science and Technology of China,
Chengdu, China
dove@uestc.edu.cn

Abstract. Wide attention has been recently paid to academic performance prediction, due to its potentials of early warning and subsequent in-time intervention. However, there are few studies to consider the effect of social influence at predicting academic performance. The major challenge comes from the difficulty of collecting a precise friend list for students. To this end, we first construct students' social relationship based on their campus behavior, and then predicts academic performance using constructed social network by semi-supervised learning. We evaluate the proposed algorithm on over 5,000 students with more than 14M behavior records. The evaluation results show the potential value of campus social network for predicting academic performance and the effectiveness of the proposed algorithm.

1 Introduction

Higher education management is centered at students, aiming to cultivate them to become contributors and leaders in every walk of life. One important factor in higher education management is academic performance prediction, which provides educators with references of their decisions. For example, if educators know students' academic performance in advance, they are able to intervene in time to provide students with guidance, so that course failure could be high probably prevented. This is important for educational management since course failure largely affects students' graduation, job seeking and even future development. There have been some methods to predict students' academic performance based on the information from different sources, such as self-report of examinees [14], automatic sensing behavior data obtained from smartphones [13], etc. However, there are few studies to consider the effect of social influence at predicting academic performance while such studies could be useful for providing guidance for educators to develop effective intervention strategies

The most important challenge lies in the difficulty to collect campus social relationship for students. Nowadays, many colleges and universities have built a variety of advanced information management and monitor systems to improve the effectiveness and convenience of students' life. When students continuously

S. Candan et al. (Eds.): DASFAA 2017, Part II, LNCS 10178, pp. 597–609, 2017.
DOI: 10.1007/978-3-319-55699-4_37

interact within a cyber-physical space, their activities in and around the campus are accumulated and collected. As suggested in [2], these behavior data provide us some potential to construct students' social relationship from their behavior data, since they find even a very small number of location co-occurrences can result in a high empirical likelihood of a social tie. Different from their work, there is no information of real friendship available, and thus it is impossible to construct social network in a supervised way. Instead, we leverage a shuffling test method for evaluating the significance of each detected social tie. In particular, based on students' behavior data, we first calculate the number of co-occurrence at a specific location; and then build a null model by shuffling behavior data for each student multiple times and re-calculating their respective co-occurrence frequency. By comparing the co-occurrence frequency between two students in these two cases, we can identify the significance of each detected social tie. Intuitively, significant co-occurrence at different locations could play a different role in identifying a social tie, so that the co-occurrence frequency at different locations could not be added together directly. Instead, we describe it as an additive fusion problem of multiple social networks, where each location implies a social network and each network is assigned a distinct weight. However, since there is no ground friendship between students used as training data, it is impossible to determine the weights of different locations via supervised learning, but these weights are considered as parameters for following optimization.

Based on the constructed campus social network with identical weights, we first investigate the significance of social influence for academic performance, and find that friends (two frequently co-occurring student) tend to have similar academic performance. Such an observation motivates a semi-supervised algorithm for academic performance prediction. Therefore, we propose a novel Label Propagation algorithm on Multiple Networks (LPMN) to predict students' academic performance, which optimizes the weights of different locations at the same time of label propagation.

Finally, we evaluate the proposed label propagation on multiple networks (LPMN) algorithm over around 5,000 students with more than 14M behavior records. Here, out of privacy concerns, the academic performance is not directly associated with educational outcomes, such as Grade Point Average(GPA). Instead, only four performance levels are considered, where each level contains 25% students. In this evaluation, 20% randomly selected students are regarded as testing users for predicting their academic performance, and among the remaining 80% students, we randomly select a fixed percentage (20%, 40%, 60%, 80%, 100%) of training data to be labeled. The results show the potential value of campus social network for predicting academic performance and the effectiveness of the proposed algorithm.

2   Related Work

2.1   Social Tie Inferring

In the community of network mining and complex network, many efforts have been devoted to inferring social ties from users' daily activities. For example, making use of the co-occurrence records in geographical space, Crandall et al. inferred social ties among Flickr users [2]. They proposed a probabilistic framework to demonstrate that the probability of social tie increases when the frequency of co-occurrence increases. Based on some specific characteristic patterns, Sadilek et al. exploited a probabilistic model to predict people's social ties at the same time of forecasting future locations [9]. Mobile behavior data [3] is used for mining several patterns. These patterns are leveraged for inferring friendship and analyzing the satisfaction of everyone. Based on the cellphone network data, a classification model is built for identifying two types of relation: family and colleague [15]. A dynamic model was proposed in [12] to understand community gathering process in Whrrl and Meetup, which is used for social ties prediction and recommendation. A generative model [5] is exploited for inferring rare links which seems to occur accidentally. An entropy-based model is proposed in [8] for inferring social ties and measuring the strength of social ties by analyzing the spatial information of users. However, there is still little study about inferring interpersonal relation in campus, the complete ecological system, especially based on large-scale behavior data from many students.

2.2   Academic Performance Prediction

Recently, a variety of machine learning models have been used to predict students' academic performance. For example, a neural network model is used for predicting performance from their placement test score [4]. Multiple instance learning is used in [16] to predict students' performance in an e-learning environment. Longitudinal data analysis is leveraged in [10] for predicting whether a student is at risk of getting poor assessment performance based on previous test performance and course history. Tensor factorization [11] between students, skills and tasks is used in predicting student performance based on the log files of solving problems in the tutoring system. PLSA [1], is used for academic performance prediction based on all available information about the educational content and users/students in intelligent tutoring systems. A fuzzy cognitive diagnosis framework is proposed in [14] for discover the latent characteristics of examinees for predicting the performance on each problem. To the best of our knowledge, there are few researches about inferring student's academic performance based on their campus friends.

3   Campus Social Network Construction

In modern university, taking advantage of the existing information management and operating systems, students use campus smartcards and generate spatial-temporal digital records in the information center during daily activities every day, such as fetching boiled water in the teaching buildings, entering the library, paying for meals in canteen, etc. Based on millions of students' behavior records, in this section, we design a framework to build the campus social network.

3.1   Social Network Construction

As illustrated in Fig. 1, *co-occurrence* means two students generate records at the same location within a short time interval, where time interval is empirically set as one minute. In this example, student A and B co-occur four times in classroom, library, canteen and campus shuttle, respectively. Thus, in our work, we collect these co-occurrence records at seven distinct locations, including canteen, library, classroom, campus shuttle, campus-to-campus bus, supermarket, bathroom.

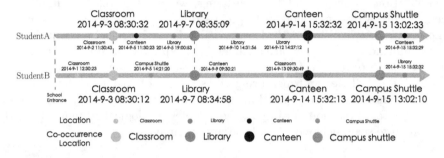

Fig. 1. Illustration of co-occurrence between two students

As suggested in [2], with the increase of co-occurrence frequency between two students, it is more evident that they are close friends. Different from their work, however, there is no any available information of real friendship information in our case, so it is impossible for us to construct social network in a supervised way. To this end, we make use of a shuffling test method to detect significant social ties among students. In particular, we first construct a null model to simulate the random case of location co-occurrences by randomly shuffling the timestamps of activity records, and then get the co-occurrence frequency in the random case. We conduct 20 rounds of permutations to approximately estimate the co-occurrence frequency distribution in the random case, and then compute the mean and standard deviation. The comparison of the co-occurrence frequency between the null model and the real case at three locations are shown in Fig. 2(a)–(c) respectively. The threshold should be set to keep the co-occurrence frequency of real case above the mean co-occurrence frequency plus two times of standard deviation of random case. For example, in canteen, when co-occurrence frequency of two students more than 110, they are likely to be friends. The thresholds of these three locations are 110, 17 and 25, respectively. It is worth to note that the threshold of dining hall is significantly larger than the other two locations. This is because the temporal distribution of the meal behavior among students almost aligns with each other so that there is a comparatively large randomly co-occurred probability. Similarly, the threshold of library is the smaller than others, because the temporal distribution of library entrance is based on the real friendship. Based on the derived thresholds, we drop those co-occurrence frequencies below the thresholds at each location since they correspond to the randomly co-occurring case.

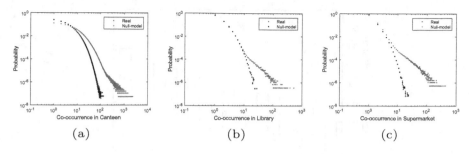

Fig. 2. (a), (b), (c) represent co-occurrence distribution of real/null model at canteen, library, supermarket respectively.

To finally obtain social ties between students, co-occurrence frequency at different locations should be combined together. But it is unreasonable to aggregate them directly, since significant co-occurrences at different locations could make distinct contribution for implying their social ties. Thus, we assume co-occurrence frequency of a location l is weighted by α_l when estimating the strength of each social tie, and these weights will be learned via the following optimization algorithm. In addition, it is intuitive that in our definition, the strength (weight) of social tie from i to j does not equal to the strength from j to i because of the bidirectional character of social influence. For example, campus stars, such as the president of a student union, have extensive social circle. Benefit from their excellent performance, stars frequently co-occur with different students and influence their behavior. However, most of students who co-occur with campus stars are unclubable, they only affect campus stars a little. The weight of the social tie from student j to student i is represented as:

$$w_{ij} = \sum_{k=1}^{L} \frac{\alpha^k O_{ij}^k}{\max_{j' \in N(i)} O_{ij'}^k},\qquad(1)$$

where L means the number of locations, $N(i)$ is defined as the set of student i's friends at location k and O_{ij}^k is the frequency of co-occurrence between student i and j at location k.

4   Social Influence

When two students become friends, they will influence each other with regard to study and manifest homophily phenomenon [7] in terms of academic performance. In other words, each student's academic performance should be close to their friends'. Below, we investigate the significance of such a conjecture. But first of all, we need to make some preprocessing. First, by simply assuming the weight of each location equals to each other, we construct a campus social network with 5,388 students and 57,994 social ties. This is reasonable to some extent since there is neither evidence on the importance of co-occurrence at different

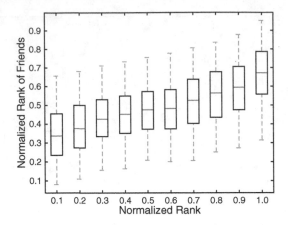

Fig. 3. Correlation between students' performance and the average of their friends' performance. X axis represents the rank of each student and Y axis represents the average rank of their friends. Normalized rank close to zero means better performance.

locations nor real friendship information used for training data. Then, out of protecting privacy, we convert each student's grade point average (GPA) into the rank among the students of the same major. For the sake of comparability of the same rank among different majors, we need to normalize the rank by the number of students of the corresponding major since the number of students varies from major to major. The smaller the normalized rank is, the better the academic performance is.

After preprocessing, we test the significant effect of social influence with regard to academic performance by comparing the similarity of academic performance between friends with that between non-friends. According to [6], we first need to define the similarity of academic performance between a students i and a student group G as follows:

$$S_G(i) = \frac{\sum_{j \in G} \text{sim}(i,j)}{|G|}. \tag{2}$$

where $\text{sim}(i,j) = |p_i - p_j|$ measures the similarity of academic performance between student i and student j. For each student i, we calculate two similarities $S_F(i)$ and $S_{NF}(i)$. $S_F(i)$ is the average similarity between student i and his friends, and $S_{NF}(i)$ is the average similarity between student i and bootstrap-sampled students from his non-friend list. A two-sample t-test on the vectors $\mathbf{S}_F$ and $\mathbf{S}_{NF}$ are conducted in the following. The null hypothesis is $H_0 : \mathbf{S}_{UF} \geq \mathbf{S}_F$ and the alternative hypothesis is $H_1 : \mathbf{S}_{UF} < \mathbf{S}_F$. In our experiment, the null hypothesis is rejected at significant level $= 0.001$ with p-value < 0.0001, indicating that students with friendship have closer academic performance than those without. Then, we further illustrate the correlation of academic performance between students' and their friends', and show them in Fig. 3. This figure clearly confirms the effect of social influence with regard to academic performance.

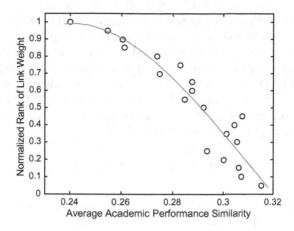

Fig. 4. Correlation the strength of social ties and the similarity of academic performance. X axis represents the average academic performance similarity and Y axis represents the normalized rank of social tie strength. The black dots mean the average academic performance of each level of strength, and the blue line is a fitting curve. (Color figure online)

Furthermore, intuitively, students' academic performance should be closer to their friends with strong tie (more co-occurrences) than to that with weak tie. Therefore, we analyze the relationship between the similarity of academic performance and the strength (weight) of social tie. We first rank all social ties based on their strength and normalize their order by subtracting the number of social ties. Then, we discretize the rank of link weight to 20 levels, so that the first level value in 0.00−0.05 means the strongest social ties and the twenty level value in 0.95−1.00 indicates the weakest social ties. Finally, we calculate the average similarity of academic performance between students and their friends in every level, and show its correlation with the strength of social ties in Fig. 4. The results confirm the intuition on the correlation between the strength of social ties and the similarity of academic performance.

5 Academic Performance Prediction

In the previous section, we have found that friends tend to have close academic performance. This phenomenon motivates us to predict academic performance using social influence based on constructed campus social network. Therefore, in this section, we introduce a novel label propagation algorithm on multiple networks to predict students' performance.

5.1 Label Propagation

To further protect students' privacy, we discretize the academic performance rank of each student into four levels (I, II, III, IV), where each level contains

25% students. We assume the constructed campus social network is represented as a $N \times N$ affinity matrix $\mathbf{W}$, where w_{ij} means the strength of link from student j to student i. And denote students' initial academic level matrix as $\mathbf{T}$ of size $N \times 4$, whose element at the position (i, k) is 1 if student i belongs to the level k, $k \in \{1, 2, 3, 4\}$. The rows with all zero entries in this matrix indicate the unknown academic performance of the corresponding students. Finally, denote the prediction score matrix as $\mathbf{F}$ of the same size as $\mathbf{T}$. Then, the loss function of label propagation (LP) is represented as follows:

$$\mathcal{L} = \lambda \|\mathbf{F} - \mathbf{T}\|_F^2 + (1 - \lambda) \sum_{i,j=1}^{N} w_{ij} (\mathbf{f}_i - \mathbf{f}_j)^2, \tag{3}$$

where $\mathbf{f}_i$ and $\mathbf{f}_j$ represent the predicting academic performance vector of student i and student j respectively. The first term of this objective function minimizes the error between the prediction and the given performance, and the second term is to capture of social influence. The parameter λ controls the balance between these two terms. Let $\frac{\partial \mathcal{L}}{\partial \mathbf{F}} = 0$, the optimal value of $\mathbf{F}$ is

$$\mathbf{F} = 2\lambda((1 - \lambda)(\mathbf{D} + \mathbf{D}^T - \mathbf{W} - \mathbf{W}^T) + 2\lambda \mathbf{I})^{-1} \mathbf{T}, \tag{4}$$

where $\mathbf{D}$ is the diagonal matrix with degrees of each vertex in social network, i.e., $d_{ii} = \sum_{j=1}^{N} w_{ij}$. Because the loss function in Eq. (3) is convex, the solution in Eq. (4) is globally optimal. Then the predicting level p_i of academic performance for the student i can be determined as $p_i = \arg\max_k f_{ik}$, since the entry f_{ik} in the matrix $\mathbf{F}$ means the final prediction score of student i belongs to academic level k.

5.2   Label Propagation on Multiple Networks

We have mentioned that different locations make distinct contribution for inferring social ties between students, and their individual contribution α_l can not be determined via supervised learning. Based on the label propagation algorithm, we are supposed to learn the location contribution, i.e., the value of vector $\boldsymbol{\alpha}$. Since the data are mainly collected from seven locations, we revise affinity matrix $\mathbf{W}$ as $\hat{\mathbf{W}} = [\hat{w}_{ij}]$, where $\hat{w}_{ij} = \sum_{k=1}^{L} \alpha_k u_{ijk}$ while $u_{ijk} = O_{ij}^k / \max_{j' \in N(i)} O_{ij'}^k$ means the weight of link from student j to student i at location k. Then, the loss function of this label propagation algorithm is represented as:

$$\mathcal{L} = \lambda \|\mathbf{F} - \mathbf{T}\|_F^2 + (1 - \lambda) \sum_{i,j=1}^{N} \hat{w}_{ij} (\mathbf{f}_i - \mathbf{f}_j)^2 + \mu \|\boldsymbol{\alpha}\|_2^2,$$

$$s.t., \quad \sum_{k=1}^{L} \alpha_k = 1, \quad \forall k, \ \alpha_k \geq 0, \tag{5}$$

where $\hat{\mathbf{D}}$ is similar to $\mathbf{D}$, i.e., $\hat{d}_{ii} = \sum_{j=1}^{N} \hat{w}_{ij}$. μ is a parameter for regularization. Since the final network can be regarded as the combination of several networks corresponding to different locations, we call this algorithm as label propagation on multiple networks (LPMN).

5.3   Optimization

We use alternative minimization to learn all parameters:

- **Optimization of F**

When α is fixed, then the value of matrix $\hat{\mathbf{W}}$ is fixed. The optimization problem is equivalent to the minimization of the following objective function:

$$\min_{\mathbf{F}} \lambda \|\mathbf{F} - \mathbf{T}\|_F^2 + (1 - \lambda) \sum_{i,j=1}^{N} \hat{w}_{ij} (\mathbf{f}_i - \mathbf{f}_j)^2 \tag{6}$$

Then, according to the solution of label propagation, we obtain the updating rule of $\mathbf{F}$ as follows:

$$\mathbf{F} = 2\lambda((1 - \lambda)(\hat{\mathbf{D}} + \hat{\mathbf{D}}^T - \hat{\mathbf{W}} - \hat{\mathbf{W}}^T) + 2\lambda\mathbf{I})^{-1}\mathbf{T}. \tag{7}$$

- **Optimization of α**

When $\boldsymbol{F}$ is fixed, the objective function is equivalent to the minimization of the following one subject to the simplex constraint:

$$\min_{\boldsymbol{\alpha}}(1 - \lambda) \sum_{i,j=1}^{N} \sum_{k=1}^{L} \alpha_k u_{ijk}(f_i - f_j)^2 + \mu\|\alpha\|_2^2. \tag{8}$$

Because we fix $\mathbf{F}$, the value of $\sum_{i,j=1}^{N} u_{ijk}(\mathbf{f}_i - \mathbf{f}_j)^2$ is a constant. Thus, we define a new vector $\boldsymbol{\beta}$ ($\beta_k = \sum_{i,j=1}^{N} u_{ijk}(f_i - f_j)^2$), then the objective function equals to

$$\min_{\boldsymbol{\alpha}} \left\|\boldsymbol{\alpha} + \frac{1-\lambda}{2\mu}\boldsymbol{\beta}\right\|_2^2$$

$$s.t., \quad \forall k, \ \alpha_k \geq 0, \quad \sum_{k=1}^{L} \alpha_k = 1. \tag{9}$$

This is a projection problem subject to the canonical simplex constraint. The Lagrange multiplier of equality and inequality constraints are denoted as γ and ν respectively. Based on the KKT conditions, if $\frac{\lambda-1}{2\mu}\beta_i + \frac{\gamma}{2} \leq 0$, $\alpha_i = 0$; otherwise $\alpha_i = \frac{\lambda-1}{2\mu}\beta_i + \frac{\gamma}{2}$. Obviously, as the value of $\frac{\lambda-1}{2\mu}\beta_i$ becomes larger, the probability that α_i equals to $\frac{\lambda-1}{2\mu}\beta_i + \frac{\gamma}{2}$ increases. Without loss of generality, we sort the vector $\frac{\lambda-1}{2\mu}\boldsymbol{\beta}$ to be $\boldsymbol{\Psi} = (\psi_1, ..., \psi_L)$ in a descending order. Then, the Lagrange multiplier γ of the equality constraint can be solved as:

$$\gamma = \frac{1}{\phi}\left(2 - 2\sum_{i=1}^{\phi} \psi_i\right), \tag{10}$$

where $1 \leq \phi \leq L$ is the number of non-zero entries in the vector $\boldsymbol{\alpha}$. The solution of $\boldsymbol{\alpha}$ with a specific γ is

$$\boldsymbol{\alpha} = \max\left\{\frac{\lambda - 1}{2\mu}\boldsymbol{\beta} + \frac{\gamma}{2}, 0\right\} \tag{11}$$

We can compute the solution of $\boldsymbol{\alpha}$ with each ϕ. After plugging each $\boldsymbol{\alpha}$ into Eq. (9), we get the global solution.

• **Time Complexity**

The time complexity of Eqs. (6) and (9) is $\mathcal{O}(N^3)$ and $\mathcal{O}(L^2)$, respectively. Because $L \ll N$, the time complexity of the whole optimization for Eq. (5) is $\mathcal{O}(\#iter\,N^3)$, where $\#iter$ is the number of iteration.

6   Experiment

6.1   Dataset Description

According to the aforementioned description, we apply the label propagation algorithm on a constructed campus social network from 5,388 students' daily routine. Currently, these behavior records are collected at 7 different locations, including teaching building, canteen, library, bathroom, campus shuttle, campus-to-campus bus and supermarket. The records of library entry are contained in the library entrance dataset, and others are contained in the consumption dataset. The data statistics are illustrated in Table 1.

Table 1. Dataset statistics

Data type	
Num. of students	5,388
Num. of consumption records	13,786,894
Num. of library entrance	927,854

6.2   Performance Evaluation

Since students' GPAs are converted into the levels of performance ranking, we evaluate the proposed label propagation algorithm by *precision*, defined as the fraction of students whose academic performance are classified correctly. We randomly divide the students in campus social network into training and testing datasets. The training dataset contains 80% students and the rest 20% are regarded as testing data. In the training dataset, we randomly select a fixed percentage of (20%, 40%, 60%, 80%, 100%) students to be labeled. To minimize the random error, we repeat the experiment 20 times and then compute the mean precision.

6.3   Result

Comparison on Single Network. According to our assumption, social ties constructed from different locations make distinct contribution in implying social ties, thus playing different parts in predicting academic performance according to the proposed model. Thus, we test label propagation algorithm on networks constructed from each location. Each case is denoted as *LP-Location*, where

Table 2. Comparison of label propagation on single location network.

Cases	20%	40%	60%	80%	100%
LP-Library	0.305	0.314	0.326	0.341	0.349
LP-Campus Shuttle	0.299	0.317	0.321	0.331	0.341
LP-Campus-to-Campus Bus	0.316	0.325	0.342	0.349	0.363
LP-Teaching building	**0.341**	**0.351**	**0.366**	**0.369**	**0.380**
LP-Supermarket	0.285	0.319	0.327	0.336	0.351
LP-Canteen	0.280	0.306	0.324	0.346	0.356
LP-Bathroom	0.273	0.297	0.294	0.294	0.306

the term *Location* is replaced as a specific location, such as teaching building, canteen, etc. The results are reported in Table 2.

According to the results, by comparing different percentages of labeled training data, we find the accuracy increases with the growing number of label data. By comparing the results of label propagation on seven networks, we can see that their performance is greatly different from each other. The network constructed by the records in the teaching building achieves the best performance. One possible reason is that if two students always attend classes together, they may have similar learning styles, resulting in the similar performance.

Comparison on Multiple Networks. Previous results show the different effect of networks constructed from each location. In this part, we evaluate our LPMN algorithm, which jointly predicts academic performance and optimizes the weight of each location, and compare it with an aggregation strategy with equal weights, denoted as *LP-All*. Since the network constructed from teaching building performs best according to Table 2, so we also show its result for comparison. All results are listed in Table 3.

Table 3. Comparison of label propagation on network constructed by location combining.

Cases	20%	40%	60%	80%	100%
LP-Teaching building	0.341	0.351	0.366	0.369	0.380
LP-ALL	0.315	0.331	0.356	0.360	0.365
LPMN	**0.355**	**0.374**	**0.377**	**0.380**	**0.401**

According to this table, we see that LPMN performs best among them while LP-All performs worst, showing the effectiveness of learning the weights for locations at the same time of predicting academic performance. This also implies that different locations may manifest different styles of social interaction. In conclusion, the results show the value of campus social network for predicting academic performance and reflect the effectiveness of LPMN algorithm on academic performance prediction.

7   Conclusion

In this paper, we tried to construct campus social networks from more than 14M behavior records based on the co-occurrences on multiple locations, and validated the significance of social influence with regard to academic performance, showing that students' academic performance is close to their friends'. Based on the support of social influence with regard to academic performance, we proposed a novel label propagation on multiple networks algorithm to predict academic performance, and evaluated it on the constructed social networks. The evaluation results revealed the potential value of campus social network for academic performance prediction and the effectiveness of the proposed algorithm.

Acknowledgement. This work is supported by grants from the Natural Science Foundation of China (61502077, 61631005) and the Fundamental Research Funds for the Central Universities (ZYGX2014Z012).

References

1. Cetintas, S., Si, L., Xin, Y.P., Tzur, R.: Probabilistic latent class models for predicting student performance. In: Proceedings of the 22nd ACM international conference on Conference on Information and Knowledge Management, pp. 1513–1516. ACM (2013)
2. Crandall, D.J., Backstrom, L., Cosley, D., Suri, S., Huttenlocher, D., Kleinberg, J.: Inferring social ties from geographic coincidences. Proc. Natl. Acad. Sci. **107**(52), 22436–22441 (2010)
3. Eagle, N., Pentland, A.S., Lazer, D.: Inferring friendship network structure by using mobile phone data. Proc. Natl. Acad. Sci. **106**(36), 15274–15278 (2009)
4. Fausett, L., Elwasif, W.: Predicting performance from test scores using backpropagation and counterpropagation. In: 1994 IEEE World Congress and IEEE International Conference on Computational Intelligence Neural Networks, vol. 5, pp. 3398–3402. IEEE (1994)
5. Friedland, L., Jensen, D., Lavine, M.: Copy or coincidence? A model for detecting social influence and duplication events. In: Proceedings of The 30th International Conference on Machine Learning, pp. 1175–1183 (2013)
6. Gao, H., Tang, J., Liu, H.: Exploring social-historical ties on location-based social networks. In: ICWSM (2012)
7. McPherson, M., Smith-Lovin, L., Cook, J.M.: Birds of a feather: homophily in social networks. Annu. Rev. Sociol. **27**, 415–444 (2001)
8. Pham, H., Shahabi, C., Liu, Y.: EBM: an entropy-based model to infer social strength from spatiotemporal data. In: Proceedings of the 2013 ACM SIGMOD International Conference on Management of Data, pp. 265–276. ACM (2013)
9. Sadilek, A., Kautz, H., Bigham, J.P.: Finding your friends and following them to where you are. In: Proceedings of the Fifth ACM International Conference on Web Search and Data Mining, pp. 723–732. ACM (2012)
10. Tamhane, A., Ikbal, S., Sengupta, B., Duggirala, M., Appleton, J.: Predicting student risks through longitudinal analysis. In: Proceedings of the 20th ACM SIGKDD International Conference on Knowledge Discovery and Data Mining, pp. 1544–1552. ACM (2014)

11. Thai-Nghe, N., Drumond, L., Horváth, T., Schmidt-Thieme, L., et al.: Multi-relational factorization models for predicting student performance. In: KDD Workshop on Knowledge Discovery in Educational Data (KDDinED) (2011)
12. Wang, C., Ye, M., Lee, W.: From face-to-face gathering to social structure. In: Proceedings of the 21st ACM International Conference on Information and Knowledge Management, pp. 465–474. ACM (2012)
13. Wang, R., Harari, G., Hao, P., Zhou, X., Campbell, A.T.: SmartGPA: how smartphones can assess and predict academic performance of college students. In: Proceedings of the 2015 ACM International Joint Conference on Pervasive and Ubiquitous Computing (UbiComp). ACM (2015)
14. Wu, R., Liu, Q., Liu, Y., Chen, E., Su, Y., Chen, Z., Hu, G.: Cognitive modelling for predicting examinee performance. In: Proceedings of the 24th International Conference on Artificial Intelligence, pp. 1017–1024. AAAI Press (2015)
15. Yu, M., Si, W., Song, G., Li, Z., Yen, J.: Who were you talking to-mining interpersonal relationships from cellphone network data. In: 2014 IEEE/ACM International Conference on Advances in Social Networks Analysis and Mining (ASONAM), pp. 485–490. IEEE (2014)
16. Zafra, A., Romero, C., Ventura, S.: Multiple instance learning for classifying students in learning management systems. Expert Syst. Appl. **38**(12), 15020–15031 (2011)

Social User Profiling: A Social-Aware Topic Modeling Perspective

Chao Ma[1,2], Chen Zhu[2], Yanjie Fu[3], Hengshu Zhu[2(✉)], Guiquan Liu[1], and Enhong Chen[1]

[1] University of Science and Technology of China, Hefei, China
{gqliu,cheneh}@ustc.edu.cn
[2] Baidu Inc., Beijing, China
{machao13,zhuchen02,zhuhengshu}@baidu.com
[3] Missouri University of Science and Technology, Rolla, USA
fuyan@mst.edu

Abstract. Social user profiling is an analytical process that delivers an in-depth blueprint of users' personal characteristics in social networks, which can enable a wide range of applications, such as personalized recommendation and targeted marketing. While social user profiling has attracted a lot of attention in the past few years, it is still very challenging to collaboratively model both user-centric information and social network structure. To this end, in this paper we develop an analytic framework for solving the social user profiling problem. Specifically, we first propose a novel social-aware semi-supervised topic model, i.e., User Profiling based Topic Model (UPTM), which can reconcile the observed user characteristics and social network structure for discovering the latent reasons behind social connections and further extracting users' potential profiles. In addition, to improve the profiling performance, we further develop a label propagation strategy for refining the profiling results of UPTM. Finally, we conduct extensive evaluations with a variety of real-world data, where experimental results demonstrate the effectiveness of our proposed modeling method.

Keywords: User profiling · Topic model · Social network

1 Introduction

With the rapid development and increasing prevalence of online social networks, a huge amount of user information has been accumulated. Along this line, a critical challenge is how to effectively infer the unobserved personal characteristics of social users, such as affiliation and education background, which is known as the problem of *social user profiling*. Indeed, social user profiling is an analytical process that delivers an in-depth blueprint of users personal characteristics in social networks, which can enable a wide range of applications, such as personalized recommendation and targeted marketing.

S. Candan et al. (Eds.): DASFAA 2017, Part II, LNCS 10178, pp. 610–622, 2017.
DOI: 10.1007/978-3-319-55699-4_38

In the past few years, social user profiling has attracted a lot of attention from both academia and industry. In the literature, social user profiling is usually regarded as a label prediction task, where most of studies focus on the explicit estimation by exploiting a variety of user-centric data, such as blog entries [6], query logs [11], and tweets [9]. However, there are several limits for inferring user characteristics from user-centric data: (i) the high cost of collecting annotated data; (ii) the inherent bias of annotation; and (iii) the high dependence on observed user characteristics with low generalization capability. For instance, the work in [10] extracts features from network structure and applies traditional regression algorithm to solve the social user profiling problem, where feature construction plays the most crucial role. However, these predefined features are usually effective for one particular dataset, and are insufficient to be generalized to other datasets. In contrast, some researchers have developed unsupervised and semi-supervised methods for inferring user characteristics from social network connections [12,22]. While these methods do not lie in observed user characteristics and have some potential to achieve good generalization, these methods neglect the user-centric information and thus the profiling performance highly depends on the definition of similarity between nodes that varies over different user characteristics.

All the above evidences suggest that it is highly appealing to investigate how to combine both observed user-centric information and social network structure for social user profiling, and moreover, to navigate both predictive accuracy and model generalization in a meaningful way. To this end, in this paper, we develop an analytic framework for solving the social user profiling problem. Specifically, we first propose a novel social-aware semi-supervised topic model, named User Profiling based Topic Model (UPTM), which can reconcile the observed user characteristics and social network structure for discovering the latent reasons behind social connections and further extracting users' potential profiles. In UPTM, a user in a social network is regarded as a document; neighboring friends are regarded as tokens; and topic distributions are regarded as the probabilities of reasons behind users' social connections. Indeed, UPTM is able to exploit the latent reasons behind social connections by unobserved characteristics of users, observed characteristics of users, the network structure of neighboring friends via the strategic analogies of documents, topics, and words. In particular, we have identified an interesting observation: the observed user characteristics can be served as a prior of UPTM for inferring unobserved user characteristics. For instance, if the observed characteristics show that a user is a New Yorker, this user is less likely to be associated with non-NYC topics and words. We propose to incorporate the impacts of observed user characteristics on topics and words as two regularization terms into UPTM. Meanwhile, we develop a Gibbs sampling based method to solve the optimization problem for parameter estimation. Thus, we can extract the inferred probabilities of regularized latent topics as an implicit estimation of user profile. However, UPTM cannot effectively capture the relationship between connected users who do not have enough common friends. Therefore, to improve the profiling performance, we further develop a label propagation strategy for refining the profiling results of UPTM. Finally, we present extensive experimental results with the real-world data collected from Facebook to demonstrate the effectiveness of our proposed method for social user profiling.

2   Related Work

In the past few years, social user profiling has attracted much attention as it is important in many areas like personalized search [21], target advertisement [1], and urban computing [23]. The task of user profiling is to infer users' personalized characteristics, such gender [15], location of interests [8,13], and age [2].

In the literature, social user profiling is usually regarded as a label prediction task, where most of studies are based on the explicit estimation by exploiting a variety of user-centric data, such as blog entries [6], query logs [11], social media posts [5] and other types of user generated data [14]. Indeed, the focus of the above methods is designing attribute-specific features with off-the-shelf supervised classifiers. For example, [20] detected Twitter user attributes by using a mixture of sociolinguistic features as well as n-gram models. [18] attempted to classify users by employing a large set of aggregate features including profile features, tweeting behavior features, linguistic content features, and social network features. However, there are several limits for inferring user characteristics from user-centric data: (i) the high cost of annotated data collection; (ii) the inherent bias of annotation; and (iii) the high dependence on observed user characteristics with low generalization capability.

In another aspect, some researchers tried to explore the social information for user profiling task [10,17,24]. For example, [17] proposed to infer the departments of university students based on their friendships on Facebook, where the assumption is that students with the same department are likely to be friends and form a community. [18] applied heuristics to directly propagate Twitter users' interests through all their following connections. [4] designed a probabilistic model to propagate Facebook users' locations via all their friendships. [10] used website traffic data to infer Twitter users' demographics by applying multi task elastic net on features extracted from original social network structure. While these methods do not lie in observed user characteristics and have some potential to achieve good generalization, these methods neglect the user-centric information and thus the profiling performance highly depends on the definition of node similarity that varies over different characteristics.

Different from the above work, we propose a novel social-aware semi-supervised topic model, which can reconcile the observed user characteristics and social network structure for user profiling. Moreover, we also develop a label propagation strategy for refining the profiling results.

3   Social User Profiling

In this section, we first describe some preliminaries of social user profile, and then introduce the technical details of our proposed method.

3.1   Preliminary and Overview

When inferring users' characteristics in a social network, the first problem is why they have connections between each other. Generally, we think the reasons behind these connections can be classified into two categories. The first one is that users with the same interests are more likely to make friends with each other, since birds of a feather flock together. The second one is that users want to make a friend with whom they are interested in. Under the second situation, maybe only little or even no preference similarity exists between users and their friends (i.e., neighbors in a social network), but the friends of a specific user are likely to share the some of the same interests or characteristic. For example, most of followers of Lionel Messi in Twitter may not have professional skills as Messi has, but only are the fans of Football Club Barcelona. Besides, in a social community with more than three users who are friends with each others, the first reason can also be regarded as a special case of the second reason. For example, all of John, Alice, and Mary like rock music and follow the others in Twitter. We can explain the reason behind these connections by the same interest of rock music. But we can also claim that why John and Alice make friends with the same one Mary, is both of them are interested in rock music. In other words, the origin assumption about formation of social connection can be simplified as the second reason in most cases.

Based on the above assumption, to make social user profiling in a general manner, we first follow the second reason about formation of social connection to propose a novel semi-supervised topic model, i.e., User Profiling based Topic Model (UPTM), which combines characteristics of users and social network structure to unveil the latent reasons behind social connections. With the latent information extracted from UPTM, we can further infer unobserved users' characteristics. However, UPTM cannot effectively capture the relationship between connected users who do not have enough common friends. To this end, we further propose a label propagation strategy to fit the origin assumption about formation of connections, and thus can help refine the profiling results of UPTM. An intuitive description of our approach is shown in Fig. 1.

In this paper, a social network is noted as $G = \{N, E, T\}$, in which N is the set of nodes (i.e., users), E is the set of edges in the network, and T represents the characteristic set of nodes. For example, a social network contains three nodes $N = \{John, Alice, Mary\}$, two edges $E = \{(John, Mary), (Alice, Mary)\}$, and an observed incomplete characteristic set $T = \{(John, male), (Alice, female), (Mary, female), (John, student), (Alice, student), (Mary, professor)\}$. To model social network by topic models, we treat each node as a document, and its neighbors in this social network as tokens in this document. For example, in the documented version of the above social network, the document of John contains only one token, $\{Mary\}$.

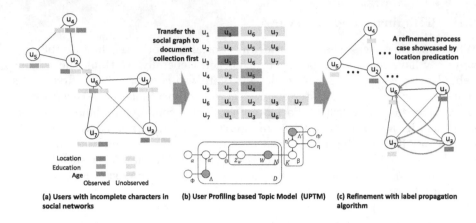

Location ▨▨ ▨▨
Education ▨▨ ▨▨
Age ▨▨ ▨▨
Observed Unobserved

(a) Users with incomplete characters in social networks   **(b) User Profiling based Topic Model (UPTM)**   **(c) Refinement with label propagation algorithm**

Fig. 1. The overview of our social user profiling framework.

3.2   User Profiling Based Topic Model

UPTM is a semi-supervised topic model, which aims to find latent reasons behind social connections from observed characteristics. Actually, UPTM can be regarded as an extension of L-LDA [19], which is a supervised topic model for labeled document collections. In L-LDA, each topic is related to a label, and the assigned topics of any tokens in a document must be selected from the related labels of this document. Here, we follow the ideas in L-LDA for incorporating observed user characteristics as constraints. Specifically, we treat each user characteristic as a label, and assume the reason behind each social connection can be attributed to one of user's characteristics. For example, if Bob is a student at Stanford, it is possible for him to have a connection with any student or faculty of Stanford in a social network. Under this situation, we attribute such social connections to his education background, i.e., Stanford. As mentioned above, in our model, each user is treated as a document and its neighbors are the corresponding tokens. Based on this analogy, in this example, Bob is a document and his friends in Stanford correspond to a part of tokens in this document.

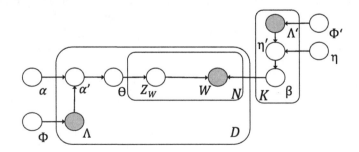

Fig. 2. Graphical model of User Profiling based Topic Model

Table 1. The notations and the corresponding description in UPTM

Notations	Description
K	Number of topics
w_i	The i-th token in a document
z_i	Topic assigned to the i-th token in document d
N_d	Number of tokens in document d
D	Number of documents
V	Length of the vocabulary
$\Lambda'^{(k)}$	Presence/absence indicators of words in topic k
Φ'_v	Prior of word v for Λ'
η	Dirichlet prior distribution of topic over words
$\eta'^{(k)}$	Dirichlet prior distribution of topic k over words
$L^{(k)}$	Matrix representation of $\Lambda'^{(k)}$ for generating $\eta'^{(k)}$
$\beta^{(k)}$	Multinomial distribution of topic k over words
$\Lambda^{(d)}$	Presence/absence indicators of topics in document d
Φ_k	Prior of topic k for Λ
α	Dirichlet prior distribution of document over topics
$\alpha'^{(d)}$	Dirichlet prior distribution of document d over topics
$L^{(d)}$	Matrix representation of $\Lambda^{(d)}$ for generating α'
$\theta^{(d)}$	Multinomial distribution of document d over topics

Furthermore, the topics of these tokens are the one related to his education background. In addition, to incorporate the unobserved characteristics, tokens can also be assigned to topics whose labels do not contradict the observed labels in UPTM. Next, we will explain UPTM from a technical perspective. Table 1 lists the notations and descriptions in UPTM.

Specifically, we describe UPTM with a document generative process following the tradition of topic modeling. Each document is represented by a tuple consisting of a list of word indices $w^{(d)} = (w_1, ..., w_{N_d})$ and a list of binary topic presence/absence indicators $\Lambda^{(d)} = (l_1, ..., l_K)$ where $w_i \in \{1, ..., V\}$ and $l_k \in \{0, 1\}$. The word assignment of topics are also controlled by a list of binary topic presence/absence indicators $\Lambda'(w) = (l_1, ..., l_V)$, where l_i equals zero means that the topic is excluded by this node observed labels. In particular, the topics that can be assigned to users are different. If a user's location is observed as New York, we will not assign Seattle to her/him. In contrary, if the user's location is unobserved, any city can be assigned to him according to the probability learned by UPTM. Therefore, UPTM is a semi-supervised model for social user profiling.

The detailed generative process is shown in Algorithm 1. There are some noticeable points in the process. Firstly, the $\beta^{(k)}$ is defined to only cover the words that can be assigned to the topic k by $\Lambda'^{(k)}$. We get $\Lambda'^{(k)}$ by observed data and domain knowledge. And $\theta^{(d)}$ is also restricted by $\Lambda^{(d)}$ in the same way. Another point should be explained here is because we can not have a friend twice at the same time, any word will not repeat in a document of a documented social network. Thus in our application, when generating the word under the topic we select, UPTM should not obey the multinomial distribution but a hyper geometric distribution. However, according to [7], when the number of samples is enough, the hyper geometric distribution will converge in a multinomial distribution. In other words, in the network with enough nodes, such settings in fact will not produce too much influence on the performance of UPTM. Here, we use the collapsed Gibbs sampling algorithm [3] to learn UPTM. Specifically, the neighbors of a user who is labeled according to the pre-established topics is represented as

$$P(d(u_i, u_j) = z|.) \propto \frac{N_{u_j z} + \eta^{(u_j)}}{\sum_{k=1}^{|N|}(N_{u_k z} + \eta_z^{(u_k)})} \times \frac{M_{u_i z} + \alpha_z^{(u_i)}}{\sum_{k=1}^{|Z|}(M_{u_i z_k} + \alpha_{z_k}^{(u_i)})}, \quad (1)$$

Algorithm 1. Generative process of UPTM

---

for *each topic* $k \in \{1, ..., K\}$ **do**
  for *each word* $v \in \{1, ..., V\}$ **do**
    $\lfloor$ Generate $\Lambda_v'^{(k)} \in \{0, 1\} \sim Bernoulli(.|\Phi_v')$
  Generate $\eta'^{(k)} = L^{(k)} \times \eta$
  Generate $\beta^{(k)} = (\beta_{l_1}, ..., \beta_{l_{N_k}})^{\mathrm{T}} \sim Dir(.|\eta'^{(k)})$
for *each document* $d \in \{1, ..., D\}$ **do**
  for *each topic* $k \in \{1, ..., K\}$ **do**
    $\lfloor$ Generate $\Lambda_k^{(d)} \in \{0, 1\} \sim Bernoulli(.|\Phi_k)$
  Generate $\alpha'^{(d)} = L^{(d)} \times \alpha$
  Generate $\theta^{(d)} = (\theta_{l_1}, ..., \theta_{l_{M_d}})^{\mathrm{T}} \sim Dir(.|\alpha'^{(d)})$
  for *each* $i \in \{1, ..., N_d\}$ **do**
    Generate $z_i \in (\lambda_1^{(d)}, ..., \lambda_{M_d}^{(d)}) \sim Multi(.|\theta^{(d)})$
    $\lfloor$ Generate $w_i \in (1, ..., V) \sim Multi(.|\beta^{(z_i)})$

---

where $d(u_i, u_j)$ represents the user u_j is a neighbor of user u_i; α and β represent the hyper parameter of Dirichlet distribution of document-topic multinomial distribution and the hyper parameter of Dirichlet distribution of topic-word multinomial distribution, respectively. $M_{u_i z}$ represents the number of topic z assigned to neighbors of the user u_i, and $N_{u_j z}$ represents the number of that the word corresponds to user u_i labeled with topic z.

After the sampling, we can get the probability distribution over topics for each user, which is also the probability distribution over user characteristics. Then, we can infer a user's certain characteristic, e.g., residential location, according to probability obtained by UPTM. If we have strong confidence to the result that the probability of some labels (e.g., New York, Seattle, Washington) can be assigned to the user is bigger than a threshold ρ, we will assign the label with biggest probability to him. Otherwise, we will not make profile inference on the user with UPTM, and implement label propagation for further profiling.

3.3 Label Propagation for User Profiling Refinement

Although UPTM is an effective approach for user profiling, it cannot capture the relationships between connected users who do not have enough common friends. Meanwhile, the user characteristics inferred with low confidence (probability) should be further refined. Therefore, here we extend the label propagation strategy [25] to refine the profiling results from UPTM. Before introducing how our algorithm works, we first define some new notations. For a label, the set of observed nodes is N_o and the set of unobserved nodes is N_u. The total node set is defined as $N = N_o \cup N_u$. We also define the distance between node i and node j as $d_{i,j}$, which in this paper is the minimum hop between them in the social network. And T is the probabilistic transition matrix of N, of which the element, $T_{i,j}$, is transition probability from node i to j. The $T_{i,j}$ can be calculated by

$$T_{i,j} = \frac{1/d_{i,j}}{\sum_{k=1}^{|N|} 1/d_{i,k}}. \tag{2}$$

Algorithm 2. Label Propagation:

---
input : N, T, τ
output: Y

for *Y does not converge* do
    Propagate $Y \leftarrow TY$
    Calculate σ, ε
    Clamp nodes by $Y \leftarrow \sigma Y + (1 - \sigma)\varepsilon$

Besides, vector Y, whose length is $|N|$, is used to represent label probability of nodes. In Y, all of observed nodes and nodes inferred by UPTM are set to 1 first. Unobserved nodes are initialized to 0.

The details of our algorithm is shown in Algorithm 1. Specifically, in each step, labels are first propagated by the transition matrix T. Then to clamp those observed nodes and nodes inferred by UPTM, we set Y as $\sigma Y + (1 - \sigma)\epsilon$, where σ and ϵ are selected to meet the following objectives: (i) the L1-norm of Y must is equal to that of ϵ, which guarantee the convergence of the algorithm; (ii) each observed node will be set to 1; (iii) each node inferred by UPTM, N_i, is equal to $(1 - \tau)Y_i + \tau$. τ is a hyper-parameter and represents the confidence level in results of UPTM. Finally, after our algorithm converges, Y is the refined label prediction result.

4 Evaluation

In this section we empirically evaluate the performance of our proposed method for social user profiling. All experiments were performed on real-world social network data, collected from Facebook, where three user profile labels (i.e., age, location, and education), were used for evaluation.

4.1 Experimental Setup

We conducted experiments with a real-world dataset, Facebook Dataset [16], in which friend relationship is represented by links. In other words, two users are linked if one is among the friend list of the other. Facebook users are annotated by a variety of profile labels, including location, education background, gender, hometown, and language. Table 2 shows the statistics of the availability of these profile labels.

The Facebook Dataset[1] contains 4,039 users and 88,234 links. Each user has 43 friends on average. Figure 3 shows some basic statistics of the Facebook dataset. Figure 3(1) illustrates the distribution of users with respect of the number of their neighbors. From the results we can observe that the distribution roughly follows power law, which indicates that only a few users have large number of neighbors while most of users only have limited neighbors. Figure 3(2) demonstrates the percentage of missing label with respect of the number of their neighbors.

To improve the quality of training data for effectively evaluating the proposed method, we first filtered out those users that have only few friends and are associated with limited social activities in Facebook. Here, we selected three major labels for evaluation, i.e., "Location", "Education", and "Age". In the experiments, we evaluated the performance of profile label prediction by several classic classification metrics, including *Precision*, *Recall*, *F1 − score*, and *Accuracy*. What should be noticed is that label prediction in our experiments is

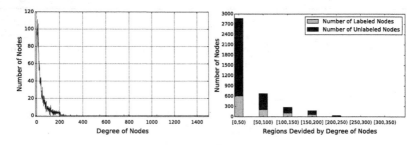

Fig. 3. The statistics of Facebook Dataset. (1) The number of users with respect of the number of their neighbors; (2) The percentage of missing label with respect of the number of their neighbors.

---

[1] http://snap.stanford.edu/data/.

Table 2. The proportion of labels given in Facebook Dataset

Attribute	Number of labeled users	Fraction
Gender	606	15.0%
Location	1,659	41.1%
Education	2,696	66.7%
Language	748	18.5%
Age	1,577	39.0%
Work-location	606	15.0%
Hometown	1,066	26.4%

a Multiclass Classification problem. Thus, *Precision* is the average of those for all classes. So are *Recall* and *F1 − score*.

4.2 Baseline Algorithms

To demonstrate the effectiveness of the proposed method, we compared our method with the following baseline algorithms.

Community Detection-Based Algorithm: This baseline is developed based on community detection, which grows a community starting with an induced subgraph until maximum normalized conductance is achieved, and then infers user attributes by setting the majority value in the community. This method is proposed in [17] and have become a seminal study in user profiling problem.

Traditional Classification Algorithms: As mentioned above, the most popular approach in user profiling is to model it by supervised learning approaches. Here, we selected two classic classification methods, Naive Bayes and SVM, as baselines.

L-LDA: The standard L-LDA [19] was also selected as a baseline. L-LDA has similar settings as UPTM for social user profiling, while it only uses the observation information as labels. Thus we aim to validate the importance of incorporating unobserved nodes by this baseline.

4.3 Results and Discussion

Figure 4 shows the overall performance comparison of our method and baseline algorithms for predicting location, education and age labels. The X-axis in Fig. 4 is the proportion of training data. We used ten-fold cross validation for obtaining the results.

It is obvious that our method outperforms all of the baselines in most cases. Specially, in education prediction task, UPTM outperforms the other baselines by at least 50%. It strongly supports the effectiveness of our method. Besides,

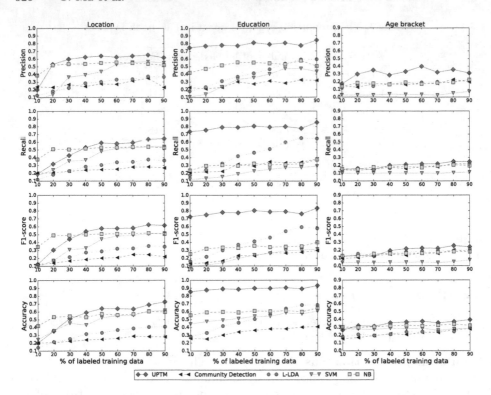

Fig. 4. The performance of UPTM and baselines in "Location", "Education", and "Age" prediction

although increasing the proportion of training data generally improves the performance of all methods, its importance varies from task to task. In "Location" and "Education" prediction tasks, we can easily observe the trend. But when predicting "Age", more training data cannot benefit a lot. And all of UPTM and baselines did not get good enough performances in "Age" prediction task. We think it may be attributed to two reasons: (i) "Age" is not an important factor when users make friends in a social network; (ii) the discretization method used for "Age" is not appropriate in this task. However, due to the anonymity of the Facebook Dataset, it is hard to prove our assumptions and we hope for further study from other researchers. To prove the robustness of the proposed approach, we will conduct experiments on more datasets in future work.

5   Concluding Remarks

In this paper, we developed an analytic framework for social user profiling. Specifically, we first proposed a novel social-aware semi-supervised topic model, i.e., User Profiling based Topic Model (UPTM), which can reconcile the observed user characteristics and social network structure for discovering the latent reasons behind social connections and further extracting users' potential profiles. Then, to improve the profiling performance, we further introduced a label propagation strategy for refining the profiling results of UPTM. Finally, we presented extensive experimental results based on the real-world data collected from Facebook, which clearly demonstrate the effectiveness of our proposed method for social user profiling.

Acknowledgments. This research was partially supported by grants from the National Natural Science Foundation of China (NSFC, Grant No. U1605251), the National Science Foundation for Distinguished Young Scholars of China (Grant No. 61325010), and the NSFC Major research program (Grant No. 91546103).

References

1. Ahmed, A., Low, Y., Aly, M., Josifovski, V., Smola, A.J.: Scalable distributed inference of dynamic user interests for behavioral targeting. In: Proceedings of the 17th ACM SIGKDD International Conference on Knowledge Discovery and Data Mining, pp. 114–122. ACM (2011)
2. Al Zamal, F., Liu, W., Ruths, D.: Homophily and latent attribute inference: inferring latent attributes of twitter users from neighbors. In: ICWSM, vol. 270 (2012)
3. Andrieu, C., Freitas, N., Doucet, A., Jordan, M.I.: An introduction to MCMC for machine learning. Mach. Learn. **50**(1–2), 5–43 (2003)
4. Backstrom, L., Sun, E., Marlow, C.: Find me if you can: improving geographical prediction with social and spatial proximity. In: Proceedings of the 19th International Conference on World Wide Web, pp. 61–70. ACM (2010)
5. Bergsma, S., Van Durme, B.: Using conceptual class attributes to characterize social media users. In: ACL, no. 1, pp. 710–720 (2013)
6. Burger, J.D., Henderson, J.C.: An exploration of observable features related to blogger age. In: AAAI Spring Symposium: Computational Approaches to Analyzing Weblogs, pp. 15–20 (2006)
7. Cha, Y., Bi, B., Hsieh, C.C., Cho, J.: Incorporating popularity in topic models for social network analysis. In: Proceedings of the 36th International ACM SIGIR Conference on Research and Development in Information Retrieval, pp. 223–232. ACM (2013)
8. Chen, X., Wang, Y., Agichtein, E., Wang, F.: A comparative study of demographic attribute inference in twitter. In: Proceedings of ICWSM (2015)
9. Cheng, Z., Caverlee, J., Lee, K.: You are where you tweet: a content-based approach to geo-locating twitter users. In: Proceedings of the 19th ACM International Conference on Information and Knowledge Management, pp. 759–768. ACM (2010)
10. Culotta, A., Kumar, N.R., Cutler, J.: Predicting the demographics of twitter users from website traffic data. In: AAAI, pp. 72–78 (2015)
11. Jones, R., Kumar, R., Pang, B., Tomkins, A.: I know what you did last summer: query logs and user privacy. In: Proceedings of the Sixteenth ACM Conference on Conference on Information and Knowledge Management, pp. 909–914. ACM (2007)
12. Jurgens, D.: That's what friends are for: inferring location in online social media platforms based on social relationships. ICWSM **13**, 273–282 (2013)

13. Li, R., Wang, C., Chang, K.C.C.: User profiling in an ego network: co-profiling attributes and relationships. In: Proceedings of the 23rd International Conference on World Wide Web, pp. 819–830. ACM (2014)

14. Lin, H., Zhu, H., Zuo, Y., Zhu, C., Wu, J., Xiong, H.: Collaborative company profiling: insights from an employee's perspective. In: AAAI (2017)

15. Liu, W., Ruths, D.: What's in a name? Using first names as features for gender inference in twitter. In: AAAI Spring Symposium: Analyzing Microtext, vol. 13, p. 01 (2013)

16. Mcauley, J., Leskovec, J.: Discovering social circles in ego networks. ACM Trans. Knowl. Discov. Data (TKDD) 8(1), 4 (2014)

17. Mislove, A., Viswanath, B., Gummadi, K.P., Druschel, P.: You are who you know: inferring user profiles in online social networks. In: Proceedings of the Third ACM International Conference on Web Search and Data Mining, pp. 251–260. ACM (2010)

18. Pennacchiotti, M., Popescu, A.M.: Democrats, republicans and starbucks afficionados: user classification in twitter. In: Proceedings of the 17th ACM SIGKDD International Conference on Knowledge Discovery and Data Mining, pp. 430–438. ACM (2011)

19. Ramage, D., Hall, D., Nallapati, R., Manning, C.D.: Labeled lDA: a supervised topic model for credit attribution in multi-labeled corpora. In: Proceedings of the 2009 Conference on Empirical Methods in Natural Language Processing: Volume 1-Volume 1, pp. 248–256. Association for Computational Linguistics (2009)

20. Rao, D., Yarowsky, D., Shreevats, A., Gupta, M.: Classifying latent user attributes in twitter. In: Proceedings of the 2nd International Workshop on Search and Mining User-generated Contents, pp. 37–44. ACM (2010)

21. Xu, S., Bao, S., Fei, B., Su, Z., Yu, Y.: Exploring folksonomy for personalized search. In: Proceedings of the 31st Annual International ACM SIGIR Conference on Research and Development in Information Retrieval, pp. 155–162. ACM (2008)

22. Xu, T., Zhu, H., Chen, E., Huai, B., Xiong, H., Tian, J.: Learning to annotate via social interaction analytics. Knowl. Inf. Syst. 41(2), 251–276 (2014)

23. Yuan, N.J., Zhang, F., Lian, D., Zheng, K., Yu, S., Xie, X.: We know how you live: exploring the spectrum of urban lifestyles. In: Proceedings of the First ACM Conference on Online Social Networks, pp. 3–14. ACM (2013)

24. Zhu, C., Zhu, H., Ge, Y., Chen, E., Liu, Q.: Tracking the evolution of social emotions: a time-aware topic modeling perspective. In: 2014 IEEE International Conference on Data Mining, pp. 697–706. IEEE (2014)

25. Zhu, X., Ghahramani, Z.: Learning from labeled and unlabeled data with label propagation (2002)

Cost-Effective Data Partition for Distributed Stream Processing System

Xiaotong Wang[1,2], Junhua Fang[1,2], Yuming Li[1,2], Rong Zhang[1,2(⊠)], and Aoying Zhou[1,3]

[1] Shanghai Key Laboratory of Trustworthy Computing, School of Data Science and Engineering, East China Normal University, Shanghai 200062, China
{xt.wang,jh.fang,ym.li}@ecnu.cn, {rzhang,ayzhou}@sei.ecnu.edu.cn
[2] International Joint Lab of Trustworthy Software,
East China Normal University, Shanghai, China
[3] School of Data Science and Engineering,
East China Normal University, Shanghai, China

Abstract. Data skew and dynamics greatly affect throughput of stream processing system. It requires to design a high-efficient partition method to evenly distribute workload in a distributed and parallel. Previous research mainly focuses on load balancing adjustment based on key-as-granularity or tuple-as-granularity, both of which have their own limitations such as clumsy balance activities or expensive network cost. In this paper, we present a comprehensive cost model for partitioning method, which makes a synthesis estimation of memory, CPU and network resource utilization. Based on cost model, we propose a novel load balancing adjustment algorithm, which adopts the idea of "Split keys on demand and Merge keys as far as possible", and is adaptive to different skewed workload. Our evaluation demonstrates that our method outperforms the state-of-the-art partitioning schemes while maintaining high throughput and resource utilization.

1 Introduction

Online real-time analysis on stream data is essential for an increasing number of applications, such as stock trading aggregating, hot topic detection and network information monitoring. However the explosive growth of data exposes great challenges to traditional centralized processing architecture. Then distributed stream processing systems (DSPSs) have gained much attention. In order to achieve high efficiency, one of the important topics is to balance the workload among parallel processing tasks to realize high throughput and low latency.

Generally there are two kinds of load balance strategies, namely operator-based [3,14,15] and data-based [4,8,10–13]. For operator-based strategy, its basic load distribution units are operators. Due to the disparity of execution time and complexity among operations, some execution plans may delegate multiple complicated operations to a single machine, which directly leads to load imbalance. Data-based strategy uses individual data as the basic load distribution units.

© Springer International Publishing AG 2017
S. Candan et al. (Eds.): DASFAA 2017, Part II, LNCS 10178, pp. 623–635, 2017.
DOI: 10.1007/978-3-319-55699-4_39

Since operations rely on data status of the same keys and data-based partitions demonstrate more scalability than operator-based ones. Then it is preferred to partition workloads via data. According to distribution granularities, the existing solutions can be divided into the following two types: (1) **Key-as-granularity**. During load balancing adjustment, it needs to transmit all the status related to a key, such as *Readj* [8]. It can maintain the semantics of key-based operations well, but limit the flexibility of load balancing adjustment. (2) **Tuple-as-granularity**. Random distribution can solve load imbalance problem caused by data skew. However, it breaks the semantics of key-based operations and leads to operational limitations. In work [4], they use random method for data distributions to solve the problem of join. Though they can promise load balance among task (to some extend), they may meet great data transmission.

But data-based strategies face the following challenges: (1) key-as-granularity could maintain key-based semantics to the utmost, but may not perform well for system equilibrium since the skewed distribution of data on keys may break the balance status among the parallel tasks. (2) tuple-as-granularity is in favor of system equilibrium, but it pays extra cost for maintaining the correctness of key-based semantics.

To achieve the goal that both balance and resource consumption should be optimized, it requires to design a cost model which synthetically analyzes various factors on system performance. In this paper, we focus on solving load imbalance problem caused by data skew in DSPSs and propose a cost model for designing a flexible and adaptive load balancing adjustment strategy. In particular, we make the following main contributions:

1. We are the first to present a comprehensive cost model for guiding designing balancing strategies, which considers memory, CPU, and network.
2. We design a novel adaptive load balancing algorithm which consists of a decision-making function and two types of load adjustment algorithms, based on the idea of "Split keys on demand and merge keys as far as possible".
3. We implement our solution over Apache Storm and conduct extensive experiments to demonstrate our advantages by comparing with the stat-of-art other techniques.

The rest of this paper is organized as follows. In Sect. 2 we introduce the background knowledge of the paper. Sect. 3 analyzes our cost model. Sect. 4 proposes our load balancing algorithms and Sect. 5 evaluates performance. At last, Sect. 6 discusses related work and Sect. 7 concludes the paper.

2   Preliminaries

2.1   Key Grouping

Stream is an unbounded sequence of data items (tuples) [9] of the form $<k, v, t>$ ordered by the timestamp t, where k is the tuple's key and v is the value. To improve the effectiveness of stream processing, it trends to partition a stream into

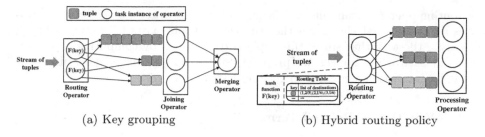

(a) Key grouping (b) Hybrid routing policy

Fig. 1. Two different partitioning schemes.

disjoint sub-streams and assign them to parallel task instances. Key Grouping (KG) has gained much attention and in most cases it maps key domain to tasks using a global hash function on keys. KG ensures tuples with the same key can be distributed to the same task and doesn't need to keep track of distribution paths of keys. If we have uniform workload distribution among tasks, KG works well. However, when workloads are skewed on keys, KG may result in high load imbalance among parallel processing tasks. In Fig. 1(a), we give an example to illustrate the potential problem of KG. Tuples from data source are distributed by a *Routing Operator* to *Joining Operator* using a hash function $F(key)$ and join results are collected in *Merging Operator*. Due to data skew(high frequency of some keys), the first task instance in *Joining Operator* is delegated workload twice more than the others and is overloaded. As a result, the first task instance in *Joining Operator* be a procrastinator and *Merging Operator* may be suspended to wait for its completion, slowing all the processing down.

2.2 Hybrid Routing Policy

When load imbalance happens, it needs to migrate a certain quantity of workload from overloaded tasks to underloaded ones. As illustrated in Fig. 1(b), we unload two tuples from the first task in *Joining Operator* and migrate them to the second and third tasks respectively. To ensure the computation correctness, it is necessary to keep track of keys that have been moved. Hence, we adopt a hybrid scheme that combines an routing table and a basic hash function. Basic hash function can be uniform hashing, consistent hashing and so on. Routing Table is used for keys that have been migrated and each task instance which is responsible for distributing keys should maintain a routing table.

Different to [8], our routing table has more direction information for those migrated keys. Specifically, we define routing item in the table as $<k, list_k : \{(d, quan_d)\}>$, where k is key, and $list_k : \{(d, quan_d)\}$ is a list of entries $(d, quan_d)$ which represents that the quantity of tuples with k in task d is $quan_d$. Keys in routing table will not be distributed by the basic hash function. Figure 1(b) shows the overall operator structure. There are two kinds of operators: *Routing Operator* is responsible for distributing and routing tuples, while *Processing Operator* is in charge of actual computation. Each task instance of routing operator maintains a routing table. Before *Routing Operator* distributes a tuple, it first checks whether the key of tuple exists in routing table. If it exists, the task that stores this tuple is determined by computing a pseudo-random hash whose ranges are proportional to the quantity of key; otherwise basic hash function is applied to get the destination task.

2.3   Definition of Related Terms

If there exists no entry for key k in routing table and it is distributed to task $d = F(k)$ via basic hash function, then task d is called **basic task** of key k. To facilitate the description of load adjustment strategies in the rest of paper, several operations during load adjustment are defined as following: (1) **Migrate Back**. This operation migrates a key which is not stored in its basic task back to its basic task; (2) **Migrate To**. This operation migrates a key from one task to another. Neither these two tasks are its basic task; (3) **Migrate Out**. This operation migrates a key from its basic task to some other task.

Among the three operations above, *Migrate Back* can reduce the number of mapping entries in routing table, whereas *Migrate Out* adds new entries to routing table and increases table size. MT has no influence on routing table size.

During load balancing adjustment, the impacts of these operations on the key status can be classified into three types: (1) **Split**. The operation only migrates a certain quantity of tuples with key k to other tasks; (2) **Merge**. The operation makes tuples with the same key that are separately stored in different tasks assembled into one task; (3) **Whole Move**. Contrary to Split, the operation migrates the entire tuples with a key from one task to the other.

3   Cost Model Analysis

Once load imbalance happens, load adjustment will be an inevitable action to improve system performance. A high-efficient migration strategy need to minimize system cost while guaranteeing load balancing.

3.1   System Resource Usage for Load Balance

System resource generally includes CPU, memory and network. It is discovered that routing table size(*routingtable*), migration volume(*migration*) and broadcast volume(*broadcast*) have great impacts on resource utilization. Here we use notation C_δ^τ to represent resource usage uniformly, where $\tau \in \{$routing table, migration, broadcast$\}$ and $\delta \in \{$CPU, Memory, Network$\}$.

CPU and Memory Cost. To ensure the correctness of processing, each task of routing operator must maintain the same routing table in memory; for each tuple distributing to processing operator, routing operator checks if there exists an entry related to the key of tuple. Obviously, the larger routing table will require more memory and cost more look-up time. The CPU cost C_{cpu}^r is defined as $C_{cpu}^r = T \cdot |RT|$, where $|RT|$ is routing table size and equals to $\sum_{k \in RT} |list_k|$, i.e.,$|list_k|$ is the size of all entries for key k. And the memory consumption C_{mem}^r of routing table is defined as $C_{mem}^r = N_r \cdot s \cdot |RT|$, where s represents the size of each entry in $list_k$. Then we get that $C_{cpu}^r \propto C_{mem}^r$.

Broadcast Cost. The broadcast cost is produced by two phases during system running, namely **(1) subsequent input tuples** while process going on and **(2) migration states** when load adjustment occurs. Let's take join processing for example to show the impact from input tuples. For each key k separately stored in different tasks, incoming tuples with k must be broadcast to all these tasks to ensure result completeness. Hence, the broadcast operation puts pressure on network. Then network cost C_{net}^b for broadcasting can be defined as $C_{net}^b = \sum_{k \in RT} D_k \cdot G_k$. Without loss of generality, during load adjustment, we only unload a certain quantity of keys out of overloaded tasks and migrate them to underloaded ones, not involving move-out operations on underloaded tasks. Hence, once load imbalance happens, no matter which key to choose for migration, the total migration volume is the same (based on key split method) and can be calculated as $C_{net}^m = \sum_{d \in UD} (L(d) - \frac{\sum_{d' \in D} L(d')}{N_d})$.

3.2   Our Idea

Our primarily goal is to improve system performance under workload imbalance, but minimize resource usages. As described above, we can draw a conclusion that routing table size and subsequent broadcast volume are two predominant factors on system resource usages. Without loss of generality, we use throughput to measure the influence of these two factors on system performance.

To achieve our goal, we propose to estimate the influence of these two factors on throughput first, which can be achieved by conducting massive experiments on varying routing table size and broadcast volume. Then we draw mapping relationships between them and throughput respectively. Based on these two mapping relationships, we expect to get a fitting function considering both routing table size and broadcast volume, guiding to select keys to migrate under current system environment. The details are elaborated in Sect. 5.

4   Load Balancing Adjustment

In this section, we first introduce the overall load balancing adjustment framework. Then we apply a novel decision-making algorithm to select a strategy to achieve our goal. At last we present two types of adjustment strategies.

4.1   Algorithm Framework

The algorithm of overall load balancing adjustment is described in Algorithm 1. First, average workload BL and maximized non-balance workload UL of downstream tasks are calculated in line 1. The parameter τ is the maximized imbalance degree of workload defined as $\frac{|L(d) - BL|}{BL}$. Once the workload of one task is more than UL, it is identified to be overloaded and need to unload extra

---

Algorithm 1. Overall Load Balancing Adjustment Framework

input: processing tasks D, broadcast volume $\sum_{u \in U} broadcast$, routing table size RTS
output: Migration plan MP and routing table RT
1: $BL \leftarrow \frac{\sum_{d \in D} L(d)}{N_d}$; $UL \leftarrow BL \cdot (1 + \tau_{max})$
2: **if** splitOrWhole($\sum_{u \in U} broadcast, RTS$) equals to *Split* **then**
3:      **foreach** each task d in D **do**
4:          **if** $L(d) > UL$ **then** splitAtFirstAdjustment(d)
5: **else**
6:      **foreach** each task d in D **do**
7:          **if** $L(d) > UL$ **then** wholeAtFirstAdjustment(d)
8: **return** MP and RT

---

workload to underloaded tasks as in line 4 and 7. Then, as described as in line 2, a decision-making algorithm is applied Finally, the corresponding adjustment strategy will be executed.

As analyzed in Sect. 3, broadcast volume and routing table size are two main factors influencing system performance. Therefore, decision-making algorithm combines both factors. It takes a combine function which represents a mapping relationship between broadcast volume and routing table size. The function takes the accumulated broadcast volume since last adjustment as input and returns a corresponding routing table size. If the returned value is bigger than current routing table size, it applies *split-keys-at-first* strategy; otherwise, it applies *whole-move-keys-at-first* strategy. The generation of this function will be introduced in Sect. 5.

4.2 Split-keys-at-first Load Balancing Strategy

A migration plan defines how to migrate data among tasks when load imbalance happens, which is formalized as $MP = <k, d_{from}, d_{to}, qty>$. It migrates tuples with key k from task d_{from} to d_{to} with the quantity of qty. Usually the adjustment procedure can be decomposed into two steps, shown in Algorithm 2:

step-1: Data Unload. It first unloads partial data to an temporary storage buffer C until the load of overloaded task d is lower than average workload BL as in line 3–10 in Algorithm 2. The keys on overloaded tasks are arranged in decreasing order based on their loads in line 1. When starting unloading, if the load of key k is lower than difference between task load $L(d)$ and average load BL, the entire tuple with k will be buffered in C and deleted from task d. Besides, if $<k, (d, \#)>$ entry exists in routing table, it must be deleted as in line 5; or else, partial tuples with k will be buffered and update the load of k in task d. Similarly, if $<k, (d, \#)>$ entry exists in routing table, it must be updated; otherwise, a new entry will be added as in line 9.

step-2: Data Load. After unloading operation, it loads the data in C to each underloaded task instance as in line 11–16. If load of key k in temporary storage C is lower than difference between load of an underloaded task $L(d')$

Algorithm 2. splitAtFirstAdjustment()

---

input: The key load set $KL = \{(key, load)\}$ of overloaded task d
output: Migration Plan MP and routing table RT
1: Arrange KL in decreasing order on key load; $C \leftarrow \emptyset$
2: **foreach** each key load entry $(k, l) \in KL$ **do**
3:     **if** $L(d) - BL > l$ **then** Add (k, l) to C; Delete (k, l) from KL
4:         **if** $< k, (d, \#) >$ exists in RT **then** // # is the proportion of data for k
5:             Delete $< k, (d, \#) >$ from RT
6:     **else**
7:         $\delta \leftarrow L(d) - BL$; Add (k, δ) to C; Replace (k, l) as $(k, l - \delta)$ in set KL
8:         **if** $<k, (d, \#)>$ exists in RT **then** Replace $<k, (d, l - \delta)>$ in RT
9:         **else** Add $<k, (d, l - \delta)>$ to RT
10:         **break**
11: **foreach** each temporary storage entry $(k, l) \in C$ **do**
12:     Choose a task d' which is underloaded
13:     **if** $BL - L(d') > l$ **then** $\delta \leftarrow l$; Delete (k, l) from C
14:     **else** $\delta \leftarrow BL - L(d')$; Replace (k, l) as $(k, l - \delta)$ in set C
15:     **if** key k exists in task d' **then** $\delta \leftarrow \delta + l'$ // l' is the load of k in d'
16:     Add $<k, d, d', \delta>$ to MP; Add $<k, (d', \delta)>$ to RT
17: **return** MP and RT

---

and average load BL, then it will be *whole-moved* to task d' and deleted from C as in line 14; otherwise it still needs to split k and move partial tuples with k in line 16.

4.3   Whole-move-keys-at-first Load Balancing Strategy

As described in Sect. 4.2, frequent key splitting may lead to massive subsequent network cost (e.g. for join). In this section, we present a whole-move-keys-at-first strategy which tries to migrate the entire tuples for a key of small granularity. The algorithm of this strategy is described in Algorithm 3 and can be divided into three steps as following:

step-1: Whole move keys in routing table. As shown in line 2–10, it first assigns priorities to keys existing in the routing table. IIf the basic task $F(k)$ of key k exists in the list $list_k$ in routing table, and that the basic task $F(k)$ still has enough storage, then the *Migrate Back* operation is executed as in line 3. Besides, for current overloaded task d, the routing entry $(d, \#)$ is deleted from $list_k$. If routing table doesn't contain entry for k, it invokes *Migrate Out* operation. If k is stored in more than one tasks, it prefers to merge tuples to one task containing k with enough free space; otherwise it will be migrated to one task not containing k.

step-2: Whole move keys not in routing table. If task d is still overloaded by $step - 1$, it then cope with keys that are not in routing tables as in line 14–15. For each underloaded task d', it selects a subset of keys and the total

---

Algorithm 3. wholeAtFirstAdjustment()

---

input: The key load set $KL = \{(key, load)\}$ of overloaded task d

output: Migration Plan MP and routing table RT

1: Differentiate the key load set $KIT = \{(key, load)\}$ in which keys exist in the routing table

2: **foreach** each key load entry $(k, l) \in KIT$ **do**

3:     **if** $<k, (F(k), \#)>$ exists in RT **and** $L(F(k)) < BL$ **then** Add $< k, d, F(k), l >$ to MP;Delete $<k, (d, \#)>$ from RT; Update $<k, (F(k), \#)>$ in RT

4:     **else if** $|list_k| > 1$ **then** //more than one entry in the list of key k in RT

5:         Choose a underloaded task d' that contains k

6:         Add $<k, d, d', l>$ to MP; Delete (k, l) from KIT

7:         Add $<k, (d', l + l')>$ to RT // l' is the load of k in d'

8:     **else**

9:         Choose a underloaded task d' that does not contain k

10:        Add $< k, d, d', l >$ to MP; Add $<k, (d', l)>$ to RT; Delete (k, l) from KIT

11: **foreach** each underloaded task d' **do**

12:     Arrange $KL - KIT$ in decreasing order on key load

13:     Choose a subset $SubKeys$ from $KL - KIT$ that $\sum_{k \in SubKeys} load_k \approx BL - L(d')$

14:     **foreach** each key load entry $(k, l) \in SubKeys$ **do**

15:         Add $<k, d, d', l>$ to MP; Add $<k, (d', \delta)>$ to RT; Delete (k, l) from $KL - KIT$

16: **if** $L(d) > BL$ **then**

17:     call splitAtFirstAdjustment() in Algorithm 2

18: **return** MP and RT

---

tuples of these keys are equal to the free storage of d'. The goal is to minimize the number of keys in such a subset. We take the approximate solutions for this problem. Before selecting operation, we arrange keys that don't exist in routing table in decreasing order on key loads. Then we use greedy algorithm to find out such a subset. Once a subset is returned, it needs to add or update the corresponding entries in the routing table and generate migration plans.

step-3: Split the remaining keys. If necessary, the split operation will be executed similar to Sect. 4.2 as in line 17.

The whole-move-keys-at-first adjustment strategy focuses on migrate tuples with the same key as a whole. Though the routing table size may grow rapidly with more keys of smaller granularity to be migrated, we take the moving back or merging operations to control the rapid growth of routing table size to a certain extent.

5 Evaluation

Environment: We implement and run all the approaches on top of *Apache Storm*[1]. The Storm system(version 0.10.0) is deployed on a 21-instance HP blade cluster, each of which has two Intel Xeon E5535 at 2.00 GHz and runs on CentOS 6.5 operating system.

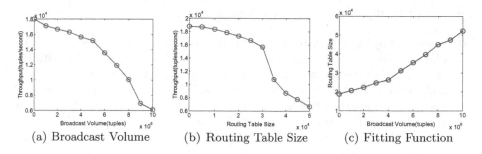

Fig. 2. Effects of broadcast volume, routing table size on throughput.

Data Sets and Workload: We conduct experiments using TPC-H generator *dbgen* proposed in [2] to generate data. We adjust parameter z for different skewness, default value is 0.8. We run equi-join query E_{Q_5} from TPC-H benchmark as our workload.

Baseline Approaches: We compare our method *SAM* with *Dynamic*[4], *MFM*[5], B_i[10] and *Readj*[8]. *Readj* runs full pairing of load-task. *Dynamic* and *MFM* are based on join-matrix model but *MFM* supports irregular shape of matrix and eliminates the restriction on task number. B_i organizes tasks as a complete bipartite graph, inside of which it is still matrix model.

Evaluation Metrics: *Throughput* is the average number of tuples processed per second in system; *Task Number* is the total number of tasks used in *Processing Operator* and each task is equipped with a constant quota of memory V; *Migration Volume* is the total number of tuples migrated to other tasks during system scaling out; *Adjustment Plan Time* is the average time spent on generating a new migration plan and routing table during load adjustment; *Load Imbalance Degree* is the maximal load imbalance degree among all the tasks of *Processing Operator*.

5.1 Decision-Making Analysis

To estimate the influence of broadcast volume on throughput, we continuously load all the 10^7 tuples with 10^4 unique keys into our system while executing query E_{Q_5}. Once load imbalance happens, decision-making method takes the accumulated broadcast volume since last adjustment as input and returns an intermediate routing table size irs. Then we compare irs with current routing table size since last adjustment. If irs is larger, we choose *whole-move-keys-at-first* for the next adjustment; otherwise we choose *split-keys-at-first*. According to the experimental results of Fig. 2(a) and (b), we then generate a polynomial relation between broadcast volume bv and routing table size $func(bv)$ as shown in Fig. 2(c). In our implementation, we set *func()* as following:

$$func(bv) = 2e - 10 \cdot bv^2 + 1.7e - 3 \cdot bv + 18331 \tag{1}$$

5.2   Load Balancing Capability (LBC)

LBC measured by the degree of system imbalance, defined as $\tau = \frac{L(d)-BL}{BL}$ in [6]. For balanced statues, each task has load lower than $UL = (1+\tau_{max})\cdot BL$ with the maximum imbalance tolerance τ_{max}. For this experiment, we set $\tau_{max} = 0.01$.

As shown in Fig. 3(a), when using 50 task instances, system imbalance of *Readj* increases dramatically when the degree of skewness is higher than 0.5. As skewness becomes more severe, the load of some key may become so large that even exceeds the average workload of tasks. No matter how *Readj* migrates keys across tasks, it can never get balanced in that *Readj* doesn't support splitting operation. On the contrary, our proposed algorithm *SAM* can migrate extra data from overloaded tasks via splitting keys on demand, hence it always can meet load balancing. *MFM*, *Bi* and *Dynamic* take random distribution as routing policy regardless of the changing of skewness. However those two methods may meet with high broadcasting cost for some specific operations, such as join in Fig. 3(c) and (d). In Fig. 3(b), we expand the task parallelism to 100. What's worse, *Readj* leads to much higher system imbalance [7].

5.3   Scalability

We continue load 12 GB data into Storm system and perform full-history join on E_{Q_5}. Figure 3(c) and (d) demonstrate the task consumption and migration volume during system scaling out. The maximum input rate is set by calling the method *setSpoutPending()* in Storm to consume all the computing power of each task.

In Fig. 3(c), it illustrates the changes of task requirements when we increase the stream volume. As data loading in, *Dynamic* meets sharp increase in task number because it has a strict requirement that the number of tasks must be a power of two and leads to high resource waste. Though *MFM* eliminates the limit on the task number, it is based on the join-matrix and consumes more tasks. Since B_i is designed especially for memory optimization, it is obvious that B_i uses less quantity of tasks. Contrarily, our algorithms performs best among all the methods. It applies for resource on real demand and especially supports to join task one by one. Figure 3(d) illustrates the changes of migration cost during scheme scaling out. Due to the massive data replication to maintain matrix structure, *Dynamic* leads to the highest migration volume than all other methods. *MFM* has less migration volume as involving less tasks than *Dynamic*. *SAM* yields low migration volume in that it adopts consistent hashing to do task deployment. Once a new task is added, only a few quantity of data need to be migrated owing to the inherit characteristics of consistent hashing.

5.4   Dynamics

In order to verify the dynamics and adaptivity of algorithms under different degrees of workload skewness, we conduct query E_{Q_5} execution using window-based model with window size as 3 min. The average input rate is about

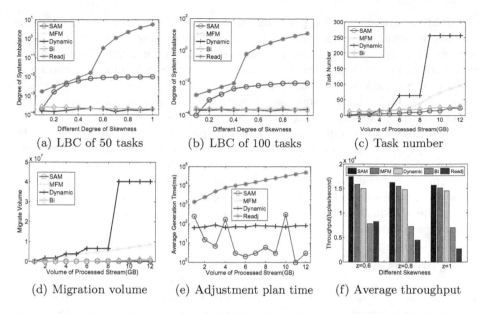

(a) LBC of 50 tasks   (b) LBC of 100 tasks   (c) Task number

(d) Migration volume   (e) Adjustment plan time   (f) Average throughput

Fig. 3. The balance capability, scalability and adaptivity of different methods.

$1.2 \cdot 10^4$ tuples per second to make full use of CPU resource. We examine the latency of adjustment plan generation and throughput as shown in Fig. 3(e) and (f). *Readj* meets much higher latency than other algorithms because of its full mapping processing as in Fig. 3(e). Both *Dynamic* and *MFM* take random distribution as routing policy, there is no need for balancing scheduling calculation. The migration plan generation latency of *SAM* varies a lot according to the choice of adjustment strategy. The higher generation latency corresponds to *whole-move-keys-at-first* strategy in that the selection of key subsets in step-2 of Algorithm 3 is an NP-hard problem; while the lower generation latency represents *split-keys-at-first* strategy because it split keys of large granularity at first, without complicated key selection. Figure 3(f) draws the throughput of each algorithm under different skewness. Throughput of *Readj* decreases as the skewness becomes more severe, because it spends more time generating migration plan. B_i represents decreasing throughput as well due to data broadcasting among groups. On the contrary, *SAM* has the highest throughput as it controls the growth of broadcast volume and routing table size, and uses a decision-making strategy to build a check-and-balance relationship between these two factors. Owing to the random distribution policy, throughputs of *MFM* and *Dynamic* change a little under different skewness.

6   Related Work

There have been two types of data-based strategies, namely tuple-based and key-based data distribution. Tuple-as-granularity strategies take tuple as distribution granularity to partition workload. Join-matrix model [13] has recently be adopted in a distributed stream join processing system *Squall* [4]. It models a join operation between two input streams as a matrix, where each side corresponds to one stream. When a tuple arrives, it is randomly distributed to tasks in the same row or column. This model can support arbitrary join predicates, but it incurs high memory consumption due to redundant storage, and it also limits the flexibility during system scaling out or down.

Analogously, key-as granularity strategies partition workload based on the key of tuple. [8] is the most similar to our proposal in this paper, which consists of a basic hash function and a routing table. However, for load imbalance adjustment, it trends to keep track of the most frequent keys and always moves back the keys of small granularity to their original destination by hash function. When the workload of keys varies dramatically, it may take too long time to generate a migration.

7 Conclusion

In this paper, we focus on designing an adaptive and cost-effective partitioning methods to handle load imbalance problem in distributed stream systems. Inspired by the idea of "Split keys on demand and Merge keys as far as possible", we propose a novel cost model to guide designing of balance schedules. Based on this cost model, our load balancing adjustment algorithms includes two type adjustment strategies, with the aim to have high throughput and low latency while using less resource. In the future, we will continue seeking for a more comprehensive cost model and more flexible and adaptive load adjustment strategies to improve system performance.

Acknowledgments. This work is partially supported by National High Technology Research and Development Program of China (863 Project) No. 2015AA0 15307, National Science Foundation of China under grant (No. 61232002, No. 61672233 and No. 61572194).

References

1. Apache storm. http://storm.apache.org/
2. The TPC-H benchmark. http://www.tpc.org/tpch
3. Abadi, D.J., Ahmad, Y., Balazinska, M., Zdonik, S.B. et al.: The design of the Borealis stream processing engine. In: CIDR, pp. 277–289 (2005)
4. Elseidy, M., Elguindy, A., Vitorovic, A., Koch, C.: Scalable and adaptive online joins. VLDB **7**(6), 441–452 (2014)
5. Fang, J., Wang, X., Zhang, R., Zhou, A.: Flexible and adaptive stream join algorithm. In: APWEB, pp. 3–16 (2016)
6. Fang, J., Wang, X., Zhang, R., Zhou, A.: High-performance data distribution algorithm on distributed stream systems. J. Softw. **28**(3), 563–578 (2017)
7. Fang, J., Zhang, R., Fu, T.Z.J., Zhang, Z., Zhou, A., Zhu, J.: Parallel stream processing against workload skewness and variance. arXiv preprint, arXiv:1610.05121 (2016)

8. Gedik, B.: Partitioning functions for stateful data parallelism in stream processing. VLDBJ **23**(4), 517–539 (2014)
9. Henzinger, M.R., Raghavan, P., Rajagopalan, S.: Computing on data streams. In: External Memory Algorithms, pp. 107–118 (1998)
10. Lin, Q., Ooi, B.C., Wang, Z., Yu, C.: Scalable distributed stream join processing. In: SIGMOD, pp. 811–825 (2015)
11. Nasir, M.A.U., Gianmarco, D.F.M. et al.: The power of both choices: practical load balancing for distributed stream processing engines. In: ICDE, pp. 137–148 (2015)
12. Shah, M., Hellerstein, J., Chandrasekaran, S., Franklin, M.: Flux: an adaptive partitioning operator for continuous query systems. In: ICDE, pp. 25–36 (2003)
13. Stamos, J.W., Young, H.C.: A symmetric and replicate algorithm for distributed joins. TPDS **4**(12), 1345–1354 (1993)
14. Xing, Y., Hwang, J., Çetintemel, U., Zdonik, S.B.: Providing resiliency to load variations in distributed stream processing. In: VLDB, pp. 775–786 (2006)
15. Xing, Y., Zdonik, S.B., Hwang, J.: Dynamic load distribution in the Borealis stream processor. In: ICDE, pp. 791–802 (2005)

A Graph-Based Push Service Platform

Huifeng Guo[1], Ruiming Tang[2], Yunming Ye[1(✉)],
Zhenguo Li[2], and Xiuqiang He[2]

[1] Shenzhen Graduate School, Harbin Institute of Technology, Shenzhen, China
huifengguo@yeah.net, yeyunming@hit.edu.cn
[2] Noah's Ark Lab, Huawei, China
{tangruiming,li.zhenguo,hexiuqiang}@huawei.com

Abstract. Learning users' preference and making recommendations is critical in information-exploded environment. There are two typical modes for recommendation, known as *pull* and *push*, which respectively account for recommendation inside and outside the item market. While previously most recommender systems adopt only pull-mode, push-mode becomes popular in today's mobile environment. This paper presents a push recommendation platform successfully deployed for Huawei App Store, which has reached 0.3 billion registered users and 1.2 million Apps by 2016. Among the various modules in developing this push platform, we recognized the task of target user group discovery to be most essential in terms of CTR. We explored various algorithmic choices for mining target user group, and highlighted one based on recent advance in graph mining, the Partially Absorbing Random Walk [13], which leads to substantial improvement for our push recommendation, compared to the state-of-the-art including the popular PageRank. We also covered our practice in deploying our push platform in both single server and distributed cluster.

Keywords: Partially Absorbing Random Walk · Push recommendation

1 Introduction

With the rapid development of the Internet and mobile devices, our daily life connects closely to online services, such as online shopping, online news and videos, online social networks, and many more. In such highly dynamic, information-exploded environment, it is crucial to learn the preference of users and make recommendations accordingly.

Recommendation often comes in one of the two modes, the *pull-mode* and *push-mode*. The pull-mode recommends items to users *after* users enter the item market. The push-mode pushes items to users proactively *before* the users enter the item market. Compared to pull recommendation, push recommendation can offer two unique advantages: to rebuild connection with users for the service

The work is done when Huifeng Guo works as an intern in Noah's Ark Lab, Huawei.

S. Candan et al. (Eds.): DASFAA 2017, Part II, LNCS 10178, pp. 636–648, 2017.
DOI: 10.1007/978-3-319-55699-4_40

(a) Push message   (b) Book listening   (c) Music   (d) Photo editor

Fig. 1. Push Services for Huawei App Store

provider and to enhance experience for the users – a user can be informed of relevant items anytime, without entering the item market. Unlike pull-mode that selects items for all users who are visiting the item market, the key in push recommendation is to identify a relatively small set of potential users for a given set of items. Unfortunately, techniques developed for pull recommendation such as matrix factorization are no longer suitable for the push-mode scenario, due to the following "cold-start" challenges: (1) the items to be pushed are usually new, with limited information available; and (2) "semi-active" and "inactive" users[1] rarely interact with the item market, and therefore not much of their information is available. While pull recommendation has been studied extensively, push recommendation is a new research area, especially to the academic community. In this paper, we present a Push Service Platform for Huawei App Store, one of the most large-scale and influential App markets in the world.

Figure 1 shows three push activities in Huawei App Store, book listening, music, and photo editor. Through the messages from the notification center (Fig. 1a), semi-active or inactive users are well informed of potentially relevant Apps without entering Huawei App Store. They can download their favorite Apps in the display pages by just clicking on the push message (Figs. 1b, c, d). Behind such convenience in connecting services to users, what is the key enabling technology?

During our extensive practice in establishing the push service for Huawei App Store, we found that identifying the right users to target is the most challenging task because too many unrelated messages could disturb users and degrade the experience. Another challenge comes from the large scale of the problem. With the versatility of smart phone and various needs from our daily life, a large number of Apps are being created by developers and installed by users. In Huawei

---

[1] In Application Market, "active users" refers to the users who visit frequently, "inactive users" are those who do not visit recently, and "semi-active users" are those who do not visit often recently.

App Store, there are 0.3 billion registered users and 1.2 million Apps by 2016. For a service (App) to be push, how to identify the users of interest from such a web-scale user pool? Especially, on average based on our statistics, for each service, there are less than 1% of the population relevant to the service, making the target user discovery extremely difficult.

In response to these practical challenges, we have established a Push Service Platform (*PSP*) for Huawei App Store, which mainly consists of three layers: distributed storage layer, application layer, and evaluation layer. The contributions of this paper are summarized as follows:

- We present a Push Service Platform for Huawei App Store. Particularly, we identify the target user discovery problem as the most significant task for the push service.
- We carefully compare different choices of algorithms for mining target user group, and highlight in details one based on recent advance in graph mining, namely Partially Absorbing Random Walk [13], which has been adopted by our push service. Particularly, we propose and implement an approximate partially absorbing random walk algorithm (A-PARW) for both single server and distributed cluster that can support very large-scale problems and can efficiently respond to a multitude of push services simultaneously.
- We conduct off-line and on-line experiments in Huawei App Store which shows that A-PARW leads to more than 27% and 16% improvement in online *CTR* and *DTR*, compared to the predecessor [7], which uses Personalized PageRank[2] in discovering target users.

In what follows, we present the full details of Huawei *PSP*. We first overview *PSP* in Sect. 2. Then we give the work flow of the Application Layer and presents the motivation, principle and implementation of A-PARW in Sect. 3. After that, we apply our system on several real marketing tasks in Huawei App Store, and carry out detailed off-line and on-line evaluation in Sect. 4. Finally, we discuss some related works in Sect. 5 and conclude the paper in Sect. 6.

2   Platform Overview

The architecture of Huawei Push Service Platform (PSP) is shown in Fig. 2, and includes *Distributed Storage Layer*, *Application Layer*, and *Evaluation Layer*.

The Distributed Storage Layer maintains two database systems for historical data storage and on-line caching. The HDFS (short for Hadoop Distributed File System) stores historical data, including users' download, click, and payment log data, which is the source data for our User-App bipartite graph (discussed later). The HBase (short for Hadoop Database) caches on-line data and users' feedback, which is critical for on-line monitoring and algorithm evaluation, and updates the historical data in HDFS periodically. In addition, this layer incorporates a Hadoop cluster to store large-scale datasets and provides parallel data processing.

---

[2] xRank, proposed in [7], is exactly Personalized PageRank (PPR) and is equivalent to the D mode of PARW. More details are presented in [3].

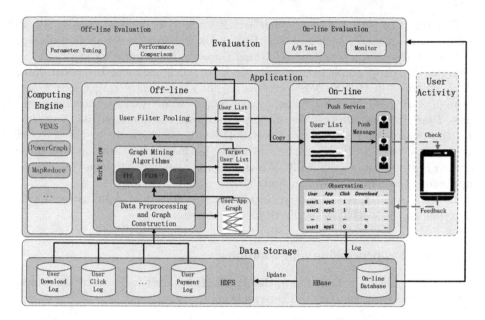

Fig. 2. PSP Architecture

The Application Layer consist of the major components (i.e., off-line target users mining and on-line pushing) of the platform. For different demands in practice, the Application Layer supports different Computing Engines, including graph engine and distributed computing engine. We will give more details of the Application Layer in Sect. 3 as it is the most challenging and important constituent of our platform.

The Evaluation Layer evaluates both off-line and on-line results. Off-line evaluation compares pre-defined off-line metrics of the results by different algorithms, which helps us to tune the parameters of the algorithms. On-line evaluation, such as A/B test, compares the performance of the algorithms which are carefully selected by off-line evaluation. The details of evaluation methods and metrics will be presented in Sect. 4.1.

3   Application Layer

The Application Layer of our PSP can be described as the following work flow:

$$History\ Data\ \xrightarrow[Graph\ Construction]{Pre-processing}\ User\text{-}APP\ Graph\ \xrightarrow{Graph\ Mining}\ Target$$

$$User\ List\ \xrightarrow{User\ Filtering}\ User\ List\ \xrightarrow[Observation]{On\text{-}line\ Pushing}\ User\ Feedback\ \xrightarrow{Logging}$$

$$Online\ Log\ \xrightarrow{Updating\ pediodly}\ Historical\ Data.$$

The input of PSP is a topic push activity, which is denoted as *seed Apps*[3] under a certain topic, such as music fans, cook lovers, etc. As an initial step, we provide *Data Preprocessing and Graph Construction* operation (Sect. 3.1) to generate User-App graph from users' download/click/payment historical data, which is stored in HDFS. Based on this graph, we can mine target user group through *A-PARW* (Sect. 3.2). Then, according to some domain knowledge and rules, *User Filtering* filters irrelevant users to obtain the final user list (Sect. 3.3) and the module of *Push Service* (Sect. 3.4) sends the message to the selected users. After that, PSP will cache the users' feedback data and update these data to History data periodically. We will review various computing engines in supporting needs in this layer in Sect. 3.5.

In the rest of this section, we introduce the details of Application Layer according to the work flow briefly described above.

3.1 Data Preprocessing and Modeling

On server log data, the first step in the Application Layer includes two modules, Data Preprocessing and Graph Construction.

Data Preprocessing: PSP can set a series of rules according to demands, such as removing pre-installed or very popular Apps from raw data before graph construction, because installing such Apps will not reflect users' interests.

Graph Construction: In this module, PSP constructs an undirected graph $\mathcal{G} = (\mathcal{U}, \mathcal{A}, \xi)$ based on the preprocessed data, where $\mathcal{U}$ denotes the set of *users* vertices, $\mathcal{A}$ is the set of *Apps* vertices, and ξ is the set of edges. Since we only use historical information to record the interaction between users and Apps, the constructed graph $\mathcal{G}$ is a bipartite graph. For instance, in Fig. 3, User vertices are on the left-hand-side and App vertices are on the right-hand-side. There exists an edge connecting U_i and A_i if user U_i installs App A_i. For example from Fig. 3, U_1 installs three Apps A_1, A_2 and A_3.

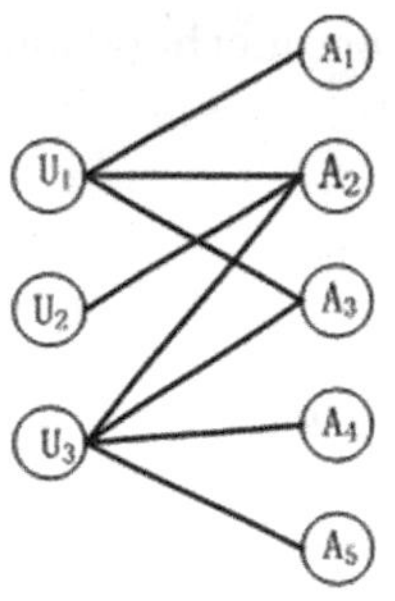

vertex	degree	PARW-I				PPR			
		iter:0	iter:1	iter:3	iter:20	iter:0	iter:1	iter:3	iter:20
v_1	3	0	0.00011104	0.0002389	0.0011157	0.00	0.28	0.18	0.28
v_2	1	0	0.000333	0.000508	0.001396	0.00	0.28	0.09	0.15
v_3	4	0	8.33E-05	0.00019306	0.0010534	0.00	0.28	0.22	0.31
v_4	1	0	0	0.00011093	0.0011146	0.00	0.00	0.05	0.08
v_5	3	0.0003	0.000333	0.000509	0.001521	1.00	0.15	0.33	0.53
v_6	2	0	0	9.71E-05	0.001084	0.00	0.00	0.10	0.14
v_7	1	0	0	8.32E-05	0.0010524	0.00	0.00	0.05	0.06
v_8	1	0	0	8.32E-05	0.0010524	0.00	0.00	0.05	0.06

Fig. 3. An example User-App bipartite graph and A-PARW-I vs PPR running case

---

[3] Seed Apps are a small set of manually-labeled Apps.

In addition, we assign uniform IDs to vertices in $\mathcal{A}$ and $\mathcal{U}$. More specifically, vertices in $\mathcal{U}$ are assigned with IDs from 1 to $|\mathcal{U}|$, and vertices in $\mathcal{A}$ are assigned from $|\mathcal{U}| + 1$ to $|\mathcal{U}| + |\mathcal{A}|$. For simplicity, we use v_i to denote vertex with ID i. For example, we use v_1 to v_3 to denote U_1 to U_3 and v_4 to v_8 to denote A_1 to A_5 respectively. We denote the adjacency matrix of $\mathcal{G}$ as $\mathcal{W}$, let $\mathcal{D} = diag(d_1, d_2, ..., d_N)$ with $d_i = \sum_j w_{ij}$ as the degree of vertex i, and define the Laplacian of $\mathcal{G}$ as $\mathcal{L} = \mathcal{D} - \mathcal{W}$.

In some push scenarios, such as online news recommendation, the graph has to be updated frequently as the hot spots are changing at any time. While in some other scenarios, such as the recommendation in application market like the one considered in this paper, the graph does not need to be updated so frequently because the interest of a user is unlikely to vary much from time to time. Hence, in our application scenario, we re-construct graph weekly with the most up-to-date information, which usually takes a few hours.

3.2   User Discovery via Graph Mining

After preprocessing, the most significant task for the push service is the problem of target user discovery. In this subsection, we carefully compare different choices.

As it is expensive to manually label all Apps and the result has no ranking information, the predecessor of PSP [7] applied PPR to mine target user group from some *seed Apps*. However, in our practice, PPR favors active users with high degree, who will download Apps with high probability no matter receiving push messages or not, and is likely to ignore the majority of inactive and semi-active users. Therefore, we need an algorithm that can mine relevant inactive and semi-active users. Below, we present the approximate Partially Absorbing Random Walk algorithm, *A-PARW*, which we found quite effective in mining target users from the seed Apps and has been adopted in our PSP push platform for Huawei App Store. (Due to space limit, the details are presented in [3].)

A-PARW: To mine target user group from a small number (e.g., 10) of *seed Apps*, we propose A-PARW by extending Partially Absorbing Random Walk (PARW) [13] for billion scale problems encountered in Huawei App Store. The formulation of PARW is $R^\top = (In)^\top \cdot (\Lambda + \mathcal{L})^{-1} \cdot \Lambda$, where R is the rank score vector, $\Lambda = diag(\lambda_1, \lambda_2, ..., \lambda_N)$ is a diagonal matrix with $\lambda_1, \lambda_2, ..., \lambda_N$ being arbitrary non-negative numbers, and In is a vector of $In(v)$ with $In(v) = 1/|seed\,Apps|$ if $v \in seed\,Apps$ and 0 otherwise.

In PARW, a random walk is absorbed at state i with probability p_i, and is transferred via a random edge of state i with probability $1 - p_i$. It is proved in [13] that a random walk starting from a set of low conductance vertices (referred as SP) is most likely absorbed in SP if $\Lambda = \alpha \cdot I$ (PARW-I), I is an identity matrix and α is a small positive value. One property of PARW-I is that the absorption probability varies slowly within SP, and drops sharply outside SP. This property suggests that PARW-I can effectively capture the underlying community structure of the graph.

Algorithm 1. dry = A-PARW(s, Λ, γ) ▷ APPROXIMATE ALGORITHM OF PARW

Input: S: seeds, $\Lambda = \{\lambda_1, \lambda_2, ..., \lambda_n\}$: regularization parameter, γ: tolerance threshold
Output: A-PARW vector dry
1: Initialize $dry = 0$ and $run = \{(s, 1/|S|)\}$
2: **while** run is not empty **do**
3:     pop a queue run element (i, w) and $dry_i = dry_i + \frac{\lambda_i}{\lambda_i + d_i} \cdot w$
4:     **if** $w > \gamma \cdot d_i$ **then**
5:         **for** all links $(i, j) \in \xi$ **do**
6:             **if** pair $(j, s) \in run$ **then**
7:                 $s = s + \frac{w}{\lambda_i + d_i}$
8:             **else**
9:                 add a new pair $(j, \frac{w}{\lambda_i + d_i})$ to run
10:             **end if**
11:         **end for**
12:     **end if**
13: **end while**

As Fig. 3 shows, low degree yet relevant vertex v_2 (the user represented by v_2 only installs A_2 (vertex v_5), which means he is more interested in A_2) absorbs higher score than v_3, which has high degree but with no deterministic preference, for A-PARW-I. In contrast, v_2 could not get good score for PPR. (Due to space limit, the details are presented in [3]).

However, to the best of our knowledge, there is no scalable implementation of PARW. Therefore we propose A-PARW in Algorithm 1 motivated by [1]. Our algorithm maintains a pair of vectors run and dry, starting with $dry = \mathbf{0}$ and $run = In$ (Line 1), then applies a series of push operations which transfer probabilities from run to dry while keeping no transfer out of dry. At vertex v_i, a push operation transfers $\lambda_i/(\lambda_i + d_i)$ fraction of run_i to dry_i (Line 3), then evenly distributes the remaining $d_i/(\lambda_i + d_i)$ fraction of run_i to v_i's neighbours (Line 5–11). We can control the precision through a strategy that A-PARW performs push operations only when $run_i \geq \gamma \cdot d_i$ (Line 4). As a result, we set γ to be 10^{-8} and select a limited number of iterations (i.e., 20) as A-PARW-I's stop condition. It is demonstrated to be good enough in our scenario.

3.3   Filtering Rule

In practice, the target user list, mined through A-PARW, includes some users who are not suitable to send push messages. Therefore we can define some practical filtering rules to filter them out. For instance, we should not select users who have turned off the function of receiving push messages, or we may not want to send messages to the users who visit Huawei App Store every day, etc.

3.4   On-Line Pushing

First of all, the module of *Push Service* will deliver messages as an alert on the notification center to the selected users' phone, whose push service is enabled and which is connected to the Internet. After that, these users will receive messages in their phone as shown in Fig. 1a, and may choose to neglect it or click on. After clicking this message, users will enter the specific page of Huawei App Store or even download the Apps contained in this page. For example, when a music fan receives an alert about music Apps on her phone's notification center (e.g., as the second message shown in Fig. 1a), she takes a look at music Apps in the display page (e.g., as in Fig. 1c), after clicking the alert message. In the end, we utilize user's feedback to generate market strategies and filtering rules. For example, we are more likely to send a message of a music-like activity to a user who has clicked the music push activity before, and we are less likely to send a push message to a user who has never clicked any push message before.

3.5   Computing Engine

As presented in Sect. 3.1 to Sect. 3.4, there are two kinds of computing tasks in the Application Layer, raw data extraction and graph mining. The former is easy to parallelize while the latter is difficult due to the heavy dependencies between vertices in a graph. Therefore, we choose MapReduce as general computing engine, but use graph engines, including VENUS [10] and Power-Graph [2], for graph mining. Specifically, we use a disk-based system–VENUS when push activity is not so urgent and the memory is limited; and we choose a memory-based distributed system–PowerGraph when push activity is highly urgent and resource is enough.

Table 1. Running Time (in seconds) of A-PARW-I on PowerGraph and VENUS

No. of push activities	1	10	100
PowerGraph	6.79	13.19	46.69
VENUS	2307.00	22588.90	N.A

In order to compare the efficiency of VENUS and PowerGraph, we ran experiments on twitter-graph [8], which contains 41,652,230 vertices and 1,468,364,884 edges. To compare fairly, the experiments are conducted on the same machine and the parameters are set to be the same as stated in the previous section. As Table 1 presents, PowerGraph needs only 6.79 s to process one push activity, while VENUS needs around 40 min. For 10 and 100 push tasks, PowerGraph uses around 13 s and less than 50 s respectively. While VENUS has to run more than 6 h when pushing 10 tasks. We don't test the case of pushing 100 tasks on VENUS as it needs several days to get the precise timing. But it's easy to estimate the time since it runs these tasks one by one.

4   Experimental Results

In this section, we first describe the data sets and evaluation metrics that are used in our experiments. Then, we compare the experimental results of A-PARW-I and PPR on real data set. Moreover, the details of experimental results on public data set are presented in [3] due to the space limit.

4.1   Data Set Description and Experiment Setting

We evaluated A-PARW algorithms on two data sets, MovieLens [4] and *APPData*, where *APPData* is collected from Huawei App Store. The difference between A-PARW-I and PPR is verified on both data sets, while evaluation on the real-life data sets furthermore confirmed the remarkable effectiveness of A-PARW-I.

Real-Life Dataset and Experiment Setting. We performed experiments on real data set–*APPData*. It is the complete user downloading log from 2015/03/01 to 2015/08/31 in Huawei App Store, and includes 96,324,654 users, 487,649 Apps, and 1,778,160,959 edges.

We first conducted experiments to evaluate the effectiveness of A-PARW-I, and then we verify the property of A-PARW-I and PPR. Our experiments on *APPData* include both off-line and on-line evaluation.

For off-line evaluation purpose, we collected users' feedback of nine push activities from Huawei App Store, for which the selected user lists were generated by PPR. The nine push activities are: **1:music; 2:camera; 3:instrument; 4:ticket; 5:listen book; 6:travel; 7:goodnight; 8:read; 9:internet**. We referred the users' feedback as the ground truth to compare the effectiveness of PPR and A-PARW-I. After receiving a push message of Apps, an interested user may click on it or even download this Apps. Therefore, we can distinguish the cases of click (or download) as follows: we refer a sample that the user clicked (or downloaded) the push message as positive, and the opposite case as negative. We ran PPR and A-PARW-I on *APPData*, and got top 1 million users as well as their ranking scores, respectively. In the off-line experiment, we adopt AUC as the evaluation metric.

In the on-line evaluation, we sent push messages to the same number of top users ranked by A-PARW-I and PPR, respectively. Making sure to receive the feedback from the majority of the users after two days, we compared the number of users who clicked (or downloaded) the recommended Apps, across the two sets of users picked by the two algorithms. We use CTR (click-through ratio) [7] and also DTR (download-through ratio) as the online evaluation metrics, where $CTR = |Users\ who\ clicked\ advertised\ app|/|Users\ who\ received\ ads|$ and $DTR = |Users\ who\ downloaded\ advertised\ apps|/|Users\ who\ received\ ads|$.

Public Dataset. As we discussed in Sect. 3, A-PARW-I is able to identify more semi-active users than PPR because it capitalizes on community structure. In order to verify this property, we ran PPR, A-PARW-I on MovieLens data set and compared degree trends of their ranking lists. The results implied PPR favors high degree vertices and ignores the semi-active users. In contrast, A-PARW-I can discover semi-active users. Due to space limit, the details are presented in [3].

---

[4] http://grouplens.org/datasets/movielens/.

4.2   Evaluation on Real-Life Data Set

Off-Line Evaluation. In off-line experiment, we used AUC to compare the ranking accuracy of A-PARW-I's and PPR's result list (we get top 1 million users from their result list). Table 2 presents click and download AUC improvement of A-PARW-I over PPR on the 9 push activities. As we can see, the performance of A-PARW-I is better than PPR at all the 9 push activities, on both click and download cases. Moreover, the improvement of A-PARW-I over PPR is more significant on download case. The improvement of A-PARW-I comes from the fact that A-PARW-I pays more attention to graph community information while PPR tends to high degree nodes. It means that A-PARW-I could find more semi-active user-nodes, which are of low degree but highly relevant to push activities.

Table 2. Off-line improvement of A-PARW-I over PPR in 9 different push activities

	1	2	3	4	5	6	7	8	9
Click AUC+	6.5%	10.9%	4.8%	3.1%	6.0%	37.7%	31.1%	39.9%	11.8%
Download AUC+	7.6%	10.9%	7.9%	9.2%	3.4%	12.0%	8.3%	12.4%	5.4%

On-Line Evaluation. In this subsection, we performed A-PARW-I and PPR algorithms on two different on-line push activities through Huawei's push service. Two lists of users are obtained by the two algorithms respectively, and the activity messages are push to these users [5]. Interested users may click the message to see the details of the Apps or perform further actions.

We calculated CTR and DTR from the log data of user feedbacks. For comparison, we define $CTR+$ and $DTR+$ as the improvement of A-PARW-I over PPR. As we can see in Table 3, the performance of A-PARW-I is significantly higher than PPR for both click and download on the two online push activities.

Table 3. On-line improvement of A-PARW-I over PPR

	Ticket	Music
$CTR+$	27%	82%
$DTR+$	16%	85%

Property of PARW. In order to analyse the property of A-PARW-I and PPR, we study the tendency of CTR/DTR and the degree of the user vertices selected by the two algorithms and sorted by their scores.

Tendency of CTR/DTR in sorted-steady-distribution. Figure 4a, b, c and d represent tendency of activity ticket's (music's) CTR and DTR. In the

---

[5] Duplicated users in the two lists are only push the message once.

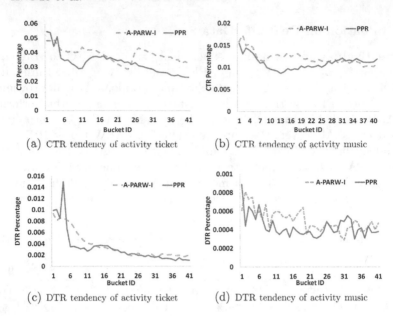

Fig. 4. Tendency of CTR/DTR.

figures, x-axis is the bucket identifier (each bucket includes 100,000 users) and y-axis is the CTR (respectively DTR) of the buckets. As we see, the curves of PPR and A-PARW-I have high CTR and DTR value at the beginning. However, PPR's curve drops more dramatically than A-PARW-I's; moreover, PPR's CTR and DTR sometimes increase at the tail of the curve. So A-PARW-I, which is steady and stable, selects the users who are more relevant to the push activity than PPR.

Change of degree in sorted-steady-distribution. Figure 5a, b presents the degree's tendency of A-PARW-I and PPR in the activity of ticket (music), where x-axis is same as Fig. 4 and y-axis is the total degree of the users in the buckets. As we see, the tendency of both red and blue curves are similar across the

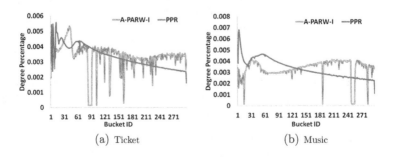

Fig. 5. Tendency of users' degree.

two figures. The PPR's curve is higher than the A-PARW-I's at first but drops rapidly and A-PARW-I's curve is more smooth and steady. It can be concluded from these two figures that PPR prefers high degree nodes.

5 Related Work

Compared to pull-mode, which gives recommendations to user within item market, push-mode pushes specific messages to users according to their characteristics even when users are not in item market. So push recommendation is able to rebuild or strengthen the connection between item market and users.

The crucial part of push-mode is target user group discovery, which can be solved by *rule-based* [5], *CF-based* [12] and *graph-based* [7,9] approaches. However, the rule-based methods can not take advantage of collaborative information among users, because it needs a set of rules that are defined so that the accuracy rate is low and not flexible enough. The CF-based methods tend to be ineffective in real-world scenarios because it requires a great deal of interactions of users and tags, which is problematic due to the sparsity of data. In graph-based approach, PageRank [11] is a well-known link analysis algorithm used by Google to rank websites according to their importance. There are many variants of PageRank, such as sensitive PageRank [6], xRank [7], and WTF [4]. However, PageRank based approach is biased to high-degree vertices. The authors of [13] propose a unified framework of graph mining and a new algorithm PARW-I, which can capture the community structure to overcome the weakness of PageRank. So we design approximate PARW, namely A-PARW, in our system.

6 Conclusions and Future Works

In this paper, we introduced Huawei Push Service Platform (PSP) to perform push recommendation by selecting target user group for a given push message. PSP includes potential users mining, online pushing, feedback caching and evaluation. In addition, we proposed A-PARW for target user group discovery on large scale data and presented a detail analysis among different choices of algorithms theoretically and empirically. We highlighted that A-PARW-I is able to discover the most relevant potential users and improve the performance of push service. As a live system, PSP supports the push recommendation in Huawei App Store and leads to a significant improvement over PPR.

Acknowledgement. This research was supported in part by NSFC under Grant No. 61572158 and Shenzhen Science and Technology Program under Grant No. JCYJ20160330163900579. We thank Dr. Qin Liu, Dr. Junbo Zhang and Chenzi Zhang for the help in scaling up PARW. We also thank Dr. Zhenhua Dong, Zhirong Liu and Benwei Gong for the valuable discussion and feedback.

References

1. Andersen, R., Chung, F., Lang, K.: Local graph partitioning using pagerank vectors. In: FOCS (2006)
2. Gonzalez, J.E., Low, Y., Gu, H., Bickson, D., Guestrin, C.: Powergraph: distributed graph-parallel computation on natural graphs. In: OSDI (2012)
3. Guo, H., Tang, R., Ye, Y., Li, Z., He, X.: A graph-based push service platform. https://arXiv.org/abs/1611.09496 (2016)
4. Gupta, P., Goel, A., Lin, J., Sharma, A., Wang, D., Zadeh, R.: WTF: the who to follow service at twitter. In: WWW (2013)
5. Han, J., Kamber, M.: Data Mining: Concepts and Techniques. Morgan Kaufmann (2000)
6. Haveliwala, T.H.: Topic-sensitive pagerank. In: WWW (2002)
7. He, X., Dai, W., Cao, G., Tang, R., Yuan, M., Yang, Q.: Mining target users for online marketing based on app store data. In: IEEE International Conference on Big Data (2015)
8. Kwak, H., Lee, C., Park, H., Moon, S.: What is twitter, a social network or a news media? In: WWW (2010)
9. Li, X., Wang, Y.Y., Acero, A.: Learning query intent from regularized click graphs. In: Proceedings of the 31st Annual International ACM SIGIR Conference on Research and Development in Information Retrieval (2008)
10. Liu, Q., Cheng, J., Li, Z., Lui, J.: VENUS: a system for streamlined graph computation on a single PC. TKDE **28**, 2230–2245 (2016)
11. Page, L., Brin, S., Motwani, R., Winograd, T.: The pagerank citation ranking: bringing order to the web. In: WWW (1999)
12. Su, X., Khoshgoftaar, T.M.: A survey of collaborative filtering techniques. Adv. Artif. Intell. **2009**, 421425:1–421425:19 (2009)
13. Wu, X.M., Li, Z., So, A.M., Wright, J., Chang, S.F.: Learning with partially absorbing random walks. In: NIPS (2012)

Edge Influence Computation in Dynamic Graphs

Yongrui Qin[1]([✉]), Quan Z. Sheng[2], Simon Parkinson[1],
and Nickolas J.G. Falkner[3]

[1] University of Huddersfield, Huddersfield, UK
{y.qin2,s.parkinson}@hud.ac.uk
[2] Macquarie University, Sydney, Australia
michael.sheng@mq.edu.au
[3] University of Adelaide, Adelaide, Australia
nickolas.falkner@adelaide.edu.au

Abstract. Reachability queries are of great importance in many research and application areas, including general graph mining, social network analysis and so on. Many approaches have been proposed to compute whether there exists one path from one node to another node in a graph. Most of these approaches focus on static graphs, however in practice dynamic graphs are more common. In this paper, we focus on handling graph reachability queries in dynamic graphs. Specifically we investigate the influence of a given edge in the graph, aiming to study the overall reachability changes in the graph brought by the possible failure/deletion of the edge. To this end, we firstly develop an efficient update algorithm for handling edge deletions. We then define the edge influence concept and put forward a novel computation algorithm to accelerate the computation of edge influence. We evaluate our approach using several real world datasets. The experimental results show that our approach outperforms traditional approaches significantly.

Keywords: Graph reachability · Dynamic graph · Edge influence

1 Introduction

Nowadays, graph structured data plays a more important role in various fields. As a foundational operation of a graph, reachability has a wide range of applications in many areas such as web data mining, biological research, social networks and computer programming, etc. It is noteworthy that, in these areas, the structure of graph is large and dynamic. To illustrate, Facebook has 1.79 billion active users monthly in the third quarter of 2016, increased from 1.55 billion active users monthly in the same period in 2015[1]. They may have different characters such as age, gender, hobbies, and may have complicated relationship with existing users.

---

[1] https://www.statista.com/statistics/264810/number-of-monthly-active-facebook-users-worldwide/, retrieved December 2016.

© Springer International Publishing AG 2017
S. Candan et al. (Eds.): DASFAA 2017, Part II, LNCS 10178, pp. 649–660, 2017.
DOI: 10.1007/978-3-319-55699-4_41

A large body of indexing techniques have been recently proposed to process reachability queries in graphs [2,3,5,11–13]. Among them, a significant portion of indexes are based on 2-hop labeling, which is originally proposed by Cohen et al. [7]. Most of the above mentioned approaches generally make the assumption that graphs are static. Some approaches investigate reachability in dynamic graphs, but they mainly focus on updating the overall indexes and supporting reachability queries in dynamic graphs [1,8,9,14].

One important question remains open: How can we evaluate the impact of an individual edge in a large graph in terms of reachability aspect? To the best of our knowledge, there is little work available in literature about the analysis of potential impact caused by changes in a large graph like edge deletions. How to evaluate the reachability influence of an edge is still an open problem. State-of-the-art approach TOL proposed in [1], which focuses on handling reachability queries in large dynamic graphs, does not provide an efficient way to compute the reachability difference caused by the failure or deletion of an edge. Therefore it remains challenging to evaluate the impact of the deletion of a given edge efficiently.

In this work, we firstly develop a decremental maintenance algorithm to efficiently update labeling index for edge deletions. We then define edge influence to indicate the impact of an edge on the reachability of the whole graph and put forward a novel computation algorithm to calculate edge influences efficiently. Experimental results show that our method outperforms state-of-the-art approach TOL in updating indexes on edge deletions and our edge influence computation algorithm is very efficient and can scale well. Potential applications of our algorithm include finding the most influencing edges in a given network, looking to building up some most important connections in an existing network, and so on.

The remainder of this paper is organized as follows. Section 2 reviews related work. Section 3 describes the details of our approach. Section 4 presents the experimental results and analysis. Section 5 concludes the paper and discusses future work.

2   Related Work

There is a large body of work on handling reachability queries in large graphs. *Tree Cover* approach is proposed by Agrawal et al. in [4], which searches a path based on a tree cover. The principle of this method is to encode each node by multiple intervals in a graph. However, even though it searches the graph efficiently, Tree Cover method can only guarantee its efficiency on static graphs. This is mainly because in order to achieve the optimal tree cover, it has to firstly establish a spanning tree when a graph changes. *Dual-Labeling* approach proposed by Wang et al. in [3] answers reachability queries by a Dual-Labeling encoding scheme. As a tree encoding method, Dual-Labeling method also encodes each node as a tree structure. Similar to Tree Cover method, it is suited for handling static graphs. *Chain Cover* approach proposed by Jagadish [6] uses a chain cover scheme to compute a reachability query. In this method, it divides a graph into several chains, which forms a chain cover for this graph. If we want to answer a reachability query, we just search whether there exists a pair in the chain cover. The disadvantage of this method is that if the graph is dynamic, for each update operation, there are numerous pairs to be modified, which reduces the efficiency especially when the graph is large. *Path-Tree Cover* approach proposed by Jin et al. [5] answers reachability queries by using a path-tree cover scheme. Its principle is similar to the Chain Cover method and the Tree Cover method. The difference is that Path-Tree Cover uses an extra scheme to deal with non-edge in the tree structure like Dual-Labeling approach.

Meanwhile, a few approaches have been proposed for handling reachability queries in dynamic graphs [8,9,14]; however, these approaches cannot scale well. State-of-the-art approach in this direction is TOL, proposed by Zhu et al. [1]. TOL uses a total order labeling scheme to answer reachability queries. It encodes a level order to each node in a graph. According to this order, TOL can compute a labeling table. Comparing with other schemes mentioned above, the advantage of TOL is that it simplifies the process of table construction. TOL has an advantage in dealing with large dynamic graphs. Surprisingly, it also outperforms most existing approaches on static graphs [1]. However, the main drawback of TOL is that it can only handle node deletions but cannot handle edge deletions.

3   Methodology

In this section, we present our approach in detail. Since our approach is inspired by state-of-the-art approach TOL [1], we firstly introduce TOL index briefly, then we develop our decremental maintenance algorithm for handling edge deletions. After that, we further define the concept of edge influence to investigate the impact of an edge in the overall reachability of a graph. We also put forward an efficient computation algorithm to compute edge influence on top of the updated labeling index of the graph.

3.1   TOL Index

The TOL Index [1] of graph $\mathcal{G}$ in Fig. 1 is shown in Fig. 2. There are three columns, n (denoting nodes), $\mathcal{L}_{in}$ and $\mathcal{L}_{out}$. Note that, TOL labeling is very similar to the *2-HOP Cover* approach proposed by Cohen et al. [7].

Example 1. According to Fig. 2, for node C, Column $\mathcal{L}_{out}$ contains $\{C\}$. It means that node C can reach all nodes that contains C in Column $\mathcal{L}_{in}$, including nodes C, D, E, and F. Similarly, for node E, Column $\mathcal{L}_{in}$ contains $\{C, E\}$. Hence node E can be reached by all nodes in Column $\mathcal{L}_{out}$ that contains C or E, including nodes A, B, C, E, and F.

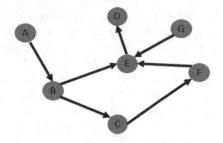

Fig. 1. Graph $\mathcal{G}$ (DAG)

n	L_{out}	L_{in}
A	{A, C}	{A}
B	{B,C}	{A, B}
C	{C}	{C}
D	{D}	{C,D,E}
E	{E}	{C,E}
F	{E,F}	{C,F}
G	{E,G}	{G}

Fig. 2. TOL Index $\mathcal{L}$

3.2 Handling Edge Deletions

Suppose the TOL index of a given graph $\mathcal{G}$ is $\mathcal{L}$. Assuming after deleting edge e from $\mathcal{G}$, we have $\mathcal{G}'$ and its corresponding index $\mathcal{L}'$. Using TOL approach, we have to compute $\mathcal{L}'$ from scratch for $\mathcal{G}'$. It is obvious that such process is not efficient. To improve efficiency, we devise a new approach that only calculates the difference between labeling indexes $\mathcal{L}$ and $\mathcal{L}'$, if an edge deletion occurs.

Given a DAG $\mathcal{G} = (V, E)$, if deleting edge $e = N_S \rightarrow N_E \in E$ from it, we can construct a new graph $\mathcal{G}_r \in \mathcal{G}$, which includes only nodes that can reach or be reached by either node N_S or node N_E. The structure of $\mathcal{G}_r$ is presented in Fig. 3.

It should be noted that, we only need to deal with those nodes whose reachability status to another node might be changed due to the edge deletion. All such nodes can be divided into six sets, N_{toSO}, N_{toEO}, N_{fromSO}, N_{fromEO}, N_{toSE} and N_{fromSE}. Details of these node sets are as follows:

N_{toSO} contains nodes that can reach node N_S but cannot reach node N_E, defined as a special ancestor node set of N_S.
N_{toEO} contains nodes that can reach node N_E but cannot reach node N_S, defined as a special ancestor node set of N_E.
N_{fromSO} contains nodes that can be reached by node N_S but cannot be reached by node N_E, defined as a special descendant node set of N_S.

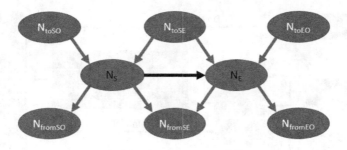

Fig. 3. Graph $\mathcal{G}_r$

N_{fromEO} contains nodes that can be reached by node N_E but cannot be reached by node N_S, defined as a special descendant node set of N_E.

N_{toSE} contains nodes that can reach both node N_S and node N_E, which form the common ancestor node set of both N_S and N_E.

N_{fromSE} contains nodes that can be reached by both node N_S and node N_E, which form the common descendant node set of both N_S and N_E.

The next process is to classify all relevant nodes into the corresponding sets, which is described in the following.

1. Find all ancestor and descendant node sets of N_S and N_E using BFS (Breadth-First-Search).
2. Compare the ancestor node set of N_S with the ancestor node set of N_E; put all nodes that appear in both ancestor node sets together and form N_{toSE}. Then N_{toSO} is the set containing the rest ancestor nodes of N_S and N_{toEO} is the set containing the rest ancestor nodes of N_E.
3. Following a similar step to the above Step 2, we can construct all the defined descendant node sets, including N_{fromSE}, N_{fromSO} and N_{fromEO}.

After completing the nodes classification, we can start to handle the edge deletion process. When deleting edge $e = N_S \rightarrow N_E \in E$, changes of each node in the labeling index can be divided into three situations as shown in Fig. 4.

Situation 1: For node sets N_{toEO} and N_{fromSO}, no changes are needed in the original labeling index. As shown in Fig. 3, a path from a node of N_{toEO} to any other node of $\mathcal{G}_r$ or a path from any node of $\mathcal{G}_r$ to a node in N_{fromSO}, will not contain edge e. In other words, reachability of any node in N_{toEO} and N_{fromSO} will not be affected by deletion of edge e.

Situation 2: For node sets N_{toSE} and N_{fromSE}, although the reachability related to the nodes in these sets is the same as before, the path of reachability may be different. This creates impact on the labeling index. We use the following steps to modify the labeling index to ensure the reachability of these nodes remains the same as before when computed from the updated labeling index.

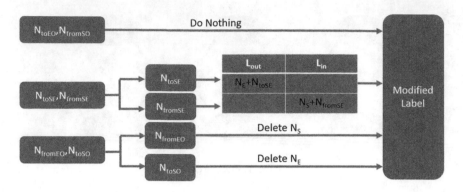

Fig. 4. Edge deletion flow chart

1. Denote node set of $\mathcal{L}_{out}$ column of node N_E in the labeling index as N_{Eout}.
2. Add N_{Eout} to $\mathcal{L}_{out}$ column of each node in N_{toSE} in the labeling index.
3. Denote node set of $\mathcal{L}_{in}$ column of N_S in the labeling index as N_{Sin}.
4. Add N_{Sin} to $\mathcal{L}_{in}$ column of each node in N_{fromSE} in the labeling index.

For the three node sets that N_{toSE} can reach (including N_{fromSO}, N_{fromSE} and N_{fromEO}), the reachability status from set N_{toSE} to set N_{fromSO} is not affected by edge e (see Fig. 3). For N_{fromSE} and N_{fromEO}, they are the only sets that can be reached by node N_E. Therefore, we only need to add N_{Eout} to the column of each node of N_{fromSE} and N_{fromEO} in the new labeling index.

Situation 3: For node sets N_{fromEO} and N_{toSO}, whether edge e is deleted or not has significant influence on their reachability status. After deleting edge e, all nodes in N_{toSO} cannot reach node N_E any more. When deleting edge e, the update of the labeling index can be achieved by deleting node N_E in the column of $\mathcal{L}_{in}$ of each node in N_{toSO}. It is similar to N_{toSO}, for N_{fromEO}, we only need to delete node N_S in the column of $\mathcal{L}_{out}$ of each node of N_{fromEO}.

Similar to deleting a node in TOL approach [1], the key step of updating index in the above situation is to modify the labeling index as follows.

For N_{toSO} and N_S:

1. Check the column $\mathcal{L}_{out}$ of node N_S, if there is node N_E, delete it.
2. Find out all $\mathcal{L}_{out}$ of child nodes of N_S except N_E, and denote them as set Set_S. Add all nodes in Set_S into $\mathcal{L}_{out}$ of N_S except the case that this node is already in $\mathcal{L}_{out}$.
3. Find out all nodes in N_{toSO}, and apply steps similar to Steps 1 to 2.

For N_{toEO} and N_E:

1. Check the column $\mathcal{L}_{in}$ of node N_E, if there is node N_S, delete it.
2. Find out all $\mathcal{L}_{in}$ of father nodes of N_E except N_S, and denote them as set Set_E. Add all nodes in Set_E into $\mathcal{L}_{in}$ of N_E except the case that this node is already in $\mathcal{L}_{in}$.
3. Find out all nodes in N_{toEO}, and apply steps similar to Steps 1 to 2.

Example 2. Given a DAG $\mathcal{G}$ and its corresponding labeling index $\mathcal{L}$ as shown in Fig. 5, if deleting edge $e = C \rightarrow F$, the process of edge deletion is as follows.

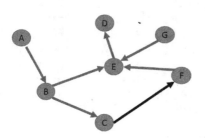

n	L_out	L_in
A	{A, C}	{A}
B	{B,C}	{A, B}
C	{C}	{C}
D	{D}	{C,D,E}
E	{E}	{C,E}
F	{E,F}	{C,F}
G	{E,G}	{G}

Fig. 5. Graph $\mathcal{G}$ and labeling index $\mathcal{L}$

Firstly, we divide all nodes of $\mathcal{G}$ into six sets, where:

N_{toSO} contains nodes A and B.
N_{fromEO} contains nodes E and D.
N_{toEO}, N_{toSE}, N_{fromSO}, N_{fromSE} All are empty sets.

It is obvious that we need to handle N_{toSO} and N_{fromEO}, which is Situation 3.

1. After checking column $\mathcal{L}_{out}$ of node C, we find that there is no F, so we do not need to delete anything.
2. We need to find out the child node of C, which is node F. Because F is the terminus node of edge e, we do not need to add anything to $\mathcal{L}_{out}$.
3. We deal with the father node of C, which is B. Because F is not in $\mathcal{L}_{out}$ of B, we just need to check all child nodes of B, E and C. For child node C, the $\mathcal{L}_{out}$ of C has already been included in $\mathcal{L}_{out}$ of B, which we do not need deal with. For child node E, the $\mathcal{L}_{out}$ of E is not in $\mathcal{L}_{out}$ of B, so we need to add E to $\mathcal{L}_{out}$ of B.
   The new $\mathcal{L}'_{out}$ of $B = \mathcal{L}_{out}$ of $B + \mathcal{L}_{out}$ of $E = \{B, E, C\}$.
4. We handle with the father node of B, which is A. Similar to B, we can compute the new $\mathcal{L}'_{out}$ of $A = \{A, B, C, E\}$. Because A does not have any father node, the process of dealing with C finishes.
5. Similar to C, after handling F, we can compute the new $\mathcal{L}'_{in}$ of $F = \{F\}$.
6. Similar to nodes B and A, we have $\mathcal{L}'_{in}$ of $E = \{E\}$ and $\mathcal{L}'_{in}$ of $D = \{D, E\}$. Because D does not have any child node, the process of dealing with F finishes.

Finally we can successfully compute the new $\mathcal{L}'$ of $\mathcal{G}'$ as shown in Fig. 6.

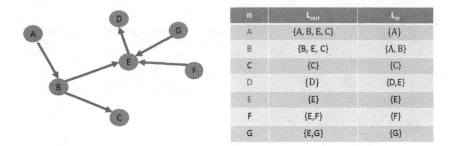

n	L_out	L_in
A	{A, B, E, C}	{A}
B	{B, E, C}	{A, B}
C	{C}	{C}
D	{D}	{D,E}
E	{E}	{E}
F	{E,F}	{F}
G	{E,G}	{G}

Fig. 6. New graph $\mathcal{G}'$ and new labeling index $\mathcal{L}$

3.3   Edge Influence

Next, we define edge influence and show how to calculate influence of a given edge efficiently. The influence provides a measure of how important an edge is to a graph. In other words, without this edge, its influence shows how greatly the reachability of all pairs of nodes of this graph will change.

Definition 1. *When one edge is deleted, the number of pairs of nodes whose reachability has been changed due to the deletion stands for the absolute influence of this edge (denoted as $\overline{Inf_e}$).*

According to Definition 1, we can calculate the maximum absolute influence of an edge in a given graph according to the following theorem.

Theorem 1. *Whatever the structure of the graph is, provided the amount of nodes is n, the maximum absolute influence (denoted as $\overline{Inf_{max}}$) of any edge in the graph is:*

$$\overline{Inf_{max}} = \begin{cases} \left(\frac{n}{2}\right)^2 & n \text{ is even.} \\ \frac{(n-1)}{2} \cdot \frac{(n+1)}{2} & n \text{ is odd.} \end{cases} \tag{1}$$

Proof. Given a graph $\mathcal{G} = (V, E)$ with n nodes, there must exist an edge $e \in E$ which divides V into three node sets A, B and C. Sets A and B meet two requirements: First, each node in A can find a path to each node in B; Second, every such path contains e. Set C contains all other nodes that do not belong to sets A and B.

If denoting the number of nodes in sets A, B and C as a, b and c, we have formula (2).

$$n = a + b + c \tag{2}$$

If deleting edge e, any path from one node in A to another node in B may become disconnected. And the amount of these paths are $a \cdot b$. Then according to Definition 1, we have formula (3).

$$\overline{Inf_e} = ab \tag{3}$$

Combining (2) and (3), we have formula (4) as follows.

$$\overline{Inf_e} = a(n - a - c) \tag{4}$$

It is obvious that when c increases, the value of Inf_e decreases. In order to achieve the maximum value of Inf_e, c should be 0. Then, we can transform (4) to (5).

$$\overline{Inf_e} = a(n - a) \tag{5}$$

Obviously, Inf_e is maximum when $a = \frac{n}{2}$. Since $n \in \mathbb{Z}^+$, if n is even, $a = \frac{n}{2}$, and if n is odd, $a = \frac{n-1}{2}$. Combining (2), we arrive at formula (1). This completes the proof. □

Once $\overline{Inf_{max}}$ of a graph is known, the normalized influence of an edge in this graph can be calculated as follows.

Definition 2. *Given a graph $\mathcal{G}' = (V, E)$ with n nodes, the influence of an edge $e \in E$ is denoted as Inf_e, which can be computed as follows:*

$$Inf_e = \frac{\overline{Inf_e}}{Inf_{max}} \tag{6}$$

Calculation. Given a DAG $\mathcal{G}$, if deleting an edge $e = N_S \rightarrow N_E$, we can apply the following steps to calculate the influence of edge c.

- According to Theorem 1, we calculate $\overline{Inf_{max}}$
- Find out the node set where nodes can be reached by node N_E including node N_E itself, denoted as Set_E
- Find out nodes that belong to Set_E and can be reached by node N_S before deleting edge e, denoted as Set_{SP}
- Find out nodes that belong to Set_E and can be reached by node N_S after deleting edge e, denoted as Set_{SL}. And then we can get node set $Set_S = Set_{SP} - Set_{SL}$. The amount of nodes in Set_S equals to the amount of bode pairs with changed reachability status that is related to node N_S after deleting edge e
- For each ancestor node of N_S, we can calculate the corresponding Set_i. It should be noted that during the calculation, once we have $Set_i = \varnothing$, we can stop calculating ancestors of node N_i
- Assuming n_i is the amount of nodes in Set_i, the absolute influence of edge e is $\overline{Inf_e} = \sum n_i$. Then we can compute the influence of edge e according to formula (6)

Example 3. Given a DAG $\mathcal{G}$, we can calculate the influence of edge $e = C \rightarrow F$ in the following.

- Calculate the maximum absolute influence $\overline{Inf_{max}}$ of graph $\mathcal{G}$.

$$\overline{Inf_{max}} = \frac{(7 - 1)}{2} \times \frac{(7 + 1)}{2} = 12$$

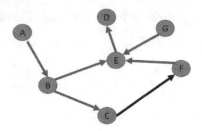

Fig. 7. Graph $\mathcal{G}$

- There are two other nodes that F can reach. Then $Set_F = \{D, E, F\}$
- $Set_{CP} = \{D, E, F\}$
- $Set_{CL} = \varnothing$ Then $Set_C = \{D, E, F\} - \varnothing = \{D, E, F\}$. Set_C has three nodes D, E and F, so $n_C = 3$
- For node C's ancestor node B, $Set_B = \{D, E, F\} - \{D, E\} = \{F\}$. There is only one node F in Set_B, so $n_B = 1$. For the second ancestor node A of node C, $Set_A = \{F\} - \varnothing = \{F\}$, which means $n_A = 1$
- $\overline{Inf_e} = \sum n_i = n_C + n_B + n_A = 3 + 1 + 1 = 5$. Then

$$Inf_e = \frac{\overline{Inf_e}}{Inf_{max}} = \frac{5}{12}$$

Therefore, the influence of edge e is $\dfrac{5}{12}$ in graph $\mathcal{G}$ in Fig. 7.

4   Experiments

We use the following five real-world datasets in our experiments: p2p-Gnutella08 (*Gnu08*, 6.3K nodes, 21K edges), p2p-Gnutella06 (*Gnu06*, 8.7K nodes, 32K edges), Wiki-Vote (*Wiki*, 7.1K nodes, 104K edges), p2p-Gnutella31 (*Gnu31*, 63K nodes, 148K edges) and soc-Epinions1 (*Epi1*, 76K nodes, 509K edges) [10]. We use these datasets to conduct two sets of experiments. The first set is to delete 100 edges generated by the graph transformation module randomly. We compare labeling index method (TOL method) with our method by performing 100 edge deletions and record the average time cost and the average index size (the changed part). The second set is to validate our edge influence algorithm also by performing 100 edge deletions and record the average time cost. All experiments were performed on a PC with 64-bit Windows 7, 8 GB RAM and 2.40 GHZ Intel i7-3630QM CPU.

It should be noted that these bar charts use log-scale plotting features. From Fig. 8(a), we can see that our method performs deletion nearly at a speed of an order of magnitude faster than TOL method among all the datasets. The updated index size of our method is also much smaller than TOL method (see Fig. 8(b)). The main reason for these two experimental results is that our method can

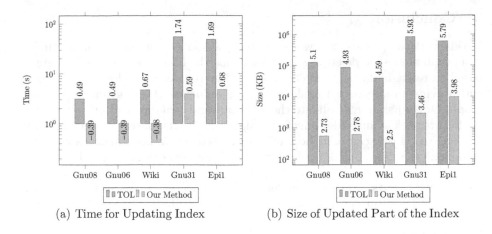

(a) Time for Updating Index   (b) Size of Updated Part of the Index

Fig. 8. Comparison with TOL approach

incrementally compute updated index for edge deletions, while TOL method can only support node deletions. For edge deletions, TOL method has to recompute the whole index from scratch. Here, we do not compare indexing time with other approaches for static graphs, as TOL method has been shown to outperform most existing approaches for static graphs [1].

Figure 9 shows the average calculation time of edge influences for the 100 deleted edges. Since this is the first attempt on the calculation of edge influence, we compare our method with a modified BFS & DFS method[2]. From the figure we can see that our method can compute edge influences orders of magnitude faster than the modified BFS & DFS method. For dataset *Epi1*, the modified BFS & DFS method cannot complete the calculation within 24 h.

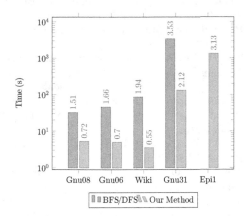

Fig. 9. Calculation time of edge influence

---

[2] BFS & DFS refers to Breadth-First-Search and Depth-First-Search. Both our method and BFS & DFS were performed on top of the updated labeling index.

5   Conclusion

In this work, we propose a new approach to calculate an updated labeling index of a graph after edge deletions. Then we define the influence of edges in a graph based on reachability between nodes affected by the potential deletions of edges, and provide an approach to calculating the influence of any given edge in the graph. Our experiments validate the efficiency and scalability of our approach.

Future work includes devising faster algorithms to handle even larger dynamic graphs (e.g., at the scale of millions of nodes and edges) and design new algorithms to rank all the edges based on our defined edge influence in large graphs.

Acknowledgments. Authors would like to thank Xiaorong Liang for the implementation of the algorithms and thank anonymous reviewers for their valuable comments.

References

1. Zhu, A.D., Lin, W., Wang, S., Xiao, X.: Reachability queries on large dynamic graphs: a total order approach. In: Dyreson, C.E., Li, F., Tamer Özsu, M. (eds.) SIGMOD Conference 2014, Snowbird, UT, USA, pp. 1323–1334, 22-27 June 2014. doi:10.1145/2588555.2612181. http://dl.acm.org/citation.cfm?id=2588555
2. Cheng, J., et al.: TF-Label: a topological-folding labeling scheme for reachability querying in a large graph. In: Proceedings of the International Conference on Management of Data. ACM (2013)
3. Wang, H., et al.: Dual labeling: answering graph reachability queries in constant time. In: Proceedings of the 22nd International Conference on Data Engineering, ICDE 2006. IEEE (2006)
4. Agrawal, R., Borgida, A., Jagadish, H.V.: Efficient management of transitive relationships in large data and knowledge bases, vol. 18, no. 2. ACM (1989)
5. Jin, R., et al.: Efficiently answering reachability queries on very large directed graphs. In: Proceedings of the ACM SIGMOD International Conference on Management of Data. ACM (2008)
6. Jagadish, H.V.: A compression technique to materialize transitive closure. ACM Trans. Database Syst. (TODS) **15**(4), 558–598 (1990)
7. Cohen, E., et al.: Reachability, distance queries via 2-hop labels. SIAM J. Comput. **32**(5), 1338–1355 (2003)
8. Bramandia, R., Choi, B., Ng, W.K.: Incremental maintenance of 2-hop labeling of large graphs. IEEE Trans. Knowl. Data Eng. **22**(5), 682–698 (2010)
9. Yildirim, H., Chaoji, V., Zaki, M.J.: Dagger: a scalable index for reachability queries in large dynamic graphs. arXiv preprint arXiv:1301.0977 (2013)
10. Leskovec, J.: Stanford large network dataset collection (2014). http://snap.stanford.edu/data/index.html
11. Yildirim, H., Chaoji, V., Zaki, M.J.: GRAIL: scalable reachability index for large graphs. PVLDB **3**(1), 276–284 (2010)
12. van Schaik, S.J., de Moor, O.: A memory efficient reachability data structure through bit vector compression. In: SIGMOD, pp. 913–924 (2011)
13. Seufert, S., Anand, A., Bedathur, S.J., Weikum, G.: Ferrari: flexible and efficient reachability range assignment for graph indexing. In: ICDE, pp. 1009–1020 (2013)
14. Schenkel, R., Theobald, A., Weikum, G.: Efficient creation and incremental maintenance of the HOPI index for complex XML document collections. In: ICDE, pp. 360–371 (2005)

Demos

DKGBuilder: An Architecture for Building a Domain Knowledge Graph from Scratch

Yan Fan[1], Chengyu Wang[1], Guomin Zhou[2], and Xiaofeng He[1(✉)]

[1] Shanghai Key Laboratory of Trustworthy Computing,
School of Computer Science and Software Engineering,
East China Normal University, Shanghai, China
eileen940531@gmail.com, chywang2013@gmail.com, xfhe@sei.ecnu.edu.cn
[2] Department of Computer and Information Technology,
Zhejiang Police College, Hangzhou, China
zhouguomin@zjjcxy.cn

Abstract. In recent years, we have witnessed the technical advances in general knowledge graph construction. However, for a specific domain, harvesting precise and fine-grained knowledge is still difficult due to the long-tail property of entities and relations, together with the lack of high-quality, wide-coverage data sources. In this paper, a domain knowledge graph construction system DKGBuilder is presented. It utilizes a template-based approach to extract seed knowledge from semi-structured data. A word embedding based projection model is proposed to extract relations from text under the framework of distant supervision. We further employ an is-a relation classifier to learn a domain taxonomy using a bottom-up strategy. For demonstration, we construct a Chinese entertainment knowledge graph from Wikipedia to support several knowledge service functionalities, containing over 0.7M facts with 93.1% accuracy.

Keywords: Knowledge graph · Taxonomy learning · Relation extraction

1 Introduction

A domain knowledge graph (DKG) is a special type of knowledge graphs that focuses on modeling relations between entities in a specific domain. It plays an important role in providing knowledge service for special-purpose applications, such as medical diagnosis, movie recommendation, etc.

Although abundant research has been conducted on general-purpose knowledge graph construction, entities and relations in a specific domain are still hard to obtain in a large quantity and a high accuracy. The difficulties mostly lie in three aspects: (i) knowledge in existing manually-built expert systems or domain relational databases usually has the low coverage; (ii) it is difficult to harvest domain facts from semi-structured/unstructured data, especially for long-tail entities and relations; and (iii) taxonomies of DKGs are often designed by experts, and constructing them is a tedious and time-consuming process.

© Springer International Publishing AG 2017
S. Candan et al. (Eds.): DASFAA 2017, Part II, LNCS 10178, pp. 663–667, 2017.
DOI: 10.1007/978-3-319-55699-4_42

In this paper, we introduce DKGBuilder, a general framework to construct a DKG solely from semi-structured and unstructured data sources. In the implementation, it takes Wikipedia pages related to a specific domain as input and extracts entities, classes, attributes and relations in a weakly supervised manner. It first constructs an initial DKG by template-based extractors over Wikipedia infoboxes and categories. We design an is-a relation classifier to build the domain taxonomy based on the Wikipedia category system in a bottom-up strategy. In order to extract long-tail relations from plain texts, a word embedding based projection model is proposed to identify new relations in the embedding space.

For demonstration, we present a transparent process of constructing a Chinese entertainment DKG from scratch. The system also supports several online tasks of knowledge service and analysis. The DKG we constructed consists of over 0.1M entities and 0.7M facts related to the entertainment industry in China. The average accuracy of these facts (i.e., attributes and relations) is 93.1%.

2   System Overview and Key Techniques

As Fig. 1 shows, DKGBuilder consists of offline and online parts. The offline system contains three modules: (i) seed knowledge graph constructor, (ii) taxonomy learner and (iii) plain-text relation extractor. The online part supports entity and class tagging, semantic search and statistical knowledge data analysis.

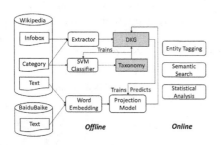

Fig. 1. Framework of DKGBuilder

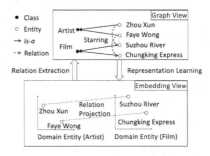

Fig. 2. Knowledge representation

The seed knowledge graph constructor builds an initial DKG with high accuracy, which employs template-based methods to extract domain entities, attributes and relations from Wikipedia infoboxes and categories with minimal human intervention. The extracted facts are later employed as training data for the latter two modules. In traditional domain databases, entities are categorized into a few coarse-grained classes (e.g. patients, diseases, symptoms and medicines in the medical domain). In DKGBuilder, the taxonomy learner classifies entities into a large number of fine-grained classes to construct the taxonomy. It employs several novel features and trains an SVM-based is-a relation classifier based on

Wikipedia category system. The plain-text relation extractor harvests long-tail knowledge by learning new relations from unstructured text. Because a limited, domain-specific corpus is usually sparse in terms of entity contexts, we propose a word embedding based projection model to learn relation representations and extract relations from text using the framework of distant supervision.

We now discuss our implementation details coping with several technical challenges in DKGBuilder.

Seed Knowledge Graph Construction. The first step is to find the most important entities in a specific domain, which minimizes the risk of extracting unrelated entities from knowledge sources. In our system, DKGBuilder takes a couple of human-defined template names from Wikipedia as input and selects entities whose infobox-template name matches one of the pre-defined names. After that, it extracts "seed" relations by mapping frequent attribute names to relations. For each attribute, we design a mapping function which converts an attribute to one/many relations. Based on the semantics of attributes, we categorize the mappings into three types: i.e., *direct, multiple* and *indirect*, inspired by the Wikipedia-based ontology YAGO [1]. We additionally use domain-specific filters to improve the precision of the initial DKG.

Fine-Grained Entity Categorization. The construction of domain taxonomy can be modeled as the *fine-grained entity categorization* problem. Because the template names in Wikipedia are relatively coarse-grained (e.g., actor, movie, etc.), we extend our prior work [2] to derive the entertainment taxonomy. For each entity, an SVM classifier is trained to predict whether there is a hyponymy-hypernym (i.e., "is-a") relation between the entity and each of its category names in Wikipedia. For example, "Hong Kong actor" is a hypernym of "Tony Leung Chiu-wai" while "1962 births" only provides relational facts about the actor. A set of features are engineered for accurate is-a relation prediction, such as the number of words in the category name, the POS tag of the head word of the category name, the common sequence of the entity and category names, the existence of specific language patterns, etc. Finally, the top-level of the domain taxonomy is constructed based on the rule mining algorithm proposed in [2].

Representation Learning and Relation Extraction. The limited coverage in the initial DKG prompts us to identify new relations from plain text to cover more long-tail facts. In a domain-specific corpus, especially for Chinese, robust relation extraction is challenging due to the flexibility of language expression and the text sparsity issue of entity contexts [3]. In DKGBuilder, rather than applying syntactic and/or lexical pattern matching methods, we learn entity and relation representations in the embedding space to support relation extraction in a semantic level. We first crawl a large-scale Chinese text corpus from *Baidu Baike*[1], consisting of 1.2M articles and 1.088B Chinese words after word segmentation. A skip-gram model [4] is trained over the text corpus to obtain the 100-dimensional embedding vector $\boldsymbol{v}(e)$ for each entity e.

---

[1] http://baike.baidu.com/.

For relation representation, similar to [6], we combine two previous relation representation approaches (i.e., vector offsets in [4] and linear projection in [5]) together in the embedding space. For an entity pair (e_i, e_j) that has a certain relation R_k, we assume there is a projection matrix M_k and an offset vector b_k such that $M_k \cdot v(e_i) + b_k \approx v(e_j)$. For ease of implementation, we learn relation representation using the distant supervision framework. We randomly sample relation instances from initial DKG, then learn the parameters by minimizing the following objective function via Stochastic Gradient Descent: $J(M_k, b_k; R_k) = \frac{1}{2} \sum_{(e_i, e_j) \in R_k} \|M_k \cdot v(e_i) + b_k - v(e_j)\|^2$.

After obtaining values of parameters, we make a single pass over the corpus. We first extract entities (e_i and e_j) that co-occur in the same sentence and pair those which have close syntax and lexical distances into candidate relation instances. For each candidate pair (e_i, e_j) and the relation R_k, the model predicts $(e_i, e_j) \in R_k$ iff $\|M_k \cdot v(e_i) + b_k - v(e_j)\| < \delta$ where δ is a pre-defined threshold.

In Fig. 2, we illustrate the two knowledge representations and their connections. In the graph view, entities and relations are expressed explicitly in the form of a directed graph. By representation learning, we can map the DKG into the embedding space. New relations are extracted or inferred from free text based on entity and relation representations, which are again added to the DKG.

3   Demonstration and Evaluation

We will demonstrate the knowledge graph construction process in DKGBuilder and showcase its semantic service functionality. Specifically, the Chinese entertainment DKG consists of 13K classes, 100K entities, 250K attribute-value pairs, 46 relation types and 480K relation instances. By random sampling, we analyze the overall accuracy of facts from all modules, summarized in Table 1.

The online system is developed in Java and uses Tomcat as the Web server. The knowledge data of DKGBuilder is managed by the Neo4j graph database. Besides the basic analysis tasks of entertainment knowledge data, the system supports semantic search of entities and relations, using a search engine-like interface. It also provides knowledge service for deep reading, which tags key entities and classes in documents. Screenshots are shown in Figs. 3 and 4.

Table 1. Accuracy Evaluation

Module	Accuracy
Seed Constructor	99.6%
Taxonomy Leaner	98.7%
Relation Extractor	71.4%
Overall	**93.1%**

Fig. 3. Screenshot I   **Fig. 4.** Screenshot II

Acknowledgements. This work is partially supported by the National Key Research and Development Program of China under Grant No. 2016YFB1000904 and NSFC-Zhejiang Joint Fund for the Integration of Industrialization and Informatization under Grant No. U1509219.

References

1. Suchanek, F.M., Kasneci, G., Weikum, G.: Yago: a core of semantic knowledge. In: WWW, pp. 697–706 (2007)
2. Li, J., Wang, C., He, X., Zhang, R., Gao, M.: User generated content oriented Chinese taxonomy construction. In: Cheng, R., Cui, B., Zhang, Z., Cai, R., Xu, J. (eds.) APWeb 2015. LNCS, vol. 9313, pp. 623–634. Springer, Cham (2015). doi:10. 1007/978-3-319-25255-1_51
3. Wang, C., Gao, M., He, X., Zhang, R.: Challenges in Chinese knowledge graph construction. In: ICDE Workshops, pp. 59–61 (2015)
4. Mikolov, T., Chen, K., Corrado, G., Dean, J.: Efficient estimation of word representations in vector space. CoRR abs/1301.3781 (2013)
5. Fu, R., Guo, J., Qin, B., Che, W., Wang, H., Liu, T.: Learning semantic hierarchies via word embeddings. In: ACL, pp. 1199–1209 (2014)
6. Wang, C., He, X.: Chinese hypernym-hyponym extraction from user generated categories. In: COLING, pp. 1350–1361 (2016)

CLTR: Collectively Linking Temporal Records Across Heterogeneous Sources

Yanyan Zou[✉] and Kasun S. Perera

Singapore University of Technology and Design, Singapore, Singapore
yanyan_zou@mymail.sutd.edu.sg, baruhupolage@sutd.edu.sg

Abstract. A huge volume of data on the Web are continually made available, which provides users rich amount of information to learn more about entities. In addition to attribute values of entities, there is often additional relational information, such as friendship on social networks, coauthorship of papers. However, to understand how these facts across heterogeneous data sources are related is challenging for users due to entity evolution over time. In this paper, we propose a novel system to help users find how records are temporally related and understand how entity profiles evolve over time. Our system is able to Collectively Link Temporal Records (CLTR) by taking advantage of evidence from both attribute and relational information on multiple sources. We demonstrate how CLTR allows users to explore time-varying history of targeted entities and visualizes multi-type relations among entities.

Keywords: Record linkage · Temporal data · Collective clustering · Heterogeneous web sources

1 Introduction

Heterogeneous Web sources provide abundant information to describe entities from different aspects over a long period of time. In addition to attribute values of entities, there is often additional relational information on the Web, such as friendship on social networks, coauthorship of papers. Understanding how facts across heterogeneous data sources are temporally related is paramount and inevitable to many applications. It also poses new challenges due to evolving information of entities where attribute values may vary over time (e.g., location, age and organization). In order to deal with time-varying property of temporal records, Li et.al. first presented a time-decay model [1] which learns the probability that an entity will change its attribute values within a time period. CHRONOS [4], the closest to our work, is implemented based on this technique. It collects data only from bibliography domain. This work was then extended by [2], where a mutation model was proposed to capture how likely an entity will return to its previous values in a given time period. Another framework, called MAROON [3], was designed to observe temporal patterns of attribute value

© Springer International Publishing AG 2017
S. Candan et al. (Eds.): DASFAA 2017, Part II, LNCS 10178, pp. 668–671, 2017.
DOI: 10.1007/978-3-319-55699-4_43

transitions. None of the models emphasizes relational information among entities, which actually provides us additional evidence to address temporal linkage across heterogeneous Web sources.

In this paper, we present a novel framework, CLTR, to help users understand how facts are temporally related across sources. The major contributions of CLTR can be summarized as follows:

1. CLTR collects data from multiple Web sources to enrich entity profiles.
2. The system combines evidence from both attribute and relational information to reconcile entities across heterogeneous web sources. It employs collective clustering mechanism. The algorithm fully utilizes relations to jointly reconcile entities which co-occur on the Web, rather than independently.
3. It enables users to search targeted entities by different domain keywords, such as name and organization. The tool presents entity history by timelines and also visualizes entity relation graph to help users further understand how entities are temporally related across sources over time.

2 Methodology

The architecture of our CLTR system is illustrated in Fig. 1. It takes as input data crawled from heterogeneous web sources and visualizes history profiles and relations of entities. The system consists of three key components: input and data preprocessing, temporal record linkage and interactive interface.

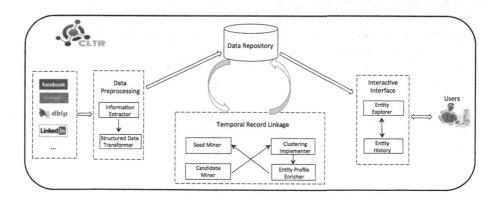

Fig. 1. Architecture of CLTR system

Data Preprocessing takes as input data provided by web sources, e.g., Facebook, Linkedin, DBLP, and personal homepages. This component transforms facts about entities into records associated with timestamps, including name, title, organization, interests, social connections and so on. It stores records in the data repository.

Temporal Record Linkage implements a collective clustering algorithm to determine related entities jointly. Initially, it discovers reconciled entities as seeds via linked web sources. Starting with seed entities, the component applies relations to further reconcile more entities which are related to them. It searches for potentially mergeable pairs from related records and pushes them into the clustering algorithm to determine whether they should be merged. Once new merging decisions are made, profiles of the corresponding entities are further enriched. On the other hand, these newly-merged records are considered as new seeds to enlarge seed entity set and to discover more candidates. The clustering results are stored in the data repository.

Relational information is beneficial to propagate evidence among records. It can also contribute to making merging decisions. To combine attribute information with relational evidence, we define a new metric:

$$Match(r_1, r_2) = \alpha \Phi(r_1, r_2) + (1 - \alpha)\Psi(N(r_1), N(r_2)) \tag{1}$$

where r_1, r_2 represent two records, $\Phi(r_1, r_2)$ denotes temporal similarity of attribute values, $\Psi(N(r_1), N(r_2))$ captures evidence from two records' relations, $N(r)$ denotes the neighbor of record r via relations, α is a balancing factor to assign weights for temporal and relational components.

Interactive Interface offers users the explorer to search by different attribute keywords, such as name, organization, and the timelines to trace the complete history for each entity.

3 Demonstration Scenario

Using a running example, we demonstrate how users can interact with our system to search for a specific entity.

Consider a user who would like to find a person named "Meng Jiang". He selects keyword domain and searches by name "meng jiang". Figure 2 depicts

Fig. 2. Entity explorer interface

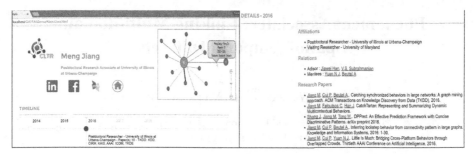

(a) Entiy timeline and relation graph   (b) Entity's details of the selected time point

Fig. 3. Detailed entity's history and relations depiction

possible search results for this. Each result is shown with latest profile summary (i.e. title and affiliation) of fuzzy matched "meng jiang" in our datasets. The user can select one of them to trace more details.

Suppose the user has selected the second result, then the system switches to profiling page of this entity. As illustrated in Fig. 3a, it shows various aspects of this person, such as his most recent affiliation, title, linked websites according to data sources. The interactive timeline summarizes profiles of the entity over years, e.g., in 2016, "Meng Jiang" is a postdoctoral researcher at "University of Illionois at Urbana-Champaign", and has published 10 papers in 7 conferences. It is accessible for users to scroll over the timeline to switch time points. If the user wants to know more details in 2016, he can click the year node, as shown in Fig. 3b, "Meng Jiang" visited "University of Maryland" this year. Related entities to the selected one are shown on an interactive graph, depicted next to entity summary. For example, we show the relation graph of "Meng Jiang", related by his coworkers, social friends, etc. The thickness of edges is proportional to the strength of the relations. The color of edges represents relations from different sources (e.g., Facebook, DBLP, etc.). Clicking on each edge, user can observe more information about the entity relationship.

References

1. Li, P., Dong, X.L., Maurino, A., Srivastava, D.: Linking temporal records. Proc. VLDB Endow. 4(11), 956–967 (2011)
2. Chiang, Y.H., Doan, A.H., Naughton, J.F.: Modeling entity evolution for temporal record matching. In: Proceedings of the 2014 ACM SIGMOD International Conference on Management of Data, pp. 1175–1186. ACM (2014)
3. Li, F., Lee, M.L., Hsu, W., Tan, W.: Linking temporal records for profiling entities. In: Proceedings of the 2015 ACM SIGMOD International Conference on Management of Data, pp. 593–605. ACM (2015)
4. Li, P., Tzviskou, C., Wang, H., Dong, X.L., Liu, X., Maurino, A., Srivastava, D.: Chronos: facilitating history discovery by linking temporal records. Proc. VLDB Endow. 5(12), 2006–2009 (2015)

PhenomenaAssociater: Linking Multi-domain Spatio-Temporal Datasets

Prathamesh Walkikar[✉] and Vandana P. Janeja[✉]

University of Maryland Baltimore County, Baltimore, USA
{prath1,vjaneja}@umbc.edu

Abstract. This paper focuses on the demonstration of an analytics dashboard application for analyzing interesting spatio-temporal associations between anomalies across multiple spatio-temporal datasets, potentially from disparate domains, to find interesting hidden relationships. The proposed system is intended to analyze spatio-temporal data across multiple phenomena from disparate domains (for example traffic and weather) to identify interesting phenomena relationships by linking anomalies from each of these domain datasets. This web-based dashboard application developed in R Shiny [1] provides interactive visualizations to quantify the multi-domain associations. The application uses a novel framework of algorithms and quantification metrics to associate these anomalies across multiple domains using spatial and temporal proximity and influence metrics.

1 Introduction

In today's world, it is not a surprise to find that almost everything in this world is inter-related particularly nearby things. These relationships also become fundamentally true across multiple application domains. For example, (a) Weather condition at a location will impact traffic [2], (b) Oil spills in oceans will adversely impact underlying aquatic animal population [3], (c) pollution in a location can affect disease spread and many more [4]. The common link across phenomenon is the underlying geographical space, which can help associate phenomena in a particular region to link with other inter-related phenomenon in the same region. This can be in the form of spatio-temporal associations. This preposition however faces numerous obstacles in the form of tremendous amount of single domain data and difficulty in combining data across different domains due to data heterogeneity issues.

These data integration and heterogeneity issues can be avoided by looking at extracted knowledge from individual phenomena and then look for potential associations across the discovered knowledge capturing data for a phenomenon, instead of looking at raw data form distinct domains. The extracted knowledge in each domain in our case is the anomalous window comprising of a set of contiguous points in a region that are unusual with respect to the rest of the points in the region. One example of multi-domain anomaly detection is discussed in [5] where circle based scan windows from single domains are linked using traditional spatial associations. In this paper, we propose a novel dashboard to illustrate the discovery of such multi-domain anomalies through

© Springer International Publishing AG 2017
S. Candan et al. (Eds.): DASFAA 2017, Part II, LNCS 10178, pp. 672–676, 2017.
DOI: 10.1007/978-3-319-55699-4_44

novel influence metrics which quantify the associations between phenomena in both spatial and spatio-temporal datasets. Our web-based dashboard application discovers these single domain anomalies across individual domain datasets and associates them to derive interesting relationships between different domains capturing multiple phenomena associations.

2   Demonstration

Developed in R, this dashboard application can assist domain experts in deriving potential knowledge from multi-domain spatio-temporal datasets and thus, facilitate researchers studying impact analysis of one phenomena over others in research fields like epidemiology, traffic accident analysis, impact of wildfires [6] to name a few. Figure 1 shows a sample of association results in the form of detected anomalous windows overlayed on a map visualization for a real world spatio-temporal health-ranking outcome dataset where we discover interesting spatio-temporal associations using influence indicators between child poverty rates and unemployment rates in the State of Maryland, USA.

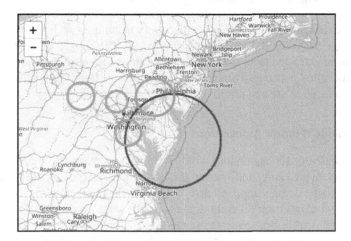

Fig. 1. Multi-domain anomalous association results

In the demo, we will show step by step how PhenomenaAssociater discovers interesting associations between such multi-domain spatio-temporal datasets.

2.1   Discovering Single Domain Spatio-Temporal Anomalies

Spatio-temporal data handling and management forms an important component in our analysis process. For handling the temporal elements of our data, we discretize the data by time and then perform anomaly detection in each of the temporal bins created. Our application is equipped with three different categories of data discretization techniques which include - equal frequency binning, equal width binning and hierarchical time series clustering based binning strategies. Data Modelling component handles the generation of specific input files from individual discretized instances of datasets, which are inputs for associations, such as specific formats required for SaTScan [7]. Figure 2 depicts this spatio-temporal data modelling and management view of our application.

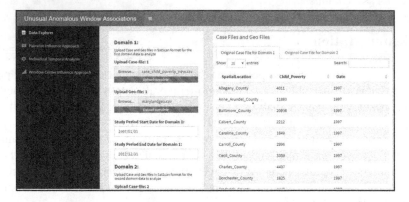

Fig. 2. Data pre-processing and management component

By using existing anomalous window discovery techniques such as space-time scan statistics using SatScan, SSLIP [8] and RWSCAN [9], single domain anomalous windows can be extracted from spatio-temporal datasets for each domain such as child poverty or unemployment data. Our interface is much more tightly coupled with SatScan mainly due to its wide use and intuitive findings.

2.2   Anomalous Window Associations

After these single-domain anomalous windows have been extracted, we associate these windows based on proximity and overlap patterns of these anomalous windows discovered from the single domain anomaly detection methods. In other words, co-occurrence of anomalous windows from different domains in the same geographical areas of proximity over time determines spatio-temporal association. We propose the concept of influence distance, which quantifies the overlap across the phenomena both in terms of spatial and non-spatial attributes. We also propose and utilize influence score - a novel metric for measurement of spatio-temporal associations as well as variations of spatio-temporal confidence and support and lift measures which quantify these interesting associations. Due to limited scope of this paper we mainly define influence distance:

Definition 1 (*Influence distance*). *Let v be the given phenomenon, and p and q be two spatial objects. We define $d^v_{p \to q}$ as the **influence distance** from spatial object p to q for the phenomenon v. $d^v_{p \to q}$ is the sum of the weights of the constituent edges of the shortest path from p to q in the network of v.*

Hence, if p and q are one spatial object, then $d^v_{p \to q} = 0$. If p and q are not connected, then we get the influence distance $d^v_{p \to q} = \infty$.

Here network of phenomena is derived using a spatial neighborhood [10] approach. Based on influence distance, we calculate the influence score, which quantifies the proximity and overlap of anomalous windows, where we take the aggregate influence distance between the spatial nodes present within each of the distinct domain windows. When a user wants to analyze a set of two distinct inter-related domains, after running anomalous window discovery methods, the user gets a series of single domain anomalous windows with respect to time for each distinct domain. To further associate these anomalies, our application provides the user to choose from a set of distinct approaches to effectively associate these domain anomalies to find interesting phenomena associations. These links are quantified using the distinct set of influence indicators which are plotted in Fig. 3.

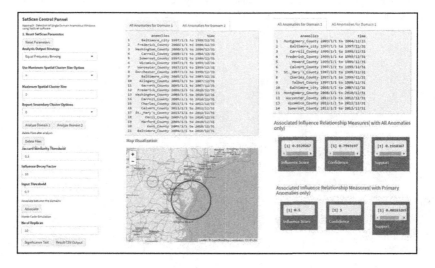

Fig. 3. Influence relationship results

These results can act as a supplementary information for domain experts in order to obtain useful phenomena linkages to further study rare phenomena relationships.

Acknowledgement. This work is supported in part by the US Army Corps of Engineers, agreement number: W9132V -15-C-0004.

References

1. Chang, W., Cheng, J., Allaire, J., Xie, Y., McPherson, J.: Shiny: web application framework for R. (2015)
2. Nookala, L.S.: Weather impact on traffic conditions and travel time prediction. Thesis (2006)
3. Barron, M.: Ecological impacts of the deepwater horizon oil spill: implications for immunotoxicity. Toxicol. Pathol. **40**, 315–320 (2011)
4. Barbara, J.: The impact of climate change on human health. In: Impact of Climate Change on Water and Health, pp. 75–105 (2012)

5. Costa, R., Pereira, M., Caramelo, L., Vega Orozco, C., Kanevski, M.: Assessing SaTScan ability to detect space-time clusters in wildfires. In: EGU General Assembly 2013 (2013)

6. Janeja, V.P., Adam, N., Atluri, V., Vaidya, J.S.: Spatial neighborhood based anomaly detection in sensor datasets. Data Min. Knowl. Disc. **20**(2), 221–258 (2010). Springer, Special issue on outlier detection

7. Kulldorff M., Information Management Services, Inc.: SaTScanTM v8.0: software for the spatial and space-time scan statistics (2009). http://www.satscan.org/

8. Shi, L., Janeja, V.P.: Anomalous window discovery through scan statistics for linear intersecting paths (SSLIP). In: Proceedings of the 15th International Conference on Knowledge Discovery and Data Mining - KDD 2009 (2009)

9. Janeja, V.P., Atluri, V.: Random walks to identify anomalous free-form spatial scan windows. IEEE Trans. Knowl. Data Eng. **20**(10), 1378–1392 (2008)

10. Janeja, V.P., Palanisamy, R.: Multi domain anomaly detection in spatial datasets. Knowl. Inf. Syst. J. **36**, 749–788 (2012). doi:10.1007/s10115-012-0534-5

VisDM–A Data Stream Visualization Platform

Lars Melander[✉], Kjell Orsborn, Tore Risch, and Daniel Wedlund

Uppsala Database Laboratory, IT Department, Uppsala University, Uppsala, Sweden
{lars.melander,kjell.orsborn,tore.risch}@it.uu.se,
daniel.wedlund@sandvik.com

Abstract. Visual Data stream Monitor (VisDM) is a new approach
to integrate a visual programming language with a data stream management system (DSMS) to support the construction, configuration, and
visualization of data stream applications through a set of building blocks,
Visual Data Flow Components (VDFCs). This functionality is provided
by extending the LabVIEW visual programming platform to support
easy declarative specification of continuous visualizations of continuous
query results. With actor-based data flows, visualization of data stream
output becomes more accessible.

Keywords: Data stream management · Data stream visualization ·
Visual data flow programming

1 Visualization of Data Streams

The capability to efficiently handling data streams in industrial processes is
becoming critical for transforming the current manufacturing industry. In an
industrial system, large volumes of sensor data are produced in the form of
continuous data streams from industrial processes and products equipped with
sensor installations. To make the output data streams intelligible by an analyst
they should be visualized in real-time.

As data stream management is becoming increasingly complex, methods that
counter-balance the complexity and make it more accessible are needed. The approach presented in this paper enables easy analysis and visualization of streaming
data, by proposing a flexible visual specification and deployment of visualizations of data stream analyses produced by a data stream management system
(DSMS) [6], where *continuous queries* (CQs) filter, transform, and combine data
streams.

Visual data flow programming [3,4] is an increasingly popular way of constructing applications, thanks to the rapid and robust prototyping this type of
programming enables. Data stream management is conceptually similar to data
flow programming, and with data flows the step between specification and implementation is eliminated; the program specification becomes the program. Development time decreases, and programming tasks can be moved closer towards the
end user.

© Springer International Publishing AG 2017
S. Candan et al. (Eds.): DASFAA 2017, Part II, LNCS 10178, pp. 677–680, 2017.
DOI: 10.1007/978-3-319-55699-4_45

Our approach, *Visual Data stream Monitor* (VisDM), utilizes the existing state-of-the-art visual programming environment in *LabVIEW* [8] to enable high-level visualization for engineering and scientific DSMS applications. LabVIEW offers a visual programming environment that is comprehensive, yet has a flat learning curve and a user interface that many find attractive [2,5,10].

A visual data flow is a program specified using graphic building blocks called *function nodes* [4,9] where each node consumes one or several input data flows and produces output data flows or visualizations. The function nodes are implicitly driven by the flow of data, rather than by explicit control structures as in regular programming. The sources of the visualized data flows are function nodes connected to CQs through a stream-oriented client-server API. It is fairly straight-forward to design a data flow environment using actors [1,7]; each actor becomes a function node, and each entity in a data flow becomes a message that is sent from one actor to another.

The prototype system provides an integrated visualization and scalable data stream analysis platform, by interfacing LabVIEW with the *SVALI* (Stream Validator) DSMS [11].

2   VisDM Data Flow Programming

Data stream processing with DSMSs usually requires custom visualization of the result from CQs. In VisDM, a library of common controls provides the basic primitives for building the visualizations, forming the foundation for the VDFCs. The visualization primitives are highly customizable using a point-and-click interface and forms, providing the support for continuous visualization of external data streams that LabVIEW does not have. The VDFCs also provide the framework for data-flow oriented specifications.

Figure 1 illustrates how VisDM is used for monitoring two machine data streams, routing data through intermediate *Corenet* servers.

A simple VisDM data flow specification is shown in Fig. 2. In the example, a CQ is executed on a SVALI server named "Mill1". The RUN QUERY VDFC node is a *producer*, a VisDM function node that is the source of a data flow, sending a CQ to the SVALI server and receiving a stream of tuples that constitutes the output data flow. The output of RUN QUERY becomes the input to a VDFC node labelled "Mill Power" that represents the output diagram. It is a *consumer* node that visualizes a stream using a LabVIEW graphical object.

The VisDM data flow specification in Fig. 3 is an extended example that has another data stream and update functionality added to the specification.

3   Summary

VisDM is a prototype system aimed a making data stream management more accessible while improving user productivity, by offering a visual data flow programming environment and automating programming tasks. It allows users to create and maintain easy data stream visualization without losing versatility, by

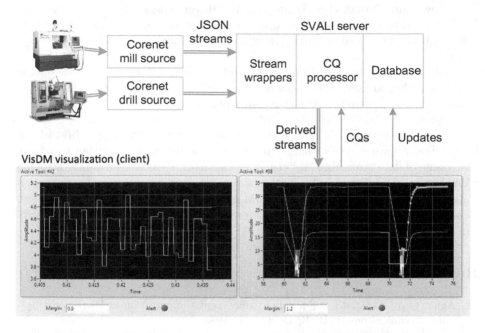

Fig. 1. Machining equipment monitoring with VisDM.

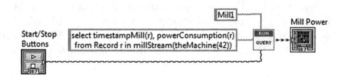

Fig. 2. Visual data flow specification.

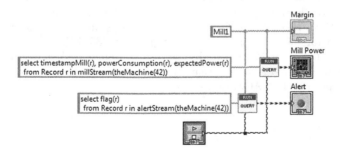

Fig. 3. Adding update functionality to a program specification with two CQ visualizations.

providing a form-based, visually programmed development environment through LabVIEW. It makes data stream application programming more user centric, by raising abstractions and providing intuitive, visual application-oriented development.

VisDM applications are built from a set of Visual Data Flow Components (VDFCs). They are based on a newly developed data flow framework that interconnects a DSMS with visualization components, providing easy specification of continuous visualizations of CQ results.

A unique strength of VisDM is its complete extensibility. To the best of our knowledge, no other system offers the same capabilities for adapting to such a vast range of data streaming and visualization tasks. VisDM is currently utilizing SVALI as the DSMS and LabVIEW as the programming and visualization environment. However, the design of VisDM is not restricted to this configuration. Other DSMSs and visualization environments that provide similar functionality can also be integrated.

References

1. Agha, G.A.: ACTORS: A Model of Concurrent Computation in Distributed Systems, Technical report, DTIC Document (1985)
2. Baroth, E., Hartsough, C.: Experience Report: Visual Programming in the Real World (1994)
3. Culler, A., Culler, D.E.: Dataflow architectures. Ann. Rev. Comput. Sci. **1**, 225–253 (1986)
4. Davis, A.L., Keller, R.M.: Data flow program graphs. Computer **2**, 26–41 (1982)
5. Ertugul, N.: Towards virtual laboratories: a survey of LabVIEW-based teaching/learning tools and future trends. Int. J. Eng. Educ. **16**, 171–180 (2000)
6. Golab, L., Öszu, M.T.: Data Stream Management. Morgan & Claypool, Williston (2010)
7. Hewitt, C., Zenil, H.: What is computation? actor model versus Turing's model. In: A Computable Universe: Understanding and Exploring Nature as Computation, pp. 159–185 (2013)
8. National Instruments, White Papers. http://ni.com/white-papers/
9. Sowa, M., Murata, T.: A data flow computer architecture with program and token memories. IEEE Trans. Comput. **100**, 820–824 (1982)
10. Whitley, K.N., Blackwell, A.F.: Visual programming in the wild: a survey of LabVIEW programmers. J. Visual Lang. Comput. **12**, 435–472 (2001)
11. Xu, C., Wedlund, D., Helgoson, M., Risch, T.: Model-based validation of streaming data. In: The 7th ACM International Conference on Distributed Event-Based Systems, pp. 107–114. ACM (2013)

Author Index

Printed in the United States
By Bookmasters